HANDBOOK OF TEEN AND NOVICE DRIVERS

Research, Practice, Policy, and Directions

HANDBOOK OF TEEN AND NOVICE DRIVERS

Research, Practice, Policy, and Directions

EDITED BY

Donald L. Fisher | Jeff K. Caird
William J. Horrey | Lana M. Trick

CRC Press
Taylor & Francis Group
Boca Raton London New York

CRC Press is an imprint of the
Taylor & Francis Group, an **informa** business

CRC Press
Taylor & Francis Group
6000 Broken Sound Parkway NW, Suite 300
Boca Raton, FL 33487-2742

Printed on acid-free paper
Version Date: 20160428

International Standard Book Number-13: 978-1-4665-8700-7 (Hardback)

Library of Congress Cataloging-in-Publication Data

Names: Fisher, Donald L., editor.
Title: Handbook of teen and novice drivers : research, practice, policy, and directions / [edited by] Donald L. Fisher [and 3 others].
Description: Boca Raton : CRC Press, 2017.
Identifiers: LCCN 2016007415 | ISBN 9781466587007
Subjects: LCSH: Automobile driver education (Secondary) | Teenage automobile drivers--Psychology. | Traffic safety--Study and teaching (Secondary)
Classification: LCC TL152.65 .H36 2017 | DDC 629.28/30835--dc23
LC record available at https://lccn.loc.gov/2016007415

Visit the Taylor & Francis Web site at
http://www.taylorandfrancis.com

and the CRC Press Web site at
http://www.crcpress.com

Printed and bound in the United States of America by Publishers Graphics, LLC on sustainably sourced paper.

In the spring of 1987, my younger brother Chris died in a traffic crash in San Diego. He was 22 and a month from being discharged from the US Navy. He planned to study geophysics at the University of Colorado. After his death, I decided to change my career direction and earned a PhD in human factors at the University of Minnesota. A fundamental motivation of my research in transportation and health care is to reduce injuries and fatalities. Occasionally, I tell this short story to students to explain why I research what I do. However, this is the first time I have written it down. I dedicate this handbook to my brother Chris.

Jeff K. Caird

I had e-mailed Hannah Frilot, a student in one of my classes, sometime in the middle of July to ask how she liked her internship. She was so excited, bubbling over with stories of just how much she was learning. And then just 2 weeks later, she was hit from behind by an automobile while walking her bicycle and died shortly thereafter. Alison Lustig, another student at UMass Amherst, was killed at a four-way signalized intersection when she was hit in a crosswalk by an automobile, a little over 100 yards from my office. I drive by that crosswalk every day on my way to work. The memory of these students' short lives reverberates in me and finds expression in the many professional talks that I give. It is to them and to the thousands of others whose lives were cut short in traffic crashes that I dedicate this handbook.

Donald L. Fisher

Contents

SECTION III Developmental Differences

SECTION IV Impairment

SECTION V Licensing, Training, and Education

SECTION VI International Perspective

SECTION VII Methods

SECTION VIII The Way Forward

Acknowledgments

The editors begin by acknowledging the very generous efforts of all of the many authors who contributed to this handbook, especially their patience with the glacial pace at which things might have seemed to progress. The handbook authors are literally among the world's leading experts in the area of teen and novice driving, and the handbook is truly an international compendium because of them. We also want to thank Tracy Zafian at UMass Amherst for her help with the final stages of the handbook and Cindy Carelli, Jill Jurgensen, and Kyra Lindholm at CRC Press and Taylor & Francis, who have helped us throughout the production process.

Jeff Caird is grateful to his numerous undergraduate and graduate students who taught him about many driving research topics, including teen and novice drivers.

Donald Fisher, first and foremost, thanks the many undergraduates, summer interns, graduate students, postdocs, visiting scientists, and colleagues who have again and again worked long hours into the night in order to keep the experiments moving forward. Their teamwork, goodwill, and extraordinary dedication to their work are evidenced in the many training programs that have proven results in decreasing the risky behaviors of novice teen drivers. Sandy Pollatsek has served as the intellectual force behind many of the studies that have been published, and without his constant critical intervention, the studies would have been much less solid. Finally, Don thanks Annie James Duffy Fisher and Jennifer Margaret Duffy Fisher for their willingness to coauthor a chapter with their father. It may be the first time ever they have been willing to appear in public with him.

William Horrey is grateful to his colleagues who have helped and supported his work over the past 15-plus years, including postdoctoral fellows, research assistants, students, and many, many intellectual contributors. William is also very appreciative of the guidance, leadership, and support of Marvin Dainoff, Ian Noy, and other directors from the Liberty Mutual Research Institute for Safety. In addition to their support for the time commitment required of this handbook, they provided the direct financial backing, on behalf of Liberty Mutual, for the beautiful cover artwork by Denis Gingras.

Lana M. Trick is grateful to her mentors, colleagues, and collaborators, who taught her so much and encouraged and supported her transition from laboratory research to applied research on driving. She also thanks the many graduate and undergraduate students, research interns, and technicians, whose hard work and enthusiasm are invaluable to the daily functioning of the University of Guelph driving simulation lab.

We also acknowledge the many agencies without whose support this handbook would not have been possible. Grants to Jeff Caird and Lana M. Trick from the Canadian Foundation for Innovation and AUTO21 Network of Centres of Excellence supported infrastructure and research, respectively, which made the University of Calgary Driving Simulator and the University of Guelph DRiVE Lab the Swiss Army knife for several generations of students at different locations across Canada. Grants from the National Institutes of Health, National Science Foundation, National Highway Traffic Safety Administration, Centers for Disease Control, New England University Transportation Center, Department of Transportation (DOT) University Transportation Centers Program, American

Automobile Association (AAA) Foundation for Traffic Safety, Arbella Insurance Charitable Foundation, State Farm, Liberty Mutual Research Institute for Safety, and Link Foundation for Simulation and Training have provided partial support to Donald Fisher for his research with novice teen drivers.

Finally, for any readers who may want to view more of the works by Denis Gingras, we would like to point you to his website (http://www.dgingras-artist.com). We are truly indebted to him for having provided us with the artwork for the cover, which captures the heart of the book, our hopes, and our fears.

It should be noted that the views expressed in this Handbook by Donald Fisher are his and do not necessarily reflect the position or policy of the Department of Transportation or the United States government.

Editors

Donald L. Fisher, PhD, is a principal technical advisor at the Volpe National Transportation Systems Center in Cambridge, Massachusetts, a professor emeritus in the Department of Mechanical and Industrial Engineering at the University of Massachusetts Amherst, and the director of the Arbella Insurance Human Performance Laboratory in the College of Engineering. He has published more than 250 technical papers, including recent ones in the major journals in transportation, human factors, and psychology. He has served as a principal or co-principal investigator on over $30 million worth of research and training grants, including awards from the National Science Foundation, the National Institutes of Health, the National Highway Traffic Safety Administration, Massachusetts Department of Transportation (MassDOT), the Arbella Insurance Group Charitable Foundation, State Farm Mutual Insurance Automobile Insurance Company (State Farm), the Liberty Mutual Research Institute for Safety, and the New England University Transportation Center. He is an associate editor of *Human Factors*, editor of the recently published *Handbook of Driving Simulation for Engineering, Medicine, and Psychology*, and co-chair of the Transportation Research Board (TRB) committee on Simulation and Measurement of Vehicle and Operator Performance. He has chaired or co-chaired a number of TRB workshops and served as a member of the National Academy of Sciences Human Factors Committee, the TRB Younger Driver Subcommittee, the joint National Research Council and Institute of Medicine Committee on the Contributions from the Behavioral and Social Sciences in Reducing and Preventing Teen Motor Crashes, and State Farm and Children's Hospital of Philadelphia Young Driver Research Initiative. Over the past 15 years, Dr. Fisher has made fundamental contributions to the understanding of driving, including the identification of those factors that increase the crash risk of novice and older drivers; impact the effectiveness of signs, signals, and pavement markings; improve the interface to in-vehicle equipment, such as forward collision warning systems, back-over collision warning systems, and music retrieval systems; and influence drivers' understanding of advanced parking management systems, advanced traveler information systems, and dynamic message signs. In addition, he has pioneered the development of both PC-based hazard anticipation training (Risk Awareness and Perception Training [RAPT]) and PC-based attention maintenance training (FOcussed Concentration and Attention Learning [FOCAL]) programs, showing that novice drivers so trained actually anticipate hazards more often and maintain attention better on the open road and in a driving simulator. This program of research has been made possible by the acquisition in 1994 of more than half a million dollars of equipment, supported in part by a grant from the National Science Foundation. He has often spoken about his results, including participating in a congressional science briefing on the novice driver research sponsored several years previous. The Arbella Insurance Human Performance Laboratory was recognized by the Ergonomics Society, receiving the best paper award for articles that appeared in the journal *Ergonomics* throughout 2009. The paper described the work in the Arbella Insurance Human Performance Laboratory on hazard anticipation. Dr. Fisher earned an AB from Bowdoin College in 1971 (philosophy), an EdM from Harvard University in 1973 (human development), and a PhD from the University of Michigan in 1982 (mathematical psychology).

Jeff K. Caird, PhD, is a professor in the Department of Psychology and an adjunct professor in the Departments of Anesthesia and Community Health Sciences at the University of Calgary. He earned his PhD from the University of Minnesota and MS from the University of Colorado. He previously directed the Cognitive Ergonomics Research Laboratory and the Canadian Foundation for Innovation (CFI)-funded University of Calgary Driving Simulator (UCDS). He currently directs the Healthcare Human Factors and Simulation Laboratory in the W21C Research and Innovation Centre (http://www.w21c.org), which is in the O'Brien Institute of Public Health and Cumming Faculty of Medicine. He has coedited a number of books on human–machine systems, including the *Handbook of Driving Simulation for Engineering, Medicine, and Psychology*. He was the coleader of the Convergent Evidence from Naturalistic, Epidemiological and Simulation Data (CENSED) Network that was part of the AUTO21 Network of Centres of Excellence (NCE). He is a member of a number of national and international committees in Canada and the United States. His broad areas of research are in transportation and health care human factors. The design of devices, environments, and systems that reduce injuries and fatalities and improve the quality of life are integral research goals.

William J. Horrey, PhD, is a principal research scientist in the Center for Behavioral Sciences at the Liberty Mutual Research Institute for Safety. He earned his PhD in engineering psychology from the University of Illinois at Urbana–Champaign in 2005. He has published numerous papers that focus on visual (selective) attention and divided attention, driver behavior, and distractions from in-vehicle devices, including cell phones and visual displays. His more recent work has examined the concept of driver calibration and how driver risk perceptions differ from actual hazards as well as how overconfidence and optimism in self-evaluations can impact driver decision making. Recently, he has also carried out studies on the adverse impact of extended and night-shift work on driver drowsiness and safety. An active member of the Human Factors and Ergonomics Society, he is a past chair of the Surface Transportation Technical Group, serves on the society's Publications Committee, and is currently an associate editor for the organization's flagship journal *Human Factors*. He recently served as lead guest editor on a special issue of the journal *Accident Analysis and Prevention* on the future directions of fatigue and safety research. He is the current chair of the Transportation Research Board standing committee on Vehicle User Characteristics (AND10). He serves or has served on several national and international committees related to transportation safety and human factors.

Lana M. Trick, PhD, is a professor at the University of Guelph in the Department of Psychology and codirector of the University of Guelph DRiVE Lab, which houses the Canadian Foundation of Innovation (CFI)-funded driving simulation facility (http://www.uoguelph.ca/drive/). She did her graduate training in the Department of Psychology of the University of Western Ontario, where she did her PhD thesis with cognitive scientist and philosopher Zenon Pylyshyn. Her career began in basic research, studying subitizing, counting, and multiple-object tracking, laboratory tasks relevant to understanding attentional performance in complex dynamic environments (*Psychological Review*), and she continues to do basic research on life span changes in attention. Her research on driving began in a collaboration with Peter Cooper and John Vavrik from the *Insurance Corporation of British Columbia*, who challenged her to write a large-scale theoretical review that incorporated the basic research on attention with the applied research on driving (*Theoretical Issues in Ergonomic Science*). During that time, she met Jeff Caird and toured his driving simulation facility, an experience that inspired her to apply for a simulator of her own when she came to work in the Psychology Department at the University of Guelph. Although she continues to balance basic and applied research, in the 10 years since the opening of the University of Guelph DRiVE lab, she has had a large number of driving-related projects, including investigations of the way that older and younger drivers react to fleeting emotions, driving challenges such as difficult driving conditions (heavy traffic, fog, wayfinding), and distractions such as in-vehicle technologies and passengers. She has done a number of studies on drivers at risk, including seniors with diminished attentional performance and teens and adults with attention deficit hyperactivity disorder, and has also done methodological work, developing a technique for measuring the dynamic ability to keep track of the positions of multiple moving vehicles at once (the multiple-vehicle tracking task) and investigating ways to reduce simulator adaptation syndrome, a common methodological problem that causes data loss in driving simulation studies. She has been a member or coleader of a variety of different multidisciplinary teams investigating safety and driving performance and, most recently, has served as coleader with Jeff Caird on the Convergent Evidence from Naturalistic, Epidemiological and Simulation Data (CENSED) Network as part of the AUTO21 Network of Centres of Excellence (NCE).

Contributors

Shelley Aggarwal
Division of Adolescent Medicine
Department of Pediatrics
Stanford University School of Medicine
Palo Alto, California

Mark Asbridge
Community Health and Epidemiology
Dalhousie University
Halifax, Nova Scotia, Canada

Paul Boase
Transport Canada
Ottawa, Ontario, Canada

Jeff K. Caird
Department of Psychology
and
Department of Community Health Sciences
University of Calgary
Calgary, Alberta, Canada

Jennifer Cartwright
Community Health and Epidemiology
Dalhousie University
Halifax, Nova Scotia, Canada

David Crundall
School of Psychology
Nottingham Trent University
Nottingham, United Kingdom

Allison E. Curry
Center for Injury Research and Prevention
The Children's Hospital of Philadelphia
Philadelphia, Pennsylvania

Drew Dawson
CQUniversity
Appleton Institute
Adelaide, South Australia, Australia

Joost C. F. de Winter
Department of Biomechanical Engineering
Delft University of Technology
Delft, Netherlands

Gautam Divekar
Battelle
Seattle, Washington

Lisa Dorn
School of Aerospace, Transport and Manufacturing
Cranfield University
Bedfordshire, United Kingdom

Johnathon Ehsani
Health Behavior Branch
Division of Intramural Population Health Research
National Institute of Child Health and Human Development
National Institutes of Health
Bethesda, Maryland

Tracy Exley
Division of Adolescent Medicine
Department of Pediatrics
Stanford University School of Medicine
Palo Alto, California

Anna James Duffy Fisher
Bristol, Rhode Island

Donald L. Fisher
Volpe National Transportation Systems Center
Cambridge, Massachusetts

Jennifer Margaret Duffy Fisher
Chicago, Illinois

Robert D. Foss
University of North Carolina Highway Safety Research Center
Chapel Hill, North Carolina

Bonnie Halpern-Felsher
Division of Adolescent Medicine
Department of Pediatrics
Stanford University School of Medicine
Palo Alto, California

William J. Horrey
Liberty Mutual Research Institute for Safety
Hopkinton, Massachusetts

Mark S. Horswill
School of Psychology
The University of Queensland
St. Lucia, Queensland, Australia

Adnan A. Hyder
International Injury Research Unit
Department of International Health
Johns Hopkins Bloomberg School of Public Health
Baltimore, Maryland

Brian A. Jonah
Road Safety Canada Consulting
Ottawa, Ontario, Canada

Sheila G. Klauer
Virginia Tech Transportation Institute
Blacksburg, Virginia

Natália Kovácsová
Department of Biomechanical Engineering
Delft University of Technology
Delft, Netherlands

Michael G. Lenné
Monash University Accident Research Centre
Monash University
Clayton, Victoria, Australia

Tsippy Lotan
Or Yarok
and
Ran Naor Research Foundation
Hod Hasharon, Israel

Jeffrey C. Lunnen
International Injury Research Unit
Department of International Health
Johns Hopkins Bloomberg School of Public Health
Baltimore, Maryland

Robert Mann
Centre for Addiction and Mental Health
Toronto, Ontario, Canada

Catherine C. McDonald
University of Pennsylvania School of Nursing
Philadelphia, Pennsylvania

Jessica H. Mirman
Craig-Dalsimer Division of Adolescent Medicine and Center for Injury Research and Prevention
The Children's Hospital of Philadelphia
Philadelphia, Pennsylvania

Eve Mitsopoulos-Rubens
VicRoads
Melbourne, Victoria, Australia

Christine Mulvihill
Monash University Accident Research Centre
Monash University
Clayton, Victoria, Australia

Jeffrey W. Muttart
Crash Safety Research Center, LLC.
East Hampton, Connecticut

John Nepomuceno
Auto Technology Research, Strategic Resources
State Farm Mutual Automobile Insurance Company
Bloomington, Illinois

Marie Claude Ouimet
Faculty of Medicine and Health Sciences
University of Sherbrooke
Longueuil, Quebec, Canada

Jessica L. Paterson
CQUniversity
Appleton Institute
Adelaide, South Australia, Australia

Anuj K. Pradhan
University of Michigan Transportation Research Institute
Ann Arbor, Michigan

Malena Ramos
Division of Adolescent Medicine
Department of Pediatrics
Stanford University School of Medicine
Palo Alto, California

Bryan Reimer
MIT AgeLab and New England University Transportation Center
Massachusetts Institute of Technology
Cambridge, Massachusetts

Bridie Scott-Parker
Adolescent Risk Research Unit
School of Social Sciences
Faculty of Arts, and Law Business
University of the Sunshine Coast
Sippy Downs, Queensland, Australia

Jean T. Shope
University of Michigan Transportation Research Institute
Ann Arbor, Michigan

Sarah Simmons
Department of Psychology
University of Calgary
Calgary, Alberta, Canada

Bruce Simons-Morton
Health Behavior Branch
Division of Intramural Population Health Research
National Institute of Child Health and Human Development
National Institutes of Health
Bethesda, Maryland

Marilyn S. Sommers
University of Pennsylvania School of Nursing
Philadelphia, Pennsylvania

Rosemary Tannock
Neurosciences and Mental Health Research
Toronto Hospital for Sick Children
Toronto, Ontario, Canada

Tomer Toledo
Faculty of Civil and Environmental Engineering
Technion—Israel Institute of Technology
Haifa, Israel

Lana M. Trick
Psychology Department
University of Guelph
Guelph, Ontario, Canada

Willem Vlakveld
Stichting Wetenschappelijk Onderzoek Verkeersveiligheid (SWOV) Institute for Road Safety Research
The Hague, Netherlands

Barry Watson
Global Road Safety Partnership
Geneva, Switzerland

Christine Wickens
Centre for Addiction and Mental Health
Toronto, Ontario, Canada

Allan F. Williams
LLC
Bethesda, Maryland

Flaura K. Winston
Center for Injury Research and Prevention
The Children's Hospital of Philadelphia
and
Division of General Pediatrics
Department of Pediatrics
Perelman School of Medicine at the University of Pennsylvania
Philadelphia, Pennsylvania

I

Introduction

1

Introduction to the Handbook of Teen and Novice Drivers

Donald L. Fisher

William J. Horrey

Jeff K. Caird

Lana M. Trick

1.1 Introduction

Let us begin by welcoming what we hope is a broad range of readers to this handbook: teens and their parents; driving instructors; legislators; staff in various state agencies charged with regulating the licensing of novice teen drivers; program officers at the different agencies, institutes, and foundations that sponsor research on novice and teen drivers; and academics from a broad spectrum of disciplines. You all have a part to play here. We hope that the material herein can inform and educate, showcase, and even inspire.

Despite a growing body of research and targeted remediation, novice teen drivers continue to be overrepresented in fatal and injurious motor vehicle crashes. In the United States, only 6.4% of the total licensed drivers in 2008 were young drivers aged 15–20 years, yet they accounted for nearly 11% of driver fatalities (National Highway Traffic Safety Administration, 2011). The World Health Organization (2007) reports that road traffic injuries are the leading cause of death globally among 15- to 19-year-olds. In the first month or two of driving, teens are some six to nine times more likely to die in a crash than they are just several years later (Insurance Institute for Highway Safety, 2004).

A book like this one would have been radically different 15 years ago. In the early 1970s, the National Highway Traffic Safety Administration reported the results of a large study in DeKalb County, Georgia, of the effects of novice driver training on crashes (National Highway Traffic Safety Administration, 1974). The results were very disappointing. It appeared that nothing could be done about novice drivers in the way of education in the classroom. Graduated driver licensing (GDL) programs had not been thoroughly evaluated. The general stereotype was that novice drivers were careless, only too willing to risk their lives and the lives of others.

But a great deal has changed over the last 15 years. We know so much more about the behaviors of drivers in general (Klauer et al., 2006, 2014), and teens in particular (Lee et al., 2011), from large-scale naturalistic studies. GDL programs have been shown to be tremendously successful (Masten and Foss, 2010). Skills training programs have now proven that they can both change behaviors related to crashes (Crundall et al.,

2010; Pollatsek et al., 2006), programs whose effects last over a year after training (Taylor et al., 2011), as well as reduce crashes among certain subpopulations of teen drivers (Thomas et al., 2016). And training programs that address higher-level skills such as self-awareness and calibration, based in part on the work the European Union did with Goals for Driver Education (Hatakka et al., 2003), have proven effects.

These developments have led to rapid growth in the number of studies of novice drivers both nationally and internationally, using everything from advanced driving simulators with motion to single desktop simulators with game controls, from naturalistic studies on the open road to field trials on the open road, and from epidemiological studies using long-available databases to those using much more detailed and sophisticated databases. This research has contributed to our enhanced understanding of differences in novice drivers' attention maintenance, hazard anticipation, and crash mitigation skills, not to mention the myriad other factors that contribute to increased risk for this vulnerable group of road users. The expansion of our knowledge in turn has led to an increasing demand from government funding and regulatory agencies that something be done to solve the novice driver problem, through engineering, education, or enforcement. It has led both to large-scale efforts by insurance companies to deliver training, over the web or on mobile simulators, to novice drivers and to smaller-scale efforts by private driver education companies to incorporate the latest elements of training into their curricula. This growth has also led to an increasing interest in the medical community in studies of the effects of hyperactivity and attention deficit disorders on novice drivers. All of these currents have relevance globally in terms of moving toward strategies for mitigating the risk that novice and teen drivers are exposed to—in developed countries as well as developing countries, where the ultimate success of any remediation needs to consider political, infrastructure, and cultural differences around the world.

But, there still remains much that can be done. Despite the long period of supervised driving in GDL programs, there appears to be little effect of this supervision on the rate at which critical skills are learned once the teen driver obtains a solo license (Foss et al., 2011). These rates start out high and, despite many hours of supervised practice, decline rapidly over the course of the first year. And despite the obvious effect that skills and self-awareness training have on behaviors that lead to crashes, few evaluations suggest that these programs actually reduce crashes.

1.2 Chapters of Interest to Our Different Sets of Readers

Before describing the contents of this handbook, we thought it useful to point our various different sets of readers toward the chapters that we think might most interest them.

1.2.1 For Parents and Caregivers

Parents and caregivers, we understand your concerns and your fears. We too have children, many of whom we taught to drive ourselves (also see Chapter 2). The worry never disappears. But there are things we can do as parents that can decrease the risks for our teens when they start driving. That said, the need for parental involvement does not disappear once your teen has a license. During the restricted license phase, it is important to engage your new teen driver in a commitment to follow the restrictions in place in your state. We hope that many of the chapters in this handbook will inform you of the various ways in which you can help improve the outlook for your teen driver(s).

1.2.2 For Teens

Hello and welcome! If there is one thing we have learned in our research with novice and teen drivers over the almost 100 years of our cumulative experience, it is that preaching safety by itself is not always enough. In this regard, teens are no different than other drivers (Romoser & Fisher, 2009). The best lessons are learned the hard way—by making mistakes, learning how to avoid the mistakes in the future, and then mastering the behaviors that led to the mistakes. With this in mind, Chapter 3 in this handbook was written specifically for you. We have kept it short (five pages total) and, we hope, to the point.

Today, there are many programs and technologies available that let you learn firsthand what types of errors you can make while driving that can get you into serious trouble. Most teens we have encountered are surprised by the errors that they make in these programs. Hopefully, you can then use this experience on the open road. Five pages is all we ask that you read. We hope you can find the time to sit down with Chapter 3 and try to digest all that it has to say.

1.2.3 For Researchers

Presumably, we do not need to convince our colleagues that research on novice teen drivers is something in which they should take an interest. Most, if not all, have presumably undertaken or are currently undertaking research in the area and know, like us, just how exciting a time it is to be doing such research. Research in driving, but especially novice teen driving, has captured our attention for many reasons. As empiricists and theoreticians, it presents us with extraordinary challenges and opportunities, whether we are interested in developing detailed quantitative or computer models of a given, very specific driving behavior or in evaluating more qualitative models of the influence of teen passengers on teen drivers (Chapters 9, 10, and 16). As data analysts, it presents us with unparalleled opportunities in terms of the sheer volume of data available to summarize and analyze (collected, say, on a driving simulator—Chapter 25, with an eye tracker—Chapter 27, or in naturalistic studies—Chapter 26), but also in terms of the incredibly challenging nature of the data analysis. As engineers and scientists, it presents us with opportunities to develop, evaluate, and deploy interventions (e.g., training programs—Chapter 18, in-vehicle safety systems—Chapter 20) that can potentially reduce on a large scale both morbidity and mortality around the world in both developed (Chapter 23) and low- and middle-income countries (Chapter 24). The list goes on and on, and we don't want to overplay our hand. After all, there are other exciting areas in which to do research. But, for those of you considering research on novice teen drivers, the area presents an array of challenges that promise to stretch far into the future.

1.2.4 For Instructors and Trainers

You, the tens of thousands of driving instructors around the world, are perhaps the unsung heroes of the effort to reduce crashes among novice drivers (see also Chapter 4). As is often the case, you may have been drawn to the field because you have seen up close and personally just how devastating a crash can be. But for far too many years, you as driving instructors have been given tools that are only partially effective. Although these tools do help novice drivers obtain a restricted license, they do less than they could to provide novice drivers with the skills that they need to reduce crashes. Such tools have simply not been available up through the turn of the century. However, as noted previously, this is now changing rapidly with the development of training programs that can be used by you both in the classroom and in the field to provide novice drivers with the skills that they need to change their behaviors in the scenarios in which they are most at risk. We encourage you to read Chapters 18 and 20 in order to gain a better understanding of what is available today that can help you provide skills training, which has been shown to reduce the behaviors that lead to crashes. Many of the training programs are freely available over the Internet (see Chapter 3; also see http://www.ecs.umass.edu/hpl/ for links to these and other training programs). We would love to hear back from you as to just how useful you (and the teens you instruct) find such programs.

1.2.5 For Federal and State Legislators

As federal and state legislators, you can play a key role in the effort to reduce novice driver crashes. Laws have already been passed in many states that apply to teens during the restricted phase of licensing, which include bans on some of the most significant risk factors (e.g., the use of cell phones and restrictions on the presence of other teen passengers—Chapter 12). However, there still remains much that could be done by you, our legislators. For example, at least here in the United States, the tests required to obtain a license contain very few, if any, questions that probe novice drivers' knowledge of those skills such as hazard anticipation that are known to be critical to the avoidance of crashes (Chapter 28). Thus, there exists broad room for improvement in the content of the tests that are required for licensure, changes that have been introduced in other countries for some time (Chapters 23 and 28). Additionally, again here in the United States, very little has been done to require that parents themselves learn something about the behaviors that are most likely to lead to crashes and the types of crashes in which their teen drivers are most likely to be involved (Chapter 5). Perhaps one of the reasons that supervised driving has not had the desired effect on the skills of novice drivers once they first obtain their solo license (Foss et al., 2011) may be that parents themselves need to better understand how best to intervene during the learner's permit phase. In short, you can help pass laws that increase the involvement of parents in their teens' driving during all phases of the licensing process (Chapter 19).

1.2.6 Motor Vehicle Departments

Motor vehicle departments can play a critical role in the development of future efforts to determine whether various interventions administered during the learner's permit phase or at the time that teens obtain their restricted license have an effect on crash rates. For either set of interventions, researchers have an acute need for information on when a driver first obtained his or her solo (restricted or unrestricted) license. Crash records, which often are available, simply do not contain this information. Researchers need to link the information in crash databases with the information at motor vehicle departments in order to determine how long an individual has been driving at the time a crash occurred. This is critical to assessing the effectiveness of programs that are targeted at reducing the crashes of teen drivers during the first 6 months of driving. Your participation in potentiating this linkage is crucial.

1.2.7 Federal and State Research Agencies, Institutes, Foundations, and Insurance Companies

The handbook can provide program officers at state and federal agencies, institutes and foundations, and insurance companies with a broad sampling of the ongoing efforts and an informed discussion of what additional research is needed. Before the turn of the century, here in the United States and around the world in developed countries, federal and state agencies, along with institutes and foundations interested in reducing novice driver crashes, had focused almost exclusively on funding the development and evaluation of GDL programs, which have now been shown to have real benefits (Chapters 17 and 23). With the turn of the century, there has been an explosion of interest in novice teen drivers by sponsors at all levels, including the funding of large naturalistic driving studies (Chapter 26), the funding of studies to evaluate training programs (Chapter 18) and in-vehicle technologies (Chapter 20), the funding of studies to determine the increasing role that distraction plays in the crashes of novice drivers (Chapter 12), and the funding of studies to identify the effect parents (Chapter 19) and teens (Chapter 16) have on novice driver crashes. And there is much more to be done. As just a start, too little is known about what teens learn during the first 6 months of solo driving that so radically increases their crashes across a broad spectrum of scenarios (Foss et al., 2011) and, related to this, whether anything can be done to train teens in the learner's permit phase to acquire the skills that would protect them during the first, very risky 6 months (Chapter 29). You as potential sponsors of research will hopefully find these opportunities, and the many other opportunities described in this handbook (also see Chapter 30), as ones where real progress can be made by the undertaking of basic and applied research.

1.3 The Plan for This Book: An Overview

Because this book is targeted toward a diverse audience, we endeavor to cover a number of different topics, in research, policy, and practice. The different topics are arranged into eight different sections.

Section I, the introduction, includes five chapters that attempt to set the stage for our readers. In addition to this broad overview, we include short chapters targeting parents (Chapter 2), their teen drivers (Chapter 3), their driving instructors (Chapter 4), and, more generally, what we know about the crash patterns among teens (Chapter 5). The major research questions addressed in the introduction and of interest to the entire audience are threefold: In what types of crashes are novice teen drivers involved? What types of behaviors are most likely to lead to these crashes? And do novice drivers differ from experienced drivers in their relative involvement in the most likely crashes?

Section II examines in much more detail the three types of skill deficiencies that are most likely to lead to crashes: deficiencies in hazard anticipation (Chapter 6), attention allocation and maintenance (Chapter 7), and hazard mitigation (Chapter 8). One of the main points from these chapters is that novice and teen drivers scan and process information in the environment differently from more experienced drivers, and these variations are most often relevant or critical when one is faced with potential (as opposed to actual or materialized) threats—from other vehicles, road users, and animals, among others. These chapters thus pick up on and amplify some of the questions addressed in the introduction.

Section III takes a more in-depth look at the developmental processes that contribute to risky behaviors in novice teen drivers. In particular, it is described how developmental changes in the brains of teens (Chapter 9) can explain, in part, the general cognitive, social, and emotional differences (Chapters 9 and 10) between teens and older drivers that lead teens to take more risks and, in particular, be willing to speed (Chapter 11).

Section IV elaborates on a number of issues concerning driver impairment, including driver distraction (Chapter 12), alcohol and tetrahydrocannabinol (THC) (Chapter 13), attention deficit disorder (Chapter 14), driver fatigue (Chapter 15), and teen passengers (Chapter 16). All of these types of impairment can have a significant impact on a teen driver's ability to safely control a motor vehicle. While their adverse impact is not solely reserved for novice and teen drivers, the presence of other factors (developmental changes, skill level, etc.) can lead to dangerous interactions. Building on the lessons from Sections II and III, these chapters contribute further to our knowledge and understanding of teen drivers—more accurately, of teens in general—and also offer some potential inroads to real-world solutions.

Section V endeavors to lay the groundwork for some countermeasures for combatting the observed (and discussed) risks. This section focuses on licensing, training, and education. Specifically, it examines the role of licensing and, in particular, GDL programs (Chapter 17), the new methods of training and education and their impact on behavior (Chapter 18), the role of parents and peers (Chapter 19), the new technologies that can monitor and coach young drivers as well as keep their parents informed (Chapter 20), and the use of simulator-based training for novice car drivers and motorcycle riders (Chapter 21). The section concludes with an overall perspective on how the road from research to practice (countermeasures) is traveled (Chapter 22). Not surprisingly, it is often filled with potholes, but hopefully, there are some general insights that lessen the impact of the inevitable bumps in the road.

Much of the research discussed in the handbook occurred in the United States. Section VI strives to provide more of an international backdrop. Chapter 23 describes key research and policy issues in developed nations around the world, including some of the approaches and relative successes of various countermeasures. In contrast, Chapter 24 describes the picture for low- and middle-income countries—those developing nations that often face staggering numbers of road traffic fatalities each year yet also face a number of political, cultural, and economic barriers to implementing effective countermeasures.

While the preceding sections describe the issues and research surrounding novice and teen drivers, Section VII focuses on various methods that have contributed to our understanding of the aforesaid issues. That is, much of the research reported in the handbook would not have been possible without the introduction of new platforms for research, new tools for assessment, and new techniques for data analysis. This section includes a discussion of driving simulators and the various issues associated with their use (Chapter 25), naturalistic (on-road) study approaches (Chapter 26), eye tracking and the windows it has opened on the causes of novice teen driver crashes (Chapter 27), and the use of hazard perception evaluation in licensing (Chapter 28).

Finally, the book concludes with a look forward (Section VIII). Much of the research on novice drivers in the United States has been undertaken by members of the Transportation Research Board Younger Driver Subcommittee. In the summer of 2015, they met to discuss the need to increase greatly the research that is being done to understand why, even after many hours of supervised practice, novice teen drivers still show a radical and rapid increase in crashes after they obtain their solo license (Chapter 29). This critical 6- to 12-month period still remains an enigma. In the concluding chapter (Chapter 30), the editors take a best guess as to the larger spectrum of research issues that need to be addressed in the context of what will be a future with automated and semiautomated vehicles (Will all drivers then become novices given so little exposure?) and vehicle-to-vehicle communication, which is estimated to reduce crashes by up to 80% (Will teens require special education in order to benefit from these technologies?).

Some of us readers have lost loved ones way too early in their lives. Some of us readers have lost children in automobile crashes. The loss is forever, and the pain never goes away. But we, the editors and the authors, believe that something can be done to reduce this loss. We hope this handbook goes a little way toward speeding us in the direction of this goal.

References

Crundall, D., Andrews, B., van Loon, E., & Chapman, P. (2010). Commentary training improves responsiveness to hazards in a driving simulator. *Accident Analysis and Prevention, 42*, 2117–2124.

Foss, R., Martell, C., Goodwin, A., O'Brien, N., & UNC Highway Research Center. (2011). *Measuring Changes in Teenage Driver Characteristics During the Early Months of Driving.* Washington, DC: AAA Foundation for Traffic Safety.

Hatakka, M., Keskinen, E., Buaghan, C., Goldenbeld, C., Gergersen, N., Groot, H., Siegrist, S., Willmes-Lenz, G., & Winkelbauer, M. (2003). *Basic Driver Training: New Models, EU-Project* (Final Report). Turku, Finland: Department of Psychology, University of Turku.

Insurance Institute for Highway Safety. (2004). *Fatality Facts 2004: Older People.* Arlington, VA: Insurance Institute for Highway Safety.

Klauer, S. G., Dingus, T. A., Neale, V. L., Sudweeks, J. D., & Ramsey, D. J. (2006). *The Impact of Driver Inattention on Near-Crash/Crash Risk: An Analysis Using the 100-Car Naturalistic Driving Study Data* (Rep. DOT HS 810 594). Washington, DC: National Highway Traffic Safety Administration.

Klauer, S., Guo, F., Simons-Morton, B., Ouimet, M., Lee, S., & Dingus, T. (2014). Distracted driving and risk of road crashes among novice and experienced drivers. *New England Journal of Medicine, 370*, 54–59.

Lee, S., Simons-Morton, B., Klauer, S., Ouimet, M., & Dingus, T. (2011). Naturalistic assessment of novice teenage crash experience. *Accident Analysis and Prevention, 43*, 1472–1479.

Masten, S., & Foss, R. (2010). Long-term effect of the North Carolina graduated driver licensing system on licensed driver crash incidence: A 5-year survival analysis. *Accident Analysis and Prevention, 42*, 1647–1652.

National Highway Traffic Safety Administration. (1974). *Driver Education Curriculums for Secondary Schools: User Guidelines. Safe Performance Curriculum and Pre-Driver Licensing Course* (Final Rep. DOT-HS-003-2-427). Washington, DC: Office of Driver and Pedestrian Research, US Department of Transportation.

National Highway Traffic Safety Administration. (2011). *Traffic Safety Facts 2009 Data: Young Drivers* (Report DOT HS 811 400). Washington, DC: NHTSA.

Pollatsek, A., Narayanan, V., Pradhan, A., & Fisher, D. L. (2006). The use of eye movements to evaluate the effect of PC-based risk awareness training on an advanced driving simulator. *Human Factors, 48*, 447–464.

Romoser, M., & Fisher, D. L. (2009). The effect of active versus passive training strategies on improving older drivers' scanning for hazards while negotiating intersections. *Human Factors, 51*, 652–668.

Taylor, T., Masserang, K., Divekar, G., Samuel, S., Muttart, J., Pollatsek, A., & Fisher, D. (2011). Long term effects of hazard anticipation training on novice drivers measured on the road. *Proceedings of the 6th International Driving Symposium on Human Factors in Driver Assessment, Training, and Vehicle Design 2011*. Iowa City: Public Policy Center, University of Iowa.

Thomas, F. D., Rilea, S., Blomberg, R. D., Peck, R. C., & Korbelak, K. T. (2016). *Evaluation of the Safety Benefits of the Risk Awareness and Perception Training Program for Novice Teen Drivers* (DOT HS 812 235). Washington, DC: National Highway Traffic Safety Administration.

World Health Organization (2007). *Youth and Road Safety*. Geneva: WHO.

2

Ten Things Parents Need to Know When Their Teens Start to Drive

Sarah Simmons

Jeff K. Caird

Donald L. Fisher

Abstract

The Chapter. Many of those reading this chapter are soon to be parents of novice teen drivers or will someday be parents of novice teen drivers. You know only too well how frightening it can be to think about handing over the keys to your children, let alone actually watch them drive away for that first time on their own. And you may be encountering material about novice teen driving for the first time. This chapter is written especially for you so that you do not need to wade through the entire *Handbook of Teen and Novice Drivers* to learn the nuggets that are buried within. Here, we outline the top 10 things that we believe parents should know when their teens start to drive. For each point, we briefly elaborate on the scientific evidence supporting our suggestions. For a more thorough treatment of each topic, we refer readers to other chapters in the *Handbook*. A number of additional resources for parents on teen driving are listed at the end of the chapter.

2.1 Introduction

If you are the parent of a current or prospective novice teen driver, this chapter is written specifically for you. Here, we describe what we believe are the 10 most important things that parents need to know when their teen begins to drive. A number of these recommendations are based on common questions that parents have when faced with their teens need to begin to drive (Mulvihill et al., 2005). For example, how do I supervise my teen when he or she begins to drive on his or her own? We focus primarily on graduated driver licensing (GDL) topics, which are highly important for parents to understand so that the risk of crash involvement for their teen can be reduced. Reviews of GDL programs have repeatedly shown that overall, they lead to reductions in crashes and fatalities (Masten et al., 2011; Russell et al., 2011). We also make suggestions about how to establish rules related to driving that make sense and how to choose the best vehicle for a young driver. A variety of additional online and publication resources are provided at the end of the chapter for those who would like to know more about a number of topics that are discussed.

2.2 The 10 Things Parents Should Know

We list here the 10 things that we believe a parent should know. They stand as a separate list for four related reasons. First, we believe that parents may want to print this list and take it with them, in the car and elsewhere. Hopefully, we have made this easy for you by grouping together these 10 recommendations. Second, we believe that these are so important that even if you don't read any further, you should take what you can from just this list. For example, you should model how your teen should drive. The chances are good that if you are following other vehicles too closely or using your smartphone while you drive, so too will your teen. Third, as researchers who talk to parents and teens, the number of parents who asked for additional information for their teen driver struck us as an unmet need. We have tried to put information into this chapter that would help and be useful to parents. Lastly, we have structured the list hoping that it provides enough in the way of interest that you do decide to read further into the chapter. For example, do you know the GDL restrictions? If you are like most parents, you may not be completely familiar with them. These restrictions save lives.

Here is the list we have assembled. The list is based on research studies that have looked at crashes or behaviors that lead to crashes. There are other, perhaps equally good, recommendations that have yet to be tested.

1. Know that teen and novice drivers are at increased risk of crashes compared to other age groups, and learn why they are at risk.
2. Take advantage of the supervised phase of the GDL program.
3. Provide an appropriate level of guidance while supervising your teen.
4. Understand and use the GDL restrictions.
5. Enforce the GDL restrictions, and closely monitor and manage your teen's independent driving.
6. Clearly communicate about rules and expectations.
7. Let your teen know that he or she should never drive or ride when alcohol, drugs, distraction, or sleep deprivation is involved.
8. Model the way that you want your teen to drive.
9. Select the safest vehicles for your teens (newer midsize or full-size cars).
10. Keep in mind that your teen may want to delay licensure, and know what to do if this is the case.

2.3 The 10 Things Parents Should Know, Explained

2.3.1 Teen and Novice Drivers Are at Increased Risk of Crashes Compared to Other Age Groups

In most countries around the world, motor vehicle crashes are highest among teen and novice drivers. Traffic crashes are the leading cause of injury and death among 15- to 19-year-old teenagers in the

United States (NHTSA, 2014) and in the world (WHO, 2015). During the first 6 months of driving, the crash rates are the highest of any age group (Mayhew et al., 2003; McCartt et al., 2003). There are many reasons for this. By definition novices are inexperienced, and inexperienced drivers, like inexperienced basketball players or pilots, are not very good. Expertise in any activity is thought to require 10,000 hours of deliberate practice (Ericsson, 1996). Fortunately, a great deal of learning occurs in the first 1000 miles of driving, but during those first 1000 miles, teens are at very high crash risk, which continues to decline for many years until reaching normal adult levels in the early 20s.

Novice drivers are known to be poor at hazard detection (Chapter 6), maintaining attention on the forward roadway (Chapters 7 and 12), and vehicle handling in routine and emergency conditions (Chapter 8). They may also be inclined to drive too fast for conditions (Chapter 11). Insufficient experience underlies many of the crashes in which teen drivers are involved; that is, they are primarily *clueless*, not *careless* (Curry et al., 2011; McKnight & McKnight, 2003). Perhaps because of this, they take more risks and are overconfident in their driving abilities. Novice drivers may also lack the judgment about how to manage risk that only comes with experience.

Parents should be aware of their teen's increased crash risk in order to fully appreciate the advice and suggestions we present in this chapter. However, we also want parents to know that reducing their teen's amount of practice driving in order to minimize the number of on-road opportunities for their teen to crash will not serve to protect them in the long term. Instead, parents should allow their teen as much practice driving as possible in a variety of driving conditions, with measures taken to keep risk levels in check. However, no matter how much supervised practice driving teens obtain, there is a great deal they can learn only by driving on their own. Hence, it is best to limit the conditions under which they drive immediately after licensure and gradually grant greater privileges as they gain experience. We expand on this as we go forward.

2.3.2 Take Advantage of the Supervised Phase of the GDL Program

GDL programs, which are now implemented in most states and provinces in the United States, Canada, and a number of other countries, reduce exposure to high-risk activities and increase parental involvement while teens are learning to drive (Chapter 17). They typically apply only to drivers under 18 years of age (we describe more about what to do with teen novice drivers older than 18 later in this chapter). GDL programs help young novice drivers increase their level of driving experience while keeping their level of risk in check. Most GDL programs follow a similar three-stage structure. During the first stage (the learner phase), novice drivers require a supervisor, which is usually a parent or guardian; during the second stage, they may drive independently as long as they comply with their GDL rules and restrictions; and during the third stage, they have an unrestricted license.

Parents should know that the learner phase is actually the *safest* period to gain driving experience in. In fact, learner drivers across the world have the lowest crash risk of any driver age group (Gregersen et al., 2003). So, parents should make sure that they allow their teen to get as much practice as possible during this time. The more experience teens receive during the supervised phase, the more prepared they'll be for the unsupervised phase (Gregersen et al., 2000). Some researchers have emphasized that, ideally, parents should aim for over 100 hours of varied practice driving, rather than tens of hours (Gregersen et al., 2001; cited in Mulvihill et al., 2005). The Insurance Institute for Highway Safety predicts that compared with novice drivers who receive no hours of supervised practice, novice drivers who receive 70 hours of supervised practice have, on average, 17% fewer crashes with insurance claims and 4% fewer fatalities (Highway Loss Data Institute, 2015).

Having said this, a number of studies have found no relationship between the number of kilometers or miles of supervised practice and postlicensure crashes (McCartt et al., 2003; Sagberg & Bjørnskau, 2006; Simons-Morton & Ouimet, 2006). However, we feel that there is a ready explanation. Parents are not always aware of exactly what skills are the most critical to teach. And why should they be? As a result, we describe in the sections that follow what we believe are the most important skills that you as parents can teach novice teen drivers during the supervised driving phase. These are already skills with which you are very familiar, but the research suggests that teens are not. And, they are the very skills that are associated with crashes. We emphasize that it is extremely important for your teen to practice driving while being supervised. As with learning to ride a bicycle, the transition to independent driving may not necessarily be smooth.

We also want to emphasize that it is important to vary the types of driving experiences that a teen receives because it appears that supervised driving practice only helps young novice drivers develop skills to the extent that these practice experiences provide them with a broad range of driving experiences (Simons-Morton & Ouimet, 2006). In other words, the quality of practice driving appears to matter more than the quantity: parents should try to provide their teen with a variety of experiences, including city, suburban, rural, and highway driving, during practice drives.

2.3.3 Provide an Appropriate Level of Guidance during the Supervised Phase of the GDL Program

During early supervised driving, parents must serve as vigilant passengers, taking time to observe and think about what aspects of their teen's driving need feedback. Parents should focus on correcting and providing feedback for one problem at a time, whether it be accelerating too fast, braking too hard, or turning too sharply. Skills should be practiced until they are correct (Ericsson, 1996). For every improvement, parents should praise their teens for their accomplishment.

Over time, teen drivers need to become less reliant on parental feedback, as they need to learn to operate the vehicle on their own (Simons-Morton & Ouimet, 2006). In general, novice drivers tend to learn basic maneuvering or vehicle handling skills and move on to higher-order skills such as identifying and avoiding hazards. A number of more advanced skills such as scanning blind spots and anticipating the poor driving habits of other drivers may not develop sufficiently if the parent is always intervening with verbal feedback (Groeger, 2000). Parents should ask their teen what they think they need to work on so that self-reliance begins to develop.

Limited research is available about how much practice and what types of practice teens require to reduce the long-term risk of crash involvement. Therefore, information about how parents can improve practice sessions with their teens as they gain experience is also limited at this time. However, there are a number of novice driver training programs that have reduced behaviors that are linked to crashes (see Chapter 18). Hidden threats or hazards can prove most deadly (e.g., angle crashes at intersections). A simple example can serve here to explain what researchers mean by *hidden* or *hidden threats*. Suppose that a marked midblock crosswalk is up ahead. There is a parking lane on either side and two travel lanes, one in each direction. You, the driver, are approaching the crosswalk. A large SUV is parked right in front of the crosswalk. You as a driver should look for a pedestrian that might be entering the crosswalk but is hidden by the SUV in the parking lane. Experienced drivers are some six times more likely to look than novice drivers (Pradhan et al., 2005). This example is intended to give parents a feel for the types of hazards that novice teen drivers have a particularly hard time identifying. Other hazards that are problematic include blind bends and intersections (see Chapters 6, 18, and 28).

Teen drivers need to learn the types of hazards that may appear but also where and when to look inside and outside the vehicle (see Chapter 7). For example, teen drivers should keep their glances inside the vehicle as infrequent as possible and, when necessary, as short as possible. While it is never safe to take one's eyes off the forward roadway, some tasks require the driver to do so. Examples include turning on the defroster, adjusting the headlights, and looking for approaching cars in the rearview and side-view mirrors, just to name a few. The current literature suggests that if one must take one's eyes off the road for whatever reason, one should ideally do so for no longer than about 2 seconds at any one time (Simons-Morton et al., 2014). Parents could emphasize the importance of short, infrequent glances inside the vehicle during the supervised driving phase.

2.3.4 Emphasize the GDL Provisions in the Restricted and Unrestricted Phases

In addition to requirements related to driver supervision, GDL programs impose a number of other restrictions on novice drivers in both the restricted and the unrestricted phase. It is important for parents to know the restrictions because most often, parents become the primary monitors and enforcers of these restrictions. Moreover, there is evidence that teens are more likely to obey the restrictions if these are enforced by their parents (Simons-Morton et al., 2004).

First, consider the provisions during the restricted phase. These provisions are in place because they have been found to save lives (Chen et al., 2006; Masten et al., 2011; Russell et al., 2011; Vanlaar et al., 2009). The restrictions differ from country to country, state to state, and province to province, so parents should make sure they know which restrictions apply in their region (AAA, 2015). Three common rules during the restricted phase include not driving at night (Chapter 15; Williams et al., 2012), not driving with other teen passengers (Chapter 16; Chen et al., 2000; Ouimet et al., 2015), and not driving while intoxicated or under the influence of drugs (Chapter 13; Elvik, 2013). The Insurance Institute for Highway Safety (Highway Loss Data Institute, 2015) estimates that fatal crashes would be reduced by 9% if parents restricted driving after 8:00 p.m. (compared to midnight) and by 16% if parents were to allow no other teen passengers in the car (compared with just one other teen passenger).

Second, consider the one restriction that applies to drivers under 21 years of age, whether they hold a restricted or unrestricted license. In particular, for drivers under 21 years of age, there is a prohibition on driving after drinking any alcohol, which applies in all 50 states and the District of Columbia. For any given blood alcohol content (BAC), younger drivers are much more likely to be in a crash than are older drivers (Peck et al., 2008.)

2.3.5 How to Enforce the Restrictions in the Various GDL Phases

You might ask why we have devoted a separate section to enforcing the GDL phases, since we have already provided evidence that these phases lead to large reductions in crashes and that parents are essential for this enforcement. However, as most parents who have tried to enforce the restrictions know, this can be difficult. Fortunately, there are ways to help increase the likelihood that the behaviors appropriate to a given phase of the GDL program will be obtained.

First, consider the supervised driving phase. Parents may find that this phase is largely a confrontational one. Adolescent drivers desire independence from their parents and also want to be accepted by their peers (Senserrick & Howath, 2003). As a result, teens may feel as though their GDL program represents a form of parental control or oppression, and this could make for a negative experience. If learning to drive is perceived negatively, it makes sense that supervised driving would be avoided by either the parent or the teen, or both. So, what can be done? It has been shown that being mutually supportive of one another during the process of learning to drive is key to keeping the experience positive (Mirman et al., 2014). Thus, parents can do something to make supervised driving more enjoyable for both themselves and their teens.

Second, consider the restricted license phase. Research suggests that many parents do not appear to be aware of, or to appreciate, the risks of allowing teen passengers in the car of a novice teen driver (Williams et al., 2006). Much the same may be true of other parents' awareness of the additional restrictions. So, although we have emphasized that the risks of being in a crash are often doubled or tripled if these restrictions are not kept in place, you may find that other parents claim that the risks are small, acceptable ones. However, with this chapter, you now have the evidence you need to argue just how important the restrictions are. Moreover, in states where certain of the restrictions are not in place, you also now have the evidence you need to make a case to your teen that such restrictions are important.

But, this still does not mean that it will be easy for you to enforce these restrictions. Fortunately, several parental programs are available online that can assist with rule setting and GDL programs for teen drivers. We outline one here, which is the *Checkpoints* Program. The *Checkpoints* Program is designed to increase parental limits on young novice independent driving, especially under high-risk conditions. The *Checkpoints* Program makes use of a video, newsletters, and a parent–teen driving contract (Simons-Morton & Ouimet, 2006). Parents and teens in the *Checkpoints* Program were found more likely to report using a driving contract and to communicate about driving rules, and parents reported imposing significantly more restrictions on teens at licensure, and also 4 and 9 months after licensure. In a large trial conducted to test the efficacy of the *Checkpoints* Program, no effect of intervention was observed on crashes, but there was an indirect effect of the program in that parents imposed more restrictions that would be expected to reduce risky driving, violations, and crashes over the longer term.

2.3.6 Clearly Communicate about Rules and Expectations

In addition to GDL restrictions, parents may consider implementing additional rules related to driving (also see Chapter 19). Here are some examples of rules that might be negotiated with teens: getting permission to use the car; reporting where they are going and when they will be home; calling to report if they will be on time or late (but not while the car is in motion); not driving at night or with other passengers (when this is not part of the GDL program for a particular state or even in some cases when it is); restricting where and how far they are allowed to drive and what to do in difficult weather (e.g., snow and ice); restricting the use of cell phones, smartphones, and associated applications such as social media and games while driving; and not driving while using alcohol or drugs. Some parents put these rules in writing, and others discuss and set the ground rules. Parents may want to have a set of questions that they go through before their teen heads out or is given the keys and another set of questions when he or she returns the keys.

Regardless of the rules themselves, clear communication of the rules is essential. Not surprisingly, research indicates that parents and teens differ in terms of their understanding of what the rules are (Hartos et al., 2004). Only a little over half (37) of the 72 driving rules reported by teens and their parents showed agreement. A number of rules are "understood," and others are written down or "absolute." Making sure that the teen understands the rules is important; teens and parents may forget rules, have different understandings of which rules were agreed upon, or have different interpretations of which rules are absolute and which are flexible. Again, not surprisingly, there may sometimes be differences in opinion. Teens may view some rules as flexible while their parents do not, and vice versa (Hartos et al., 2004).

Another source of contention between parents and teens is the consequences of breaking rules and the fairness of those consequences (Hartos et al., 2004). It is important to discuss these consequences when these rules are first established, that is, whether the teen receives a warning, a restriction on future driving, a total loss of driving privileges, or grounding. Latitude for application of consequences depending on circumstances is the nature of administering parental justice; however, teens may interpret lenient consequences or your failure to follow through on restrictions as an indication that they will be able to get away with future rule infractions. To be effective, it is important that parents enforce the driving rules consistently while at the same time being open to special circumstances and willing to listen.

The I Promise Program (IPP), like *Checkpoints*, is one program that attempts to codify a set of rules and consequences. However, it consists of a broader set of rules than just those comprised by the GDL program. In particular, it consists of a parent–youth contract and a rear-window decal with a toll-free number connected to a community monitoring service, which allows other road users to call in to a professional call center to report on the driver's behavior. Items included in the contract cover driving sober, refraining from using a cell phone, and limiting the number of passengers. Driver behavior reports are mailed to the vehicle's registered owner after community members call in to report (Votta & MacKay, 2005). In an evaluation of the program, novice drivers reported feeling targeted, indicating that adults drive irresponsibly as well. However, many novice drivers and parents thought the contract in the IPP was a useful communication tool. In particular, contract restrictions on cell phone use, eating, drug/alcohol use while driving, vehicle maintenance, emergency materials, and number of passengers were found to be important. Many novice drivers disliked the presence of the decal because they thought it conveyed the impression that they were not good drivers. However, for both programs, it seems that the process of negotiating a parent–teen driving contract is helpful for parents trying to enforce driving rules. Based on these findings, we suggest that parents consider drafting a contract of sorts with their teen. Doing so ensures that parents and teens participate in a discussion about rules, their rigidity, and their consequences.

There are ways to monitor compliance with rules during unsupervised driving through the use of devices such as in-vehicle cameras or cell phone apps (Chapters 18 and 20). However, privacy and trust need to be discussed before imposing these systems on the teen driver. Parents must try their best to delicately balance trust, privacy, fairness, and firmness when discussing these rules with their teen (McCartt et al., 2007). We provide additional resources to the programs discussed at the end of this chapter that parents may find useful in aiding compliance with driving rules (please see Internet Information Links for Parents at the end of this chapter).

2.3.7 Let Your Teen Know That They Should Never Drive or Ride When Alcohol, Drugs, Distraction, or Sleep Deprivation Is Involved

Alcohol, drugs, distraction, and fatigue kill many drivers each year, including teens. For everyone, the general rule should be, never drive or ride when impaired, and don't let your friends do so either. As a ground rule, teens should be able to get a ride home from Mom or Dad with no questions asked at the time. An in-depth discussion of alcohol and drugs, distraction, and fatigue as crash contributors can be found in Chapters 13, 12, and 15, respectively. These crash contributors also relate in obvious ways to the restrictions in the GDL program at all stages.

First, consider alcohol. In 2013, 10,076 people were killed in crashes involving alcohol-impaired drivers in the United States, which translates into an average of one fatality every 52 minutes (NHTSA, 2014). Alcohol is perhaps the best-known contributor to crashes. According to a case–control study, crash risk begins to increase at a breath alcohol concentration or BrAC of 0.03, doubles at 0.05, triples at 0.07, and nearly quadruples at 0.08 (Compton & Berning, 2015). However, alcohol increases crash risk for young drivers more than for drivers of other age groups (Williams, 2003). As well, alcohol affects confidence, leading some people to underestimate their intoxication level after two or three drinks (Ronen et al., 2008), which could be particularly problematic given that teen drivers already tend to overestimate their driving ability.

It is important for parents to discuss the dangers of impaired driving (or riding with an impaired driver) with their teens, especially because teen drivers are already at an increased level of crash risk. However, if parents suspect that there is already an alcohol problem, there are resources available to them. The National Institute on Alcohol Abuse and Alcoholism has a website for parents that describes a number of these resources as well as the way to structure conversations with your teen (see Internet Information Links for Parents at the end of this chapter for the website).

Other than alcohol, marijuana is the most frequently detected drug in drivers involved in crashes (Chapter 13; Compton & Berning, 2015). The effect of marijuana on crash risk is difficult to evaluate using crash data because testing positive for THC may not indicate current impairment; it can take days (infrequent users) or weeks (frequent users) for THC to clear the body after ingestion (Cary, 2006). Studies show that marijuana impairment affects psychomotor skills, divided attention, lane tracking, and cognitive functions (Chapter 13; Ronen et al., 2008). As with alcohol, parents should discuss the dangers of drug-impaired driving (or riding with a drug-impaired driver) with their teens. Again, the

US federal government is an initial place to turn to for resources, in particular, the National Institute on Drug Abuse (see Internet Information Links for Parents at the end of this chapter for the website).

Distraction is another well-known contributor to crash risk. In 2013, 3,152 people were killed in crashes involving driver distraction, and an estimated 424,000 were injured (NHTSA, 2015). Distracted driving, particularly from handheld phone use and activities requiring the driver's eyes to be diverted from the roadway, also presents a high level of risk to drivers, particularly to novice and teen drivers (WHO, 2011, 2015). In 2013, 10% of all drivers 15–19 years old who were involved in fatal crashes were reported to be distracted when the crash occurred (NHTSA, 2015). Text messaging while driving is a particularly dangerous example of distracted driving; it has been shown to increase the odds of a safety-critical event by 10 times (Klauer et al., 2014; Simmons et al., 2016). Texting while driving prolongs and increases the frequency of glances away from the roadway and leads to slower responses to hazards, higher frequencies of crash involvement, and decreased lane-keeping accuracy (Caird et al., 2014). Survey data show that although drivers appear to be aware of the dangers of distracted driving and perceive other distracted road users as threats to personal safety, drivers still appear willing to continue engaging in their *own* distracted driving behaviors (Chapter 12). Teens should be encouraged to put their phones away while driving and only use them when the vehicle is safely parked. Again, parents need to discuss the dangers of distracted driving with their teens. Teens should be able to be reached by phone but not while driving. Teens should never answer, call, or text while driving. Instead, they should let the passenger do the communication or stop the car and call back. Additional information about distracted driving is available in the Internet Information Links for Parents at the end of the chapter.

Finally, drowsy or sleepy driving contributes to crash risk. In 2009, drowsy drivers were implicated in 832 fatal crashes (NHTSA, 2011). In one study, participants deprived of sleep for 24 hours suffered impairments in reaction time equivalent to participants who were rested but had a blood alcohol concentration of 0.05 (Maruff et al., 2005). Drowsy or sleepy driving is more frequent among young drivers than older drivers (Dahl, 2008; Shope & Bingham, 2008), as most adolescents do not sleep enough, particularly on school nights (Dahl, 2008). Consequently, risk may be elevated due to lapses in attention, falling asleep at the wheel, impaired judgment and decision making, amplified effects of alcohol due to sleep deprivation, and increased reactive aggression (Chapter 15; Dahl, 2008).

It is important for parents to be aware of their teen's sleeping patterns and prevent their teen from driving when they are drowsy or sleep deprived. This may be more easily said than done. There are a number of factors that cause insufficient sleep in adolescents, many of which can be controlled. Teens' bedtimes may be delayed due to competing demands on their time: extracurricular activities, school, work, and social demands, such as spending time with peers in the evening or communicating electronically (Moore & Meltzer, 2008). Early school starting times may compound the issue. As well, adolescents may maintain poor sleep hygiene, such as the inability to maintain a consistent sleep schedule, consumption of caffeine in the late afternoon or evening, and electronics before bedtime (Moore & Meltzer, 2008). Frequently, teens text friends at all hours of the night. Parents should have their teens turn off their phones so they can get uninterrupted sleep. Parents can encourage their teens to maintain a consistent bedtime to achieve good sleep hygiene, and to avoid overloading their schedules during the day. As well, parents can become involved in local, state, and national efforts to introduce later school start times, a measure that initially has proven very promising. The Centers for Disease Control (an agency of the US government) has more information on its website (see Internet Information Links for Parents at the end of this chapter).

2.3.8 Model the Way That You Want Your Teen to Drive

Teens generally adopt the driving styles of their parents. For example, when parents drive while distracted, their teens are likely to adopt similar behaviors (Prato et al., 2010). A number of studies have also found that parents' driving styles and attitudes are related to their children's styles (Bianchi & Summala, 2005). So, it is important for parents to know that teens model their parents' behavior—if parents set an example and reduce their own risk, their teen may reduce his or her own risk as well.

In addition, parents should be aware that their teens probably will not appreciate being required to follow parental rules that their parents violate themselves. In one study, teens became dissatisfied when their parents did not follow the rules that they expected their teens to follow (Votta & MacKay, 2005). So, it is important for parents to lead by example—failing to do so may not only impact their teen's driving habits but also lead to perceptions of unfairness. If necessary, have your teen help you with your own driving shortcomings, such as using your cell phone while driving.

2.3.9 Newer Midsize or Full-Size Cars Are the Best Vehicles for Your Teen

Parents should choose the vehicle for their teen with safety under consideration. Parents frequently state a preference for airbags (driver, passenger, side), antilock brakes, and electronic stability control (Hellinga et al., 2007). Ideally, a midsize, large, or full-size car would be chosen because generally, these provide teen drivers with more safety features. Although SUVs and pickups are large, they can be difficult to handle and may roll over. Small cars, while economical, do not necessarily provide as much safety as midsize or large cars even if the small car is newer (McCartt & Teoh, 2015).

However, parents often face practical considerations, such as price, insurance and maintenance costs, fuel economy, and availability of the family car (Hellinga et al., 2007). These can limit teen vehicle choices. Research shows that most often the teen in the family is assigned to drive an older-model vehicle that is already owned by the family. In one study, only about one-quarter of parents purchased an additional vehicle specifically for their teen, and 90% of these vehicles were used.

2.3.10 Keep in Mind That Your Teen May Want to Delay Licensure

Most teens want to drive once they are of legal age. However, these days, more teens are choosing not to for a number of reasons. Common reasons for choosing to delay licensing include the following: the ease of public transit, ridesharing with peers, the costs of having a car (e.g., vehicle maintenance, insurance, and operation), not getting enough practice to be licensed, and not finishing educational requirements (McCartt et al., 2007; Tefft et al., 2014).

The later your teen is licensed, the smaller the probability that he or she will be injured or die in an automobile crash while in his or her teen years. However, parents should not forget that this delay comes with a potential cost. There is a great deal that parents can do during the supervised driving phase of the GDL program, as described earlier in this chapter. This supervised driving phase is mandated by law for young teens. However, for teens 18 years or older, there is often no supervised driving phase, though the decrease in crash risk in the first 6 months due to delayed licensing is only marginal (Vlakveld, 2005). Thus, parents lose all control (legally) over the supervision of their children during the learner's permit phase since none exists. We would encourage parents to consider some modification of the activities that are described here in the permit phase for their older teen drivers. Again, we want to emphasize that most teens are clueless, not careless (McKnight & McKnight, 2003). There is every reason to involve yourselves as parents as much as possible in the education process.

2.4 Summary

In this chapter, we present 10 practical recommendations based on research on teen and novice drivers that parents should know and implement when their teen starts driving. In summary, parents should know that teen and novice drivers are at increased risk of crashes compared to other age groups. This is largely due to driver inexperience, which is best remedied by parents ensuring that their teens receive as much varied driving experience as possible (ideally 70-plus hours). Parents should know that they should take full advantage of the learner period for this purpose because it is the safest period for their teen drivers to gain the experience they need. Parents should also be mindful of their level of guidance—though feedback is useful, it should focus on improving one driving skill at a time. Based on scientific evidence from skill acquisition studies in other domains, it is our opinion that too much intervention can inhibit teens from developing the skills they need for independent driving.

Parents should also become familiar with GDL program requirements because, largely, it is up to parents to enforce these rules and restrictions. Parents should appreciate that these rules and restrictions are designed to mitigate risk for teen and novice drivers as they develop driving experience, so parents can help their teens remain safe by ensuring compliance with the requisite restrictions. If parents want to implement other rules, they should ensure that these rules (and their consequences) are clearly communicated. In all cases, rules need to be enforced in an authoritative manner in order to be most effective (although the ways that rules are enforced must also take into account personality variations that may have an impact on the characteristics of the teen–parent relationship).

Parents should know that alcohol, drugs, distraction, and fatigue are all significant contributors to crashes. Discussing the dangers of these behaviors and restricting teens from engaging in them is one way to address these issues. However, parents should also ensure that they set an example. For example, if parents do not want their teen to text while driving, they should keep from texting while driving as well. To reduce the risk of harm in the event of a crash, parents should select their teen's vehicle with safety in mind, ideally a larger car with more safety features.

Finally, parents should know that some teens opt to delay getting their license. This is something that parents should discuss with their teen. But, parents should realize that delaying their teen's licensure, while clearly decreasing the crash risk while the teen is unlicensed, will not lead to large decreases in the crash risk when the teen first becomes licensed. Parents need to be as involved in teaching their teens to drive at later licensure ages as they are at earlier licensure ages. This can be especially challenging given that there are no legal requirements for such supervision and given that their teens are well on their way to complete independence.

Acknowledgments

We would like to thank Lana M. Trick, Bill Horrey, and Bruce Simons-Morton for their extensive edits and comments, which strengthened the chapter. We would also like to acknowledge the AUTO21 Network of Centres of Excellence (NCE), which funded aspects of the research on which this chapter is based.

Internet Information Links for Parents

AAA Digest of Motor Laws: Information about graduated licensing laws in the United States and Canada, http://drivinglaws.aaa.com/tag/graduated-drivers-licensing/

Centers for Disease Control: Information about sleep and school, http://www.cdc.gov/media/releases/2015/p0806-school-sleep.html

Checkpoints Program: A parent-oriented program for teen drivers, http://www.injurycenter.umich.edu/programs/checkpoints-program%E2%84%A2

Distraction.gov: Information about distracted driving, http://www.distraction.gov/

Insurance Institute for Highway Safety: Information about teen drivers, http://www.iihs.org/iihs/topics/t/teenagers/topicoverview

I Promise Program: A parent-youth driving contract, http://www.ipromiseprogram.com/

National Highway Traffic Safety Administration: Information about teen drivers, http://www.nhtsa.gov/Teen-Drivers

National Institute on Alcohol Abuse and Alcoholism: Information about alcohol use, http://pubs.niaaa.nih.gov/publications/MakeADiff_HTML/makediff.htm

National Institute on Drug Abuse: Information about talking to your kids about drugs, http://www.drugabuse.gov/publications/marijuana-facts-parents-need-to-know/talking-to-your-kids-communicating risks

References

AAA. (2015). Digest of motor laws. Graduated driver's licensing. Accessed December 29, 2015, retrieved from http://drivinglaws.aaa.com/tag/graduated-drivers-licensing/

Bianchi, A., & Summala, H. (2005). The "genetics" of driving behavior: Parents' driving style predicts their children's driving style. *Accident Analysis & Prevention, 36*, 655–659.

Caird, J.K., Johnston, K., Willness, C., Asbridge, M., & Steel, P. (2014). A meta-analysis of the effects of texting on driving. *Accident Analysis & Prevention, 71*, 311–318.

Cary, P.L. (2006). The Marihuana Detection Window: Determining the Length of Time Cannibinoids Will Remain Detectable in Urine Following Smoking: A Critical Review of Relevant Research and Cannibinoid Detection Guidance for Drug Courts. National Drug Court Institute, Volume IV, Number 2. Accessed May 5, 2016, retrieved from http://www.ndci.org/sites/default/files/ndci/THC_Detection_Window_0.pdf

Chen L., Baker S.P., Braver E.R., & Li G. (2000). Carrying passengers as a risk factor for crashes fatal to 16- and 17-year-old drivers. *JAMA, 283*(12), 1578–1618.

Chen, L-H., Baker, S.P., & Li, G. (2006). Graduated driver licensing programs and fatal crashes of 16-year-old drivers. *Pediatrics, 118*, 56–62.

Compton, R.P., & Berning, A. (2015). *Traffic Safety Factors Research Note: Drug and Alcohol Crash Risk (Rep. No. DOT-HS-812-117).* Washington, DC: NHTSA.

Curry, A.E., Hafetz, J., Kallan, M.J., Winston, F.K., & Durbin, D.R. (2011). Prevalence of teen driving errors leading to serious motor vehicle crashes. *Accident Analysis & Prevention, 43*, 1285–1290.

Dahl, R.E. (2008). Biological, developmental, and neurobehavioral factors relevant to adolescent driving risks. *American Journal of Preventive Medicine, 35*(3), S278–S284.

Elvik, R. (2013). Risk of road accident associated with the use of drugs: A systematic review and meta-analysis of evidence from epidemiological studies. *Accident Analysis & Prevention, 60*, 254–67.

Ericsson, K.A. (Eds.) (1996). *The Road to Excellence: The Acquisition of Expert Performance in the Arts Sciences, Sports, and Games.* Mahwah, NJ: Erlbaum.

Gregersen, N., Berg, H., Engstrom, I, Noeln, S., Nyberg, A., & Rimmo, P. (2000). Sixteen years of age limit for Learner drivers in Sweden—An evaluation of safety effects. *Accident Analysis & Prevention, 32*, 25–35.

Gregersen N., Nyberg A., & Berg, H. (2003). Accident involvement among learner drivers—An analysis of the consequences of supervised practice. *Accident Analysis & Prevention, 35*, 725–730.

Groeger, J.A. (2000). *Understanding Driving: Applying Cognitive Psychology to a Complex Everyday Task.* London: Taylor and Francis.

Hartos, J.L., Shattuck, T., Simons-Morton, B.G., & Beck, K.H. (2004). An in-depth look at parent-imposed driving rules: Their strengths and weaknesses. *Journal of Safety Research, 35*, 547–555.

Hellinga, L.A., McCartt, A.T., & Haire, E.R. (2007). Choice of teenagers' vehicles and views on vehicle safety: Survey of parents of novice teenage drivers. *Journal of Safety Research, 38*, 707–713.

Highway Loss Data Institute. (2015). Teenagers. State laws. GDL crash reduction calculator. Ruckersville, VA: Insurance Institute for Highway Safety. Accessed December 29, 2015, retrieved from http://www.iihs.org/iihs/topics/laws/gdl_calculator?topicName=teenagers

Klauer, S.G., Guo, F., Simons-Morton, B.G., Ouimet, M.C., Lee, S.E., & Dingus, T.A. (2014). Distracted driving and risk of road crashes among novice and experienced drivers. *New England Journal of Medicine, 370*, 54–59.

Maruff, P., Felleti, M.G., Collie, A., Darby, D., & McStephen, M. (2005). Fatigue-related impairment in the speed, accuracy and variability of psychomotor performance: Comparison with blood alcohol levels. *Journal of Sleep Research, 14*(1), 21–27.

Masten, S.V., Foss, R.D., & Marshall, S.W. (2011). Graduated driver licensing and fatal crashes involving 16- to 19-year-old drivers. *JAMA, 306*(10), 1098–1103.

Mayhew, D.R., Simpson, H.M., & Pak, A. (2003). Changes in collision rates among novice drivers during the first months of driving. *Accident Analysis & Prevention, 35,* 683–691.

McCartt, A.T., & Teoh, E.R. (2015). Type, size and age of vehicles driven by teenage drivers killed in crashes during 2008–2012. *BMJ, 21,* 133–136.

McCartt, A.T., Shabanova, V.I., & Leaf, W.A. (2003). Driving experience, crashes and traffic citations of teenage beginning drivers. *Accident Analysis & Prevention, 35,* 311–320.

McCartt, A.T., Hellinga, L.A., & Haire, E.R. (2007). Age of licensure and monitoring teenagers' driving: Survey of parents of novice teenage drivers. *Journal of Safety Research, 38,* 697–706.

McKnight, A.J., & McKnight, A.S. (2003). Young novice drivers: Careless or clueless? *Accident Analysis & Prevention, 35,* 921–925.

Mirman, J.H., Curry, A.E., Wang, W., Fisher Thiel, M.C., & Durbin, D.R. (2014). It takes two: A brief report examining mutual support between parents and teens learning to drive. *Accident Analysis & Prevention, 60,* 23–29.

Moore, M., & Meltzer, L.J. (2008). The sleepy adolescent: Causes and consequences of sleepiness in teens. *Paediatric Respiratory Reviews, 9,* 114–121.

Mulvihill, C., Senserrick, T., & Haworth, N. (2005). *Development of a Model Resource for Parents as Supervisory Drivers (MUARC Rep. No. 243).* Clayton, Victoria, Australia: Monash University Accident Research Centre.

National Highway Traffic Safety Administration (NHTSA) (2011). *Traffic Safety Facts: Drowsy Driving (Rep. No. DOT-HS-811-449).* Washington, DC: NHTSA.

National Highway Traffic Safety Administration (NHTSA) (2014). *Traffic Safety Facts, 2012 Data: Young Drivers (Rep. No. DOT-HS-812-019).* Washington, DC: NHTSA.

National Highway Traffic Safety Administration (NHTSA) (2014). *Traffic Safety Facts, 2013 Data: Alcohol-Impaired Driving (Rep. No. DOT-HS-812-102).* Washington, DC: NHTSA.

National Highway Traffic Safety Administration (NHTSA) (2015). *Traffic Safety Facts: Distracted Driving 2013 (Rep. No. DOT-HS-812-132).* Washington, DC: NHTSA.

Ouimet, M.C., Pradhan, A.K., Brooks-Russell, A., Ehsani, J.P., Berbiche, D., & Simons-Morton, B.G. (2015). Young drivers and their passengers: A systematic review of epidemiological studies on crash risk. *Journal of Adolescent Health, 57,* S24–S35.

Peck, R.C., Gebers, M.A., Voas, R.B., & Romano, E. 2008. The relationship between blood alcohol concentration (BAC), age, and crash risk. *Journal of Safety Research, 39,* 311–319.

Pradhan, A.K., Hammel, K.R., DeRamus, R., Pollatsek, A., Noyce, D.A., & Fisher, D.L. (2005). The use of eye movements to evaluate the effects of driver age on risk perception in an advanced driving simulator. *Human Factors, 47,* 840–852.

Prato, C.G., Toledo, T., Lotan, T., & Taubman-Ben-Ari, O. (2010). Modeling the behavior of novice young drivers during the first year after licensure. *Accident Analysis & Prevention, 42,* 480–486.

Ronen, A., Gershon, P., Drobiner, H., Rabinovich, A., Bar-Hamburger, R., Mechoulam, R., Cassuto, Y., & Shinar, D. (2008). Effects of THC on driving performance, physiological state and subjective feelings relative to alcohol. *Accident Analysis & Prevention, 30*(3), 926–934.

Russell, K.F., Vandermer, B., & Hartling, L. (2011). Graduated driver licensing for reducing motor vehicle crashes among young drivers (Review). *The Cochrane Library, 10.*

Sagberg, F., & Bjørnskau, T. (2006). Hazard perception and driving experience among novice drivers. *Accident Analysis & Prevention, 382,* 407–414.

Senserrick, T., & Haworth, N. (2003). *Review of Literature regarding National and International Young Driver Training, Licensing and Regulatory Systems (Rep. No. 239).* Clayton, Victoria, Australia: Monash University Accident Research Centre.

Shope, J.T., & Bingham, R. (2008). Teen driving: Motor-vehicle crashes and factors that contribute. *American Journal of Preventive Medicine, 35*(3S), S261–S271.

Simmons, S., Hicks, A., & Caird, J.K. (2016). Safety-critical events associated with cell phone tasks measured through naturalistic driving studies: A systematic review and meta-analysis. *Accident Analysis & Prevention, 87,* 161–169.

Simons-Morton, B., & Ouimet M.C. (2006). Parent involvement in novice teen driving: A review of the literature. *Injury Prevention, 12,* i30–i37.

Simons-Morton, B.G., Hartos, J.L., & Beck, K.H. (2004). Increased parent limits on teen driving: Positive effects from a brief intervention administered at the motor vehicle administration. *Prevention Science, 5*(2), 101–111.

Simons-Morton, B.G., Guo. F., Klauer, S.G., Ehsani, J.P., & Pradhan, A.K. (2014). Keep your eyes on the road: Young driver crash risk increases according to the duration of distraction. *Journal of Adolescent Health, 54,* S61–S67.

Tefft, B.C., Williams, A.F., & Grabowski, J.G. (2014). Driver licensing and reasons for delaying licensure among young adults ages 18–20, United States, 2012. *Injury Epidemiology, 1,* 4.

Vanlaar, W., Mayhew, D., Marcoux, K., Wets, G., Brijs, T., & Shope, J. (2009). An evaluation of graduated driver licensing programs in North America using a meta-analytic approach. *Accident Analysis & Prevention, 41,* 1104–1111.

Vlakveld, W. (2005). *Young, Novice Motorists, Their Crash Rates, and Measures to Reduce Them: A Literature Study.* Leidschendam, The Netherlands: SWOV.

Votta, E., & MackKay, M. (2005). Evaluating the acceptability and feasibility of the I Promise Program: A driving program for families with young new drivers. *Injury Prevention, 11,* 369–372.

Williams, A.F. (2003). Teenage drivers: Patterns of risk. *Journal of Safety Research, 34,* 5–15.

Williams, A.F., Leaf, W.A., Simons-Morton, B.G., & Hartos, J.L. (2006). Parents' views of teen driving risks, the role of parents, and how they plan to manage the risks. *Journal of Safety Research, 37,* 221–226.

Williams, A.G., West, B.A., & Shults, R.A. (2012). Fatal crashes of 16- to 17-year-old drivers involving alcohol, nighttime driving, and passengers. *Traffic Injury Prevention, 13,* 1–6.

World Health Organization (WHO). (2011). *Mobile Phone Use: A Growing Problem of Driver Distraction.* Geneva, Switzerland: World Health Organization.

World Health Organization (WHO). (2015). *Global Status Report on Road Safety 2015.* Geneva, Switzerland: World Health Organization.

3 For Teens

Donald L. Fisher

Anna James Duffy Fisher

Jennifer Margaret Duffy Fisher

Abstract

It is not often that a few brief seconds are of any consequence. But in an automobile, as you know all too well, just a few seconds can mean the difference between life and death. What can we possibly tell you that you don't already know? Wear your seat belts? But you know this. Don't speed? But you know this. We will skip almost all of what we think you know. But we believe that there are a few critical things that you may not know and so take this opportunity to address them. This knowledge can give you the few seconds you need to save your life or someone else's.

3.1 Don Fisher

3.1.1 Fall 1968

Nineteen years old. I was traveling 60 miles an hour, upside down, in a metallic blue Volkswagen Fastback. Above my head, I could see sparks flying where the road had scraped through the roof. Then all went black. I was wearing my seat belt. I walked away with only a chipped fingernail. The car was not so lucky. Although I was traveling under the speed limit, I was going too fast for conditions (fog). The road just disappeared. And then it reappeared, only it curved sharply to the left. I steered, oversteered, and flipped. Oversteering is an all too common occurrence among novice drivers. But no one told me, let alone taught me what to do in the situation I faced.

3.1.2 Fall 1986 and Fall 1999

My wife and I were blessed with two beautiful daughters, first Jennifer and then Annie (who are now coauthors of this chapter). For the entire first 6 months of their lives, I think I knew to the day how old each one was—3 months and 21 days, 3 months and 22 days, and so on.

3.1.3 Fall 1994

I was a professor in the Department of Industrial Engineering and Operations Research at the University of Massachusetts Amherst. Together with some colleagues, we raised money to purchase a driving simulator costing in the neighborhood of half a million dollars, thanks in no small measure to help from the National Science Foundation. Driving simulators were rare back then.

The press was on campus the day it was installed. The room was abuzz with activity. In the middle of everything, I received a telephone call from a woman whom I did not know. She said something that sounded like "3 months and 21 days." I assumed that she was talking about a newborn, but I could not understand why the woman was already interested in driving. And then, 2 minutes into the conversation, I understood. It had been 3 months and 21 days since her 16-year-old daughter had died in an automobile crash.

3.1.4 Fall 1997

I received a grant from the AAA Foundation for Traffic Safety (AAA FTS) to evaluate a program available on a CD that, it was hoped, would ultimately reduce novice driver crashes. We could use the driving simulator to evaluate the effectiveness of the training. Although we can't determine what will happen when the trained drivers are out on the open road, we can determine on the driving simulator whether the training decreases those behaviors that lead to the spike in crashes just after teens get their solo (restricted) license. The AAA FTS program works, the first one in the history of driver education to impact the behaviors that lead to crashes (Fisher et al., 2002). That program is still available on the AAA FTS website (DriverZED) as a DVD (https://www.aaafoundation.org/content/driver-zed%C2%AE-30-computer-game). Annie actually got higher scores on this than her father, but I will never admit to that.

3.1.5 Spring 2000

My wife received a call after having an x-ray of her femur. There was a mass that needed checking. Our two daughters, Annie and Jenny, were respectively 10 and 15 at the time. She died of cancer 20 months later, on Valentine's Day. The loss is forever. She has missed so much. And she was 52.

3.1.6 Spring 2009—Hazards I Did Not Anticipate

By the spring of 2009, there were a number of researchers around the world interested in why it was that novice drivers are six times more likely to crash during the first month of solo driving than they are just 1 year later. Before the turn of the century, the general impression among adults (and maybe even teens) was that the crashes were largely due to teens being careless, willing to take risks that older drivers would not take. But that attitude was changing rapidly. It was becoming more and more clear that teens did not know where to look in order to avoid the most likely crashes and then take actions to decrease the likelihood that a crash would occur (McKnight & McKnight, 2003). They weren't necessarily careless, simply untrained. The fault lay with us, the adults, not the teens. We don't throw a child into the water who can't swim. But we were putting our children on the road without giving them the skills that they needed to avoid crashes.

The results from a number of studies confirmed that what are now called hazard anticipation skills were closely linked to the likelihood of crashing (Horswill & McKenna, 2004). Moreover, the results from other studies had long confirmed that novice drivers—teens—are much less likely to anticipate hazards than experienced drivers (Borowsky et al., 2007; McKenna & Crick, 1991). The only question now was whether novice teen drivers could be trained in a short period of time (say 30 minutes) to anticipate hazards better than they did without training. As you can imagine, no one was sure that

FIGURE 3.1 (See color insert.) Amity St., Amherst, Massachusetts. (A multithreat scenario where a minivan parked in front of a library obscures a driver's view of pedestrians that might enter the crosswalk.)

something like hazard anticipation, which takes well over a year to be learned when a teen is behind the wheel, could be learned in anything like 30 minutes on a PC and then retained for a year.

But it turns out that a short computer training program actually can help teen drivers predict and react to dangers. For example, consider the scenario in Figure 3.1. Several large vehicles are parked close to a marked crosswalk that leads from the town library in Amherst (our hometown) to shops across the street. Pedestrians are hidden by these vehicles to drivers approaching the crosswalk. A research team created the same scenario on a driving simulator as appears in Figure 3.1 and compared how likely an experienced driver was to look for a pedestrian with how likely a novice driver was to look for a pedestrian (Pollatsek et al., 2006). In that experiment, experienced drivers were six times more likely to look for the pedestrian than were novice drivers (we used an eye tracker to determine where the driver was glancing). There were many other hidden hazards embedded in the simulation where, as in the example I just gave, novice drivers glanced less frequently than experienced drivers. As already noted, researchers believe that this is largely because novice drivers were not taught to do such.

So, a team of researchers developed and evaluated a training program that taught novice drivers to anticipate hazards, which takes only 30 minutes to administer. It turns out that the training produces better hazard anticipation not only in a driving simulator (Pollatsek et al., 2006; Pradhan et al., 2009) but also on the open road up to a year after training (Taylor et al., 2011). This program is available to download for free (http://www.ecs.umass.edu/hpl/software.html#RAPT). Other programs based on it are available and free as well. The one developed by State Farm has been evaluated and proves equally effective at improving hazard anticipation skills (http://teendriving.statefarm.com/road-aware).

3.1.7 Fall 2011—Hazards I Did Not See

I thought I knew it all. Hazard anticipation was the one skill that clearly could be taught. What else did I need to know? What other skill figured critically in crashes and could possibly be learned in a matter of minutes? As we just learned, it is certainly the case that if you are looking at the road and don't predict a danger—say, a car pulling out from behind a truck idling by the side of the road—you're more likely to get into a crash. That's not surprising. What is surprising is just how little time you can spend looking away from the road before putting yourself at greatly inflated risk for a crash.

A landmark study was reported in 2006 in which 100 cars were instrumented with cameras and sensors to detect speed, acceleration, braking, and a whole host of other variables (Klauer et al., 2006). The 200 drivers of these vehicles were recorded for over 2,000,000 vehicle miles, totaling almost 43,000 hours of data. It was determined that glances inside the vehicle over 2 seconds in length were associated with an increase in crash risk by a factor of three. In other words, spending 2 seconds to glance at an incoming text message could make you three times more likely to get in a car crash. It can take you that long just to find the location of the text on your smartphone display! Other studies on a driving simulator have shown that the crash risk is greatly inflated for glances greater than just 1.6 seconds (Horrey & Wickens, 2007). While there is no such thing as a safe glance away from the roadway (Liang et al., 2012), the existing evidence suggests that glances longer than 2 seconds are unacceptably risky (National Highway Traffic Safety Administration, 2012).

Again, researchers asked whether a simple PC-based training program could be developed that would teach drivers how to behave more safely, in this case, how to limit their glances to less than 2 seconds. The answer is yes, that such a training program can be developed, and it has effects both on a driving simulator (Divekar et al., 2013) and in the field (Pradhan et al., 2011) for up to 3 months. The program is available as a free download (http://www.ecs.umass.edu/hpl/software.html#RAPT). And, if you are lucky enough to live in the Commonwealth of Massachusetts, Arbella Insurance's Distractology 101 program lets drivers experience for themselves the dangers of especially long glances inside the vehicle.

3.2 Annie Fisher—Winter 2010

I learned how to drive in the snow. I used to make my dad take me to the high school parking lot after a big snowstorm and test the brakes. I wanted to know how far the car would go on snow. The first few times, we were barely going fast enough to need to stop short. By the third and fourth times, I was going faster and was startled as five, six, seven parking spaces whizzed by before we came to a complete halt. Stopping distances, speeds—all the things my dad had talked about were starting to become more of a reality.

It wasn't until I hit a patch of black ice merging onto I-94 that I understood what it felt like to be truly out of control of the car. My friend and I had left early in the morning to catch the first flight out of Detroit. Michigan is not known for plowing their roads, and it had only just begun snowing. I knew that the roads would be bad, so I was going slower than I generally would. As we rounded the bend on the entrance ramp, we hit the black ice. We started sliding off the road to the right. Instead of turning into the skid, like I'd been lectured on, I overcorrected and steered toward the merging lane. We spun around once and then went head first into the Jersey barrier blocking the merging lane with the four-lane highway. We had been going maybe 25 miles per hour at the time, fast enough so that I wasn't able to control the car but not fast enough to deploy the airbags. When the cop came, he checked to make sure the two of us were alright. Neither of us was injured, just stunned. I had just totaled someone else's car.

The cop drove the car to the side of the road, and the EMTs checked us out. As we were discharged from the ambulance, the cop called out to the EMTs, "See you at the next one." The EMTs looked at him and said, "We'll be there."

(Don: Do you know what it means to steer into a skid? And do you know what to do as soon as the tires begin to gain traction? Life and death—just a few seconds separate them.)

3.3 Jenny Fisher—Winter 2016

We are very glad that my father decided to wear his seat belt that day nearly 50 years ago. It's hard to imagine that he was once a teenager—and that I will have a teenager myself one day, too. As I write this, my son is 4 months old, sleeping in an elephant onesie in his crib. He's just learning to roll over, and we're captivated by his beautiful smile. One day, though, I'll teach him to drive—like my father once taught me.

At 16, I may not have been the easiest pupil. I didn't want to listen to my father's advice to watch for pedestrians at crosswalks, turn off the music as I merged onto the highway, and slow down on the curves (yes, I clipped a mailbox or two). Trying to learn stick shift at the same time as I learned to drive made it even tougher. Fifteen years later, I still don't know how to drive a car with a stick shift, but I can't back out of my driveway without hearing my father's voice in my head, telling me to watch for hidden dangers and put down the phone. With my infant son strapped in the back, it suddenly seems much more important.

3.4 Annie, Jenny, and Don

If you are intrigued by what you have read here, you might want to glance at Chapter 18, where other advances in driver training and education are discussed, or look at other chapters where methods and information are described that can help you understand the risk of traveling with teen passengers or at night, of driving while fatigued or even with very low levels of alcohol, or of speeding and being overconfident. Or go online to test your skills on any of the driving programs we mentioned.

Acknowledgments

Don would like to thank Volkswagen for making three-point harnesses a standard feature in their cars back in 1968 for idiots like him. A day doesn't pass when he does not look back and wonder what if...

References

Borowsky, A., Shinar, D., & Oron-Gilad, T. (2007). Age, skill, and hazard perception in driving. *Proceedings of the Fourth International Driving Symposium on Human Factors in Driver Assessment, Training and Vehicle Design*. Iowa City, IW: University of Iowa.

Divekar, G., Pradhan, A. K., Masserang, K. M., Pollatsek, A., & Fisher, D. L. (2013). A simulator evaluation of the effects of attention maintenance training on glance distributions of younger novice drivers. *Transportation Research F*, 20, 154–159.

Fisher, D. L., Laurie, N. E., Glaser, R., Connerney, K., Pollatsek, A., Duffy, S. A., & Brock, J. (2002). The use of an advanced driving simulator to evaluate the effects of training and experience on drivers' behavior in risky traffic scenarios. *Human Factors, 44*, 287–302.

Horrey, W., & Wickens, C. (2007). In-vehicle glance duration. *Transportation Research Record, 2018*, 22–28.

Horswill, M., & McKenna, F. (2004). Drivers' hazard perception ability: Situation awareness on the road. In S. Banbury & S. Tremblay (Eds.), *A Cognitive Approach to Situation Awareness* (pp. 155–175). Aldersot, UK: Ashgate.

Klauer, S., Dingus, T., Neale, V., Sudweeks, J., & Ramsey, D. (2006). *The Impact of Driver Inattention on Near-Crash/Crash Risk: An Analysis Using the 100-Car Naturalistic Driving Study Data*. Washington, D.C.: National Highway Traffic Safety Administration.

Liang, Y., Lee, J., & Yekhsatyan, L. (2012). How dangerous is looking away from the forward roadway: Algorithms predict crash risk from glance patterns in naturalistic driving. *Human Factors, 54*, 1104–1116.

McKenna, F., & Crick, J. (1991). Experience and expertise in hazard perception. In G. Grayson and J. Lester (Eds.), *Behavioural Research in Road Safety 1990* (pp. 39–46). Crowthorne, England: Transport Research Laboratory.

McKnight, A., & McKnight, S. (2003). Young novice drivers: Careless or clueless? *Accident Analysis & Prevention, 35*, 921–925.

National Highway Traffic Safety Administration. (2012). *Visual-Manual NHTSA Driver Distraction Guidelines for In-Vehicle Devices*. Washington, DC: National Highway Traffic Safety Administration. Retrieved January 1, 2015, from http://www.distraction.gov/downloads/pdfs/visual-manual-nhtsa-driver-distraction-guidelines-for-in-vehicle-electronic-devices.pdf.

Pollatsek, A., Narayannan, V., Pradhan, A., & Fisher, D. (2006). The use of eye movements to evaluate the effect of PC-based risk awareness training on an advanced driving simulator. *Human Factors, 48*, 447–464.

Pradhan, A., Pollatsek, A., Knodler, M., & Fisher, D. (2009). Can younger drivers be trained to scan for information that will reduce their risk in roadway traffic scenarios that are hard to identify as hazardous? *Ergonomics, 52*, 657–673.

Pradhan, A. K., Divekar, G., Masserang, K., Romoser, M., Zafian, T., Blomberg, R. D., Thomas, F. D., Reagan, I., Knodler, M., Pollatsek, A., & Fisher, D. L. (2011). The effects of focused attention training (FOCAL) on the duration of novice drivers' glances inside the vehicle. *Ergonomics, 54*, 917–931.

Taylor, T., Masserang, K., Divekar, G., Samuel, S., Muttart, J., Pollatsek, A., & Fisher, D. (2011). Long term effects of hazard anticipation training on novice drivers measured on the road. *Proceedings of the 6th International Driving Symposium on Human Factors in Driver Assessment, Training, and Vehicle Design 2011.* Iowa City: Public Policy Center, University of Iowa.

4

How Science Informs Engineering, Education, and Enforcement: A Message for Driving Instructors

Joost C. F. de Winter

Natália Kovácsová

Abstract

The aim of this chapter is to illustrate to driving instructors how science contributes to cumulative knowledge on road safety. We do this by reviewing a scientific study for each of the three classical *Es* of road safety: (1) education, (2) enforcement, and (3) engineering.

Regarding education, we review the DeKalb experiment from the 1980s, which was a large-sample randomized controlled trial that studied the effect of driver education on postlicense crash rates. The DeKalb experiment showed that participants who were assigned to a state-of-the-art driver education program performed better on theory and road tests, and became licensed sooner than control participants who did not receive formal driving instruction. Although the state-of-the-art education improved these target outcomes, there is no consistent evidence that it reduced crash risk. The recent consensus is that theoretical knowledge and skillful maneuvering alone are not sufficient for safe driving. Drivers should also have postlicense on-road experience and the lifestyle and attitudes that contribute to a safe driving style.

Regarding enforcement, we describe a UK study from the late 1990s on the statistical reliability of the formal road test. In this study, driving test candidates were asked to retake the test with a different examiner. The results showed surprisingly low consistency between the two tests, indicating that an assessment of a 30-minute drive might not be trustworthy. We provide several recommendations (such as increasing the test duration and implementing standardized routes and checklists) for improving the reliability of road testing. Furthermore, the value of computerized testing (e.g., hazard perception testing) and long-term data collection (e.g., in-vehicle driver state monitoring) is addressed.

Regarding engineering, the growing prevalence of active safety systems in vehicles has raised the question of how to treat such technologies in driver education curricula. A study on electronic stability control (ESC) was reviewed to illustrate how advances in technology improve road safety and affect elements of on-road training. In the case of ESC, skid training has become less relevant, but it is unknown whether learner drivers should experience critical driving situations during which the ESC gets activated. This may foster their overconfidence.

4.1 Introduction

Worldwide, 1.3 million fatal road traffic crashes occur on a yearly basis, making road injuries the eighth leading cause of death (Lozano et al., 2013). Young drivers are overrepresented, with 20–30% of the traffic fatalities resulting from crashes involving a driver under the age of 25 (Organisation for Economic Co-operation and Development [OECD], 2006). Fortunately, the high-income countries are making great strides in improving road safety (for more information, see Chapter 23). The ongoing implementation of road safety measures allows the setting of strict safety targets, with the long-term goal of zero fatalities in traffic (Rosencrantz et al., 2007).

4.1.1 The Three *E*s: Education, Enforcement, and Engineering

Road safety measures are traditionally categorized into the three *E*s: education, enforcement, and engineering (Learoyd, 1950; McKenna, 2012; Rothengatter, 1982). We define education as those mechanisms that intend to improve the knowledge and behavior of road users. This includes on-road practice, classroom courses, and mass media road safety campaigns (Beanland et al., 20113; Wakefield et al., 2010). Emerging methods such as simulator-based training (e.g., De Winter et al., 2009; Park et al., 2015) and in-vehicle monitoring systems that allow for real-time or postdrive feedback (e.g., Musicant & Lampel, 2010) also belong to the category of education (for further information, turn to Chapters 18 and 20). In North America and Australia, the term *driver education* is often used in reference to formal in-class and in-vehicle training prior to licensed driving (e.g., Mayhew & Simpson, 2002). Thus, driver education encompasses, and has a broader meaning than, driver training (see also Beanland et al., 2013). However, McKenna (2010) argued that in practice, people do not recognize the difference between the words *training* and *education*. In the present chapter, we use the term *education* for both classroom teaching and on-road instruction.

Enforcement includes the development and application of laws and regulations that aim to eliminate undesirable behaviors. Enforcement concerns not only such salient measures as police patrolling and speed cameras, but also driver testing, restricted driving in graduated driver licensing, breath alcohol testing, traffic regulations, vehicle safety standards and regulations, and laws regarding road design (Groeger & Banks, 2007; Zaal, 1994).

Engineering refers to the invention, design, construction, and modification of physical systems. Examples are modifications in road design such as black-spot treatments and traffic calming measures

(Elvik et al., 2009); the introduction of passive safety systems such as airbags and crumple zones; and, more recently, the introduction of active safety systems such as driver assistance and automation technology (e.g., Lee, 2007).

4.1.2 The Three *E*s and Driving Instructors

Among the three *E*s, driving instructors are probably most familiar with the first *E*, education. It is important that instructors know the scientific consensus and apply evidence-based education, not unlike clinicians who practice evidence-based medicine. However, education cannot be understood in isolation from the other two *E*s. After all, drivers drive in *engineered* vehicles and have to pass a formal driving test before being allowed to drive independently. Another example of the interaction between the three *E*s concerns the safety effectiveness of seat belts. Research has shown that the mere legislation of this technology in the 1970s (mandating that seat belts are installed in new cars and that it is compulsory to wear them) had limited effectiveness. It required substantial further investments in publicity campaigns and enforcement to ensure that people actually started wearing seat belts (Jonah et al., 1982; Mäkinen & Hagenzieker, 1991; Williams & Wells, 2004). Thus, driving instructors need to be familiar not only with the science behind education but also with issues of enforcement and engineering.

4.1.3 Aim of This Chapter

The aim of this chapter is to illustrate to driving instructors and other practitioners how the scientific method contributes to the development of road safety knowledge. We do this by describing three example scientific studies, one each in the areas of education (Stock et al., 1983), enforcement (Baughan & Simpson, 1999), and engineering (Farmer, 2006). For each of the three studies, we show the main results and explain the relevance for driving instructors. Furthermore, we discuss the limitations of these studies in an attempt to shed light on the limits of the acquired knowledge.

4.2 Education: Why Driver Education Sometimes Fails to Reduce Crashes

4.2.1 Prelicense Driver Education

One of the measures aiming to reduce novice driver crashes is prelicense driver education. The assumption that driver education produces safe drivers led to the introduction of formal driver education as a part of the licensing process in the first half of the twentieth century. The popularity of driver education grew in the 1950s and 1960s, stimulated by evaluation studies reporting that driver education was effective in reducing novice drivers' crash risk (see Mayhew, 2007 for a review). However, most of the early studies suffered from serious methodological weaknesses (e.g., no randomized controlled designs, small sample sizes), which means that the validity of their results is questionable.

4.2.2 Evaluation of Driver Education Effectiveness: The DeKalb Study

As a response to the growing popularity of driver education but ongoing concerns about its effectiveness, the National Highway Traffic Safety Administration (NHTSA) designed a state-of-art education program and a corresponding experiment to determine the effect of this program on road safety (Stock et al., 1983). This study took place between December 1977 and June 1981 in DeKalb County, Georgia. Herein, we report the results of the NHTSA final report (Stock et al., 1983) and reanalyses conducted by the Insurance Institute for Highway Safety (Lund et al., 1986) and by R. C. Peck & Associates (Peck, 2011).

We selected the DeKalb study as an illustration of a well-designed experiment. It had a large sample size and used a stratified randomization procedure for assigning participants to groups. Random assignment is considered to be a gold standard for investigating cause–effect relationships by ensuring that each participant has an equal chance of being placed in any group. Thus, at the end of the study, differences between groups can confidently be attributed to the effects of the experimental treatment (i.e., the type of driver education) on the dependent variables (i.e., indices of the effect of training). Another strength of the DeKalb study was that it evaluated educational effectiveness on measures of actual safety (i.e., crash and violation records from the Georgia Department of Administrative Services).

Students who had reached the age of 15 years (i.e., the legal licensing age), who did not already have a driver's license, who were not already participating in driver education, and who were motivated to obtain their driver's license as soon as possible could apply to participate in the DeKalb study (Stock et al., 1983). Over 16,000 secondary school students were randomly assigned to either of two educational groups or one control group while they were matched for sex, socioeconomic status, and grade point average (GPA). Students assigned to the first educational group participated in an advanced driver education program called the Safe Performance Curriculum (SPC). The SPC "was developed in such a way that it represented the best that the driver education community and its supporting scientific and technical resources had to offer as an accident countermeasure" (Riley & McBride, 1974, p. 5). Specifically, the SPC group received about 70 hours of formal education, consisting of three modes of formal instruction: (1) classroom instruction, including film-based driving simulation instruction, (2) instruction on a driving range, focusing on the initial development of vehicle control skills, skills in interacting with various roadway configurations, and emergency skills, and (3) on-road training focusing on the enhancement of the skills required in actual traffic. These types of formal instruction were complemented by practice-with-parents sessions and by guided learning designed to respond to individual needs (Riley & McBride, 1974; Weaver, 1978). In guided learning, the students could interact with an instructor during waiting intervals (e.g., when another group of students received the film-based driving simulation instruction). The duration of the in-vehicle instruction provided by the DeKalb study (range instruction and on-road training) was approximately one-third of the total time of formal instruction, whereas the remaining two-thirds was devoted to in-class education. The second group received a 20-hour education, which was called the predriver licensing curriculum (PDL) (Riley & McBride, 1974; Stock et al., 1983). The PDL aimed to develop only those skills and knowledge necessary for passing the driving test and covered less safety content. For example, the modules on hazard perception, alcohol and drugs, and skid control were not treated in the PDL. The control group did not receive any education provided by the DeKalb study (Stock et al., 1983; Weaver, 1978). It was expected that students assigned to this group were taught to drive by their parents or friends, or in commercial driving schools.

The results reported by Stock et al. (1983) were as follows:

- *Crashes and violations per assigned student.* About one year after the completion of the project, there were no statistically significant differences in the number of violations and crashes per student between the educational groups and the control group (Table 4.1 and Figure 4.1).
- *Crashes and violations per student who completed the course and obtained a driver's license.* During the first 6 months of licensed driving, there were slightly fewer crashes (average of 0.1021 [n = 3545], 0.1010 [n = 3375], and 0.1221 [n = 4135] for the SPC, PDL, and control groups, respectively) and violations (average of 0.1391, 0.1425, and 0.1753, respectively) for students in the SPC and PDL groups than for students in the control group, when analyzing only those students who had completed the SPC/PDL course and subsequently became licensed. These results are in line with the work of Peck (2011), who similarly concluded that the DeKalb study showed evidence of a small short-term crash and violation reduction *per licensed driver.* However, one limitation of these statistics is that not all assigned students actually completed the SPC/PDL course. The possibility that the more motivated/competent students completed the course, and hence skewed the results, cannot be ruled out. Stock et al. (1983) explained that "the percent of high GPA students among

TABLE 4.1 Crashes and Violations of All Assigned Students

		Crashes		Violations	
	Number of Assigned Students	% of Students with at Least One Crash	Mean Crashes per Student	% of Students with at Least One Violation	Mean Violations per Student
SPC	5464	28.61	.3776	45.59	.9771
PDL	5430	26.46	.3611	44.51	.9565
Control	5444	26.75	.3643	43.37	.9772

Source: Data from Tables II-7, II-8 and II-12 in Stock, J. R. et al., *Evaluation of Safe Performance Secondary School Driver Education Curriculum Demonstration Project* (Final Report DOT-HS-6-01462), National Highway Traffic Safety Administration, Washington, DC, 1983.

Note: PDL, predriver licensing; SPC, Safe Performance Curriculum. The crash and violation data were current as of December 1981 and December 1982, respectively.

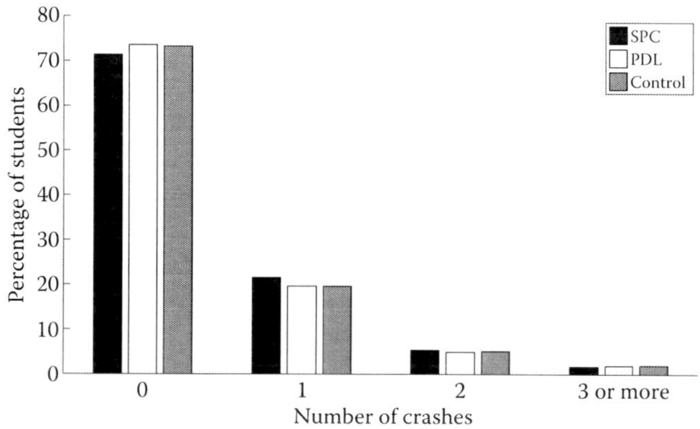

FIGURE 4.1 Distribution of the number of crashes for all assigned students (n = 5464 for Safe Performance Curriculum [SPC], n = 5430 for predriver licensing [PDL], n = 5444 for the control group). The crash and violation data were current as of December 1981 and December 1982, respectively. (Data from Table C-1 in Stock, J. R. et al., *Evaluation of Safe Performance Secondary School Driver Education Curriculum Demonstration Project* [Final Report DOT-HS-6-01462], National Highway Traffic Safety Administration, Washington, DC, 1983.)

the SPC group, 65.3 percent high GPA, and the PDL group, 65.8 percent high GPA, is somewhat higher than among the Control group, 59.6 percent high GPA. This difference probably reflects a self-selection factor in completing the SPC and PDL programs" (p. II-19).

- *Licensing rates.* Students assigned to SPC and PDL groups became licensed at greater rates compared to students assigned to the control group. Specifically, 70.6%, 66.7%, and 58.8% of students assigned to the SPC, PDL, and control groups, respectively, were licensed within 6 months of course completion or their 16th birthday, whichever was later.
- *Driving tests.* A subset of students completed additional tests of driving knowledge and skills. The SPC students scored higher than PDL students on a 56-item driving knowledge test administered on the last day of the quarter in which the student took driver education (the mean scores were 48.18 [n = 955] and 44.43 [n = 994], respectively). Furthermore, SPC students scored higher than the PDL and control groups on a standardized 30-minute on-road performance test, which was administered after the students were already licensed (mean percentages of correct behaviors were 68.75% [n = 100], 64.82% [n = 117], and 62.10% [n = 242], respectively).
- *Mileage.* By means of telephone surveys, it was determined that students in the control group had a higher driving exposure per licensed driver (the mean miles driven the day before the survey were 21.05 [n = 500] for SPC, 22.82 [n = 517] for PDL, and 24.93 [n = 498] for the control group, excluding 73, 73, and 80 students who reported they did not drive the previous day, respectively).

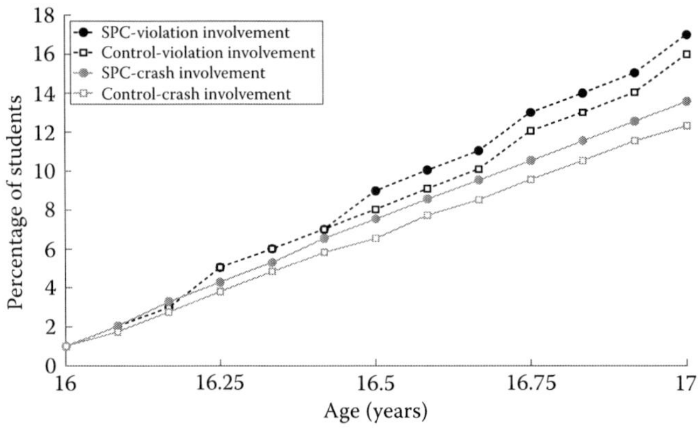

FIGURE 4.2 Estimated percentage of students having received a traffic violation and estimated percentage of students involved in a crash, per age group. The PDL data were omitted for clarity. (Data extracted from Figures 2 and 3 in Lund, A. K. et al., *Accid. Anal. Prev.*, 18, 349–357, 1986.)

Lund et al. (1986) reanalyzed the DeKalb data and applied a statistical model that controlled for students' GPA, parental education, parental occupation, sex, and the period during which they received the education. According to the statistical model by Lund et al. (1986), students assigned to the SPC group were 16% more likely to be licensed than students assigned to the control group. Furthermore, SPC students were 11% more likely to have crashed and 8% more likely to have received a traffic violation than the control group (see Figure 4.2).

4.2.3 Implications for Driving Instructors

The results of the DeKalb experiment yielded no consistent evidence that SPC and PDL programs reduced crash risk. A small crash reduction was observed per licensed driver, an effect that was detectable up to 18 months after licensure (Peck, 2011). However, if one wishes to express the effectiveness of the DeKalb study from a public-health per-capita point of view, the inescapable conclusion is that the "state-of-the-art" SPC program increased the likelihood of crashing compared to the control group. Whether one should adopt the per-licensed-driver (only students who completed the SPC course) or the per-capita (all students assigned to the course whether they completed it or not) perspective remains debatable (e.g., Peck, 2011).

One potential cause behind the limited safety effectiveness of the DeKalb program is that the SPC group in particular focused extensively on maneuvering at the driving range and on classroom instruction. Students in the SPC group indeed performed significantly better than the other two groups in a road test and a theory test. However, basic driving skills and knowledge about traffic rules are not sufficient for safe driving. The recent consensus is that drivers should acquire at least several months of independent postlicense driving experience in order to be safe drivers (Foss, 2011; Maycock & Lockwood, 1993). Appropriate lifestyle, attitudes, and skills for self-control are important prerequisites for safe driving as well (Hatakka et al., 2002; Jessor, 1987). In other words, although drivers clearly become more skillful and safe simply through learning by doing, risky driving attitudes are resistant to change. There is evidence that deliberate traffic violations, such as drunk driving, even *increase* with licensure (De Winter et al., 2015; Foss, 2011).

The DeKalb study demonstrated that students assigned to a driver education program became licensed sooner than students assigned to the control group. Although this is a positive outcome, it also means that the educational programs stimulated getting young people onto the roads who otherwise would not be driving, hence increasing the overall risk exposure. An additional issue is that young persons, males in particular, have riskier driving styles than older persons, due to their neurobiological immaturity (Dahl, 2008; Evans, 2006; Steinberg, 2008). To address these risks, current licensing systems aim to increase the licensing age and the amount of on-road driving experience prior to solo driving (OECD, 2006). For example, graduate licensing systems and multiphase driver educational programs worldwide aim to decrease fatalities by increasing the time period for achieving a full license and by letting novice drivers practice in protective conditions (OECD, 2006; Waller, 2003; Williams et al., 2012).

In the past decades, the effectiveness of driver education has been investigated in a number of studies (see Beanland et al., 2013 and Kardamanidis et al., 2010 for reviews on car driver education and motorcycle riding education, respectively). Unfortunately, many of these studies suffered from methodological weaknesses, such as attrition bias and a lack of randomized assignment (Beanland et al., 2013; Kardamanidis et al., 2010). Nonetheless, the available high-quality research indicates that driver education is useful for becoming skillful at the tasks that are the actual focus of the education. Examples of such target skills are to score highly on a road test, to perform well on a computerized test of a safety-relevant driving skill such as hazard anticipation, and to improve habits of wearing seat belts or helmets (e.g., Boele-Vos & De Craen, 2015; Horswill et al., 2015; Pradhan et al., 2009; Underwood et al., 2011). For example, in one randomized controlled trial in Thailand, it was found that driver education was successful in raising the proportion of motorcyclists who always wore helmets from 20.5% in the control group to 46.5% in the intervention group (Swaddiwudhipong et al., 1998). Furthermore, there is increasing evidence that safe driving skills can be acquired before licensure in simulator-based and PC-based training programs that target complex driving skills such as hazard anticipation, hazard mitigation, and attention, along with driver attitudes and motivation (Chapter 18; see also Chapters 21 and 28). In short, it would appear that drivers before licensure can develop important target skills that transfer to the open road, behaviors that are related to crash risk. However, the effect of education on actual crashes remains uncertain.

Despite the absence of consistent evidence that formal driver education reduces road traffic crashes, driver education continues to remain popular among the instructors who deliver it as well as among those who receive it (McKenna, 2012). In the last decades, driving instructors have rightly started to recognize that safe driving involves more than just theoretical knowledge of safe driving practices and skillful maneuvering at the driving range (Hatakka et al., 2002). It should be emphasized here that it is the research community that bears full responsibility for not having identified training programs that have been proven effective on actual measures of crash involvement. Driving instructors cannot be expected to develop and evaluate different training programs on their own. The driver education community is doing the very best it can with what researchers have given them as tools.

4.3 Enforcement: On the Statistical Reliability of On-Road Driver Testing

4.3.1 Driver Testing

In most countries, learner drivers have to pass a driving test in order to obtain their driver's license (Twisk & Stacey, 2007). Not only novice drivers but also professional and older persons with medical conditions have to participate in road tests (Siren & Haustein, 2015). Despite substantial advances in computerized visual and psychometric testing, the road test is still regarded as the gold standard of driver fitness (e.g., Dickerson et al., 2014; Rizzo et al., 2002). However, a study conducted in 1998 cast some doubt on the presumption that the outcome of the road test is particularly informative about the competence of a driver. Although road tests are closely tied to education, we treat driver testing as being in the category of enforcement because the driver's license indicates whether one is legally allowed to drive.

4.3.2 The Reliability of the Road Test

In November and December 1998, a study was undertaken at 20 test centers in the United Kingdom (Baughan & Simpson, 1999). Test candidates were asked whether they would like to take a second driving

TABLE 4.2 Number of Candidates Who Passed and Failed the Driving Tests

	Second Test	
First Test	Pass	Fail
Pass	80	57
Fail	75	154

Source: Baughan, C., & Simpson, B., Consistency of Driving Performance at the Time of the L-Test, and Implications for Driver Testing, in G. B. Grayson (ed.), *Behavioural Research in Road Safety IX*, Crowthorne: Transport Research Laboratory, pp. 206–214, 1999.

Note: The pass rate in the first test was 37.4% ([80 + 57]/366). The pass rate in the second test was 42.3% ([80 + 75]/366). This slight improvement in pass rates could indicate a learning effect.

test a few days later free of charge. The candidates were given a pass certificate if they passed the first test, the second, or both. Neither the candidate nor the examiner of the second driving test were provided with feedback about how the candidate had done in the first driving test until after the candidate had completed the second. A total of 366 candidates took part in the study.

The results revealed low consistency between the two tests (Table 4.2). Only in 64% of the driving tests were the results of the first and second tests the same. When expressed as a correlation coefficient, the test–retest reliability was $r = 0.25$. This is a weak association, especially when considering that the two driving tests were conducted at the same test center, thereby not incorporating regional differences in test difficulty.

4.3.3 Implications for Driving Instructors

In order to understand the implications of low test–retest reliability for driving instructors, it is useful to analyze where disagreement between the two driving tests could have arisen. Four sources of unreliability can be identified. First, there is the issue of interrater reliability. That is, even if two examiners independently assess the same driving test, they do not necessarily assign the same rating to this test, because humans differ regarding their perceptions and valuations (e.g., Boele-Vos & De Craen, 2015). Second, the capacities of the examiners as well as the candidates vary across time, because of momentary distractions as well as fluctuations in alertness, fatigue, and emotion. Third, the traffic conditions vary from one driving test to the other. That is, whether a candidate makes a mistake during a driving test depends on the behavior of other vehicles, weather conditions, and the route driven. Fourth, as explained by Baughan and Simpson (1999), it is likely that learner drivers apply for the driving test only when they are just sufficiently competent to pass the test (see also Baughan et al., 2005). A very poor driver will probably not apply for the road test but will continue practicing to increase the likelihood of passing. Therefore, driving test candidates are probably a homogeneous group, and no strong reliabilities are to be expected. Among statisticians, this phenomenon is known as *restriction of range*, whereby the association between two traits cannot be strong if all people are very much alike (see Kirkegaard, 2015 for an intuitive online demonstration).

Several recommendations can be put forward to improve the reliability of driving tests. First, it is possible to make the driving test longer. In the Baughan and Simpson (1999) study, the test lasted 35 minutes. Making the driving test longer will increase the amount of data (e.g., assessments, faults) that are collected and hence will increase test–retest reliability (Baughan & Simpson, 1999). Reliability can also be improved by using highly standardized routes and checklists, and by retraining the examiners such that they apply more homogeneous norming.

Another solution is to use computerized testing, such as video-based hazard perception tests and simulator-based testing (e.g., Horswill et al., 2015; Vlakveld, 2014; Chapter 28). The major advantage of computerized testing is that objective scoring is possible and that exactly the same traffic situations can be offered to all test candidates, guaranteeing a higher reliability than road testing. The disadvantage of computerized testing is the issue of validity. For example, it is known that people underestimate distance in driving simulators (e.g., Saffarian et al., 2015) and drive faster than they normally do in a car (Boer et al., 2000; De Groot et al., 2011). In addition, simulators are known to induce simulator sickness in a portion of the population, which means that they probably cannot be used for testing sensitive groups such as older drivers (e.g., Carsten & Jamson, 2011; also see Chapter 25).

A final lesson learned from the study by Baughan and Simpson (1999) is the fact that the situations we encounter, and our judgments thereof, are poorly replicable. Schmidt and Hunter (1999) explained that "the human central nervous system contains considerable noise at any given moment. This 'neural noise' can, for example, cause a person to answer two semantically identical questions differently, because of misreading a single word, because of a stray worry that popped up, etc." (p. 193). In order to obtain a statistically reliable assessment, driver behavior has to be recorded across long periods, and the collected data have to be aggregated across multiple measurement instances. In the near future, lifelong assessment and learning may indeed become a possibility. For example, driver state monitoring devices could be used for providing real-time alerts on risky driving behaviors and to keep track of one's driving style in the long term (Lee et al., 2015; Musicant & Lampel, 2010; also see Chapters 18 and 20). Furthermore, with such technology, parents can monitor their children's driving behavior via the Internet (Farmer et al., 2010).

4.4 Engineering: Electronic Stability Control Reduces Single-Vehicle Crashes by 40%

4.4.1 What Is Electronic Stability Control (ESC)?

Electronic stability control (ESC) is an active safety technology that aims to prevent skidding. The ESC system continuously compares the desired state of the vehicle (determined from the steering wheel angle and wheel speeds) with its current state (determined from the yaw rate and lateral acceleration). When the ESC detects that the vehicle is not traveling in the direction that it should be, it automatically applies the brakes of the individual wheels. For example, if the ESC detects that the yaw rate is smaller than the target yaw rate (*understeer*), it can brake the inner rear wheel in order to generate a corrective yaw moment. The ESC typically operates in conjunction with the engine and drivetrain systems, and can have additional functionalities such as rollover mitigation (Liebemann et al., 2004). In normal driving conditions, the driver cannot notice the presence of the ESC, because it is continuously analyzing sensor data but not implementing any corrective action. Only when the tires approach the maximum forces they can generate, the ESC applies a corrective braking action, in which case the driver may notice that an intervention has taken place.

4.4.2 Evaluation of the Safety Effectiveness of ESC

ESC was first introduced in 1995 and is now required for all passenger cars manufactured after September 2011 for sale in the United States (NHTSA, 2007). In the European Union, ESC is required in all new car models manufactured after November 2011 and in all newly registered cars from November 2014 onward (European Parliament and the Council for the European Union, 2009). The adoption of ESC and the subsequent requirement by various federal regulatory agencies that it be included in all manufactured vehicles is a consequence of accumulated scientific evidence supporting its safety effectiveness. With extensive test-track (e.g., Breuer, 1998) and driving simulator (e.g., Papelis et al., 2010) experiments, it has been shown that ESC has the potential to reduce crashes, in particular, loss-of-control and rollover crashes. However, the decisive scientific evidence came from actual on-road crash statistics.

There have been at least a dozen scientific publications on the on-road safety effectiveness of ESC (see Høye, 2011 for a review). We selected the work by Farmer (2006) as an exemplar because this is a representative study that features a large sample size and a straightforward method. Specifically, Farmer (2006) collected information on all police-reported crashes from 10 states for the years 2001–2003. He then extracted the number of crashes across 41 vehicle models having ESC as standard equipment and compared it to the same 41 vehicle models without ESC (or with ESC as option).

A total of 867 single-vehicle crashes were observed among the 41 ESC-equipped vehicles, while 1477 single-vehicle crashes were expected assuming that ESC-equipped vehicles had the same crash risk per registered vehicle as vehicles without ESC. Thus, because of ESC, single-vehicle crashes were reduced by 41% (i.e., 100% * [1477 – 867]/1477). The calculation of the expected crash risk included a correction factor (between 2% and 8%) to account for vehicle age. This correction factor was applied because it is known that older vehicles are more likely to be involved in car crashes, for example, because the quality of the vehicle has deteriorated or because older vehicles are driven by people who adopt riskier driving styles (e.g., teen drivers driving second-hand cars).

Additionally, Farmer (2006) found that ESC reduced injury crashes by 45% (337 observed versus 617 expected crashes) and fatal crashes by 56% (89 observed versus 204 expected crashes). The safety gains of ESC were even greater for rollover crashes, where 39 crashes were observed and 163 expected, an impressive reduction of 76%. Based on these numbers, it is clear why ESC has been called "the greatest safety innovation since the safety belt" (Nason, 2006). The safety effectiveness of ESC is especially good news for male novice drivers, who are known to be overinvolved in single-vehicle crashes (Laapotti & Keskinen, 1998).

These promising results must be somewhat tempered because single-vehicle crashes accounted for only 12% of all police-reported crashes (Farmer, 2006). Because ESC is designed to prevent loss-of-control crashes, it is perhaps not surprising that ESC had no statistically significant effect on multiple-vehicle crashes (Farmer, 2006). Several other studies have found that ESC even slightly *increases* certain types of multiple-vehicle crashes, such as rear-end collisions (Høye, 2011). A possible explanation is a phenomenon called *behavioral adaptation*. When drivers know that ESC is present in their cars, they may feel more confident and adopt riskier driving styles (Kulmala & Rämä, 2013). On the other hand, self-selection and police-reporting bias cannot be ruled out. For example, ESC-equipped car crashes may be more likely to be entered into the police records for the simple reason that equipped cars are more expensive or used by different types of drivers than nonequipped cars (Scully & Newstead, 2008). This could mean that the crash reduction potential of ESC is actually underestimated.

4.4.3 Implications for Driving Instructors

The growing prevalence of ESC has clear implications for driver education. One evident example is skid training, which becomes less important as ESC becomes more prevalent (Barker & Woodcock, 2011). An important question is whether learner drivers should experience the functionality of ESC, for example, by means of a skid pad or high-speed cornering exercise. Although learning by experiencing seems a sensible thing to do, there are potential downsides. Letting learner drivers experience the limits of the vehicle may indeed improve their handling skills but could also lead to behavioral adaptation and overconfidence (e.g., Beanland et al., 2013; Katila et al., 1996; McKenna, 2012).

ESC as well as other types of technologies, such as route navigation devices, blind-spot monitors, and advanced emergency braking systems (AEBS), are gradually finding their way into consumer vehicles. Ongoing research is trying to determine how to treat such technologies in driver education curricula (Hedlund, 2007; Panou et al., 2010). It is currently possible for a student to be trained in a car with automatic transmission and to take the driving test in such a car (in which case, in some jurisdictions, the driver's license does not permit driving a vehicle with manual transmission). In the future, driver education and licensing procedures will have to be adjusted to include highly automated driving and the use of in-vehicle interfaces (see Hancock & Parasuraman, 1992 for an early discussion on this topic).

Of course, not all technology is beneficial for road safety. Cell phones and infotainment devices can seriously undermine safety, especially in teen and novice drivers who like to stay in contact with peers and have little spare mental capacity for performing secondary tasks (Lee, 2007; Young & Stanton, 2007; also see Chapter 12). It has been recommended that driver education should improve learner drivers' awareness of their risky habits (Hatakka et al., 2002).

4.5 Discussion and Conclusion

The aim of the present chapter is to illustrate to driving instructors and other stakeholders how science contributes to the expansion of knowledge on road safety. We provided three examples, one for each *E*: education (the DeKalb driver education study by Stock et al., 1983), enforcement (the study on the reliability of the road test by Baughan & Simpson, 1999), and engineering (the study on the effectiveness of ESC by Farmer, 2006). These examples provide an illustration of how research has contributed to cumulative knowledge.

The three selected papers rely on a number of scientific methods, such as a randomized controlled trial, where it is only the effect of the treatment, not some other factor, that can explain why the treatment produces whatever results are observed (Stock et al., 1983); the blinding of experimental conditions to the individuals involved in the evaluation so that bias the candidate or examiner might have is removed from the assessment (Baughan & Simpson, 1999); and systematic archiving and analysis of crash data (Farmer, 2006). In essence, these methods are intended to protect scientists from self-deception. This is important because humans all have certain ideas and conceptions of how the world works, and this may bias their observations. As explained by Wolpert (1994) in his book *The Unnatural Nature of Science*, "ordinary, day-to-day common sense—will never give an understanding about the nature of science" (p. xi).

Although the authors of the present handbook write about novice and teen drivers, they do not necessarily have firsthand experience in automotive engineering, police enforcement, or driver education. In fact, an author of a chapter in this handbook and a leading authority on the value of hazard perception testing in the licensing process openly admits he does not have a driver's license, and he had the following proposition in his PhD thesis: "It is an advantage to study driver behaviour without having a driving licence" (Vlakveld, 2011). Vlakveld's position is not strange or absurd. Considering the wide array of biases and predispositions toward driving (Vanderbilt, 2008), it seems reasonable that scientists—in their quest for objectivity—dissociate themselves from the activity of driving and devote their attention to science.

In this chapter, we showed several things: (1) driver education is known to improve target skills (e.g., obtaining a driver's license), but whether it actually reduces crashes compared to informal education remains unproven; (2) a subjective assessment of a 30-minute drive is statistically unreliable; and (3) ongoing technological innovations, including ESC, have a major positive impact on road safety. We argue that future driving will look different from today. Most likely, there will be more in-vehicle technologies, more automated driving systems, more data on driver and vehicle state, and more vehicle-to-vehicle and vehicle-to-infrastructure communication than exist today. These developments will allow us to predict, prevent, and mitigate crashes with ever-greater effectiveness. The need for driver education is not likely to disappear. It is true that automatically driving cars may one day be the norm. However, just as pilots need to interpret a large number of displays in the cockpit and to take over control when automation fails, so too will drivers need to know how to take over control when the automated driving suite fails or reaches its functional limitations. Thus, driver education may become even more critical with the emergence of technology.

Acknowledgments

This work has been supported in part by the Marie Curie Actions of the European Union's Seventh Framework Programme FP7/2007–2013 under REA grant agreement number 608092.

References

Barker, P., & Woodcock, A. (2011). Driving skills, education and in-vehicle technology. *International Journal of Vehicle Design, 55*, 189–207.

Baughan, C., & Simpson, B. (1999). Consistency of driving performance at the time of the L-test, and implications for driver testing. In G. B. Grayson (Ed.), *Behavioural Research in Road Safety IX* (pp. 206–214). Crowthorne: Transport Research Laboratory.

Baughan, C. J., Gregersen, N. P., Hendrix, M., & Keskinen, E. (2005). *Towards European Standards for Testing* (Final Report). Brussels: Commission Internationale des Examens de Conduite Automobile (CIECA). Retrieved from http://www.cieca.eu/sites/default/files/documents/projects_and_studies/EU_TEST_Project_Final_Report.pdf.

Beanland, V., Goode, N., Salmon, P. M., & Lenné, M. G. (2013). Is there a case for driver training? A review of the efficacy of pre- and post-licence driver training. *Safety Science, 51*, 127–137.

Boele-Vos, M. J., & De Craen, S. (2015). A randomized controlled evaluation study of the effects of a one-day advanced rider training course. *Accident Analysis & Prevention, 79*, 152–159.

Boer, E. R., Yamamura, T., Kuge, N., & Girshick, A. (2000). Experiencing the same road twice: A driver centered comparison between simulation and reality. *Proceedings of Driving Simulation Conference*, Paris, France, 33–55.

Breuer, J. J. (1998). Analysis of driver–vehicle interactions in an evasive manoeuvre—Results of "moose test" studies. *16th International Technical Conference on the Enhanced Safety of Vehicles*, Windsor, Ontario, 620–627. Retrieved from http://www-nrd.nhtsa.dot.gov/pdf/nrd-01/Esv/esv16/98S2W35.PDF.

Carsten, O., & Jamson, A. H. (2011). Driving simulators as research tools in traffic psychology. In B. E. Porter (Ed.), *Handbook of Traffic Psychology* (pp. 87–96). London: Academic Press.

Dahl, R. E. (2008). Biological, developmental, and neurobehavioral factors relevant to adolescent driving risks. *American Journal of Preventive Medicine, 35*, S278–S284.

De Groot, S., De Winter, J. C. F., Mulder, M., & Wieringa, P. A. (2011). Nonvestibular motion cueing in a fixed-base driving simulator: Effects on driver braking and cornering performance. *Presence: Teleoperators and Virtual Environments, 20*, 117–142.

De Winter, J. C. F., De Groot, S., Mulder, M., Wieringa, P. A., Dankelman, J., & Mulder, J. A. (2009). Relationships between driving simulator performance and driving test results. *Ergonomics, 52*, 137–153.

De Winter, J. C. F., Dodou, D., & Stanton, N. A. (2015). A quarter of a century of the DBQ: Some supplementary notes on its validity with regard to accidents. *Ergonomics, 58*, 1745–1769.

Dickerson, A. E., Brown Meuel, D., Ridenour, C. D., & Cooper, K. (2014). Assessment tools predicting fitness to drive in older adults: A systematic review. *The American Journal of Occupational Therapy, 68*, 670–680.

Elvik, R., Høye, A., Vaa, T., & Sørensen, M. (2009). *The Handbook of Road Safety Measures*. Bingley: Emerald Group Publishing.

European Parliament and the Council for the European Union (2009). Regulation (EC) No 661/2009. Concerning type-approval requirements for the general safety of motor vehicles, their trailers and systems, components and separate technical units intended therefor. Retrieved from http://eur-lex.europa.eu/legal-content/EN/TXT/?uri=celex:32009R0661.

Evans, L. (2006). Innate sex differences supported by untypical traffic fatalities. *Chance, 19*, 10–15.

Farmer, C. M. (2006). Effects of electronic stability control: An update. *Traffic Injury Prevention, 7*, 319–324.

Farmer, C. M., Kirley, B. B., & McCartt, A. T. (2010). Effects of in-vehicle monitoring on the driving behavior of teenagers. *Journal of Safety Research, 41*, 39–45.

Foss, R. D. (2011). *Measuring Changes in Teenage Driver Crash Characteristics during the Early Months of Driving*. Washington, DC: AAA Foundation for Traffic Safety.

Groeger, J. A., & Banks, A. P. (2007). Anticipating the content and circumstances of skill transfer: Unrealistic expectations of driver training and graduated licensing? *Ergonomics, 50*, 1250–1263.

Hancock, P. A., & Parasuraman, R. (1992). Human factors and safety in the design of intelligent vehicle-highway systems (IVHS). *Journal of Safety Research, 23*, 181–198.

Hatakka, M., Keskinen, E., Gregersen, N. P., Glad, A., & Hernetkoski, K. (2002). From control of the vehicle to personal self-control; broadening the perspectives to driver education. *Transportation Research Part F: Traffic Psychology and Behaviour, 5*, 201–215.

Hedlund, J. (2007). Novice teen driving: GDL and beyond. *Journal of Safety Research, 38*, 259–266.

Horswill, M. S., Hill, A., & Wetton, M. (2015). Can a video-based hazard perception test used for driver licensing predict crash involvement? *Accident Analysis & Prevention, 82*, 213–219.

Høye, A. (2011). The effects of Electronic Stability Control (ESC) on crashes—An update. *Accident Analysis & Prevention, 43*, 1148–1159.

Jessor, R. (1987). Risky driving and adolescent problem behavior: An extension of problem-behavior theory. *Alcohol, Drugs, & Driving, 3*, 1–11.

Jonah, B. A., Dawson, N. E., & Smith, G. A. (1982). Effects of a selective traffic enforcement program on seat belt usage. *Journal of Applied Psychology, 67*, 89–96.

Kardamanidis, K., Martiniuk, A., Ivers, R. Q., Stevenson, M. R., & Thistlethwaite, K. (2010). Motorcycle rider training for the prevention of road traffic crashes. *Cochrane Database of Systematic Reviews, 10*, CD005240.

Katila, A., Keskinen, E., & Hatakka, M. (1996). Conflicting goals of skid training. *Accident Analysis & Prevention, 28*, 785–789.

Kirkegaard, E. (2015). Understanding restriction of range. Retrieved from https://emilkirkegaard.shinyapps.io/Understanding_restriction_of_range.

Kulmala, R., & Rämä, P. (2013). Definition of behavioural adaptation. In C. M. Rudin-Brown, & S. L. Jamson (Eds.), *Behavioural Adaptation and Road Safety: Theory, Evidence and Action* (pp. 11–21). Boca Raton: CRC Press.

Laapotti, S., & Keskinen, E. (1998). Differences in fatal loss-of-control accidents between young male and female drivers. *Accident Analysis & Prevention, 30*, 435–442.

Learoyd, C. G. (1950). The carnage on the roads. *The Lancet, 255*, 367–369.

Lee, J. D. (2007). Technology and teen drivers. *Journal of Safety Research, 38*, 203–213.

Lee, Y.-C., Belwadi, A., Bonfiglio, D., Malm, L., & Tiedeken, M. (2015). Techniques for reducing speeding beyond licensure: Young drivers' preferences. *Proceedings of the Eighth International Driving Symposium on Human Factors in Driver Assessment, Training and Vehicle Design*, Salt Lake City, Utah, 168–174. Retrieved from http://drivingassessment.uiowa.edu/sites/default/files/DA2015/papers/026.pdf.

Liebemann, E. K., Meder, K., Schuh, J., & Nenninger, G. (2004). Safety and performance enhancement: The Bosch electronic stability control (ESP). *SAE Paper*, 2004-21-0060. Retrieved from http://www-nrd.nhtsa.dot.gov/pdf/nrd-01/esv/esv19/05-0471-O.pdf.

Lozano, R., Naghavi, M., Foreman, K., Lim, S., Shibuya, K., Aboyans, V., & Cross, M. (2013). Global and regional mortality from 235 causes of death for 20 age groups in 1990 and 2010: A systematic analysis for the Global Burden of Disease Study 2010. *The Lancet, 380*, 2095–2128.

Lund, A. K., Williams, A. F., & Zador, P. (1986). High school driver education: Further evaluation of the DeKalb County study. *Accident Analysis & Prevention, 18*, 349–357.

Mäkinen, T., & Hagenzieker, M. P. (1991). Strategies to increase the use of restraint systems—Background paper. *Proceedings of a workshop organized by SWOV and VTT at the VTI-TRB International Conference Traffic Safety on Two Continents.* Leidschendam: SWOV, 2–5.

Maycock, G., & Lockwood, C. R. (1993). The accident liability of British car drivers. *Transport Reviews, 13*, 231–245.

Mayhew, D. R. (2007). Driver education and graduated licensing in North America: Past, present, and future. *Journal of Safety Research, 38*, 229–235.

Mayhew, D. R., & Simpson, H. M. (2002). The safety value of driver education and training. *Injury Prevention, 8*, ii3–ii8.

McKenna, F. (2010). *Education in Road Safety: Are We Getting Right?* (Report No. 10/113). London: RAC Foundation.

McKenna, F. P. (2012). How should we think about the three E's: education, engineering and enforcement? *The 5th International Conference on Traffic and Transport Psychology*, Groningen, the Netherlands, August 29–31, 2012. Retrieved from http://www.icttp2012.com/images/stories/presentations/29th/Keynote/McKenna_ICTTP2012_Keynote_How_should_we_think.pdf.

Musicant, O., & Lampel, L. (2010). When technology tells novice drivers how to drive. *Transportation Research Record: The Journal of the Transportation Research Board, 2182*, 8–15.

Nason, N. R. (2006). Remarks of Nicole R. Nason, Administrator, National Highway Traffic Safety Administration U.S.A. Made at the *140th Session of the United Nations' World Forum for Harmonization of Vehicle Registration* (WP.29). Geneva, November.

NHTSA. (2007). 49 Code of Federal Regulations (CFR) Parts 571 and 585. US Government Printing Office, Washington, DC.

Organisation for Economic Co-operation and Development. (2006). *Young Drivers: The Road to Safety.* Paris: OECD.

Panou, M. C., Bekiaris, E. D., & Touliou, A. (2010). ADAS module in driving simulation for training young drivers. *13th International IEEE Conference on Intelligent Transportation Systems (ITSC)*, 1582–1587.

Papelis, Y. E., Watson, G. S., & Brown, T. L. (2010). An empirical study of the effectiveness of electronic stability control system in reducing loss of vehicle control. *Accident Analysis & Prevention, 42*, 929–934.

Park, G. D., Allen, R. W., & Rosenthal, T. J. (2015). Novice driver simulation training potential for improving hazard perception and self-confidence while lowering speeding risk attitudes for young males. *Proceedings of the Eighth International Driving Symposium on Human Factors in Driver Assessment, Training and Vehicle Design*, 253–259. Retrieved from http://drivingassessment.uiowa.edu/sites/default/files/DA2015/papers/039.pdf.

Peck, R. C. (2011). Do driver training programs reduce crashes and traffic violations?—A critical examination of the literature. *IATSS Research, 34*, 63–71.

Pradhan, A. K., Pollatsek, A., Knodler, M., & Fisher, D. L. (2009). Can younger drivers be trained to scan for information that will reduce their risk in roadway traffic scenarios that are hard to identify as hazardous? *Ergonomics, 52*, 657–673.

Riley, M. C., & McBride, R. S. (1974). *Safe Performance Curriculum for Secondary School Driver Education: Program Development, Implementation, and Technical Findings* (Final Report. DOT-HS-003-2-427). Washington, DC: National Highway Traffic Safety Administration.

Rizzo, M., Jermeland, J., & Severson, J. (2002). Instrumented vehicles and driving simulators. *Gerontechnology, 1*, 291–296.

Rosencrantz, H., Edvardsson, K., & Hansson, S. O. (2007). Vision zero–Is it irrational? *Transportation Research Part A: Policy and Practice, 41*, 559–567.

Rothengatter, T. (1982). The effects of police surveillance and law enforcement on driver behaviour. *Current Psychological Reviews, 2*, 349–358.

Saffarian, M., De Winter, J. C. F., & Senders, J. W. (2015). Measuring drivers' visual information needs during braking: A simulator study using a screen-occlusion method. *Transportation Research Part F: Traffic Psychology and Behaviour, 33*, 48–65.

Schmidt, F. L., & Hunter, J. E. (1999). Theory testing and measurement error. *Intelligence, 27*, 183–198.

Scully, J., & Newstead, S. (2008). Evaluation of electronic stability control effectiveness in Australalasia. *Accident Analysis & Prevention, 40*, 2050–2057.

Siren, A., & Haustein, S. (2015). Driving licences and medical screening in old age: Review of literature and European licensing policies. *Journal of Transport & Health, 2*, 68–78.

Steinberg, L. (2008). A social neuroscience perspective on adolescent risk-taking. *Developmental Review, 28*, 78–106.

Stock, J. R., Weaver, J. K., Ray, H. W., Brink, J. R., & Sadof, M. G. (1983). *Evaluation of Safe Performance Secondary School Driver Education Curriculum Demonstration Project* (Final Report DOT-HS-6-01462). Washington, DC: National Highway Traffic Safety Administration.

Swaddiwudhipong, W., Boonmak, C., Nguntra, P., & Mahasakpan, P. (1998). Effect of motorcycle rider education on changes in risk behaviours and motorcycle-related injuries in rural Thailand. *Tropical Medicine & International Health, 3*, 767–770.

Twisk, D. A. M., & Stacey, C. (2007). Trends in young driver risk and countermeasures in European countries. *Journal of Safety Research, 38*, 245–257.

Underwood, G., Crundall, D., & Chapman, P. (2011). Driving simulator validation with hazard perception. *Transportation Research Part F: Traffic Psychology and Behaviour, 14*, 435–446.

Vanderbilt, T. (2008). *Traffic. Why We Drive the Way We Do (and What It Says about Us)*. London: Allen Lane.

Vlakveld, W. P. (2011). Hazard anticipation of young novice drivers: Assessing and enhancing the capabilities of young novice drivers to anticipate latent hazards in road and traffic situations. Doctoral Dissertation, University of Groningen, the Netherlands.

Vlakveld, W. P. (2014). A comparative study of two desktop hazard perception tasks suitable for mass testing in which scores are not based on response latencies. *Transportation Research Part F: Traffic Psychology and Behaviour, 22*, 218–231.

Wakefield, M. A., Loken, B., & Hornik, R. C. (2010). Use of mass media campaigns to change health behaviour. *The Lancet, 376*, 1261–1271.

Waller, P. F. (2003). The genesis of GDL. *Journal of Safety Research, 34*, 17–23.

Weaver, J. K. (1978). Does quality driver education produce safer drivers? *Traffic Safety, 78*, 18–20.

Williams, A. F., & Wells, J. K. (2004). The role of enforcement programs in increasing seat belt use. *Journal of Safety Research, 35*, 175–180.

Williams, A. F., Tefft, B. C., & Grabowski, J. G. (2012). Graduated driver licensing research, 2010–present. *Journal of Safety Research, 43*, 195–203.

Wolpert, L. (1994). *The Unnatural Nature of Science*. Cambridge: Harvard University Press.

Young, M. S., & Stanton, N. A. (2007). What's skill got to do with it? Vehicle automation and driver mental workload. *Ergonomics, 50*, 1324–1339.

Zaal, D. (1994). *Traffic Law Enforcement: A Review of the Literature* (Report No. 53). Leidschendam: SWOV.

5

Novice Teen Driver Crash Patterns

Catherine C. McDonald

Marilyn S. Sommers

Flaura K. Winston

Contents

Abstract

Teen driver crashes are a threat to public health. Novice teen driver crash rates and deaths in the United States and other developed countries have consistently dropped in the last 30 years, indicating the success of multipronged prevention efforts through behavior change (e.g., seat belt use), policy (e.g., graduated driver licensure [GDL], drinking–driving laws), and technology (e.g., safer vehicles). Experience behind the wheel contributes to skill acquisition for newly licensed drivers and has a strong influence on crash reduction in the first year of driving. Scenarios where teens crash include situations where they must integrate complex decision making, such as turning at intersections, following lead vehicles, and running off the road. In these and other types of scenarios, teens often commit an error or lack the skill to effectively avoid the crash. However, hazardous behaviors also contribute to crash risk, and the intersection between skill and adolescent risk taking needs to be better delineated in future research. Methodological challenges in deconstructing the factors contributing to teen driver crash patterns include difficulties with prospective data collection strategies and the lack of detail available in some crash data. Further research is also needed to bolster existing crash prevention efforts and develop new ones in order to prevent future deaths and injuries related to novice teen driver motor vehicle crashes.

5.1 Introduction

Novice teen drivers are at their highest risk for motor vehicle crashes in their first year of licensure, and this risk declines almost all the way to adult levels over the first 2–3 years of licensure (Mayhew et al., 2003; McCartt et al., 2003; Williams & Tefft, 2014). The consequences of novice teen driver crashes have a profound, negative impact on the public's health, as evidenced by the high morbidity and mortality for adolescents and young adults (Centers for Disease Control and Prevention, 2014b). In 2011, almost 2650 teens died, and another 292,000 were treated for injuries from motor vehicle crashes in the United States (Centers for Disease Control and Prevention, 2014b). The economic impact associated with teen driver crashes is burdensome, with over 35% of lifetime medical cost for crash injuries in 2012 ($6.5 billion out of $18 billion) attributed to young drivers (ages 15–29) (Centers for Disease Control and Prevention, 2014a; Sommers et al., 2011).

Teen driver crash rates have dropped dramatically in the last 40 years. A number of factors have contributed to the historical decreases in crash rates in all ages of drivers, but disentangling the reasons for decreased novice teen crash rates is complex. In the months after a teen exits the learner phase and enters independent licensure, there is a sharp decrease in crash rates, which continue to fall throughout the first years of licensure (Foss et al., 2011). This decline can be attributed to factors such as an increase in experience and skill (Guo et al., 2013) as well as evolving social and emotional maturity that can influence risk-taking tendencies (Keating, 2007). In this chapter, we will discuss the historical trends in teen crashes in the United States, the influence of experience and hazardous driving, common situations in which teens crash, errors that contribute to crashes, and the challenges associated with the research in novice teen driver crash patterns. Given the substantial global impact of motor vehicle crashes, we briefly address the global impact among novice teen drivers on crash epidemiology (also see Chapter 24).

5.2 Patterns of Crashes

5.2.1 Historical Trends in Crash Rates

Since the 1970s in the United States, for teens and adults alike, the rate of passenger-vehicle drivers involved in fatal crashes per 100,000 people has declined (Insurance Institute for Highway Safety, 2015a; Shope & Bingham, 2008). From 1975 to 2013, there has been a 71% reduction in teen deaths due to motor vehicle crashes (Insurance Institute for Highway Safety, 2015a), with substantial drops from 2005 to 2011 (The Center for Injury Research and Prevention at the Children's Hospital of Philadelphia, 2013). This drop in fatal crashes has been consistent across time for young drivers: even from 2011 to 2012 when there was a slight increase in fatal crashes in the general population, young driver fatal crash rates continued to fall (Insurance Institute for Highway Safety, 2015a). Factors that contributed to the reductions over the past 40 years include advances in vehicle and road safety, increased seat belt use, decreased driving after drinking, and the implementation of graduated driver licensure (GDL) provisions for teen drivers (Centers for Disease Control and Prevention, 2010; Williams et al., 2012).

Across the United States, geographic differences in the driving environment can influence teen driver crash rates. From 2004 to 2008, rates of fatal crashes involving 16- to 17-year-old drivers fell across the United States, but the 5-year annualized rate for crashes ranged from 9.7 per 100,000 population in New York and New Jersey to 59.6 in Wyoming (Centers for Disease Control and Prevention, 2010). Factors such as weather, urbanization, speed limit restrictions, and laws determine the environment in which teens drive. For newly licensed teens, the state in which they are first licensed can have important implications for when and how they learn to drive, in particular, the state restrictions in place when they become independent drivers. Local, regional, and state geographic driving environments, as well as the overall traffic safety culture of the region (Ward, 2007), influence teen driving behaviors and crash rates. For example, in an examination of teen driver crashes from 2002 to 2008 identified from the Iowa Department of Transportation, Peek-Asa et al. (2010) found that younger teen drivers had higher overall crash rates in more rural as compared to urban areas; for older teen drivers, the overall crash rates were lower for rural areas. In addition, rural teen crashes were about five times more likely to lead to a fatal or severe-injury crash than urban teen crashes. These differences in crashes by geographic location and characteristic are important considerations for prevention efforts.

5.2.2 Licensure and Experience

As already noted, during the first years of licensure, teen driver crash rates decline. An analysis of data from Nova Scotia by Mayhew et al. (2003) showed that the largest decline in crashes occurs in the first 6 months of licensure. Research has shown that the decline continues throughout the first 2–3 years of independent licensure (Foss et al., 2011). These findings from crash and licensure data have also been supported by naturalistic driving data in the United States. In a longitudinal naturalistic driving study with newly licensed teens from the Virginia area, the decline in the first 6 months of licensure of crash/near-crash (CNC) events mirrored archival crash data; the decline continued for an additional 12 months (Lee et al., 2011). As expected, teens' crash rates were higher during the first 18 months than those of their parents. Given the nature of the crash rate decline in the first few years of independent driving, the change is largely attributed to the notion that as teens gain experience behind the wheel, they have fewer crashes (McKnight & McKnight, 2003).

The identification of skill acquisition as a contributor to crash reduction has influenced the implementation and strengthening of GDL, an important protective factor for newly licensed drivers that provides guidelines on learner driver requirements and restrictions on passengers, nighttime driving, and secondary task engagement (Williams, 2013). (Note: While GDL is covered in depth in Chapter 17, we briefly describe how it relates to novice teen driver crash patterns.) GDL provisions were designed to help novice drivers gain experience in safer driver situations before transitioning to an unrestricted license. Generally, GDL allows novice drivers to move through a learner phase where they must drive with an adult, a restricted phase where the teen has his or her license and can drive solo but has restrictions in place that limit driving activity, and then an unrestricted phase. For example, under GDL in California, teens can acquire their learner's permit at 15.5 years and are required to hold it for a minimum of 6 months and complete at least 50 hours of supervised practice driving; once licensed, teens have nighttime restrictions (cannot drive between 11 p.m. and 5 a.m.) and are not allowed to have passengers in the first 12 months of driving (Governors Highway Safety Association, 2014). Each state has different components in their GDL provisions that can influence resultant novice teen driver crashes.

Evidence indicates that GDL provisions have helped to decrease, per capita, crash rates of 16- and 17-year-old drivers (Curry et al., 2013; Ehsani et al., 2013a,b; Shope, 2007). However, GDL provisions do not necessarily decrease the per capita crash rates of teens ages 18 and 19, and the reasons why and how this happens are not yet entirely clear (Masten et al., 2011). In many states, individuals 18 or older without a license are not required to proceed through GDL provisions. Therefore, they likely miss out on the benefits of GDL and do not have the protective effects against crash risk. Further research is needed to better understand the needs of older novice teen drivers (Chapman et al., 2014; Masten et al., 2011).

The GDL provisions of restrictions on nighttime driving and passengers have been shown to be effective in reducing crash rates in teens. For example, McCartt et al. (2010) found estimated reductions of fatal crashes for 15- to 17-year-old drivers for restrictions both on nighttime driving (each additional hour indicated further reductions: 12% at midnight, 14% at 11 p.m., 16% at 10 p.m.) and on the number of passengers, with the greatest reduction when no peer passenger was allowed (21%) and a smaller reduction when one peer passenger was allowed (7%). Although typical in countries like Australia (Senserrick & Haworth, 2005), decal provisions as part of GDL are infrequent in the United States. A decal provision requires newly licensed drivers to display a sticker—typically on the front and rear license plates—that identifies them as drivers in the provisional stage of licensure. Failure to display the decal would result in a fine. New Jersey included a decal provision in their GDL, which requires drivers 21 or younger who are learner or intermediate drivers to place a highly visible decal on their vehicle. The intent is to make learner or intermediate drivers identifiable to law enforcement so they can more easily monitor if these drivers are following the provisions (such as passenger or nighttime driving restrictions). Curry et al. (2015) estimated that in New Jersey, more than 3000 crashes were prevented in the 2 years postimplementation of the decal. Although there are challenges and issues of noncompliance associated with the decals (McCartt et al., 2012), the impact of the decal in New Jersey on crash reduction presents important information on the specific components of GDL and what states can do to implement stronger GDL provisions.

5.2.3 Crash Scenarios and Types

An accumulation of both archival crash data and interviews with teens helps identify and describe situations associated with novice teen driver crashes. Together, these data can help target where existing prevention efforts can be improved or new ones developed. Braitman et al. (2008) conducted a study on novice teen driver crashes (16-year olds in the first 8 months of licensure) using Connecticut police reports of nonfatal crashes and interviews with the teens. They found that crashes primarily occurred when the teen ran off the road, rear-ended another vehicle, or collided with another vehicle that had the right-of-way. In addition, for the crashes investigated, 75% of the novice teen drivers were at fault.

Also using state crash data, Foss et al. (2011) conducted an in-depth examination of crashes for newly licensed teen drivers in North Carolina in their first 36 months of licensure. In this sample of data from 2001 to 2008, they found that newly licensed teen drivers are predominately involved in two-vehicle crashes (68.7%) and in crashes on roads with posted speed limits between 35 and 54 mph (65.3%). They examined a few key variables related to the crashes, notably the movement of the vehicle and the first harmful events in the crash. In the next two paragraphs, we will discuss both of these variables, as the analysis by Foss et al. (2011) provides important information about how teens crash during this initial period of licensure.

With respect to vehicular movement, the crashes involving vehicles going straight (e.g., index vehicle was going straight and rear-ended a lead vehicle) accounted for almost 60% of the crashes during the time period (12.33 per 1000 licensed drivers in the 1st month and 5.72 per 1000 licensed drivers in the 36th month) (Foss et al., 2011). The investigators found that the crashes where the driver was making a left turn, making a right turn, or entering the roadway declined quickly with new licensure, although their rates varied (e.g., left turn, 3.88 per 1000 licensed drivers in 1st month and 0.87 per 1000 licensed drivers in 36th month; entering a roadway, 0.61 per 1000 licensed drivers in 1st month and 0.18 per 1000 licensed drivers in 36th month). The authors note that experience and skill acquisition, such as handling the mechanics of the vehicle and judgment, likely contribute to this quick decline in specific crashes (Foss et al., 2011).

For the *first harmful event type* (the first injury- or damage-producing event in the crash), Foss et al. (2011) found that the most frequent types in the first month were rear end/slowing stopped (5.54 per 1000 licensed drivers) and angle collisions (3.98 per 1000 licensed drivers; vehicles hitting at or near right angles, e.g., front of one striking side of another). They found a decline from the 1st to the 36th month for these two types of crashes, to 3.09 per 1000 licensed drivers for rear end/slowing stopped and 1.51 per 1000 licensed drivers for angle collision. Each first harmful event type accounted for 33% and 15% of all crashes, respectively, for the 36 months of crash data. Overall, they found a decline in almost all first harmful event crash types. Some first harmful event types declined more rapidly than others, such as rollovers, running off the road on the right side, colliding with another vehicle on an intersection while making a left, and hitting a fixed object or parked car. Again, these declines may be linked to experience and skill acquisition in the early months of driving. Around the time when the provisional licensure ended (at about the seventh month), a slight uptick occurred in all crashes in the North Carolina data. Foss et al. (2011) suggested that this increase in crashes could be attributed to the lifting of restrictions related to nighttime driving and passengers that occurs in North Carolina after 6 months of an intermediate (or restricted) license.

The previous studies provide important information that can be used to identify situations and changes in novice teen drivers crashes. However, comparison of novice teen crashes to experienced adult driver crashes can help characterize the differences between inexperience and experience when considering skill and hazardous behaviors. Teens crash at higher rates than adult drivers (Insurance Institute for Highway Safety, 2015b). However, for those teens and adults that do crash, if they do so in in similar situations, it may point to certain hazardous behaviors, skills, or situations that do not differ by age or experience. Alternatively, if there are scenarios in which teens' and adults' crashes differ from one another, those differences can help identify areas for skill development or targeted interventions. However, comparing teen crashes that include length of licensure with adult crashes is difficult. National databases might not contain specifics on length of licensure that state-level data may be able to address. A few studies have compared teen and adult crash scenarios, but the findings are limited because the length of licensure for the teens was unknown in the data (Bingham & Ehsani, 2012; McDonald et al., 2014). While limitations to the data exist, these findings are useful in helping to understand teen driver crash scenarios.

Using the Fatality Analysis Reporting System (FARS) and the General Estimates System (GES) data from 2005 to 2009, Bingham and Ehsani (2012) compared crash types for adolescents (age, 15–19) and adults (age, 45–64). They examined both fatal and nonfatal crashes for the teens and adults, as well as differences between male and female crash-involved teen and adult drivers. For all groups (male adolescents, female adolescents, male adults, and female adults), the largest proportion of crashes was single-vehicle crashes. However, adolescent males had the greatest likelihood of being involved in a single-vehicle crash (both fatal and nonfatal; e.g., odds ratio and 95% confidence interval [CI] of adolescent males compared to adolescent females in nonfatal single-vehicle crashes: 1.57 [1.47–1.68]). They found a general age difference in nonfatal front-to-rear crashes (i.e., index vehicle striking another vehicle from behind), with both male and female adolescents more likely than the male and female adults to be involved in that crash type. Adolescent males were more likely than adolescent females (1.16 [1.05–1.28]) and both adult groups (compared to male adults, 1.29 [1.21–1.37], and female adults, 1.10 [1.02–1.19]) to be involved in fatal head-on collisions. Adults were more likely to be involved in fatal rear-end crashes (i.e., index vehicle struck from behind) than adolescents.

McDonald et al. (2014) compared serious crash scenarios of teens (age, 16–19) and adults (35–55) using data from the National Motor Vehicle Crash Causation Survey (NMVCCS). In their examination of NMVCCS data that included serious crashes, the top five crash scenarios among teen drivers accounted for 37.3% of their crashes. The scenarios, with the prevalence ratio (PR) and 95% CI comparing teens to adults, included the following: (1) going straight, other vehicle stopped, rear end (1.08 [0.64–1.83]); (2) stopped in traffic lane, turning left at intersection, turn into path of other vehicle (1.74 [0.96–3.15]); (3) negotiating curve, off right edge of road, right roadside departure (2.03 [0.93–4.44]); (4) going straight, off right edge of road, right roadside departure (1.09 [0.53–2.24]); and (5) stopped in lane, turning left at intersection, turn across path of other vehicle (0.91 [0.52–1.62]). The top five crash scenarios among adult drivers, accounting for 33.9% of their crashes, included the same scenarios as the teen drivers with the exception of scenario (3) and the addition of going straight, crossing over an intersection, and continuing on a straight path (0.38 [0.19–0.75]). Comparisons were also made between male and female adolescents and male and female adults: female teens were more likely than male teens to be involved in going straight, other vehicle stopped, rear end (3.45 [1.56–7.59]).

Taken together, these data from state-level and national crash reports provide an important picture of the role of crash types or scenarios in patterns of teen driver crashes. In particular, the work of Foss et al. (2011) provides details about crashes in the first 36 months of driving, where rates of almost all crash types (movement of vehicle and first harmful event type) decrease in the first 36 months. Not all rates decrease with the same slope, however. It would be beneficial to replicate their analysis in other states with different GDL provisions and driving environments to see if the findings can be generalized to other populations of teens. Bingham and Eshani (2012) and McDonald et al. (2014) found that many adults and teens are in similar crash scenarios, but it is still not clear if these crashes occur for the same reasons. For example, similar crash scenarios may happen for teens and adults because both groups engage in risky behaviors such as speeding or both groups lack skill in a certain domain of driving. Alternatively, the same crash scenario may occur in teens because they lack skills such as hazard anticipation, whereas for adults, alcohol may contribute to the crash (Centers for Disease Control and Prevention, 2015; McKnight & McKnight, 2003). A limitation of the analyses was that length of licensure was not a variable reported in the findings (Bingham & Ehsani, 2012; McDonald et al., 2014). Further comparisons between teens and adults can help to uncover patterns.

Overall, these studies indicate that there are a number of common crash scenarios that are important to delineate teen driving risk. These include single-vehicle crashes (which could include running off the road), rear-end crashes (both where they are the striking vehicle and the vehicle being struck), as well as making turns at intersections (both left and right). While teen drivers are involved in a multitude of crash types, a targeted understanding of how these specific more frequent crash types occur is important to the advancement of crash prevention efforts.

5.2.4 Contributors to Crashes

Although it is helpful to understand the situations in which teens crash as we described previously, it is also crucial to identify the contributing factors to the crashes in order to develop interventions to mitigate these causes. The decrease in crash rates that occurs in the first few years of licensure can, in part, be attributed to skill acquisition (Masten & Foss, 2010). However, with a teen's developmental stage, risk-taking or hazardous driving behaviors (such as speeding, reckless driving, alcohol involvement, and secondary task engagement) (Steinberg, 2007) can also contribute to their high crash rates (also see Chapter 18). Given the importance of both skill and hazardous behaviors to teen driver crashes, we will address both here in this section.

5.2.4.1 Skill Errors

McKnight and McKnight (2003) concluded from their data that for teens, ages 16–19, a majority of crashes were the result of poor skill—errors in hazard anticipation, emergency maneuvers, visual search and attending to the roadway, and speed relative to road conditions—rather than risky driving behaviors (such as alcohol-impaired driving). In the investigation of crashes of newly licensed teens in Connecticut, three factors contributed about equally to their crashes: failing to detect another vehicle or traffic control device, speeding, and losing control of the vehicle or sliding (Braitman et al., 2008). Curry et al. (2011) also found that in serious crashes, recognition errors (such as inadequate surveillance and distraction), decision errors (such as following too closely or travelling too fast for conditions), and performance errors (such as loss of control) were the main types of errors that teens made. Aggressive driving, impaired driving, and drowsy driving were infrequently the critical reason for a serious crash (Curry et al., 2011).

In the comparison of crash scenarios of teens and adults, McDonald et al. (2014) also examined the driver critical errors to which the crashes are attributed or the main driver-behavior contributing factor (i.e., not weather or external environmental factors). In the NMVCCS data, driver critical errors included nonperformance errors (e.g., sleeping, medical conditions); recognition errors (e.g., inattention, inadequate surveillance, distraction); decision errors (e.g., too fast for road conditions, following too closely, misjudgment of gap/others' speed); and performance errors (e.g., overcompensation, poor directional control). In their sample of crash-involved teen and adult drivers who were assigned a driver-related critical reason, 71% of teen drivers versus 44% of adults drivers were assigned a driver-related critical reason; teens were more often assigned a driver-related behavior that was the immediate reason for the events leading up to the crash. For two crash scenarios (going straight, other vehicle stopped, rear end; negotiating curve, off right edge of road, right roadside departure), teens were more likely than adults to make a critical decision error (e.g., traveling too fast for conditions). For the remainder of the scenarios examined, the estimated PRs of driver critical errors did not indicate differences between teens and adults (McDonald et al., 2014).

5.2.4.2 Hazardous Driving Behaviors

Even though there is evidence that errors are a large contributor to teen crashes, the relationship between novice teen driver skill and hazardous behaviors as they contribute to crashes is complex. Other factors related to adolescent development that also influence skill acquisition and execution are relevant to prevention efforts. In an 18-month naturalistic study of novice teen drivers, Guo et al. (2013) delineated three groups of teen drivers by CNC rates—low (2.1 CNC per 10,000 kilometers), moderate (8.3 CNC per 10,000 kilometers), and high-risk (21.8 CNC per 10,000 kilometers). They found that the moderate-risk group was the only group to have a significantly decreasing CNC rate over the 18 months of driving. The high- and low-risk teens did not demonstrate any changes in their CNC rates over the 18 months. Low-risk teens had a stable low CNC rate, and that of high-risk teens remained high. In particular, the authors concluded that the high-risk teens were insensitive to experience—their CNC risk remained high throughout the 18 months in the study. However, there were not any differences among the three groups of teens on characteristics such as sensation seeking and personality factors (e.g., extraversion, agreeableness, and conscientiousness). These data are important in helping identify teen crash patterns, but more information is needed to further identify what contributes to these patterns. For example, it is difficult to determine if the high-risk group lacks skill even while gaining more experience or if risk engagement increases (even while skill may be increasing) and contributes to the propensity for a crash. Alternatively, with the low-risk group, further understanding of how their skill changes with experience is also needed.

Ouimet et al. (2014) found in this same sample that a higher cortisol response among teens during a stress-inducing task at baseline was associated with lower CNC rates at follow-up as well as a faster decline in CNC rates. They determined that given the relationship between low cortisol response and other risk behaviors such as aggressiveness, alcohol use, and asocial behavior, there is the potential to further explore how neurobiological markers influence teen driver crash risk. Building on the knowledge gained from analyses of CNC, naturalistic studies—even though they tend to have CNC on the lower end of severity—provide insight into what differentiates typical versus CNC-involved trips. In addition, these naturalistic studies with prospective data collection with teen drivers in early stages of licensure provide important information about teen risk contributions to crashes.

Distracted-driving crashes are an area of great public attention (also see Chapters 7 and 12 in this handbook). Use of technology in the current generation of teens is exploding: 93% of teens use the Internet; of those, 63% use it on a daily basis, and 75% have a cell phone (Lenhart et al., 2010). In-vehicle distractions, such as cell phone calls, texting, and social media app use—also known as secondary task engagement—draw attention away from the roadway and increase teen crash likelihood (Caird et al., 2014; Klauer et al., 2014). Teens' engagement in cell phone use while driving is concerning: almost half of teens report texting while driving in the previous month (Olsen et al., 2013). In a naturalistic driving study of 1691 crashes of drivers ages 16–19 years, cell phone use was identified in 12% of crashes, and a driver operating or looking at a cell phone spent an average of 4.1 seconds (out of 6) before the crash looking away (Carney et al., 2015). In an 18-month naturalistic study of novice teen drivers, Klauer et al. (2014) found that in teens, texting or accessing the internet is associated with a 3.87 (95% CI, 1.62–9.25) increase in CNC risk. They also observed during the course of 18 months that novice drivers increased their secondary task engagement over time (including mobile technology, HVAC [heating, ventilation, and air-conditioning], eating, looking at roadside objects, and reaching for objects).

These results and others about secondary task engagement have led to the passing of laws designed to reduce in-vehicle distractions among teens. Among these laws are cell phone restrictions, but they are largely ineffective in changing behaviors of teens behind the wheel (Foss et al., 2009; Goodwin et al., 2012; McCartt et al., 2014). Further, there have been few studies examining the effectiveness of mobile phone restrictions on novice teen driver crashes. Ehsani et al. (2014) found that text messaging bans in Michigan did not decrease crash rates of 16- or 17-year-old drivers, as well as in the age groups of 18, 19, 20–24, and 25–50. However, Ferdinand et al. (2014) found that primary enforcement of texting bans for younger drivers had the strongest effect (incidence rate ratio [IRR], 0.88; 95% CI, 0.79–0.98) on the youngest drivers (ages 15–21).

Impaired driving in teens includes situations in which alcohol, other drugs, and sleep deficit may affect driving (also see Chapters 13 and 15 in this handbook). The effects of alcohol on driving behaviors are damaging—underage drinking and driving are associated with higher relative crash risk (Peck et al., 2008; also see Chapter 13). Reductions in teen drunk driving have emerged. Between 1991 and 2012, the percentage of teens in high school who drink and drive has decreased by more than half (Centers for Disease Control and Prevention, 2012). Sleep deficits and drowsy driving are other factors that contribute to impaired driving in novice teen drivers and have been linked to motor vehicle crashes (Martiniuk et al., 2013).

5.2.5 Challenges Associated with Examining Novice Teen Driver Crash Patterns

Many challenges exist to identifying patterns in novice teen driver crashes, including elements associated with collecting retrospective and prospective data. Retrospective data sources, such as police reports or national crash databases, are often limited in the amount and detail of information that can be obtained for a given crash. Reliance on police reports, use of data that are collected from just one state or jurisdiction, or lack of information available from the individuals involved in the crash can limit conclusions about the crash scenarios. We know that the crash risk drops in the first year, but more information on what it is about the experience gained related to driving exposure is needed. However, not all is lost. The data available from state-level crash records and nationally representative crash databases have helped identify the patterns of how teens crash as a function of year of licensure, time since licensure, level of experience, and crash scenario. In the future, in-vehicle monitoring devices have the potential to contribute to our knowledge about teen driver crashes, collecting rich prospective data on vehicle information, driver behavior, and environmental factors (Farmer et al., 2010; Horrey et al., 2012; McGehee et al., 2007). However, the length of time, intensity, and financial implications of long-term in-vehicle monitoring limit the availability of these types of data (see Chapter 20). Problems with using near crashes as surrogates for crashes are also becoming ever more apparent, a serious problem for naturalistic studies since there aren't often enough crashes to be analyzed by themselves (Knipling, 2015).

More generally, teen crashes are a significant public health problem around the world. Globally, there are great variations in how young people are injured in motor vehicle crashes (see Chapters 23 and 24 in this handbook), but sometimes, it can be difficult to get reliable international crash data about novice drivers. Some explanation of global variation in teen crash injury is enlightening, however. For example, in low-income countries, pedestrian injuries predominate as young people do not have vehicles to drive (World Health Organization, 2013); in middle-income countries, young people have cars, but there is no infrastructure for training, licensure, and enforcement, and cars may be older, with a lack of safety features (Winston et al., 1999); and in high-income countries, efforts have been aimed at improving existing programs, but safety concerns related to issues such as distracted driving, seat belt use, speeding, and helping new drivers gain experience behind the wheel still exist. The global outlook for reductions in motor vehicle crashes shows promise but also areas where more work is needed (see Chapters 23 and 24). The overarching advances in vehicle safety and road safety can be seen not only in the United States but also in the global context.

5.3 Conclusions

Motor vehicle crashes are the leading cause of death in teens, though teen driver crash rates have dropped substantially in the last three decades. The decline can be attributed to advances in vehicle safety, policy, and behavior change for novice teen drivers. Crashes for novice teen drivers include scenarios where they are integrating complex decision making, and teens often commit an error or lack the skill to effectively avoid a crash. These scenarios include car-following scenarios that lead to rear-end collision, negotiating intersections, and curves in the road that result in running off the road. Hazardous driving behaviors also contribute to crash risk, and the intersection between skill and risk taking needs to be better delineated in future research. Further research is also needed to bolster existing crash prevention efforts and develop new ones in order to prevent future deaths and injuries related to novice teen driver motor vehicle crashes.

Acknowledgments

We would like to acknowledge Jeff Caird, Donald Fisher, and William Horrey for their thoughtful review of the drafts of this chapter. During the time this chapter was written, Catherine C. McDonald was supported by the National Institute of Nursing Research of the National Institutes of Health under award numbers K99NR013548 and R00 NR013548. The content is solely the responsibility of the authors and does not necessarily represent the official views of the National Institutes of Health. In addition, we acknowledge the Center for Global Women's Health (M. S. Sommers, center director) at the University of Pennsylvania School of Nursing and the Center for Injury Research and Prevention (F. K. Winston, scientific director) at the Children's Hospital of Philadelphia and Perelman School of Medicine.

References

Bingham, C.R., & Ehsani, J.P. (2012). The relative odds of involvement in seven crash configurations by driver age and sex. *Journal of Adolescent Health, 51*(5), 484–490. doi: http://dx.doi.org/10.1016/j.jadohealth.2012.02.012.

Braitman, K.A., Kirley, B.B., McCartt, A.T., & Chaudhary, N.K. (2008). Crashes of novice teenage drivers: Characteristics and contributing factors. *Journal of Safety Research, 39*(1), 47–54. doi: 10.1016/j.jsr.2007.12.002.

Caird, J.K., Johnston, K.A., Willness, C.R., Asbridge, M., & Steel, P. (2014). A meta-analysis of the effects of texting on driving. *Accident Analysis & Prevention, 71*(0), 311–318. doi: http://dx.doi.org/10.1016/j.aap.2014.06.005.

Carney, C., McGehee, D.V., Harland, K., Weiss, M., & Raby, M. (2015). *Using Naturalistic Driving Data to Assess the Prevalence of Environmental Factors and Driver Behaviors in Teen Driver Crashes.* Washington, DC: AAA Foundation for Traffic Safety.

Centers for Disease Control and Prevention. (2010). Drivers Aged 16 or 17 Years Involved in Fatal Crashes—United States, 2004–2008. *Morbidity and Mortality Weekly Report, 59*(41), 1329–1334.

Centers for Disease Control and Prevention. (2012). Teen Drinking and Driving. Accessed November 26, 2014, retrieved from http://www.cdc.gov/vitalsigns/teendrinkinganddriving/index.html

Centers for Disease Control and Prevention. (2014a). Motor vehicle crash injuries: Costly but preventable. Accessed August 17, 2015, retrieved from http://www.cdc.gov/vitalsigns/crash-injuries/index.html.

Centers for Disease Control and Prevention. (2014b). Teen Drivers: Fact Sheet. Accessed August 17, 2015, retrieved from http://www.cdc.gov/Motorvehiclesafety/Teen_Drivers/teendrivers_factsheet.html.

Centers for Disease Control and Prevention. (2015). Impaired driving: Get the Facts. Accessed September 10, 2015, retrieved from http://www.cdc.gov/motorvehiclesafety/impaired_driving/impaired-drv_factsheet.html.

Chapman, E.A., Masten, S.V., & Browning, K.K. (2014). Crash and traffic violation rates before and after licensure for novice California drivers subject to different driver licensing requirements. *Journal of Safety Research, 50*, 125–138. doi: http://dx.doi.org/10.1016/j.jsr.2014.05.005.

Curry, A.E., Hafetz, J., Kallan, M.J., Winston, F.K., & Durbin, D.R. (2011). Prevalence of teen driver errors leading to serious motor vehicle crashes. *Accident Analysis & Prevention, 43*(4), 1285–1290. doi: 10.1016/j.aap.2010.10.019.

Curry, A.E., Pfeiffer, M.R., Localio, R., & Durbin, D.R. (2013). Graduated driver licensing decal law: Effect on young probationary drivers. *American Journal of Preventive Medicine, 44*(1), 1–7. doi: http://dx.doi.org/10.1016/j.amepre.2012.09.041.

Curry, A.E., Elliott, M.R., Pfeiffer, M.R., Kim, K.H., & Durbin, D.R. (2015). Long-term changes in crash rates after introduction of a graduated driver licensing decal provision. *American Journal of Preventive Medicine, 48*(2), 121–127. doi: http://dx.doi.org/10.1016/j.amepre.2014.08.024.

Ehsani, J.P., Raymond Bingham, C., & Shope, J.T. (2013a). The effect of the learner license Graduated Driver Licensing components on teen drivers' crashes. *Accident Analysis & Prevention, 59*, 327–336. doi: http://dx.doi.org/10.1016/j.aap.2013.06.001.

Ehsani, J.P., Raymond Bingham, C., & Shope, J.T. (2013b). Graduated driver licensing for new drivers: Effects of three states' policies on crash rates among teenagers. *American Journal of Preventive Medicine, 45*(1), 9–18. doi: http://dx.doi.org/10.1016/j.amepre.2013.03.005.

Ehsani, J.P., Bingham, C.R., Ionides, E., & Childers, D. (2014). The impact of Michigan's text messaging restriction on motor vehicle crashes. *Journal of Adolescent Health, 54*(S5), S68–S74. doi: http://dx.doi.org/10.1016/j.jadohealth.2014.01.003.

Farmer, C.M., Kirley, B.B., & McCartt, A.T. (2010). Effects of in-vehicle monitoring on the driving behavior of teenagers. *Journal of Safety Research, 41*(1), 39–45. doi: http://dx.doi.org/10.1016/j.jsr.2009.12.002.

Ferdinand, A.O., Menachemi, N., Sen, B., Blackburn, J.L., Morrisey, M., & Nelson, L. (2014). Impact of texting laws on motor vehicular fatalities in the United States. *American Journal of Public Health, 104*(8), 1370–1377. doi: 10.2105/AJPH.2014.301894.

Foss, R.D., Goodwin, A.H., McCartt, A.T., & Hellinga, L.A. (2009). Short-term effects of a teenage driver cell phone restriction. *Accident Analysis & Prevention, 41*(3), 419–424. doi: http://dx.doi.org/10.1016/j.aap.2009.01.004.

Foss, R.D., Martell, C.A., Goodwin, A.H., O'Brien, N.P., & Center, U.H.S.R. (2011). *Measuring Changes in Teenage Driver Crash Characteristics during the Early Months of Driving.* Washington, DC: AAA Foundation for Traffic Safety.

Goodwin, A.H., O'Brien, N.P., & Foss, R.D. (2012). Effect of North Carolina's restriction on teenage driver cell phone use two years after implementation. *Accident Analysis & Prevention, 48*, 363–367. doi: http://dx.doi.org/10.1016/j.aap.2012.02.006.

Governors Highway Safety Association. (2014). Graduated Driver Licensing (GDL) Laws. Accessed November 21, 2014, retrieved from http://www.ghsa.org/html/stateinfo/laws/license_laws.html.

Guo, F., Simons-Morton, B.G., Klauer, S.E., Ouimet, M.C., Dingus, T.A., & Lee, S.E. (2013). Variability in crash and near-crash risk among novice teenage drivers: A naturalistic study. *The Journal of Pediatrics, 163*(6), 1670–1676. doi: http://dx.doi.org/10.1016/j.jpeds.2013.07.025.

Horrey, W.J., Lesch, M.F., Dainoff, M.J., Robertson, M.M., & Noy, Y.I. (2012). On-board safety monitoring systems for driving: Review, knowledge gaps, and framework. *Journal of Safety Research, 43*(1), 49–58. doi: http://dx.doi.org/10.1016/j.jsr.2011.11.004.

Insurance Institute for Highway Safety. (2015a). Fatality Facts 2013: Teenagers. Accessed September 10, 2015, retrieved from http://www.iihs.org/iihs/topics/t/teenagers/fatalityfacts/teenagers.

Insurance Institute for Highway Safety. (2015b). Teenagers: Overview. Accessed September 10, 2015, retrieved from http://www.iihs.org/iihs/topics/t/teenagers/topicoverview.

Keating, D.P. (2007). Understanding adolescent development: Implications for driving safety. *Journal of Safety Research, 38*(2), 147–157. doi: http://dx.doi.org/10.1016/j.jsr.2007.02.002.

Klauer, S.G., Guo, F., Simons-Morton, B.G., Ouimet, M.C., Lee, S.E., & Dingus, T.A. (2014). Distracted driving and risk of road crashes among novice and experienced drivers. *New England Journal of Medicine, 370*(1), 54–59. doi: doi:10.1056/NEJMsa1204142.

Knipling, R.R. (2015). Naturalistic driving events: No harm, no foul, no validity. Paper presented at the Eighth International Driving Symposium on Human Factors in Driver Assessment, Training and Vehicle Design, Salt Lake City, Utah.

Lee, S.E., Simons-Morton, B.G., Klauer, S.E., Ouimet, M.C., & Dingus, T.A. (2011). Naturalistic assessment of novice teenage crash experience. *Accident Analysis & Prevention, 43*(4), 1472–1479. doi: http://dx.doi.org/10.1016/j.aap.2011.02.026.

Lenhart, A., Purcell, K., Smith, A., & Zickuhr, K. (2010). Social media and mobile internet use among teens and young adults. *The Millenials.* Accessed November 6, 2014, retrieved from http://pewinternet.org/~/media//Files/Reports/2010/PIP_Social_Media_and_Young_Adults_Report_Final_with_toplines.pdf.

Martiniuk, A.C., Senserrick, T., Lo, S., Williamson, A., Du, W., Grunstein, R.R., Woodward, M., Glozier, N., Stevenson, M., Norton, R., & Ivers, R.Q. (2013). Sleep-deprived young drivers and the risk for crash: The drive prospective cohort study. *JAMA Pediatrics, 167*(7), 647–655. doi: 10.1001/jamapediatrics.2013.1429.

Masten, S.V., & Foss, R.D. (2010). Long-term effect of the North Carolina graduated driver licensing system on licensed driver crash incidence: A 5-year survival analysis. *Accident Analysis & Prevention, 42*(6), 1647–1652. doi: http://dx.doi.org/10.1016/j.aap.2010.04.002.

Masten, S.V., Foss, R.D., & Marshall, S.W. (2011). Graduated driver licensing and fatal crashes involving 16 to 19 year old drivers. *JAMA, 306*(10), 1098 1103. doi: 10.1001/jama.2011.1277.

Mayhew, D.R., Simpson, H.M., & Pak, A. (2003). Changes in collision rates among novice drivers during the first months of driving. *Accident Analysis & Prevention, 35*(5), 683–691. doi: http://dx.doi.org/10.1016/S0001-4575(02)00047-7.

McCartt, A.T., Shabanova, V.I., & Leaf, W.A. (2003). Driving experience, crashes and traffic citations of teenage beginning drivers. *Accident Analysis & Prevention, 35*(3), 311–320. doi: http://dx.doi.org/10.1016/S0001-4575(02)00006-4.

McCartt, A.T., Teoh, E.R., Fields, M., Braitman, K.A., & Hellinga, L.A. (2010). Graduated licensing laws and fatal crashes of teenage drivers: A national study. *Traffic Injury Prevention, 11*(3), 240–248. doi: 10.1080/15389580903578854.

McCartt, A.T., Oesch, N.J., Williams, A.F., & Powell, T.C. (2012). New Jersey's license plate decal requirement for graduated driver licenses: Attitudes of parents and teenagers, observed decal use, and citations for teenage driving violations. *Traffic Injury Prevention, 14*(3), 244–258. doi: 10.1080/15389588.2012.701786.

McCartt, A.T., Kidd, D.G., & Teoh, E.R. (2014). Driver cellphone and texting bans in the United States: Evidence of effectiveness. *Annals of the Advancement of Automative Medicine, 58*, 99–114.

McDonald, C.C., Curry, A.E., Kandadai, V., Sommers, M.S., & Winston, F.K. (2014). Comparison of teen and adult driver crash scenarios in a nationally-representative sample of serious crashes. *Accident Analysis & Prevention, 72*, 302–308. doi: 10.1016/j.aap.2014.07.016.

McGehee, D.V., Raby, M., Carney, C., Lee, J.D., & Reyes, M.L. (2007). Extending parental mentoring using an event-triggered video intervention in rural teen drivers. *Journal of Safety Research, 38*(2), 215–227. doi: http://dx.doi.org/10.1016/j.jsr.2007.02.009.

McKnight, A.J., & McKnight, A.S. (2003). Young novice drivers: Careless or clueless? *Accident Analysis & Prevention, 35*(6), 921–925. doi: http://dx.doi.org/10.1016/S0001-4575(02)00100-8.

Olsen, E.O.M., Shults, R.A., & Eaton, D.K. (2013). Texting while driving and other risky motor vehicle behaviors among US high school students. *Pediatrics, 131*(6), e1708–e1715.

Ouimet, M., Brown, T.G., Guo, F., Klauer, S.G., Simons-Morton, B.G., Fang, Y., Lee, S.E., Gianoulakis, C., & Dingus, T.A. (2014). Higher crash and near-crash rates in teenaged drivers with lower cortisol response: An 18-month longitudinal, naturalistic study. *JAMA Pediatrics, 168*(6), 517–522. doi: 10.1001/jamapediatrics.2013.5387.

Peck, R.C., Gebers, M.A., Voas, R.B., & Romano, E. (2008). The relationship between blood alcohol concentration (BAC), age, and crash risk. *Journal of Safety Research, 39*(3), 311–319.

Peek-Asa, C., Britton, C., Young, T., Pawlovich, M., & Falb, S. (2010). Teenage driver crash incidence and factors influencing crash injury by rurality. *Journal of Safety Research, 41*(6), 487–492. doi: http://dx.doi.org/10.1016/j.jsr.2010.10.002.

Senserrick, T., & Haworth, N. (2005). *Review of Literature regarding National and International Young Driver Training, Licensing and Regulatory Systems.* Victoria, Australia: Monash University Accident Research Center.

Shope, J.T. (2007). Graduated driver licensing: Review of evaluation results since 2002. *Journal of Safety Research, 38*(2), 165–175. doi: http://dx.doi.org/10.1016/j.jsr.2007.02.004.

Shope, J.T., & Bingham, C.R. (2008). Teen driving: Motor-vehicle crashes and factors that contribute. *American Journal of Preventive Medicine, 35*(3, Supplement 1), S261–S271. doi: 10.1016/j.amepre.2008.06.022.

Sommers, B.D., Fargo, J.D., Lyons, M.S., Shope, J.T., & Sommers, M.S. (2011). Societal costs of risky driving: An economic analysis of high-risk patients visiting an urban emergency department. *Traffic Injury Prevention, 12*(2), 149–158. doi: 10.1080/15389588.2010.536599.

Steinberg, L. (2007). Risk taking in adolescence. *Current Directions in Psychological Science, 16*(2), 55–59. doi: 10.1111/j.1467-8721.2007.00475.x.

The Center for Injury Research and Prevention at the Children's Hospital of Philadelphia. (2013). Miles to go: Focusing on risks for teen driver crashes. Accessed November 18, 2014, retrieved from https://teendriving.statefarm.com/system/article_downloads/2013_miles_to_go_report.pdf.

Ward, N. (2007). *The Culture of Traffic Safety in Rural America* (pp. 241–256). Washington, DC: AAA Foundation for Traffic Safety.

Williams, A.F. (2013). Commentary: Teenage driver fatal crash rate trends: What do they reveal? *Traffic Injury Prevention, 15*(7), 663–665. doi: 10.1080/15389588.2013.878802.

Williams, A.F., & Tefft, B.C. (2014). Characteristics of teens-with-teens fatal crashes in the United States, 2005–2010. *Journal of Safety Research, 48*, 37–42. doi: http://dx.doi.org/10.1016/j.jsr.2013.11.001.

Williams, A.F., Tefft, B.C., & Grabowski, J.G. (2012). Graduated driver licensing research, 2010–present. *Journal of Safety Research, 43*(3), 195–203. doi: http://dx.doi.org/10.1016/j.jsr.2012.07.004.

Winston, F.K., Rineer, C., Menon, R., & Baker, S. P. (1999). The carnage wrought by major economic change: Ecological study of traffic related mortality and the reunification of Germany. *British Medical Journal, 318*(7199), 1647–1649. doi: 10.2307/25184983.

World Health Organization. (2013). *Global Status Report on Road Safety 2013: Supporting a Decade of Action*. Geneva, Switzerland: WHO.

II

Skill Differences

6

Hazard Avoidance in Young Novice Drivers: Definitions and a Framework

Anuj K. Pradhan

David Crundall

Abstract

There are a number of elements of driving skills and abilities of young novice drivers that have been identified as critical to their safe operation of motor vehicles. These play a significant role in the overrepresentation of that cohort in roadway injuries and fatalities. The ability to efficiently and accurately detect, predict, recognize, plan for, and mitigate overt and developing hazards have been recognized to be of prime importance, with performance in these skills generally discriminating between novice and experienced drivers. The consensus from research groups and institutions across the globe is that these skills are critical to the safety of novice drivers. There is also acknowledgement that these skills can be trained in an accelerated fashion without months of behind-the-wheel exposure. However, despite, or perhaps because of, the diversity of research on the subject matter, there is little agreement about the terminology and definitions of constructs and concepts in the field. Additionally, the field lacks a theory to explain how drivers detect and respond to a roadway hazard. In this chapter, we put forward definitions and terminology for critical concepts and constructs related to roadway hazards. We then assess the current theoretical understanding of hazard perception and propose a working framework of hazard avoidance to help us understand why young novice drivers crash.

6.1 Introduction: Young Drivers and Roadway Hazards

Young novice drivers are overrepresented in crashes as compared to experienced drivers, especially in the first 6–12 months after licensure (Foss et al., 2011; McCartt et al., 2003; Chapter 5). This statistic holds true in the United States and globally (Christie, 2001; Mayhew et al., 2003). A variety of diverse factors may play a role, e.g., deficits in vehicle-handling skills, lack of driving experience, immaturity, distraction, or the dangerous practice of driving under the influence of alcohol or drugs. However, safe driving also requires that novice drivers efficiently, effectively, and accurately predict, detect, recognize, plan for, and mitigate against overt or developing hazards in the driving environment (McKnight & McKnight, 2003). This is the focus of this chapter.

A significant amount of research has shown that this particular skill set (termed variously as *hazard perception*, *hazard anticipation*, *hazard recognition*, etc.) is clearly deficient in young novice drivers, as compared to experienced drivers (McKenna & Crick, 1994; Quimby & Watts, 1981). This deficiency arises due to a combination of inexperience (e.g., lack of exposure to a comprehensive set of driving situations with hazards) (Braitman et al., 2008; McCartt et al., 2009), lack of training (e.g., not enough emphasis on the role of higher-order skills such as anticipating and predicting hazards during driver education) (Mayhew & Simpson, 2002), and the resulting miscalibration of abilities (e.g., the young driver thinks himself more capable of safe driving than he actually is) (De Craen et al., 2011; Horrey et al., 2015). The breadth of research conducted on the subject matter further underscores the general scientific consensus on the importance of this skill set. In addition to the pioneering research that showed these skills to discriminate between young novice drivers and experienced drivers, there have been a significant number of research efforts aimed at identifying and understanding the finer distinctions of these skills. This includes work on precursors of hazards (Crundall et al., 2012; Sagberg & Bjørnskau, 2006) and work on the tactical and strategic aspects of this skill set (Borowsky et al., 2009; Pradhan et al., 2005).

Collectively, all of this research has produced important tangible outcomes in terms of translational research on development and training (Chapter 22; Fisher et al., 2006; McDonald et al., 2015; Pradhan et al., 2009). Chapter 18 in this handbook offers a comprehensive look at training programs for novice drivers related to roadway hazards, and although this section will not expand more on that research, it is still of relevance given the aims of this chapter.

Significant advances have been achieved in the hazard perception field, and a diverse and important body of literature has been generated. However, the diversity of the literature has produced a lack of cohesion and scientific standardization in terminology, methods, approaches, metrics, participant cohorts, and analytical methods. Such shortcomings do a disservice to the field in terms of communication and replication, especially as efforts converge in translational research that has the potential to make a significant and real impact on the safety of young novice drivers. In recognition of these issues, this chapter will take a first step toward establishing common terminology and definitions, and then use these definitions to develop a conceptual framework for understanding the processes involved in safely and efficiently dealing with roadway hazards.

6.2 Aims: Terminologies and Framework

The ability to predict or detect driving-related hazards in sufficient time to avoid a collision has been given many different names. Early UK pioneers termed it *hazard perception* (e.g., McKenna & Crick, 1994; Quimby & Watts, 1981). This terminology has persisted in some research groups (cf. Crundall, 2016; Horswill et al., 2015) and has been enshrined in official tests that now form part of the official driver-licensing procedure in the United Kingdom and certain states of Australia. These tests require participants to watch a series of videos that present the point of view of a driver. Participants have to press a button as quickly as possible to acknowledge any hazards they perceive, with quicker responses reflecting safer drivers, who are then allowed to progress to their on-road test (see Chapter 28 for an in-depth discussion on hazard perception tests).

Over time, however, different research groups around the world have adopted other terms, such as *hazard recognition*, *detection*, *processing*, *avoidance*, and *anticipation* (e.g., Doherty et al., 1998; McDonald et al., 2015; Underwood et al., 2013; Vidotto et al., 2011). Some of these definitions are used interchangeably, though all have slightly different connotations. Even between the chapters in this handbook, there is a lack of standardization of terminology associated with the science and research on hazard perception. Unfortunately, this means that many researchers now use different terms to define similar, if not identical, processes. Not only does this create confusion within the research community; it acts as a barrier to end users, such as practitioners and legislators, and potentially limits the impact that hazard-based testing and training could have on global crash statistics.

For instance, McDonald et al. (2015) have rejected *hazard perception* in favor of *hazard anticipation* to represent "the broad number of terms used in the literature depicting the multiple components related to constructs of cognitive awareness, visual perception, and experiential and schema-based recognition." In contrast, Crundall (2016) has retained the overarching term *hazard perception* but has defined *hazard prediction* (the ability to predict an imminent hazard before it is considered hazardous) as a subcomponent of hazard perception. The terms *prediction* and *anticipation* appear to have a good deal of overlap, yet these authors use these words to refer to very different concepts (either a very small part of the hazard response process or the whole process itself). This lack of agreement in the literature is perhaps most apparent when two authors, representing different terminological camps, come together to write a chapter such as this one.

In order to provide some clarity in definition and terminology for the field and thus ensure better communication between researchers, we decided that a logical and useful *first aim of the chapter* would be to generate common definitions that are as precise as possible. These definitions will serve to help the reader understand the different processes involved in how novice drivers tackle on-road hazards. If comprehensive and noncontroversial, we hope this effort may help other authors adopt these definitions, with the field eventually progressing toward a universal terminology.

The other problem in the field of hazard perception is the lack of a robust theory to explain how the act of responding in a timely manner to an on-road hazard is performed. This theoretical lacuna at the heart of the field has not gone unnoticed (Crundall, 2016; Vlakveld, 2008), but the only theoretical framework that has been put forward is that of *situation awareness* (SA) (e.g., Horswill & McKenna, 2004). At best we have a loose association between hazard perception and SA, though many of the sub-processes that we suspect are involved in responding to hazards are not explained through this methodological approach.

Certainly, further work is required in identifying a theory of hazard perception. An additional benefit of defining the various terms in the field (beyond improving the current authors' interpersonal communication) is that the creation and consolidation of definitions requires scrutiny directed at the hypothesized processes that underlie hazard responses. Clear definitions will make it easier to discuss theoretical models and identify gaps in our understanding, while the process of defining terms may even lead to new theoretical insights. This is *the second aim of this chapter*: to assess the current theoretical understanding of the hazard perception process in regard to young and novice drivers and propose a novel conceptual framework for hazard avoidance while driving.

Finally, we will discuss the future work and logical extensions of this effort, including the need to relate the literature on young novice drivers' crash-related behaviors to the proposed framework. Given the wealth of studies examining the deficiencies of young novice drivers in these skills, we anticipate that the proposed definitions and theoretical structures will allow the parsing of current evidence to identify the aspects of hazard avoidance that are most critical for this cohort. Such an exercise will provide directions for the development of additional training interventions, within an overarching framework, for those subskills necessary for efficient hazard avoidance behaviors in young novice drivers.

6.3 Defining the Concepts, Constructs, and Components of the Hazard Avoidance Process

The first definition that we require is one to describe the whole process, from detecting a hazard, though the stages of processing and appraisal, before one makes a behavioral response in order to avoid a collision. The traditional term *hazard perception* is perhaps inadequate, as the word *perception* does not reflect the postperceptive processes of appraisal and decision making. There have been many definitions of *hazard perception* (cf. Jackson et al., 2009) that vary in the number of subcomponents included, but even those that include an element of responding in their definition (e.g., *hazard latency* [Deery, 2000]; *perception-reaction time* [Sagberg & Bjørnskau, 2006]) imply a simple response, such as a button press, rather than the selection of the most appropriate evasive maneuver. We argue that a new term (or at least a reimagined term) is required to describe the whole process from initially being vigilant to the possibility of a hazard; through to spotting the hazard, processing the hazard, and evaluating the risk that it poses; and finally selecting the appropriate response and successfully avoiding the hazard.

We thus propose that an appropriate overarching term to describe the whole process is *hazard avoidance*. Within *hazard avoidance* lie a number of subprocesses. Performance differences in some of these subprocesses can be found to discriminate between young novice drivers and safer and more experienced drivers. We have listed initial definitions of these subprocesses in Table 6.1, along with some other terms of note. The remainder of this section will attempt to better define and elucidate these terms and subprocesses, while drawing on examples from the literature on novice drivers to support these definitions.

TABLE 6.1 Suggested Definitions for the Processes Involved in Avoiding On-Road Hazards

Term	Definition
Hazard perception	A collection of hazard avoidance subprocesses, which variably include hazard searching, hazard prediction, precursor prioritization, hazard fixation, hazard processing, hazard appraisal, and hazard reaction.
Hazard avoidance	The overarching term to describe the process of avoiding a collision with a hazard from initial searching for hazards through to the successful selection of an appropriate response.
Hazard salience	The ability of a hazard to draw attention to itself through bottom-up features including sudden movement, looming, color, luminance, etc.
Hazard searching	The direction of overt attention (i.e., eye movements) to areas of the visual scene that are most likely to produce hazards.
Hazard precursor	The clues to an upcoming hazard (e.g., a pedestrian on the pavement is a precursor prior to stepping into the road and becoming a hazard).
Hazard evidence	The visual evidence contained within an object or location that the driver uses to judge immediate danger (a hazard) or a strong potential for danger (a precursor).
Abrupt hazard	A hazard that appears without (or ostensibly without) any precursor.
Precursor prioritization	Prioritizing and labeling precursors most likely to produce hazards for continued monitoring through overt and covert attention.
Hazard prediction	Predicting a hazard before it appears on the basis of the hazard evidence present in precursors.
Hazard onset	When a precursor turns into a hazard, e.g., a pedestrian steps off the sidewalk into the road or from behind a parked vehicle, an oncoming car begins to turn across your path, etc.
Hazard fixation latency	The time taken to first fixate the hazard. This can be zero if the driver if fixating the precursor at the time it becomes a hazard.
Hazard processing	The time taken to identify the object as a hazard, primarily reflected in fixation durations and dwell time upon the hazard.
Hazard appraisal	Assessing the level of risk posed by the hazard both in absolute terms and in relation to one's own self-perceived skills.
Hazard mitigation	The act of reducing the probability of a collision with a future hazard by changing one's driving behavior (e.g., changing lane position, headway) on the basis of prioritized precursors.
Hazard reaction	Any behavioral outcome from identifying a hazard. This could be positive (braking) or negative (freezing).
Hazard response	A subsection of hazard reactions composed of deliberate actions (e.g., intentional braking). While more likely to be positive, a poorly chosen intentional response can still be negative (e.g., trying to overtake a braking car when oncoming traffic is too close).

Visual search for on-road hazards is driven by both *hazard salience* and *hazard searching*. Salience refers to the ability of external objects to attract attention to themselves via their conspicuity. Much research has been done on modeling the way that the attentional system devotes attention to natural scenes on the basis of *bottom-up* factors such as color, luminance, movement, contrast etc. (Kinchla & Wolfe, 1979). It is generally accepted that sudden changes in conspicuity are especially good at attracting attention (e.g., a flashing light, the sudden emergence of a car from a side road, etc.). However, many studies have also demonstrated that models of bottom-up salience do not always capture visual search in naturalistic settings. This is primarily due to the goals of the viewer when looking at the scene. For instance, a driver searching for a particular side road is more likely to seek out road signs than another driver who has no intention to turn. There is good reason to believe that many highly salient objects in the driving scene may be relevant to safety (McCarley et al., 2014). Brake lights, traffic signals, and warning signs are designed to catch attention via salience, while sudden movement and looming can also capture attention. However, as McCarley et al., point out, some safety-critical information may only be found in low-salient areas.

Such top-down influences on visual search were reported nearly half a century ago (Yarbus, 1967) but are only recently being added to bottom-up models of salience (e.g., Torralba et al., 2007). We believe that these two concepts, captured in the terms *hazard salience* and *hazard searching*, are highly useful when discussing young novice drivers' hazard avoidance skills. Although it may be a slight oversimplification, one typically might expect top-down factors reflected in *hazard searching* to benefit much more from training and experience than the bottom-up factors that contribute to *hazard salience*.

Many researchers have found that young novice drivers' general visual search patterns differ from those of more experienced drivers (Chapman & Underwood, 1998; Crundall & Underwood, 1998; Falkmer & Gregersen, 2005; Konstantopoulos et al., 2010; Lehtonen et al., 2014; Mourant & Rockwell, 1970, 1972; Underwood et al., 2002, 2003). In simulators, video-based tests, and naturalistic driving studies, researchers consistently find that young inexperienced drivers have a narrower search pattern, focusing on the road ahead, while more experienced drivers tend to search a wider area of the visual scene. Novices also often focus closer to the car hood than more experienced drivers. Underwood et al. (2002) argued that finding such differences even when drivers simply watch videos of driving suggests that this effect is not solely due to the cognitive demand of controlling the vehicle (which may be understandably greater for novices) but that this reflects a lack of the appropriate mental model for how to search. Without knowledge of where to look to spot potential hazards, novice drivers are less likely to employ the most appropriate visual strategies. See Chapter 7 for further discussion of the interaction of top-down and bottom-up factors in the guidance of drivers' visual attention.

While both *searching* and *salience* may direct a driver's attention to an actual hazard, it is more likely that one will be first attracted to a *hazard precursor*. The *hazard precursor* is different than the hazard itself, in that it provides clues to upcoming hazards. There are different types of precursors (which again suffer from different definitions from different authors), though two of the most studied include the behavioral precursor and the environmental precursor (e.g., Crundall, 2016; Crundall et al., 2012). A behavioral precursor is the same object as the hazard but has yet to become hazardous. Pedestrians on the sidewalk may be a precursor, and when they step into the road, they become the hazard. Environmental precursors often identify locations of the scene that may hide hazards, such as a high-sided parked vehicle, the brow of a hill, or a blind curve. In such instances, the high-sided vehicle could be considered an environmental precursor to a pedestrian hazard, with learned statistical co-occurrences leading more experienced drivers to better anticipate the possibility that a parked bus might hide a small child who is about the cross the road. The *hazard precursor* is a logical precondition of any hazard, and the ability to identify and recognize these precursors discriminates between safer, more experienced drivers and less safe inexperienced drivers.

Without a precursor, there is no *hazard evidence* that will allow for an experienced driver to plan an appropriate response prior to the appearance of the actual hazard. Thus, *abrupt hazards* (defined as having no precursor—e.g., a soccer ball suddenly enters the road from the side) will most likely favor younger drivers, as they will have faster simple reaction times to abrupt onsets (Salthouse, 1996). These faster reaction times may mask the effects of experience (see also Chapter 8 for a discussion of hazard mitigation). For instance, Yeung and Wong (2015) compared three groups of drivers (young inexperienced, young experienced, and old experienced) viewing abrupt hazards and found no significant differences in response times, though older drivers were slower to first fixate the hazard. They argue that deterioration in detection speed with age masked the effect of experience.

Fixation of these precursors has been found to be particularly difficult for novice drivers. Crundall et al. (2012) found that drivers were slower to fixate behavioral precursors when compared to more experienced drivers; however, Crundall (2016) found that hazards preceded by an environmental hazard were better discriminators than hazards with behavioral precursors, and many studies have shown that novices are less likely to look at environmental precursors than experienced drivers (e.g., Borowsky et al., 2010; Pradhan et al., 2005). If novice drivers fail to fixate precursors, or fail to comprehend their importance even following fixation (Crundall et al., 2012), then they are less likely to spot a subsequent hazard in time to avoid it.

Assuming that one or more hazard precursors are indeed identified, they will then be *prioritized* for subsequent monitoring. The experienced driver should be able to *predict* whether a particular hazard is likely to occur on the strength of the *hazard evidence* in these precursors and will continuously monitor the most likely sources of danger. Thus, if a precursor should develop into a hazard, the experienced driver should spot it first.

The transition from precursor to actual hazard is termed the *hazard onset*, but can this be defined with any temporal accuracy? Let us consider the pedestrian on the sidewalk. If the pedestrian makes frequent over-the-shoulder glances into the roadway, this increases the *hazard evidence* in this precursor (and therefore increases prioritization of the precursor for future monitoring). Finally, pedestrians become a hazard when they begin on a course of action that will inevitably bring them into conflict with the participant's car (e.g., stepping into the road). The precise definition of the hazard onset will, of course, vary from hazard to hazard: stepping into the road may be an obvious onset in one particular situation but less so in another (where the road may be quite wide and the pedestrian steps into the lane but is still obviously waiting for the car to pass). Thus, while some hazard onsets may be very easy to define temporally, other—highly similar—hazards may be more difficult. For this reason, each hazard onset must be considered individually. This creates particular problems for defining temporal scoring windows in traditional hazard perception tests (see Chapter 28).

Following onset, drivers will tend to *fixate* on the hazard in order to process it. *Hazard processing* involves identifying what the hazard is and whether it is actually dangerous. Thus, the sudden brake lights of the car ahead must first be identified as indicative of a rapid change in speed of the car ahead and then be related to one's current headway and the likely duration and extent of the braking of the car ahead (which is inevitably related to the cause of the braking). While many of these cues (current headway, severity of the braking, etc.) can be accessed simply by looking at the car ahead, some of them may require looking elsewhere (i.e., ahead of the braking car to ascertain the cause of the sudden deceleration and whether it will continue). Unfortunately, hazards have a tendency to capture the eyes of the viewer, often preventing cues from beyond the immediate vicinity from being considered. This is especially the case in young novice drivers, who have been noted to have longer fixation durations on hazards while watching hazard perception clips (e.g., Chapman & Underwood, 1998). Fortunately, a successful prediction of the hazard should reduce the required processing time, perhaps in a similar way to the effects of semantic priming, with cues to the hazard lowering the threshold for accepting the hazard once it appears (cf. Crundall & Underwood, 2001).

Hazard appraisal overlaps with hazard processing in that it requires drivers to assess the absolute danger posed by the event, yet it goes beyond the processing of the hazardous object in that it also requires the driver to relate this perceived level of danger to his or her self-perceived driving skills (and an understanding of what the car is capable of). Assuming a rational driver, if the perceived absolute danger is greater than self-perceived skill, then the driver will take appropriate evasive action, though this may be modified according to risk propensity and sensation-seeking traits. For example, imagine a situation where a young driver notices a car emerging from a side road ahead into her lane. Experienced drivers may decelerate to ensure that they do not crash into the emerging car. Novice drivers, however, may not perceive the distance between the car ahead and themselves to be dangerous, as they believe that the braking distance is within their skill range and the capacity of their car.

We know that novice drivers tend to have problems processing and appraising hazards, even if they have spotted them. First, we have evidence that novice drivers' fixations on hazards tend to be longer than those of more experienced drivers (Chapman & Underwood, 1998). Secondly, there is evidence that inexperienced drivers demonstrate fewer physiological responses to obvious hazards. Kinnear et al. (2013) reported that the probability of a hazard evoking a skin conductance response (SCR) increased with driver experience, which they argue reflects inappropriate appraisal of the hazard and events leading up to the hazard (precursors). Finally, Crundall et al. (2012) found that learner drivers did fixate some environmental precursors but without realizing their importance. Thus, these precursor fixations did not help them respond to the hazard as fast as the more experienced drivers. Even if novice drivers do successfully process the event or object as a developing hazard, they must then relate it to their own skill and the capability of the car. Unfortunately, inexperienced drivers are likely to be inadequately calibrated for this task (Horrey et al., 2015) and may inadvertently underappraise the level of risk posed.

Finally, in Table 6.1, we have made a distinction between three different types of behavior. *Hazard mitigation* refers to preemptive behaviors that reduce the possibility of a collision based on the evidence present in a precursor. For instance, if there is a car in a side road that may pull out, it might be appropriate to change one's lane position or speed to better handle the hazard should it occur. Successful mitigation can prepare the driver to better respond should the predicted hazard occur or may even render the precursor inert. For example, if one accelerates past a freeway slip lane, this removes the possibility that a parallel car attempting to join the freeway will come into conflict with him or her (see also Chapter 8). Without perception of the precursors, followed by appropriate processing and appraisal, novice drivers are unlikely to be able to mitigate the threat of a hazard. This was noted in a recent simulator study, which found that a group of novice drivers who were given training in producing a live commentary slowed down and adjusted their lane position in a simulator in the presence of precursors. Untrained novices did not (Crundall et al., 2010).

If a hazard does manifest, the driver may make a *hazard reaction* or a *hazard response*. A reaction is any unplanned behavior as a result of noticing the hazard, which can be either positive (e.g., automatically braking) or negative (e.g., stepping on the accelerator by mistake, freezing, steering toward the hazard). A response, however, is a deliberate action that is selected (e.g., intentional braking or swerving). A response is more likely to lead to a positive outcome than a reaction, especially if the response is primed by successful hazard prediction and is preceded by hazard mitigation.

Taken together, these subprocesses can be used to describe a number of scenarios. For instance, the conscious search for a hazard can be explained as *searching > precursor prioritization > hazard prediction > fixation > processing > appraisal > response*. Alternatively, an automatic reaction to a late-appearing abrupt hazard might be described as *salience > hazard fixation > reaction*, while a driver who mitigates the danger evident in a precursor might pass through the following stages: *searching > precursor prioritization > hazard prediction > mitigation > precursor prioritization* and so on, in an iterative loop.

As described in the previous paragraphs, there is evidence that the problem of young novice drivers manifests in several of the subprocesses. Novices do not have the knowledge of where to look in the scene and have inflexible scan patterns that focus too narrowly and too close to the car ahead (see also Chapter 7). Thus, precursors are missed, and even when spotted, they may be misunderstood or incorrectly appraised. When the hazard finally appears, these novices are ill prepared to respond and are therefore more likely crash.

6.4 Theories Relevant to Hazard Perception and a Proposed Framework

Endsley's model of SA suggests that people develop situational models of their environment in relation to themselves, moving through an iterative linear process of *perception, comprehension, and prediction* (Endsley, 1988, 1995). While this sounds like an intuitively appealing way to explain how people avoid hazards on the road, some contend that the level of empirical evidence for SA is below the threshold for developing a coherent theory (Durso & Gronlund, 1999). Certainly, the number of studies linking SA to a variety of applied settings has increased considerably over the last decade, but Salmon and Stanton's recent editorial (2013) makes it clear that theoretical consensus has yet to surface.

While SA has undoubtedly served a purpose in the hazard avoidance literature, it is both too limited and too vague to account for hazard avoidance behavior. The three levels do not explain how one perceives the elements of the scene in the first place, what processing is involved in the comprehension level, or how one chooses the correct response on the basis of SA. Endsley (2000) explicitly stated that SA was an antecedent to decision making, and that it was possible to successfully predict the upcoming situation yet still choose a suboptimum course of action. Thus, it appears that SA is not sufficient to explain hazard avoidance and even has trouble explaining the hazard perception subprocesses of hazard avoidance. It is not clear that SA provides a better framework for hazard avoidance than our table of subprocesses, though we acknowledge the fact that SA has contributed to the development of these definitions.

Michon's review of models of driver behavior also identified a three-level framework that included *strategic, tactical,* and *operational* levels (Michon, 1985). As this framework was intended to cover all driving behavior, the three levels were extremely broad. The strategic level, for instance, included navigation planning prior to setting off on the journey, while the tactical level was concerned with making decisions while driving. The operational level was reserved for the most immediate behaviors, such as maneuvers. In regard to hazard avoidance, while one can argue that planning a trip to avoid certain roads (at Michon's strategic level) could be considered part of hazard management, if we restrict the concept of *hazard avoidance* to how one survives driving from instant to instant (as most researchers do), then Michon's model is somewhat limited for the current topic. All hazard-related perception, processing, and response will occur at Michon's tactical level, which leaves the model with limited explanatory power for comprehensively understanding hazard avoidance.

However, if Michon's categories are recalibrated to the activity of driving along a single road, then the model becomes more relevant. Similarly, the levels within Endsley's SA model offer a process flow and relevance to environmental factors that can contribute to any modeling of the hazard avoidance processes. These joint relevancies allow for the conceptualization of an adapted framework for hazard avoidance based on constructs of cognition, perception, processing, and action.

In addition to the adaptation of relevant elements from Michon's and Endsley's models, any conceptual framework for hazard avoidance will benefit from a consideration of the concepts of environmental perception in terms of distance and space. Space can be categorized by relative proximity to the perceiver, with category boundaries determined by one's ability to interact with items at varying distances. One such distinction is between personal, peripersonal, and extrapersonal space (e.g., Spence & Ho, 2008; Previc, 1998). Personal space is the area that one's body currently occupies. Peripersonal space is the area beyond the body in which one can still reach and manipulate objects, while extrapersonal space refers to those locations that fall beyond one's reach. These three representations of space may be relevant to the driving context. Personal space might still reflect the location of the body in the driver's seat, while peripersonal space might include all space within the car that one could reach, with everything else being classed as extrapersonal. However, tool use can ostensibly extend peripersonal space. Berti and Frassinetti (2000) found that a patient with left visuospatial neglect had a high level of error when bisecting a nearby (peripersonal) line with a finger but a low level of error when bisecting a far line (extrapersonal) with a laser pointer. When the patient used a stick (i.e., a tool) to bisect the far line, the level of error rose to a comparable level with that of near bisection. This suggests that tool use extends peripersonal space, resulting in previously extrapersonal locations being encoded more in line with peripersonal representations. If one considers the car as a tool, then peripersonal space could be hypothesized to extend to those objects that one can immediately interact with via the tool, rather than by hand. Thus, nearby traffic and pedestrians could be considered to reflect peripersonal space, while the whole interior of the car becomes personal space (e.g., Blakeslee & Blakeslee, 2007). Traffic beyond the immediate space would be encoded as extrapersonal space, though as Spence and Ho (2008) note, there is no clear borderline that separates the extrapersonal from the peripersonal. Recent research on distance evaluations from the perspective of drivers and pedestrians (Moeller et al., 2015) also concluded that cars altered the perception of far distances because they modulated the driver's perception in the same way as tools typically do, changing the driver's perceived potential for action. Thus, compared to a nondriver, the environmental perception of the driver is somewhat skewed as it relates to distance perception and, hence, the potential for action.

Therefore, the driver's environmental perception and potential for action should be taken into account when conceptualizing a framework for hazard avoidance. Our proposed framework involves zones of perceived distance where the classification of visual objects and scenes takes place and hazards and their precursors are identified. There are also zones where hazard processing occurs and mitigation strategies are formed. The framework can be illustrated by imagining that the road ahead is divided into different zones of functional importance to the driver according to distance ahead of the vehicle (Figure 6.1). The notion of segregating the path of travel by distance is not new. Indeed, Gibson and Crooks (1938) discussed the role of distance-based zones ahead of the vehicle based on one's ability to interact with objects at distance (e.g., the minimum stopping zone). This current framework proposes four functional zones rather than the three in Michon's model: a *vigilance* zone, a *strategic* zone, a *tactical* zone, and an *operational* zone.

The *vigilance* zone typically incorporates the focus of expansion and is the furthest that one can see (the green zone in Figure 6.1). In this zone, distantly located roadway elements or traffic does not pose an immediate threat and is probably too small for the driver to visually interrogate (i.e., the *hazard evidence* is too low to be even classed as a precursor). However, appropriate vigilance in this zone can help timely identification of elements that can be marked for further inspection when the driver gets into the strategic zone.

In the *strategic zone* (yellow in Figure 6.1), potential precursors to hazards become salient, though they may not necessarily require an immediate response. Drivers may use a strategic model of prediction based on experience, risk schemas, or heuristics and prioritize the hazards accordingly. If the *hazard evidence* for a precursor reaches a particular threshold, this may induce hazard mitigation strategies (e.g., moving toward the edge of the lane, slowing down, etc.) even before the precursor evolves into a hazard. Hazards will generally not occur in the strategic zone. For example, if a pedestrian steps into the road within a driver's strategic zone, he or she should be sufficiently far away to pose little threat.

The *tactical zone* (red in Figure 6.1) is closer to the car than the strategic zone. It is in this zone that objects become hazardous. For instance, a pedestrian stepping in front of the driver's car in the tactical zone poses an immediate threat that may require an evasive maneuver. If a hazard appears in the tactical zone but the driver did not spot a corresponding precursor in the strategic zone, then the hazard will appear abruptly. *Abrupt hazards* are the most difficult to avoid, as there is no opportunity to mitigate the hazard in the strategic zone and no opportunity to plan for an evasive tactical maneuver. If the driver has prepared for the possibility of a tactical hazard, he or she will make an intentional *hazard response* that should hopefully avoid a collision. If the driver was unprepared for the tactical hazard, then he or she may make no response at all (simply colliding with the hazard, or luckily passing by without realizing how close he or she came), or he or she will make an instinctive *hazard reaction*. While a hazard response is intentional and most likely to lead to a successful avoidance, a hazard reaction is a more automatic behavior that could actually increase the level of danger (e.g., swerving into oncoming traffic).

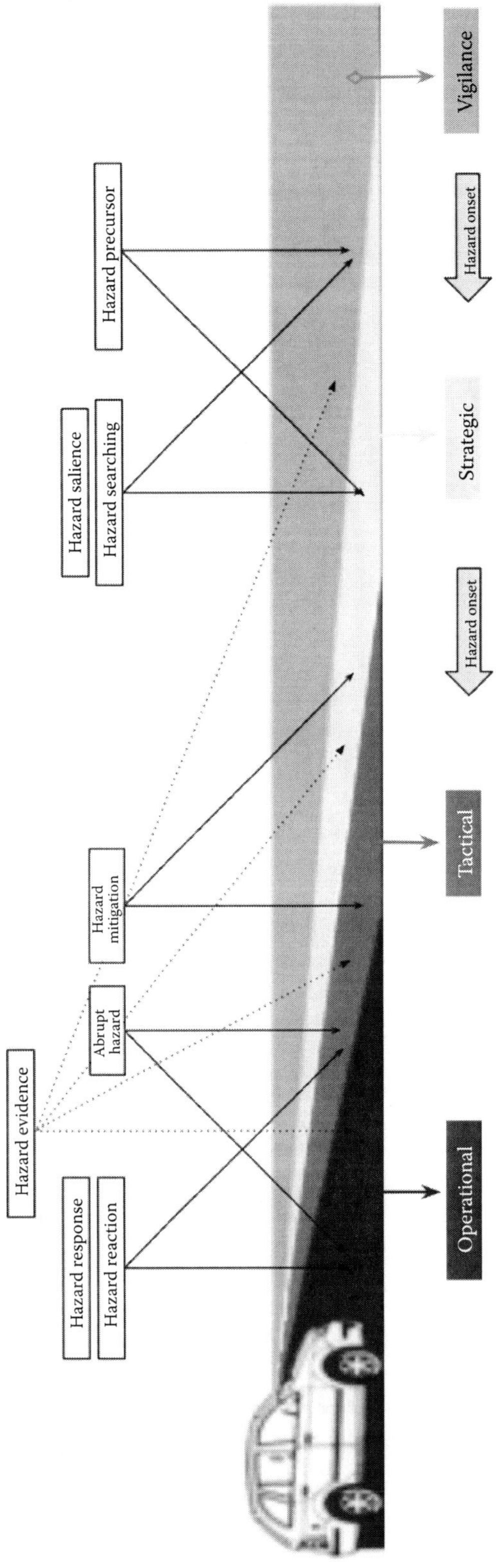

FIGURE 6.1 (See color insert.) The four visual zones based on distance from the driver's car.

Finally, the *operational zone* is closest to the car and is a zone that will rarely require overt fixation. Hazards that appear in this zone are unlikely to be avoided if they have not already been fixated, mitigated, and responded to already. The information in this zone is typically acquired via peripheral vision for lane maintenance (e.g., Summala et al., 1996).

6.5 Future Work, Extensions, and Discussion

The primary aims for this chapter were to propose a unified terminology and a novel framework for approaching the research on young novice drivers and their hazard avoidance skills. The hope was that this would help tie together the scientific literature on age- and experience-related differences in skills, thus facilitating development of programs for targeted training, evaluation, and testing of these skills. Future work will be needed to determine the extent to which novice and experienced drivers differ in terms of the various components of this framework, particularly as it relates to crash-related behaviors. There are many studies that have demonstrated young novice drivers to have deficiencies in a variety of skills and subskills that may relate to their probability of being involved in a collision, and many of these have been linked to their ability to detect and respond to on-road hazards (see Chapters 7, 8, and 28 for an excellent review of this and other related evidence). Our hope is that the definitions and theoretical structures will permit the current evidence to be parsed in such a way that it is possible to identify the specific hazard perception processes where novice drivers are lacking. This will facilitate the creation of targeted training programs that help novice drivers develop these skills.

References

Berti, A., & Frassinetti, F. (2000). When far becomes near: Remapping of space by tool use. *Journal of Cognitive Neuroscience, 12*(3), 415–42.

Blakeslee, S., & Blakeslee, M. (2007). *The Body Has a Mind of Its Own: How Body Maps in Your Brain Help You Do (Almost) Everything Better*. New York: Random House.

Borowsky, A., Oron-Gilad, T., & Parmet, Y. (2009). Age and skill differences in classifying hazardous traffic scenes. *Transportation Research Part F: Traffic Psychology and Behaviour, 12*(4), 277–287.

Borowsky, A., Shinar, D., & Oron-Gilad, T. (2010). Age, skill, and hazard perception in driving. *Accident Analysis & Prevention, 42*(4), 1240–1249.

Braitman, K.A., Kirley, B.B., McCartt, A.T., & Chaudhary, N.K. (2008). Crashes of novice teenage drivers: Characteristics and contributing factors. *Journal of Safety Research, 39*(1), 47–54.

Chapman, P.R., & Underwood, G. (1998). Visual search of driving situations: Danger and experience. *Perception-London, 27*, 951–964.

Christie, R. (2001). The effectiveness of driver training as a road safety measure: An international review of the literature. In *Road Safety Research, Policing and Education* Conference, Melbourne, Victoria, Australia. Canberra, Australia: Australasian College of Road Safety.

Crundall, D. (2016). Hazard prediction discriminates between novice and experienced drivers. *Accident Analysis & Prevention, 86*, 47–58.

Crundall, D., Andrews, B., Van Loon, E., & Chapman, P. (2010). Commentary training improves responsiveness to hazards in a driving simulator. *Accident Analysis & Prevention, 42*(6), 2117–2124.

Crundall, D., & Underwood, G. (1998). The effects of experience and processing demands on visual information acquisition in drivers. *Ergonomics, 41*, 448–458.

Crundall, D., & Underwood, G. (2001). The priming function of road signs. *Transportation Research Part F: Traffic Psychology and Behaviour, 4*(3), 187–200.

Crundall, D., Chapman, P., Trawley, S., Collins, L., Van Loon, E., Andrews, B., & Underwood, G. (2012). Some hazards are more attractive than others: Drivers of varying experience respond differently to different types of hazard. *Accident Analysis & Prevention, 45*, 600–609.

De Craen, S., Twisk, D.A.M., Hagenzieker, M.P., Elffers, H., & Brookhuis, K.A. (2011). Do young novice drivers overestimate their driving skills more than experienced drivers? Different methods lead to different conclusions. *Accident Analysis & Prevention, 43*(5), 1660–1665.

Deery, H.A. (2000). Hazard and risk perception among young novice drivers. *Journal of Safety Research, 30*(4), 225–236.

Doherty, S.T., Andrey, J.C., & MacGregor, C. (1998). The situational risks of young drivers: The influence of passengers, time of day and day of week on accident rates. *Accident Analysis & Prevention, 30*(1), 45–52.

Durso, F.T., & Gronlund, S.D. (1999). Situation awareness. In: F.T. Durso (Ed.), *Handbook of Applied Cognition* (pp. 283–314). Chichester: John Wiley & Sons.

Endsley, M.R. (1988). Design and evaluation for situation awareness enhancement. In *Proceedings of the Human Factors and Ergonomics Society Annual Meeting* (Vol. 32, No. 2, pp. 97–101). Thousand Oaks, CA: Sage Publications.

Endsley, M.R. (1995). Toward a theory of situation awareness in dynamic systems. *Human Factors: The Journal of the Human Factors and Ergonomics Society, 37*(1), 32–64.

Endsley, M.R. (2000). Theoretical underpinnings of situation awareness: A critical review. *Situation Awareness Analysis and Measurement*, 3–32.

Falkmer, T., & Gregersen, N.P. (2005). A comparison of eye movement behavior of inexperienced and experienced drivers in real traffic environments. *Optometry & Vision Science, 82*(8), 732–739.

Fisher, D.L., Pollatsek, A.P., & Pradhan, A. (2006). Can novice drivers be trained to scan for information that will reduce their likelihood of a crash? *Injury Prevention, 12*(Suppl 1), i25–i29.

Foss, R.D., Martell, C.A., Goodwin, A.H., & O'Brien, N.P. (2011). *Measuring Changes in Teenage Driver Crash Characteristics during the Early Months of Driving.* Washington, DC: AAA Foundation for Traffic Safety.

Gibson, J.J., & Crooks, L.E. (1938). A theoretical field-analysis of automobile-driving. *The American Journal of Psychology, 51*(3), 453–471.

Horrey, W.J., Lesch, M.F., Mitsopoulos-Rubens, E., & Lee, J.D. (2015). Calibration of skill and judgment in driving: Development of a conceptual framework and the implications for road safety. *Accident Analysis & Prevention, 76*, 25–33.

Horswill, M.S., & McKenna, F.P. (2004). Drivers' hazard perception ability: Situation awareness on the road. In: S. Banbury & S. Tremblay (Eds.), *A Cognitive Approach to Situation Awareness: Theory and Application.* Farnham: Ashgate Publishing.

Horswill, M.S., Hill, A., & Wetton, M. (2015). Can a video-based hazard perception test used for driver licensing predict crash involvement? *Accident Analysis & Prevention, 82*, 213–219.

Jackson, L., Chapman, P., & Crundall, D. (2009). What happens next? Predicting other road users' behaviour as a function of driving experience and processing time. *Ergonomics, 52*(2), 154–164.

Kinchla, R.A., & Wolfe, J.M. (1979). The order of visual processing: "Top-down," "bottom-up," or "middle-out." *Perception & Psychophysics, 25*(3), 225–231.

Kinnear, N., Kelly, S.W., Stradling, S., & Thomson, J. (2013). Understanding how drivers learn to anticipate risk on the road: A laboratory experiment of affective anticipation of road hazards. *Accident Analysis & Prevention, 50*, 1025–1033.

Konstantopoulos, P., Crundall, D., & Chapman, P. (2010). Driver's visual attention as a function of driving experience and visibility. Using a driving simulator to explore visual search in day, night and rain driving. *Accident Analysis & Prevention, 42*, 827–834.

Lehtonen, E., Lappi, O., Koirikivi, I., and Summala, H. (2014). Effect of driving experience on anticipatory look-ahead fixations in real curve driving. *Accident Analysis & Prevention, 70*, 195–208.

Mayhew, D.R., & Simpson, H.M. (2002). The safety value of driver education and training. *Injury Prevention, 8*(Suppl 2), ii3–ii8.

Mayhew, D.R., Simpson, H.M., & Pak, A. (2003). Changes in collision rates among novice drivers during the first months of driving. *Accident Analysis & Prevention, 35*(5), 683–691.

McCarley, J.S., Steelman, K.S., & Horrey, W.J. (2014). The view from the driver's seat: What good is salience? *Applied Cognitive Psychology, 28*(1), 47–54.

McCartt, A.T., Shabanova, V.I., & Leaf, W.A. (2003). Driving experience, crashes and traffic citations of teenage beginning drivers. *Accident Analysis & Prevention, 35*(3), 311–320.

McCartt, A.T., Mayhew, D.R., Braitman, K.A., Ferguson, S.A., & Simpson, H.M. (2009). Effects of age and experience on young driver crashes: Review of recent literature. *Traffic Injury Prevention, 10*(3), 209–219.

McDonald, C.C., Goodwin, A.H., Pradhan, A.K., Romoser, M.R., & Williams, A.F. (2015). A review of hazard anticipation training programs for young drivers. *Journal of Adolescent Health, 57*(1), S15–S23.

McKenna, F.P., & Crick, J.L. (1994). Hazard perception in drivers: A methodology for testing and training. TRL Contractor Report 313. Wokingham: Transport Research Laboratory.

McKnight, A.J., & McKnight, A.S. (2003). Young novice drivers: Careless or clueless? *Accident Analysis & Prevention, 35*(6), 921–925.

Michon, J.A. (1985). A critical view of driver behavior models: What do we know, what should we do? In *Human Behavior and Traffic Safety* (pp. 485–524). New York: Springer.

Moeller, B., Zoppke, H., & Frings, C. (2015). What a car does to your perception: Distance evaluations differ from within and outside of a car. *Psychonomic Bulletin & Review*, 1–8.

Mourant, R.R., & Rockwell, T.H. (1970). Mapping eye-movement patterns to the visual scene in driving: An exploratory study. *Human Factors: The Journal of the Human Factors and Ergonomics Society, 12*(1), 81–87.

Mourant, R.R., & Rockwell, T.H. (1972). Strategies of visual search by novice and experienced drivers. *Human Factors: The Journal of the Human Factors and Ergonomics Society, 14*(4), 325–335.

Pradhan, A.K., Hammel, K.R., DeRamus, R., Pollatsek, A., Noyce, D.A., & Fisher, D.L. (2005). The use of eye movements to evaluate the effects of driver age on risk perception in an advanced driving simulator. *Human Factors, 47*, 840–852.

Pradhan, A.K., Pollatsek, A., Knodler, M., & Fisher, D.L. (2009). Can younger drivers be trained to scan for information that will reduce their risk in roadway traffic scenarios that are hard to identify as hazardous? *Ergonomics, 52*, 657–673.

Previc, F.H. (1998). The neuropsychology of 3-D space. *Psychological Bulletin, 124*(2), 123.

Quimby, A.R., & Watts, G.R. (1981). Human factors and driving performance. TRRL, No. LR 1004 Monograph.

Sagberg, F., & Bjørnskau, T. (2006). Hazard perception and driving experience among novice drivers. *Accident Analysis & Prevention, 38*(2), 407–414.

Salmon, P.M., & Stanton, N.A. (2013). Situation awareness and safety: Contribution or confusion? Situation awareness and safety editorial. *Safety Science, 56*, 1–5.

Salthouse, T.A. (1996). The processing-speed theory of adult age differences in cognition. *Psychological Review, 103*(3), 403–428.

Spence, C., & Ho, C. (2008). Multisensory warning signals for event perception and safe driving. *Theoretical Issues in Ergonomics Science, 9*(6), 523–554.

Summala, H., Nieminen, T., & Punto, M. (1996). Maintaining lane position with peripheral vision during in-vehicle tasks. *Human Factors: The Journal of the Human Factors and Ergonomics Society, 38*(3), 442–451.

Torralba, A., Murphy, K.P., & Freeman, W.T. (2007). Sharing visual features for multiclass and multiview object detection. *Pattern Analysis and Machine Intelligence, IEEE Transactions on, 29*(5), 854–869.

Underwood, G., Crundall, D., & Chapman, P. (2002). Selective searching while driving: The role of experience in hazard detection and general surveillance. *Ergonomics, 45*, 1–12.

Underwood, G., Chapman, P., Brocklehurst, N., Underwood, J., & Crundall, D. (2003). Visual attention while driving: Sequences of eye fixations made by experienced and novice drivers. *Ergonomics, 46*, 629–646.

Underwood, G., Ngai, A., & Underwood, J. (2013). Driving experience and situation awareness in hazard detection. *Safety Science, 56*, 29–35.

Vidotto, G., Bastianelli, A., Spoto, A., & Sergeys, F. (2011). Enhancing hazard avoidance in teen-novice riders. *Accident Analysis & Prevention*, *43*(1), 247–252.

Vlakveld, W.P. (2008). Testing and training of hazard perception: Study of the testability and trainability of hazard perception in young novice drivers in 2007. Report No. D-2008-2. The Hague: Dutch Institute for Road Safety Research (SWOV).

Yarbus, A.L. (1967). Eye movements during perception of complex objects. In: *Eye Movements & Vision* (pp. 171–211). New York: Plenum Press.

Yeung, J.S., & Wong, Y.D. (2015). Effects of driver age and experience in abrupt-onset hazards. *Accident Analysis & Prevention*, *78*, 110–117.

7

Attention Allocation and Maintenance in Novice and Teen Drivers

William J. Horrey

Gautam Divekar

Abstract

The Problem. The driving environment is dynamic and rich with critical information to help drivers navigate safely. Timely and appropriate allocation of attention toward this information is paramount, and the relative inexperience of young and new drivers carries implications for how they sample, prioritize, and process available information. ***Scope of the Chapter***. We describe attention allocation in the context of models of supervisory control and highlight the important role of experience (or inexperience) in determining sampling behavior as well as potential attentional failures. We also consider failures of maintenance of attention on the driving task where the driver has elected to divert attention away from the driving task in favor of another task (e.g., driver distraction). Lastly, we discuss briefly visual attention allocation and maintenance in the context of the training approaches aimed at increasing or enhancing driving performance, risk awareness, or attention maintenance on the driving task. ***Limitations and Recommendations***. Given the space constraints, the current chapter is not intended to represent an exhaustive review of the extant literature on visual attention and scanning behaviors of novice and experienced drivers. Rather, we aim to highlight some important theoretical constructs and models and illustrate how they relate to some of the observed experimental findings.

7.1 Introduction

Teenage years are rife with change. During adolescence, individuals experience many developmental, emotional, and social changes (see, e.g., Chapters 9 and 10), which are often manifest in the alteration or emergence of new behaviors, such as risk taking, sensation seeking, and impulsivity. Naturally, these behaviors are also reflected in driving; however, in this domain, teens face additional challenges as a result of their level of inexperience.

The driving context is a dynamic and information-rich environment, and the importance of the complete and timely allocation of attention to relevant information is paramount. Although the definition and categorization of attentional failures can vary across study and data sources, attentional errors and inattention are often cited as significant causal factors in crashes for young driver age groups (e.g., Chapter 5; Braitman et al., 2008; Curry et al., 2011; McKnight & McKnight, 2003). The relative inexperience of young and new drivers carries implications for how they sample, prioritize, and process available information. This chapter focuses predominantly on sampling and prioritization—concepts that underlie attentional control.

We primarily consider the allocation of visual attention towards information that supports the driving task (i.e., how drivers sample the environment and where they seek information in support of safe driving). In doing so, we describe attention allocation in the context of models of supervisory control and highlight the important role of experience (or inexperience) in determining sampling behavior. It follows from the approaches that attentional failures can be realized by inappropriate, misdirected, or misprioritized attention—even when drivers remain engaged in the driving task. For example, novice drivers might scan areas that are less informative vis-à-vis potential hazards, or they might overmonitor some aspects of vehicle control, such as lane keeping, at the expense of a more balanced approach to information gathering (e.g., Pradhan et al., 2009; Romer et al., 2014). We consider hazard anticipation and awareness to be a subset of the full range of attention allocation tasks (see, e.g., Chapter 6 for a more focused treatment of hazard anticipation). In this vein, we review select findings from the extant literature on young drivers' scanning behavior (see also Chapter 27 on eye movements).

We also consider the maintenance of attention on the driving task. Whereas the discussion of the allocation of attention tends towards strategies of sampling task-relevant information, attention maintenance generally refers to the more global assignment of resources to the driving task. In fact, it is the failure of attention maintenance, or the complete withdrawal of attention from driving-related tasks, that is of most concern. This captures situations where, for some reason, the driver has elected to divert attention away in favor of another task (e.g., glancing away from the roadway towards a smartphone). This is often implicated in discussions of driver distraction (e.g., Chapter 12; Regan et al., 2008)—a noteworthy issue for young drivers especially as they tend to be early adopters and frequent users of new and mobile technology (Foss & Goodwin, 2014; Lee, 2007) or may have increased difficulties in the concurrent performance of these activities (e.g., Dingus, 2014; Klauer et al., 2014; Simons-Morton et al., 2014).

Finally, given their relevance to how drivers process the world around them, we discuss briefly different training approaches based on visual attention allocation and maintenance. These aim to increase or enhance driving performance, risk awareness, or attention maintenance on the driving task.

7.2 Attention Allocation

In a visually rich setting, such as the driving environment, how do we choose what information to attend to and what information to ignore or disregard? Visual attention has been studied extensively in countless basic and applied settings, and these studies have informed the development of many models and theories of attention allocation and orientation. Oftentimes, terms such as *covert/overt*, *exogenous/endogenous*, and *stimulus-driven (bottom up)/goal-directed (top-down)* are implicated in these models (e.g., Bundesen, 1990; Itti & Koch, 2001; Posner, 1980; Trick et al., 2004; Wolfe, 2007). Many of these relate to discrete visual search tasks; however, in many applied settings, it is important to know not just where to look but also when to look where. This is a hallmark of supervisory control, which is generally related to a situation or process where a human operator is continuously receiving system feedback and information from different information sources and must make intermittent responses to preserve system function (e.g., Sheridan, 1970). Driving is essentially a supervisory control task in which the driver monitors information from different channels (e.g., different parts of the roadway, instrument panel, mirrors) and acts upon this information whenever necessary (e.g., steer to correct lateral lane position, brake in response to traffic ahead). In the following section, we review some models of supervisory control and their relevance for teen and novice drivers.

7.2.1 Models of Attention and Supervisory Control

Given the constraints and restricted range of focal vision—often the area of the retina most strongly associated with visual attention—we can arguably only look at one location or thing at a time. Thus, the eye is a single-server queue (i.e., locations and objects to be scanned are placed in a queue), and eye movements are a means of servicing this queue (e.g., Moray, 1986; Senders, 1964 [https://youtu.be/kOguslSPpqo]; although see Fisher, 1982). Models of supervisory control differ from visual search models in that the observer, in this case, the driver, is tasked with monitoring a set of dynamic and changing areas of interest rather than searching for a static target among a field of nontarget information. As noted, the emphasis is more on the observer's knowledge of when to look where as opposed to knowing simply where to look. For example, scanning to the left and right while driving is a prudent activity in most circumstances; however, scanning to the left and right precisely when and where potential hazards could emerge is paramount. Knowing when to look where is not an automatic process and is one in which experience and exposure are important elements. For example, novice drivers might not fully appreciate the importance of diligent scanning towards a parked vehicle with a driver seated in it—that is, until they encounter a situation where the door is swung open into their path as they approach. It follows that teens and novice drivers need time and practice to develop and refine their understanding of the most effective strategy for sampling the environment (see Chapter 6 for a more detailed discussion of hazard anticipation, just one of many cases where it is critical to know where and when to look).

Early work in supervisory control highlighted the importance of information bandwidth in guiding visual attention (e.g., Senders, 1964). While the common definition of bandwidth relates to data transfer or processing capacity, here it is used to characterize information available in the environment. For example, drivers might glance to the forward roadway or to the lane markings in order to get information regarding their current position in the lane (i.e., "Am I still centered in the lane?"). Whenever the eyes are directed elsewhere, such as the instrument panel, the rearview mirrors, or the driver's smartphone, uncertainty about lane position increases to the point where the driver will need to update himself or herself by scanning back to the roadway. This uncertainty can be influenced by many factors, such as the degree of road curvature, the presence of lateral wind gusts, and the alignment of the wheels on the driver's car, among many other things. When task-relevant information occurs more frequently at a given location (uncertainty increases more rapidly), observers tend to scan toward this location more frequently than locations where information occurs less frequently (or never). When the occurrence (or bandwidth) of lane-related information is high, such as on a very curvy road, with lots of wind gusts and in a vehicle with poor alignment, the frequency of scans to the road should likewise be high. On a straight road in wind-free conditions and with perfect alignment (low bandwidth), one might expect the frequency of glances to be decreased. Put another way, information bandwidth characterizes the frequency with which task-relevant information occurs, and observers tend to sample places where they expect to find this information (e.g., Theeuwes & Hagenzieker, 1993).

Knowledge of (or calibration to) information bandwidth is linked to experience—a function that might explain some of the discrepancies in the scanning of novice drivers, compared to more experienced ones (e.g., Underwood et al., 2002). In particular, there are two important aspects of information bandwidth and novice drivers that merit further dwelling. First, and as noted previously, novice drivers generally have not built up sufficient experience in order to have a precise understanding of where and how often relevant road information occurs. For example, they might scan less frequently to the sides of the road where latent hazards appear, because they do not appreciate that objects can intrude from these areas (e.g., Crundall & Underwood, 1998; Pradhan et al., 2009). Secondly, novice drivers have an increased need for visual information in support of driving tasks that eventually become more automated with more practice. For example, inexperienced drivers often direct more glances toward lane information, in support of the lateral control task—a task that, later in driving, does not command as much attention or can be supported by more ambient/peripheral processing (e.g., Summala et al., 1996).

Visual attention is not determined by information bandwidth alone. The relative value of seeing or processing the information (or the costs associated with missing it) is also an important factor determinant of visual scanning (e.g., Carbonell, 1966; Horrey et al., 2006; Wickens et al., 2003). For example, the value of perceiving a pedestrian who has walked into a driver's path is much greater for the task of driving safety than correctly perceiving the sale items on display in a shop (however, this latter information would have more value for the task of finding good deals). Unfortunately, the information that has the highest bandwidth is not always the most valuable information. It follows that drivers should oversample areas of high risk, even though events might be rare. Oftentimes, (hopefully) infrequent events such as the aforementioned pedestrian example carry the most value in terms of crash-free (safe) driving.

Information bandwidth and value are considered to be top-down determinants of visual scanning (although this list is not exhaustive), and as noted, the role of experience and expertise is an important consideration in this context. However, bottom-up factors can also impact visual scanning. For example, observers are often inclined to scan toward highly salient information (e.g., Itti & Koch, 2001)—even when this information is not task relevant. A highly conspicuous and dynamic billboard is much more likely to draw the eyes (and for longer) than a static inconspicuous one (e.g., Belyusar et al., 2014; Samuel & Fisher, 2015).

Theoretical and experimental approaches have offered support for the role of information bandwidth and value as well as bottom-up saliency-driven factors in determining momentary control and allocation of visual attention. While these factors are useful in articulating differences between novice and experienced drivers, it is important to note that there are many other factors that can impact the manner in which attention is regulated, such as information access effort (Wickens et al., 2003) and arousal or fatigue level (e.g., Kahneman, 1973; Chapter 15), to name a few. Also, to this point, the characteristics of information have been largely articulated in terms of task- or driving-related information occurring in the traffic environment (i.e., outside the vehicle). However, these same properties readily apply to in-vehicle tasks, both those related and those unrelated to driving, such as mirror checks and reading text messages. Naturally, different in-vehicle tasks lend themselves to varying degrees of value, and each task itself can vary considerably across drivers and contexts. Information presented on a navigation display will be highly valued when the driver is in an unfamiliar traffic environment; however, this information will be less valuable when the driver is in a very familiar environment (even though the bandwidth of information across the two contexts may be similar). Likewise, the value afforded to activities that allow drivers to be productive (i.e., the "mobile office") will vary across drivers and circumstances. This shifting prioritization of in-vehicle tasks remains a challenge in our understanding and modeling of task value. Ideally, driving should always be the top priority; however, as crash statistics, experimental data, and casual observation suggest, this may not always be the case.

7.2.2 Experimental Findings

While the previous section outlines some of the theoretical issues concerning the governance of visual attention allocation in the driving context, here we review some of the published research concerning the glance behavior of novice and teen drivers. Where possible, this is contrasted to the glance behavior of more experienced drivers. Given space constraints, we do not conduct an exhaustive review of this literature but, rather, use a few studies as illustrative examples. In addition to the papers cited, the interested reader is directed to other sources for more details (e.g., Fisher et al., 2011; Green, 2002).

One of the earliest studies of novice and experienced drivers was carried out by Mourant and Rockwell (1972). They showed that novice drivers, defined here as drivers with virtually no driving experience, scanned the world less broadly than experienced drivers. That is, they tended to look predominantly in a small region of space ahead of the vehicle, making few glances to the periphery, fewer glances to the rearview mirrors, and fewer downstream glances (i.e., they tended to make glances closer in front of the vehicle). Crundall and Underwood (1998) found that differences in the breadth of scanning might be context specific, with experienced drivers showing broader scan patterns on freeways than novice drivers—scans that were presumed to be in search of merging or exiting traffic. Other road situations showed more consistent patterns between the two groups. Also, experienced drivers' fixations tended to be longer in information-poor areas, whereas novice drivers' scans tended to be longer in information-rich areas—potentially an unsafe pattern (also, Chapman & Underwood [1998] found that novice driver scans tended to be prolonged in dangerous situations).

Subsequent work by Underwood et al. (2002) suggested that the different patterns of scanning observed in novice drivers were due to an impoverished mental model of the road and traffic environment and the potential for hazards. Collectively, these results highlight some important aspects of scanning behavior as it relates to novice drivers. Information needs are different across the different driver groups because of their relative skill set. For example, novice drivers, who are less skilled at controlling the vehicle, need to scan information related to the current vehicle lane position. Relatedly, they are less adept at using peripheral or ambient vision in support of lane keeping, another reason that they need to fixate on this information (e.g., Horrey & Wickens, 2004; Summala et al., 1996). Moreover, these studies suggest that novice drivers' expectations regarding potential hazards might not be sufficiently realized. It is now known that this is indeed the case (Chapter 6). It follows that their sampling strategy likely yields a picture of the roadway that is degraded or incomplete (i.e., reduced situation awareness).

In addition to differences in where novice and experienced drivers tended to scan, Underwood et al. (2003) found differences in the sequencing of glances and the transitions from one area to another. In particular, novice drivers tended to transition more frequently to the road ahead from other areas, whereas experienced drivers would often transition to other areas of relevance before returning to the forward road. This is suggestive of different approaches to managing the priority of glances in the single-server queue (per the supervisory control framework) and also underscores the demanding role of vehicle control for novice and inexperienced drivers (see also Scott et al., 2013).

There have also been several studies that have highlighted the severe and safety-critical deficits that novice drivers have in their anticipation (expectation) of roadway hazards. For example, Pradhan et al. (Pollatsek et al., 2006; Pradhan et al., 2005) found pronounced differences between novice and experienced drivers in the frequency of anticipatory glances to areas where hazards could materialize. This study, and a number of related follow-on studies, nicely illustrates the role of expectations in driving visual attention allocation. Per the supervisory control framework, novice drivers might not be sensitive to the information bandwidth associated with potential (or actual) hazards, or they might be misprioritizing their glances to other parts of the driving scene because of perceived demands.

In this section, we have used a supervisory control framework to describe how drivers' visual attention might be allocated to different areas of interest within the traffic environment. By sampling these areas, the driver should establish and maintain an accurate awareness of the road and related hazards and, by extension, make necessary control inputs. Several factors can influence how attention is allocated; here we described only a few: information bandwidth, value, and salience. While they might not always lead to optimal scanning and situation awareness, experience and expertise influence a driver's calibration to the rate and importance of information occurring in the vicinity of one's own vehicle. As drivers gain more and more experience, especially in shedding the demands associated with vehicle control, they are better able to distribute their attention to areas of importance, including locations where real and potential hazards reside.

7.3 Attention Maintenance

Although not independent from the discussion of attention allocation, attention maintenance refers more generally to continuous engagement in the driving task (i.e., eyes on the road or on driving-related information). When drivers fail to maintain attention on the driving task, usually, this signifies that they have intentionally or unintentionally diverted their attention toward a competing activity, such as the case for driver distraction.*

In early work, Wierwille (1993) found that drivers generally followed a deterministic pattern of sampling when looking away from the road and towards in-vehicle displays. When the information required for a particular task cannot be completely processed in under 1.6 seconds, typically, the driver will return his or her attention to the roadway before resuming the in-vehicle task. Unfortunately, there are many instances where drivers can and do take longer glances away from the roadway—especially given the advent of new and more complex embedded and portable technologies (e.g., smartphones, in-dash information systems). Several studies have documented the importance of especially long glances in contributing to motor vehicle crash risk (e.g., Horrey & Wickens, 2007; Klauer et al., 2006; Liang et al., 2014). One obvious implication is that the driver who is looking away from the roadway for 2-plus seconds is at risk of missing critical information that has changed in the interim, such as a pedestrian stepping out in front or a lead vehicle abruptly braking. However, these prolonged glances toward nondriving tasks can also impact awareness in less catastrophic ways as well. Following from the models espoused in Section 7.2, the eye is a single server, and therefore, long disruptions in sampling behavior degrade not only one's awareness of the critical space on the road ahead but also one's awareness of all road-related information. Some recent experimental work has shown some of the detrimental effects that alternating glances inside and outside the vehicle have on a driver's situation awareness in anticipating hazards (e.g., Borowsky et al., 2015; Samuel & Fisher, 2015).

With respect to novice drivers, recent investigations have revealed some important trends concerning long off-road glances. In both field and simulator studies, researchers have shown that inexperienced drivers make frequent prolonged (over 2.5 seconds) off-road glances—over three times as many as compared to experienced drivers (Chan et al., 2008; Wikman et al., 1998). The results again underscore the absence of a refined strategy of attentional allocation vis-à-vis the information demands of the driving task (e.g., knowing when to look where) or a failure on the part of novice drivers to adhere to their scan policy—more so than experienced drivers (e.g., allowing the momentary demands of a secondary activity to disrupt their prioritization policy). In light of results such as these, one might wonder whether there are any inroads for facilitating and improving the development of good attentional control, apart from that gleaned through practice and exposure.

7.3.1 Training and Attention Maintenance

Several attempts have been made to promote the development of superior strategies for attentional allocation and maintenance in young and novice drivers. For example, Pradhan et al. (2006) showed that a risk awareness and perception training (RAPT) program could train novice drivers to scan potential hazard areas, and these effects could persist for several days posttraining. Other subsequent work demonstrated that these training benefits could translate to real-world situations as well (e.g., Pradhan et al., 2009) and last up to a year after training (Taylor et al., 2011). The premise of the training is well grounded in the models of supervisory control as they represent a means to teach novice drivers where and when they might expect to find safety-relevant (i.e., valuable) information—most importantly, in a safe setting where errors and failures do not have life-threatening consequences.

Others have employed alternate approaches to changing novice drivers' scan patterns. For example, Chapman et al. (2002) trained scanning behavior using a commentary drive approach, where the driver verbally described what he or she was looking at moment to moment. This intervention led to changes

* We note that the withdrawal of attention can be due to other factors, such as fatigue (Chapter 15); however, we focus our discussion here on secondary, nondriving activities.

in scanning behavior while viewing subsequent risky road situations compared to a control group who did not receive any intervention (see also Underwood, 2007), thus underscoring the potential utility of eye behaviors and scanning in promoting novice and teen driver development.

With respect to attention maintenance, a number of studies have also examined potential inroads for training. For example, Pradhan et al. (2011) applied a training technique, the Forward Concentration and Attention Learning (FOCAL) approach, designed to teach novice drivers to avoid long glances away from the roadway. They showed a training advantage for drivers on the open road compared to a placebo-trained group with respect to the number of long glances made while driving. Divekar et al. (2013) likewise found that novice drivers could benefit from the FOCAL approach; however, though these benefits were realized for a variety of in-vehicle activities, they did not appear to translate to especially long out-of-vehicle glances away from the roadway (e.g., to roadside signs). While more work is merited in this area, these and other studies show promise for teaching novice drivers to prioritize the driving task and not to divert too much attention away from it.

7.4 Conclusion

Although a theme that continues to emerge in the various chapters of this handbook, it is important to again underscore the risks that teenagers face, as drivers. The current chapter endeavors to describe some of the important and theoretically motivated issues for our understanding of how drivers, including novice and teen drivers, sample their visual surroundings. Given the dynamic and information-rich driving environment, complete and timely allocation of attention to relevant information is of critical importance, and attentional errors are often attributed to motor vehicle crashes in young age groups (e.g., Chapter 5; McKnight & McKnight, 2003).

Models of supervisory control help to articulate some of the important factors that drive attention but also highlight some of the areas where novice and teen drivers deviate from more experienced drivers. For example, novice drivers' scanning with respect to potential hazards can be related to underdeveloped or miscalibrated mental models concerning the frequency and importance of certain information. This also showcases how the increased demands of certain driving subtasks (e.g., lane maintenance) can spill over into other aspects of driving. Fortunately, training approaches aimed at increasing or enhancing driving performance, risk awareness, or attention maintenance on the driving task show some promise in improving novice drivers' behaviors. However, as a message of critical importance carried throughout this handbook, there is still much more work to do. The most obvious place to find future advances is in real-time driver-state models, which will be able to capture how attentive the driver is and when the level of attention falls below some threshold (Liang et al., 2007).

We should note before ending just how much more challenging the issues will become as autonomous and semiautonomous vehicles become a larger and larger part of the driving experience. The driver not only will need to be aware of everything while in full control of the vehicle but will need to make sure when the vehicle is in semiautonomous mode that he or she keeps aware of the situation enough so that control can be assumed if the automated driving suite reaches a boundary condition. How to keep drivers in the loop when they are out of the loop has long been a concern in aviation (e.g., Bainbridge, 1983). It is only recently becoming a concern in surface transportation (e.g., Cunningham & Regan, 2015; Merat et al., 2014). The problems are likely to be exacerbated for the novice driver, who lacks the knowledge necessary to be situationally aware even when completely in the loop.

Acknowledgments

We are very grateful to Don Fisher, Jeff Caird, Lana M. Trick, and Marvin Dainoff for their comments and suggestions on early drafts of this chapter.

References

Bainbridge, L. (1983). Ironies of automation. *Automatica, 19*(6), 775–779.

Belyusar, D., Reimer, B., Shoup, A., Jokubaitis, B., Pugh, B., Mehler, B., & Coughlin, J.F. (2014). A preliminary report on the effects of digital billboards on glance behavior during highway driving. In *Transportation Research Board 93rd Annual Meeting* (No. 14-4670).

Borowsky, A., Horrey, W.J., Liang, Y., Garabet, A., Simmons, L., & Fisher, D.L. (2015). The effects of momentary visual disruption on hazard anticipation and awareness in driving. *Traffic Injury Prevention, 16*(2), 133–139.

Braitman, K.A., Kirley, B.B., McCartt, A.T., & Chaudhary, N.K. (2008). Crashes of a novice teenage drivers: Characteristics and contributing factors. *Journal of Safety Research, 39*(1), 47–51.

Bundesen, C. (1990). A theory of visual attention. *Psychological Review, 97*(4), 523–547.

Carbonell, J.R. (1966). A queueing model of many-instrument visual sampling. *Human Factors in Electronics, IEEE Transactions on*, (4), 157–164.

Chan, E., Pradhan, A.K., Knodler Jr, M.A., Pollatsek, A., & Fisher, D.L. (2008). Empirical evaluation on driving simulator of effect of distractions inside and outside the vehicle on drivers' eye behavior. In *Transportation Research Board 87th Annual Meeting* (No. 08-2910).

Chapman, P.R., & Underwood, G. (1998). Visual search of driving situations: Danger and experience. *Perception, 27*(8), 951–964.

Chapman, P., Underwood, G., & Roberts, K. (2002). Visual search patterns in trained and untrained novice drivers. *Transportation Research Part F: Traffic Psychology and Behaviour, 5*(2), 157–167.

Crundall, D.E., & Underwood, G. (1998). Effects of experience and processing demands on visual information acquisition in drivers. *Ergonomics, 41*(4), 448–458.

Cunningham, M., & Regan, M.A. (2015). Autonomous vehicles: Human factors issues and future research. *Proceedings of the 2015 Australasian Road Safety Conference.* Gold Coast, Australia: Australasian College of Road Safety.

Curry, A.E., Hafetz, J., Kallan, M.J., Winston, F.K., & Durbin, D.R. (2011). Prevalence of teen driver errors leading to serious motor vehicle crashes. *Accident Analysis & Prevention, 43*(4), 1285–1290.

Dingus, T.A. (2014). Estimates of prevalence and risk associated with inattention and distraction based upon naturalistic data. *Annals of Advances in Automotive Medicine, 58*, 60–68.

Divekar, G., Pradhan, A.K., Masserang, K.M., Reagan, I., Pollastek, A., & Fisher, D.L. (2013). A simulator evaluation of the effects of attention maintenance training on glance distributions of younger novice drivers inside and outside the vehicle. *Transportation Research Part F: Traffic Psychology and Behavior, 20*, 154–169.

Fisher, D.L. (1982). Limited channel models of automatic detection: Capacity and scanning in visual search. *Psychological Review, 89*, 662–692.

Fisher, D.L., Pollatsek, A., & Horrey, W.J. (2011). Eye behaviors: How driving simulators can expand their role in science and engineering. In Fisher, D., Rizzo, M., Caird, J., & Lee, J. (Eds.), *Handbook of Driving Simulation for Engineering, Medicine and Psychology*. Boca Raton, FL: CRC Press.

Foss, R.D., & Goodwin, A.H. (2014). Distracted driver behaviors and distracting conditions among adolescent drivers: Findings from a naturalistic driving study. *Journal of Adolescent Health, 54*(5 Suppl), S50–S60.

Green, P. (2002). Where do drivers look while driving (and for how long). In Dewar, R.E., & Olson, P.L. (Eds.), *Human Factors in Traffic Safety* (pp. 77–110). Tucson, AZ: Lawyers & Judges.

Horrey, W.J., & Wickens, C.D. (2004). Focal and ambient visual contributions and driver visual scanning in lane keeping and hazard detection. *Proceedings of the 2004 Human Factors and Ergonomics Society Annual Meeting* (pp. 2325–2329). Santa Monica, CA: HFES.

Horrey, W., & Wickens, C. (2007). In-vehicle glance duration: Distributions, tails, and model of crash risk. *Transportation Research Record, 2018*, 22–28.

Horrey, W.J., Wickens, C.D., & Consalus, K.P. (2006). Modeling drivers' visual attention allocation while interacting with in-vehicle technologies. *Journal of Experimental Psychology: Applied, 12*(2), 67–78.

Itti, L., & Koch, C. (2001). Computational modelling of visual attention. *Nature Reviews Neuroscience, 2*(3), 194–203.

Kahneman, D. (1973). *Attention and Effort*. Englewood Cliffs, NJ: Prentice-Hall.

Klauer, S.G., Dingus, T.A., Neale, V.L., Sudweeks, J.D., & Ramsey, D.J. (2006). *The Impact of Driver Inattention on Near-Crash/Crash Risk: An Analysis Using the 100-Car Naturalistic Driving Study Data* (No. HS-810 594).

Klauer, S.G., Guo, F., Simons-Morton, B.G., Ouimet, M.C., Lee, S.E., & Dingus, T.A. (2014). Distracted driving and risk of road crashes among novice and experienced drivers. *New England Journal of Medicine, 370*(1), 54–59. Washington, DC: National Highway Traffic Safety Administration.

Lee, J.D. (2007). Technology and teen drivers. *Journal of Safety Research, 38*(2), 203–213.

Liang, Y., Reyes, M.L., & Lee, J.D. (2007). Real-time detection of driver cognitive distraction using support vector machines. *IEEE Transactions on Intelligent Transportation Systems, 8*, 340–350.

Liang, Y., Lee, J.D., & Horrey, W.J. (2014). A looming crisis the distribution of off-road glance duration in moments leading up to crashes/near-crashes in naturalistic driving. *Proceedings of the 2014 Human Factors and Ergonomics Society Annual Meeting* (pp. 2102–2106). Santa Monica, CA: HFES.

McKnight, A.J., & McKnight, A.S. (2003). Young novice drivers: Careless or clueless? *Accident Analysis & Prevention, 35*(6), 921–925.

Merat, N., Jamson, A.H., Lai, F.C., Daly, M., & Carsten, O.M. (2014). Transition to manual: Driver behaviour when resuming control from a highly automated vehicle. *Transportation Research Part F: Traffic Psychology and Behaviour, 27*, 274–282.

Moray, N. (1986). Monitoring behavior and supervisory control. In Boff, K.R., Kaufman, L., & Thomas, J.P. (Eds.), *Handbook of Perception and Human Performance, Vol. 2: Cognitive Processes and Performance* (pp. 1–51). Chichester: John Wiley & Sons.

Mourant, R.R., & Rockwell, T.H. (1972). Strategies of visual search by novice and experienced drivers. *Human Factors, 14*(4), 325–335.

Pollatsek, A., Fisher, D.L., & Pradhan, A. (2006). Identifying and remedying failures of selective attention in younger drivers. *Current Directions in Psychological Science, 15*(5), 255–259.

Posner, M.I. (1980). Orienting of attention. *Quarterly Journal of Experimental Psychology, 32*(1), 3–25.

Pradhan, A.K., Hammel, K.R., DeRamus, R., Pollatsek, A., Noyce, D.A., & Fisher, D.L. (2005). Using eye movements to evaluate effects of driver age on risk perception in a driving simulator. *Human Factors, 47*(4), 840–852.

Pradhan, A., Fisher, D., & Pollatsek, A. (2006). Risk perception training for novice drivers: Evaluating duration of effects of training on a driving simulator. *Transportation Research Record, 1969*, 58–64.

Pradhan, A.K., Pollatsek, A., Knodler, M., & Fisher, D.L. (2009). Can younger drivers be trained to scan for information that will reduce their risk in roadway traffic scenarios that are hard to identify as hazardous? *Ergonomics, 52*, 657–673.

Pradhan, A.K., Divekar, G., Masserang, K. et al. (2011). The effects of focused attention training on the duration of novice drivers' glances inside the vehicle. *Ergonomics, 54*(10), 917–931.

Regan, M.A., Lee, J.D., & Young, K. (Eds.). (2008). *Driver Distraction: Theory, Effects, and Mitigation*. Boca Raton, FL: CRC Press.

Romer, D., Lee, Y.C., McDonald, C.C., & Winston, F.K. (2014). Adolescence, attention allocation, and driving safety. *Journal of Adolescent Health, 54*(5), S6–S15.

Samuel, S., & Fisher, D.L. (2015). Evaluation of the minimum forward roadway glance duration critical to latent hazard detection. *Transportation Research Record, 1707*, 9–17.

Scott, H., Hall, L., Litchfield, D., & Westwood, D. (2013). Visual information search in simulated junction negotiation: Gaze transitions of young novice, young experienced and older experienced drivers. *Journal of Safety Research, 45*, 111–116.

Senders, J.W. (1964). The human operator as a monitor and controller of multidegree of freedom systems. *Human Factors in Electronics, IEEE Transactions on*, (1), 2–5.

Sheridan, T. (1970). On how often the supervisor should sample. *IEEE Transactions on Systems Science and Cybernetics*, *2*(6), 140–145.

Simons-Morton, B.C., Guo, F., Klauer, S.G., Ehsani, J.P., & Pradhan, A.K. (2014). Keep your eyes on the road: Young driver crash risk increases according to duration of distraction. *Journal of Adolescent Health*, *54*(5 Suppl), S61–S67.

Summala, H., Nieminen, T., & Punto, M. (1996). Maintaining lane position with peripheral vision during in-vehicle tasks. *Human Factors: The Journal of the Human Factors and Ergonomics Society*, *38*(3), 442–451.

Taylor, T.G.G., Masserang, K.M., Divekar, G., Samuel, S., Muttart, J.W., Pollatsek, A., & Fisher, D.L. (2011). Long term effects of hazard anticipation training on novice drivers measured on the open road. *Proceedings of the 6th International Driving Symposium on Human Factors in Driver Assessment, Training, and Vehicle Design 2011*. Iowa City, IA: Public Policy Center, University of Iowa.

Theeuwes, J., & Hagenzieker, M.P. (1993). Visual search of traffic scenes: On the effect of location expectations. *Vision in Vehicles*, *4*, 149–158.

Trick, L.M., Enns, J.T., Mills, J., & Vavrik, J. (2004). Paying attention behind the wheel: A framework for studying the role of attention in driving. *Theoretical Issues in Ergonomics Science*, *5*(5), 385–424.

Underwood, G. (2007). Visual attention and the transition from novice to advanced driver. *Ergonomics*, *50*(8), 1235–1249.

Underwood, G., Chapman, P., Bowden, K., & Crundall, D. (2002). Visual search while driving: Skill and awareness during inspection of the scene. *Transportation Research Part F: Traffic Psychology and Behaviour*, *5*(2), 87–97.

Underwood, G., Chapman, P., Brocklehurst, N., Underwood, J., & Crundall, D. (2003). Visual attention while driving: Sequences of eye fixations made by experienced and novice drivers. *Ergonomics*, *46*(6), 629–646.

Wierwille, W.W. (1993). Visual and manual demands of in-car controls and displays. In Peacock, B. & Karwowski, W. (Eds.), *Automotive Ergonomics* (pp. 299–320). Washington, DC: Taylor & Francis.

Wickens, C.D., Goh, J., Helleberg, J., Horrey, W.J., & Talleur, D.A. (2003). Attentional models of multitask pilot performance using advanced display technology. *Human Factors*, *45*(3), 360–380.

Wikman, A.S., Nieminen, T., & Summala, H. (1998). Driving experience and time-sharing during in-car tasks on roads of different width. *Ergonomics*, *41*(3), 358–372.

Wolfe, J.M. (2007). Guided search 4.0. In Gray, W.D. (Ed.), *Integrated Models of Cognitive Systems*, (pp. 99–119). Oxford: Oxford University Press.

8

The Differences in Hazard Mitigation Responses Implemented by Novice and Experienced Drivers

Jeffrey W. Muttart

Donald L. Fisher

Abstract

The Problem. The three most common crash types for drivers under age 18 are run-off-road crashes (primarily on curves), intersection crashes (primarily left turns), and rear-end crashes (primarily on straight segments), accounting for some 80% of the total crashes. We will focus on just these three crash types, the former two being among the deadliest for teen and novice drivers. ***Novice Drivers Need to Mitigate More Often.*** It is well known that novice drivers fail to anticipate hazards and maintain attention on the forward roadway. Thus novice drivers will need to mitigate hazards more frequently than experienced drivers. ***Novice Drivers Mitigate More Poorly.*** Not only do novice drivers mitigate hazards less often than experienced drivers, but their mitigation behaviors are often less safe than experienced drivers. We will discuss what types of mitigation behaviors should be initiated before a hazard materializes (the potential hazard phase) and, where research is available, the differences between the types of mitigation behaviors of novice and experienced drivers. And we will discuss what types of mitigation behavior should be initiated when a hazard materializes (the immediate hazard phase) and the differences that exist between the mitigation behaviors of novice and experienced drivers during this phase irrespective of whether the driver anticipated the hazard. ***Failures of Hazard Mitigation Given Hazard Was Anticipated***. Finally, we will ask whether it is the case, given that both experienced and novice drivers anticipate a hazard during the potential hazard phase, they are equally likely to mitigate the hazard. Put slightly differently, we are asking here whether the conditional likelihood of various mitigation behaviors, given a driver has anticipated a hazard, varies between experienced and novice drivers.

8.1 Introduction

Chapter 6 examined the differences in the hazard anticipation behaviors of novice and experienced drivers, it generally being the case that novice drivers anticipated hazards, especially latent hazards, less well than experienced drivers. Chapter 7 analyzed the differences in attention allocation and maintenance behaviors of novice and experienced drivers. Again, it was generally the case that novice drivers were less safe, in this case withdrawing attention from the forward roadway for longer periods of time than more experienced drivers. The consequence of failing to anticipate a hazard while maintaining attention, or worse still, not maintaining attention, is that on occasion, this will require the driver to initiate a series of responses in order to mitigate the hazard.

In the remainder of the introduction, we describe briefly the three most frequent types of crashes since these are the ones in which mitigation is most involved. We then define various terms and conventions. Finally, in the main body of the chapter, we describe in some detail the ways in which novice and experienced drivers differ in their responses to the mitigation of each of these three most common crashes.

To begin, consider the most frequent crash types among novice and teen drivers. Using the National Motor Vehicle Crash Causation Survey (NMVVCS) a list was compiled of all serious (tow-away) crashes involving drivers ages 16–18 in 2005 and 2006 (Winston, 2011). All drivers had restricted or unrestricted licenses. The most frequent crash type for this age group included left-turn crashes, followed by road departures and rear-end crashes. For teenage drivers, these three crash types account for more than 80% of the 677 serious teen-involved crashes. Not only are these crashes the most frequent among teen and novice drivers, but also, they are among the deadliest—left-turn and run-off-road (ROR) crashes (Abdel-Aty et al., 1999; Braitman et al., 2008; Mayhew et al., 2003).

These crash types are also the ones that decrease rapidly in frequency over the first 12 months of driving (Chapter 29). For example, Mayhew et al. (2003) pointed out that novice driver crash involvement decreased by approximately 70% after 11–12 months of licensure. Similar findings were reported more recently by Foss et al. (2011).

Because they are frequent and deadly and decrease rapidly with experience, we concentrate on these three crash types in this chapter. Before discussing the causes of such crashes, we want to briefly differentiate between two categories of hazard response phases. The two hazard response phases are related to the time to impact, which is the calculated time necessary for the subject vehicle to travel from its current position at the onset of a hazard to the impact location, assuming a constant speed for the subject vehicle. Responses in the *potential hazard phase* typically occur 5 seconds or more before the time to impact. The typical responses in this case are to slow down moderately and/or move slightly in the lane, along with increased eye glances to the area where a potential hazard may occur and toward the area where information may exist that would cause the driver to mitigate in some specific fashion. Responses in the *immediate hazard phase* occur at a much shorter time to impact (between 0 and 5 seconds before time to impact). The typical responses in this case are hard steering and braking.

In this chapter, we discuss the appropriate (safe) mitigation behaviors during the potential and immediate hazard phases for each of the three crash types. Where research is available, we describe the differences between novice and experienced drivers in the frequencies and types of mitigation behaviors in these phases. Finally, we describe briefly the results of those studies that have controlled for anticipation in novice and experienced drivers and looked for possible differences in their mitigation behaviors even when they are equally likely to anticipate a particular hazard.

8.2 Run-Off-Road Crashes: Rollovers, Off-Road Impacts, and Crashes at Curves

We begin each discussion of a crash type with the relevant crash statistics. Briefly, younger drivers have been overrepresented in steering-related crashes such as ROR crashes (Braitman et al., 2008; Mayhew et al., 2003; Muttart, 2015). A query of the Fatal Accident Reporting System (FARS, 2012) showed that drivers ages 16–19 were much more likely than drivers ages 35–38 to be involved in ROR crashes that involved rollovers (590 versus 270) and off-road impacts (1861 versus 813). In particular, 1000 more 16- to 19-year-olds in the United States died in a crash that involved striking an off-road obstacle than did drivers aged 35–38, even though the number of drivers in the latter age group is much higher. Curves have also been linked to ROR events among younger drivers (McElheny et al., 2006; McLaughlin et al., 2009; Muttart, 2013; Muttart et al., 2013).

ROR crashes might stem from failures during the potential hazard phase, immediate hazard phase, or both. Examples of problems that lead to ROR crashes during the potential hazard phase include being distracted (Chapter 7), being fatigued (Chapter 15), or failing to anticipate a hazard, as in the case of a driver who fails to slow before a sharp curve (Chapter 6). For the purposes of this chapter, each one of these causes has much the same effect on mitigation during the potential hazard phase: the driver, for whatever reason, does not initiate the mitigation behaviors that could reduce the likelihood of a crash. Thus, we can leave the discussion of exactly why a driver might have failed to engage in mitigation behaviors to earlier chapters and focus here exclusively on the mitigation behaviors appropriate to the potential and immediate hazard phases. Finally, we need to ask whether, given that drivers anticipate a hazard in the potential hazard phase, novice drivers are as likely to mitigate the hazard as experienced drivers.

8.2.1 Potential Hazard Phase

First, consider what behaviors are important during the potential hazard phase of ROR crashes, in particular, ROR crashes associated with curves. First, lane shifts are critical as a driver enters a curve. These lane shifts elongate the radius of the curve and make the curve easier to negotiate at the same speed due to reduced lateral forces. The potential hazard phase occurs before a driver arrives at the apex of the curve. Drivers who make the curve sharper by failing to move to the outside of the curve within the lane will experience greater lateral forces and have fewer available frictional resources. Such lane shifts among experienced drivers have been observed in a field study reported by Bonneson and Pratt (2009). In their study, most drivers moved laterally within their lane by an average of 3 feet when approaching the apex of a curve. Specifically, most drivers start near the outside of the curve and move to the inside of the curve at the apex. Second, slowing overall and slowing early are important. Finally, taking anticipatory glances to the extent of the sight line is important because without such glances, one cannot slow or shift in the lane in time (Land and Lee, 1994; Suh et al., 2006).

So, do novice and experienced drivers differ in the way they respond during the potential hazard phase on curves on each of the aforementioned factors? First, consider a simulator study in which each driver navigated through three curves, where each curve had a different radius (Muttart et al., 2013). One of the scenarios included a gentle curve left, and one included a tightening (spiral) curve right where a stopped vehicle was positioned partially in the driver's lane after the driver rounded the curve. Crashes were possible in this latter scenario. Glance behaviors, speed loss, accelerator pedal movements, and braking movements were recorded for each driver beginning 10 seconds before the hazard location.

The study showed that the critical region where anticipatory behaviors were needed was 3–5 seconds before the curve (also see Glennon et al., 1986; McElheny et al., 2006). Experienced drivers exhibited more anticipatory slowing when within 5 seconds of a sharp curve and within 3 seconds of a gradual curve. Experienced drivers also began slowing significantly earlier in the curves than did novice drivers. Additionally, the experienced drivers made more anticipatory glances to the extent of the sight line.

Finally, the experienced drivers were more likely to shift within their lane (Muttart et al., 2013). The net result was that novice drivers were more likely to leave the road or crash after driving through the apex of the curve than were the drivers over age 25 (Muttart, 2013; Muttart et al., 2013).

8.2.2 Immediate Hazard Phase

Failures to mitigate hazards in the immediate hazard phase also lead to ROR crashes on curves and other roadway segments. Steering is typically the major problem for novice drivers. From the 2nd Strategic Highway Research Program (SHRP-2) naturalistic database, drivers under the age of 25 years accounted for more than 68% of the 73 drivers who steered to avoid a rear-end collision and drivers in the 16- to 19-year-old age group accounted for 34% of the drivers who swerved (hard steering), both proportions larger than the proportion of drivers in the associated age groups (Muttart, 2015). These results are consistent with those reported by Araki and Matsuura (1990), who found that young drivers steered 38% more often than the experienced drivers in their research (also see Muttart, 2015).

But what about ROR crashes on straight segments and curves? After all, steering and swerving are not by themselves bad. Unfortunately, not only do young drivers steer more often; they steer farther and harder, and they use steering in riskier conditions like curves, wet roads and rain, and nighttime (Muttart, 2015). Additionally, they are more likely to oversteer or understeer (Maeda et al., 1977; McGehee et al., 1999; Muttart, 2015; Muttart et al., 2013) and more likely to countersteer at a greater rate (Maeda et al., 1977).

This overreliance on steering can cause problems in a number of situations, which potentially lead to ROR crashes. For example, consider a tire blowout, pothole, or road (edge) drop-off situation. Before a sudden tire blowout or a tire dropping into a rut, a driver might not receive any information related to the hazard or the appropriate response. After a blowout or lane drop-off, drivers have been known to steer more than necessary (Blythe et al., 1998; Hallmark et al., 2006). Novice drivers' steering response is compromised, as noted, and so they handle these situations more poorly, running off the road being one such consequence.

Frequently, it is often the second steering action that leads to the loss of control. Consider a driver who comes upon a slow-moving vehicle ahead on the highway and in the right lane. The subject driver realizes very late that the slow-moving vehicle is not traveling slightly slower than the speed limit but instead is traveling much slower. A common response to this situation has been to steer (Muttart, 2015). Suppose the driver steers left and avoids the crash but either steers too hard initially or is slow to countersteer (straighten), as Maeda et al. (1977) revealed in the case of young drivers. After the driver commits too far left, countersteering right is called for. In a countersteer situation, a crash might result from the attempted recovery steering (e.g., McGehee et al., 1999).

Similar steering problems occur in curves. As noted, anticipating, slowing, and shifting toward the outside of a lane before a curve are the optimal solutions to steering problems that can arise in a curve (Glennon et al., 1986; Mikolajetz et al., 2009). This is what experienced drivers do. However, when this does not occur, as is often the case with novice drivers, and a driver is traveling too fast when negotiating a curve, hard braking or hard steering might become necessary to keep the car in the appropriate lane. For instance, drivers who approach a curve at speeds significantly greater than the design speed were two to four times more likely to touch or cross the centerline when at the curve (Hallmark et al., 2006). If the driver brakes hard, the potential occurs for understeering. When a driver brakes hard and locks the front wheels, the steering functionality is lost unless the vehicle is equipped with advanced safety systems (Anti-lock Braking System [ABS], Electronic Stability Control [ESC], or traction control). As a result, the driver might not be able to steer adequately to maintain the curvature of the road, which results in understeering (Zegeer et al., 1990). On the other hand, a driver who is traveling too fast for the curve might steer to get back in the lane but oversteer (Glennon et al., 1986; Maeda et al., 1977), producing a turn that is sharper than the highway curve and further reducing the available friction, leading on occasion to an ROR crash.

8.2.3 Relation between Anticipation and Mitigation

Only one study of which we are aware has tried to determine the exact relationship between hazard anticipation and hazard mitigation (Muttart et al., 2013). It is well known that novice drivers are less likely to mitigate hazards at curves, because they fail to anticipate the hazards. For example, in a driving simulator study, when bushes obscured a stop sign at the end of a curve to the right, the novice drivers braked much harder on average when approaching the stop sign than did the experienced drivers, despite a "stop ahead" sign presented earlier in the approach (Fisher et al., 2002). Also, experienced drivers, but not novice drivers, reduced their speed by the amount predicted by the Mikolajetz et al. (2009) or Bonneson and Pratt (2009) speed loss models. Unfortunately, their research cannot be used to determine whether novice drivers failed to anticipate the hazard or, instead, anticipated the hazard but failed to mitigate it.

Muttart et al. (2013) examined the relationship between hazard anticipation and hazard mitigation by examining the eye and vehicle behaviors of experienced and novice drivers. The study was undertaken on a driving simulator. It showed that slowing is associated with glancing: the likelihood of a driver slowing, whether novice or experienced, was higher given that the driver had glanced to the far extent of a curve (Figure 8.1). And it was clear that experience had a role to play as well: the likelihood that a driver slows, given that he or she has glanced, is higher for experienced drivers than it is for novice drivers. This answers the question set forth at the beginning. In particular, even when novice drivers noticed that a curve was ahead (as evidenced by a glance to the far extent), only 60% of the novice drivers slowed as compared to 90% of the experienced drivers.

In summary, drivers who fail to anticipate a curve by slowing and shifting within the lane will likely have a much more difficult time remaining in their lane as they approach the apex of the curve. We hypothesize that novice drivers are much less likely to respond to curves with appropriate steering and braking actions in the potential hazard phase, because they do not glance toward the extent of the sight line. As noted, there was almost a one-to-one relation between glances toward the extent of the sight line and anticipatory slowing and lane shifting for the experienced drivers, but not the novice drivers. Together, at least to us, these results suggest that novice drivers are largely clueless (not careless) when it comes to hazard mitigation at curves (McKnight & McKnight, 2003). Quite simply, they are not making the anticipatory glances that are correlated with actions that lead to mitigation within the potential hazard phase. Instead, they are having to deal with their failure to mitigate the hazard early on by taking actions later that require of them skills they would appear not yet to have developed.

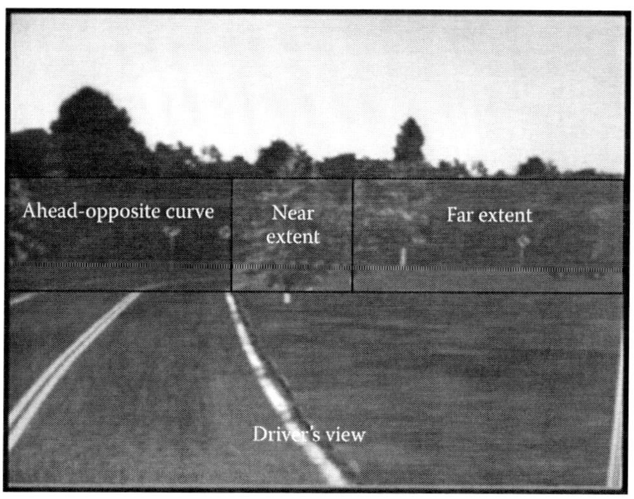

FIGURE 8.1 **(See color insert.)** Near and far extent when approaching a curve to the right.

8.3 Left-Turn-at-Intersection Crashes

Whereas the FARS database was used to identify the frequency, type, and severity of ROR crashes, the NMVCCS database was used to identify the frequency, severity, and type of intersection crashes. Within the NMVCCS database, 677 crashes were identified that were serious and involved teen drivers (Winston, 2011). A serious crash was defined as a crash that was associated with at least one vehicle being towed from the crash scene. Among these crashes, the left-turn-across-path (LTAP) crash type (the driver is struck by a vehicle in the opposing lane that is turning left) was the most common serious crash among the teen drivers (Figure 8.2, right panel). The LTAP crash type accounted for 16% of teen tow-away crashes in 2005–2006. The typical intersection left-turn maneuver was the second-leading crash type (the driver is struck by a vehicle in cross traffic that is moving perpendicular to the driver's path of travel) and accounted for 13% of the teen tow-away crashes (Figure 8.2, left panel). The FARS database showed that 500 more teen drivers (ages 16–19) died in angular crashes than did drivers ages 35–38 (1479 versus 940). Furthermore, according to the American Automobile Association (AAA), drivers ages 16–19 travel only half the annual miles of drivers ages 35–38.

The next question is why novice drivers are especially likely to crash in the aforementioned two types of intersection crashes. Using the NMVCCS database, Choi (2010) found that the cause of the crashes for novice drivers at these two intersections differed as a function of their type and whether the driver was the turning or the through driver. The causes included obstructed views, inadequate surveillance, and poor judgments of the gap size (bottom of Figure 8.2). Notably, Choi (2010) reported different failures for drivers under age 24 when compared to drivers ages 25 and older.

8.3.1 Left Turn across the Path of Opposite-Direction Traffic

Consider first the scenario on the right-hand side of Figure 8.2. We need to consider both the turning driver and the through driver during the potential hazard phase.

8.3.1.1 The Turning Driver

8.3.1.1.1 Potential Hazard Phase—Approaching the Turn and Deciding Whether to Yield or Not

To begin, assume that we are focusing on the turning driver. What are the mitigating behaviors that help the turning driving during the potential hazard phase? Sometimes, the best action is no action at

Left turn **Left turn across path**

Turning driver:
- Failure to adequately look for approaching traffic (inadequate surveillance)
- Failure to estimate closing speed of approaching traffic (misjudged gaps)

Through driver:
- Speed choice (does not slow or anticipate conflicting traffic)
- Obstructions: through drivers fail to account for what might be obstructed (enter too fast)

FIGURE 8.2 **(See color insert.)** The two left-turn configurations associated with the most fatalities among young drivers and the typical failures by drivers in those crashes.

all. Such is the case with drivers before attempting a turn in the LTAP scenario. Moving farther into the intersection before turning increases the time and distance that a driver can survey traffic when approaching the turn. Hough (2008) referred to the extension of forward movement as *delayed apexing*. When a driver begins a turn earlier (cuts the turn short), he or she is exposed to oncoming traffic for a longer distance and has less opportunity to make a secondary glance toward oncoming traffic. Furthermore, moving farther into the intersection is also associated with reduced speed (Happer et al., 2009). Figure 8.3 shows a proper turn, where a driver delays the turn (delayed apexing), and an improper turn, where the driver cuts the corner. In summary, when drivers delay the turn, they must slow more, and they give themselves more time to glance for threatening traffic. If that traffic should appear, they should remain relatively straight, thus allowing themselves to respond to late-appearing traffic while still remaining in their lane. Of course, in delaying a turn, drivers must still look for oncoming traffic and judge the appropriate gap size.

Of the three factors described that make an LTAP maneuver safer for the turning driver (delayed apexing, glances toward oncoming traffic, gap judgment), differences between novice and experienced drivers have only been examined for the latter two factors. Perhaps, not surprisingly, novice drivers are much less likely to look for oncoming traffic as they are taking the turn, especially if their view of oncoming traffic is obstructed. For example, in one simulator study, 70% of the experienced drivers made a secondary glance toward the direction of possible oncoming traffic, while only 14% of the novice drivers did so (Pradhan et al., 2005). In this case, the oncoming vehicle was hidden during the turn by a vehicle in the opposing lane that was making a left turn. Novice drivers are also less able to judge safe gap sizes than are more experienced drivers (Choi, 2010). This is somewhat surprising given that aging has a detrimental effect on the perceptual abilities that govern performance in a gap acceptance task. That said, the difference could be entirely due to novice drivers' willingness to accept smaller gap sizes as safe that, indeed, are not safe.

More specifically, novice drivers were more willing to accept gaps within what is referred to as the dilemma zone. The dilemma zone is the time period where nearly 50% of drivers will attempt a turn and 50% choose to yield. Novice drivers were as likely to attempt a turn with an oncoming vehicle 6.3 seconds away as an experienced driver was willing to accept a gap with an approaching vehicle 7.4 seconds away. Also, as turn time increased, fewer experienced drivers attempted the turn, while novice drivers did not adapt as well to the changing task demands (Mitsopoulos-Rubens et al., 2009). Accepting smaller gaps and failing to adapt to increased demands leaves much less margin for the correct interpretation of speed and requires more skill and greater acceleration. When the

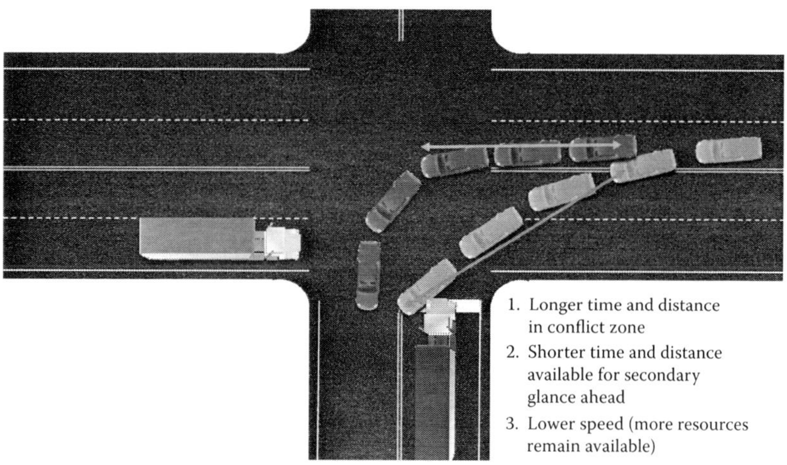

FIGURE 8.3 **(See color insert.)** An explanation of delayed apexing or proper turning when turning across opposing traffic.

approaching vehicle is traveling too fast or the turning car hesitates, a crash or near crash becomes more likely.

8.3.1.1.2 Immediate Hazard Phase—Responding to the Through Driver after Starting the Turn

Once the turning driver initiates the turn, his or her vehicle will usually be in the oncoming driver's path within 2 seconds. Given that this driver has made a decision to enter the intersection, it is unlikely that the turning driver will be able to perceive the hazard, react, and implement an avoidance maneuver in the short time available. Undoubtedly, the LTAP turning scenario is an unforgiving one, which is why it is the second-leading crash scenario for novice drivers.

To reiterate, it is really only possible to avoid a potential conflict in this scenario when the driver glances just as he or she starts turning if the driver (1) delays the start of the turn by traveling well into the intersection, (2) slows, and (3) remains aligned somewhat straight ahead. The unsuspecting driver (i.e., the driver who looks at the very last second) who does this then has time to abort the turn before entering the path of the opposing vehicle.

8.3.1.2 The Through Driver

8.3.1.2.1 Potential Hazard Phase—Approaching an Intersection Where a Vehicle Might Turn into the Drivers' Paths

Consider next the through driver. In particular, consider a through driver approaching the scenario that is shown in Figure 8.4 (perspective view; plan view in Figure 8.2, right panel). (For the turning driver, the scenario is an LTAP crash scenario.) The crosshairs show where this driver is looking. Note that as a driver approaches this intersection, there is a large truck blocking the view of the intersection. Despite the traffic signal displaying a green ball, the truck remains stopped with its left direction signal activated.

What are the mitigating behaviors needed in this type of scenario? Clearly, the through driver needs to glance to the left front of the truck. Recall that Choi (2010) found that an obstructed view was a primary factor in LTAP crashes for the turning driver. But this is true as well for the through driver in this case. It is also the case that the driver should slow some and move to the right.

A simulator study was designed to look at these differences. In the simulator study, teenage drivers with less than 1 year of licensure were twice as likely to crash in the scenario shown in Figure 8.4 than drivers over age 26 (Muttart, 2013). There were several differences noted between the age groups.

First, the experienced drivers made frequent glances to the extent of the sight line. Specifically, the experienced drivers were much more likely to glance toward the front of the truck that created the obstruction when 3–9 seconds from the stop line. Second, following the increased glances came the increased speed loss of the experienced drivers. Between 8 and 9 seconds before the stop line, the speed

FIGURE 8.4 (See color insert.) Through driver's view when approaching a left-turn-across-path scenario in a driving simulator.

of the two groups was similar. However, 7 seconds from the stop line, the experienced drivers began to slow, while the novice teen drivers continued to accelerate. When 1–4 seconds before the stop line, the experienced group were traveling between 6.5 and 10 mph slower than the novice teen drivers at the same locations. Finally, as a result, while driving on a 40 mph road, the novice teen drivers continued at an average speed of 44 mph, which is the same speed they traveled on the straight road segment with no obstructions (Muttart, 2013). Thus, the speed choice by the novice drivers did not vary due to the circumstances. As a comparison, the experienced drivers traveled through the intersection at an average speed of 34 mph despite facing a green traffic signal.

8.3.1.2.2 Immediate Hazard Phase—When Opposing Traffic Turns into Drivers' Paths with Less than 4 Seconds to Impact

The primary reason that LTAP scenarios are so dangerous is that there is little option for the through driver to avoid the crash. In the LTAP scenario, the turning driver usually does not stop but is traveling at a greater average speed than a driver who recently started from a stop. Many times, the turning driver moves from his or her proper lane to the impact location within 2 seconds. When the time to implement an avoidance maneuver is greater than the time to impact, the crash situation is unavoidable.

As time to contact approaches and exceeds 3 seconds, the probability of avoidance increases as well. Usually, LTAP scenarios involve a time to contact of less than 3 seconds. Given that a discussion of perception-response time applies more often to intersection intrusions, please refer to that section for further discussion of perception (emergency) response times (Section 8.3.2.2.2).

8.3.2 Left Turns at Intersections

Next, consider a left turn at an intersection where the threat vehicles are crossing traffic, not the opposing traffic, and the driver is starting the turn from a stop (Figure 8.2, left panel).

8.3.2.1 The Turning Driver, or the Driver Emerging from the Side Road

8.3.2.1.1 Potential Hazard Phase—Before the Turning Driver Enters the Intersection

Here we are addressing the side-road (turning) driver. A left turn at an intersection involves several small acts, each of which is necessary to be able to successfully negotiate the turn. At a higher level, these acts fall into two categories: (1) drivers must look for what they might hit, and then (2) they must watch out for what might hit them. Once a decision is made to pull out, the driver must fit a 6-foot peg (the car) into a 12-foot hole (the lane). Each of these tasks might seem trivial, but failures at any stage have catastrophic consequences.

Although the driver is starting the turn from a stop at a stop line, yield marking, or flashing red traffic signal, proper mitigation needs to begin before the stop. One intersection in particular has been addressed in several studies and will be used to explain the concepts involved in good mitigation (Chan et al., 2010; Muttart et al., 2014; Pradhan et al., 2005, 2006a,b). The intersection in question can be seen in Figure 8.5. Traffic moving east and west (left and right) has no a priori obligation to stop, while traffic north and south must stop at a stop sign. Notice the sidewalk to the right that is obstructed by a tall hedge. This intersection offers an environment that allows for the examination of preacceleration driver behavior.

As the driver approaches and then attempts to turn into an intersection, there are several subtasks. Each of these subtasks will be addressed in the order they occur. First, a driver must stop or be prepared to stop at the stop bar. Unfortunately, many traffic engineers paint the stop bar at a location considerably in advance of where drivers customarily stop. When drivers become accustomed to stopping beyond the stop bar, they learn to dismiss its presence. Unfortunately, sometimes the stop bar means stop, particularly when it is placed immediately before a crosswalk. This is particularly true of the intersection shown in Figure 8.5. The stop bar is far upstream of the crossroad. Due to the hedge to the right, a pedestrian or bicycle could emerge from behind the hedge and enter the road mere seconds before a car rolls through

FIGURE 8.5 **(See color insert.)** Field test intersection of the actions of through-movement drivers. Lincoln Avenue and Amity Street in Amherst, Massachusetts.

the area. Since the driver recognizes that the stop line does not afford an adequate view of cross traffic, many drivers move to a position approximately 2 meters from the road edge (Harwood et al., 1996; Kosaka et al., 2007), which is often beyond the stop line. Both pedestrian (Agran et al., 1994) and bicycle crashes (Summala et al., 1996) are a problem at intersections like the one shown in Figure 8.5. Research on simulators has shown that novice drivers in particular are unlikely to glance at the stop bar for potential bicyclists or pedestrians when it is located far upstream of the cross street (Pradhan et al., 2005).

Second, once a driver stops or at least neutralizes the potential that a pedestrian or bicyclist will emerge from the side when a crosswalk is present, the driver can move forward closer to the cross street and stop once again. The driver must now identify vehicles that might hit him or her (not pedestrians or bicyclists that they might hit). Most drivers make a glance to the left and the right (Muttart, 2013; Muttart et al., 2013; Pradhan et al., 2005). However, much fewer make a secondary glance as they move into the intersection. A secondary glance is a glance toward an area from which a threat might emerge at a time after the foot moves from the brake, or after the start of acceleration into the intersection when there is not a stop. A secondary glance is the last best chance to abort the movement into the intersection. In research by Pradhan et al. (2005), at the through intersection with the hidden sidewalk, experienced drivers made over five times more anticipatory glances.

The timing of the glances relative to the vehicle's movements shows how dangerous it might be to fail to make a secondary glance. Overall, the research indicates that it takes 1.8–2.9 seconds to perform glances and then start moving forward (Hostetter et al., 1986). Imagine a driver who glances left, toward the direction of an approaching vehicle, while at the stop line. Yet, the approaching through vehicle is not visible at that point (suppose that the approaching through vehicle is beyond the sight line or that the turning driver simply failed to see the approaching vehicle with the first glance). Next, consider that sometime between when the driver glanced to his or her left and when the driver started into the intersection, the approaching vehicle became visible to the driver who was turning. When the turning driver does not take a secondary glance to the left immediately before entering the road (when the through vehicle would be visible in this example), an approaching vehicle traveling 50 mph will close 140–210 feet before the turning driver ever enters the intersection. Clearly, failing to monitor an approaching vehicle for this amount of time is not optimal (see also Chapters 6 and 7).

8.3.2.1.2 Immediate Hazard Phase—After the Turning Driver Enters the Intersection

Next, suppose a driver fails to make a secondary glance and moves into the intersection and into the path of a vehicle that had no obligation to stop (a through driver). The intruding driver has already made any side-to-side glances he or she was planning to make and is now concentrating on the intended destination. Therefore, the approaching hazard is up to 90° to the driver's left or right. Considering the large angle and the burden of the turning task, novice drivers engaged in left and right turning movements take, on average, over 2 seconds to respond to an approaching vehicle (Muttart, 2005) (this is also true of experienced drivers [Muttart, 2003]). Consequently, a driver who starts into an intersection and accelerates 2–3 seconds into the intersection might be able to first apply the brakes at the instant of impact. However, to avoid a crash, a driver must respond well before the impact. Obviously, there is inadequate time to avoid the crash in such situations, which is why a secondary glance becomes so important. A secondary glance is the last chance to abort the turn and therefore avoid the need to respond to the immediate hazard.

8.3.2.2 The Through Driver or the Driver Who Has No Obligation

Next, consider the through driver. Once a side-road vehicle starts into a through driver's path, the through driver must respond quickly and reasonably. In this case, depending upon the time remaining when the intruder enters the intersection, there might little chance to mitigate a crash. However, at other times, there are foreshadowing events (Garay-Vega & Fisher, 2005; Muttart, 2013) that give the driver the opportunity to avoid a crash. It is to this potential hazard phase that we want to address our comments.

8.3.2.2.1 Potential Hazard Phase—Anticipation That a Roadside Obstacle Might Enter the Driver's Path

Of most immediate relevance here to the understanding of the through driver's behavior in the potential hazard phase is a study that was undertaken both on a driving simulator (Fisher et al., 2007) and in the field (Pradhan et al., 2009). There were no experienced drivers; however, there were novice drivers who were trained and those who were not trained to scan for traffic when a sign appeared that indicated that a fork was ahead where traffic was entering from the side. The trained drivers on the simulator glanced toward the fork 58% of the time; the untrained drivers, 17% of the time. Similar results were found in the field. The trained drivers glanced toward the fork 50% of the time; the untrained drivers, 18% of the time. Training typically brings novice drivers' level of hazard anticipation halfway to the level of experienced drivers (Taylor et al., 2011). Thus, we can infer that experienced drivers, had they been evaluated in these experiments, would have glanced about 80% of the time for a potentially intruding vehicle at the fork. Being prepared to stop (perhaps by hovering the foot over the brake) can greatly decrease the likelihood of a crash, and therefore, once again, experienced drivers are advantaged at intersections during the potential hazard phase.

8.3.2.2.2 Immediate Hazard Phase—Responding with Emergency Braking or Steering in an Attempt to Avoid a Crash

There are times when an intruding vehicle enters a driver's path without cueing or a foreshadowing event. Similarly, there are instances when due to a failure to respond during the potential hazard phase, a situation materializes into an immediate hazard event. In these situations, the outcome depends in part on the perception-response time and in part on the action taken by the driver.

Here, perception-response time is the period necessary for drivers to respond to an immediate path intrusion. After an intruding driver crosses the stop line or similar intersection location, through drivers begin to brake hard approximately 1.3 seconds later in the daytime and within 1.6 seconds in darkness (Muttart, 2004). Nearly 90% of all drivers have perception-response times within 2 seconds after a vehicle intrudes at an intersection. These responses include field studies (Mazzae et al., 1999b), simulator studies (Hankey et al., 1996; Lechner & Malaterre, 1991; Mazzae et al., 1999a; Perron et al., 1998), and naturalistic studies (Muttart, 2003, 2005).

We can ask whether the response times of novice and experienced drivers differ in such circumstances. Studies that measured drivers' response times in emergency response events have concluded that young drivers have similar response times to the more experienced driver groups (McGehee et al., 1999; Olson & Sivak, 1986). Even when there was a difference between age groups in average response times, the 85th-percentile response times were similar (Lerner et al., 1996). And when examined at a more fine-grained level, older and younger drivers have similar leg movement times (Olson & Sivak, 1986), and generally, most drivers take similar lengths of time from the onset of braking to full braking (Ising et al., 2012; Mazzae et al., 1999b).

However, there are three common instances when drivers' perception-response times increase, and it is in these instances that novice drivers are more likely to be involved than experienced drivers. The first instance is longer leg movement and braking times by distracted and overburdened drivers (Chang et al., 2008; Fisher et al., 2009; Muttart, 2004). And we know that young drivers, particularly novice drivers, are more susceptible to becoming overburdened and distracted (Muttart, 2015).

The second instance where perception-response times are likely to increase is when drivers are looking away from hazards. A meta-analysis of 18 path intrusion response time studies showed that drivers' responses increased by 0.3 seconds for every 10° visual angle away from the hazard that the driver was glancing. Therefore, if a driver is looking 45° away when a vehicle enters his or her path, the average response time increases by more than 1.2 seconds (Muttart, 2003, 2005). Novice drivers generally scan horizontally (side to side) less than do experienced drivers, so they are differentially affected here as well (Mourant & Rockwell, 1972).

Finally, there is a third common way that perception-response times increase. This is when the driver is responding to a concurrent task. This is sometimes referred to as a visual–manual distraction. When a driver is engaged in texting, reaching, typing, or some other task involving manual action, response times increase by an average of 0.7 seconds compared to drivers who are not engaged in a secondary task (Muttart, 2003; Chapter 12). We know from numerous studies that novice drivers are engaged in secondary tasks more frequently than experienced drivers.

Reaction time is not the entire story. When measuring reaction times, it is not possible to measure the reaction time for drivers who fail to respond. Over 20% of drivers involved in police-reportable or severe crashes in the SHRP-2 database failed to respond before a crash. For drivers in age categories less than 25 years old, 29 of 148 failed to react at all. Also, 27 of 67 crashes involving drivers aged 26–69 involved no reaction. However, when examining the effectiveness of the response, we see a difference. Drivers aged 16–25 did not lock the brakes 47 of 148 times. Yet, only 15 of the 67 drivers aged 26–69 failed to lock the brakes (Transportation Research Board, 2013). This is consistent with the finding that novice drivers have been known to brake with less force than experienced drivers (Loeb et al., 2015).

There are other incorrect responses to an immediate hazard. For example, when comparing steering responses, drivers under age 25 were more likely to swerve, more likely to swerve out of the travel lane, and more likely to swerve hard. While more steering might seem to be a favorable result, as already noted earlier, younger drivers are inclined to oversteer (Maeda et al., 1977), lose control (McGehee et al., 1999), and leave the road (Mayhew et al., 2003). Therefore, once a driver initiates a response, the vulnerability of younger drivers might not cease.

8.3.3 Relation between Hazard Anticipation and Hazard Mitigation

The next and final question we want to answer is whether, given equal levels of anticipation, novice drivers are as likely as experienced drivers to mitigate a hazard. The two turn locations reveal the nuances involved in the failures of novice drivers. Before left turns at busy intersections, all hazards are usually evident. However, suppose that there is an obstructed view (Figure 8.2, right panel). Experienced drivers who anticipated the hazard were only slightly more likely slow to a target speed of 0 mph earlier than were novice drivers (the driver should stop before turning), although the average speed between the two groups did not differ significantly (Muttart, 2013). When the experienced and novice drivers who glanced were compared with those who failed to glance, there was a 20% increase in slowing for those who glanced versus those who did not.

Now consider an intersection where a driver is taking a left turn across the path of opposing drivers (Figure 8.2, right panel). And suppose that there is an obstructed view of the opposing through traffic.

Given that the driver anticipated the hazard (i.e., given that the driver glanced toward the potential threat), experienced drivers are 1.92 times more likely to slow than were novice drivers. Those who do anticipate the hazard (glance), whether novice or experienced drivers, are more likely to mitigate the hazard (slow). But more notably, experienced drivers are more likely to anticipate and mitigate hazards that are not initially visible.

8.4 Rear-End Crashes

Intersection and ROR crashes are among the deadliest. Fortunately, they are not the most frequent. Rear-end crashes take this distinction. In a sample of 629,144 North Carolina drivers between 16 and 17 years old, over an 8-year period, a total of 256,975 crashes were recorded for the 3 years after a driver obtained his or her intermediate license (Foss et al., 2011). The most frequent crash type, here defined by the first harmful event in the crash, is the rear-end crash. Over a 36-month period, 33.2% of the crashes were classified as ones where the first harmful event was a rear-end collision with a slowing or stopped vehicle, more than double the percentage of events that fell into the next classification (angle collision, 15.1%). Perhaps the statistic of most relevance is the conditional probability that a driver is in the striking vehicle in a rear-end crash given his or her age. Studies show that drivers younger than 18 are almost 100 times more likely to be the striking vehicle given that they are in a rear-end crash (the conditional probability is equal to 0.082) than are drivers between 50 and 69 (0.005) (Singh, 2003).

For purposes of our analysis of hazard mitigation behaviors in rear-end crashes, these crashes fall into two general categories based upon the distance between the lead vehicle and the following vehicle at the onset of recognized closing. There are rear-end crashes where the lead vehicle suddenly slows and is struck in the rear by the following vehicle, and there are rear-end crashes where the lead vehicle is stopped or moving slowly and is struck by the vehicle approaching from the rear. The factors that influence novice drivers' responses for these two crash scenarios in the potential hazard phase are very different. We do not go on to discuss the novice drivers' responses in the immediate hazard phase that distinguish them from experienced drivers, since these responses differ in ways that have been discussed previously: failing to apply enough force to the brakes.

8.4.1 Vehicles Ahead That Slow Suddenly

In the typical rear-end crash scenario, there are several visual cues available for the approaching driver. The brake lights flash on; there is an immediate change in following distance and perhaps a pitching of the front of the lead vehicle downward due to hard braking. Drivers, even novice teen drivers, have been able to easily interpret what they see when a vehicle ahead slows suddenly (Muttart et al., 2005). The average perception-response time for these crash types has been near 1 second (standard deviation [SD] = 0.3 second) (Chang et al., 2008; Chisholm et al., 2005; Muttart, 2003; Muttart et al., 2005). This is more than enough time to keep from hitting the lead car when the driver is following 2 seconds behind the lead vehicle (assuming that the driver starts braking as hard as the lead vehicle).

So what failures to mitigate a rear-end collision in the potential hazard phase could account for the increased involvement of novice drivers in rear-end collisions? As we noted previously, and has been discussed throughout the chapters in the handbook, distraction is an obvious one (Chapters 7 and 12). To repeat, long glances away from the forward roadway place the driver at a greatly increased risk of crashing (Horrey & Wickens, 2007). Another obvious place to look is the following distance. In one simulator study where the average age of the younger sample was 20 years old and the average age of the older sample of was 70 years old, the mean following distance at 65 mph distance was 286 feet for younger drivers and 473 for older drivers (i.e., 60% smaller for the younger drivers) (Yoo & Green, 1999). In a field study where the speeds varied between 50 and 100 kph, there was a positive association between driver age and the minimum and comfortable time headways (Taieb-Maimon & Shinar, 2001).

8.4.2 Stopped or Slow-Moving Vehicle on a High-Speed Multilane Road

Perhaps surprising to many, most rear-end crashes do not involve one car following closely behind another. Instead, most rear-end crashes involve a lead vehicle that is stopped and being struck by an approaching vehicle. Several studies have shown that more than 70% of rear-end crashes involve a vehicle that is stopped or traveling less than 14 mph (20 kph) (Knipling et al., 1993; McGehee et al., 1997; Sorock et al., 1996). Furthermore, rear-end crashes have been shown to increase when there are large differences in traffic speed (Brehmer, 1990; Soloman, 1964; Taylor et al., 2000), as is often the case when there are stopped and slow-moving vehicles in free-flow traffic. Regardless, as noted previously, younger drivers are much more likely to be the striking driver in rear-end crashes than are more experienced drivers.

Why might this be the case? Again, distraction could be almost the entire explanation (Chapter 12). But we also need to ask whether there is something about identifying a stopped or slowing vehicle that varies as a function of age. Unlike the driver of a slowing vehicle, a driver of a stopped or slow-moving vehicle may not apply the brakes. The driver approaching the stopped or slow-moving vehicle needs to rely on other clues.

Of relevance here, Summala et al. (1998) showed that driver experience did not influence a driver's ability to detect closing speed in the periphery. In their experiment, young novice drivers and experienced drivers looked away from the forward roadway at visual eccentricities of up to 50° from straight ahead. The more eccentric was the vehicle, the more difficult it was to detect the closing speed. This indicates that the distribution of the glance locations of novice and experienced drivers with respect to the forward roadway will determine just how likely the driver of a given age is to detect the closing speed. Earlier studies have indicated that novice drivers have a restricted scanning range left to right (Mourant & Rockwell, 1972). This would suggest that novice drivers might actually be quicker overall to detect closing speed. But we know that novice drivers are more distracted (and therefore more likely to be looking inside the vehicle or at a nomadic device) (Lerner & Boyd, 2005). Thus, it is difficult to predict just what will be the net effect.

Finally, when looking away, a driver closing in on a lead vehicle has to extrapolate the motion of other traffic. Drivers with an average age of 58 did not perform as well with motion extrapolations when compared to a group of drivers with an average age of 22 (Delucia & Mather, 2006). Delucia and Mather did not examine the ability of novice drivers to extrapolate motion, but such a study would be interesting given the research mentioned previously that young drivers have had difficulty estimating closing speed in left-turning crashes (Choi, 2010)—a situation that is similar if the driver is not directly glancing at the oncoming traffic.

8.4.3 Relation between Hazard Anticipation and Hazard Mitigation

In the one study of which we are aware that controlled for hazard anticipation, the novice drivers again failed to execute the proper mitigating behaviors even though they anticipated the hazard (Muttart, 2013). Specifically, in the scenario in question, a bus in the right lane ahead slowed to a stop as the participant driver and a lead vehicle approached the bus. The lead vehicle moved out of the right lane and began to pass the bus, but a pedestrian emerged from in front of the bus, which caused the car (lead vehicle) to stop suddenly. Among the experienced drivers, the probability of a crash given that the driver slowed and glanced toward the pedestrian was only 5%. It was four times higher in the novice drivers (21%). Again, we see that the correct behaviors, slowing and glancing, do not translate into equivalent reductions in crashes.

8.5 Conclusions

In the three crash types discussed here, in both the potential and immediate hazard phases, novice drivers' hazard mitigation behaviors are compromised. Because they do not anticipate a hazard in the potential hazard phase, they do not execute the steering and braking maneuvers that keep the hazard from developing into one that requires a response in the immediate hazard phase. This is particularly problematic for novice drivers because they are much less adept than experienced drivers at steering and braking appropriately in the immediate hazard phase. And even those novice drivers who anticipate a potential hazard are less likely to engage in behaviors that mitigate the hazard, and when they do engage in such behaviors, they still are at risk because the mitigation, although present, is not optimal. Clearly, better hazard anticipation skills would decrease the reliance of novice drivers on the need to deploy hazard mitigation skills (Chapter 6). In general, the data support the idea that novice drivers are not as prepared to mitigate hazards as experienced drivers. This may be because novice drivers fail to anticipate the hazards. However, even when they do anticipate the hazards, they may not react in a way that reduces their risk of crashing as much as experienced drivers.

Acknowledgments

We would like to thank the members of the Human Performance Laboratory at the University of Massachusetts.

References

Abdel-Aty, M., Chen, C., & Radwan, A. (1999). Using conditional probability to find driver age effect in crashes. *Journal of Transportation Engineering, 125*(6), 502–507.

Agran, P.F., Winn, D.G., & Anderson, C.L. (1994). Differences in child pedestrian injury events by location. *Pediatrics, 93*, 284–288.

Araki, K., & Matsuura, Y. (1990). *Driver's Response and Behavior on Being Confronted with a Pedestrian or a Vehicle Suddenly Darting across the Road* (SAE Technical paper #900144). Warrendale, PA: Society of Automotive Engineers.

Blythe, W., Day, T., & Grimes, W. (1998), *3-Dimensional Simulation of Vehicle Response to Tire Blow Outs.* (Technical paper no. 980221). Warrendale, PA: Society of Automotive Engineers.

Bonneson, J.A., Pratt, M.P. (2009). A model for predicting speed along horizontal curves on two-lane highways, Paper No. 09-1419, Washington, DC: Transportation Research Board.

Braitman, K.A., Kirley, B.B., McCartt, A.T., & Chaudhary, N.K. (2008). Crashes of novice teenage drivers: Characteristics and contributing factors. *Journal of Safety Research, 39*(1), 47–54.

Brehmer, B. (1990). Variable errors set a limit to adaptation. *Ergonomics*, 33(10/11), 1231–1239.

Chan, E., Pradhan, A.K., Knodler, M.A, Pollatsek, A. & Fisher, D.L. (2010). Are driving simulators effective tools for evaluating novice drivers' hazard anticipation, speed management, and attention maintenance skills? *Transportation Research F, 13*, 343–353.

Chang, S.H., Lin, C.Y., Fung, C.P., Hwang, J.R., & Doong, J.L. (2008). Driving performance assessment: Effects of traffic accident location and alarm content. *Accident Analysis & Prevention, 40*, 1637–1643.

Chisholm, S.L., Caird, J.K., Lockhart, J.A., Vacha, N.H., & Edwards, E.J. (2005) Do in-vehicle advance signs benefit older and younger driver intersection performance? *Proceedings of the Third International Driving Symposium on Human Factors in Driving Assessment, Training, and Vehicle Design*. Rockport, ME.

Choi, E.-H. (2010). *Crash Factors in Intersection-Related Crashes: An On-Scene Perspective,* DOT HS 811 366. Washington, DC: National Highway Traffic Safety Administration.

Delucia, P.R., & Mather, R.D. (2006). Motion extrapolation of car-following scenes in younger and older drivers. *Human Factors, 48*, 666–674.

Fatal Accident Reporting System (FARS). (2012). National Highway Traffic Safety Administration. Accessed November 12, 2012, retrieved from http://www.nhtsa.gov/people/ncsa/fars.html

Fisher, D.L., Laurie, N.E., Glaser, R., Connerney, K., Pollatsek, A., Duffy, S.A., & Brock, J. (2002). The use of an advanced driving simulator to evaluate the effects of training and experience on drivers' behavior in risky traffic scenarios. *Human Factors, 44*, 287–302.

Fisher, D.L., Pradhan, A.K., Pollatsek, A., & Knodler, M.A. Jr. (2007). Empirical evaluation of hazard anticipation behaviors in the field and on a driving simulator using an eye tracker. *Proceedings of the 86th Transportation Research Board Annual Meeting*. Washington, DC: National Research Council.

Fisher, D.L., Knodler, M., & Muttart, J. (2009). Driver-Eye-Movement-Based Investigation for Improving Work-Zone Safety. Project NETC 04-2, 2009, The New England Transportation Consortium. Retrieved from http://www.netc.umassd.edu/netcr71_04-2.pdf

Foss, R.D., Martell, C.A., Goodwin, A.H., & O'Brien, N.P. (2011). *Measuring Changes in Teenage Driver Crash Characteristics during the Early Months of Driving*. Washington, DC: AAA Foundation for Traffic Safety. Retrieved from https://www.aaafoundation.org/sites/default/files/2011MeasuringCharacteristicsOfTeenCrashes_0.pdf

Garay-Vega, L. & Fisher, D.L. (2005). Can novice drivers recognize foreshadowing risks as easily as experienced drivers? *Proceedings of the Third International Driving Symposium on Human Factors in Driver Assessment, Training and Vehicle Design*, pp. 471–477.

Glennon, J.C., Neuman, T.R., & Leisch, J.T. (1986). *Safety and Operational Considerations for Design of Rural Highway Curves, (DOT-FH-11-9575)*. McLean, VA: Federal Highway Administration.

Hallmark, S.L., Veneziano, D., McDonald, T., Graham, J., Bauer, K.M., Patel, R., & Council, F.M. (2006). *Safety Impacts of Pavement Edge Drop-offs*. Washington, DC: AAA Foundation for Traffic Safety.

Hankey, J.M., McGehee, D.V., Dingus, T.A., Mazzae, E.N., & Garrott, W.R. (1996). Initial driver avoidance behavior and reaction time to an unalerted intersection incursion. *Proceedings of the Human Factors and Ergonomics Society 40th Annual Meeting*. Philadelphia, PA, September 2–6, 1996.

Happer, A.J., Peck, M.D., & Hughes, M.C. (2009). *Analysis of Left-Turning Vehicles at a 4-way Medium-Sized Signalized Intersection*. Warrendale, PA: Society of Automotive Engineers.

Harwood, D.W., Mason, J.M., Brydia, R.E., Pietrucha, M.T., & Gittings, G.L. (1996). *Intersection Sight Distance, NCHRP Report 383*. Washington, DC: Transportation Research Board.

Horrey, W., & Wickens, C. (2007). In-vehicle glance duration. *Transportation Research Record, 2018*, 22–28.

Hostetter, R.S., McGee, H.W., Crowley, K.W., Sequin, E.L., & Dauber, G.W. (1986). *Improved Perception-Reaction Time Information for Intersection Sight Distance*, Report No. FHWA/RD-87/015. McLean, VA: Federal Highway Administration.

Hough, D.L. (2008). *Proficient Motorcycling: The Ultimate Guide to Riding Well*, 2nd Ed. Irvine, CA: BowTie Press.

Ising, K.W., Droll, J.A., Kroeker, S.G., D'Addario, P.M., & Goulet, J.-F. (2012). Driver-related delays in emergency braking response to a laterally incurring hazard. *Proceedings of the Human Factors and Ergonomics Society*.

Knipling, R., Mironer, M., Hendricks, D., Allen, J., Tijerina, L., Everson, J., & Wilson, C. (1993). *Assessment of IVHS Countermeasures for Collision Avoidance: Rear-End Crashes. Final Report (Report No. DOT HS 807 995)*. Washington, DC: National Highway Traffic Safety Administration.

Kosaka, H., Hashikawa, T., Highashikawa, N., Noda, M., Nishitani, H., Uechi, M., & Sasaki, K. (2007). *On-the-Stop Investigation of Negotiation Patterns of Passing Cars without Right of Way at Non-Signalized Intersections* (Technical paper 2007-01-3599). Warrendale, PA: Society of Automotive Engineers.

Land, M.F., & Lee, D.N. (1994). Where we look when we steer. *Nature, 369*(6483), 742–744.

Lechner, D., & Malaterre, G. (1991). *Emergency Maneuver Experimentation Using a Driving Simulator. (SAE Paper No. 910016)*. Warrendale, PA: Society of Automotive Engineers.

Lerner, N., & Boyd, S. (2005). *On-Road Study of Willingness to Engage in Distracting Tasks* (DOT HS 809 863). Washington, DC: National Highway Traffic Safety Administration.

Lerner, N.D., Huey, R.W., McGee, H., & Sullivan, A. (1996). Older driver PRT for intersection sight distance and object detection. *Accident Investigation Quarterly, 9*, 20–36.

Loeb, H., Kandadai, V., McDoinald, C.C., & Winston, F.K. (2015). Emergency braking in adults versus teen drivers. *Transportation Research Record, Journal of the Transportation Research Board*, (2516), 8–14. Transportation Research Board, Washington, DC, doi: 10.3141/2516-02

Maeda, T., Irie, N., Hidaka, K., & Nishimura, H. (1977). *Performance of Driver-Vehicle System in Emergency Avoidance (Technical paper 770130)*. Warrendale, PA: Society of Automotive Engineers.

Mayhew, D.R., Simpson, H.M., & Pak, A. (2003). Changes in collision rates among novice drivers during the first months of driving. *Accident Analysis & Prevention, 3*, 683–691.

Mazzae, E.N., Baldwin, G.H.S., & McGehee, D.V. (1999a). *Driver Crash Avoidance Behavior with ABS in an Intersection Incursion Scenario on the Iowa Driving Simulator* (SAE paper no. 1999-01-1290). Warrendale, PA: Society of Automotive Engineers.

Mazzae, E.N., Barickman, F., Baldwin, G.H.S., & Forkenbrock, G. (1999b). *Driver Crash Avoidance Behavior with ABS in an Intersection Incursion Scenario on Dry versus Wet Pavement* (SAE Paper no. 1999-01-1288). Warrendale, PA: Society of Automotive Engineers.

McElheny, M., Blanco, M., & Hankey, J. (2006). On-road evaluation of an in-vehicle curve warning device. *Proceedings of the Human Factors and Ergonomic Society 50th Annual Meeting*. San Francisco, CA. October 1–5, 2007.

McGehee, D.V., Brown, T., & Wilson, T. (1997). Examination of drivers' collision avoidance behavior in a stationary lead vehicle situation using a front-to-rear-end collision warning system. USDOT/NHTSA Office of Crash Avoidance Research Technical Report. Contract DTNH22-93-C-07326.

McGehee, D., Mazzae, E., & Scott Baldwin, G. (1999), 'Examination of drivers' collision avoidance behavior using conventional and antilock brake systems on the Iowa driving simulator'. NHTSA Light Vehicle Antilock Brake Research Program, Task 5. National Technical Information Service, Springfield, Virginia.

McKnight, J.A., & McKnight, S.A. (2003). Young novice drivers: Careless or clueless. *Accident Analysis & Prevention, 35*, 921–925.

McLaughlin, S.B., Hankey, J.M., Klauer, S.G., & Dingus, T.A. (2009). *Contributing Factors to Run-Off-the-Road Crashes and Near Crashes*, DOT HS 811 079. Washington, DC: National Highway Traffic Safety Administration.

Mikolajetz, A., Henning, M.J., Tenzer, A., Zobel, R., Krems, J.F., & Petzoldt, T. (2009). Curve negotiation: Identifying driver behavior around curves with the driver performance database. *Proceedings of the fifth International Driving Symposium on Human Factors in Driver Assessment, Training and Vehicle Design*. San Francisco, CA. October 16–20, 2006, pp. 391–397.

Mitsopoulos-Rubens, E., Triggs, T., & Regan, M. (2009). Comparing the gap acceptance and turn time patterns of novice with experienced drivers for turns across traffic. *Proceedings of the fifth International Driving Symposium on Human Factors in Driver Assessment, Training, and Vehicle Design*. Big Sky, Montana. June 22–25, 2009.

Mourant, R., & Rockwell, T. (1972). Strategies of visual search by novice and experienced drivers. *Human Factors, 14*, 325–335.

Muttart, J. W. (2003). *Development and Evaluation of Driver Perception-Response Equations Based upon Meta-analysis* (Technical paper no. 2003-01-0885). Warrendale, PA: Society of Automotive Engineers.

Muttart, J.W. (2005). Quantifying driver response times based upon research and real life data. *3rd International Driving Symposium on Human Factors in Driver Assessment, Training, and Vehicle Design*, Rockport, Maine, June 27–30, 2005. 3, 8–29. http://drivingassessment.uiowa.edu/DA2005/PDF/03_Muttartformat.pdf

Muttart, J.W. (2013). Identifying Hazard Mitigation Behaviors That Lead to Differences in The Crash Risk between Experienced and Novice Drivers (a dissertation submitted in partial fulfillment of the doctoral degree). University of Massachusetts, Amherst.

Muttart, J.W. (2015). *Influence of Age, Secondary Tasks and Other Factors on Drivers' Swerving Responses before Crash or Near-Crash Events* (Technical paper no. 2015-01-1417). Warrendale, PA: Society of Automotive Engineers.

Muttart, J.W., Messerschmidt, W., & Gillen, L. (2005). *Relationship between Relative Velocity Detection and Driver Response Times in Vehicle Following Situations* (Technical paper no. 2005-01-0427). Warrendale, PA: Society of Automotive Engineers.

Muttart, J.W., Fisher, D.L., Pollatsek, A.P., & Marquard, J. (2013). Hazard mitigation when approaching curves. *7th International Driving Symposium on Human Factors in Driver Assessment, Training, and Vehicle Design*, Lake George, NY.

Muttart, J.W., Fisher, D.L., & Pollatsek, A.P. (2014). *Hazard Mitigation When Approaching Obstructed Intersections in a Simulator.* Washington, DC: Transportation Research Board of the National Academies.

Olson, P.L., & Sivak, M. (1986). Perception-response time to unexpected roadway hazards. *Human Factors, 28*, 91–96.

Perron, T., Chevennement, J., Damville, A., Mautuit, C., Thomas, C., & Le Coz, J.Y. (1998). Pilot Study of Accident Scenarios on a Driving Simulator, *Proceedings of the 16th ESV Conference*, June 1–4, 1998, Windsor, Canada, Paper no. 98-S2-O-02, pp. 374–385.

Pradhan, A.K., Fisher, D.L., & Pollatsek, A. (2006a). Risk perception training for novice drivers: Evaluating duration of effects on a driving simulator. *Transportation Research Record, 1969*, 58–64.

Pradhan, A.K., Fisher, D.L., Pollatsek, A., Knodler, M., & Langone, M. (2006b). Field evaluation of a risk awareness and perception training program for younger drivers. *Proceedings of the Human Factors and Ergonomics Society 50th Annual Meeting*, San Francisco, 2388–2391.

Pradhan, A.K., Hammel, K.R., DeRamus, R., Pollatsek, A., Noyce, D.A., & Fisher, D.L. (2005). The use of eye movements to evaluate the effects of driver age on risk perception in an advanced driving simulator. *Human Factors, 47*, 840–852.

Pradhan, A.K., Pollatsek, A., Nodler, M., & Fisher, D.L. (2009). Can younger drivers be trained to scan for information that will reduce their risk in roadway traffic scenarios that are hard to identify as hazards? *Ergonomics, 52*(6), 657–673.

Singh, S. (2003). *Driver Attributes and Rear-End Crash Involvement Propensity (DOT HS 809 540).* Washington, DC: National Highway Traffic Safety Administration.

Soloman, D. (1964). *Accidents on Main Rural Highways Related to Speed, Driver and Vehicle.* Washington, DC: United States Government Printing Office.

Sorock, G.S., Ranney, T.A., & Lehto, M.R. (1996). Motor vehicle crashes in roadway construction work zones: An analysis using narrative text from insurance claims. *Accident Analysis & Prevention, 28*, 131–138.

Suh, W., Park, Y.-J., Ho Park, C., & Soo Chon, K. (2006). Relationship between speed, lateral placement, and drivers' eye movement at two-lane rural highways. *Journal of Transportation Engineering, 132*(8), 649–653.

Summala, H., Pasanen, E., Raesaenen, M., & Sievaenen, J. (1996). Bicycle accidents and drivers' visual search at left and right turns. *Accident Analysis & Prevention, 28*, 147–153.

Summala, H., Lamble, D., & Laakso, M. (1998). Driving experience and perception of the lead car's braking when looking at in-car targets. *Accident Analysis & Prevention, 30*(4), 401–407.

Taieb-Maimon, M., & Shinar, D. (2001). Minimum and comfortable driving headways: Reality versus perception. *Human Factors, 43*, 159–172.

Taylor, M.C., Lyman, D.A., & Baruya, A. (2000). *The Effects of Drivers' Speed on the Frequency of Road Accidents.* TRL Report No. 421. Crowthorne, Berkshire: Transport Research laboratory.

Taylor, T.G.G., Masserang, K.M., Divekar, G., Samuel, S., Muttart, J.W, Pollatsek, A., & Fisher, D.L. (2011). Long term effects of hazard anticipation training on novice drivers measured on the open road. *Proceedings of the 6th International Driving Symposium on Human Factors in Driver Assessment, Training, and Vehicle Design 2011*, Lake Tahoe, CA. Iowa City: Public Policy Center, University of Iowa.

Transportation Research Board of the National Academy of Sciences. (2013). The 2nd Strategic Highway Research Program Naturalistic Driving Study Dataset. Available from the SHRP 2 NDS InSight Data Dissemination. Web site: https://insight.shrp2nds.us

Winston, F. (2011). *677 Teens Who Were Involved in Serious Crashes between 2005-2006: Query from the National Motor Vehicle Crash Causation Survey (NMVVCS), 2008.* DOT HS 811 051, Washington, DC: National Highway Traffic Safety Administration (email correspondence to D. L. Fisher, June 10, 2011).

Yoo, H., & Green, P. (1999). *Driver Behavior While Following Cars, Trucks, and Buses* (UMTRI-99-14). Ann Arbor, MI: University of Michigan.

Zegeer, C., Stewart, R., Reinfurt, D., Council, F.M., Neuman, T., Hamilton, E., Miller, T., & Hunter, W. (1990). *Cost Effective Geometric Improvements for Safety Upgrading of Horizontal Curves*. Publication No. FHWA-RD-90-021. Chapel Hill, NC: University of North Carolina, Highway Safety Research Center.

III

Developmental Differences

9

Developmental Factors in Driving Risk

Bonnie Halpern-Felsher

Malena Ramos

Tracy Exley

Shelley Aggarwal

Abstract

Developmental factors juxtaposed with contextual changes and influences that take place during the period of time when adolescents learn to drive result in greater vulnerabilities to risky driving and crashes. The developmental factors related to adolescent driving include physical, cognitive, and psychosocial maturation. The well-known risk factors for adolescents' involvement in automobile crashes include lack of driver experience; risk taking (including speeding, reckless driving, and running red lights); distractions from road rules and safety concerns (e.g., influence of passengers, cell phones, texting, or radio use); substance use; night driving and driving when fatigued; and improper use or nonuse of restraint devices. Understanding how developmental factors contribute to adolescent driving risk will help us better understand and promote safe driving, including understanding adolescent crash risk. Further, there is a need for reasonable cultural and social norms to combine to set a standard of expectation that leads to safe driving habits and practices. In this chapter, we describe the physical development, including brain maturation, that occurs throughout adolescence and into young adulthood. We then discuss adolescents' cognitive and psychosocial maturity, both of which influence and are influenced by adolescents' physical development. We end with a discussion of social factors, including peers and parents, that influence development and influence the relationship between development and driving-related decision making.

9.1 Introduction

Kathy, age 17, is driving her friend Priya (age 16) to volleyball practice. As the music is blasting and they are intermittently singing and laughing, Kathy gets a text from another friend. With one hand on the wheel, she fumbles to find her phone to read the text, and Priya hurriedly grabs the falling smartphone that has now slipped from Kathy's hand. As both girls are looking at the falling phone, Kathy hears a honk and realizes that she has veered into the other lane.

Eighteen-year-old Lionel and his friend Taylor are on their way to a high school football game to meet some friends. As Lionel drives, Taylor turns up the radio and is demonstrating his air guitar skills. Lionel approaches an intersection as the traffic light is turning from yellow to red, and Taylor yells, "Step on it!" Lionel steps on the accelerator, speeding through the intersection, just as another car is about to cross.

Linda just got her license. Her parents have allowed her to drive to and from school each day. She is not allowed to have friends in the car unless accompanied by an adult chaperone, and Linda has respected her parents' request. During her routine drive to school at 8 a.m., road construction has begun along her route, and the usual road signs are blocked by heavy equipment. The noise of drill work is distracting, and Linda is frantically scanning the street trying to identify her usual turnoff. Simultaneously, a speeding driver is attempting to make a hurried left turn as Linda is deciding which direction to go.

These scenarios are only a few examples of what teenage drivers may face when getting behind the wheel of a vehicle. While experienced adult drivers can also encounter similar situations, developmental factors juxtaposed with contextual changes and influences that take place during the period of time when adolescents learn to drive result in greater vulnerabilities to risky driving and crashes. The well-known risk factors for adolescents' involvement in automobile crashes include lack of driver experience; risk taking (including speeding, reckless driving, and running red lights); distractions from road rules and safety concerns (e.g., influence of passengers, cell phones, texting, or radio use); substance use (see also Chapter 13); night driving and driving when fatigued; and improper or nonuse of restraint devices. These risks rarely occur in isolation; there is frequent overlap among many of them.

The developmental factors related to adolescent driving include physical, cognitive, and psychosocial maturation. Understanding how these developmental factors contribute to adolescent driving risk will help us better understand and promote safe driving, including understanding adolescent crash risk. Throughout the chapter, it is important to note that adolescents of similar chronological age may be at varying stages of physical, cognitive, or psychosocial development, and possess differing abilities to manage complex tasks and situations. More and more is being learned about adolescent social risks, such as driving safety and the relationship between psychosocial development and biological development, specifically brain maturation. In the following sections, we will first describe adolescents' physical development, including brain maturation, that occurs throughout adolescence and into young adulthood. We then discuss adolescents' cognitive and psychosocial maturity, both of which influence and are influenced by adolescents' physical development. We end with a discussion of social factors, including peers and parents, that influence development and influence the relationship between development and driving-related decision making (see also Chapters 10, 16, and 19).

9.2 Adolescent Development

9.2.1 Physical Development

Brain development is a complex process that involves a variety of changes that advance throughout childhood and into adolescence and young adulthood. Brain maturation occurs with a back-to-front progression. The areas of the brain involved with more primitive processes, like motor and sensory activities, mature before areas involved with more high-level processing, such as executive functioning and planning (Casey et al., 2008; Galvan et al., 2007). The amygdala and limbic system, responsible for emotions and the reward center, respectively, mature earlier than the prefrontal cortex, or the area of the brain responsible for executive functions such as decision making and higher-level reasoning (Casey et al., 2008; Steinberg, 2007). The prefrontal cortex is one of the last areas to develop and generally does not fully mature until early adulthood (approximately age 25). Along with the maturation of various parts of the brain, connections between neurons in the brain are also evolving. Before puberty, there is a period of time when the number of connections between neurons in the brain greatly increases. As puberty progresses, there is a selective elimination of some of these connections and an "editing" of neuronal pathways, better known as synaptic pruning, thus eliminating connections that are used less often in order to maximize efficiency (Blakemore & Choudhury, 2006; Gogtay et al., 2004).

Concurrently, the remaining neuronal connections are myelinated, and together, the pruning and myelination processes optimize neuronal processing; as the adolescent brain continues to mature, decision making, selective attention, and processing speed are expected to improve over time (Blakemore & Choudhury, 2006; Luna et al., 2004). The processes of pruning and myelination also occur in a back-to-front manner, with the frontal lobes, involved more in executive functioning, being pruned and myelinated last (Gogtay et al., 2004). Planned behaviors require integrating multiple complex processes such as working memory and response inhibition, and executive tasks in younger individuals may require increased attentional effort (Luna et al., 2004, 2010). As a result, the adolescent years, prior to the completion of the various maturation processes, could be a higher-risk interval for initiating driving, related to immature biological development, as the ability to organize certain tasks may not be fully evolved.

Adolescents and adults are similarly able to distinguish between risky and safe choices, but due to the maturation differences between the two, adolescents may be less likely to make a safe choice and more likely to make a risky one (Giedd & Rapoport, 2010). As noted, areas of the brain involved with emotions (the amygdala) and the reward centers (limbic system) mature before the prefrontal cortex, responsible for critical thinking and decision making, with connections continuing to mature into early adulthood (Casey et al., 2008). This pattern of maturation may lead the adolescent to act more impulsively, choose more risky behaviors, and make decisions influenced by the emotional and reward centers of their brains (Casey et al., 2008; Galvan et al., 2007; Steinberg, 2007). This would suggest that rational thinking may be overridden by emotions or the desire for rewards. As the adolescent driver matures into adulthood, so does the brain, and risky and impulsive decisions begin to decrease concurrently (Casey et al., 2008).

It is important to remember that there are also individual and gender differences in the maturation of the brain. Galvan et al. (2007) conducted a study looking at individual differences in risk-taking behaviors, and their findings suggest that due to developmental neural changes, some adolescents are predisposed to engage in risky behaviors, while others are not. This may explain why some adolescents participate in risky behaviors, while others do not.

Gender differences in maturation of the brain help explain the gender differences that have been found between males and females with regard to their driving behaviors. Horvath et al. (2012) looked at the beliefs that motivate young drivers to speed and found that high intenders among males were significantly more likely to speed than females. Another study by Møller and Haustein (2013) looked at young male drivers, ages 18 and 28, and found that perceptions of friends' speeding were the most important predictor of speeding. Males tended to voluntarily engage in high-risk behavior, were found to be less compliant toward traffic rules, and generally perceived traffic situations as less risky compared to females (Oltedal & Rundmo, 2006).

9.2.2 Cognitive Development

Are adolescents equipped to handle and understand the rules and responsibilities of operating a motor vehicle? Although there is certainly within-age-group variation, many studies show that by age 16, adolescents' general cognitive abilities, such as the ability to understand consequences including the risks and benefits of their actions and decisions, process information, and reason, are very similar to adults' cognitive abilities (Steinberg et al., 2009). During adolescence, abstract cognition and less concrete thinking allow adolescents the ability to consider many components necessary for competent decision making at one time and consider potential positive and negative outcomes associated with each decision. However, despite adolescents and adults having similar cognitive abilities, motor vehicle accidents remain the leading cause of death and a leading cause of nonfatal injury among adolescents aged 16–20 years in the United States (Carter et al., 2014; WISQARS, 2011). This could be due to several factors, such as personality traits more commonly found among young adult male drivers (e.g., sensation seeking and/or impulsivity) as well as adolescents' and young adults' tendency to overestimate their abilities while underestimating danger (Constantinou et al., 2011). In addition to general increases in cognitive ability over time, two additional aspects of adolescent development are important to consider in influencing driving ability and crash rates: the role of experience and perceptions of driving-related risks and benefits. These are described next.

9.2.2.1 Importance of Experience

Gaining experience behind the wheel has a critical role in adolescents' ability to identify risk and in how adolescents react under certain driving situations and conditions. Multiple factors influence adolescent driving behaviors, and inexperience is a core issue that must be differentiated from other developmental factors. It is important to distinguish here between just having knowledge and having actual behind-the-wheel driving experience. While it is certainly important for adolescents to gain knowledge about driving, including driving rules, potential obstacles, and driving policies, it is as important for novice drivers to gain actual driving experience, whereby they can learn how to translate knowledge into on-the-road driving, increase reaction times, and so on. A less experienced or novice driver may have a lower comprehension of driving safety and the associated risks as well as consequences (Keating, 2007; McKnight & McKnight, 2003; Romer et al., 2014). Young drivers tend to identify fewer potential hazards compared to adults (Borowsky & Oron-Gilad, 2013; Borowsky et al., 2010), and they tend to overestimate their abilities and underestimate danger (Constantinou et al., 2011). There is a steady decline in crash rates with each year of driving, with rates of crashes being highest amongst adolescents within the first 6 months after licensure, implicating the importance of experience related to driving cues and situational awareness (Mayhew et al., 2003; Romer et al., 2014). In their comprehensive review, Romer et al. (2014) divided attention to driving tasks into two categories: (1) task inattention, including failure to allocate attention to the road, and (2) hazard inattention, or failure to attend and respond to hazards even when attention is given to driving tasks. They suggest that novice drivers may be unable to recognize and respond to hazards, due to a lack of skill and experience. While inexperience is a key influencer for adolescent driving risk, driver distraction is also of notable concern and may be more linked to brain development and immaturity (Keating & Halpern-Felsher, 2008).

9.2.2.2 The Role of Perceptions of Driving-Related Risks and Benefits

A hallmark of cognitive development is the ability to identify and understand consequences associated with a particular behavior. Perceptions of social, physical, and/or health risks associated with any given behavior as well as perceived social and physical benefits are key components of any competent decision. Adolescents and adults are generally similar in their ability to identify and consider positive and negative consequences of their decision. In some cases, adolescents actually perceive greater risks than adults.

Driving-related risk perceptions are the subjective evaluation of how well the driver thinks he or she is able to handle the road and apply appropriate actions and precautions (Borowsky & Oron-Gilad, 2013). Studies have shown that skills-based processes relate to the drivers' ability to anticipate hazards and evaluate risks associated with driving. In general, young drivers identify fewer potential hazards than older, more experienced drivers (Borowsky & Oron-Gilad, 2013).

Rhodes and Pivik (2011) examined the roles of positive affect and risk perception in age and gender differences in risky driving. They found that positive affect more strongly predicted risky driving among adolescent males than adolescent females. That is, when adolescents, particularly males, are put in more emotionally charged situations, they are more likely to engage in risky driving. This might explain why adolescents, especially males, are more likely to get into accidents when driving with friends. Further, Borowsky and Oron-Gilad (2013) and Borowsky et al. (2010) examined hazard awareness and risk perception skills concurrently by looking at the effects of age and driving experience on the ability to detect road hazards. They found that drivers' experience improved their awareness of potential hazards.

Despite young drivers' understanding and knowledge of the risks, and the role that experience plays in driving competency and perceptions, other aspects of development, including desire for independence and autonomy, influence from peers, desire for rewards and sensation seeking, limited or nascent impulse control, and brain development, all help explain why adolescents still take driving risks that can and do result in crashes. These psychosocial factors are described next.

9.2.3 Psychosocial Maturity

While adolescent cognitive abilities are largely forged by about age 16, psychosocial maturity continues to develop and contribute to the elevated driving crash risk in the adolescent age group. Psychosocial maturity includes social and peer comparison, impulsivity, peer affiliation, susceptibility to peer pressure, ability to understand and plan for the future, and ability to consider and acknowledge other people's perspectives. With psychosocial maturation, one can see an increase in self-control, stronger resistance to peer influences, and the willingness to forsake immediate gratification in order to achieve future goals (Monahan et al., 2009), of which all are important components in safe driving practices.

Critical aspects of psychosocial development, such as those associated with peer pressure, sensation seeking, reward seeking, and impulse control, are much less developed during adolescence and into young adulthood, compared to adults. It is these areas of immaturity that most affect adolescents' decision to engage in driving-related risk behaviors that often lead to the crash rates seen for this age group.

9.2.3.1 Role of Sensation Seeking and Impulsivity

Sensation seeking, or the desire for exciting, new, thrilling, and rewarding experiences, increases during adolescence and continues into the mid-20s. While health-relevant models suggest a negative relationship between risky driving and perceived risk of outcomes, Hatfield et al. (2014) suggest that high sensation-seekers may value the "thrill" of the risk, thus accounting for the positive associations between sensation seeking and risky driving. They concluded that an emphasis on risk might be of limited benefit to sensation-seekers. Thus, subsequent road safety laws and campaigns should consider these factors in their messaging. Carter et al. (2014) used the theory of normative social behavior to provide a framework for understanding how adolescent risk taking and sensation seeking combine with parent and peer influences to shape adolescent distracted driving behavior. They found that adolescent risk perception and descriptive norms (perception of the extent to which others were engaging in the same behavior) were important predictors of adolescent distracted driving, with 92% of their adolescent sample reporting regularly engaging in distracted driving.

In their study looking at predictors of distracted driving behavior, Carter et al. (2014) found that male participants showed higher sensation seeking and lower risk perception when compared with female participants. Adolescents are able to think about and understand the risks associated with behaviors, but ultimately, their decisions may be influenced by emotion and reward centers (Steinberg, 2007). In addition, multitasking, such as speaking on the phone, reduces attention and response time to braking (Strayer et al., 2003), and dialing a cell phone significantly increases the risk of crash or near crash, particularly in novice drivers.

Impulsivity refers to making decisions in a quick fashion, with limited foresight or information. Impulsivity steadily declines from age 10 on (Steinberg et al., 2008). Becoming competent to make decisions requires adolescents to be able to control their desire to act impulsively.

Studies have highlighted the complex relationship between impulsivity, peer pressure, and delinquent behavior, where individuals with low impulsivity are actually more vulnerable to delinquent peer influences than those with high impulsivity. Those with high impulsivity are likely to engage in risky behavior, whereas those with low impulsivity are particularly sensitive to peer pressure that may also result in engaging in risky behavior (Vitulano et al., 2010).

Thus, adolescence is a time of low impulse control coupled with high rates of sensation seeking, resulting in a higher likelihood of distracted driving and crashing. The coupling of low impulse control and high sensation seeking is especially harmful in more emotionally charged situations, in which adolescents are seeking rewards and pleasure yet don't have the ability to control these desires, as often occurs when driving with peers or in situations where it is more difficult to control their desires.

9.3 Adolescent Decision Making

Ultimately, cognitive, psychosocial, and experiential factors combine to result in more or less competent decision making. Deliberate and analytic processes necessary for decision making are described in traditional models of decision making. According to theories such as the theory of reasoned action and the theory of planned behavior (Ajzen, 1985; Fishbein & Ajzen, 1975), applied to driving, decisions are based on cognitive processes, including the following: (1) an assessment concerning both the potential positive and negative outcomes associated with driving, and driving under specific circumstances; (2) an assessment of perceptions of vulnerability or likelihood of experiencing personal harm or harm to others, including the likelihood that each positive (benefit) and negative (risk) outcome can and would occur; (3) consideration of one's desire to engage in the behavior despite potential consequences; (4) perceptions of the extent to which similar others are engaging in the behavior (descriptive social norms); (5) perceptions of the extent to which others would accept or not accept engagement in the behavior (inductive social norms); and (6) intention to engage in the behavior.

While these decision models are used often to describe ideal decision-making processes, few adolescents actually engage in these processes to make decisions, including decisions regarding driving, in part because, as described previously, while adolescents have the cognitive capacity to make sound decisions, they are often more swayed by psychosocial and emotional factors such as affect, emotions, reward seeking, impulsivity, and peer influences.

Decision making, which in reality combines cognitive and psychosocial maturity, is especially relevant to adolescent driving, which inherently involves both the deliberate decision process as well as social influences and external factors that adolescents are not always equipped to process. It has been argued that in explaining adolescent behavior, including risky driving, we need to move beyond traditional models of rational decision making. This notion is supported by the data showing limited reduction in crash rates despite traditional driver's education (Rhodes & Pivik, 2011).

Instead of employing the traditional cognitive decision models, the dual-process models have been utilized to explain adolescent behavior. The dual-process models involve the cognitive path that include the more traditional, deliberate, reasoned, and informed aspects of the decision process. In this path, decisions utilize cognitive skills, such as weighing risks and benefits and social norms, and these attitudes are expected to predict intentions and, ultimately, behavior. The second path involves the noncognitive aspects of decision making, such as impulsivity, sensation seeking, and reward seeking. This path is based on the asynchrony observed in the adolescent and young adult brain's structure and function, and the more hypersensitive affective system that results in decisions that are more affectively based and influenced by psychosocial factors, such as impulse control, sensation seeking, and self-regulation (Albert et al., 2013; Reyna & Farley, 2006). Few studies have solely focused on the reward-seeking aspect of decision making as it pertains to young and inexperienced drivers, but this may account for the number of motor vehicle accidents this age group is known for.

9.4 Contextual Influences

In addition to focusing on cognitive and psychosocial development, it is important to consider contextual and social factors that are particularly salient for adolescent behavior and particularly relevant for adolescent risky driving. The transition to adolescence is marked by a decrease in time spent with parents and an increase in time spent either alone or with peers (Steinberg & Morris, 2001). This is a time period in which the opinions and actions of peers become increasingly important in influencing behavior (Crone & Dahl, 2012) and have been found to play a major role in the number of vehicular accidents among adolescent drivers (see also Chapters 10, 16, and 19). The role of peers and adults in adolescent risky driving is discussed next.

9.4.1 Influence of Peers on Adolescent Risky Driving

The presence of peers has been associated with risk taking as well as heightened reward activity in the adolescent brain (Chein et al., 2011; Falk et al., 2014). This is portrayed in our scenario of Lionel making the risky decision to speed through the red light when Taylor urges him to do so.

By age 16, susceptibility to peer pressure that is undesirable or goes against one's goals generally decreases, with many adolescents able to resist peer pressure by age 12. In order to make competent decisions, individuals must have the ability to resist undue pressure from others. That being said, studies also show that peers remain powerful influences on and reinforcers of behavior even into late adolescence and emerging adulthood.

Observational research indicates that adolescents are more likely to drive unsafely (Mirman et al., 2012) in the presence of peers and groups (Albert et al., 2013; Zimring, 2000). The negative effect of peer passengers on drivers' behavior is likely due to both distraction and increased intentional risk taking (Mirman et al., 2012), as peer passengers may increase cognitive load, which could reduce situational awareness, potentially increasing risks for motor vehicle accidents (Pradhan et al., 2014).

Experimental studies have also shown that adolescents are more likely to make riskier decisions when they are told that they are being observed by peers (Albert et al., 2013). This is believed to be because the presence of peers stimulates the area of the brain that anticipates potential rewards (Albert et al., 2013). Another reason adolescents may make riskier decisions with peers may simply be because they are around their peers more often than adults or because of maturational imbalances between the two brain systems (incentive processing and cognitive control systems) (Strang et al., 2013). Until these two systems mature in adulthood, adolescents continue to show a propensity toward riskier decision making in front of their peers.

9.4.2 The Role of Adults in Adolescent Risky Driving

Adolescent driving is strongly influenced by parenting styles and modeling of behaviors (Carter et al., 2014). A study of family communication patterns around driving showed that the frequency of family communication with teens was positively correlated with teen attitudes toward safe driving habits (Yang et al., 2013). Parents can also reduce their teen's engagement in risky driving by controlling car access and regulating with whom their teen is driving (Mirman et al., 2012). Health care providers can also play a role in beginning a dialogue between parents and teens concerning driving and risk taking. Anticipatory guidance regarding driving should occur at every routine visit from early teen to "driving-age" years. In addition to giving anticipatory guidance in the clinical setting, providers also have the opportunity to serve as advocates for safer driving in the community; in their legislatures; and by partnering with other government, community, and public health organizations. This is particularly true when dealing with important regulatory provisions such as graduated driver licensure (GDL) laws (see Chapter 17 for more information on GDLs). Parents can play an important role and be an active part of interventions to promote safe driving practices.

9.5 Conclusions

Adolescent risky driving results from the interrelated and complex influence of developmental, social, and environmental factors. With some individual variation, skills necessary for safe driving, such as experience, understanding of risks and benefits, resistance to peer influence, and reduced need for rewards, develop gradually and continuously over time, from adolescence and into young adulthood.

Further, the developing brain during this time provides the physical context involved in risky driving and explains the psychosocial immaturity that occurs during this age period.

Thus, reducing adolescent risky driving will require understanding the developmental factors that explain crash rates and setting expectations and policies that take these vulnerabilities into account. For example, appropriate anticipatory guidance from parents, educators (both in school and in private driving school companies), and health care professionals (pediatrician, family physician) as well as public policies such as the GDL make sense (Mirman et al., 2012; see also Chapters 16, 17, and 19). Further, there is a need for reasonable cultural and social norms to combine to set a standard of expectation that leads to safe driving habits and practices.

References

Ajzen, I. (1985). From intentions to actions. In J. Kuhl & J. Beckman (Eds.), *Action Control from Cognition to Behavior*. New York: Springer-Verlag.

Albert, D., Chein, J., & Steinberg, L. (2013). The teenage brain peer influences on adolescent decision making. *Current Directions in Psychological Science, 22*(2), 114–120.

Blakemore S.J., & Choudhury, S. (2006). Development of the adolescent brain: Implications for executive function and social cognition. *Journal of Child Psychology and Psychiatry, 47*(3–4), 296–312.

Borowsky, A., & Oron-Gilad, T. (2013). Exploring the effects of driving experience on hazard awareness and risk perception via real-time hazard identification, hazard classification, and rating tasks. *Accident Analysis & Prevention, 59*, 548–565.

Borowsky, A., Shinar, D., & Oron-Gilad, T. (2010). Age, skill, and hazard perception in driving. *Accident Analysis & Prevention, 42*, 1240–1249.

Carter, P.M., Bingham, C.R., Zakrajsek, J.S., Shope, J.T., & Sayer, T.B. (2014). Social norms and risk perception: Predictors of distracted driving behavior among novice adolescent drivers. *Journal of Adolescent Health, 54*, S32–S41.

Casey, B.J., Getz, S., & Galvan, A. (2008). The adolescent brain. *Developmental Review, 28*(1), 62–77.

Chein, J., Albert, D., O'Brien, L., Uckert, K., & Steinberg, L. (2011). Peers increase adolescent risk taking by enhaning activity in the brain's reward circuitry. *Developmental Science, 14*(2), F1–F10.

Constantinou, E., Panayioutou, G., Konstantinou, N., Loutsiou-Ladd, A., & Kapardis, A. (2011). Risky and aggressive driving in young adults: Personality matters. *Accident Analysis & Prevention, 43*, 1323–1331.

Crone, E.A., & Dahl, R.E. (2012). Understanding adolescence as a period of social–affective engagement and goal flexibility. *Nature Reviews Neuroscience, 13*, 636–650.

Falk, E.B., Cascio, C.N., O'Donnell, M.B., Carp, J., Tinney, Jr, F.J., Bingham, C.R., Shope, J.T., Ouimet, M.C., Pradhan, A.K., & Simons-Morton, B.G. (2014). Neural responses to exclusion predict susceptibility to social influence. *Journal of Adolescent Health, 54*, S22–S31.

Fishbein, M., & Ajzen, I. (1975). *Beliefs, Attitudes, Intention, and Behavior: An Introduction to Theory and Research*. Reading, MA: Addison-Wesley.

Galvan, A., Hare, T., Voss, H., Glover, G., & Casey, B.J. (2007). Risk-taking and the adolescent brain: Who is at risk? *Developmental Science, 10*(2), F8–F14.

Giedd, J., & Rapoport, J.L. (2010). Structural MRI of pediatric brain development: What have we learned and where are we going? *Neuron, 67*, 728–734.

Gogtay, N., Giedd, J.N., Lusk, L. et al. (2004). Dynamic mapping of human cortical development during childhood through early adulthood. *Proceedings of the National Academy of Sciences of the United States of America, 101*(21), 8174–8179.

Hatfield, J., Fernandes, R., & Soames Job, R.F. (2014). Thrill and adventure seeking as a modifier of the relationship of perceived risk with risky driving among young drivers. *Accident Analysis & Prevention, 62*, 223–229.

Horvath, C., Lewis, I.M., & Watson, B.C. (2012). Peer passenger identity and passenger pressure on young drivers' speeding intentions. *Transportation Research Part F: Traffic Psychology and Behaviour, 15*(1), 52–64.

Keating, D.P. (2007). Understanding adolescent development: Implications for driving safety. *Journal of Safety Research, 38*(2), 147–157.

Keating, D.P., & Halpern-Felsher, B.L. (2008). Adolescent drivers: A developmental perspective on risk, proficiency, and safety. *American Journal of Preventive Medicine, 35,* S272–S277.

Luna, B., Garver, K.E., Urban, T.A., Lazar, N.A., & Sweeney, J.A. (2004). Maturation of cognitive processes from late childhood to adulthood. *Child Development, 75,* 1357–1372.

Luna, B., Padmanabhan, A., & O'Hearn, K. (2010). What has fMRI told us about the development of cognitive control through adolescence? *Brain and Cognition, 72*(1), 101–113.

Mayhew, D.R., Simpson, H., & Pak, A. (2003). Changes in collision rates among novice drivers during the first months of driving. *Accident Analysis & Prevention, 35*(5), 683–691.

McKnight, J.A., & McKnight, A.S. (2003). Young drivers: Careless or clueless? *Accident Analysis & Prevention, 35*(6), 921–925.

Mirman, J.H., Albert, D., Jacobsohn, L.S., & Winston, F.K. (2012). Factors associated with adolescents' propensity to drive with multiple passengers and to engage in risky driving behaviors. *Journal of Adolescent Health, 50,* 634–640.

Møller, M., & Haustein, S. (2013). Peer influence on speeding behavior among male drivers aged 18 and 28. *Accident Analysis & Prevention, 64,* 92–99.

Monahan, K.C., Steinberg, L., Cauffman, E., & Mulvey, E.P. (2009). Trajectories of antisocial behavior and psychosocial maturity from adolescence to young adulthood. *Developmental Psychology, 45*(6), 1654–1668.

Oltedal, S., & Rundmo, T. (2006). The effects of personality and gender on risky driving behaviour and accident involvement. *Safety Science, 44,* 621–628.

Pradhan, A.K., Li, K., Bingham, C.R., Simons-Morton, B.G., Ouimet, M.C., & Shope, J.T. (2014). Peer passengers influence on male adolescent drivers' visual scanning behavior during simulated driving. *Journal of Adolescent Health, 54*(5 Suppl.), S42–S49.

Reyna, V., & Farley, F. (2006). Risk and rationality in adolescent decision making: Implications for theory, practice and public policy. *Psychological Science in the Public Interest, 7*(1), 1–44.

Rhodes, N., & Pivik, K. (2011). Age and gender differences in risky driving: The roles of positive affect and risk perception. *Accident Analysis & Prevention, 43,* 923–931.

Romer, D.R., Lee, Y., McDonald, C.C., & Winston, F.K. (2014). Adolescence, attention, and driving safety. *Journal of Adolescent Health, 54*(5 Suppl.), S6–S15.

Steinberg, L. (2007). Risk taking in adolescence: New perspectives from brain and behavioral science. *Current Directions in Psychological Science, 16*(2), 55–59.

Steinberg, L., & Morris, A.S. (2001). Adolescent development. *Journal of Cognitive Education and Psychology, 2*(1), 55–87.

Steinberg, L., Albert, D., Cauffman, E., Banich, M., Graham, S., & Woolard, J. (2008). Age differences in sensation seeking and impulsivity as indexed by behavior and self-report: Evidence for a dual systems model. *Developmental Psychology, 44*(6), 1764–1778.

Steinberg, L., Cauffman, E., Woolard, J., Graham, S., & Banich, M. (2009). Are adolescents less mature than adults? Minors' access to abortion, the juvenile death penalty, and the alleged APA "flip-flop." *American Psychologist, 64*(7), 583–594.

Strang, N.M., Chein, J.M., & Steinberg, L. (2013). The value of the dual systems model of adolescent risk-taking. *Frontiers in Human Neuroscience, 7,* 223.

Strayer, D.L., Drews, F.A., & Johnston, W.A. (2003). Cell phone-induced failures of visual attention during simulated driving. *Journal of Experimental Psychology Applied, 9*(1), 23–32.

Vitulano, M.L., Fite, P.J., & Rathert, J.L. (2010). Delinquent peer influence on childhood delinquency: The moderating effect of impulsivity. *Journal of Psychopathological Behavioral Assessment, 32,* 315–322.

WISQARS. (2011). *Ten Leading Causes of Death*. Cited September 2014. http://webappa.cdc.gov/sasweb/ncipc/leadcaus10_us.html.

Yang, J., Campo, S., Ramirez, M., Krapfl, J.R., Cheng, G., & Peek-Asa, C. (2013). Family communication patterns and teen drivers' attitudes toward driving safety. *Journal of Pediatric Health Care, 27*(5), 334–341.

Zimring, F.E. (2000). *American Youth Violence*. Oxford: Oxford University Press.

10

Emotional and Social Factors

Bridie Scott-Parker

Barry Watson

Abstract

The Problem. Teen drivers have an elevated risk of crashing during the earliest stages of independent driving and often carry teen passengers in their vehicle. Consequently, teen crashes are the leading cause of death and disability for adolescents around the world. Teen drivers are also adolescents, and thus, normative influences including psychobiological immaturity and vulnerability to social and emotional influences exert a strong influence on their road use behavior. ***Social and Emotional Influences.*** Teens are developmentally vulnerable to social influences, including the influence of peers, parents, police, and other drivers, and as such, they are likely to become more risky drivers in the presence of negative influences. Teens are also developmentally vulnerable to emotional influences, including personal characteristics such as sensation seeking, reward sensitivity, depression, and anxiety. ***Theories Relevant to Understanding and Thus Reducing the Negative Influence of Social and Emotional Factors.*** Theories relevant to understanding and thus reducing the negative influence of social and emotional factors include criminological theories (e.g., Akers's social learning theory), psychosocial theories (e.g., Gerrard and Gibbons's prototype willingness model), and driving frameworks (e.g., Goals for Driver Education matrix). ***Interventions Addressing Emotional and Social Influences.*** Interventions that are discussed in this chapter include graduated driver licensing, and resilience-focused interventions. ***Research Priorities.*** Research priorities include the need to develop and evaluate more effective interventions that target the role of social and emotional factors, including those involving parents, peers, and the use of technology.

10.1 Introduction

Driving that impacts on the safety of young and novice drivers (herein referred to as *teen drivers*) includes risky behaviors that increase risk involvement (e.g., driving in excess of speed limits) and protective behaviors that reduce crash involvement or crash severity (e.g., refraining from impaired driving, wearing seat belts). Central to these driving behaviors is the development of the teen driver (Bingham et al., 2008; Jessor et al., 1997) and the associated emotional and social factors influencing their driving behavior. The physical, psychological, social, and biological development of the individual appears to be inextricably linked in terms of their influence upon the behavior of the adolescent, which in turn impacts on the driving behavior of the teen driver. As such, as noted in the previous chapter (Chapter 9), developmental factors should be borne in mind when investigating teen driving behavior. Of particular interest are normative emotion-related factors such as reward sensitivity, sensation seeking, and impulsivity. In addition, adolescence can be a tumultuous period of psychosocial development, which has been found to be associated with psychological distress such as anxiety and depression (Avenevoli et al., 2008). Peers, parents, other drivers, and police can exert social influences upon adolescent behavior through a variety of mechanisms, including modeling and imitation, and being a source of actual or anticipated punishments and rewards. Each of these emotional and social factors can influence the decision making (Allen & Brown, 2008; Keating & Halpern-Felsher, 2008), and therefore the driving behavior, of the teen driver. The literature regarding emotional and social factors is derived from a variety of methodologies, which are elaborated upon further in Section VII (Chapters 25 through 28), including naturalistic driving studies, crash data studies, cohort and longitudinal studies, self-report studies, and driving diaries. This chapter will review and discuss the evidence relating to the influence of emotional and social factors on driving behavior; review and discuss theories that have been used to explain the influence of these factors on teen driving behavior; and review and discuss interventions that have historically been used to target these factors.

10.2 Emotional Factors

Teen drivers appear to be developmentally prone to emotional driving. Three-quarters of youth surveyed in an American study reported that they had seen teens driving while experiencing strong emotions (e.g., anger, excitement), with over half reporting that they had seen instances of road rage (CHOP, 2007). Strong emotions may also emerge during the journey; for example, learner drivers in Victoria, Australia, reported that they and their supervisor became upset during some of their journeys (Harrison, 2004). Becoming angry during a drive in an instrumented vehicle resulted in more speeding by greater amounts (Mesken et al., 2007), with aggression predicting crashes in longitudinal research in New Zealand (Gulliver & Begg, 2007) and high-risk driving by teen drivers in Norway (Ulleberg, 2002). Teen drivers also are vulnerable to impression management (Leary & Kowalski, 1990; Leary et al., 1994), while the psychosocial purpose of risky driving (e.g., demonstrate progression to adulthood [Møller, 2004; Møller & Gergersen, 2008; Scott-Parker et al., 2015]) and perceived "vehicle personality," in which anthropomorphized traits are ascribed to their car by the teen driver, is associated with self-reported aggressive driving (Benfield et al., 2007). In some instances, teen drivers may commence a driving trip in an emotional state, possibly using the driving task as a means of coping with their emotions. For example, teen drivers, particularly young males, report that their driving changes in response to stressful events (Lonczak et al., 2007).

Emotional factors that can influence the behavior of teen drivers can be conceptualized as different types of person-related factors that can influence the emotional state of the teen driver. First of all, there are state factors, which are transient characteristics that can be fleeting or last for longer periods of time, and include driver mood (particularly anger), driver stress, depression, and anxiety. Secondly, trait factors are also of interest and include relatively stable and enduring personality traits of the teen driver, like sensitivity to reward, sensation-seeking propensity, impulsivity, and trait anxiety. The following sections review the available evidence regarding the impact of these factors on teen driving behavior and crash risk.

10.2.1 Driver Stress

Stress is defined by the American Psychological Association as "the pattern of specific and nonspecific responses an organism makes to stimulus events that disturb its equilibrium and tax or exceed its ability to cope" (APA, 2014). Within the context of the driving environment, research demonstrates that the impact of stress upon driver behavior is multifaceted: stress intrinsic to driving arises within and in response to the driving environment, such as in response to situational factors, while stress extraneous to the driving environment includes stress that can transfer from the broader lives of the teen driver, including stress in the teen driver's life more generally and the workplace specifically (Rowden et al., 2006, 2011). Drivers who experience time pressure, which may be for various reasons, including meeting deadlines as they drive as their occupation, also exhibit greater stress due to a combination of goal and safety conflicts, time constraints, and uncertainty, evidenced as more risky driving behavior generally (Coeugnet et al., 2013).

Teen drivers report driving as a way to relieve stress in their lives more generally, and that their driving behavior becomes more risky (e.g., driving in excess of posted speed limits) in response to stress (Scott-Parker & Proffitt, 2015; Scott-Parker et al., 2010). Interestingly, stress can have a pervasive impact upon driver mood; for example, it can be experienced as anger (see Section 10.2.2 for further discussion), happiness, or fear, with simulator-based research revealing that both happiness and anger, but not fear, led to deficits in driving performance and perceived workload among teen drivers (Jeon et al., 2014).

Stress can also be evidenced as worry. Worry is defined as "mental distress or agitation resulting from concern usually for something impending or anticipated" (Roche-Cerasi et al., 2013, p. 698) and can reflect driver stress regarding possible traffic conflicts. Interestingly, cluster analysis reveals that drivers who use public transport are more worried about a road crash than those who use private transport, suggesting that an element of risk control may be important in ameliorating road user worry (Roche-Cerasi et al., 2013). Driver stress can also distract drivers, with the considerable role of distraction in teen driver road crashes being well recognized (see also Chapter 12). Distraction generated by involving drivers in an emotionally charged conversation, as shown through simulator-based studies, can lead drivers to make more driving errors and to experience visual tunneling whereby they restrict their visual fixations, compared to drivers who are not emotionally involved (Briggs et al., 2011). Similarly, contentious conversations between romantically involved drivers and passengers have been found to negatively impact upon following distance and lateral vehicle control in a simulated driving task, even when the conversation was via a mobile telephone but particularly when the romantic partner was actually present in the vehicle (Lansdown & Stephens, 2013).

Conversely, distress tolerance can be conceptualized as "an individual's capability to experience and endure negative emotional states" (Beck et al., 2013, p. 51) and thus can represent a teen driver's ability to cope with driving-related and life-related stress. Interestingly, drivers who characterize themselves as "hurried drivers," female hurried drivers in particular, exhibit lower distress tolerance, more driving frustration, and more impatience toward other drivers, which appears to demonstrate a general lack of coping with negative emotional states during the driving task (Beck et al., 2013).

10.2.2 Driver Mood

As noted previously, driver stress can impact upon driver mood, and driving anger is frequently the resulting mood, particularly if the discourteous action of another road user is perceived as intentional or disrespectful (Lonsdale, 2010; Roseborough & Wisenthal, 2014; Wickens et al., 2013b). Driving anger can be expressed as aggressive driving. It is noteworthy that there is much variation in the definitions of both driving anger and driving aggression; however, it is generally agreed that driving aggression involves an intention to exert a negative influence—a harm, whether it be physical or psychological—upon another road user (e.g., see Soole et al., 2011). Notwithstanding this, a number of driving behaviors have consistently been found to contribute to driving anger and to provoke aggressive responses among drivers. For example, content analysis of diaries reveals that driver behaviors such as cutting in, slow driving, speeding, and tailgating can generate an angry response in drivers, and these responses could be reliably categorized as driving negligence, time urgency, and retaliation (Wickens et al., 2013a). Interestingly, research also demonstrates that male drivers are more likely to report angry responses to on-road police presence, while female drivers respond angrily to traffic obstructions (Gonzalez-Iglesias et al., 2012), consistent with the findings of Berdoulat et al. (2013) that impeded progress and driver aggressiveness predicted driving violations, aggressive driving violations in particular. Furthermore, research demonstrates that instigators of aggressive driving tend to attribute their behaviors to external and temporary causes, such as driving errors, while recipients of aggressive driving attribute the behaviors to the instigators' driving deficits, with instigators reporting a greater emotional impact of aggressive driving compared to recipients (Lennon et al., 2011). It is noteworthy, however, that situation-specific factors are also important (e.g., Mesken et al., 2007), such that impeded progress generally elicits aggressive responses in teen drivers (Stephens & Groeger, 2009).

An angry driving mood can result from (1) trait anger, which reflects more stable personality patterns; (2) driving-specific anger; and (3) situation-specific anger (Nesbit et al., 2007). Research operationalizing the five-factor personality model (see Goldberg, 1981) has revealed that the personality traits of greater neuroticism and conscientiousness and less agreeableness predicted driving anger, while greater driving anger and less agreeableness and conscientiousness predicted aggressive driving among teen drivers, suggesting that the influence of personality variables upon aggressive driving are mediated by perceived driving anger (Jovanovic et al., 2011). Reckless and angry driving styles alike were found in teen drivers who reported greater extraversion and sensation seeking, and less agreeableness and conscientiousness (Taubman-Ben-Ari & Yehiel, 2012); greater extraversion and driving anger have been found to be associated with speeding and rule violations (Sarma et al., 2013). In addition, trait rumination—"a mode of responding to distress that involves repetitively and passively focusing on symptoms of distress and on the possible causes and consequences of these symptoms" (Nolen-Hoeksema et al., 2008, p. 400, cited in Suhr & Nesbit, 2013)—can contribute to driving anger, with trait rumination associated with risky and aggressive driving (Suhr & Nesbit, 2013). Drivers exhibiting greater on-road aggression report greater difficulties with managing the aggression of others (e.g., arguing with passengers/other drivers), greater trait revenge, and more physical aggression (Nesbit & Conger, 2012), highlighting the relationship between maladaptive thinking and aggressive driving behaviors.

In addition to driving diaries, teen driving-related anger is frequently measured through self-report questionnaires, such as the 33-item *Driving Anger Scale* (DAS) (Deffenbacher et al., 1994; Lennon et al., 2011; McLinton & Dollard, 2010; Nesbit & Conger, 2012) and the *Driving Anger Expression Inventory* (DAX) (e.g., Jovanovic et al., 2011; Villieux & Delhomme, 2010). Angry drivers have been found to exceed posted speed limits (Fernandes et al., 2010), to perform greater longitudinal and lateral acceleration maneuvers (e.g., Roidl et al., 2013), and to accelerate more strongly in simulator-based studies (e.g., Roidl et al., 2014). High-aggression drivers have also been found to have more driving offenses and crashes (Nesbit & Conger, 2012). Interestingly, teen drivers who exhibit risk characteristics such as being male and having a high motivation to drive also exhibit more driving anger, thereby increasing their on-road risk through emotional driving (Roidl et al., 2013). Other risk factors, coupled with driving anger, include low premeditation and perseverance, and high sensation seeking and urgency (Bachoo et al., 2013). Aggression and hostility have also been found to be associated with both prelicense (Begg et al., 2012) and unsupervised learner driving, which itself was predicted by prelicense driving (Langley et al., 2013) by teen drivers. In addition, longitudinal research has revealed a slight increase in driving anger experienced among drivers aged 18 years compared to those aged 23 years (Møller & Haustein, 2013).

Emotion regulation also appears to play an important role in managing driver mood, including driver anger and driver aggression. Emotion regulation includes such variables as emotional awareness, impulse control, and the ability to use situationally appropriate strategies to regulate emotions (e.g., see Gratz & Roemer, 2004). Drivers who exhibit difficulties in emotion regulation have been found to demonstrate anxious, angry, dissociative, and risky driving styles, as measured by the *Multidimensional Driving Style Inventory* (MDSI). Drivers who do not experience emotion regulation difficulties demonstrate a careful driving style (Trogolo et al., 2014). Also, greater frequency of texting while driving has been found for teen drivers who exhibit less mindfulness, with the impact of mindfulness upon texting frequency mediated by emotional regulation (Feldman et al., 2011).

10.2.3 Depression and Anxiety

Teen drivers, by virtue of their age, are adolescents who are undergoing physiological, cognitive, social, and behavioral development (Sprinthall & Collins, 1995). This maturation can have implications for risky behavior (Dahl, 2008), and of note for road safety is the increased incidence of psychological distress experienced by the adolescent as anxiety and depression. Depression is experienced as psychological distress and may be a transient state reactive to an adverse life event, or may be more persistent (Avenevoli et al., 2008). The vulnerability of the adolescent to depression has been found to be related to the processing of social rewards and maturation of the prefrontal cortex (Davey et al., 2008), and the effects of depression on the adolescent are pervasive (Jaycox et al., 2009). Research undertaken in the United States estimates the prevalence of depression in adolescence to be 24% (Avenevoli et al., 2008), further suggesting that 1 in 10 adolescents are depressed at any time (Strine et al., 2008). Anxiety can also be a transient reactive state to an adverse life event or may be a more stable personality trait. There are also differences in the experience of psychological distress for each gender and over the developmental period of adolescence, irrespective of ethnicity (Huang et al., 2009), with greater prevalence of depression (Paxton et al., 2007) and anxiety (Scott-Parker et al., 2012a) among females, who tend to experience symptoms earlier (Avenevoli et al., 2008). In addition, greater rates of depression and anxiety have been reported by university students compared to the general teen driver population (Scott-Parker et al., 2013a), with the experience of anxiety and depression being relatively stable throughout a 6-month period of adolescence (Scott-Parker et al., 2013a).

While road safety researchers acknowledge that psychological distress is likely to impact upon driving behavior (e.g., Lonczak et al., 2007; Schwebel et al., 2006), there has been limited research regarding the relationship between psychological distress and teen risky driving. Research has revealed that risky driving (e.g., distracted driving) by university students was related to high levels of anxiety (Ferreira et al., 2009), and frequent drunk-driving offenders were found to have significantly higher levels of depression and anxiety than the general population (Hubicka et al., 2010). As many as five decades ago, researchers suggested that single-vehicle and single-occupant car crashes may indeed include "socially acceptable suicides," chosen by depressed young males in particular (Jenkins & Sainsbury, 1980; Peck & Warner, 1995), and greater psychological distress is associated with greater suicidal ideation (Avenevoli et al., 2008). In the Australian state of Victoria, teen drivers who report high levels of risky driving (e.g., speeding, drunk driving) previously reported 1.7 and 1.4 times the levels of anxiety and depression, respectively, of low-level risky drivers (Vassallo et al., 2008). Mental health problems such as depression have been found to be significantly more common in fatally injured intoxicated (drugs and/or alcohol) drivers than sober drivers (Karjalainen et al., 2012). In addition, drivers with an anxious driving style reported they were more likely to avoid difficult driving conditions (e.g., driving in poor weather) and maneuvers (e.g., changing lanes on motorways) (Gwyther & Holland, 2012). Other research has reported no relationship between psychological distress and risky driving; however, these findings are confounded by time lags of 2 (Martiniuk et al., 2010) and 5 years (Begg et al., 2003) between measures of distress and risky driving behavior.

These mixed results suggest that the relationship between anxiety and risky driving behavior is complex. As further evidence of this, driving-related anxiety has been found to decrease with greater driving exposure (e.g., Møller & Haustein, 2013). Notwithstanding this, trait anxiety has been found to predict driving lapses, and it is suggested that the worry that emerges as a result of anxiety actually interferes with the capacity of the working memory, and the lapses result from the cognitive overload (Wong et al., 2015). Cluster analysis of teen drivers' response to the 10-item *State-Trait Anxiety Inventory* (STAI-X) revealed that more cautious driving behavior was evidenced in a group of drivers who exhibited greater anxiety. Simulator-based research that incorporated self-report and electrodermal measures has demonstrated that driver anxiety increases in response to increases in the conflict between safe driving and speedy driving (Schmidt-Daffy, 2013). Moreover, drivers who reported greater crash involvement (three or more crashes during the past 3 years) reported significantly greater trait anxiety as measured by the *Minnesota Multiphasic Personality Inventory* (MMPI) and the *California Psychological Inventory* (CPI) (Mayer & Treat, 1977).

In addition, it is noteworthy that teen drivers who exhibit anxiety may have comorbidity such as attention deficit hyperactivity disorder (ADHD; see Chapter 14), with driving diary research noting that, compared to drivers without ADHD, drivers with ADHD have difficulty sustaining attention while driving and, as a result, make driver errors (e.g., Rosenbloom & Wultz, 2011). Simulator-based research has revealed that low-stimulus driving specifically is problematic for ADHD drivers, with adolescent and young adult ADHD drivers more likely to be distracted in such environments (Reimer et al., 2010).

10.2.4 Sensitivity to Reward and Sensitivity to Punishment

Rewards and punishments are pivotal in the learning, repetition, and cessation of behavior, including risky driving. Rewards are motivating, acting as incentives to gain expected outcomes, as well as being reinforcing. Therefore, they are pivotal in learning new behaviors (Beck, 1990). Anticipated rewards such as a quicker journey, as well as those received previously for engaging in risky driving such as feelings of excitement (see Section 10.2.5) and status in the peer group (see Section 10.3.1), motivate and reinforce risky behaviors such as speeding, thereby increasing the likelihood that the teen driver will speed in the future. In contrast, punishments serve to prevent, curtail, or extinguish learned behaviors (Beck, 1990). The teen driver who is detected speeding and receives a fine and demerit points is less likely to speed in similar circumstances in the future. Therefore, driving behavior is constantly altered by its consequences (Fuller, 2002).

Besides the external administration (or lack) of rewards and punishments by significant others such as peers, parents, and police, the teen driver can be motivated by the perceived or anticipated rewards (advantages) and punishments (disadvantages) for risky driving. Sensation seeking through risky driving can be rewarding for the teen driver (Rimmo & Aberg, 1999), and behaviors that elicit pleasurable sensations are more likely to be repeated, while behaviors that do not are less likely to be repeated. This is further validated by the finding that risky drivers report more advantages and fewer disadvantages for risky driving than drivers who are not risky drivers (Deery, 1999; Horswill et al., 2004; Horvath et al., 2012). Avoiding negative consequences was also rewarding for a group of young male teen drivers who reported being involved in an excess speed–related crash without incurring personal harm, thereby reinforcing driving in excess of speed limits (Falk & Montgomery, 2007).

Individuals' responses to rewards appear to be regulated by their behavioral activation system (BAS), while their responses to punishments appear to be regulated by their behavioral inhibition system (BIS) of motivation. These neurological systems influence the individual's sensitivity to reinforcing events and also control their experiences of emotion (Torrubia et al., 2001): the BAS has been aligned with impulsivity, and the BIS has been aligned with anxiety (Sava & Sperneac, 2006). The *Sensitivity to Punishment and Sensitivity to Reward Questionnaire* (SPSRQ) created by Torrubia et al. (1995, cited in O'Connor et al., 2004) has been utilized in university populations around the world and has been found to predict risky attitudes and behaviors in drivers (e.g., in Spain, Torrubia et al., 2001; in Greece, Constantinou et al., 2011; in Taiwan, Li et al., 2007). Individuals with low reward sensitivity and high punishment sensitivity are least likely to infringe on road rules, and reward sensitivity is the greatest predictor of risky driving.

There has also been mixed evidence regarding the extent to which teen drivers are sensitive to these measures, such that reward sensitivity has been found to influence the risky driving behavior of adolescents, while punishment sensitivity did not (diCastella & Perez, 2004; Scott-Parker et al., 2012a). Young males also have been found to respond more to rewards than to punishments in comparison to young females (Li et al., 2007), and the most risky teen drivers have been found to exhibit the greatest reward sensitivity (Scott-Parker et al., 2013b). Furthermore, reward sensitivity, and thus the influence of reward sensitivity, appears relatively stable throughout adolescence (Scott-Parker et al., 2013a), while the influence of punishment sensitivity appears to be subsumed within the influence of psychological distress experienced as anxiety and depression (Scott-Parker et al., 2012a). Reward and punishment sensitivity have also been found to influence teens' responses to road safety advertising, such that drivers with greater reward sensitivity demonstrate attentional bias toward advertising incorporating incentives, with commensurate increases in the perception that the message would be effective in improving the road safety of teen drivers. Interestingly, while drivers with greater reward sensitivity demonstrate attentional bias toward advertising incorporating loss-focused messaging, they report that they are unlikely to comply with the road safety message (Kaye et al., 2013).

10.2.5 Sensation-Seeking Propensity and Impulsivity

Sensation seeking and impulsivity have also been found to influence teen driver behavior. Frequently, teen drivers report that they drive for reasons other than an economical and efficient means of travel. Of particular concern is driving to facilitate sensation seeking (see Jonah, 1997, for a review), such as expressing feelings of excitement, anger, frustration (Arnett et al., 1997; Sullman, 2006), and competitiveness (Ulleberg, 2004). Such sensation-seeking behavior is frequently impulsive (i.e., occurs in response to emotional and social cues in the immediate context). This is problematic for road safety, as teen drivers may change the way they drive so they can take risks to experience the accompanying thrill. For example, young male drivers frequently report that they like to engage in sensation-seeking behaviors (Begg & Langley, 2004; Dahlen et al., 2005; Jonah, 1997; Schwebel et al., 2006), including risky behavior such as speeding (Rhodes & Pivik, 2011). A desire for thrill seeking contributed to one-quarter of teen drivers' offenses (Ross & Guarnieri, 1994, cited in Cavallo et al., 1997), and sensation-seeking propensity has been found to be associated with risky driving, offenses, and crashes (Rimmo & Aberg, 1999). Young people who reported sensation seeking through behavior such as speeding also reported more than double the incidence of being injured in a car crash during the previous year compared to teen drivers who did not engage in such risky behavior (Blows et al., 2005).

Research repeatedly documents more favorable attitudes toward risky driving such as speeding (e.g., Waylen & McKenna, 2008) and more risky driving behaviors (Jonah, 1997; Schwebel et al., 2006; Ulleberg, 2002; Zuckerman, 1994, 2007) like drunk driving (Fernandes et al., 2010), driving offenses in particular (Rimmo & Aberg, 1999; Schwebel et al., 2006), in drivers with greater sensation-seeking propensity. Higher sensation-seeking tendencies have been found among prelicense (Begg et al., 2012) and unsupervised learner drivers (Scott-Parker et al., 2013b), and passengers traveling with an alcohol-impaired driver (Kim & Kim, 2012). Interestingly, there appears to be a relationship between sensation seeking and driver aggression: some young male drivers report that they "hoon" (an Australian term referring to a range of risky driving behaviors, often accompanied by loud music, noise, and smoke from spinning tires [Queensland Government, 2015]) and engage in more aggressive driving, such as honking horns, merging closely to the front of the vehicle being overtaken (Shinar & Compton, 2004), and using rude gestures (Sullman, 2006). It is noteworthy also that drivers who engage in hooning report significantly more offenses, including speeding (Leal et al., 2010).

Furthermore, research persistently reports that males have a higher sensation-seeking propensity than females (e.g., Arnett et al., 1997; Waylen & McKenna, 2008). In addition, sensation seeking has been found to be a predictor of adolescent risky behavior in longitudinal research, with younger adolescents with greater propensity at baseline reporting significantly more smoking and drinking after 2 years (Sargent et al., 2010). Sensation seeking via engaging in risky behaviors is normative for adolescents, and it serves many purposes, including the development of identity and autonomy (Bonino et al., 2005; Johnson & Malow-Iroff, 2008). As such, sensation seeking appears to be inherently rewarding, and in accordance with the reinforcing effects of rewarding behaviors as discussed thus far, sensation-seeking behavior is likely to be repeated. Sensation seeking by the adolescent is very risky when the similarly normative optimism bias is coupled with a powerful vehicle in an unforgiving environment (Keating, 2007). Furthermore, socializing with friends, which is often characterized by little planning and high impulsivity, is associated with risky driving behaviors such as racing other drivers and driving at high speed to impress friends (Møller & Gregersen, 2008).

10.2.5.1 General Sensation-Seeking and Impulsivity Scales

The persistent finding linking greater sensation seeking and impulsivity with risky driving behavior and crashes is a robust phenomenon. For example, research from nearly four decades ago that operationalized impulsivity-oriented items from the MMPI and CPI revealed that drivers who reported greater crash involvement (three or more crashes during the past 3 years) also reported significantly greater impulsive tendencies (Mayer & Treat, 1977). Sensation-seeking and impulsivity tendencies are measured through a number of scales (see Appendix 10.1).

10.2.5.2 Driving-Specific Sensation-Seeking and Impulsivity Scales

Driving-specific dimensions of sensation seeking as measured by the nine-item *Thrill Seeking Scale* (TSS) (Lawton et al., 1997) has also been found to predict risky driving behavior in young adults (e.g., Bates et al., 2009; Scott-Parker et al., 2009a). In contrast, a correlational study in which the eight-item *Driver Thrill Seeking Scale* (DTSS) was applied in a sample of car club and racetrack enthusiasts found that it did not predict self-reported risky driving, despite being moderately correlated with a competitive attitude ($r = .44$) and favorable attitudes to racing ($r = .35$) (Vingilis et al., 2013) (Appendix 10.1).

10.2.5.3 Functional Magnetic Resonance Imaging Studies

Perhaps unsurprisingly, functional magnetic resonance imaging (fMRI) imaging studies reveal that, when shown high-arousal images, individuals with high sensation-seeking tendencies demonstrate stronger responses in arousal- and reinforcement-related brain regions, while individuals with low sensation-seeking tendencies show greater responses in emotional regulation–related brain regions (Joseph et al., 2009). Further, it appears that sensation-seeking propensity is consistent with reward sensitivity, and research suggests that, while the two variables are highly correlated (e.g., $r = .45$ [Torrubia et al., 2001]; $r = .55$ [Scott-Parker et al., 2012a]; $r = .42$ [Scott-Parker et al., 2013a]; $r = .50$ [Scott-Parker et al., 2013c]), they are independent variables that exert a unique influence upon the risky driving behavior of the teen driver (Scott-Parker et al., 2012a, 2013a,c).

10.3 Social Factors

As noted earlier, central to teen driver behavior is that teens are also adolescents, and this has considerable implications for safe road use (Gregersen & Bjurulf, 1996). Essentially, teen drivers are still developing their personal identities and learning appropriate "adult" attitudes and behaviors through experiences, emulation, and experimentation (Bonino et al., 2003). Adolescents are raised in a social environment in which they are exposed to the attitudes and behaviors of their parent(s), and as they mature, they are exposed to the attitudes and behaviors of others such as older siblings, peers and friends, police, and other drivers. This exposure can have considerable influence upon the risky behavior of adolescents, particularly when they internalize the attitudes (e.g., perceived parental approval of drinking corresponds to more drinking by adolescents [Foley et al., 2004]) and emulate the behaviors and attitudes of these models (Andrews et al., 1997; Lahatte & Le Pape, 2008; Scott-Parker et al., 2009b; Sheehan et al., 2002). Interestingly, the nature of the social relationships also appears to be important, with romantic and friendship groups exerting more influence over the behavior of the young novice than the general peer group (Kobus, 1998). The nature of the relationships is also relevant for the parent–child dynamic: children who have a good relationship with their parents are significantly more likely to imitate alcohol and marijuana use (Andrews et al., 1997), and children who have strained relationships with parents during early adolescence report more risky driving, including multiple crashes (Smart et al., 2005). As such, it is important to review the available evidence regarding the social influences of peers, parents, other drivers, and police on teen driving behavior.

10.3.1 Peers

Consistent with the developmental milestone of adolescence, teen drivers increase their reliance on peers—and friends in particular—in forming attitudes and behaviors (Sharpley, 2003; Sigelman, 1999). As discussed more fully in Chapter 16, peers can be models to be imitated, can encourage risky—and discourage safe—driving, and can reward and punish the young novice's attitudes and behaviors. Teen drivers are vulnerable to the negative influences of their peers and are susceptible to a need for social approval from these peers (Arnett, 2002), engaging in risky behaviors that are subsequently further reinforced by intergroup playfulness (Buckley, 2005; Rhodes et al., 2005; Williams et al., 2007). Much risky driving is impacted upon by the social context in which it occurs—indeed, teen drivers cannot show off unless there is someone to whom they can show off (Harre et al., 2004). Moreover, teen drivers use their car for social purposes (Arnett, 2002; Harrison et al., 1999), and psychologically salient (Cameron, 1999) group-approved behavior can, and is expected, to occur (Harre et al., 2000). Teen drivers also report that their friends explicitly encourage them to drive in a risky manner (Buckley, 2005), including speeding (CHOP, 2007, 2009). Adolescents can deliberately engage in risky behavior to fit in and please their friends (Allen & Brown, 2008): traveling as a passenger of a drinking driver (Kim & Kim, 2012); driving after drinking in young male drivers in particular (Gonzalez-Iglesias et al., 2014); speeding as teen drivers (Cristea et al., 2013); and driving in a more reckless style (Taubman-Ben-Ari & Katz-Ben-Ami, 2012). Such behavior is likely to meet emotional needs (e.g., to reduce anxiety and to engage in sensation seeking, as noted in Section 10.2) as well as social needs (e.g., to belong).

10.3.2 Parents

Parental influence upon the behaviors and attitudes of teen drivers is apparent across the prelicense, learner, and provisional license phases (Scott-Parker, under review) (please note that parental influence on teen driver safety and behaviors is addressed in greater detail in Chapter 19). Similar to peer influence, primary mechanisms of parent influence include modeling and imitation, rewards, and punishments. For the teen driver, if parents do not punish risky behavior, such as the learner intentionally speeding while under their supervision, the teen driver is likely to continue speeding when he or she can drive independently if it is emotionally (e.g., a "fun" journey) or socially rewarding (e.g., greater social status) for them (Foss, 2007).

10.3.3 Other Drivers

Other drivers in the road environment include teen drivers as well as older, more experienced drivers. Drivers of all ages commonly report pressure to speed up and stay with the traffic flow (Fleiter et al., 2010), and the risky behavior of other drivers, including teen drivers, can also influence the driving of teen novices if they perceive that behavior favorably. The other drivers act as a form of *prototype*—a positive image that is a socially sanctioned model to imitate (Beullens & Van den Bulck, 2008)—and imitation can fill social (e.g., to fit in with the surrounding traffic) and emotional needs (e.g., to reduce anxiety in traffic). In addition, the perceived norms of the teen driver's friendship group (peers) and of other drivers can also influence driver behavior. That is, if the teen driver believes that his or her friends would speed if they were carrying passengers late at night, the teen driver is likely to engage in that behavior (see Section 10.4.1 for further discussion of the influence of norms).

10.3.4 Police

The influence of police upon teen driver behavior reflects a broader cultural influence and, as such, is discussed in greater detail in Chapter 22. It is noteworthy that interactions between police and teen drivers, however, occur within a social context, and as such, the influence of police on teen driver behavior will be discussed briefly. Legislation has been pivotal in the safety of all road users (e.g., seat belt laws [Milne 1985]; blood alcohol concentration laws [Voas et al., 2003]), and as will be discussed in more detail in Chapter 22, enforcement of legislation by Police (Travis, 2005) has been and continues to be essential (Williams, 2006; Wundersitz et al., 2010; Yannis et al., 2007). Graduated driver licensing (GDL) (see Section 10.5.1) has emerged as a particularly effective legislation, reducing the risks for the teen driver. Frequently, the efficacy of legislation is dependent upon the acceptance of the driving public (Foss et al., 2001; Huq et al., 2011; McCartt & Eichelberger, 2011). In addition, the effect of leniency by police is not well understood: qualitative research suggest that officers frequently use their discretion to verbally counsel noncompliant drivers rather than impose formal sanctions such as financial penalties and imposition of demerit points, particularly when drivers appear repentant (Schafer & Mastrofski, 2005). Relatedly, paying attention to police presence on the roads in order to avoid detection for illegal behavior appears normative for teen drivers (Scott-Parker et al., 2012b): 91% of learners and 72% of provisional drivers reported paying attention (Scott-Parker et al., 2011a). Provisional drivers also reported that they actively took steps to avoid the police, with 25% of males and 13% of females and 23% of rural drivers and 15% of urban drivers reporting such behavior (Scott-Parker et al., 2011b). In addition to avoiding on-road police presence, teen drivers report a complex web of punishment avoidance behaviors—meeting social (e.g., avoidance of peer-group ostracism) and emotional (e.g., avoidance of stress) needs—including having parents take the demerit points and associated fines on their behalf, missed consecutive offenses, and talking themselves out of a ticket (Scott-Parker & Bates, under review; Scott-Parker et al., 2011b, 2012b, 2013d).

10.4 Theories Relevant to Emotional and Social Factors

There is a plethora of theories conceptualizing human behavior that are relevant to the realm of teen driver safety, such as those from the domains of psychology, public health, sociology, criminology, and ergonomics, and a variety of more traffic-specific models that can inform intervention, including tools such as the Guarding Automobile Drivers through Guidance Education and Technology (GADGET) model (Hatakka et al., 2002). Each theory and model has its own epistemological position, with differing perspectives regarding the agent of change, and as such, each theory and model is uniquely positioned to provide insight into not only behaviors of interest, including predicting the role of a breadth of factors in the target behavior, but also how to intervene in these behaviors. Consequently, these frameworks can be used to obtain a greater insight into the factors that influence teen driver behavior and to develop, evaluate, and refine countermeasures targeting the group. The following section reviews some of the more common theoretical perspectives that have been applied in the road safety domain to examine the emotional and social influences on driver behavior. This review is not intended to be exhaustive but more illustrative of the approaches adopted, particularly to exploring teen driver behavior.

10.4.1 Theories from the Domains of Psychology and Criminology

Theories from the psychological and criminological domains that have been applied to understand the nature and extent of social and emotional influences upon teen driving include Akers's social learning theory, Gerrard and Gibbons's prototype willingness model, and Ajzen's theory of planned behavior. These psychosocial theories are concerned with explicating how behaviors are learned and the myriad of individual and social factors that contribute to behaviors and behavioral intentions. Accordingly, if we can understand *how* and *what* risky behaviors are learned and what factors contribute to risky behavior, we can also *change* these behaviors and factors by targeting the mechanisms and agents of learning and the magnitude of the influence of personal and other factors. Traffic-specific theories that have been applied in the teen driver area include the GADGET model, Matthews' transactional model of driver stress, and Shinar's application of the frustration–aggression hypothesis.

10.4.1.1 Akers' Social Learning Theory

Akers' theory was developed in the 1960s to account for the persistent finding that youth are more likely to indulge in deviant behavior if they associate with peers who are accepting of and/or promote such deviance (Akers & Sellers, 2004; Akers et al., 1979). As such, the (non) risky driving behavior of the young driver emerges from a complex combination of the *attitudes* held by the young driver toward the behavior and its alternatives; the nature and extent of *interaction* with significant others (such as parents and peers) who model driving behaviors; and the *rewards* and *punishments* (social, nonsocial, and instrumental; Scott-Parker et al., 2013c) for performing the (non-) risky driving behavior. Within the realm of road safety, the following findings have emerged from applications of Akers' social learning theory:

- Differential association with peers, differential reinforcements, modeling, and peer attitudes favoring risky behavior significantly predicted whether a sample of American youths aged less than 15 years traveled as passengers of drinking drivers (DiBlasio, 1987).
- Parents and peers have been found to be influential upon the seatbelt use of Spanish young drivers (Gras et al., 2007).
- Differential association was found to be the principal psychosocial influence upon the intentions of 309 suspended and disqualified adult Australians to drive while unlicensed, with prediction based on Akers' social learning variables being superior to prediction based on deterrence theory for the most noncompliant participants (Watson, 2004).
- Akers' social learning variables significantly explained self-reported speeding behavior over and above the explanatory contribution of deterrence theory in 320 Australian adults aged 17 to 79 years (Fleiter et al., 2006).
- A study of the drug driving behavior of university students in Queensland who were aged 17 to 56 years reported that drug driving was positively correlated with social rewards and negatively correlated with social punishments (Armstrong et al., 2005).
- While the 16–24 year old young driver's sociodemographic variables of age, gender, and driving exposure explained 19% of the variance in their risky behavior, Akers' social learning variables explained a further 42% of the variance of this risky behavior (Scott-Parker et al., 2009a). Furthermore, when the influences of parents and peers upon the risky behavior of the young drivers were separated for the variables of anticipated rewards, anticipated punishments, and norms through the creation of separate (rather than composite) subscales, the linear combination of peer and parent factors accounted for 59% of the variance in the risky driving behavior. Parents and peers were also found to differ in their influence on the risky behavior of young drivers: overall, the significant predictors of young driver risky behavior were found to be peer norms and peer and parent rewards, but exploratory analyses revealed that these influences differed by the gender

of the young driver. To illustrate, for young female drivers, parent rewards and peer punishment were significant predictors of risky behavior, while for young male drivers, peer and parent punishment and parent norms were approaching significance (Scott-Parker et al., 2009b).

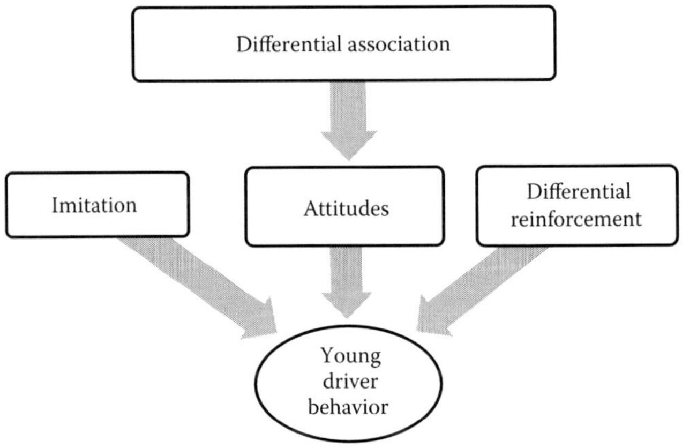

From Akers, R. L. (2009). A social learning theory of crime. In F. R. Scarpatti, A. L. Nielsen, and J. M. Miller (Eds.), *Crime and Criminals: Contemporary and Classic Reading in Criminology* (2nd ed.), pp. 214–225, New York: Oxford University Press.

10.4.1.2 Gerrard and Gibbons' Prototype Willingness Model

Extending the notions of the importance of rewards and the friendship network as incorporated within Akers' SLT, the prototype willingness model asserts that the majority of adolescent health risk is as a consequence of risky prototype appraisal, a willingness to be risky, prior risky behavior, and intentions to engage in risky behavior in the future (Gibbons & Gerrard, 1995; Ouellette et al., 1999). In addition to these constructs, personal attitudes and normative beliefs of parents and peers are also considered in the prototype willingness model, and it is noteworthy that these constructs are also captured within Akers' SLT. The temporal relationship between prototype willingness model variables and risky behavior is essential, with constructs measured at Time 1 predicting risky driving measured at Time 2. As can be seen in the figure, Time 1 measures of previous risky driving behavior, attitudes, subjective norms, prototypes, behavioral intention, and willingness influence Time 2 risky driving behavior.

Scott-Parker et al. (2013c) applied Akers' social learning theory, augmented by the prototype willingness model, to understand the factors influential in self-reported speeding of 378 young drivers who had held a provisional license for a period of 6 months. Significant predictors in the final model, in decreasing beta size, included Learner speeding, personal attitudes, depression, reward sensitivity, gender, and car ownership. It is also noteworthy that separate gender-based analyses (notwithstanding the small male sample size [$n = 113$], which prevented definitive conclusions being drawn in this instance) suggested that influential variables differ considerably across genders, with willingness to speed and associated rewards important for males, and sensation seeking propensity and age influential for females. It is also worth noting that this application of theory provided insight into who should be targeted in a speeding intervention and when this intervention should occur. The consideration of personal characteristics (such as depression and car ownership), the application of Akers' SLT (such as the role of attitudes), and the augmentation of Akers' social learning theory with the prototype willingness model variables (which incorporated the role of prior behaviors) revealed that intervention needs to consider not only the characteristics of the individual but also the role, and actions, of the supervisor during the Learner phase. Young driver attitudes have been found to be formed long before licensure (Berg, 2006), suggesting early and repeated intervention merits consideration.

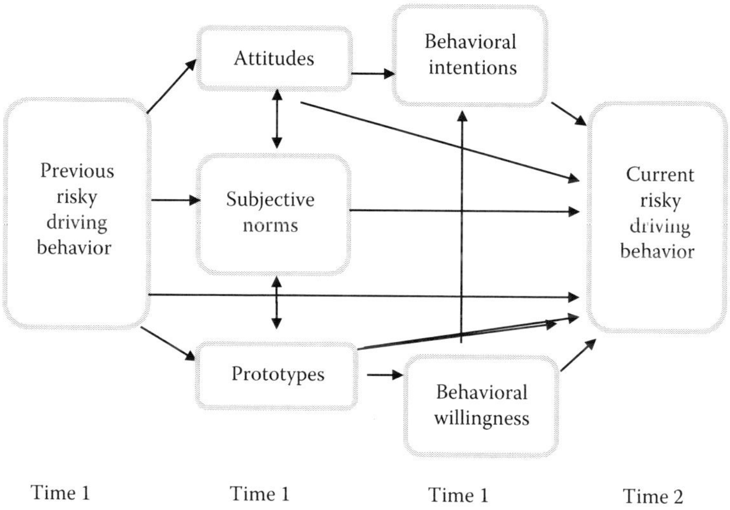

Adapted from Gerrard, M. et al., *Developmental Review, 28*, 29–61, 2008.

10.4.1.3 Ajzen's Theory of Planned Behavior

The theory of planned behavior has been one of the dominant models in social psychology over the last 20 years and has been used to examine a range of health behaviors including road user behavior (e.g., Armitage & Conner, 2001). Horvath et al. (2012) applied Ajzen's (1991) theory with some additions to explicate the peer passenger factors influential in self-reported speeding intentions of 398 university students who were young drivers with either a provisional or open (unrestricted) license. Theory of planned behavior measures *attitudes* toward speeding, *subjective norms* (perceived dis/approval of behaviors by persons important to the young driver), and *perceived behavioral control* (incorporating controllability [the ability to control the driving behavior] and self-efficacy [belief that they can control the driving behavior]). Hierarchical multiple regression analyses revealed that the theoretical constructs explained an additional 23%–32% of variance in speeding intentions, over and above sociodemographic characteristics of gender, age, exposure, sensations seeking, self-esteem, and past behavior. Predictors of speeding intentions included attitudes, self-efficacy and past behavior, controllability, and sensation seeking. Similarly, group norms—the young driver's perceptions regarding what their friends' support for texting while driving—have been found to predict intentions to text while driving (Nemme & White, 2010).

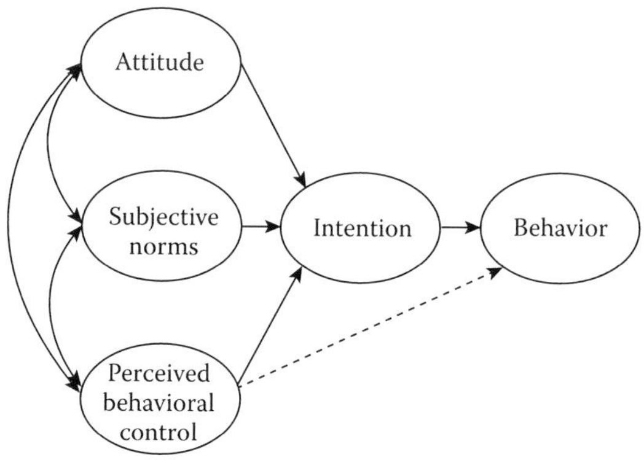

From Ajzen, I., *Organizational Behavior and Human Decision Processes, 50*, 179–211, p. 182, 1991.

10.4.1.4 GADGET Model

GADGET is a hierarchical model of the driving behavior (and thus, the training, education, and skill requirements of the young driver), which would be expected of a *safe* young driver derived from the Michon Model of driver behavior. Skills and abilities at lower levels of the matrix (e.g., vehicle maneuvering and mastery of traffic situations) influence the skills and abilities at higher levels (e.g., driving goals and goals for life). Essential learning and teaching curricula are categorized as *knowledge and skills* (fundamental to the driving tasks at each level of the hierarchy), *risk-increasing factors* (which derive from a combination of the various behavior levels, knowledge and skills, and self-evaluation), and *self-evaluation* (particularly, self-calibration). Interestingly, GADGET can similarly provide unique insight regarding understanding, and effectively intervening in, young driver road safety.

Within the context of the GADGET model, the social, and potentially negative, influence of peers (irrespective of presence in the vehicle) and other young drivers can be captured within the matrix "goals for life and skills for living/risk-increasing factors" (complying with social pressure). Similarly, emotional influences can also be captured within this cell (e.g., high level of sensation seeking) and the adjoining cell "goals for life and skills for living/self-evaluation" (e.g., risky tendencies). Somewhat relatedly, Ouimet et al. (2014) found in a longitudinal study that young drivers who had a more pronounced affective response to stress (incorporated within the behavioral aspects level of GADGET), in this case measured by cortisol levels observed immediately following a nondriving task designed to be stressful, had both lower baseline crash risk upon licensure and a faster rate of decrease in crash risk over time subsequently, compared with drivers who had a lesser cortisol response to stress.

Hierarchical Levels of Behavior	Essential Curriculum		
	Knowledge and Skills	Risk-Increasing Factors	Self-Evaluation
Goals for life and skills for living (general) This level will further be called *Behavioral aspects.*	**Knowledge about/control over how life goals and personal tendencies affect driving behavior** • Lifestyle/life situation • Peer group norms • Motives • Self-control, other characteristics • Personal values	**Risky tendencies** • Acceptance of risks • Self-enhancement through driving • High level of sensation seeking • Complying with social pressure • Use of alcohol and drugs • Values, attitudes toward society...	**Self-evaluation/ awareness of** • Personal skills for impulse control • Risky tendencies • Safety-negative motives • Personal risky habits...
Driving goals and context (journey-related) This level refers to the Michon-level *Strategic tasks* and will further be called this way.	**Knowledge and skills concerning** • Effects of journey goals on driving • Planning and choosing routes • Evaluation of requested driving time • Effects of social pressure inside the car • Evaluation of necessity of the journey	**Risks connected with** • Driver's condition (mood, BAC, etc.) • Purpose of driving • Driving environment (rural/urban) • Social context and company • Additional motives (competitive, etc.)...	**Self-evaluation/ awareness of** • Personal planning skills • Typical driving goals • Typical risky driving motives...
Mastery of traffic situations This level refers to the Michon-level *Maneuvering tasks* and will further be called this way.	**Knowledge and skills concerning** • Traffic regulations • Observation/selection of signals • Anticipation of the development of situations	**Risks caused by** • Wrong expectations • Risk-increasing driving style (e.g., aggressive) • Unsuitable speed adjustment • Vulnerable road users	**Self-evaluation/ awareness of** • Strong and weak points of basic traffic skills • Personal driving style

(*Continued*)

Hierarchical Levels of Behavior	Essential Curriculum		
	Knowledge and Skills	Risk-Increasing Factors	Self-Evaluation
	• Speed adjustment • Communication • Driving path • Driving order • Distance to others/safety margins...	• Not obeying regulations/ unpredictable behavior • Information overload • Difficult conditions (darkness, etc.) • Insufficient automatism or skills	• Personal safety margins • Strong and weak points for hazard situations • Realistic self-evaluation...
Vehicle maneuvering This level refers to the Michon-level *Control task* and will further be called this way.	**Knowledge and skills concerning** • Control of direction and position • Tire grip and friction • Vehicle properties • Physical phenomena...	**Risks connected with** • Insufficient automatism or skills • Unsuitable speed adjustment • Difficult conditions (low friction, etc.)...	**Awareness of** • Strong and weak points of basic maneuvering skills • Strong and weak points of skills for hazard situations • Realistic self-evaluation

Source: Adapted from Hatakka, M. et al., *Transportation Research, Part F*, 5, 201–215, p. 209, 2002.

10.4.1.5 Matthews' Transactional Model of Driver Stress

Matthews' transactional model of driver stress emphasizes the stress–strain relationship, which can impact upon driver behavior, such that environmental stressors like congested traffic spark cognitive stress processes (such as the driver appraising their ability to cope with the demands of the environmental stressors), which are also impacted upon by personality and self-awareness. Cognitive stress processes such as appraisal and regulation of coping strategies, in turn, impact upon driving performance (evidenced, for example, as distraction from the driving task) and subjective emotional symptoms such as anxiety. In this way, stress from the external environment becomes internalized by the teen driver. Interestingly, it is noteworthy that the subjective response can be further impacted upon by the emotional states and traits of the teen driver; for example, it is reasonable to expect that the strain experienced by the environmental stressor may be exacerbated for the teen driver who is already anxious and depressed. While the transactional model has not featured extensively in teen driver research, research suggests that external stressors do indeed impact upon driver performance (Rowden et al., 2006, 2011).

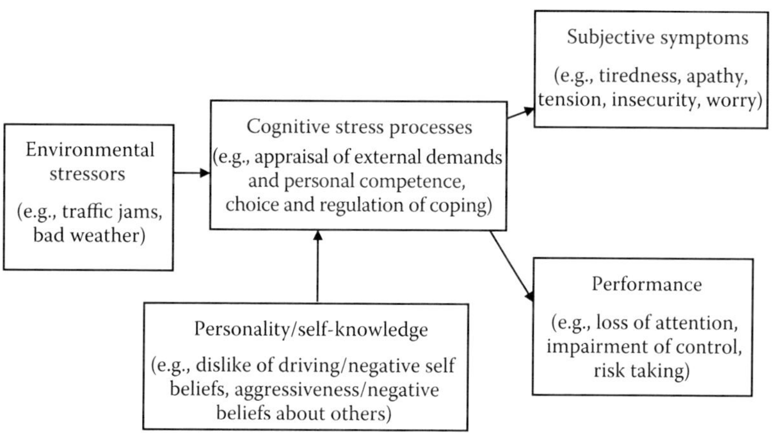

Adapted from Matthews, G. (2001). In Rowden, P., Watson, B., & Biggs, B. (2006). The transfer of stress from daily hassles to the driving environment in a fleet sample. *Proceedings of the Australasian Road Safety Research, Policing and Education Conference*, Gold Coast, Queensland.

10.4.1.6 Shinar's Application of the Frustration–Aggression Hypothesis

The frustration–aggression hypothesis (Dollard et al., 1939, cited in Soole et al., 2011) asserts that aggression is the result of frustration. Extending the frustration–aggression hypothesis, Shinar's model (1998) suggests that the emergence of aggression depends on environmental factors (e.g., degree of driver anonymity), personality factors (e.g., trait anger), and situation appraisal factors (e.g., intent on behalf of the other driver). An integration of the general aggression model and Shinar's model has resulted in a revised driver aggression model, which recognizes important limitations of earlier models, which include the following: environmental triggers do not always evoke frustration—rather other emotional reactions such as anxiety may be evoked; frustration does not always lead to aggression; and a complex web of cognitive and emotional factors are involved in appraising the environmental trigger (Soole et al., 2011).

10.5 Interventions Addressing Emotional and Social Influences upon Teen Driver Behavior

Historically, a range of approaches have been developed to manage the risky driving behavior of teen drivers. Interventions principally are designed to ameliorate and manage on-road risky driving behavior. Interventions include such measures as driver training and education; in-car technology; media campaigns; incentive programs; peer and parent monitoring and support programs; medical professionals; legislation and enforcement; and rehabilitation, remediation, and retraining programs. They have been applied in various formats in motorized jurisdictions throughout the world with varying degrees of success in modifying the behavior and crash involvement of teen drivers. Section V of the book, "Licensing, Training, and Education," explores many of these interventions in greater depth. Therefore, only a brief summary of the emotional and social factors related to GDL and resilience programs will be provided at this time.

10.5.1 Graduated Driver Licensing

As noted in Chapter 17, GDL refers to a system of licensing for teen drivers that traditionally has at least three levels of driving privileges and restrictions, with a growing body of evidence confirming its effectiveness evidenced through evaluations (e.g., New Zealand [Lewis-Evans & Lukkien, 2007]). Programs with night-driving and passenger restrictions appear to be the most effective (e.g., Masten et al., 2013), suggesting that social and emotional influences upon teen driving can be ameliorated to some extent through GDL program restrictions (e.g., delaying licensure to allow psychosocial and physiological maturity, and reducing the capacity for negative peer pressure impacting upon social and emotional needs through passenger restrictions).

10.5.2 Resilience Programs

Historically, the most common approach adopted to address the high crash risk of teen drivers has been driver education and training. However, there is little evidence that conventional driver education and training programs are effective in reducing crash involvement rates among teen drivers (Beanland et al., 2013). A newer approach, with particular relevance to the role of emotional and social factors, that appears promising are programs designed to build resilience among teens. The research findings suggest that social skills and mental health interventions may be effective in ameliorating the risky driving behavior of teen drivers. As noted earlier, teen drivers report peer pressure to engage in risky driving. Accordingly, fostering the development of social and resilience skills to resist negative peer influences are important in ameliorating the risky behavior of teen drivers. Thus, peer support (e.g., Ulleberg, 2004) and resilience training (Botvin et al., 2001; Senserrick et al., 2009;

Siegrist, 1999) programs merit further application and evaluation. Recently, education programs have considered the nature and extent of psychosocial contributors to crashes and the role of protective behavioral strategies in risky behavior (e.g., Borden et al., 2011; Botvin et al., 2001; Liourta & van Empelen, 2008). While no difference in offenses were detected between groups of teen drivers, participants of extended resilience-focused education in New South Wales, Australia, were found to have a 44% lower relative risk of crashing than drivers who participated in a 1-day driving-risk workshop (Senserrick et al., 2009).

10.6 Research Priorities

In light of the previous discussions, there are a number of research priorities relating to the influence of emotional and social factors on teen driver behavior.

10.6.1 Priority 1: Understanding the Influence of Emotional and Social Factors on Behavior

While much has been learned regarding the influence of emotional and social factors on teen driver behavior, more needs to be learned if we are to reduce the negative influence of social and emotional factors in particular. To illustrate in the emotional factor context, while sensation-seeking propensity has been recognized for some time as an important emotional risk-related factor, reward sensitivity has only recently been identified as playing an important and distinct role in risky behavior (Scott-Parker & Weston, under review). In addition, the literature regarding the impact of psychological distress has only recently recognized the influence of these emotional states, despite the well-recognized pervasive impact of depression and anxiety upon other health- and risk-related behaviors. Further, the role of self-reflection in behavior regulation remains relatively unexplored and merits further investigation. Within the social factor context, the influence of parents and peers is well understood; however, the influence of police and the broader road user community is less well understood. The influence of police and disparate enforcement practices upon teen and novice driving behavior in particular has only recently begun to receive attention. Importantly, a number of emerging methodologies can provide unique insight into emotional and social factors in teen driving (for example, naturalistic driving studies [Chapter 26] and enhanced simulation techniques [Chapter 25]), complementing findings from traditional epidemiological approaches of crash and data analysis, and experimental research approaches of self-report, observation, interview, and focus group methodologies.

10.6.2 Priority 2: Use of Theory to Understand the Influence of Factors

It is highly likely that teen driver road safety could benefit considerably from applications of theory, and it is noteworthy that many theories that have shaped other fields of behavior change, behavior regulation, and skill acquisition can inform these applications (see web references in Section 10.4.1). Perhaps most importantly, theory can guide the development of effective interventions that can address both emotional and social factors that influence teen driver behavior, such as elucidating the important role of parents and peers, beyond the well-recognized influence of age and gender (see Section 10.4).

10.6.3 Priority 3: The Development of Effective Interventions

While all of the mechanisms behind the increased risk experienced by teen drivers in psychological distress remain unknown at this time, the recurring finding that teen drivers in psychological distress engage in more risky driving behavior, including speeding in particular, merits further consideration. Psychological distress should be addressed not only for social justice and a right to the improvement of life quality for all persons, but also because the depressed and anxious adolescent is likely to become the depressed and anxious adult (Avenevoli et al., 2006). Further, medical professionals treating teen drivers injured through road crashes need to be aware that the drivers could also be depressed and anxious. More generally, mental health practitioners treating depressed or anxious adolescents need to be aware that their patient could be at additional risk of harm through more risky driving. Peer and parent interventions appear warranted, in addition to efforts addressing all road users, and police as enforcers of GDL-specific and general road rules. Peer intervention programs such as resilience training and passengers intervening in the case of risky behavior by their romantic partner (e.g., Buckley & Foss, 2012) require consideration, as do interventions addressing speeding and aggressive driving behaviors, particularly as such risky driving behaviors are more likely to result in a fatal crash if the teen driver is carrying peer passengers (Lambert-Belanger et al., 2012). Interventions also need to be rigorously evaluated, with methodological difficulties plaguing much research on teen drivers (e.g., Beanland et al., 2011; Scott-Parker & Senserrick, 2013, under review). Parents need to be aware of the considerable role they play in the road safety of their teen driver child. Public education campaigns can highlight the significant social influence of parents, similar to the Australian Drinking Nightmare campaign targeting youth binge drinking (Department of Health and Ageing, 2008). Efforts targeting parents will also need to encourage them to explicitly sanction (Preston & Goodfellow, 2006; Winfree & Bernat, 1998) and to not explicitly or implicitly reward risky driving behaviors, while rewarding safe driving behaviors (King & Vidourek, 2010).

10.7 Conclusions

This chapter has highlighted the role of emotional and social influences on the driving behavior of teen drivers. Influential emotional factors include driver mood, driver stress, anxiety and depression, reward sensitivity, sensation seeking, and impulsivity. In addition, social sources of influence range from parents and peers to police and other drivers. Numerous areas requiring additional research were identified in relation to better understanding the influence of emotional and social factors on teen driver behavior, especially as it is informed by the application of theory, and the development of effective interventions. Teen driver road safety interventions may need to more fully incorporate elements addressing social and emotional influence to effect real improvements in teen road safety outcomes such as crashes and the fatalities and injuries arising from those crashes.

Appendix 10.1 Scales Used to Measure Sensation-Seeking and Impulsivity Tendencies

- **20-item Arnett Inventory of Sensation Seeking (AISS):** Found to be a significant predictor of driving while intoxicated (Gonzalez-Iglesias et al., 2014).
- **19-item Zuckerman–Kuhlman Impulsivity–Sensation Seeking Scale (ISSS) (Zuckerman et al., 1993):** Found to be predictive of risky behavior (e.g., illicit drug use, alcohol consumption [Zuckerman & Kuhlman, 2000]), including risky driving (Jonah et al., 2001; Zuckerman & Kuhlman, 2000) and the driving violations subscale of the Driver Behaviour Questionnaire (DBQ) (Rimmo & Aberg, 1999).
- **45-item Urgency Premeditation Perseverance Sensation Seeking Impulsive Behaviour Scale (UPPS) (Whiteside & Lynam, 2001):** Drivers (postgraduate university students) exhibiting greater sensation-seeking tendencies have been found to report more risky driving behaviors (e.g., speeding, following too closely [Bachoo et al., 2013]), more aggressive driving, and more frequent driving transgressions (Berdoulat et al., 2013).
- **10-item thrill and adventure seeking (TAS) subscale of Zuckerman's 40-item Sensation Seeking Scale—Form V (SSSV):** The TAS was found to be a significant predictor for not wearing seat belts (Hatfield et al., 2014); scores on the SSSV have shown that higher sensation seekers

were significantly more likely to report speeding intentions (Fernandes et al., 2010) and to choose higher traveling speeds when posed with a time-pressured driving scenario (Peer & Rosenbloom, 2013).

- **30-item Barratt Impulsiveness Scale (BaImSc):** Young drivers caught for a traffic offense exhibited greater impulsiveness than nonoffenders (O'Brien & Gormley, 2013).
- **8-item Brief Sensation Seeking Scale (BSSS) (Palmgreen et al., 2001):** Found to predict risky driving behavior including speeding (Scott-Parker et al., 2012a, 2013a,b).
- **10-item dangerous thrill seeking and 10-item impulsivity thrill seeking subscales of the sensation seeking facet element of the International Personality Item Pool:** The most risky subgroup of moped riders aged 14–15 years were found to exhibit greater impulsivity and sensation seeking; they experienced significantly more crashes and more self-reported risky driving behaviors, including noncompliance with road rules (Marengo et al., 2012).
- **7-item Driving Related Sensation Seeking (DRSS) scale:** Sensation seeking was a better predictor of speeding intentions and self-reported behaviors than driving anger for a sample of French young drivers (Delhomme et al., 2012).

References

Akers, R. L., & Sellers, C. S. (2004). *Criminological Theories: Introduction, Evaluation, and Application.* Los Angeles, CA: Roxbury Publications.

Akers, R. L., Krohn, M. D., Lanza-Kaduce, L., & Radosevich, M. (1979). Social learning and deviant behaviour: A specific test of a general learning theory. *American Sociological Review, 44*, 636–655.

Allen, J. P., & Brown, B. B. (2008). Adolescents, peers, and motor vehicles. The perfect storm? *American Journal of Preventive Medicine, 35*(3S), S289–S293.

American Psychological Association. (2014). Glossary of psychological terms. Accessed October 1, 2014, retrieved from http://www.apa.org/research/action/glossary.aspx.

Andrews, J. A. A., Hops, H., & Duncan, S. C. (1997). Adolescent modeling of parent substance use: The moderating effect of the relationship with the parent. *Journal of Family Psychology, 11*, 259–270.

Armitage, C. J., & Conner, M. (2001). Efficacy of the theory of planned behaviour: A meta-analytic review. *British Journal of Social Psychology, 40*, 471–499.

Armstrong, K., Wills, A., & Watson, B. (2005). Psychosocial influences on drug driving in young Australian drivers. Accessed March 23, 2007, from the Register of Australian and Drug Alcohol Research, http://www.radar.org.au.

Arnett, J. J. (2002). Developmental sources of crash risk in young drivers. *Injury Prevention, 8*(SII), ii17–ii23.

Arnett, J. J., Offer, D., & Fine, M. A. (1997). Reckless driving in adolescence: 'State' and 'trait' factors. *Accident Analysis & Prevention, 29*, 57–63.

Avenevoli, S., Knight, E., Kessler, R. C., & Merikangas, K. R. (2008). Epidemiology of depression in children and adolescents. In J. R. Z. Abela, and B. L. Hankin (Eds.), *Handbook of Depression in Children and Adolescents* (pp. 6–32). New York: Guilford Press.

Bachoo, S., Bhagwanjee, A., & Govender, K. (2013). The influence of anger, impulsivity, sensation seeking and driver attitudes on risky driving behaviour among post-graduate university students in Durban, South Africa. *Accident Analysis & Prevention, 55*, 67–76.

Bates, L., Watson, B., & King, M. J. (2009). *Factors Influencing Learner Driver Experiences.* Canberra: Australian Transport Safety Bureau.

Beanland, V., Goode, N., Salmon, P. M., & Lenne, M. G. (2013). Is there a case for driver training? A review of the efficacy of pre- and post-licence driver training. *Safety Science, 51*(1), 127–137.

Beck, K. H., Daughters, S. B., & Ali, B. (2013). Hurried driving: Relationship to distress tolerance, driver anger, aggressive and risky driving in college students. *Accident Analysis & Prevention, 51*, 51–55.

Beck, R. C. (1990). *Motivation: Theories and Principles* (3rd ed.). Englewood Cliffs, NJ: Prentice-Hall.

Begg, D. J., & Langley, J. D. (2004). Identifying predictors of persistent non-alcohol or drug-related risky driving behaviours among a cohort of young adults. *Accident Analysis & Prevention, 26*, 1067–1071.

Begg, D., Langley, J., & Stephenson, S. (2003). Identifying factors that predict persistent driving after drinking, unsafe driving after drinking, and driving after using cannabis among young adults. *Accident Analysis & Prevention, 35*, 669–675.

Begg, D., Sullman, M., & Samaranayaka, A. (2012). The characteristics of young pre-licensed drivers: Evidence from the New Zealand Drivers Study. *Accident Analysis & Prevention, 45*, 539–546.

Benfield, J. A., Szlemko, W. J., & Bell, P. A. (2007). Driver personality and anthropomorphic attributions of vehicle personality relate to reported aggressive driving tendencies. *Personality and Individual Differences, 42*, 247–258.

Berdoulat, E., Vavassori, D., & Munoz Sastre, M. T. (2013). Driving anger, emotional and instrumental aggressiveness, and impulsiveness in the prediction of aggressive and transgressive driving. *Accident Analysis & Prevention, 50*, 758–767.

Beullens, K., & Van den Bulck, J. (2008). News, music videos and action movie exposure and adolescents' intentions to take risks in traffic. *Accident Analysis & Prevention, 40*, 349–356.

Bingham, C. R., Shope, J. T., Zakrajsek, J., & Raghunathan, T. E. (2008). Problem driving behaviour and psychosocial maturation in young adulthood. *Accident Analysis & Prevention, 40*, 1758–1764.

Blows, S., Ameratunga, S., Ivers, R. Q., Lo, S. K., & Norton, R. (2005). Risky driving habits and motor vehicle driver injury. *Accident Analysis & Prevention, 37*, 619–624.

Bonino, S., Cattelino, E., & Ciairano, S. (2005). *Adolescents and Risk. Behavior, Functions, and Protective Factors*. Milan: Springer-Verlag Italia.

Borden, L. A., Martens, M. P., McBride, M. A., Sheline, K. T., Bloch, K. K., & Dude, K. (2011). The role of college students' use of protective behavioral strategies in the relation between binge drinking and alcohol-related problems. *Psychology of Addictive Behaviors, 25*(2), 346–351.

Botvin, G., Griffin, K. W., Diaz, T., & Ifill-Williams, M. (2001). Preventing binge drinking during early adolescence: One- and two-year follow-up of a school-based preventive intervention. *Psychology of Addictive Behaviors, 15*, 360–365.

Briggs, G. F., Hole, G. J., & Land, M. F. (2011). Emotionally involving telephone conversations lead to driver error and visual tunneling. *Transportation Research Part F: Traffic Psychology and Behaviour, 14*, 313–323.

Buckley, L. (2005). Adolescents' perspective of transport related risk-taking and injury: Definitions, consequences, and risk and protective factors. In *Proceedings of Australasian Road Safety Research, Policing and Education Conference*, 14–16 November, Wellington, New Zealand.

Buckley, L., & Foss, M. S. (2012). Protective factors associated with young passenger intervening in risky driving situations. *Journal of Safety Research, 43*, 351–356.

Cameron, J. E. (1999). Social identity and the pursuit of possible selves: Implications for the psychological well-being of university students. *Group Dynamics: Theory, Research, and Practice, 3*, 179–189.

Cavallo, A., Montero, K., Sangster, J., & Maunders, D. (1997). Youth work with high risk young drivers VICRoads and RMIT peer education program. In *Staysafe 39, Young Drivers* (pp. 209–217). Sydney: Parliament of NSW.

CHOP. (2007). Driving: Through the eyes of teens. Accessed October 10, 2009, retrieved from http://www/chop/edu/youngdrivers.

CHOP. (2009). Driving: Through the eyes of teens. A closer look. Accessed October 10, 2009, retrieved from http://www/chop/edu/youngdrivers.

Coeugnet, S., Navetuer, J., Antoine, P., & Anceaux, F. (2013). Time pressure and driving: Work, emotions and risks. *Transportation Research Part F: Traffic Psychology and Behaviour, 20*, 39–51.

Constantinou, E., Panayioutou, G., Konstantinou, N., Loutsiou-Ladd, A., & Kapardis, A. (2011). Risky and aggressive driving in young adults: Personality matters. *Accident Analysis & Prevention, 43*, 1323–1331.

Cristea, M., Paran, F., & Delhomme, P. (2013). Extending the theory of planned behavior: The role of behavioral options and additional factors in predicting speeding behavior. *Transportation Research Part F: Traffic Psychology and Behaviour, 21*, 122–132.

Dahl, R. E. (2008). Biological, developmental, and neurobehavioral factors relevant to adolescent driving risks. *American Journal of Preventive Medicine, 35*, S278–S284.

Dahlen, E. R., Martin, R. C., Ragan, K., & Kuhlman, M. M. (2005). Driving anger, sensation seeking, impulsiveness, and boredom proneness in the prediction of unsafe driving. *Accident Analysis & Prevention, 37*, 341–348.

Davey, C. G., Yucel, M., & Allen, N. B. (2008). The emergence of depression in adolescence: Development of the prefrontal cortex and the representation of reward. *Neuroscience and Biobehavioural Reviews, 32*, 1–19.

Deery, H. A. (1999). Hazard and risk perception among young novice drivers. *Journal of Safety Research, 30*, 225–236.

Deffenbacher, J. L., Oetting, E. R., & Lynch, R. S. (1994). Development of a driving anger scale. *Psychological Reports, 74*, 83–91.

Delhomme, P., Chaurand, N., & Paran, F. (2012). Personality predictors of speeding in young drivers: Anger vs. sensation seeking. *Transportation Research Part F: Traffic Psychology and Behaviour, 15*, 654–666.

Department of Health and Ageing. (2009). *National binge drinking campaign—Evaluation survey April 2009*. Accessed June 1, 2009, retrieved from http://www.health.gov.au/internet/drinkingnightmare/publishing.nsf/Content/3F34473572CF15F2CA257679007C3A7A/$File/eval.pdf.

DiBlasio, F. A. (1987). Predriving riders and drinking drivers. *Journal of Studies on Alcohol, 49*, 11–15.

di Castella, J., & Perez, J. (2004). Sensitivity to punishment and sensitivity to reward and traffic violations. *Accident Analysis & Prevention, 36*, 947–952.

Falk, B., & Montgomery, H. (2007). Developing traffic safety interventions from conceptions of risks and accidents. *Transportation Research Part F, 10*, 414–427.

Feldman, G., Greeson, J., Renna, M., & Robbins-Monteith, K. (2011). Mindfulness predicts less texting while driving among young adults: Examining attention- and emotion-regulation motives as potential mediators. *Personality and Individual Differences, 51*, 856–861.

Fernandes, R., Hatfield, J., & Job, R. F. S. (2010). A systematic investigation of the differential predictors for speeding, drink-driving, driving while fatigued, and not wearing a seat belt, among young drivers. *Transportation Research Part F: Traffic Psychology and Behaviour, 13*, 179–196.

Ferreira, A. I., Martinez, L. F., & Guisande, M. A. (2009). Risky behavior, personality traits and road accidents among university students. *European Journal of Education and Psychology, 2*(2), 79–98.

Fleiter, J. J., Lennon, A., & Watson, B. (2010). How do other people influence your driving speed? Exploring the "who" and the "how" of social influences on speeding from a qualitative perspective. *Transportation Research Part F: Traffic Psychology and Behaviour, 13*, 49–62.

Foley, K. L., Altman, D., Durant, R., & Wolfson, M. (2004). Adults' approval and adolescents' alcohol use. *Journal of Adolescent Health, 34*, 345–354.

Foss, R. D. (2007). Improving graduated driver licensing systems: A conceptual approach and its implications. *Journal of Safety Research, 38*, 185–192.

Foss, R. D., Stewart, J. R., & Reinfurt, D. W. (2001). Evaluation of the effects of North Carolina's 0.08% BAC law. *Accident Analysis & Prevention, 33*, 507–517.

Fuller, R. (2002). The psychology of the young driver. In R. Fuller, and J. A. Santos (Eds.), *Human Factors for Highway Engineers* (pp. 241–253). Oxford: Elsevier.

Gibbons, F. X., & Gerrard, M. (1995). Predicting young adults' health risk behaviour. *Journal of Personality and Social Psychology, 69*, 505–517.

Goldberg, L. R. (1981). Language and individual differences: The search for universals in personality lexicons. In L. Wheeler (Ed.), *Review of Personality and Social Psychology, Vol. 2*. Beverly Hills, CA: Sage.

Gonzalez-Iglesias, B., Gomez-Fraguela, J. A., & Luengo-Martin, M. A. (2012). Driving anger and traffic violations: Gender differences. *Transportation Research Part F: Traffic Psychology and Behaviour*, 15, 404–412.

Gonzalez-Iglesias, B., Gomez-Fraguela, J. A., & Luengo, M. A. (2014). Sensation seeking and drunk driving: The mediational role of social norms and self-efficacy. *Accident Analysis & Prevention, 71*, 22–28.

Gras, M. E., Cunill, M., Sullman, M. J. M., Planes, M., & Font-Mayolas, S. (2007). Predictors of seat belt use amongst Spanish drivers. *Transportation Research Part F, 10*, 263–269.

Gratz, K., & Roemer, L. (2004). Multidimensional assessment of emotion regulation and dysregulation: Development, factor structure, and initial validation of the difficulties in emotion regulation scale. *Journal of Psychopathology, 26*(1), 41–54.

Gregersen, N. P., & Bjurulf, P. (1996). Young novice drivers: Towards a model of their accident involvement. *Accident Analysis & Prevention, 28*, 229–241.

Gulliver, P., & Begg, D. (2007). Personality factors as predictors of risky driving behavior and crash involvement among young drivers. *Injury Prevention, 13*, 376–381.

Gwyther, H., & Holland, C. (2012). The effect of age, gender and attitudes on self-regulation in driving. *Accident Analysis & Prevention, 45*, 19–28.

Harre, N., Brandt, T., & Dawe, M. (2000). The development of risky driving in adolescence. *Journal of Safety Research, 31*, 185–194.

Harre, N., Brandt, T., & Houkamau, C. (2004). An examination of the actor-observer effect in young drivers' attributions for their own and their friends' risky driving. *Journal of Applied Social Psychology, 34*, 806–824.

Harrison, W. A. (2004). Investigation of the driving experience of a sample of Victorian learner drivers. *Accident Analysis & Prevention, 36*, 885–891.

Harrison, W. A., Triggs, T. J., & Pronk, N. J. (1999). *Speed and young drivers: Developing countermeasures to target excessive speed behaviours amongst young drivers.* Monash University Accident Research Centre Report No. 159. Melbourne: MUARC.

Hatakka, M., Keskinen, E., Gregersen, N. P., Glad, A., & Hernetkoski, K. (2002). From control of the vehicle to personal self-control: Broadening the perspectives to driver education. *Transportation Research, Part F, 5*, 201–215.

Hatfield, J., Fernandes, R., & Job, R. F. S. (2014). Thrill and adventure seeking as a modifier of the relationship of perceived risk with risky driving among young drivers. *Accident Analysis & Prevention, 62*, 223–229.

Horswill, M. S., Waylen, A. E., & Tofield, M. I. (2004). Drivers' ratings of different components of their own driving skill: A greater illusion of superiority for skills that relate to accident involvement. *Journal of Applied Social Psychology, 34*, 177–195.

Horvath, C., Lewis, I., & Watson, B. (2012). The beliefs which motivate young male and female drivers to speed: A comparison of low and high intenders. *Accident Analysis & Prevention, 45*, 334–341.

Huang, J. P., Xia, W., Sun, C. H., Zhang, H. Z., & Wu, L. J. (2009). Psychological distress and its correlates in Chinese adolescents. *Australian and New Zealand Journal of Psychiatry, 43*, 674–681.

Hubicka, B., Kallmen, H., Hiltunen, A., & Bergman, H. (2010). Personality traits and mental health of severe drunk drivers in Sweden. *Social Psychiatry and Psychiatric Epidemiology, 45*(7), 723–731.

Huq, A., Tyler, T. R., & Schulhofer, S. J. (2011). Why does the public cooperate with law enforcement? The influence of the purposes and targets of policing. *Psychology, Public Policy, and Law, 17*(3), 419–450.

Jaycox, L. H., Stein, B. D., Paddock, S., Miles, J. N. V., Chandra, A., Meredith, L. S., Tanielian, T., Hickey, S., & Burnam, M. A. (2009). Impact of teen depression on academic, social, and physical functioning. *American Academy of Pediatrics, 124*(4), e596–e605.

Jenkins, J., & Sainsbury, P. (1980). Single-car road deaths—Disguised suicides? *British Medical Journal, 281*, 1041.

Jeon, M., Walker, B. N., & Yim, J. B. (2014). Effects of specific emotions on subjective judgment, driving performance, and perceived workload. *Transportation Research Part F: Traffic Psychology and Behaviour, 24*, 197–209.

Jessor, R., Turbin, M. S., & Costa, F. M. (1997). Predicting developmental change in risky driving: The transition to young adulthood. *Applied Developmental Science, 1*, 4–16.

Johnson, P. B., & Malow-Iroff, M. S. (2008). *Adolescents and Risk: Making Sense of Adolescent Psychology.* Westport, CT: Praeger Publishers.

Jonah, B. A. (1997). Sensation seeking and risky driving: A review and synthesis of the literature. *Accident Analysis & Prevention, 29*, 651–665.

Jonah, B. A., Thiessen, R., & Au-Yeung, E. (2001). Sensation seeking, risky driving and behavioural adaptation. *Accident Analysis & Prevention, 33*, 679–684.

Joseph, J. E., Liu, X., Jiang, Y., Lynam, D., & Kelly, T. H. (2009). Neural correlates of emotional reactivity in sensation seeking. *Psychological Science, 20*, 215–223.

Jovanovic, D., Lipovac, K., Stanojevic, P., & Stanojevic, D. (2011). The effects of personality traits on driving-related anger and aggressive behaviour in traffic among Serbian drivers. *Transportation Research Part F: Traffic Psychology and Behaviour, 14*, 43–53.

Karjalainen, K., Blencowe, T., & Lillsunde, P. (2012). Substance use and social, health and safety-related factors among fatally injured drivers. *Accident Analysis & Prevention, 45*, 731–736.

Kaye, S. A., White, M. J., Lewis, I. M. (2013). Individual differences in drivers' cognitive processing of road safety messages. *Accident Analysis & Prevention, 50*, 272–281.

Keating, D. P. (2007). Understanding adolescent development: Implications for driving safety. *Journal of Safety Research, 38*, 147–157.

Keating, D. P., & Halpern-Felsher, B. L. (2008). Adolescent drivers. A developmental perspective on risk, proficiency, and safety. *American Journal of Preventive Medicine, 35*, S272–S312.

Kim, J. H., & Kim, K. S. (2012). The role of sensation seeking, perceived peer pressure, and harmful alcohol use in riding with an alcohol-impaired driver. *Accident Analysis & Prevention, 48*, 326–334.

Kobus, K. (1998). Peers and adolescent smoking. *Addiction, 98*(S1), 37–55.

Lahatte, A., & Le Pape, M. C. (2008). Is the way young people drive a reflection of the way their parents drive? An econometric study of the relation between parental risk and their children's risk. *Risk Analysis, 28*(3), 627–634.

Lambert-Belanger, A., Dubois, S., Weaver, B., Mullen, B., & Bedard, M. (2012). Aggressive driving behaviour in young drivers (aged 16 through 25) involved in fatal crashes. *Journal of Safety Research, 43*, 333–338.

Lansdown, T. C., & Stephens, A. N. (2013). Couples, contentious conversations, mobile telephone use and driving. *Accident Analysis & Prevention, 50*, 416–422.

Lawton, R. J., Parker, D., Manstead, A. S. R., & Stradling, S. G. (1997). The role of affect in predicting social behaviour: The case of road traffic violations. *Journal of Applied Social Psychology, 27*, 1258–1276.

Leal, N., Watson, B., & Armstrong, K. (2010). Risky driving or risky drivers? Exploring driving and crash histories of illegal street racing offenders. *Transportation Research Record: Journal of the Transportation Research Board, 2182*, 16–23.

Leary, M. R., & Kowalski, R. M. (1990). Impression management: A literature review and two-component model. *Psychological Bulletin, 107*(1), 34–47.

Leary, M. R., Tchividjian, L. R., & Kraxberger, B. E. (1994). Self-presentation can be hazardous for your health: Impression management and health risk. *Health Psychology, 13*, 461–470.

Lennon, A., Watson, B., Arlidge, C., & Fraine, G. (2011). "You're a bad driver but I just made a mistake": Attribution differences between the "victims" and "perpetrators" of scenario-based aggressive driving incidents. *Transportation Research Part F: Traffic Psychology and Behaviour, 14*, 209–221.

Lewis-Evans, B., & Lukkien, C. (2007). Crash profile of New Zealand novice drivers. In *Proceedings of the Australasian Road Safety, Research, Policing and Education Conference*, 17–19 October, Melbourne, Australia.

Li, C. S. R., Huang, C. Y., Lin, W. Y., & Sun, C. W. V. (2007). Gender differences in punishment and reward sensitivity in a sample of Taiwanese college students. *Personality and Individual Differences, 43*, 475–483.

Liourta, E., & van Empelen, P. (2008). The importance of self-regulatory and goal-conflicting processes in the avoidance of drunk driving among Greek young drivers. *Accident Analysis & Prevention, 40*, 1191–1199.

Lonczak, H. S., Neighbors, C., & Donovan, D. M. (2007). Predicting risky and angry driving as a function of gender. *Accident Analysis & Prevention, 39*, 536–545.

Marengo, D., Settanni, M., & Vidotto, G. (2012). Drivers' subtypes in a sample of Italian adolescents: Relationship between personality measures and driving behaviors. *Transportation Research Part F, 15*, 480–490.

Martiniuk, A. L. C., Ivers, R. Q., Glozier, N. et al. (2010). Does psychological distress increase the risk for motor vehicle crashes in young people? Findings from the DRIVE study. *Journal of Adolescent Health, 47*(5), 488–495.

Masten, S. V., Foss, R. D., & Marshall, S. W. (2013). Graduated driver licensing program component calibrations and their association with fatal crash involvement. *Accident Analysis & Prevention, 57*, 105–113.

Mayer, R. E., & Treat, J. R. (1977). Psychological, social and cognitive characteristics of high-risk drivers: A pilot study. *Accident Analysis & Prevention, 9*, 1–8.

McCartt, A. T., & Eichelberger, A. (2011). *Attitudes toward Red Light Camera Enforcement in Cities with Camera Programs.* Arlington, VA: Insurance Institute for Highway Safety.

McLinton, S. S., & Dollard, M. F. (2010). Work stress and driving anger in Japan. *Accident Analysis & Prevention, 42*, 174–181.

Mesken, J., Hagenzieker, M. P., Rothengatter, T., & de Waard, D. (2007). Frequency, determinants, and consequences of different drivers' emotions: An on-road study using self-reports, (observed) behaviour, and physiology. *Transportation Research Part F: Traffic Psychology and Behaviour, 10*, 458–475.

Milne, P. W. (1985). *Fitting and Wearing of Seat Belts in Australia. The History of a Successful Countermeasure.* Canberra: AGPS.

Møller, M. (2004). An explorative study of the relationship between lifestyle and driving behaviour among young drivers. *Accident Analysis & Prevention, 36*, 1081–1088.

Møller, M., & Gergersen, N. P. (2008). Psychosocial function of driving as predictor of risk-taking. *Accident Analysis & Prevention, 40*, 209–215.

Møller, M., & Haustein, S. (2013). Keep on cruising: Changes in lifestyle and driving style among male drivers between the age of 18 and 23. *Transportation Research. Part F: Traffic Psychology and Behaviour*, 20, 59–69.

Nemme, H. E., & White, K. M. (2010). Texting while driving: Psychosocial influences on young people's texting intentions and behaviour. *Accident Analysis & Prevention, 42*, 1257–1265.

Nesbit, S. M., & Conger, J. C. (2012). Predicting aggressive driving behavior from anger and negative cognitions. *Transportation Research Part F: Traffic Psychology and Behaviour, 15*, 710–718.

Nesbit, S. M., Conger, J. C., & Conger, A. J. (2007). A quantitative review of the relationship between anger and aggressive driving. *Aggression and Violent Behavior: A Review Journal, 12*, 156–176.

O'Brien, F., & Gormley, M. (2013). The contribution of inhibitory deficits to dangerous driving among young people. *Accident Analysis & Prevention, 51*, 238–242.

O'Connor, R. M., Colder, C. R., & Hawk, L. W. Jr. (2004). Confirmatory factor analysis of the sensitivity to punishment and sensitivity to reward questionnaire. *Personality and Individual Differences, 37*, 985–1002.

Ouellette, J. A., Gerrard, M., Gibbons, F. X., & Reis-Bergan, M. (1999). Parents, peers, and prototypes: Antecedents of adolescent alcohol expectancies, alcohol consumption, and alcohol-related life problems and rural youth. *Psychology of Addictive Behaviours, 13*, 183–197.

Palmgreen, P., Donohew, L., Lorch, E. P., Hoyle, R. H., & Stephenson, M. T. (2001). Television campaigns and adolescent marijuana use: Tests of sensation seeking targeting. *American Journal of Public Health, 91*, 292–295.

Paxton, R. J., Valois, R. F., Watkins, K. W., Huebner, E. S., & Drane, J. W. (2007). Associations between depressed mood and clusters of health risk behaviors. *American Journal of Health Behavior, 31*(3), 272–283.

Peer, E., & Rosenbloom, T. (2013). When two motivations race: The effects of time-saving bias and sensation-seeking on driving speed choices. *Accident Analysis & Prevention, 70*, 282–292.

Preston, P., & Goodfellow, M. (2006). Cohort comparisons: Social learning explanations for alcohol use among adolescents and older adults. *Addictive Behaviors, 31*, 2268–2283.

Queensland Government. (2015). *Hooning*. Retrieved at https://www.qld.gov.au/law/crime-and-police/types-of-crime/hooning/.

Reimer, B., Mehler, B., D'Ambrosio, L. A., & Fried, R. (2010). The impact of distractions on young adult drivers with attention deficit hyperactivity disorder (ADHD). *Accident Analysis & Prevention, 42*, 842–851.

Rhodes, N., & Pivik, K. (2011). Age and gender differences in risky driving: The roles of positive affect and risk perception. *Accident Analysis & Prevention, 43*(3), 923–931.

Rhodes, N., Brown, D., & Edison, A. (2005). Approaches to understanding young driver risk taking. *Journal of Safety Research—Traffic Records Forum Proceedings, 36*, 497–499.

Rimmo, P. A., & Aberg, L. (1999). On the distinction between violations and errors: Sensation seeking associations. *Transportation Research Part F, 2*, 151–166.

Roche-Cerasi, I., Rundmo, T., Sigurdson, J. F., & Moe, D. (2013). Transport mode preferences, risk perception and worry in a Norwegian urban population. *Accident Analysis & Prevention, 50*, 698–704.

Roidl, E., Siebert, F. W., Oehl, M., & Hoger, R. (2013). Introducing a multivariate model for predicting driving performance: The role of driving anger and personal characteristics. *Journal of Safety Research, 47*, 47–56.

Roidl, E., Frehse, B., & Hoger, R. (2014). Emotional states of drivers and the impact on speed, acceleration and traffic violations—A simulator study. *Accident Analysis & Prevention, 70*, 282–292.

Roseborough, J., & Wiesenthal, D. (2014). Roadway justice—Making angry drivers, happy drivers. *Transportation Research Part F, 24*, 1–7.

Rosenbloom, T., & Wultz, B. (2011). Thirty-day self-reported risky driving behaviours of ADHD and non-ADHD drivers. *Accident Analysis & Prevention, 43*, 128–133.

Rowden, P., Watson, B., & Biggs, B. (2006). The transfer of stress from daily hassles to the driving environment in a fleet sample. In *Proceedings of the Australasian Road Safety Research, Policing and Education Conference*, Gold Coast, Queensland.

Rowden, P., Matthews, G., Watson, B., & Biggs, H. (2011). The relative impact of work-related stress, life stress and driving environment stress on driving outcomes. *Accident Analysis & Prevention, 43*, 1332–1340.

Sargent, J. D., Tanski, S. Stoolmiller, M., & Hanewinkel, R. (2010). Using sensation seeking to target adolescents for substance use interventions. *Addiction, 105*, 506–514.

Sarma, K. M., Carey, R. N., Kervick, A. A., & Bimpeh, Y. (2013). Psychological factors associated with indices of risky, reckless and cautious driving in a national sample of drivers in the Republic of Ireland. *Accident Analysis & Prevention, 50*, 1226–1235.

Sava, F. A., & Sperneac, A. M. (2006). Sensitivity to reward and sensitivity to punishment rating scales: A validation study on the Romanian population. *Personality and Individual Differences, 41*, 1445–1456.

Schafer, J. A., & Mastrofski, S. D. (2005). Police leniency in traffic enforcement encounters: Exploratory findings from observations and interviews. *Journal of Criminal Justice, 33*, 225–238.

Schmidt-Daffy, M. (2013). Fear and anxiety while driving: Differential impact of task demands, speed and motivation. *Transportation Research Part F: Traffic Psychology and Behaviour, 16*, 14–28.

Schwebel, D. C., Severson, J., Ball, K. K., & Rizzo, M. (2006). Individual difference factors in risky driving: The roles of anger/hostility, conscientiousness, and sensation-seeking. *Accident Analysis & Prevention, 38*, 801–810.

Scott-Parker, B. (under review). "*You're so used to having someone tell you what to do*": Experiences of young drivers during the Provisional licence phase. *Transportation Research Part F.*

Scott-Parker, B., & Bates, L. (under review). "*...it just feels like you are a suspect for everything*": Investigating young drivers' perceptions regarding interactions with Police. *Transportation Research Part F.*

Scott-Parker, B., & Proffitt, C. (2015). Validation of the behaviour of young novice drivers scale in a New Zealand young driver population. *Accident Analysis & Prevention, 77*, 62–71.

Scott-Parker, B., & Senserrick, T. (2013). Methodology and broader implications of young driver research published in traffic injury prevention in the past five years. *Australasian College of Road Safety Conference*. Adelaide, October.

Scott-Parker, B., & Senserrick, T. (under review). Brief report: Improving young driver research methodology and reporting: A review of selected 2006–2013 publications. *Injury*.

Scott-Parker, B., & Weston, L. (submitted 23 December 2014). Sensitivity to reward and risky driving, risky decision making, and risky health behaviour: A literature review. *Transportation Research Part F.*

Scott-Parker, B., Watson, B., & King, M. J. (2009a). Understanding the psychosocial factors influencing the risky behaviour of young drivers. *Transportation Research Part F: Traffic Psychology and Behaviour, 12*, 470–482.

Scott-Parker, B., Watson, B., & King, M. J. (2009b). Exploring how parents and peers influence the behaviour of young drivers. *2009 Australasian Road Safety Research, Policing and Education Conference*, 10–12 November, Sydney, New South Wales, Australia.

Scott-Parker, B., Watson, B., & King, M. J. (2010). The risky behaviour of young drivers: Developing a measurement tool. *Proceedings of the 20th Canadian Multidisciplinary Road Safety Conference, Niagara Falls, Canada, June 6–9, 2010.*

Scott-Parker, B., Bates, L., Watson, B., King, M. J., & Hyde, M. K. (2011a). The impact of changes to the graduated driver licensing program in Queensland, Australia on the experiences of Learner drivers. *Accident Analysis & Prevention, 43*(4), 1301–1308.

Scott-Parker, B., Watson, B., King, M. J., & Hyde, M. K. (2011b). Mileage, car ownership, experience of punishment avoidance and the risky driving of young drivers. *Traffic Injury Prevention, 12*(6), 559–567.

Scott-Parker, B., Watson, B., King, M. J., & Hyde, M. K. (2012a). The influence of sensitivity to reward and punishment, propensity for sensation seeking, depression and anxiety on the risk behaviour of novice drivers: A path model. *British Journal of Psychology, 103*(2), 248–267.

Scott-Parker, B., Watson, B., King, M. J., & Hyde, M. K. (2012b). "They're lunatics on the road": Exploring the normative influences of parents, friends, and police on young novice's risky driving decisions. *Safety Science, 50*, 1917–1928.

Scott-Parker, B., Watson, B., King, M. J., & Hyde, M. K. (2013a). A further exploration of sensation seeking propensity, reward sensitivity, depression, anxiety and the risky behaviour of young novice drivers in a structural model. *Accident Analysis & Prevention, 50*, 465–471.

Scott-Parker, B., Watson, B., King, M. J., & Hyde, M. K. (2013b). Revisiting the concept of the "problem young driver" within the context of the "young driver problem." *Accident Analysis & Prevention, 59*, 144–152.

Scott-Parker, B., Hyde, M. K., Watson, B., & King, M. J. (2013c). Speeding by young novice drivers: What can personal characteristics and psychosocial theory add to our understanding? *Accident Analysis & Prevention, 50*, 242–250.

Scott-Parker, B., Watson, B., King, M. J., & Hyde, M. K. (2013d). Punishment avoidance and intentional risky driving behaviour: What are the implications for 'getting away with it'? In N. Castro (Ed.), *Psychology of Punishment: New Research* (pp. 55–77). New York: Nova.

Scott-Parker, B., King, M., & Watson, B. (2015). The psychosocial purpose of driving and its relationship with the risky driving behaviour of young novice drivers. *Transportation Research Part F: Traffic Psychology and Behaviour, 33*, 16–26.

Senserrick, T., Ivers, R., Boufous, S., Chen, H. Y., Norton, R., Stevenson, M., van Beurden, E., & Zask, A. (2009). Young driver education programs that build resilience have the potential to reduce road crashes. *Pediatrics, 124*, 1287–1292.

Sharpley, D. (2003). Driver behaviour and the wider social context. In L. Dorn (Ed.), *Driver Behaviour and Training* (pp. 381–387). Burlington, VT: Ashgate.

Sheehan, M., Siskind, V., & Greenslade, J. (2002). Social and psychological predictors of young people's involvement in fatal and serious injury crashes. In *Proceedings 2002 Australasian Road Safety, Research, Policing and Education Conference*, 4–5 November, Adelaide, South Australia.

Shinar, D., & Compton, R. (2004). Aggressive driving: An observational study of driver, vehicle, and situational variables. *Accident Analysis and Prevention, 36*, 429–437.

Siegrist, S. (Ed.). (1999). *Driver Training, Testing and Licensing—Towards Theory-Based Management of Young Drivers' Injury Risk in Road Traffic. Results of EU-Project GADGET, work package 3*. Bern: BFU.

Sigelman, C. K. (1999). *Life-Span Human Development* (3rd ed.). Melbourne: Brooks/Cole.

Smart, D., Vassallo, S., Sanson, A., Cockfield, S., Harris, A., Harrison, W., & McIntyre, A. (2005). *In the Drivers' Seat: Understanding Young Adults' Driving Behaviour. Research Report No. 12 2005*. Canberra: Australian Institute of Family Studies.

Soole, D., Lennon, A., Watson, B., & Bingham, R. C. (2011). Towards a comprehensive model of driver aggression: A review of the literature and directions for the future. In Ferraro, C. N. (Ed.), *Traffic Safety*. New York: Nova Science Publishers.

Sprinthall, N. A., & Collins, W. A. (1995). *Adolescent Psychology. A Developmental View* (3rd ed.). Sydney: McGraw-Hill.

Stephens, A. N., & Groeger, J. A. (2009). Situational specificity of trait influences on drivers' evaluations and driving behaviour. *Transportation Research Part F: Traffic Psychology and Behaviour, 12*, 29–39.

Strine, T. W., Mokdad, A. H., Balluz, L. S., Gonzalez, O., Crider, R., Berry, J. T., & Kroenke, K. (2008). Depression and anxiety in the United States: Findings from the 2006 Behavioral Risk Factor Surveillance System. *Psychiatric Services, 59*(12), 1383–1390.

Suhr, K. A., & Nesbit, S. M. (2013). Dwelling on "road rage": The effects of trait rumination on aggressive driving. *Transportation Research Part F: Traffic Psychology and Behavior, 21*, 207–218.

Sullman, M. J. M. (2006). Anger amongst New Zealand drivers. *Transportation Research Part F, 9*, 173–184.

Taubman-Ben-Ari, O., & Katz-Ben-Ami, L. (2012). The contribution of family climate for road safety and social environment to the reported driving behaviour of young drivers. *Accident Analysis & Prevention, 47*, 1–10.

Taubman-Ben-Ari, O., & Yehiel, D. (2012). Driving styles and their associations with personality and motivation. *Accident Analysis & Prevention, 45*, 416–422.

Torrubia, R., Avila, C., Molto, J., & Caseras, X. (2001). The Sensitivity to Punishment and Sensitivity to Reward Questionnaire (SPSRQ) as a measure of Gray's anxiety and impulsivity dimensions. *Personality and Individual Differences, 31*, 837–862.

Travis, L. E. (2005). *Introduction to Criminal Justice* (5th ed.). Cincinnati, OH: Matthew Bender & Co.

Trogolo, M. A., Melchior, F., & Medrano, L. A. (2014). The role of difficulties in emotion regulation on driving behaviour. *Journal of Behavior, Health and Social Issues, 6*, 107–117.

Ulleberg, P. (2002). Personality subtypes of young drivers. Relationship to risk-taking preferences, accident involvement, and response to a traffic safety campaign. *Transportation Research Part F, 4*, 279–297.

Ulleberg, P. (2004). Social influence from the back seat: Factors related to adolescent passengers' willingness to address unsafe drivers. *Transportation Research Part F, 7*, 17–30.

Vassallo, S., Smart, D., Sanson, A., Cockfield, S., Harris, A., McIntyre, A., & Harrison, W. (2008). Risky driving among young Australian drivers II: Co-occurrence with other problem behaviours. *Accident Analysis & Prevention, 40*, 376–386.

Villieux, A., & Delhomme, P. (2010). Driving anger and its expressions: Further evidence of validity and reliability for the Driving Anger Expression Inventory French adaptation. *Journal of Safety Research, 41*, 417–422.

Vingilis, E., Seeley, J., Wiesenthal, D. L., Wickens, C. M., Fischer, P., & Mann, R. E. (2013). Street racing video games and risk-taking driving: An Internet survey of automobile enthusiasts. *Accident Analysis & Prevention, 50*, 1–7.

Voas, R. B., Tippetts, A. S., & Fell, J. C. (2003). Assessing the effectiveness of minimum legal drinking age and zero tolerance laws in the United States. *Accident Analysis & Prevention, 35*, 579–587.

Waylen, A. E., & McKenna, F. P. (2008). Risky attitudes towards road use in pre-drivers. *Accident Analysis & Prevention, 40*, 905–911.

Whiteside, S. P., & Lynam, D. R. (2001). The five factor model and impulsivity: Using a structural model of personality to understand impulsivity. *Personality and Individual Differences, 30*, 669–689.

Wickens, C. M., Roseborough, J. E. W., Hall, A., & Wiesenthal, D. L. (2013a). Anger-provoking events in driving diaries: A content analysis. *Transportation Research Part F: Traffic Psychology and Behaviour, 19*, 108–120.

Wickens, C. M., Wiesenthal, D., Hall, R., & Roseborough, J. (2013b). Driver anger on the information superhighway: A content analysis of online complaints of offensive driver behaviour. *Accident Analysis & Prevention, 51*, 84–92.

Williams, A. F. (2006). Young driver risk factors: Successful and unsuccessful approaches for dealing with them and an agenda for the future. *Injury Prevention, 12*, i4–i8.

Williams, A. F., Ferguson, S. A., & McCartt, A. T. (2007). Passenger effects on teenage driving and opportunities for reducing the risks of such travel. *Journal of Safety Research, 38*, 381–390.

Winfree Jr., L. T., & Bernat, F. P. (1998). Social learning, self-control, and substance abuse by eighth grade students: A tale of two cities. *Journal of Drug Issues, 28*(2), 539–558.

Wong, I. Y., Mahar, D., & Titchener, K. (2015). Driven by distraction: Investigating the effects of anxiety on driving performance using the attentional control theory. *Journal of Risk Research, 18*(10), 1293–1306.

Wundersitz, L., Doecke, S.D., & Baldock, M. R. J. (2010). *Annual performance indicators of enforced driver behaviours in South Australia, 2008* (CASR073), Centre for Automotive Safety Research, Adelaide.

Yannis, G., Papadimitriou, E., & Antoniou, C. (2007). Impact of enforcement on traffic accidents and fatalities: A multivariate multilevel analysis. *Safety Science, 46*(5), 738–750.

Zuckerman, M. (1994). *Behavioural Expressions and Biosocial Bases of Sensation Seeking.* New York: Cambridge University Press.

Zuckerman, M. (2007). *Sensation Seeking and Risky Behavior.* Washington, DC: American Psychological Association.

Zuckerman, M., & Kuhlman, D. M. (2000). Personality and risk-taking: Common biosocial factors. *Journal of Personality, 68*(6), 999–1029.

Zuckerman, M., Kuhlman, D. M., Joireman, J., Teta, P., & Kraft, M. (1993). A comparison of three structural models for personality: The big three, the big five, and the alternative five. *Journal of Personality and Social Psychology, 65*(4), 757–768.

11

Speeding and Other Risky Driving Behavior among Young Drivers

Brian A. Jonah

Paul Boase

Abstract

This chapter reviews research on risky driving among young drivers aged 16–24 that has been conducted mostly during the last 5 years. The main focus in the chapter is on speeding among young drivers, but other risky behaviors are examined as well (e.g., alcohol- and drug-impaired driving, distracted driving). Young drivers are more likely to be involved in fatal and injury collisions compared to older drivers. There is also considerable evidence that young drivers are more likely than older drivers to be involved in speed-related fatal and injury collisions. In observational surveys, young drivers are more often found to be drinking and driving, using drugs and driving, and driving while distracted. They are also more likely to indicate in self-report surveys that they speed as well as engage in other risky behavior such as alcohol- and drug-impaired driving and distracted driving, particularly the use of electronic communication devices. Many risky driving behaviors have been found to be correlated, thereby supporting the concept of a risky driving syndrome. Effective interventions to reduce speeding include legislation (e.g., antiracing laws), automated enforcement (e.g., speed cameras, point-to-point speed measurement), and the use of incentives and feedback.

11.1 Introduction

It has been known for decades that young drivers (16–24 years of age) take greater risks while driving and that this behavior may be due to inexperience (i.e., lack of knowledge, poor danger perception, poor driving skills) or lack of maturity (i.e., thrill seeking, belief in immortality, reacting to peer pressure) or both. There have been a number of reviews of the literature on risky driving by youth in the past, including those by Jonah (1986), the Organization for Economic Cooperation and Development (OECD, 2006), and the World Health Organization (Toroyan and Peden, 2007). More recently, Palarma et al. (2012) reviewed the literature and determined that young drivers are more likely to speed, not use seat belts, drive while distracted or fatigued, drive after drinking, and drive without an appropriate license and that the young people who engage in these risky driving behaviors also engage in other risky behaviors such as binge drinking, drug use, unsafe sexual activity, etc. These reviews have clearly demonstrated that drivers aged 16–24 have a greater risk of being killed in a collision per distance traveled than any other age group and that this higher risk of death is in large part due to their risky driving (e.g., speeding, failure to use seat belts, alcohol-impaired driving, etc.).

The purpose of this chapter is to review some of the recent research literature (generally over the last 5 years) on risky driving by youth (alcohol- and drug-impaired driving, distracted driving), with an emphasis on speeding, to present some of the data on age differences in risky driving in fatal and serious-injury collisions, and then to present some recent self-reported risky driving data based on a national public opinion survey on road safety conducted in Canada. Young drivers are defined as those aged 16–24. Where available, differences between younger (16–20) and older youth (21–24) are shown. This chapter also reviews effective interventions that have been implemented to reduce speeding.

11.2 Age Differences in Collision Involvement in Canada

Analyses of Canada's National Collision Database (NCDB) for the years 2007–2012 were conducted for several age groups for fatal crashes involving light-duty vehicles (i.e., passenger cars, vans, SUVs, pickup trucks). While drivers aged 16–20 represented 6.3% of all licensed drivers, they accounted for 13.4% of the fatalities (i.e., relative risk ratio [RRR] of 2.1). Drivers aged 21–24 represented 6.9% of licensed drivers, but they accounted for 10.9% of the fatalities (i.e., RRR of 1.6). Drivers aged 25–44 represented 35.4% of licensed drivers and 34.1% of fatalities (RRR of 0.9), and older drivers (45+) were underrepresented (RRR of 0.8) in fatalities compared to the number of licensed drivers in this group.

Similar analyses were performed on injury data from NCDB. Drivers aged 16–20 accounted for 12.1% of injury collisions (RRR of 1.9), and those aged 21–24 accounted for 10.3% (RRR of 1.5). Drivers aged 25–44 accounted for 39.3% of injury collisions (RRR of 1.1). Older drivers (45+) were underrepresented in injury collisions (RRR of 0.7).

These collision data indicate that Canadian drivers aged 16–24 are overrepresented in fatal and injury collisions relative to the number of licensed drivers in each age group (see Chapter 5 on novice driver crash patterns).

11.3 Speeding

11.3.1 Prevalence of Speeding

A review of research on speeding behavior has been conducted by the OECD (2007), and a manual on speed management has been prepared by the Global Road Safety Partnership (2008). There have been several recent studies on speeding among young drivers using collision data or survey results.

11.3.1.1 Collision Data

Analysis of Canadian fatality data from the NCDB for the years 2007–2012 indicates that speed was a contributing factor to fatal collisions more often when the drivers involved were younger. Police identified speed as a factor in 21% of fatal collisions involving drivers aged 16–20 and 17% of those involving drivers aged 21–24. The percentages for older drivers declined from 13% for those aged 25–44 to 5% for those 65 and older. Similarly, speed was identified as a contributing factor in injury collisions more often when a driver was 16–20 years old (10%) or 21–24 (8%) compared to older drivers, with percentages declining from 5% for drivers aged 25–44 to 3% for those 65 years of age or older.

The US National Highway Traffic Safety Administration (NHTSA, 2014a) issued a Traffic Safety Facts sheet based on 2011 Fatality Analysis Reporting System (FARS) data that showed that 37% of male drivers aged 15–20 and 37% of male drivers aged 21–24 involved in fatal collisions were considered to have been speeding (i.e., racing, driving too fast for conditions, or exceeding the speed limit) at the time of the crash. These percentages were higher than those for older males, which decreased from 31% for drivers aged 25–34 to 8% for drivers 75 and over. Similarly, for female drivers involved in fatal crashes, 24% of drivers aged 15–20 and 19% of drivers aged 21–24 had been speeding, higher than that for older drivers, which decreased from 17% for drivers aged 25–34 to 7% for drivers 75 and over.

Williams (2013) analyzed data from FARS for the years 2000–2011, and Ferguson (2013) analyzed data for the years 2007–2011 and found the same results as those in the NHTSA fact sheet. Williams reported that the number of collisions involving drivers aged 16–19 that were considered to be speeding-related declined by about 50% during that time period. Ferguson also noted that drivers aged 16–19 were much more likely to be involved in single-vehicle collisions and that speeding was much more likely a contributing factor to these types of collisions. Speed-related fatal collisions involving drivers aged 16–19 were also more likely if drivers were accompanied by teenage passengers compared to when they were driving alone.

Swedler et al. (2012) examined FARS data on collisions occurring during 2007–2009 that involved teen drivers and noted that the percentage of females who were speeding dropped from 38% to 22% between 15 and 19 years of age, but for males, the percentage remained stable at 38%. These results suggest that young women mature earlier than young men regarding speeding behavior.

Lambert-Belanger et al. (2012) analyzed FARS data for the period 1991–2008 and noted that drivers aged 16 were more likely to have been speeding and engaging in other aggressive driving if there were passengers present in the vehicle compared to older drivers (odds ratios increased from 1.14 to 1.95 as the number of passengers increased from one to three).

Curry et al. (2014) examined collision data in New Jersey to determine the role of several driver actions where a moving violation was issued by the police. They found that drivers 21 and older were more likely to have been cited for the action of unsafe speed as a result of a collision than drivers under 21 and that this finding was observed for both males and females. While this finding appears to contradict those of NHTSA, Williams, Ferguson, and the Canadian data noted previously, it is based on total collisions, while the other studies are based on fatal and serious-injury collisions, and it used only data from New Jersey rather than national data. It is also noteworthy that among drivers under 21 who were cited for unsafe speed, about 60% either were unlicensed or had a suspended license.

11.3.1.2 Survey Data

Beck et al. (2013) examined the relationship between hurried driving ("I get in a hurry when I drive") and a number of other variables using a web-based survey conducted in the Washington, DC, area with a group of university undergraduate students aged mostly between 18 and 22. They found that as age increased, reported hurried driving decreased. However, once variables such as aggressive driving, driver anger, risky driving (e.g., cell phone use, driving after few drinks, etc.) were included in the regression analysis, age was no longer a significant predictor, suggesting that hurried driving covaries with other risky driving behaviors.

Hassan and Abdel-Aty (2013) conducted a survey with 680 novice drivers in Florida and noted that those who were involved in at-fault crashes or violations when they were aged 16–17 were more likely to be involved in such collisions when they were 18–24. Moreover, positive attitudes toward speeding, self-reported speeding, and in-vehicle distractions also predicted at-fault collisions among the 18- to 24-year-old drivers.

Scott-Parker et al. (2013) asked 1170 novice drivers aged 17–25 in Queensland, Australia, to indicate how often they engaged in speeding using the speeding subscale from the Behaviour of Young Novice Drivers Scale (Scott-Parker et al., 2010). While simple correlations indicated that speeding during the learning phase or the provisional phase of graduated licensing decreased with age, hierarchical regression analyses indicated that age was no longer significantly related to speeding once other factors such as sensation seeking, depression, and gender were included in the model. However, separate regression analyses for males and females revealed that younger females were more likely to report speeding than older females, but for males, age was not a significant variable. These results suggest that females mature out of speeding behavior but males do not, at least up to age 25.

A French study examined factors influencing 3002 young (18–25) drivers' intentions to drive more than 110 kph on roads where the limit is 90 kph (Cestac et al., 2011). Using the theory of planned behavior (TPB) (Ajzen, 1991), they examined the role of attitudes, norms, and perceived control in speeding intentions. Drivers were asked to imagine that they were driving in various situations described in written scenarios. It was found that mens' intentions to speed were influenced by injunctive norms ("People in your circle think you should drive at 110 kph in this situation"), while women's intentions were influenced more by self-description variables (i.e., whether they believe that they are show-offs/reckless/dangerous). Sensation seeking had a greater influence on the intentions of males and those of novice (i.e., younger) drivers. Also, the intention to speed was greater among more experienced drivers.

Moller and Haustein (2014) conducted a survey among a sample of 4000 Danish male drivers drawn from driver licensing records and then compared self-reported driving behavior of those who were aged 18 with an average of 6–12 months of driving experience and those who were 28 years old with an average of about 11 years of experience. The results showed that the older drivers were more likely to report exceeding the speed limits in urban and rural areas than the younger drivers. Interestingly, the younger drivers thought that their friends were more likely to exceed the limit than did the older drivers for both urban and rural driving. Older drivers had more negative attitudes toward measures to reduce speeding (e.g., tougher penalties, higher limits) than younger drivers. While the results of this study appear inconsistent with those of Scott-Parker et al. (2013), this study looked only at males, while Scott-Parker included both males and females, suggesting that age differences should be examined separately for men and women.

Horvath et al. (2012) used the TPB to look at intention to speed and determined that among 265 female Australians aged 17–24, high intenders to speed were more likely to perceive advantages of speeding, less likely to perceive disadvantages, and more likely to be encouraged to speed on familiar and inappropriately signed roads than low intenders. Among 133 young males in this age group, high intenders were more likely to speed on familiar and inappropriately signed roads, have greater perceptions of support from their friends to speed, and perceive speeding to be more exciting than low intenders. Males were not influenced by advantages and disadvantages of speeding, and females were not influenced by support from their friends. Interestingly, intention to speed was higher if they were driving without passengers than with them. Age was not a predictor of intention to speed once these other factors were entered into regression equations.

Leandro (2012) had 210 Costa Rican drivers aged 18–30 complete a questionnaire about speed selection based on TPB while they were watching a video that depicted a real driving situation involving speeding. The drivers were asked to select the speed that they would drive in each situation. The results indicated that perceived norms and behavioral control best explained the intention to decrease speed and actual speed selection in various situations. However, in only one of the six driving scenarios where they were supposed to drive like other drivers do, were younger drivers more likely to select higher speeds than older drivers.

Jonah (2015) asked a sample of 3888 licensed Canadian drivers how often they drove 10 kph or more over the limit on the highway. The results of the survey indicated that 84% of drivers aged 16–20 admitted to speeding at least some of the time, increasing to 90% for those aged 21–24 and then dropping off as age increased. A regression analysis showed that the major predictors of speeding on the highway were how dangerous they thought speeding was, higher income, younger drivers, and traveling greater distances. Together, these variables accounted for 23% of the variance in reported speeding.

11.3.1.3 Naturalistic Driving Studies

Richard et al. (2012) conducted a study that examined the speeding behavior of drivers in their own vehicles during 3–4 weeks of naturalistic driving in urban (Seattle, Washington) and rural (College Station, Texas) settings. Data collected from 164 drivers included recordings of vehicle position and speed using a GPS receiver. A general pattern in the logistic regression models showed that young males and females aged 18–25 in Seattle were more likely than older drivers (35–55) to engage in speeding during a trip. This pattern was evident with young males in both locations, but it appeared that young females exhibited qualitatively different speeding behavior in Seattle than in Texas (i.e., young females drove more like young males in Seattle but more like older females in Texas). However, when other predictors were entered into the regression equation, such as reckless driving (i.e., frequently committing dangerous violations such as racing, driving drunk, and tailgating) and road rage (i.e., expressing hostile or angry actions toward other drivers), age no longer predicted speeding. This study supports the idea that speeding is related to other types of risky driving.

11.3.1.4 Summary

Taken together, these studies indicate that young drivers aged 16–24 are more likely than older drivers to be involved in speed-related fatal and injury collisions, be observed to be speeding in a naturalistic driving study, and report speeding in surveys. There is also some evidence indicating that young drivers aged 16–20 are more likely to be involved in speed-related fatal and injury collisions than those aged 21–24. When behavioral intentions or self-reported behavior are considered, the results are more mixed. It may be that young drivers aged 16–20 are less than truthful about reporting their speeding intentions or behavior compared to older drivers. When other risky driving behaviors are included in regression models, age does not remain a significant predictor of speeding, suggesting a positive correlation among risky driving behaviors. There is also some evidence that female drivers may mature out of speeding behavior earlier than males and that the presence of young passengers in the vehicle may encourage speeding by young drivers, particularly young male passengers.

11.3.2 Reducing Speeding

It has been argued that if speeds could be reduced by 2–5 kph, then there could be reductions in collision injuries in the range of 10%–30% (Molin and Brookhuis, 2007). There have been a number of measures adopted to try to reduce speeding by drivers in general and young drivers in particular, including legislation, speed limit enforcement, use of incentives, education/training/awareness programs, and the use of feedback technology.

11.3.2.1 Legislation

The Ministry of Transportation of Ontario introduced legislation in 2007 to reduce street racing, which was defined as driving at speeds exceeding the speed limit by 50 kph (Street Racers, Stunt and Aggressive Drivers Legislation). Given that young drivers are more likely to engage in this type of risky driving, this legislation was particularly relevant to them. For example, Vingilis et al. (2011) reported that about 20% of grade 11–12 high school students reported street racing in the last 12 months. Penalties under the legislation include an immediate 7-day roadside vehicle impoundment and driver's license suspension, prior to conviction. Upon conviction, the driver could receive a fine ranging from $2000 to $10,000, 6 demerit points, possible imprisonment for up to 6 months, and a license suspension up to 2 years for the first conviction. If there is a second conviction within 10 years of first conviction, there could be a license suspension up to 10 years.

An evaluation of the impact of this racing legislation discovered that there was a significant decrease in the observed average driving speed at three test sites in Ontario from before to after the implementation of the law (Meirambayeva et al., 2014). There were also significant reductions in speed-related injury collisions from before to after the law went into effect, but these were significant only for young (16–24) males and not for other age/gender groups. The authors estimate that the legislation resulted in a reduction of 58 injuries per month among the young male driver group, or about 700 per year. Similar legislation in other jurisdictions could have similar effects on speed-related collisions involving young drivers.

As noted in Chapter 17, all but one Canadian jurisdiction and many others around the world (e.g., United States, Australia) have implemented graduated driver licensing (GDL) programs for novice drivers, who are generally young. The rationale for these GDL programs is that it is safer to ease novices into the driving environment gradually, allowing them to gain driving experience under lower-risk conditions. In order to graduate from the novice phase to the intermediate phase, novices must keep a clean driving record. If they are caught for speeding or other traffic offenses, they will receive demerit points and as a result may not be allowed to move to the intermediate phase, which has fewer restrictions. The desire to graduate from the novice to the intermediate phase may serve as an incentive to the novice to drive within the speed limit. Furthermore, while in the novice phase, novices must have an experienced driver with them, often a parent, which may also prevent speeding. However, in some jurisdictions (e.g., Quebec), you can age out of GDL so that the restrictions no longer apply. Completing GDL may not serve as an incentive to drive safely in these jurisdictions.

GDL systems in Ontario and other jurisdictions have been shown to be effective in reducing the involvement of novice young drivers in collisions (Hedlund and Compton, 2004; Williams et al., 2012). Williams et al. note that GDL appears to be most effective with 16- and 17-year olds and less so with drivers aged 18 and 19, where the evidence is conflicting. This research suggests that passenger and nighttime restrictions are effective for young GDL drivers, although the passenger effect appears complex. However, they also note that there is some evidence that young drivers may delay getting a license to avoid going through GDL in the United States. It is not known to what extent GDL has actually resulted specifically in reduced speeding among young drivers in the GDL programs.

A study by Keall and Newstead (2013) suggests that GDL restrictions should be extended to the operation of high-performance vehicles. Using Australian and New Zealand crash data, they reported that high-performance vehicle owners under the age of 25 were 69% more likely than older owners (aged 40–59) of such vehicles to have been involved in collisions and twice as likely to have been in injury collisions. The authors estimated a potential 0.4%–2.5% reduction in injury rates if such a vehicle restriction for young owners were implemented. This study was based on owners of high-performance vehicles, but if such a restriction were applied to the operation of any high-performance vehicle by young drivers, whether owned by them or their parents, the benefits may be even greater. Given that these vehicles can accelerate quickly and attain high speeds, such a restriction may result in less speeding by young drivers.

11.3.2.2 Enforcement

Police officers have enforced speed limits for decades using radar devices and, more recently, lidar. While this method of enforcement is effective, it does require significant expenditure of time by officers in order to detect speeders. There are automated enforcement techniques that are available today that are being used in some communities.

Speed or safety cameras have been found to be effective in reducing speeding and traffic casualties in Europe (e.g., France, Sweden, United Kingdom) as well as in Australia and Canada. Wilson et al. (2006) conducted a systematic review of speed cameras for the prevention of road traffic injuries and deaths by analyzing the results from 35 studies that met their inclusion criteria. They found reductions in average speeds ranging from 1% to 15% and reductions in the proportion of vehicles speeding ranging from 14% to 65%. In addition, they reported that in the vicinity of camera sites, pre–post reductions ranged from 8% to 49% for all crashes and 11% to 44% for fatal and serious-injury crashes. The authors concluded that "the consistency of reported reductions in speed and crash outcomes across all studies show that speed cameras are a worthwhile intervention for reducing the number of road traffic injuries and deaths" (Wilson et al., 2006, p. 2).

The UK Department for Transport conducted an evaluation of the impact of speed cameras in 38 areas that were employing cameras as part of a national program between April 2000 and March 2004 (PA Consulting Group, 2005). Surveys showed that average speeds dropped at speed camera sites by 6%, and there was a 31% reduction in drivers breaking the speed limit. Fatalities and serious injuries were 42% lower at the camera sites, resulting in a benefit–cost ratio for speed cameras of 2.7:1. There was also evidence suggesting that speeds were reduced not only where the cameras were located but on other roads in the surrounding area as well.

A pilot study in three regions of Quebec (Montreal, Montérégie and Chaudière-Appalaches) in 2009–2010 evaluated the effects of fixed and mobile speed cameras (Ministère des Transports du Québec, 2013) over a 9-month period. With the speed cameras present, average speeds dropped by 9–12 kph depending on the site. Furthermore, the percentage of drivers over the speed limit declined by 63%. At sites with fixed cameras, total collisions were reduced from 10% to 16% depending on the specific control group used, but injury collisions actually increased from 3% to 9%. However, if only the last 6 months of the time when the cameras were in operation were used, the reductions in collisions ranged from 25% to 29%, and the reductions in injury collisions ranged from 27% to 32%, suggesting that it took about 3 months for the fixed cameras to have an effect. For mobile cameras, there were reductions from 4% to 10% in total collisions over 9 months and reductions of 25%–30% in injury collisions. Again, looking at just the last 6 months, the reductions were greater. While not all the reductions in collisions were statistically significant at the .05 level, the results taken together indicate the effectiveness of the speed cameras.

A recent evaluation of Winnipeg's photoenforcement program for speeding on green lights and running red lights found mixed results (Vanlaar et al., 2011). The researchers looked at the impact of photoenforcement at 48 intersections over the period 1994–2008 on collisions using time series analyses. The authors concluded, "Regarding the results from the time series analyses of speeding-related crashes, when only considering the strongest evidence (significant at 5% level), there were no increases or decreases in injury crashes or property damage only (PDO) crashes, not at the camera intersections and not at other intersections in Winnipeg. When considering all the evidence (i.e., including effects significant at the 10% level), it can be concluded that there was a 24% decrease in injury crashes at camera intersections, a 13% decrease in PDO crashes at camera intersections and a 2% increase in PDO crashes at other intersections without cameras."

These authors noted that these cameras may not be effective in reducing high levels of speeding through these intersections given that this behavior is more common among high-risk drivers, who are less affected by such measures. This raises the question about whether photoenforcement would affect the speeding behavior of young drivers, a question not yet addressed by research. This study also suggests that photoenforcement may not be as effective in reducing speeding through intersections compared to speeding elsewhere. The positive effects of photoenforcement on speeding may be site specific.

Soole et al. (2013) conducted a review of the literature on the effectiveness of average-speed (point-to-point) enforcement programs in reducing speeds and speed-related collisions. These enforcement programs are typically used on controlled-access highways, where the vehicle's license plate is scanned at entry and departure from the highway. Knowing the time of entry and departure and the distance between the two points, the average speed to travel on this stretch of highway can be determined, and a traffic violation ticket can be sent to the owner of the speeding vehicle. Having reviewed about 25 empirical studies, most conducted in the United Kingdom, the authors concluded that these average-speed enforcement programs have significantly reduced speeds, particularly excessive speeds, and fatal and serious crashes.

While not specifically targeted at young drivers, these automated speed enforcement programs may reduce speeding among this group as well as reduce speed-related collisions. It would be useful if the impact of automated enforcement on the speed of drivers of different ages were determined. While such automated enforcement techniques are effective in deterring speeding, it is recognized that these devices have not always been politically acceptable.

11.3.2.3 Training, Awareness, and Education Programs

It seems logical that driver education should instill good driving habits within young drivers that should result in less risky driving, including speeding, and consequently fewer collisions. Unfortunately, the evidence does not support this belief, since drivers who take basic driver education do not appear to be any safer than those who do not (Mayhew and Simpson, 2002; Senserrick and Haworth, 2005; Elvik et al., 2009). Indeed, it appears that driver education programs may get novice drivers on the road sooner given that novices who take driver education can move to the intermediate phase of GDL earlier in some jurisdictions, which has fewer restrictions and hence increases their exposure to risk. However, more advanced driver training and education programs have been developed.

Prabhakharan and Molesworth (2011) evaluated a training program where young drivers under 25 were actively involved in a simulated driving session where they received personalized feedback to reduce their speeding. Drivers in control groups read three collision reports involving speeding or engaged in a card sorting exercise. Results showed that the drivers receiving the personalized feedback on their driving exhibited less speeding in a subsequent driving test session.

It has been noted that young male drivers from economically and socially deprived families are more likely to be involved in collisions than other drivers. For example, Clarke et al. (2008) examined 900 collisions in the United Kingdom and determined that they were more likely to involve excessive speed, driver impairment, failure to wear seat belts, and unlicensed driving among those drivers who were from socially and economically deprived areas based on postal codes. This relationship led a group of British researchers (Tapp et al., 2013) to develop a different type of advanced driving training program. An in-vehicle data recorder (IVDR) was used to provide feedback to young drivers but also allowed the researchers to monitor "incidents," defined as aggressive braking and cornering at an unsafe speed, both indicative of high-risk driving. In addition, professional driving instructors from the Institute of Advanced Motorists took the students for driving tests and provided coaching regarding their skills and incentives for engagement in the training program (e.g., opportunity to drive go-carts on track). The program started with 42 young participants, but only 23 completed the program, illustrating one of the problems with addressing this group of drivers. However, the results demonstrated that there was a significant decline in the frequency of risky incidents as measured by the IVDR both during the program and 4 months after the students completed it.

Tapp et al. (2013) also noted that the students performed better in the drive checks conducted by the professional instructors, with scores improving from an average of 34 to 21 (lower number represents better driving). Furthermore, there were significant improvements on the Driver Behaviour Questionnaire (Parker et al., 1995) regarding self-reported tailgating, braking, and racing other drivers. While this approach to driver training for young drivers from socially and economically deprived communities holds some promise, this study had issues regarding student drop-out rate and a lack of a control group that did not take the training, which detract from the conclusion that this program was effective. Furthermore, it is not known what the longer-term impact of such programs may be on speeding. Nevertheless, given that this target group is overinvolved in serious collisions, further work on such programs should be pursued.

Paaver et al. (2013) evaluated the effectiveness of an intervention targeted at Estonian young drivers ($N = 1899$) that focused on encouraging them to acknowledge their personal risk factors, particularly impulsiveness. Based on a rationale that teaching behavioral techniques for controlling risky behavior is more effective in reducing collisions than changing attitudes, these researchers addressed the issue of impulsivity. A group of novice drivers who were taking driver education courses were allocated to either the intervention condition or the control condition. The intervention was embedded in the standard driver education course. The control group received only the standard course. The intervention comprised a 45-minute lecture on impulsivity as a personality trait and information processing style that may be biological in nature and how it may lead to risky driving, how situational factors may trigger impulsive behavior, how to recognize their impulsive tendencies, and how to regulate them. This lecture was followed by a group exercise where the students identified psychological factors involved in collisions and discussed their own risk of being involved in such collisions and how they might reduce their risk. The researchers followed the driving records of the students in the intervention and control groups for a period of 17 months using databases maintained by the police and the insurance fund. The results showed that the students who participated in the intervention had a lower incidence of speeding violations than did the control group, but there were no differences in at-fault collisions or impaired-driving offenses.

Interestingly, the effect of the intervention on speeding violations was moderated by the ADRA2A CC genotype, which affects novel information processing and attention and has been shown to be related to attention deficit hyperactivity disorder (ADHD) and impulsivity (Roman et al., 2006). The intervention was more effective among students with the ADRA2A CC genotype than among those with the CG/GG genotypes. Furthermore, the students with this CC genotype who were exposed to the intervention had lower levels of traffic collisions and violations in general compared to the controls. This study demonstrates the value of training approaches that attempt to raise self-awareness among risky drivers and teach them ways of coping with impulsiveness. Moreover, it demonstrates that this type of intervention does not work with all young drivers, and as such, education and training programs, whether they are cognitive or motivational in nature, need to be tailored to the specific needs of novice drivers.

A recent study evaluated the impact of the Risk Awareness and Perception Training (RAPT) program using a PC on young driver crashes and traffic violations (Thomas et al., 2016). Drivers 16–18 years of age were recruited immediately after they passed the on-road driving exam at six California Department of Motor Vehicles licensing offices. Participants (N = 5251) were assigned to a group that completed the RAPT program or to a comparison group that completed a pretest but did not take the RAPT. The participants in the RAPT group viewed a series of nine driving situations, (e.g., two-lane roadway approaching two-way stop; potential hazards obscured by hedge on the right corner) and were asked to point the cursor where they would look. Analyses of pretest and posttest data included in the RAPT program showed substantial improvements in trainee performance, which suggests that participants attended to the training materials. Crash analyses over a 12-month period did not show an overall main effect for the RAPT, but there was a significant treatment-by-sex interaction effect such that the males who underwent the program had about a 24% lower collision involvement, while females who took the RAPT showed an increase of about 11%, although it was not significant. Researchers used regression analysis to evaluate the number of weeks after being licensed at which each driver had his or her first crash (time to first crash), but there was no significant effect of the RAPT on this measure. While this study did not address speeding specifically, the results provide some evidence that brief, computer-based training interventions can have a positive influence on driving safety for newly licensed teen drivers.

Enhanced driver training does not always work to improve safety (Ker et al., 2005). For example, a Dutch study looked at the ability of a more intensive driving course to reduce collisions involving injury or property damage among 436 recently licensed drivers (de Craen and Vlakveld, 2013). The more intensive training addressed vehicle maneuvering; mastering traffic situations; decisions in traffic (turning left); conditions of driving (e.g., driving when too tired or impaired by alcohol, resisting peer pressure to speed); and norms and values in relation to driving (knowing your limitations). The latter two components are not typically addressed in traditional driver education. Novice drivers who took the intensive driver training or the standard training were followed for 2 years. It was found that 43% of drivers in the intensive driver training reported collisions, compared to 26% of drivers who took the standard training. However, speed-related collisions were not examined. Given that the novices chose to take the intensive rather than the standard training, the authors thought that there might have been some selection bias but could find no evidence for such a bias in comparing the two groups.

Glendon and Walker (2013) compared antispeeding messages based on protection motivation theory (PMT) with standard antispeeding messages used on Australian roads. There were two components of the PMT model: threat appraisal and coping appraisal. Threat appraisal consists of the severity of the outcome ("Kill your speed, not yourself"), vulnerability ("No one is safe from a speed camera"), and rewards ("Speeding—the thrill that kills"), while coping appraisal consists of self-efficacy ("You can keep your mates safe; you can drive the speed limit"), response efficacy ("Don't speed and you don't get a fine"), and response cost ("Better late than never"). The 18 PMT-based messages (three for each of six model components) were more effective in getting young drivers aged 18–25 to say that they would drive within the speed limit than the standard messages, and this was the case for both males and females. The severity and vulnerability messages (i.e., threat component) were most effective in changing intentions about speeding. While the use of the PMT model to develop messages to reduce speeding appears to be effective in changing intentions to speed, it is not known if they actually have an impact on speeding by young drivers.

The Transport Accident Commission (TAC) in the state of Victoria in Australia has been running speed commercials on television for a number of years. Eight "Wipe off 5" campaigns have been released by TAC, with all of them emphasizing that even a small reduction in speed of 5 kph can make the difference between life and death. Some of these campaigns focused on the consequences of speed—not just for the victims but also for the family of the driver. Others have taken a more statistical, scientific approach to demonstrate the lower-speed/lower-impact approach. Over time, TAC claims that there has been a change in community attitudes toward speeding and also in behavior. According to TAC (2002), people who reported that they speed most or all of the time dropped from 25% to 11% after the campaign.

There are a variety of web-based programs for young drivers in Canada and elsewhere. For example, there is Parachute's National Teen Driver Safety Week (http://www.parachutecanada.org/programs/topic/C281), which is using social media (Facebook, Twitter, etc.) to reach out to teen drivers and encourage them to drive safely. In the United States, there is the national IKnowEverything program (http://iknoweverything.com/), which is testing teens regarding their knowledge about impaired and distracted driving in particular and then showing them that while they know some things, they do not know everything, particularly in dealing with unsafe drivers or passengers.

The results on the effectiveness of education and training programs are mixed. It appears that the effectiveness depends very much on the nature of the program and the characteristics of the young drivers taking these programs. Also, there may be a selection bias such that the safe drivers are more likely to take these programs. There is clearly a need for more evaluation of these programs.

11.3.2.4 Use of Incentives

A number of studies have examined the use of incentives to reduce speeding.

A Dutch study (Bolderdijk et al., 2011) examined the effectiveness of a pay-as-you-speed (PAYS) program with a group of 141 young drivers under 30 years of age whose driving speed was also monitored using GPS. The incentive resulted in a significant reduction in the percentage of time speeding from 18.6% to 17.7%, a 5% reduction over a 4-month period. Speeding by the control group actually increased from 17.9% to 19.7%. The incentive program appeared to reduce speeds more on roads with 100 kph limits than on other roads.

Lahrmann et al. (2012) assessed the impact of a PAYS program and informative Intelligent Speed Adaptation (ISA) devices on driving over the speed limit in Sweden. Drivers ($N = 153$) in a field trial were offered a reduction of their vehicle insurance premiums of 30% during the trial. If they drove over the limit, they accumulated penalty points, which resulted in a loss of a part or all of their premium reduction. In addition, their vehicle was fitted with ISA equipment, which informed drivers when they exceeded the speed limit (i.e., provided feedback). There was a 2 × 2 experimental design: PAYS—Yes or No, ISA—Yes or No. It was found that both ISA and PAYS had significant effects in reducing the proportion of the distance that the driver traveled over the speed limit. However, the informative ISA reduced this proportion more than the incentive, particularly on rural roads. The combined effects of the information and incentive conditions were not significantly greater than their individual effects. The researchers compared the effect of the ISA and PAYS for younger (18–28) and older (29+) drivers and observed that they were equally effective for both age groups, although the younger age group returned to the baseline level of speeding faster than did older drivers.

Samsung in Australia conducted an interesting pilot project called the S-Drive (www.youtube.com/watch?v=u9TlxdyYNUY), a program that promotes safe driving, especially targeting youth. Drivers were encouraged to install a special app on selected Samsung Galaxy smartphones and to use a custom phone holder in the vehicle. The phone can verbalize text, indicate to drivers that they are speeding in excess of the posted speed limit, and warn them if they attempt to touch the phone while the vehicle is in motion. This is an incentive-based program where the driver gets points for appropriate behavior (not texting or speeding). Drivers can form groups and share points, increasing social pressure to drive appropriately. The points can then be traded in for items from industry partners. The effectiveness of this program in reducing speeding by young drivers is not known.

These studies indicate that providing incentives with or without feedback to drivers may result in reductions in speeding and that these measures are effective for younger drivers as well.

11.3.2.5 Use of Feedback

Several studies have used feedback to influence speeding. Chapter 20 examines the use of feedback to improve driving behavior.

There are several on-board vehicle devices that are being used to influence the behavior of drivers, some specifically intended for young drivers. The Ford Motor Company (2013) has developed the MyKey technology, which reminds young drivers with a chime to slow down if they are driving over a preset speed. It also provides a constant chime when the teen driver is not wearing a seat belt and mutes the radio until he or she does. It has a "Do Not Disturb" function, which allows parents to block phone calls and texts when teens are driving. However, this technology has not yet been evaluated for its ability to influence teen driving behavior.

There are other systems where parents can monitor the speed and hard braking of young drivers. The Aegis TeenSafer (Aegis Mobility, 2015) application disables text and phone capabilities when young drivers are operating a vehicle (except for emergency calls). Other features include the ability to monitor and receive reports on driver behavior, including drive time, speeding, fast acceleration, and hard braking; a secure parental portal, which provides reports on driving behaviors, including route-specific events displayed on maps; and notifications sent to parents via e-mail for exceeding preset thresholds.

There are also on-board devices that drivers can plug into the vehicle's On-Board Diagnostic System Version II (OBD-II) port that will give the driver a visual and/or auditory signal that he or she is speeding, or they could be used with haptic feedback such that the driver has to press harder on the accelerator to keep the speed above a preset level. An on-road trial in Sweden found a decrease in collisions of 10%–15% when informative (i.e., tells drivers they are speeding) ISA systems were used, while a decrease on the order of 10%–24% was observed for active systems that automatically reduced vehicle speed (Biding and Lind, 2002).

DeLeonardis et al. (2014) assessed the effect of an on-board speed warning system based on GPS. A group of Maryland drivers ($N = 83$) with at least three speeding violations in the past 3 years participated in the study. The speed monitoring device, called tiwiPro, was mounted on the lower left of the windshield, where it could be seen by the driver. During the 2-week baseline, the subjects drove without receiving feedback from the device. During the 4- to 8-week treatment period (length of period varied by driver), the device gave a verbal alert when the driver was traveling more than 8 mph over the limit, and it also recorded speeds during each trip. Following the treatment period, there was a follow-up period without feedback being provided. During the treatment period, the percentage of trip distances where the vehicle was traveling over the posted speed limit declined from 45% to 43% and then returned to 44% during the follow-up baseline. While younger drivers aged 21–29 were more likely than older drivers to drive over the limit, they were affected by the warnings as much as older drivers (30+). However, the feedback had an even greater impact on driving more than 8 mph over the limit for both younger (reduction from 22% to 17%) and older drivers (reduction from 15% to 11%), although the effects disappeared during the follow-up period. This study shows that such feedback devices can affect speed choice by younger and older drivers, but the challenge is getting drivers to accept the monitoring device. The researchers approached 6361 drivers to participate, but only 271 (4%) contacted them about participating in the study, raising concerns about a sample bias. However, if the installation of the devices were required for drivers with multiple speed offenses to continue to drive or young drivers in GDL, they may be effective in reducing speed.

Carney et al. (2010) assessed whether or not event-triggered video technology that provided feedback to novice 16-year-old drivers and their parents in Minnesota influenced unsafe driving behaviors. During the intervention phase, 23 young drivers were provided immediate feedback via a blinking LED on the video recording unit, and their parents were provided a report card weekly showing the cumulative event frequency along with a DVD with safety-related clips on it. The study consisted of an initial baseline period of 6 weeks, followed by a 40-week intervention period and then a 6-week return-to-baseline period. With the video intervention in place, there was a significant reduction of 62% in unsafe events, which included making turns and taking curves too quickly (i.e., speeding), which decreased by 78%. There was also a significant drop in abrupt acceleration, but unlike turns and curves, these events returned quickly to the baseline. It is noteworthy that 77% of the young drivers did not think that the video recording was an invasion of privacy, 67% said that they had conversations with their parents about the reports on their driving, 94% felt that they could drive without activating the trigger to record, and 90% were glad that they participated in the program, all of which suggest acceptance of the program.

11.3.2.6 Summary of Interventions

The research suggests that there are a number of interventions that could be used to influence the speeding behavior of young drivers. While some of these have been tested using young drivers, others have not. More research should be conducted to assess the effectiveness of these measures specifically for young drivers.

11.4 Other Risky Driving Behavior

11.4.1 Impaired Driving

Impaired driving by young drivers has been a road safety concern for decades. A general overview of impaired driving has been conducted by the World Health Organization (2007). A number of recent studies have examined impaired driving by alcohol or other drugs among youth using collision data or survey data. Note that Chapter 13 focuses on alcohol- and drug-impaired driving.

11.4.1.1 Collision Data

A number of recent studies have examined the prevalence of alcohol and other drugs in collisions involving drivers of different ages.

The Traffic Injury Research Foundation (2016) reported that drivers aged 26–35 were slightly more likely (53%) to have been drinking prior to being killed in fatal collisions in 2012 than younger drivers aged 20–25 (43%) but considerably more likely than teenaged drivers (34%), based on coroner data. Looking at the victims of fatal crashes, whether they were drivers, passengers, or pedestrians, the involvement of alcohol was highest for the 26–35 age group (53%) followed by the 20–25 age group (55%) and the 16–19 age group (39%). Using police-reported information and a surrogate measure of drinking and driving (single-vehicle nighttime crashes), it was determined that 26% of drivers aged 20–25 who were involved in a serious-injury collision (i.e., stayed in hospital overnight) had positive blood alcohol concentrations (BACs) compared to 26% of teenage drivers (16–19) and 21% of drivers aged 26–35.

Using coroner data, Beasley et al. (2011) examined the presence of alcohol and/or psychotropic drugs among fatally injured drivers for the years 2000–2008 and determined that about 37% of drivers aged 16–19 and 49% of drivers aged 20–24 had some alcohol present in their bodies compared to drivers aged 25–34, who had the highest prevalence of alcohol at 50%. About 49% of drivers aged 20–24 had some drug present, compared to 50% for drivers aged 25–34 and 16–20. All drivers aged 16–24 were more likely to have had some other drug present rather than alcohol. Among those drivers aged 16–24 who were positive for drugs, cannabis was considerably more prevalent than it was for older drivers. An earlier report by Beasley and Beirness (2011) indicated that fatally injured drivers under 35 years of age were more likely than older drivers to have both alcohol and other drugs in their body. More recently, the Traffic Injury Research Foundation (2016) reported that drug use was highest among fatally injured drivers aged 26–35 (46%), followed by drivers under 20 (45%) and those aged 20–25 (42%).

An examination of the NCDB data for all injury collisions for 2007–2012 shows that 3.8% of Canadian drivers aged 16–20 and 4.9% of drivers aged 21–24 had been drinking or were impaired according to the police, which was more frequent than the next closest age group, 25–44, at 3.1%. Drugs were involved in 0.3% of injury collisions for both age groups 16–20 and 21–24, slightly higher than that for the 25–44 age group at 0.25%.

FARS data for 2013 (NHTSA, 2014c) showed that 17% of drivers aged 16–20 involved in fatal collisions in the United States had BACs greater than the .08 legal limit, increasing to 33% among drivers aged 21–24 and then declining from 29% for those aged 25–34 to 5% for those 75 and older.

These studies using fatal and injury collision data generally indicate that alcohol-impaired driving increases through the teen years, likely due to greater access to alcohol, but is highest among drivers in the 21–24 and 25–35 age categories, which do not really differ.

11.4.1.2 Survey Data

Beirness and Beasley (2010) conducted a nighttime roadside survey in British Columbia during 2010 and found that 3% of drivers aged 16–18 had been drinking and driving, compared to 9% of drivers aged 19–24 and 13% of drivers aged 25–44. About 8% of drivers aged 16–24 were drug positive, but this did not vary significantly from the percentage for older drivers. A subsequent BC roadside survey in 2012 (Beasley and Beirness, 2012) found that no drivers aged 16–18 had been drinking and 5% of drivers aged 19–24 had positive BACs, which was lower than for drivers aged 25–44 (8%). The percentage of drug-positive drivers ranged from 5% for drivers aged 16–18 to 12% for those aged 35–44, but the differences were not significant, due to small sample sizes. The results on type of drug used showed that cannabis was the most commonly detected drug, with no significant differences by age.

A roadside survey conducted in the United States in 2007 (Lacey et al., 2009a) determined that about 16% of drivers aged 21–34 had been drinking, compared to 7% of drivers aged 16–20 and 13% of drivers aged 35–44. Another report based on the same US roadside survey (Lacey et al., 2009b) determined the use of illicit and medicinal drugs by drivers during the daytime and the nighttime. During the daytime, drivers aged 45–64 were more likely to be using drugs while driving (13%) compared to younger drivers aged 16–20 (6%) or 21–44 (11%). However, during the nighttime, 16% of drivers aged 16–20 were using drugs, and 17% of those aged 21–34 had some drug present based on oral-fluid testing, compared to drivers aged 35–44 (15%), 45–64 (10%), and 65 and older (2%).

When the class of drug was examined by age, it was found that during the day, the use of marijuana by drivers aged 21–34 was highest at 7%, followed by drivers aged 35–44 at 5% and drivers aged 16–20 at 4%. However, during the nighttime, marijuana use was higher among drivers aged 16–20 at 10% than those aged 21–34 (9%) and those aged 35–44 (4%). It is also noteworthy that drivers aged 16–20 (3%) and those aged 21–34 (3%) were more likely than older drivers to be using more than one class of drug at night.

TIRF (2011) conducted a public opinion survey regarding drinking and driving and found that 12% of drivers aged 16–24 reported driving within 2 hours of drinking any amount of alcohol compared to 25% of drivers 25 and older. Further, 8% of young drivers admitted to driving while they thought that they were over the legal limit, compared to 5% of older drivers.

A recent national survey on driving after using alcohol or drugs (Jonah, 2015) revealed that about 17% of drivers aged 20–24 and 55–64 reported drinking alcohol and then driving within 2 hours, compared to 16% of drivers aged 16–19, 25–34, and 45–54. Self-reported use of cannabis and then driving within 4 hours of using was highest for drivers aged 25–34 at 4%, followed by drivers aged 16–24 at 3%.

Pashley et al. (2014) conducted a survey on drugs and driving and reported that about 8% of drivers aged 16–24 admitted to driving within 2 hours of taking potentially impairing prescription drugs, compared to only 1.3% of drivers aged 65 and older, in 2013. This finding raises concerns about the illegal use of prescription drugs by young people.

A survey conducted in the United Kingdom asked respondents to indicate if they had driven while they thought that they were over the .08 legal limit in the last 12 months, including the morning after (Brake Organization, 2014). Overall, only 2% reported that they were sure that they had driven over the limit, and 8% said that it was possible, but 9% of young drivers (17–24) said they were sure, and 18% said it was possible.

Bingham and Shope (2014) examined the developmental trajectories of psychosocial and nondriving problem behaviors that lead to risky driving and determined through a longitudinal study of over 2000 subjects that young adult risky drivers were more likely to have had a low level of parental monitoring, greater parental permissiveness, a weaker social bond with their parents, and a rapid increase in substance use.

11.4.1.3 Summary

The research generally shows that the age group 21–24 is more likely to drive after drinking, although in some studies, this group is not much different than the 25–34 age group. Drivers aged 16–20 are generally less likely to drink and drive, likely due to less access to alcohol, given that most of them are under age. The results regarding driving after using other drugs are not clear. In some cases, young drivers are more likely to be using drugs than older drivers, particularly cannabis. As will be shown later, speeding is related to drinking and driving and, to a lesser extent, cannabis use and driving. The use of alcohol and cannabis may reduce inhibitions to risky driving and hence foster speeding.

11.4.2 Driver Distraction

The World Health Organization (2011) has underscored the growing problem of the use of mobile phones by drivers. Several recent studies have compared younger drivers with older drivers on distracted driving, particularly the use of electronic communication devices (ECDs). Note that Chapter 12 focuses on driver distraction.

11.4.2.1 Collision Data

It has been estimated using FARS data for 2014 that 10% of fatal collisions and 18% of injury crashes in the United States involved distracted driving, which is defined as a specific type of inattention that occurs when drivers divert their attention from the driving task to focus on some other activity instead (NHTSA, 2014a). Looking at age differences, it was found that drivers aged 15–19 killed in collisions were more likely to have been distracted by something (10%) compared to drivers aged 20–29 (8%) or most older drivers (6–7%), although for drivers 70 and over, the rate was 9%. However, 19% of drivers aged 15–19 who were distracted were distracted by the use of a cell phone, which was higher than for those aged 20–29 (18%) or for older age groups (ranging from 2% to 15%).

Analyses of NCDB fatal collision data for 2007–2012 indicate that distraction was a contributing factor, according to the police, in 10.1% of collisions in Canada involving drivers aged 16–20, 7.9% for those aged 21–24, 8.1% for those aged 25–44, 7.9% for those aged 45–64, and 15.1% for those 65 and older. For injury collisions, 16.7% of drivers aged 16–20 had been distracted, compared to 14.3% of those aged 21–24, 11.9% of those aged 25–44, 11.8% of those aged 45–64, and 18.5% of those 65 and older. These data suggest that distraction is a contributing factor in fatal and injury collisions more often for younger drivers aged 16–20 and for drivers 65 and older. However, these data do not distinguish between distraction by cell phones and other types of distractions, since the type of distraction is not reported in most Canadian jurisdictions.

11.4.2.2 Survey Data

Observational and public opinion surveys have been used to estimate the use of ECDs while driving.

Jonah (2014) conducted a national roadside observational survey of 104,000 drivers stopped at rural and urban intersections in Canada in 2012 and 2013 and determined that drivers estimated to be under 25 years of age were significantly more likely (7%) to have been using a handheld (ECD) for either talking or typing/texting than drivers aged 25–49 (5%) or drivers 50 and over (2%). The same age effect was observed at urban and rural sites.

A 2014 observational survey conducted in the United States among 36,000 drivers stopped at intersections found that 5.8% of drivers aged 16–24 were using a handheld electronic device compared to 4.3% of drivers aged 25–69 and 0.8% of those 70 and older (NHTSA, 2014b). It is noteworthy that the gap between younger and older drivers' use of these devices has declined since 2005 when the comparable numbers were 10%, 6%, and 1%, with the younger drivers showing the largest decline in use. However, it is quite possible that the use of hands-free ECDs has increased.

An earlier NHTSA survey (NHTSA, 2009) found that drivers aged 18–20 were more likely (13%) than all older age groups (range from 0% to 2.2%) to report that they had been using an ECD for talking or sending/reading a text message or e-mail at the time of a collision or near collision.

Tison et al. (2011) conducted a telephone survey with about 6000 drivers regarding distracted driving in the United States. Of the 6% of drivers who reported being in a crash or near crash in the past year, 13% of younger drivers aged 18–20 said that they were using a phone at the time of the event. Interestingly, drivers aged 25–34 were more likely to have been talking on the phone prior to the event than all other age groups, but drivers aged 18–20 were more likely to have been reading (3%) or sending a text message or e-mail (8%) than other age groups. Young drivers (18–24) were more likely to say that talking on the phone made no difference to their driving (62%), and 24% said that texting/e-mailing made no difference, which was lower than for drivers aged 25–34 (30%).

A survey conducted for the Centers for Disease Control (O'Malley et al., 2013) showed that self-reported texting while driving at least once while driving in the past 30 days increased from 33% among 16-year-old students to 58% among those 18 and over.

Zhao et al. (2013) asked a group of 108 drivers in the Boston area, "How often do you use a cell phone while driving?" It was found that drivers in their 20s were much more likely to say a few times a week or a few times a day (61%) compared to those who were in their 40s (47%) or in their 60s (34%). The subjects were asked to drive an instrumented vehicle on a highway for about 40 minutes while their driving behavior was recorded, although they were not using a cell phone while driving. It was observed that frequent cell phone users of all ages drove faster, and did significantly more lane changes, hard braking, and throttle accelerations, suggesting that cell phone use may be linked to other risky driving behavior.

11.4.2.3 Naturalistic Driving Study

A recent study by the Insurance Institute for Highway Safety (Farmer et al., 2014a) using data from the 100-car Naturalistic Driving Study determined that younger drivers under 21 interacted with cell phones more than older drivers. However, this greater phone usage did not translate into more frequent crashes or near crashes for young drivers while using these devices. Farmer et al. presented data that suggest that all drivers may be compensating for the increased perceived risk of cell phone use by driving more slowly. A subsequent study by Farmer et al. (2014b) further suggests that all drivers using cell phones may stop engaging in other distracting behavior such as talking to passengers, eating, etc.

11.4.2.4 Summary

These studies on distracted driving indicate that young drivers aged 16–24 are more likely to engage in such behavior than older drivers, particularly the use of ECDs, and that it is more prevalent among those aged 16–20, especially texting.

TABLE 11.1 Correlation of Self-Reported Risky Driving

	Driving while Texting	Driving while Talking on a Mobile Phone That Is Not Hands-Free	Driving after Taking Drugs Such as Cocaine or Methamphetamines	Aggressive Driving	Exceeding the Speed Limit by 10 kph on a Highway Where the Limit Is 80 kph or More	Not Wearing a Seat Belt	Driving after Consuming Two or More Drinks of Alcohol in 2 Hours	Driving while Tired	Driving after Taking Marijuana/Cannabis	Driving after Taking Prescription Drugs
Driving while Texting	1	.538	.275	.316	.244	.159	.255	.270	.217	.066
Driving while Talking on a Mobile Phone That Is Not Hands-Free	.538	1	.220	.272	.248	.215	.210	.268	.202	.093
Driving after Taking Drugs Such as Cocaine or Methamphetamines	.275	.220	1	.207	.044	.231	.302	.126	.460	.188
Aggressive Driving	.316	.272	.207	1	.294	.157	.209	.282	.208	.123
Exceeding the Speed Limit by 10 kph on a Highway Where the Limit Is 80 kph or More	.244	.248	.044	.294	1	.084	.191	.281	.110	.110
Not Wearing a Seat Belt	.159	.215	.231	.157	.084	1	.138	.113	.185	.093
Driving after Consuming Two or More Drinks of Alcohol in 2 Hours	.255	.210	.302	.209	.191	.138	1	.190	.267	.135
Driving while Tired	.270	.268	.126	.282	.281	.113	.190	1	.113	.154
Driving after Taking Marijuana/Cannabis	.217	.202	.460	.208	.110	.185	.267	.113	1	.117
Driving after Taking Prescription Drugs	.066	.093	.188	.123	.110	.093	.135	.154	.117	1

11.4.3 Risky Driving Syndrome

The risky driving behaviors from a public opinion survey on road safety (Jonah, 2015) were intercorrelated, and it can be seen in Table 11.1 that many of the behaviors are correlated, some as high as .54. Given the large sample size, correlations over .05 are statistically significant at the $P < .05$ level. However, in general, many of the correlations are in the .20–.30 range, suggesting that a driver who engages in one type of risky driving is likely to engage in other risky driving behaviors. The correlations with driving after using prescription drugs and not wearing a seat belt are generally lower.

Other studies have found that risky driving behaviors are correlated. For example, McNally and Bradley (2014) observed that among 522 Australian and New Zealand students aged 17–25, the correlations among a distracted behavior factor (e.g., texting), a substance abuse factor (e.g., driving with BAC over .05), an extreme behavior factor (e.g., racing other drivers), and a positioning behavior factor (e.g., driving 15 kph over limit) ranged from .31 to .61. Scott-Parker et al. (2014) reported that among 1076 young (aged 18–20) novice drivers, those who said that they drove after drinking or drove after using drugs like cannabis were more likely to also say that they speed, turn right in front of another vehicle, drive while using a handheld cell phone, not wear a seat belt, and exceed night passenger limits. All these studies support the notion of a risky driving syndrome. Jonah and Dawson (1987) and Jonah (1990) looked at the correlation of risky behaviors by age group and determined that as drivers age, the relationship among these behaviors weakens.

11.5 Conclusions

This chapter has reviewed recent research on risky driving among young drivers aged 16–24, with a focus on speeding. Overall, this research indicates that young drivers are more likely to be involved in fatal and injury collisions. They are also more involved in speed-related fatal and injury collisions. There is considerable evidence based on observational, naturalistic, or self-report surveys that young drivers are more likely to engage in risky driving such as speeding, driving while talking or texting on a cell phone, and driving after using drugs or alcohol. However, for some risky driving behaviors (e.g., alcohol-impaired driving), the prevalence was higher among older drivers aged 25–34, and for other behaviors, the youngest age group (16–20) had higher levels of prevalence (e.g., driving after using cannabis, speeding). Therefore, risky driving is not always more prevalent among young drivers. More research is required to better understand why young drivers are more likely to engage in some risk driving behaviors but not others.

A number of measures have been implemented to address speeding, and some of these have been shown to deter young drivers from engaging in such behavior (e.g., antiracing legislation, automated speed enforcement, monetary incentives, and feedback technology). Further research is required to determine which measures, alone and in concert with others, are most effective in deterring speeding among drivers, particularly younger ones.

Correlational data from a recent Canadian public opinion survey, as well as those from other studies, underscore the fact that risky driving behaviors are related and that drivers who speed are more likely to engage in other risky behavior, which supports the notion of a risky driving syndrome, or what Jessor (1987) calls problem behavior. Further work is required to better understand this syndrome and to identify effective measures to discourage such risky driving behavior among young drivers.

References

Aegis Mobility (2015) Free Mobile App designed to keep teens from texting while driving. https://www.youtube.com/watch?v=qzBS-y3pWV0, March 9.

Ajzen, I. (1991) The theory of planned behavior. *Organizational Behavior and Human Decision Processes*, 50, 179–211.

Beasley, E. and Beirness, D. (2011) *Drug Use by Fatally Injured Drivers in Canada (2000–2008)*, Canadian Centre on Substance Abuse, Ottawa.

Beasley, E. and Beirness, D. (2012) Alcohol and Drug Use Among Drivers Following the Introduction of Immediate Roadside Prohibitions in British Columbia: Findings from the 2012 Roadside Survey. Report prepared for British Columbia Ministry of Justice, Office of the Superintendent of Motor Vehicles.

Beasley, E.E., Beirness, D.J., and Porath-Waller, A.J. (2011) *A Comparison of Drug- and Alcohol-involved Motor Vehicle Driver Fatalities*, Canadian Centre on Substance Abuse, Ottawa.

Beck, K.H., Daughters, S.B., and Ali, B. (2013) Hurried driving: Relationship to distress tolerance, driver anger, aggressive and risky driving in college students. *Accident Analysis & Prevention*, 51, 51–55.

Beirness, D.J. and Beasley, E.E. (2010) A roadside survey of alcohol and drug use among drivers in British Columbia. *Traffic Injury Prevention*, 11, 215–221.

Beirness, D. and Beasley, E. (2011) *Alcohol and Drug Use Among Drivers*. Canadian Centre on Substance Abuse, Ottawa.

Biding, T. and Lind, G. (2002) Intelligent Speed Adaptation (ISA). Results of large-scale trials in Borlänge, Lidköping, Lund and Umeå during the period 1999–2002. 2002:89. Borlange, Sweden: Vagverket (Swedish National Road Administration).

Bingham, C.R. and Shope, J.T. (2014) Adolescent developmental antecedents of risky driving among young adults. *Journal of Studies in Alcohol*, 65, 84–94.

Bolderdijk, J.W., Knockaert, J., Steg, E.M., and Verhoef, E.T. (2011) Effects of Pay-As-You-Drive vehicle insurance on young driver's speed choice: Results of a Dutch field experiment. *Accident Analysis & Prevention*, 43(3), 1181–1186.

Brake Organization (2014) *Fit to Drive: Drink Driving*. Direct Line Survey, United Kingdom.

Carney, C., McGehee, D.V., Lee, J.D., Reyes, M.L., and Raby, M. (2010) Using an event-triggered video intervention system to expand the supervised learning of newly licensed adolescent drivers. *American Journal of Public Health*, 100, 1101–1106.

Cestac, J., Paran, F., and Delhomme, P. (2011) Young drivers' sensation seeking, subjective norms, and perceived behavioural control and their roles in predicting speeding intention: How risk-taking motivations evolve with gender and driving experience. *Safety Science*, 49, 424–432.

Clarke, D.D., Ward, P., Truman, W., and Bartle, C. (2008) A poor way to die: Social deprivation and road traffic fatalities. *Behavioural Research in Road Safety*. Available from http://dft.gov.uk/pgr/roadsafety/research/rsrr/themes/poorwaytodie.pdf.

Curry, A.E., Pfeiffer, M.R., Myers, R.K., and Durbin, D.R. (2014) Statistical implications of using moving violations to determine crash responsibility in young driver crashes. *Accident Analysis & Prevention*, 65, 28–35.

De Craen, S. and Vlakveld, W. (2013) Young drivers who obtained their licence after an intensive driving course report more incidents than drivers with traditional driver education. *Accident Analysis & Prevention*, 58, 64–69.

DeLeonardis, B., Huey, R., and Robinson, E. (2014) Investigation of the use and feasibility of speed warning systems. National Highway Traffic Safety Administration, DOT HS 811996.

Elvik, R., Høye, A., Vaa, T., and Sørensen, M. (2009) *The Handbook of Road Safety Measures*, 2nd ed. Emerald Group Publishing, Bingley.

Farmer, C.M., Klauer, S.G., McClafferty, J.A., and Guo, F. (2014a) Relationship of Near-Crash/Crash Risk to Time Spent on a Cell Phone While Driving. Insurance Institute for Highway Safety, Arlington, VA.

Farmer, C.M., Klauer, S.G., McClafferty, J.A., and Guo, F. (2014b) Secondary Behavior of Drivers on Cell Phones. Insurance Institute for Highway Safety, Arlington, VA.

Ferguson, S. (2013) Speeding-Related Fatal Crashes among Teen Drivers and Opportunities for Reducing the Risks. Governors Highway Safety Association, Washington, D.C.

Ford Motor Company, Ford Media Center (2013) Ford MyKey. Press release.

Glendon, A.I. and Walker, B.L. (2013) Can anti-speeding messages based on protection motivation theory influence reported speeding intentions? *Accident Analysis & Prevention*, 57, 67–79.

Global Road Safety Partnership (2008) Speed Management: A Road Safety Manual for Decision-Makers and Practitioners, Geneva.

Hassan, H.M. and Abdel-Aty, M.A. (2013) Exploring the safety implications of young drivers' behavior, attitudes, and perceptions. *Accident Analysis & Prevention*, 50, 361–370.

Hedlund, J. and Compton, R. (2004) Graduated driver licensing research in 2003 and beyond. *Journal of Safety Research*, 35, 5–11.

Horvath, C., Lewis, I., and Watson, B. (2012) The beliefs which motivate young male and female drivers to speed: A comparison of low and high intenders. *Accident Analysis & Prevention*, 45, 334–341.

Jessor, R. (1987) Risky driving and adolescent problem: An extension of problem behavior theory. *Alcohol, Drugs and Driving*, 3, 1–13.

Jonah, B.A. (1986) Accident risk and risk-taking behaviour among young drivers. *Accident Analysis & Prevention*, 18, 255–271.

Jonah, B.A. (1990) Age differences in risky driving. *Health Education Research*, 5, 139–149.

Jonah, B.A. (2014) Use of Electronic Communication Devices (ECD) by Canadian Drivers, Canadian Council of Motor Transport Administrators, Ottawa.

Jonah, B.A. (2015) Public Opinion Survey of Canadian Drivers on Road Safety. Paper presented at Canadian Association of Road Safety Professionals Conference, Ottawa, May.

Jonah, B.A. and Dawson, N.E. (1987) Youth and risk: Age differences in risky driving, risk perception, and risk utility. *Alcohol, Drugs and Driving*, 3, 13–29.

Keall, M.D. and Newstead, S. (2013) The potential effectiveness of young high-performance vehicle restrictions as used in Australia. *Accident Analysis & Prevention*, 52, 154–161.

Ker, K., Roberts, I., Collier, T., Beyer, F., Bunn, F., and Frost, C. (2005) Post-licence driver education for the prevention of road traffic crashes: A systematic review of randomised controlled trials. *Accident Analysis & Prevention*, 37, 305–313.

Lacey, J.H., Kelley-Baker, T., Furr-Holden, D., Voas, R.B., Romano, E., Torres, P., Tippetts, S., Ramirez, A., Brainard, K., and Berning, A. (2009a) 2007 National Roadside Survey of Alcohol and Drug Use by Drivers: Alcohol Results. NHTSA, DOT HS 811 248.

Lacey, J.H., Kelley-Baker, T., Furr-Holden, D., Voas, R.B., Romano, E., Moore, C., Torres, P., Ramirez, A., Brainard, K., and Berning, A. (2009b) 2007 National Roadside Survey of Alcohol and Drug Use by Drivers: Drug Results. NHTSA, DOT HS 811 249.

Lahrmann, H., Agerholm, N. Tradisauskas, N., Berthelsen, K.K., and Harms, L. (2012) Pays as you speed, ISA with incentive for not speeding: Results and interpretation of speed data. *Accident Analysis & Prevention*, 48, 17–28.

Lambert-Belanger, A.L., Dubois, S., Weaver, B., Mullen, N., and Bedard, M. (2012) Aggressive driving behaviour in young drivers (16 through 25) involved in fatal crashes. *Journal of Safety Research*, 43, 333–338.

Leandro, M. (2012) Young drivers and speed selection: A model guided by the Theory of Planned Behavior. *Transportation Research Part F: Traffic Psychology and Behavior*, 15, 219–232.

Mayhew, D.R. and Simpson, H.M. (2002) The safety value of driver education and training. *Injury Prevention*, 8 (Suppl. 2), ii3–ii8.

McNally, B. and Bradley, G.L. (2014) Re-conceptualizing the reckless driving behaviour of young drivers. *Accident Analysis & Prevention*, 70, 245–257.

Meirambayeva, A., Vingilis, E., McLeod, I., Elzohairy, Y., Xiao, J., Zou, G., and Lai, Y. (2014) Road safety impact of Ontario street racing and stunt driving law. *Accident Analysis & Prevention*, 71, 72–81.

Ministère des Transports du Québec (2013) Cinemometres Photographiques et Systems Photographiques de Controle de Circulation aux Feux Rouges.

Molin, E.J.E., and Brookshuis, K.A. (2007) Modelling acceptability of the intelligent speed adapter. *Transportation Research Part F: Traffic Psychology and Behaviour*, 10, 99–108.

Moller, M. and Haustein, S. (2014) Peer influence on speeding behaviour among male drivers aged 18 and 28. *Accident Analysis & Prevention*, 64, 92–99.

National Highway Traffic Safety Administration (NHTSA) (2009) Traffic Safety Facts: Distracted Driving. NHTSA DOT HS 811 737.

National Highway Traffic Safety Administration (2014a) Traffic Safety Facts 2012 Data: Speeding, NHTSA, DOT HS 812 021.

National Highway Traffic Safety Administration (2014b) Traffic Safety Facts: Driver Electronic Device Use in 2012. NHTSA, DOT HS 811 884.

National Highway Traffic Safety Administration (2014c) Traffic Safety Facts: Alcohol-impaired Driving, NHTSA, DOT HS 812 102.

O'Malley, E., Shults, R.A., and Eaton, D.K. (2013) Texting while driving and other risky motor vehicle behaviours among U.S. high school students. *Pediatrics*, 131, e1708–e1715.

Organization for Economic Cooperation and Development (2006) *Young Drivers*. Paris, France.

Organization for Economic Cooperation and Development (2007) *Speed Management*. Paris, France.

PA Consulting Group (2005) Evaluation of the national speed camera programme: Four year report. Prepared for U.K. Department for Transport.

Paaver, M., Eensoo, D., Kaasik, K., Vaht, M., Maestu, J., and Harro, J. (2013) Preventing risky driving: A novel and efficient brief intervention focusing on acknowledgement of personal risk factors. *Accident Analysis & Prevention*, 50, 430–437.

Palarma, P., Molnar, L., Eby, D., Kopinanthan, C., Langford, J., Gorman, J., and Broughton, M. (2012) Review of young driver risk taking and its association with other risk taking behaviours. Curtin Monash Accident Research Centre and Transportation Research Institute, University of Michigan.

Parker, D., Reason, J.T., Manstead, A., and Stradling, S.G. (1995) Driving errors, driving violations, and accident involvement. *Ergonomics*, 38, 1036–1048.

Pashley, C.R., Robertson, R.D., and Vanlaar, W.D. (2014) The Road Safety Monitor: Drugs and driving 2013. Traffic Injury Research Foundation, Ottawa.

Prabhakharan, P. and Molesworth, B.R. (2011) Repairing faulty scripts to reduce speeding behaviour in young drivers. *Accident Analysis & Prevention*, 43, 696–702.

Richard, C.M., Campbell, J.L., Lichty, M.G., Brown, J.L., Chrysler, S., Lee, J. D., Boyle, L., and Reagle, G. (2012) Motivations for speeding, Volume I: Summary report. NHTSA, DOT HS 811 658.

Roman, T., Polanczyk, G.V., and Zeni, C. (2006) Further evidence of the involvement of alpha 2a-adrenergic receptor gene (ADRA2A) in inattentive dimensional scores of attention-deficit/hyperactivity disorder. *Molecular Psychiatry*, 11, 8–10.

Scott-Parker, B., Hyde, M.K., Watson, B., and King, M.J. (2010) The risky driving behaviour of young drivers: Developing a measurement tool. In Proceedings of the 24th Canadian Multidisciplinary Road Safety Conference, Niagara Falls, Canada, June 6–9, 2010.

Scott-Parker, B., Hyde, M.K., Watson, B., and King, M.J. (2013) Speeding by young novice drivers: What can personal characteristics and psychosocial theory add to our understanding? *Accident Analysis & Prevention*, 50, 242–250.

Scott-Parker, B., Hyde, M.K., Watson, B., King, M.J., and Hyde, M.K. (2014) "I drove after drinking alcohol" and other risky driving behaviours reported by young novice drivers. *Accident Analysis & Prevention*, 70, 65–73.

Senserrick, T. and Haworth, N. (2005) Review of the literature regarding national and international young driver training, licensing, and regulatory systems. Monash University Accident Research Centre (MUARC), Clayton, Victoria, Australia.

Soole, D.W., Watson, B.C., and Fleiter, J.J. (2013) Effects of average speed enforcement on speed compliance and crashes: A review of the literature. *Accident Analysis & Prevention*, 54, 46–56.

Swedler, D.I., Bowman, S.M., and Baker, S. (2012) Gender and age differences among teen drivers in fatal crashes. *Association for the Advancement of Automotive Medicine*, 56, 97–106.

Tapp, A., Pressley, A., Baugh, M., and White, P. (2013) Wheels, skills, and thrills: A social marketing trial to reduce aggressive driving from young men in deprived areas. *Accident Analysis & Prevention*, 58, 148–157.

Thomas, F.D., Rilea, S., Blomberg, R.D., Peck, R.C., and Korbelak, K.T. (2016) Evaluation of the Safety Benefits of the Risk Awareness and Perception Training Program for Novice Teen Drivers. Report for National Highway Traffic Safety Administration, Washington, D.C., DOT HS 812 235.

Tison, J., Chaudhary, N., and Cosgrove, L. (2011) National Phone Survey on Distracted Driving Attitudes and Behaviour. Report for National Highway Traffic Safety Administration, Washington, D.C., DOT HS 811 555.

Toroyan, T. and Peden, M. (2007) Youth and road safety. World Health Organization, Geneva.

Traffic Injury Research Foundation (TIRF) (2011) The Road Safety Monitor: Youth Drinking and Driving.

Traffic Injury Research Foundation (2016) Alcohol-Crash Problem in Canada: 2016. Report prepared for the Canadian Council of Motor Transport Administrators.

Transport Accident Commission (2002) Wipe Off 5 Speed Campaign. TAC website, Victoria, Australia.

Vanlaar, W., Robertson, R., and Marcoux, K. (2011) Evaluation of the photo enforcement safety program of the city of Winnipeg: Final report.

Vingilis, E., Smart, R., Mann, R., Paglia-Boak, A., Stoduto, G., and Adlaf, E. (2011) Prevalence and correlates of street racing among ontario high school students, *Traffic Injury Prevention*, 12(5), 443–450.

Williams, A. (2013) Teenage Driver Fatalities by State: 2013 Preliminary Data. Governors' Highway Safety Association, Washington, D.C.

Williams, A., Tefft, B.C., and Grabowski, J.G. (2012) Graduated driver licensing research, 2010–Present. *Journal of Safety Research*, 43, 195–203.

Wilson, C., Willis, C., Hendrikz, J.K., and Bellamy, N. (2006) Speed Enforcement Detection Devices for Preventing Road Traffic Injuries (Review). *Cochrane Database of Systematic Reviews*, 2, CD004607.

World Health Organization (2007) Drinking and Driving: A Road Safety Manual for Decision Makers and Practitioners. Geneva, Switzerland.

World Health Organization (2011) Mobile Phone Use: A Growing Problem of Driver Distraction. Geneva.

Zhao, N., Mehler, B., Ambrosio, L.A., and Coughlin, J.F. (2013) Self-reported and observed risky driving behaviors among frequent and infrequent cell phone users. *Accident Analysis & Prevention*, 61, 71–77.

IV

Impairment

12

A Review of Novice and Teen Driver Distraction

Jeff K. Caird

William J. Horrey

Abstract

The Problem. Driver distractions contribute to the deaths and injuries of teen and adult drivers. ***Scope of Chapter.*** The central theme of this chapter is whether or not teen drivers are more likely to be distracted and more adversely affected while driving than other age groups. Crash databases, naturalistic studies, driver observations, surveys, and experimental research, which are concentrated within the past 5 years, are reviewed so that a number of focused questions concerning the teen driver can be addressed. In addition, potential countermeasures to reduce distracting activities and consequential outcomes in teen drivers are also discussed. ***Limitations and Recommendations.*** Many gaps in knowledge are evident with respect to a complete understanding of the effects of distracting activities on driving in general and teens in particular. Studies that determine the crash risk and performance decrements of teen and novice drivers with a range of distractions are unfortunately rare, use a limited number of measures, and are not replicated. Similarly, countermeasure effectiveness studies are also infrequent and delayed relative to immediate threats from changing technologies. Effective interventions require urgent action to prevent additional deaths and injuries from driver distraction.

12.1 Introduction

Technology use in the lives of teens is ubiquitous. Teens are intimately and constantly connected to their friends and family through their phones. In many respects, carryover communication and socialization while driving has become the norm, and those who do not drive while performing a variety of distracting tasks appears to be the exception. The hyperbolic meme that driving is a distraction from everyday distractions suggests a need for driverless vehicles.

As a result, teen driver deaths and injuries from distractions are common occurrences. Overall, motor vehicle crashes are the leading cause of death of those aged 15–20 in the United States (NHTSA, 2014b) and worldwide (WHO, 2015). In the most recent analysis of distracted driving in the United States in 2013, 10% of fatal and 18% of injury crashes involved distraction, and teens are overrepresented in fatal distraction crashes compared to other age groups (NHTSA, 2015). From a public health perspective, the proliferation of potential distractions and negative outcomes has the indications of an epidemic. Against this ominous backdrop and societal interest, we review the effects of distractions on teen and novice drivers and various countermeasures aimed at reducing crashes and use of potentially distracting technology.

Historically, driver distraction is not a new problem (see, e.g., Caird & Dewar, 2007; Regan et al., 2008). However, the media continuously thrusts the issue of smartphone use while driving into the public eye, which leaves the impression that driver distraction is a new problem. More recently, distractions and teen drivers have been highlighted by a number of researchers who have reviewed portions of the literature (e.g., Durbin et al., 2014; Klauer et al., 2015). Others have identified driver distraction in the context of discussing a variety of issues that face teen drivers (NRC, IOM, TRB, 2007; Pediatrics, 2006; Shope & Bingham, 2008; Williams et al., 2012a). Reviews of cell phone distraction have also identified young drivers as an important driver group (Cazzulino et al., 2014; Collet et al., 2010; GHSA, 2013; McCartt et al., 2006; WHO, 2011). Recent research that focuses on teen distractions, including the special issue of the *Journal of Adolescent Health* in 2014, has produced a number of interesting results and recommendations that have not been brought together with previous research. This chapter synthesizes some of these findings, incorporating previous reviews, new research, and overlooked studies. We used a variety of database search strategies and reference backtracking approaches to identify relevant research on teen and novice driver distraction and countermeasures. Study selection emphasized national sampling and high-quality studies so that stronger and generalizable conclusions could be synthesized from our review. Crash database analyses, surveys, observational studies, naturalistic studies, and driving simulator experiments are synthesized with respect to a specific set of specific research questions that are bound by the constraints of each method. Interesting results are summarized in a tabular quantitative format so that readers can compare results across studies and arrive at their own interpretations.

The focus of this chapter is on *teen driver distraction,* but it excludes impairment by drugs and alcohol (see Chapter 13) and fatigue (see Chapter 15). This chapter overlaps with a number of other chapters, including Chapter 5 (novice driver crash patterns), Chapter 7 (attention allocation and maintenance), Chapter 10 (emotional and social factors), Chapter 11 (risky driving), Chapter 14 (attention deficit hyperactivity disorder [ADHD]), and Chapter 16 (passengers). Where overlap occurs, we reference other chapters within the handbook but also retain a singular focus on teen driver distraction. In addition, vulnerable road users such as pedestrians, bicyclists, motorcyclists, and all-terrain vehicles (ATVs) are not considered in this chapter but represent an important and underinvestigated set of problems (WHO, 2015). For example distracted walking is exceedingly common among teens and adults, which likely also contributes to deaths and injuries.

12.2 Review of Teen Driver Distraction

12.2.1 Definitions and Caveats

A variety of definitions of driver *distraction* have been put forward. For example, "[d]istraction occurs when a driver voluntarily diverts [his or her] attention to something not related to driving that uses the driver's eyes, ears, or hands" GHSA (2013, p. 3; also see WHO, 2011, pp. 7–10). Lee et al. (2008, p. 34) put forth a simpler definition, "the diversion of attention away from activities critical for safe driving toward a competing activity." In contrast, *mind wandering* is when "attention drifts from its current train of thought (often an external task [such as driving]) to mental content generated by the individual rather than cued by the environment" (Smallwood & Schooler, 2015, p. 489). Self-generated thought is an internal source of distraction, whereas when the source is from the environment, it is considered an external distraction. *Inattention* is insufficient attention to the primary task of driving and may result from internal or external sources. A number of publications define distraction in ways that do not align with the aforementioned definitions, such as what is in a police report (e.g., NHTSA, 2010). Similarly, operational definitions, as opposed to conceptual distraction terms, use a variety of research approaches that require readers to understand the relationships among variables, measurement processes, experimental designs, participant instructions, and conceptual interpretations. Where appropriate, information is provided about definition variances with respect to the interpretation of results.

Novice drivers are considered those who have less than 6 months of driving experience, whereas teen drivers typically range from 14 to 19 years of age (see Chapter 1). Many studies include samples of drivers that are not ideally centered in this age range. For example, samples of drivers typically range from 18 to 25 years of age, which we refer to as "young drivers." When studies of young drivers are the only research available on a particular topic, we will comment on the strengths and weaknesses of the results relative to a sample of teen drivers.

12.2.2 Do Teens and Novice Drivers Engage in Distracting Tasks More Frequently?

A common assumption is that teens are more likely to engage in distraction-related activities and that doing so while driving affects them more due to inexperience than other age groups. In this section, clear and convincing evidence for these working hypotheses and potential effects is examined and critiqued.

12.2.2.1 Self-Reports

A number of nationwide surveys in the United States surveyed drivers who report engaging in a variety of distraction-related activities (see Table 12.1). In general, the percentage of drivers who report doing an activity while driving has increased annually as more drivers own phones and bring them into their vehicles. In 2011, young drivers appeared to text more than drivers aged 35–44, although this latter age group tended to talk more on their phones than those aged 18–20 (Tison et al., 2011). For some tasks, the year-to-year differences within the age groups are less evident. In some cases, the tasks associated with cell phone use change as technology progresses, and use patterns change with respect to talking, texting, and app-based interactions. Questions asked in more recent surveys have expanded to capture these new uses (Ehsani et al., 2015a).

Caution when comparing across columns (studies) in Table 12.1 is required. How a question is asked of respondents and how their answers are categorized can affect the interpretation of the results (see Table 12.1 footnotes). In addition, what people say they do may not always reflect what they actually do. For example, when initially asked, "Do you send text messages or e-mails while driving?," 21% of drivers said that they "always," "almost always," "sometimes," or "rarely" did so (Tison et al., 2011). When these drivers were asked a follow-up question to confirm that they texted while driving, 67% or two-thirds responded positively; however, one-third changed their answer to "no." Those who switched their answer tended to have previously responded that they rarely or sometimes sent texts or e-mail while driving. Over the course of a survey interview, negative social and public perceptions of cell phones and driving may affect responses to subsequent questions. Thus, overall patterns of increases or decreases in reported technology use should be considered in relative terms.

Several older US national surveys of teens are not included in Table 12.1 but provide interesting results in comparison. A survey by the Pew Research Center (Madden & Lenhart, 2009) found that 34%

TABLE 12.1 Percentage of Drivers Who Say They Engage in a Particular Distraction-Related Activity while Driving

Survey/Activity	Tison et al. (2011)[a]		Schroeder et al. (2013)[b]		Ehsani et al. (2015b)[c]
	18–20	35–44	16–20	All	Teens
Read or send text[d]	44.0%	19.0%	71.0%	69.1%	64.8%
Make or accept phone calls	42.6%	50.7%		61.8%	71.1%
Talk to passenger(s)	81.4%	85.3%		94%	
Eat or drink	46.5%	52.4%		72.6%	
Personal grooming	8.1%	7.0%		19.9%	
Read or send e-mail					20.3%
Adjust car radio	70.5%	74.8%		84.4%	
Frequently change music					71.6%
Use iPad, tablet, or computer				3.7%	12.8%
Check a website, social media, Facebook, Twitter					29.1%
Use map, phone, or navigation device					52.6%

[a] Based on a sample of 6002 drivers 18 years and older from all 50 states and the District of Columbia who completed a survey by December 2010. Percentage indicates those who said that they did the activity on all, most, or some of their driving trips. Text and e-mail indicates percentage of drivers indicating that they have sent a text or e-mail while driving.

[b] Based on interviews with 6016 drivers conducted from February to June 2012. Sampling of landline and cell phones approximated a population 16 years and older. Those aged 16–20 comprised 6.5% of the total sample. Percentages indicate whether a driver responded that they always, almost always, sometimes, or rarely did a particular distracting activity (i.e., "yes").

[c] Based on a sample of 1243 teen drivers in grade 11 during the 2012–2013 academic year who were licensed to drive independently. Drivers were asked if they had engaged in a secondary task (Y/N) in the previous 30 days.

[d] Category includes texting and sending an e-mail for Schroeder et al. (2013). Whether a text was read or typed is not discriminated in the three surveys.

of teens aged 16 or 17 ($N = 283$) texted while driving and 52% ($N = 242$) said they talked on a cell phone while driving. In general, younger teen drivers (aged 16–17) tended to report less use than older teens (i.e., aged 18–19). Consistent with the general annual increases observed, in 2009, teen use of technology while driving was lower than in subsequent national surveys (Ehsani et al., 2015a; Schroeder et al., 2013; Tison et al., 2011).

A number of safety-related behaviors, such as texting and not wearing a seat belt, appear to be interrelated. Olsen et al. (2013) reported similar results to Tison et al. (2011) for a sample of 8505 students aged 16 and older in US high schools who said they texted or e-mailed while driving. Texting was associated with an array of other risky behaviors, including nonuse of seat belt, drinking and driving, or being a passenger with a driver who had been drinking. Using cluster analysis, Schroeder et al. (2013) found that drivers aged 16–20 were categorized as being more "distraction prone" (58.3% of drivers; i.e., more likely to engage in more than one type of activity while driving, such as those activities listed in Table 12.1) than "distraction averse" (41.7%). With the exception of respondents aged 21–24 (64.8% distraction prone), all other age categories (i.e., 25–34 or 65+) were more averse to distracted driving than 16- to 20-year-old drivers. In a survey of 1893 Iowa teenage drivers, Westlake and Boyle (2012) found a similar clustering of infrequent, moderate, and frequent distraction engagers. Thus, the collective findings suggest that some teens seek out or engage in activities without necessarily being aware that they are distractions. Others purposively avoid distractions.

Questionnaire research has explored *why* drivers engage in distracting activities even when they know that it is unsafe to do so (Atchley et al., 2011; O'Brien et al., 2010). Young drivers rate texting and driving as dangerous but also concede that this risk had little or no effect on their behavior of texting while driving (Atchley et al., 2011). Social norms on the negative implications and perceptions of distracted driving, including texting, that results in a crash where the driver is responsible have not yet been adopted by drivers compared to norms associated with drinking and driving (Atchley et al., 2012). Others have found that adding other dimensions of social norms, such as group and moral norms, explained more variance in attitudes toward texting and driving (Nemme & White, 2010). For example, young drivers in particular may feel that communicating with friends, family, and coworkers while driving takes precedence over perceived risk and safety. Rejecting peer pressure to text and drive is difficult for younger drivers—especially if they feel left out by their friends when they do not respond immediately. Impulse control or inhibition of immediate gratification from social contact is perhaps more difficult for young drivers, whose self-regulatory capabilities are still developing (Dahl, 2008; Chapter 9). Moderating unpleasant emotional states or feeling lonely may also motivate individuals to seek out social contact or try to block out negative states while driving (Feldman et al., 2011), which also has implications for drugs and alcohol (Chapter 13). A number of complex social, emotional, and individual differences appear to moderate distraction-related activities.

12.2.2.2 Observational Studies

Although the data from self-reports provide an interesting perspective, it is important to corroborate some of the patterns using different approaches in light of potential reporting biases. Observational studies, carried out in the field, offer one such avenue. Table 12.2 presents observations of drivers who were holding a cell phone to their ear, using a hands-free (HF) setup, or manipulating a handheld device while driving from the years 2011 to 2014. The approximate age of the drivers and activity of the drivers were observed while they were stopped at controlled intersections in locations throughout the United States during the day (Pickrell, 2014; Pickrell & Kc, 2015; Pickrell & Ye, 2013). Extrapolating from these observations at any given moment in the United States in 2014, approximately 6% of young drivers were holding a cell phone to their ear, and 5% were manipulating their phone, which includes texting activities. Based on many years of National Highway Traffic Safety Administration (NHTSA) observations (also see NHTSA, 2009, 2010), young drivers are more likely to engage in talking and texting with a cell phone than other age groups. However, the use of headsets to talk appears roughly equivalent across young and middle-aged groups, although there have been some declines in recent years in the latter group. Several naturalistic studies, which reflect small regional samples, also report similar rates of cell phone talking and manipulation and age group differences (Foss & Goodwin, 2014; Klauer et al., 2014). Based on changes in smartphone technology and uses, increases in manipulation and decreases in holding a phone to the ear are expected in subsequent years—a trend that is already starting to emerge in Table 12.2. In general, distraction engagement at intersections is likely higher when stopped than while drivers are in motion.

Several regional observational studies of drivers in the United States identify a wider range of distraction types that drivers engage in but do not necessarily identify teen and young drivers. Of the 3265 drivers observed during winter, weekday mornings in Jefferson County, Alabama, 32.7% were engaged

TABLE 12.2 Percentage of Drivers Who Were Observed Engaging in Certain Activities while Stopped at Controlled Intersections during the Daytime

Survey	Age Range	2011	2012	2013	2014
Drivers holding cell phone to ear	16–24	6.5	5.9	5.9	5.8
	25–69	4.8	5.4	4.7	4.3
Drivers speaking with a visible headset	16–24	0.6	0.7	0.7	0.7
	25–69	0.7	0.7	0.4	0.4
Drivers manipulating handheld devices	16–24	3.7	3.0	2.9	4.8
	25–69	1.1	1.4	1.6	2.0

Source: Pickrell, T. M., and Kc, S., *Driver Electronic Device Use in 2014 (Traffic Safety Facts: Research. Note DOT HS 812 197)*. NHTSA, Washington, DC, 2015.

Note: Percentage of drivers observed doing a particular activity, as part of the National Occupant Protection Use Survey (NOPUS).

in distracting activities while approaching 11 intersections (Huisingh et al., 2015). The types of distraction included interacting with passengers, talking on a phone, external vehicle distraction, texting/dialing, grooming, smoking, drinking, reaching into another seat, eating, manipulating the radio, wearing headphones, singing, pets, and other. Kidd et al. (2015) observed 16,556 drivers in four Northern Virginia communities during daytime and nighttime while traveling in traffic, in a roundabout and while stopped, or moving through a signalized intersection. Engagement in different distraction types, which were similar to the preceding list, varied depending on the driving context, with higher use at intersections and lower use in roundabouts. Driver age, presence of passengers, and time of day also affected engagement. These studies illustrate the scope of distraction types, which are much broader than just using smartphones. Traffic and vehicle environment, driver demand, and individual factors also affect engagement.

In summary, over time, teens and drivers of all ages report doing more distracting activities including smartphone use, and in observational studies, young drivers are more frequently observed to be manipulating handheld (HH) devices than older drivers. Patterns of distracting and risky behavior appear to be interrelated. Thus, performing one distracting task may increase the likelihood of engaging in other distracting or risky behaviors. It is plausible that learning, feedback, consequences, and individual differences play a role in governing the continuation and expansion of in-vehicle activities. Some teen drivers are more willing to engage in distractions, while others are not; however, there is a need for further research using more sophisticated approaches, such as path analysis, structural equation modeling, and time series, in order to better elucidate these effects.

12.2.3 What Are the Sources of Interference between Distracting Tasks and Driving?

The previous section articulated some of the patterns of use with respect to teenage drivers and various distracting activities. But, given that a task is being attempted concurrently, what are the sources of interference with driving? Studies of driver distraction have examined different types of tasks (e.g., music manipulation, texting), employing different methods (e.g., naturalistic, driving simulation). In general, distracting tasks can interfere with the driving task in different ways, mapping conceptually to encoding, processing, or response functions. Distraction is sometimes described in terms of a simple trichotomy of visual (eyes off road), cognitive (mind off road), and physical (hands and feet off the wheel or pedals), which has become common language to communicate with the public about how distraction affects driving (GHSA, 2013; WHO, 2011). Similarly, within traffic safety domains, tasks are categorized as visual–manual or cognitive, which also insufficiently describes the processing requirements placed on the driver.

On closer inspection, however, different distractions affect a range of information processing capabilities in common patterns when shared or interleaved with driving. To illustrate potential interactions among task requirements and information processing, Table 12.3 lists common everyday tasks and evolving technology uses that may produce performance decrements. Intrinsic teen motives often center on communicating and being with friends and listening to music. These tasks are illustrated in Table 12.3. Other common tasks, such as eating, playing games, and accessing web-based media, can be likewise coded. In this model, drivers are conceptualized as multitaskers who coordinate driving and nondriving activities as goals and desires evolve during travel. Obviously, trying to look at the road and nondriving tasks concurrently produces visual interference (e.g., Chisholm et al., 2008), whereas other tasks must share limited attention and cognitive resources, leading to varying decrements (Horrey & Wickens, 2003; Wickens, 2002). Importantly, the intrinsic task structure, frequency of use, timing of use relative to internal and external driving demands, and overall driving exposure also affect relative safety and crash risk.

TABLE 12.3 Examples of Everyday and Technology Task Effects, along with Driving Subtasks, and Their Mapping to (Potentially Overlapping) Perceptual, Processing, and Response Modes

Activity	Encoding/Perception (e.g., Visual, Auditory, Tactile)	Cognitive Processing (e.g., Spatial, Verbal)	Response (e.g., Manual, Vocal)
Driving			
Vehicle control	• Visual (e.g., vehicle position and status, traffic environment) • Auditory (e.g., road noise) • Tactile feedback (from road)	• Integration and memory for objects and features (spatial) • Processing differences between actual and desired vehicle state (e.g., speed, lane position)	• Eye and head movements to access information • Hand (steering) and foot (pedal) inputs to correct vehicle status
Hazard detection	• Visual (objects and potential obstructions) • Auditory (e.g., sirens)	• Processing spatial locations of potential obstacles; prediction of future areas of risk	• Eye and head movements to access information
Navigation	• Visual (e.g., streets, landmarks) • Auditory (in some cases)	• Searching and processing of spatial locations; route or landmarks in working memory • Processing verbal information (e.g., street signs)	• Eye and head movements to access information • Coordinate timing of vehicle control with route selection
Nondriving			
Text messaging	• Visual (access phone, look at screen, select, read, verify responses) • Auditory (attend to notifications and alerts) • Haptic (interpret notifications and alerts)	• Processing verbal information • Working memory (WM)/ long-term memory (LTM) (translate abbreviations, contextualize sender/thread)	• Reach for and manipulate phone; type, hold phone, send
Conversing on cell phone	• Visual (find phone, answer) • Verbal (speak) • Auditory (attend to notifications and alerts) • Haptic (interpret notifications)	• Processing (attending), response selection	• Pick up phone, hold phone to ear, speak (verbal responses)
Managing music	• Visual (manipulate phone or player, scan music) • Auditory (listen to track and compare to intended song or emotion)	• Process song titles • WM/LTM (contextualize song or playlist affect)	• Manipulate phone or player; scroll and select songs, stations, or playlists
Interacting with passengers	• Visual, auditory, and (possibly) tactile inputs • Potential nonverbal (visual) information	• Processing (attending), response selection	• Verbal responses, manual gestures • Eye and head movements to areas of mutual interest

12.2.4 How Are Driving Performance, Crash Risk, and Eye Behaviors Impacted by Distracting Tasks for Teen Drivers?

While the previous section articulates conceptually how distracting tasks can interfere with the encoding, processing, and response functions required while driving, the current section focuses on findings from empirical studies of driver distraction. Given the extensive literature on the subject, we necessarily limit our scope somewhat, highlighting recent findings most relevant for teen drivers.

12.2.4.1 Driving Performance

Driving performance is multifaceted and is often described in terms of vehicle control and hazard awareness. There are many different measures of vehicle control that have been examined in studies of driving, but generally, they tend to be related to the lateral and longitudinal dimensions. (See SAE, 2015, for definitive definitions of specific measures.) For example, lateral control can be inferred from a number of measures regarding how well centered the vehicle is in its respective lane. Longitudinal control includes measures of speed control as well as the clearances and spacing with other surrounding vehicles (e.g., headway). Hazard awareness has likewise been defined in different ways; however, it often relates to the timely detection and awareness of potential road hazards as well as the appropriate and effective mitigation of these hazards (see Chapters 6 through 8 and 28). For example, in the case of a vehicle cutting sharply in front of a driver, hazard awareness, anticipation, and mitigation might be characterized by the amount of time needed to orient toward the emerging threat and the time until one is required to execute a braking or steering response.

Talking and texting—two common in-vehicle tasks—are examined in depth. To illustrate the effects on driving performance, we compare studies that have used novice or teen drivers with composite measures from two recent meta-analyses. A total of 113 experiments with 4720 participants were analyzed for a wide range of performance measures while talking on a mobile phone or with a passenger (Caird et al., in review). When talking, HF and HH phones modestly increase reaction time to critical events (RT), detection of targets, and collisions over baseline (i.e., driving without a distracting task). Adaptive safety strategies such as increasing headway, decreasing speed, or improving lateral control showed little or no effect, which indicates that drivers do not necessarily compensate while talking. Performance decrements were similar for HF and HH phones. Three studies with sufficient quality assessed novice (Chisholm et al., 2006; Kass et al., 2007) or teen drivers (Stavrinos et al., 2013) while talking on a cell phone and driving. The former two studies used HF phones, whereas the latter, an HH phone. Chisholm et al. found long perception-reaction times (PRTs) for novices (M = 17 years) compared to experienced (M = 32 years) drivers when braking to a lead vehicle, to a vehicle pullout, or to a pedestrian event. Kass et al. (2007) found that novice drivers had more infractions such as collisions and speeding compared to experienced drivers when conversing on an HF phone during simulated driving. Thirty novice (16–18 years) drivers did not differ from young adults (M = 23.3) in lane deviations or speed variability when talking on an HH phone compared to driving without doing these tasks (Stavrinos et al., 2013).

Eyes off the road from texting, as previously introduced, has a detrimental effect on processing of driving and nondriving tasks. In a meta-analysis of 28 texting studies with 977 participants, large effect sizes for eye movements while typing and reading messages were found (Caird et al., 2014). Moderate effect sizes were found for RT, headway (time, distance), and headway variance. Modest effect sizes were found for detection, collisions, lateral control, and speed. Almost every variable, with the exception of speed variability, was affected by typing texts. Reading affected performance measures but to a lesser degree. Only 1 of the 28 studies (Stavrinos et al., 2013) also contained a sample of texting teen drivers, and no differences were found between novice and young drivers for speed variability and lane deviation. Narad et al. (2013) and Rumschlag et al. (2015) published more recent studies with samples of 16- to 17- and 18- to 24-year-old drivers, respectively. Neither study found differences between teens and more experienced drivers. Overall, samples of teen drivers when talking or texting sometimes exhibited differences when compared to more experienced groups, but other studies did not necessarily replicate these results.

12.2.4.2 Crashes

Are teen drivers involved in more crashes, including fatal crashes, from distracted driving than other age groups? In the United States, the proportion of fatal distraction crashes is higher for 15- to 19-year-old drivers than any other age group, including those aged 30–39 (NHTSA, 2013, 2014b, 2015). Across 2011–2013, the percentage of fatalities for both age groups remained constant but was higher in teens (see

Table 12.4). However, from 2011 to 2013, the percentage of 15- to 19-year-old drivers using a cell phone when a fatality occurred declined from 21% to 15%. Overall, cell phone use produced 411 fatal and 34,000 injury crashes in 2013 (NHTSA, 2015). This decline may be related to the way cell phones are used by this age group (e.g., texting) or by legislation aimed at limiting cell phone use while driving; however, these inferences are speculative.

Other studies of nationally available crash databases have also highlighted the important role of distraction or inattention in contributing to motor vehicle crashes. For example, McKnight and McKnight (2003) analyzed 1000 crashes each from Maryland and California for drivers who were 16–19 years old, and they found that distractions contributed to 3.9% of crashes overall. Similarly, Braitman et al. (2008) examined crashes of 16 year-old drivers in Connecticut during the first 6 months of driving; they found that distraction and inattention were contributors. Using national-level data for the United States, Curry et al. (2011) examined the National Motor Vehicle Crash Causation Survey (NHTSA) between July 2005 and December 2007 with a single driver, vehicle, or environmental factor as the critical reason that led to a crash. A total of 822 teen drivers were involved in 795 serious crashes, which resulted in injury or property damage requiring emergency medical services (EMS). Hazard recognition errors, which include distraction and inattention as factors, were a major contributor to crashes. In this study, distraction was defined as "attention that was directed away from the driving task and instead to some event, object, person, or activity inside or outside the vehicle" (p. 1287). This study prominently identified distraction from internal and external events as a contributor to serious teen crashes.

Passengers who are travelling with a teen driver, representing a potential nontechnological source of distraction, may either increase or decrease the likelihood of a crash occurring, depending upon their age and relationship to the driver (also see Chapter 16). For example, Chen et al. (2000) found a linear increase in crash risk for 16- and 17-year-old drivers as the number of teen passengers present increased from one to three or more. In contrast, drivers who were older and had one or more passengers in their vehicle had a lower crash risk than when no passengers were present. It follows that teens transporting teens is a known crash risk problem. Reasons for increased crash risk are assumed to be associated with distractions that produced looked-but-did-not-see errors (White & Caird, 2010) or increased risky behavior such as speeding (Simons-Morton et al., 2005). The gender of the passenger and relationships among driver and passengers may directly and indirectly affect behavior and crash risk. These complex interrelationships are just beginning to be investigated. Curry et al. (2012, p. 592) found that male drivers with male or female peer passengers "were more likely to perform an aggressive act," perform "an illegal maneuver," and be "distracted by an external factor" than when driving alone. Female drivers with male passengers were "more likely to be engaged in at least one interior distracting activity" than when driving alone. Pradhan et al. (2014), using a high-fidelity simulator, found that teen drivers who were 16–19 years of age reduced their horizontal and vertical scanning in the presence of an adolescent

TABLE 12.4 Percentage of 15- to 19- and 30- to 39-Year-Old Drivers in Fatal Distracted Crashes and Percentage of Those in Distracted Fatal Crashes Who Were Using a Cell Phone in 2011–2013

Crash Type	Age	2011	2012	2013
Percentage of fatal crashes, distracted[a]	15–19	11%	10%	10%
	30–39	7%	7%	7%
Percentage of fatal distracted crashes, using cell phone[b]	15–19	21%	19%	15%
	30–39	16%	16%	16%

Note: Proportion of fatal crashes based on Fatal Accident Reporting System (FARS) 2011–2013 (see NHTSA, 2013, 2014b, 2015).

[a] Distraction is attributed to another occupant, a moving object in a vehicle, audio/climate controls, electronic device use, reaching for an object, external person/object/event, smoking, eating, and so forth. Distraction crashes are assumed to be underreported on police accident reports (PARs), and the details of crashes are frequently unknown.

[b] Using a cell phone includes talking or listening on a cell phone, dialing, and other cell phone-related actions.

confederate passenger. The results were interpreted as resulting from increased cognitive load or normative influences of the passenger.

More recently, Ouimet et al. (2015) conducted a systematic review of the contribution of passengers to crash risk based on studies screened and selected from 1989 to August 1, 2013. Eleven studies specifically compared crash risk when passengers were present with either older or younger drivers. Compared to driving alone, passenger presence produced higher fatal and nonfatal crash risk for younger drivers. Similar to the results of Chen et al. (2000), crash risk increased as the number of peer passengers increased. Overall, crash risk is higher for younger drivers with peer passengers and increases as the number of peer passengers increases.

Passenger effects have also been linked to other causal factors. For example, A.G. Williams et al. (2012) points out that many crashes associated with passengers occur in the presence of alcohol greater than 0.08 blood alcohol concentration (BAC) at night. Stronger alcohol, nighttime, and passenger restrictions address all three areas of potential intervention. Nighttime driving doubles fatal crash risk in 16- and 17-year-old drivers due to fatigue (Chapter 15), alcohol and THC (tetrahydrocannabinol) (Chapter 13), and passengers (Williams & Tison, 2012; A.F. Williams et al., 2012; Chapter 16).

Efforts have been made to elucidate the crash risk arising from different forms of distraction. For example, Neyens and Boyle (2008), using data from the 2003 General Estimates System (GES), found that teens are at higher risk for being severely injured in a motor vehicle crash if they were distracted by a cell phone or passenger than by in-vehicle devices (e.g., climate control, radio, eating) or if they were generally inattentive (e.g., appeared lost in thought, "looked but did not see"). Passengers were also more likely to be injured if the teenage driver was distracted by devices or by passengers compared to drivers who were generally inattentive or were not distracted. In some earlier work, Neyens and Boyle (2006) also found that different distractions (cognitive, cell phone, passenger, in-vehicle device) led to different likelihoods of involvement in different crash configurations. For example, at intersections, teen drivers distracted by cognitive activity or passengers were more likely to be involved in a rear-end or angular collisions. In general, cell phone distractions tended to yield higher rates of rear-end collisions, whereas in-vehicle distractions were more likely to lead to collisions with fixed objects. Donmez and Liu (2014) used the US National Automotive Sampling System (NASS) General Estimates System from 2003 to 2008 to examine the interaction of age and distraction type. Inattention, which is a more encompassing term than distraction, from in-vehicle devices, passengers, dialing or texting, talking on a cell phone, and external sources was analyzed for young (16–24 years), middle-aged (25–64 years), and older drivers (65+ years). Dialing and texting had increased odds ratios of severe crashes as age group increased from young to old. However, other patterns of odds ratios were in conflict with other research. For example, passengers and outside sources of distraction were protective for younger drivers, which is at odds with research that was previously discussed.

In light of such divergent database results, it is clear that more work is needed to enhance our understanding of different types of crash configurations with certain distraction contributors. At the same time, a frank and sober understanding of the inherent limitations of these different approaches and data sets is an important consideration. Coding of distraction into databases from police investigations, witness statements, or circumstantial inferences remains inherently unreliable. In addition, variability in police report forms and state reporting requirements biases reported statistics. The result is that crash databases likely underestimate the extent of distraction as a crash contributor.

In addition to database and epidemiological approaches, crashes and near crashes or safety-critical events (SCEs) have also been examined in the context of naturalistic driving studies (see Chapter 26). As with database crash reporting, SCEs from naturalistic studies also have an inherent set of limitations such as variance in crash and near-crash coding and crash surrogates that may not have strong relationships to actual crash contributors (see, e.g., Simmons et al., 2016, for review). Nevertheless, naturalistic studies have illuminated a number of interesting results concerning teen behaviors with distractions. For example, Guo et al. (2013) describe SCE rates of 42 teen drivers during each quarter (3 months) of their first 18 months of licensure. High-, moderate-, and low-risk groups were defined and tracked over the study period. Only the

TABLE 12.5 Odds Ratios and 95% Confidence Intervals for a Number of In-Vehicle Distractions

Activity/Study	Klauer et al. (2014)[a] OR (95% CI)	Simmons et al. (2016)[b] OR (95% CI)
Dialing	8.8 (2.8–24.4)	4.0 (2.7, 6.2)
Locate/answer phone	8.0 (3.7–17.5)	3.6 (2.5, 5.1)
Texting/browsing	3.9 (1.6–9.3)	10.3 (2.4, 44.7)
Eating	3.0 (1.3–6.9)	–
Talking	0.6 (0.2–1.6)	0.9 (0.8, 1.1)

[a] Based on a sample of 42 teen drivers (20 M, 22 F; M = 16.4, SD = 0.3) who were observed for 18 months using a case–cohort approach to determine the prevalence of crash and near-crash incidents.

[b] A stratified meta-analysis of pooled odds ratios from six naturalistic driving studies, which included adult commercial truck drivers and civilian drivers.

moderate-risk group showed decreases in SCEs. Similarly, Lee et al. (2011) described the scope of investigation of the 40-teen SCE study, and initial results indicated that teens have more crashes than adults and have more crashes during the first 6 months of driving (also see Mayhew et al., 2003).

One naturalistic study has described increases in crash risk relative to driving without a distraction for a range of distracting in-vehicle activities (Klauer et al., 2014). The odds ratios associated with various activities are contrasted with studies that have focused on adult drivers (Simmons et al., 2016). Table 12.5 contrasts teen naturalistic crash risk with a meta-analysis of all naturalistic studies that have examined the same activities. Comparisons between teen and adult drivers indicate that teens had a higher crash risk for dialing, locating a phone/answering, and texting. Differences between the results of Klauer et al. (2014) and pooled odds ratio estimates from six naturalistic studies may reflect the effects of teen inexperience and greater likelihood of distraction activity. Eating could not be compared within Table 12.5 because it was not meta-analyzed, whereas talking produced a protective effect, but less so for adults than teens. Naturalistic studies that estimate the crash risk of conversation have found odds ratios less than 1 (Fitch et al., 2013; Klauer et al., 2014). In contrast, odds ratios found in case-crossover designs are about 3–4 (McEvoy et al., 2005; Redelmeier & Tibshirani, 1997). Elvik (2011) pooled odds ratios across crash risk studies at 2.8. The nonconvergent results for talking across epidemiological, naturalistic, and driving performance methods require a close examination of the strengths and weaknesses inherent with each method. Interpretation of the crash risk data to mean that conversation is safe requires consideration that reaching for a phone incurs a higher crash risk.

Simons-Morton et al. (2011) analyzed the "40-teen" ($N = 42$) naturalistic study data of SCEs that occurred in the presence of different passengers. As expected, driving with adult passengers produced 75% fewer SCEs, whereas in the presence of "risky friends," it was 96% higher. Classification of risky friends was based on a median split of a questionnaire that examined how often teens thought their friends smoked, drank, used marijuana, drove without a seat belt, and sped (see also Chapters 10 and 11; Foss & Goodwin, 2014).

12.2.4.3 Eye Behavior

Performance decrements and crash risk due to distraction have been associated with drivers' visual (eye) scanning behavior. Generally, novice and teen drivers have different scanning strategies than more experienced drivers (see Chapters 7 and 27), and clearly, distracting tasks—especially those that involve some visual components (e.g., reading a text message)—will have an impact on visual scanning and driving performance. Since eye movements are a limited resource, such scans away from the roadway will dilute drivers' ability to monitor for critical road information. For example, while looking at an in-dash navigation system to confirm a spoken command, a driver might fail to notice that the vehicle ahead has begun to brake sharply. Other tasks that do not involve visual information (e.g., purely cognitive tasks) can also impact visual scanning behavior. For example, researchers have observed a reduction in the breadth of horizontal and vertical scanning during cognitive distraction (see, e.g., Reimer et al., 2012, and many others, such as Recarte & Nunes, 2003). However, this effect has not necessarily been explicitly tested in novice or teen age groups.

Although the overall amount of time spent looking away from the roadway is of concern, many research studies have identified the importance of especially long single-glance durations from the roadway. By glance duration, we mean the amount of time that drivers look away before returning their attention to the road ahead. For example, a number of studies have shown an association between especially long off-road glances and crash risk (e.g., Horrey & Wickens, 2007; Klauer et al., 2006, 2014; Liang et al., 2012; Pradhan et al., 2011; Simons-Morton et al., 2014). With more and increasingly sophisticated technology, there are many avenues for drivers to take especially long off-road glances to access information from HH, portable, or embedded devices.

While long glances can be problematic for drivers of all age groups, some studies have documented disproportionate glance patterns across drivers in different age groups. For example, Wikman et al. (1998) found that young, inexperienced drivers made in-vehicle glances lasting at least 2.5 seconds 46% of the time in an on-road study. In contrast, experienced drivers exhibited such long glances only 13% of the time. Likewise, Chan et al. (2008) found that novice drivers made in-vehicle glances exceeding 2.5 seconds for 46% of the tasks, while these glances only presented for 12% of the tasks for experienced drivers. Consistent with the pattern for all driver age groups, naturalistic driving studies have shown that teen drivers have a greater likelihood of SCEs based on the length of individual glances (e.g., Simons-Morton et al., 2014).

Although these findings are telling, a more targeted study of teens from 15 to 18 needs to be conducted on a driving simulator. If teens are more willing to text and keep their eyes off the road for longer periods of time (Chan et al., 2008; Olsen et al., 2007), results of a teen-specific sample would exhibit worse eyes-off-road time and degraded lateral and longitudinal vehicle control. The potential combination of increased eyes off the road, which increases crash risk, and less developed vehicle control due to inexperience would provide additional insight into why crash/near crash, injury, and fatal crashes are more likely in teenage drivers.

12.3 Countermeasures

The scientific evidence paints a clear picture that teen drivers are often engaged in distracting activities and that these can have detrimental impacts on performance and safety. A more prominent challenge is in dealing with this significant road safety concern. It is interesting to note that, although many teen (and adult) drivers recognize the safety concerns related to driver distraction, they persist in engaging in these types of activities. For example, surveys have found that respondents who have indicated that driver distraction (or a specific task) is a threat for traffic safety often indicate that they themselves sometimes or often engage in these activities.

Some researchers have framed the issue in the context of calibration, where people's general self-appraisals and situational evaluations are inaccurate; they often overestimate their own performance and ability to deal with distraction (Horrey et al., 2015). Not unrelated, Lerner & Boyd (2005) found that teens were much more willing to engage in distractions, although these were based on momentary (i.e., situational) self-reports and not on actual behavior. More work is needed to understand more of the motivational factors that contribute to such risky behavior—those arising from errors in evaluated skills and higher tolerance for risk—in order to more effectively administer countermeasures. In the following sections, we describe a sample of current and ongoing avenues toward remediation, highlighting areas that have exhibited some success as well as areas where evidence or data are lacking. It is likely that a multipronged (versus single-pronged) approach will ultimately be most effective.

Traffic safety advocates are concerned about the outcomes of teen driver distraction, and calls to action are often made with a limited understanding of countermeasure effectiveness (Pless & Pless, 2014). Legislation, education, and enforcement comprise common solutions for driver distraction.

However, a critical examination of the effectiveness of single and multiple strategies is needed and is briefly discussed in subsequent sections.

12.3.1 Are Laws That Target Cell Phone Use or Texting Effective?

The effectiveness of laws that limit HH mobile phone use and texting in the United States is reviewed by McCartt et al. (2014). In evaluations of bans, mixed results have been reported (see GHSA, 2015; McCartt et al., 2014, for specific studies). Some studies found reductions in crashes and fatalities, while others showed no reductions. High-quality national-level studies are exceptionally difficult to achieve given numerous data, methods, and statistical challenges. Results are dependent on the types of bans, the outcome measures analyzed (e.g., observed use, police-reported crashes, hospitalizations or fatalities), jurisdictional data reporting requirements, control comparisons, and statistical approaches. Based on their review, strong evidence for the effectiveness of cell phone or texting bans is lacking and requires concerted investigation.

One conclusive outcome is that all-driver HH bans in three states reduced long-term, observed use of cell phones in vehicles (McCartt et al., 2010). An increase in HF technology use is also noted in states with HH bans. High-visibility enforcement decreased HH and texting use in Connecticut and New York (Cosgrove et al., 2011).

Three studies have examined teen-specific cell phone bans (Foss et al., 2009; Goodwin et al., 2012; Lim & Chi, 2013), which were previously reviewed (Durbin et al., 2014; McCartt et al., 2014). North Carolina limited cell phone use by drivers younger than 18, and short-term (Foss et al., 2009) and long-term (Goodwin et al., 2012) findings found minimal impact on cell phone use.

Four additional studies have appeared since McCartt et al. (2014) did their review of cell phone and texting law effectiveness. Using the Fatal Accident Reporting System (FARS) to examine fatal crashes from 2000 to 2010, Ferdinand et al. (2014) found that texting laws reduced fatalities by 3% overall. The greatest reductions were in young drivers aged 15–21 in states with these bans. A similar study examined hospitalizations from crashes in the states that had all-age texting bans (Ferdinand et al., 2015). Based on these latter two studies, states that do not have texting bans may want to consider them.

Ehsani et al. (2014) examined the effects of Michigan's all-age texting restriction introduced in 2010 on fatal/disabling injury, nondisabling injury, and property-damage crashes. It is difficult to assess the effects that policy change has had on texting behavior. There is a pattern of small significant changes for some types of crash and some age groups, but after adjusting for the effects of age, gasoline prices, and unemployment, these effects are minimal. The effects of laws on behaviors and resultant crash-type changes are complex and difficult to determine. Ehsani et al. (2015b) found that after adjusting for exposure, in states where restrictions on "texting for all drivers were in place, teens were significantly less likely to report talking or texting while driving." Whether restrictions for all drivers or restrictions specific to teens were more or less effective was not necessarily established (Ehsani et al., 2015b).

Although mobile phone use while driving is considered to be increasing worldwide and a serious threat to road safety, the extent of the problem globally is not known. Implementation of any countermeasures in low- or middle-income countries has not been considered and represents a very large gap in knowledge (WHO, 2015).

12.3.2 Are Passenger Restrictions in Graduated Driver Licensing Effective in Reducing Peer Pressure and Distracting Activities?

Restrictions of peer passengers in graduated driving licensing (GDL) are intended to reduce peer pressure and distracting activities that may affect the driver. The effects of passenger restrictions in GDL appear to vary depending on the stage of licensing (Chen et al., 2006; Vanlaar et al., 2009). For example, Williams et al. (2010) analyzed FARS data for 16- and 17-year-old drivers who had passengers who were from 13 to 19 (but not older or younger) when a fatal crash occurred. A similar proportion of crashes when teen passengers were present was found in 2004 (43%) and 2008 (41%), although the number of states with passenger restrictions increased across the 5 years of data. Other studies have concluded that passenger restrictions are effective at reducing fatalities (Chen et al., 2006; see Chapter 17). Most recently, McCartt and Teoh (2015) found that teen driver fatal crashes declined from 1996 to 2012 for two or more passengers at night, which is consistent with nighttime and passenger GDL restrictions.

12.3.3 Are Media Campaigns Effective in Reducing Teen Driver Distraction?

Media and other public policy initiatives have often been used to address motor vehicle safety concerns (e.g., drunk driving, seat belt use), employing different delivery methods (e.g., television, radio, social media), levels/scopes (e.g., municipal, national), and messaging (e.g., positive/reinforcing, rationale, emotional). Recently, Phillips et al. (2011) carried out a meta-analysis of 67 studies evaluating the efficacy of road safety campaigns internationally. They showed a weighted average reduction of 9% in crashes as a result of these campaigns, suggestive that these approaches can be effective. They also found that the use of personal communication as part of the messaging as well as roadside media were associated with greater reductions. However, the campaigns used in their analysis were primarily alcohol or seat belt related. As such, it is unknown whether these findings generalize to campaigns related to driver distraction.

There have been many media campaigns related to distraction sponsored by government agencies, nonprofit organizations, and telecommunication providers, among others. One noteworthy effort is the Distraction.gov initiative sponsored by the US Department of Transportation, where they compile targeted information and public service announcements (PSAs) for teens, parents, educators, employers, and community groups. However, in spite of these various efforts, including those targeted at teenagers, the scientific evidence for the efficacy of these is generally lacking. Others, such as Sherin et al. (2014), have advocated further legislation to establish public awareness campaigns that target text messaging and driving, which is a good suggestion if sufficient evaluation for establishing effectiveness is included. It is unrealistic to expect that a generalized appeal to avoid distraction will be effective with either teens or adults when drivers hear a phone ring, ping, or buzz in the vehicle.

12.3.4 Can Technology Be Useful in Reducing In-Vehicle Distraction for Teens?

Technological innovations can target the issue of driver distraction in different ways. For example, devices themselves might offer features and functionality that render the interaction less distracting. Voice technologies may allow users to listen to and respond to text messages without having to look at or manually interact with a smartphone. There is some evidence that voice technologies can lead to better driving performance outcomes (e.g., improved lane keeping and speed control) compared to manual interfaces (e.g., manual dialing or texting). Voice technologies reduce visual demand as intended; however, compared to driving alone, performance is often worse with voice interactions (e.g., Barón & Green, 2006). Other recent studies have explored some of the workload implications for a variety of currently available embedded or mobile platforms for voice interaction, showing that different systems and tasks can have very different implications for cognitive workload (Cooper et al., 2014; Strayer et al., 2014, 2015). For example, very short tasks or simple commands incur less workload than do tasks that are more complex or prolonged; these latter tasks are often characterized by overly verbose systems and a higher number of required steps, and they are also more prone to errors (e.g., technology fails to recognize commands and/or executes an undesired action [Strayer et al., 2014]). In addition to potentially detrimental effects on workload, one could speculate that voice technologies, marketed as a safe alternative to HH or manual devices, could lead to inflated perceptions of safety and, consequently, a propensity for more frequent interactions. Many of these effects need to be explored further in controlled studies and with a broad array of performance outcomes—especially for younger drivers.

Another technological innovation involves algorithms that can selectively block certain applications or features when the vehicle starts or when the phone is estimated to be moving in a vehicle. For example, incoming calls or text messages might be silenced and hidden from view, while outgoing calls (save for emergency ones) are blocked. Sometimes, these blocking features need to be manually enabled by the driver before beginning a trip (and hence must be remembered). Recent work by Ebel et al. (2015) carried out a randomized control trial for a cell phone blocking system and found that drivers aged 15–18 decreased the frequency of calling or texting behaviors. Creaser et al. (2015) found similar results for their cell blocking system; however, they also noted that a small subset of teen drivers were motivated and were able to bypass the blocking functions. This latter outcome reinforces the need to test a wider array of use-case scenarios but also that a single approach to remediating risky distracting behaviors is not likely to succeed. The proverbial caveat is that those who need these systems are not likely to use them.

Lastly, we make one last note regarding parental involvement—a nontechnological one: researchers have elucidated the important role that parents play as a role model in the development of teenagers' later driving habits (Carter et al., 2014; Prato et al., 2010). Teenagers who are exposed to safer driving habits by their parents are more likely to engage in these behaviors when they become drivers compared to teens who are not exposed to positive role behaviors. It follows that parents should be fully aware of their own driving habits and styles—especially when driving with their progeny.

12.4 Summary and Conclusion

In this chapter, we have reviewed evidence that teen drivers often expose themselves to distracting activities and more so than other age groups. We have also examined some of the sources of interference and, consequently, some of the detrimental impacts on driving performance, crash risk, and eye behavior. A summary of technology, passenger, and countermeasure effects is listed in Appendix 12.1. As noted earlier, the current evidence presents somewhat of an ominous backdrop for our discussion. As technology and new sources of potential distraction continue their explosive growth, it is important that safety research and countermeasures remain in lockstep; otherwise, an epidemic of unfavorable outcomes will result. Given the relative pace and nature of these evolving issues, this is and will remain a challenge for researchers, policy makers, as well as parents and teens themselves.

12.4.1 Research Limitations, Gaps, and Future Research

Although there has been much research on teen drivers, distracted drivers, and distracted teen drivers, it is clear that there is much more work to do. We note a number of gaps and deficiencies in the studies reviewed. In some cases, samples of teen or novice drivers in experimental distraction studies are somewhat rare, possibly due to the institutional review board (IRB) requirement of obtaining informed consent from their parents or guardians. The cost of recruitment and running a study with teens may be prohibitive.

Another issue that is insufficiently described or controlled for in studies is whether teens are more skilled at using technology such as texting than a comparison group of adult drivers. Thus, teens and young drivers may mitigate some of the decrements produced by distractions by executing interactions with technology with greater speed and efficiency (e.g., Owens et al., 2011). Although teens and young drivers may be more skilled at using technology, they may use technology more frequently in a variety of contexts because they believe that their skills allow them to do so. Many studies do not test drivers to determine baseline device or task performance and do not assess frequency of use. As a result, group differences (or a lack thereof) may reflect proficiency differences between technology and driving experience. However, as a counterpoint, proficiency at a distracting task such as texting might lead to overconfidence in one's ability to perform this task simultaneously while driving (e.g., Horrey et al., 2015).

Age and experience are typically confounded and not easily separated or sufficiently treated as covariates (McCartt et al., 2009; Shope & Bingham, 2008). In addition, many studies report the age of participants but not the number of years a person has driven or, more importantly, the number of kilometers (or miles) driven over his or her career. This problem is compounded by teens and young drivers who do not feel compelled to get their driver's license until they are older for reasons associated with being able to get around without a car, costs, and access (Tefft et al., 2014).

A number of prominent research needs are evident. Gender differences in texting, talking, and interacting with passengers are only dimly understood, and observational, simulation, and naturalistic studies are needed. Emotional development and transient affective states as predictors of compulsive and dangerous distraction require further research (see Chapter 10). The conceptualization of novice and teen drivers as serial processors of single tasks is näive. Facile direction of attention to driving and nondriving tasks with attendant risk accumulations over a period of maturation is the greater context to which investigations need to relate. Finally, teens are a generation of multitaskers who will require support for a set of behaviors that have the least risk.

Appendix 12.1

A Summary of Teen Distraction Research by Method or Approach

Method or Approach	Technology and Passenger Distraction Results	Research Limitations and Gaps
Epidemiological	• Based on databases in the United States, teen drivers have a higher proportion of distracted fatal crashes than other age groups. • Fatal and nonfatal crash risk increases as the number of peer passengers increases and decreases with adult supervision.	Results likely underestimate the true prevalence of distraction as a contributor to fatal, injury, and property-damage crashes.
Naturalistic	• One naturalistic study of teens found higher safety-critical event risk for dialing, reaching for a phone, and texting, but not for talking on a phone. • When compared with a meta-analysis of six naturalistic studies, the teen odds ratios are greater than those from a mixed sample of drivers and vehicle types.	Although naturalistic studies provide interesting insights into driver behavior, a variety of SCE sampling and validity limitations require consideration when generalizing to other populations.
Observational	• Young drivers were observed holding a cell phone to their ear or manipulating it more often than other age groups. • The presence of male passengers with male drivers increases vehicle speed.	Determination of age and activity through windscreens lacks precision.
Survey	• Teens report performing a number of technology and daily activities while driving with or without awareness of safety implications. • The percentage of teens reporting use and distraction increased from 2010 to 2014 but did so in other age groups too. • The types of technology tasks that teens report using have changed over time as smartphone apps have emerged.	What people say they do and what they do in their own vehicles are not necessarily similar, and they require systematic verification.
Driving simulation	• Texting degrades almost all measures of driving performance for adult, young, and teen drivers. • Talking on a cell phone or with a passenger slows RT for both adults and teens. • Manipulation of nomadic music devices affects steering control and RT in adults and young drivers. • Passengers, depending on the interaction with the driver, may produce "looked but did not see" errors, increased speed, and lateral control deviations for young drivers.	Engaging in device use when desired by a participant as opposed to required by an experimenter requires alternative method protocols.
Countermeasures	• GDL passenger restrictions are effective at reducing fatalities.	Although promising, cell phone legislation, education, and a variety of technologies require further evaluations. Translation of effective countermeasures to middle- and low-income countries is a challenge.

Acknowledgments

We are grateful to Marv Dainoff, Don Fisher, David Kidd, Sandra Mish, Sarah Simmons, and Lana M. Trick for insightful comments and edits that improved our chapter. The AUTO21 Network of Centres of Excellence (NCE) provided funding through the Convergent Evidence from Naturalistic, Simulation and Epidemiological Data (CENSED) Network, which supported aspects of this research.

References

Atchley, P., Atwood, S., & Boulton, A. (2011). The choice of text and drive in younger drivers: Behavior may shape attitude. *Accident Analysis & Prevention, 43*, 134–142.

Atchley, P., Hadlock, C., & Lane, S. (2012). Stuck in the 70s: The role of social norms in distracted driving. *Accident Analysis & Prevention, 48*, 279–284.

Barón, A., & Green, P. (2006). *Safety and usability of speech interfaces for in-vehicle tasks while driving: A brief literature review (Rep. No. UMTRI-2006-5).* Ann Arbor, MI: University of Michigan Transportation Research Institute.

Braitman, K.A., Kirely, B.B., McCartt, A.T., & Chaudhary, N.K. (2008). Crashes of novice teen-aged driver: Characteristics and contributing factors. *Journal of Safety Research, 39*, 47–54.

Caird, J.K., & Dewar, R.E. (2007). Driver distraction. In R.E. Dewar & P. Olson (Eds.), *Human Factors in Traffic Safety (2nd Ed.)* (pp. 195–229). Tucson, AZ: Lawyers & Judges Publishing.

Caird, J.K., Johnston, K., Willness, C., Asbridge, M., & Steel, P. (2014). A meta-analysis of the effects of texting on driving. *Accident Analysis & Prevention, 71*, 311–318.

Caird, J.K., Wiley, K., Johnston, K., Simmons, S., & Horrey, W. (in review). Does talking on a cell phone or with a passenger affect driving performance? A systematic review and meta-analysis of experimental studies. *Accident Analysis & Prevention.*

Carter, P.M., Bingham, C.R., Zakrajsek, J.S., Shope, J.T., & Sayer, T.B. (2014). Social norms' and risk perception: Predictors of distracted driving behavior among novice adolescent drivers. *Journal of Adolescent Health, 54*(5 Suppl), S32–S41.

Cazzulino, F., Burke, R.V., Muller, V., Arbogast, H., & Upperman, J.S. (2014). Cell phones and young drivers: A systematic review regarding the association between psychological factors and prevention. *Traffic Injury Prevention, 15*, 234–242.

Chan, E., Pradhan, A.K., Knodler Jr., M.A., Pollatsek, A., & Fisher, D.L. (2008). Empirical Evaluation on Driving Simulator of Effect of Distractions Inside and Outside the Vehicle on Drivers' Eye Behavior. In *Transportation Research Board 87th Annual Meeting* (No. 08-2910).

Chen, L.-H., Baker, S.P., Braver, E.R., & Li, G. (2000). Carrying passengers as a risk factor for crashes fatal to 16- and 17-year-old drivers. *Journal of the American Medical Association, 283*, 1578–1582.

Chen, L.-H., Baker, S.P., & Li, G. (2006). Graduated driver licensing programs and fatal crashes of 16-year-old drivers. *Pediatrics, 118*, 56–62.

Chisholm, S., Caird, J.K., Teteris, E., Lockhart, J., & Smiley, A. (2006). Novice and experienced driver performance with cell phones. *Proceedings of the 50th Annual Human Factors and Ergonomics Meeting* (pp. 2354–2358). Santa Monica, CA: Human Factors and Ergonomics Society.

Chisholm, S., Caird, J.K., & Lockhart, J. (2008). The effects of practice with MP3 players on driving performance. *Accident Analysis & Prevention, 40*, 704–713.

Collet, C., Guillot, A., & Petit, C. (2010). Phoning while driving II: A review of driving conditions influence. *Ergonomics, 53*(5), 602–616.

Cooper, J.M., Ingebretsen, H., & Strayer, D.L. (2014). *Mental Workload of Common Voice-Based Vehicle Interactions across Six Different Vehicle Systems.* Washington, DC: AAA Foundation for Traffic Safety.

Cosgrove, L., Chaudhary, N., & Reagan, I. (2011). *Four High-Visibility Enforcement Demonstration Waves in Connecticut and New York Reduce Hand-Held Phone Use (Traffic Safety Facts, Research Note, Rep. No. DOT HS 811 845).* Washington, DC: NHTSA.

Creaser, J.I., Edwards, C.J., Morris, N.L., & Donath, M. (2015). Are cellular phone blocking applications effective for novice teen drivers? *Journal of Safety Research, 54*, 75–78.

Curry, A.E., Hafetz, J., Kallen, M.J., Winston, F.K., & Durbin, D.R. (2011). Prevalence of teen driving errors leading to serious motor vehicle crashes. *Accident Analysis and Prevention, 43*, 1285–1290.

Curry, A.E., Mirman, J.H., Kallan, M.J., Winston, F.K., & Durbin, D.R. (2012). Peer passengers: How do they affect teen crashes? *Journal of Adolescent Health, 50*, 588–594.

Dahl, R.E. (2008). Biological, developmental, and neurobehavioral factors relevant to adolescent driving risks. *American Journal of Preventive Medicine, 35*, S278–S284.

Donmez, B., & Liu, Z. (2014). Associations of distraction involvement and age with driver injury severities. *Journal of Safety Research, 52*, 23–28.

Durbin, D.R., McGehee, D.V., Fisher, D., & McCartt, A. (2014). Special considerations in distracted driving with teens. *Annals of Advances in Automotive Medicine, 58*, 69–83.

Ebel, B.E., Boyle, L., O'Conner, S., Qui, Q., Bresnahan, B., Maeser, J., Kernic, M., & Rowhani-Rahbar, A. (2015). Randomized trial of cell phone blocking and in-vehicle camera to reduce high-risk driving events among novice drivers (Abstract). Retrieved from http://www.abstact2view.com/pas/view/php?nu=PAS15L1_2175.2.

Ehsani, J., Bingham, C.R., Ionides, E., & Childers, D. (2014). The impact of Michigan's text messaging restriction on motor vehicle crashes. *Journal of Adolescent Health, 54*, S68–S74.

Ehsani, J., Li, K., & Simons-Morton (2015a). Teenage drivers portable electronic device use while driving. *Proceedings of the Eighth International Driving Symposium on Human Factors in Driver Assessment, Training and Vehicle Design* (pp. 218–224). Iowa City: University of Iowa.

Ehsani, J., Simons-Morton, B.G., Perlus, J., & Xie, Y. (2015b). Association between cell phone restrictions and teens' self-reported cell phone use while driving. *Proceedings of the Eighth International Driving Symposium on Human Factors in Driver Assessment, Training and Vehicle Design* (pp. 323–329). Iowa City: University of Iowa.

Elvik, R. (2011). Effects of mobile phone use on accident risk: Problems of meta-analysis when studies are few and bad. *Transportation Research Record, 2236*, 20–26.

Feldman, G., Greeson, J., Renna, M., & Robbins-Monteith, K. (2011). Mindfulness predicts less texting while driving among young adults: Examining attention- and emotion-regulation motives as potential mediators. *Personality and Individual Differences, 51*, 856–861.

Ferdinand, A.O., Menachemi, N., Sen, B., Blackburn, J.L., & Morrisey, M. (2014). The impact of texting laws on motor vehicle fatalities. *American Journal of Public Health, 104*(8), 1370–1377.

Ferdinand, A.O., Menachemi, N., Blackburn, J.L., Sen, B., Nelson, L., & Morrisey, M. (2015). The impact of texting bans on motor vehicle crash-related hospitalizations. *American Journal of Public Health, 105*(5), 859–865.

Fitch, G.M., Soccolich, S.A., Guo, F. et al. (2013). *The Impact of Hand-Held and Hands-Free Cell Phone Use on Driving Performance and Safety-Critical Event Risk (Rep. No. DOT HS 811 757).* Washington, DC: National Highway Traffic Safety Administration.

Foss, R.D., & Goodwin, A.H. (2014). Distracted driving behaviors and distracted conditions among adolescent drivers: Findings from a naturalistic study. *Journal of Adolescent Health, 54*, S50–S60.

Foss, R.D., Goodwin, A.H., McCartt, A.H., & Hellinga, L.A. (2009). Short-term effects of a teenage driver cell phone restriction. *Accident Analysis & Prevention, 41*, 419–424.

Goodwin, A.H., O'Brien, N.P., & Foss, F.D. (2012). Effect of North Carolina's restriction on teenage driver cell phone use two years after implementation. *Accident Analysis & Prevention, 48*, 363–367.

Governors Highway Safety Association (GHSA). (2013). Distracted driving: Survey of the states. Retrieved from http://www.ghsa.org/html/stateinfo/laws/cellphone_laws.html. Accessed November 5, 2013.

Governors Highway Safety Association (GHSA). (2015). Cell phone and texting laws. Retrieved from http://www.ghsa.org/html/stateinfo/laws/cellphone_laws.html. Accessed July 9, 2015.

Guo, F., Simons-Morton, B.G., Klauer, S.E., Ouimet, M.C., Dingus, T.A., & Lee, S.E. (2013). Variability in crash and near-crash risk among novice teenage drivers: A naturalistic study. *Journal of Pediatrics, 163*(6), 1670–1676.

Horrey, W.J., & Wickens, C.D. (2003). Multiple resource modeling of task interference in vehicle control, hazard awareness and in-vehicle task performance. *Proceedings of the Second International Driving Symposium on Human Factors in Driver Assessment, Training, and Vehicle Design, Park City, Utah,* 7–12.

Horrey, W.J., & Wickens, C.D. (2007). In-vehicle glance duration: Distributions, tails, and a model of crash risk. *Transportation Research Record, 2018,* 22–28.

Horrey, W.J., Lesch, M.F., Mitsopoulos-Rubens, E., & Lee, J.D. (2015). Calibration of skill and judgment in driving: Development of a conceptual framework and the implications for road safety. *Accident Analysis & Prevention, 76,* 25–33.

Huisingh, C., Griffin, R., & McGwin, G. (2015). The prevalence of distraction among passenger vehicle drivers: A roadside observational approach. *Traffic Injury Prevention, 16,* 140–146.

Kass, S.J., Cole, K.S., & Stanny, C.J. (2007). Effects of distraction and experience on situation awareness and simulated driving. *Transportation Research: Part F, 10,* 321–329.

Kidd, D.G., Tison, J., Chaudhary, N.K., McCartt, A.T., & Casanova-Powell, T.D. (2015). *The Influence of Roadway Situation, Other Contextual Factors, and Driver Characteristics on the Prevalence of Driver Secondary Behaviors.* Arlington, VA: Insurance Institute for Highway Safety.

Klauer, S.G., Dingus, T.A., Neale, T.A., Sudweeks, J., & Ramsey, D.J. (2006). *The Impact of Driver Inattention on Near-Crash/Crash Risk: An Analysis Using the 100-Car Naturalistic Study Data (Rep. No. DOT HS 810 594).* Washington, DC: National Highway Traffic Safety Administration.

Klauer, S.G., Guo, F., Simons-Morton, B.G., Ouimet, M.C., Lee, S.E., & Dingus, T.A. (2014). Distracted driving and risk of road crashes among novice and experienced drivers. *New England Journal of Medicine, 370,* 54–59.

Klauer, S.G., Ehsani, J.P., McGehee, D.V., & Manser, M. (2015). The effect of secondary task engagement on adolescents' driving performance and crash risk. *Journal of Adolescent Health, 57,* S36–S43.

Lee, J.D., Young, K.L., & Regan, M.A. (2008). Defining driver distraction. In M.A. Regan, J.D. Lee, & K.L. Young (Eds.), *Driver Distraction: Theory, Effects and Mitigation* (pp. 31–40). New York: CRC Press.

Lee, S.E., Simons-Morton, B.G., Klauer, S.E., Oimet, M.C., & Dingus, T.A. (2011). Naturalistic assessment of novice driver crash experience. *Accident Analysis & Prevention, 43,* 1472–1479.

Lerner, N., & Boyd, S. (2005). *On-road Study of Willingness to Engage in Distracting Tasks.* Washington, DC: NHTSA.

Liang, Y., Lee, J.D., & Yekhshatyan, L. (2012). How dangerous is looking away from the road? Algorithms predict crash risk from glance patterns in naturalistic driving. *Human Factors: The Journal of the Human Factors and Ergonomics Society, 54*(6), 1104–1116.

Lim, S.H., & Chi, J. (2013). Are cell phone laws in the U.S. effective in reducing fatal crashes involving young drivers. *Transportation Policy, 27,* 158–163.

Madden, M., & Lenhart, A. (2009). *Teens and Distracted Driving: Texting, Talking and Other Uses of the Cell Phone behind the Wheel.* Washington, DC: Pew Research Center.

Mayhew, D.R., Simpson, H.M., & Pak, A. (2003). Changes in collision rates among novice drivers during the first few months of driving. *Accident Analysis & Prevention, 35,* 683–691.

McCartt, A., & Teoh, E.R. (2015). Tracking progress in teenage driver crash risk in the United States since the advent of graduated licensing programs. *Journal of Safety Research, 53,* 1–9.

McCartt, A., Hellinga, L.A., & Braitman, K.A. (2006). Cell phones and driving: A review of the literature. *Traffic Injury Prevention, 7,* 89–106.

McCartt, A., Mayhew, D.R., Braitman, K.A., Ferguson, S.A., & Simpson, H.M. (2009). Effects of age and experience on young driver crashes: Review of recent literature. *Traffic Injury Prevention, 10,* 209–219.

McCartt, A., Hellinga, L.A., Strouse, L.M., & Farmer, C.M. (2010). Long-term effects of handheld cell phone laws on driver handheld cell phone use. *Traffic Injury Prevention, 11*, 133–141.

McCartt, A.T., Kidd, D.G., & Teoh, E.R. (2014). Driver cellphone and texting bans in the United States: Evidence of effectiveness. *Annals of Advances in Automotive Medicine, 58*, 99–114.

McEvoy, S.P., Stevenson, M.R., McCartt, A.T., Woodward, M., Haworth, C., Palamara, P., & Cercarelli, R. (2005). Role of mobile phones in motor vehicle crashes resulting in hospital attendance: A case-crossover study. *BMJ, 331*, 428–30.

McKnight, A.J., & McKnight, A.S. (2003). Young novice drivers: Careless or clueless? *Accident Analysis & Prevention, 35*, 921–925.

Narad, M., Garner, A.A., Brassell, A.A., Saxby, D., Antonini, T.N., O'Brien, K.M., Tamm, L., Matthews, G., & Epstein, J.N. (2013). The impact of distraction on the driving performance of adolescents with and without attention deficit hyperactivity disorder. *JAMA Pediatrics, 167*(10), 933–938.

National Highway Traffic Safety Administration (NHTSA). (2009). *Driver Electronic Device Use in 2008 (Rep. No. DOT-HS-811-184)*. Washington, DC: NHTSA.

National Highway Traffic Safety Administration (NHTSA). (2010). *Driver Electronic Device Use in 2009 (Rep. No. DOT-HS-811-372)*. Washington, DC: NHTSA.

National Highway Traffic Safety Administration (NHTSA). (2013). *Traffic Safety Facts: Distracted Driving 2011 (Rep. No. DOT-HS-811-737)*. Washington, DC: NHTSA.

National Highway Traffic Safety Administration (NHTSA). (2014a). *Traffic safety facts: Distracted driving 2012 (Rep. No. DOT-HS-812-012)*. Washington, DC: NHTSA.

National Highway Traffic Safety Administration (NHTSA). (2014b). *Traffic Safety Facts, 2012 Data: Young Drivers (Rep. No. DOT-HS-812-019)*. Washington, DC: NHTSA.

National Highway Traffic Safety Administration (NHTSA). (2015). *Traffic Safety Facts: Distracted Driving 2013 (Rep. No. DOT-HS-812-132)*. Washington, DC: NHTSA.

National Research Council, Institute of Medicine, Transportation Research Board (2007). *Preventing Teen Motor Crashes: Contributions of the Behavioral and Social Sciences, Workshop Report*. Washington, DC: National Academies Press.

Nemme, H.E., & White, K.M. (2010). Texting while driving: Psychosocial influences on young peoples' intentions and behaviour. *Accident Analysis & Prevention, 42*, 1257–1265.

Neyens, D.M., & Boyle, L.N. (2006). The effect of distractions on the crash types of teenage drivers. *Accident Analysis & Prevention, 39*, 206–212.

Neyens, D.M., & Boyle, L.N. (2008). The influence of driver distraction on the severity of injuries sustained by teenage drivers and their passengers. *Accident Analysis & Prevention, 40*, 254–259.

O'Brien, N.P., Goodwin, A.H., & Foss, R.D. (2010). Talking and texting among teenage drivers: A glass half empty or half full. *Traffic Injury Prevention, 11*, 549–554.

Olsen, E.C.B., Lee, S.E., & Simons-Morton, B.G. (2007). Eye movement patterns for novice drivers: Does 6 months of driving experience make a difference? *Transportation Research Record, 2009*, 8–14.

Olsen, E.O., Shults, R.A., & Eaton, D.K. (2013). Texting while driving and other risky motor vehicle behaviors among US high school students. *Pediatrics, 131*, e1708–15.

Ouimet, M.C., Pradhan, A.K., Brooks-Russell, A., Ehsani, J.P., Berbiche, D., & Simons-Morton, B.G. (2015). Young drivers and their passengers: A systematic review of epidemiological studies on crash risk. *Journal of Adolescent Health, 57*, S24–S35.

Owens, J.M., McLaughlin, S.B., & Sudweeks, J. (2011). Driver performance while text messaging using handheld and in-vehicle systems. *Accident Analysis & Prevention, 43*, 939–947.

Pediatrics. (2006). The teen driver. *Pediatrics, 118*(6), 2570–2581.

Phillips, R.O., Ullberg, P., & Vaa, T. (2011). Meta-analysis of the effect of road safety campaigns on accidents. *Accident Analysis & Prevention, 43*, 1204–1218.

Pickrell, T.M. (2014). *Driver Electronic Device Use in 2012 (Traffic Safety Facts: Pre. No. DOT HS 811 884)*. Washington, DC: NHTSA.

Pickrell, T.M., & Kc, S. (2015). *Driver Electronic Device Use in 2014 (Traffic Safety Facts: Research. Note DOT HS 812 197).* Washington, DC: NHTSA.

Pickrell, T.M., & Ye, T.J. (2013). *Driver Electronic Device Use in 2011 (Traffic Safety Facts: Pre. No. DOT HS 811 719).* Washington, DC: NHTSA.

Pless, C., & Pless, B. (2014). Mobile phones and driving. *BMJ, 348*, 1193.

Pradhan, A., Simons-Morton, B., Lee, S., & Klauer, S. (2011). Hazard perception and distraction in novice drivers: Effects of 12 months driving experience. In *Proceedings of the Sixth International Driving Symposium on Human Factors in Driving Assessment, Training and Vehicle Design*. Iowa City: Iowa.

Pradhan, A.K., Li, K., Bingham, C.R., Simons-Morton, B.G., Ouimet, M.C., & Shope, J.T. (2014). Peer passenger influences on male adolescent drivers' visual scanning behavior during simulated driving. *Journal of Adolescent Health, 54*, S42–S49.

Prato, C.G., Toledo, T., Lotan, T., Taubman, O., & Taubman-Ben-Ari, O. (2010). Modeling the behavior of novice young drivers during the first year after licensure. *Accident Analysis & Prevention, 42*, 480–486.

Recarte, M.A., & Nunes, L.M. (2003). Mental workload while driving: Effects on visual search, discrimination, and decision making. *Journal of Experimental Psychology: Applied, 9*(2), 119–137.

Redelmeier, D.A., & Tibshirani, R.J. (1997). Association between cellular telephone calls and motor vehicle collisions. *New England Journal of Medicine, 336*, 453–458.

Regan, M.A., Lee, J.D., & Young, K.L. (Eds.) (2008). *Driver Distraction: Theory, Effects and Mitigation*. New York: CRC Press.

Reimer, B., Mehler, B., Wang, Y., & Coughlin, J.F. (2012). The impact of systematic variation of cognitive demand on drivers' visual attention across multiple age groups. *Human Factors, 54*(3), 454–468.

Rumschlag, G., Palumbo, T., Martin, A., Head, D., George, R., & Commissaris, R.L. (2015). The effects of texting on driving performance in a driving simulator: The influence of driver age. *Accident Analysis & Prevention, 74*, 145–149.

Schroeder, P., Meyers, M., & Kostyniuk, L. (2013). *National Survey on Distracted Driving Attitudes and Behaviors—2012 (Rep. No. DOT HS 811 729).* Washington, DC: National Highway Traffic Safety Administration.

Sherin, K.M., Lowe, A.L., Harvey, B.J., Leiva, D.G., Malik, A., Matthews, S., Suh, R., & the American College of Preventive Medicine Prevention Practice Committee. (2014). Preventing texting and driving: A statement of the American College of Preventive Medicine. *American Journal of Preventive Medicine, 47*(5), 681–688.

Shope, J.T., & Bingham, R. (2008). Teen driving: Motor-vehicle crashes and factors that contribute. *American Journal of Preventive Medicine, 35*(3S), S261–S271.

Simmons, S., Hicks, A., & Caird, J.K. (2016). Safety-critical events associated with cell phone tasks measured through naturalistic driving studies: A systematic review and meta-analysis. *Accident Analysis & Prevention, 87*, 161–169.

Simons-Morton, B.G., Lerner, N., & Singer, J. (2005). The observed effects of teenage passengers on the risky behavior of teenage passengers. *Accident Analysis & Prevention, 37*, 973–982.

Simons-Morton, B.G., Oimet, M.-C., Zhang, Z., Klauer, S.E., Lee, S.E., Wang, J., Chen, R., Albert, P., & Dingus, T.A. (2011). The effect of passengers and risk-taking friends on risky driving and crashes/near crashes among novice teenagers. *Journal of Adolescent Health, 49*, 587–593.

Simons-Morton, B.G., Guo, F., Klauer, S.G., Ehsani, J.P., & Pradhan, A.K. (2014). Keep your eyes on the road: Young driver crash risk increases according to the duration of distraction. *Journal of Adolescent Health, 54*, S61–S67.

Smallwood, J., & Schooler, J.W. (2015). The science of mind wandering: Empirically navigating the stream of consciousness. *Annual Review of Psychology, 66*, 487–518.

Society of Automotive Engineering (SAE). (2015). *Operational Definitions of Driving Performance Measures and Statistics. Surface Vehicle Recommended Practice J2944.* Warrendale, PA: SAE.

Stavrinos, D., Jones, J.L., Garner, A.A., Griffin, R., Franklin, C.A., Ball, D., Welburn, S.C., Ball, K.K., Sisiopiku, V.P., & Fine, P.R. (2013). Impact of distracted driving on safety and traffic flow. *Accident, Analysis & Prevention, 61*, 63–70.

Strayer, D.L., Turrill, J., Coleman, J., Ortiz, E.V., & Cooper, J.M. (2014). *Measuring Cognitive Distraction in the Automobile II: Assessing In-Vehicle Voice-Based Interactive Technologies*. Washington, DC: AAA Foundation for Traffic Safety.

Strayer, D.L., Cooper, J.M., Turill, J., Colemen, J.R., & Hopman, R.J. (2015). *The Smartphone and the Driver's Cognitive Workload: A Comparison of Apple, Google and Microsoft's Intelligent Personal Assistants*. Washington, DC: AAA Foundation for Traffic Safety.

Tefft, B.C., Williams, A.F., & Grabowski, J.G. (2014). Driver licensing and reasons for delaying licensure among young adults ages 18–20, United States, 2012. *Injury Epidemiology, 1*, 4.

Tison, J.N., Chaudhary, N., & Cosgrove, L. (2011). *National Phone Survey on Distracted Driving Attitudes and Behaviors (Rep. No. DOT HS 811 555)*. Washington, DC: National Highway Traffic Safety Administration.

Vanlaar, W., Mayhew, D., Marcoux, K., Wets, G., Brijs, T., & Shope, J. (2009). An evaluation of graduated driver licensing programs in North America using a meta-analytic approach. *Accident Analysis & Prevention, 41*, 1104–1111.

Westlake, E.J., & Boyle, L.N. (2012). Perceptions of driver distraction among teenage drivers. *Transportation Research: Part F, 15*, 644–653.

White, C., & Caird, J.K. (2010). The blind date: The effects of change blindness, passenger conversation and gender on looked-but-failed-to-see (LBFTS) errors. *Accident Analysis & Prevention, 42*, 1822–1830.

Wickens, C.D. (2002). Multiple resources and performance prediction. *Theoretical Issues in Ergonomics Science, 3*(2), 159–177.

Wikman, A.S., Nieminen, T., & Summala, H. (1998). Driving experience and time-sharing during in-car tasks on roads of different width. *Ergonomics, 41*(3), 358–372.

Williams, A.F. (2003). Teenage drivers: Patterns of risk. *Journal of Safety Research, 34*, 5–15.

Williams, A.F., & Tison, J. (2012). Motor vehicle crash profiles of 13–15-year olds. *Journal of Safety Research, 43*, 145–149.

Williams, A.F., Ali, B., & Shults, R.A. (2010). The contribution of fatal crashes involving teens transporting teens. *Traffic Injury Prevention, 11*, 567–572.

Williams, A.F., Tefft, B.C., & Grabowski, J.G. (2012a). Graduated driver licensing research 2010–present. *Journal of Safety Research, 43*, 195–203.

Williams, A.F., West, B.A., & Shults, R.A. (2012b). Fatal crashes of 16- to 17-year-old drivers involving alcohol, nighttime driving, and passengers. *Traffic Injury Prevention, 13*, 1–6.

World Health Organization (WHO). (2011). *Mobile Phone Use: A Growing Problem of Driver Distraction*. Geneva, Switzerland: World Health Organization.

World Health Organization (WHO). (2015). *Global Status Report on Road Safety 2015*. Geneva, Switzerland: World Health Organization.

13

Alcohol, Cannabis, and New Drivers

Mark Asbridge

Christine Wickens

Robert Mann

Jennifer Cartwright

Abstract

The Problem. Learning how to drive, and drive safely, is a very complex process and one on which the lives of new drivers and others can depend. For young, new, and novice drivers, the process of learning how to drive can also occur at the same time that they are learning about the use of alcohol or other drugs like cannabis. The combination of novice driving and experimentation with alcohol and others drugs may be particularly hazardous. For example, the impact of a given amount of alcohol, including low doses, on the collision risk of young drivers is substantially greater than its impact on older drivers. ***Epidemiology of Driving after Drinking and Driving after Using Cannabis.*** For many years, researchers have examined the occurrence of driving under the influence of alcohol (DUIA), and research suggests that rates of DUIA have declined recently. Rates of driving under the influence of cannabis (DUIC) have received attention only more recently, and existing research suggests that among young drivers, rates of DUIC may equal or exceed rates of DUIA. ***Effects of DUIA and DUIC on Driving Skills.*** Many studies have assessed how alcohol affects driving skills. Results of studies examining the effects of alcohol on the performance of laboratory tasks involving psychomotor and cognitive skills, simulated and natural driving, and risk of collision among drivers converge to clearly demonstrate that DUIA impairs driving skills and increases collision risk. This effect appears to be exponential as the amount of alcohol consumed increases. Evidence on the effects of cannabis, on the other hand, is much less and has been more controversial. While some early studies suggested that DUIC did not increase collision risk, more recent evidence now indicates that it does. ***Prevention Efforts.*** Much effort has been directed to preventing DUIA among the general driving population and among young drivers, including educational, legal, and enforcement initiatives. Much is known about the effectiveness of DUIA prevention efforts, including their ability to reduce collision-related casualties, although more work to understand these effects among young and new drivers specifically is needed. On the other hand, efforts to prevent DUIC are in their infancy, and little evidence on their impact is available.

13.1 Introduction

Young drivers face greater collision risk than any other age group (Organisation for Economic Co-operation and Development, 2006; World Health Organization, 2015), making it imperative that we identify and address factors that contribute to this risk. According to the National Highway Traffic Safety Administration (2013), 31% of motor vehicle fatalities in 2012 were alcohol related, and almost 24% of fatally injured legally impaired drivers that year were between 16 and 24 years of age. Numerous studies have also reported that the prevalence of driving under the influence of cannabis is highest among younger drivers (Beirness & Davis, 2006; Couper & Peterson, 2014). Recognizing the impact of alcohol and cannabis use, alone and in combination, on collision-related injuries and fatalities among young drivers, the current chapter will (1) discuss the prevalence of driving under the influence of alcohol (DUIA) and driving under the influence of cannabis (DUIC) among young drivers; (2) address the impact of these substances on driving skills; (3) review epidemiological evidence linking use of these substances to collisions; (4) summarize prevention efforts focused on DUIA and DUIC among young drivers; and (5) identify knowledge gaps and research needs in understanding and addressing DUIA and DUIC among young drivers.

13.2 DUIA and DUIC: An Overview of the Evidence

13.2.1 DUIA and DUIC among Youth: Prevalence, Trends, and Key Correlates

Driving under the influence of alcohol or cannabis is a public health priority, especially among young and novice drivers. Considerable evidence has accumulated regarding the prevalence of drinking and driving in the developed world. Following the development of the first per se* drinking and driving law in Norway in 1936, most jurisdictions around the world have adopted per se legislation in the form of either an administrative or a criminal sanction. These laws have been coupled with active enforcement, aided by roadside breath testing technologies, and have led to considerable declines in rates of drinking and driving (Asbridge et al., 2004; Deshapriya & Iwase, 1998; Eisenberg, 2003; Foss et al., 2001; Mann et al., 2001; McCartt et al., 2013; Voas & Tippetts, 1999).

In recent years, declines in drinking and driving rates have become less pronounced in the adult population. Rates of self-reported DUIA have remained stable, with slightly fewer than one in five Canadian adults reporting having driven under the influence of alcohol at least once in the past 30 days between 1999 and 2010 (Vanlaar et al., 2012). However, evidence from a number of surveys of young people in Canada, the United States, and Europe shows that despite high rates of alcohol consumption and binge drinking, rates of drinking and driving among youth and novice drivers are declining. Data from Ontario indicate that among students in grades 10–12, self-reported rates of DUIA have decreased from 14% in 1999 to 7% in 2011 and 4% in 2013 (Boak et al., 2013). Drawing on data from a series of provincial surveys collected in 2007 and 2008, we see rates of DUIA among students in grade 12 ranging from 12% to 20% (Young et al., 2011). Similarly, self-reported past-30-day DUIA rates among a representative sample of young drivers in the United States have also decreased in recent years, from 17% in 1991 to 10% in 2013 (Kann et al., 2014). In a nationally representative sample of 11th grade students in the United States, Li and colleagues (Li et al., 2013) report a DUIA rate of 12.5%.

Observations from self-report surveys measuring adolescent rates of DUIA are corroborated by evidence from a handful of roadside surveys of drivers in Canada and the United States. Roadside surveys rely on randomly timed stops of drivers at checkpoints, during which breathalyzer and oral fluid samples (or alternatively, urine or blood samples) are collected to provide objective assessments of the presence of alcohol or drugs. In the most recent roadside surveys, alcohol was detected in 3.4% of 16- to

* *Per se* means "by itself" or "in and of itself "or " intrinsically" and is typically used in describing a matter of law.

18-year olds in British Columbia in 2010 and 0% of drivers 16–18 years of age in British Columbia in 2012. Conversely, 9.2% and 4.9% of drivers aged between 19 and 24 years tested positive for alcohol in British Columbia in 2010 and 2012, respectively. A similar pattern was observed in Ontario in 2014 (Beasley & Beirness, 2012; Beirness & Beasley, 2009, 2011; Beirness et al., 2015).

The exceptionally low rates of DUIA seen among teens in recent years offer strong support for federal and provincial laws directed at drinking and driving and, in particular, graduated driver licensing programs that are present in many jurisdictions in North America, most of which include a zero blood alcohol concentration (BAC) requirement for new and novice drivers during the first years of driving (Begg et al., 2001).

While rates of DUIA among young drivers have declined in recent years, DUIC has emerged as a serious public health issue. Despite being regulated by Canada's Controlled Drugs and Substances Act, cannabis is the most widely used psychoactive substance after alcohol (MacDonald et al., 2003; Mann et al., 2003). Provincial surveys of students across Canada estimate that between 20.9% and 36.8% of Canadian students have used cannabis at least once in their lifetime, and between 16.7% and 32.4% report use in the past year (Young et al., 2011). More recent student data from Ontario and Atlantic Canada estimate past-year use rates ranging from 23.0% to 34.7%, with higher rates among males and older students, and with rates fluctuating over time within provinces (Asbridge & Langille, 2013; Boak et al., 2013). Similarly, recent statistics from the United States Monitoring the Future study report that rates of past-year cannabis use remained stable between 2011 and 2014, at roughly 28% among grade 10 students and 26% among grade 12 students (Miech et al., 2015). Similar rates were observed in a 30-country cross-national study of cannabis use among youth in Europe and North America, where the United States and Canada reported the highest use (ter Bogt et al., 2014).

To understand rates of DUIC among young drivers, we again look to student surveys from Canada and the United States that ask students about their past-year involvement in driving within 1 hour after using cannabis. In most jurisdictions, the prevalence of driving after cannabis use has risen in recent years across the entire population, with rates disproportionately higher among younger drivers (Asbridge et al., 2005; Fergusson et al., 2008; Hall & Degenhardt, 2009; O'Malley & Johnston, 2013; Young et al., 2011). The 2012 Canadian Alcohol and Drug Use Monitoring Survey (CADUMS) revealed that 8.3% of drivers aged 18–19 and 6.5% of drivers 15–17 years of age engaged in DUIC, compared to 2.6% of drivers overall (Health Canada, 2013). Meanwhile, a sweeping view of student survey data across Canada reveals considerable provincial variation in rates of DUIC among students in grade 12. In Ontario, 13% of grade 12 students reported DUIC in 2011 and 11.6% in 2013 (Boak et al., 2013); in British Columbia, 14.9% of grade 12 students engaged in DUIC, and in Alberta, 10.6% (Young et al., 2011); in Atlantic Canada, 22% of Nova Scotia grade 12 students, 16.1% of grade 12 students in Newfoundland and Labrador, and 15.7% of grade 12 students in New Brunswick reported past-year DUIC (Asbridge & Langille, 2013). In the United States, drawing on a nationally representative sample of students in grade 12, DUIC rates were generally consistent from 2001 to 2011, ranging between 10.4% (2008) and 14.6% (2001), with a reported prevalence of 12.4% in 2011 (O'Malley & Johnston, 2013).

A review of more objective indicators confirms this trend. A 2012 British Columbia roadside survey, relying on oral fluid samples from drivers, found that 7.5% of drivers between the ages of 16 and 18 who were stopped at checkpoints tested positive for cannabis, the highest for any age group; in comparison, 6.8% of 19- to 24-year olds tested positive for cannabis (Beasley et al., 2013). Similarly, in 2014 in Ontario, 6.6% of drivers 16–18 years of age tested positive for cannabis at the roadside (Beirness et al., 2015).

When we look at key sociodemographic and lifestyle correlates of driving under the influence of alcohol or cannabis among young people, we see many similarities across studies. Engaging in DUIA and DUIC is more common among male than female adolescents, and for older compared to younger adolescents (Adlaf et al., 2003; Asbridge et al., 2005; O'Malley & Johnston, 2013; Whitehill et al., 2014). Higher rates of DUIA and DUIC have also been linked to poor academic performance, school truancy, coming from a single-parent family, and those drivers with a history of traffic violations. In terms of psychosocial factors, poor mental health, sensation seeking, low self-control, and having an aggressive or risk-taking personality are more common in young people who engage in impaired driving (Asbridge et al., 2005; Begg & Langley, 2004; O'Malley & Johnston, 2013; Young et al., 2011; see also Chapters 9 and 10). Similarly, both DUIA and DUIC are linked to a history of substance use and misuse including binge drinking, cigarette smoking, and the use of other illicit drugs, as well as dependence and addiction; impaired drivers also frequently ride as a passenger of a drinking driver (Butters et al., 2005; Leadbeater et al., 2008; Li et al., 2013; Poulin et al., 2007; Yu et al., 2004).

In looking at rates of DUIA and DUIC, two important developments are worth discussing. First, DUIA and DUIC appear to go hand in hand. Those young people who engage in DUIA typically also engage in DUIC, an overlap in impaired driving that has been noted in a number of adolescent surveys (Asbridge et al., 2005; O'Malley & Johnston, 2013; Whitehill et al., 2014; Young et al., 2011). For example, in a study of high school students in Atlantic Canada, Asbridge et al. (2005) note that the strongest observed relationship was between DUIC and DUIA, where those youth who engaged in DUIA had sixfold increased odds of DUIC. Alcohol consumption alone had no association with DUIC (Asbridge et al., 2005). Second, an emerging trend among young drivers in the past 10 years is that rates of DUIC parallel, and in many instances surpass, rates of DUIA (Fergusson et al., 2008; Young et al., 2011). O'Malley and Johnston (2013), looking at 11 years of Monitoring the Future data, note the changing patterns of impaired driving. In 2001, 16% of high school seniors reported DUIA, and 12.1%, DUIC; yet between 2009 and 2011, rates of DUIC (10.8%, 11.9%, and 12.4%) surpassed rates of DUIA (9.4%, 9.2%, and 8.7%). Similarly, in Atlantic Canada and Ontario, the prevalence of DUIC among senior students was higher than that of DUIA during the same time period, despite the higher prevalence of alcohol consumption relative to cannabis use (Adlaf et al., 2003; Asbridge et al., 2005). In a recent report on college students in the United States, despite higher numbers of students reporting drinking alcohol in the past month compared to using marijuana, rates of driving after drinking were much lower than those for cannabis and driving (Whitehill et al., 2014).

13.2.2 Effects of DUIA and DUIC on Driving Skills

13.2.2.1 Laboratory Studies

Three experimental methods have been used to further our understanding of *how* alcohol and cannabis may impact driver performance: laboratory, simulation, and naturalistic studies. Laboratory studies assess basic cognitive, sensory, perceptual, and psychomotor skills related to driving, such as tracking and attentional abilities. Hundreds of studies have examined alcohol or cannabis impairment of these elemental skills using abstract laboratory tests. This research has generally suggested that alcohol is most impairing for tasks involving controlled skills that focus attention on multiple information sources and require complex information processing (Moskowitz & Fiorentino, 2000, as cited in Creaser et al., 2011). Alcohol affects visual search patterns, processing speed, and ability to focus or switch attention (Creaser et al., 2011). Cannabis research has identified impairment of several driving-related skills including tracking, coordination, perception, and vigilance (Moskowitz, 1985).

Although highly informative, these studies of elemental driving-related skills lack external validity; the research cannot assess the simultaneous and coordinated use of basic skills necessitated by the driving task. This is particularly problematic when assessing substance-related driver impairment, given that many studies have shown progressive deterioration of the driver's ability to perform attention-related tasks with increasing BAC or tetrahydrocannabinol (THC), the primary psychoactive components in cannabis (Hartman & Huestis, 2013; Hunault et al., 2009; Leung & Starmer, 2005). Instead, simulated and naturalistic driving studies are used to assess the coordinated and simultaneous use of basic driving-related skills.

13.2.2.2 Simulation and Naturalistic Studies

Starting in the early 1980s, driving simulation became interactive (i.e., responsive operator controls) and offered accurate vehicle dynamics (Smiley, 1986, 1999). Since then, driving simulation has experienced profound technological advancements and has provided increasingly more realism and significantly more detailed data (Chapter 25). Likewise, the development of in-vehicle instrumentation has expanded greatly, improving measurement accuracy and facilitating naturalistic or in situ assessment of drivers on public roads or closed-course tracks (Chapter 26). These technologies have been used to assess the impact of cellular telephones (both talking and texting), social influence, fatigue, and both alcohol and drugs on driver performance.

13.2.2.2.1 *Alcohol Impairment Effects*

There has been extensive simulation and naturalistic study of alcohol impairment effects on driver performance. This vast literature has been fairly consistent in its identification of affected behaviors (Creaser et al., 2011), including increased speed and speed variability (Arnedt et al., 2001; Charlton & Starkey, 2015; Gawron & Ranney, 1988; Moskowitz et al., 2000; Weafer et al., 2008), steering error and instability (Roehrset al., 1994; Weiler et al., 2000), reduced following distances (Weiler et al., 2000), increased within-lane position variability and lane departures (Arnedt et al., 2001; Charlton & Starkey, 2015; Gawron & Ranney, 1988; Hartman et al., in press; Moskowitz et al., 2000; Ramaekers et al., 2000; Weafer et al., 2008), slower reaction time (Burns et al., 2002), and more crashes (Gawron & Ranney, 1988; Moskowitz et al., 2000; Roehrs et al., 1994). Performance degradation for each of these skills and behaviors tends to follow a dose-dependent pattern, demonstrating increasing impairment with increasing BAC (Arnedt et al., 2001; Mets et al., 2011).

One of the earliest and most impactful studies of alcohol-impaired driving, known as the Grand Rapids study, involved the comparison of BACs from drivers involved versus those not involved in a motor vehicle collision (Borkenstein et al., 1964). Analyses of the Grand Rapids data set suggest that alcohol impairment may be modified by age, but only among drivers younger than 18 years of age and those older than 70 years. It was theorized that age impacts alcohol impairment as the result of some developmental change in physiology or psychology. However, age is also associated with driver inexperience (Moskowitz et al., 2000), which, in addition to age differences in annual mileage and limited experience with alcohol consumption, could also account for the relationship between age and alcohol impairment.

Since the Grand Rapids research, there have been a few experimental studies examining the differential impact of age on alcohol impairment, but these have failed to find a significant effect (e.g., Jones & Neri, 1994; Moskowitz et al., 2000; Quillian et al., 1999). None of these studies included drivers younger than 21 years of age, thus neglecting a demographic group for which age had previously been identified as a moderator of alcohol impairment (Moskowitz et al., 2000). A simulation study by Lenné et al. (2010) examined the impairment effect of alcohol and cannabis on young inexperienced drivers (aged 18–21 years) versus older experienced drivers (aged 25–40 years). Only one significant interaction between alcohol and age/driver experience was found, such that speed variability under a low dose of alcohol increased among young inexperienced drivers but not among older experienced drivers. However, this difference disappeared under a high dose of alcohol. At present, evidence for or against an age interaction with alcohol effects is equivocal. None of the existing experimental research has been able to eliminate the level of driving experience as a potential confounder. Moreover, with restrictions on providing alcohol to minors (aged 21 years in most US states and aged 18 or 19 years in Canadian provinces), efforts to research this issue further are unlikely in North America and many other jurisdictions.

13.2.2.2.2 *Cannabis Impairment Effects*

Simulation and naturalistic studies of the impairing effects of cannabis have been far fewer and produced less consistent findings than studies of alcohol. There appears to be some agreement that motorists who drive under the influence of cannabis tend to engage in compensatory behavior (Ramaekers et al., 2004; Robbe, 1998; Sewell et al., 2009), driving more slowly (Downey et al., 2013; Lenné et al., 2010; Ronen et al., 2008, 2010; Sexton et al., 2000, 2002; Stein et al., 1983), with an increased following distance (Downey et al., 2013; Lenné et al., 2010; Sexton et al., 2002; Smiley et al., 1981). They also attempt fewer passes or require a wider gap in traffic before attempting to pass (Dott, 1972; Ellingstad et al., 1973;

Smiley et al., 1981). Although researchers generally agree that this type of reduced risk-taking cannot compensate for all cannabis-induced driver impairment, a clear picture of what other dangerous behaviors emerge in cannabis-impaired drivers is still in question. Previous research has demonstrated both improved (Krueger & Vollrath, 2000) and reduced lane control (Downey et al., 2013; Hartman et al., in press; Lenné et al., 2010; Papafotiou et al., 2005; Ramaekers et al., 2000; Ronen et al., 2010; Sexton et al., 2000, 2002; Smiley et al., 1981) or failed to find an effect (Anderson et al., 2010; Liguori et al., 1998; Ronen et al., 2010; Stein et al., 1983); however, standard deviation of lateral position, one measure of lane control, is becoming increasingly more recognized as a driving measure that is sensitive to driving-related cannabis impairment (e.g., Hartman et al., in press; Society of Automotive Engineering, 2015). Reaction time to unexpected roadway events has generally been greater under the influence of cannabis (Liguori et al., 1998; Sexton et al., 2000), but again, this has not been a consistent finding (Anderson et al., 2010; Downey et al., 2013; Liguori et al., 2002; Ramaekers et al., 2000; Sexton et al., 2002). Some studies have found an increase in reaction time and other behavioral effects under dual-task conditions, such as driving while verbalizing mental arithmetic tasks (e.g., addition, subtraction) (Anderson et al., 2010; Lenné et al., 2010; Ronen et al., 2008; Smiley et al., 1981). As with alcohol, the impairing effects of cannabis on driver behavior appear to follow a dose–response pattern.

To the best of our knowledge, the only simulation or naturalistic study to assess age as a modifier of the cannabis impairment effect was the simulation study by Lenné et al. (2010), which found that young inexperienced drivers (aged 18–21 years) demonstrated poorer skill and control than older experienced drivers (aged 25–40 years). Moreover, interaction effects were found for variability in steering behavior such that variability increased with increasing doses of cannabis, but only among inexperienced drivers.

Much of the existing research of driving-related cannabis impairment effects has been conducted on drivers under 30 years of age, meaning that our state of knowledge on cannabis impairment is most applicable to the youngest generation of drivers. However, if decriminalization and/or legalization of recreational cannabis use spreads to more jurisdictions, this activity may become more socially acceptable, and the average age of recreational cannabis users may rise. This type of societal change would make the potential role of age as a moderator of the cannabis impairment effect an even more relevant and vital research question. At the time this chapter was written, several US states had already legalized/decriminalized recreational cannabis use, and the Canadian government was considering this policy change. With the potential for generation of significant tax revenue and reduction in crimes associated with illegal drug trade (e.g., violence, theft), it is likely that other governments will follow suit.

13.2.2.3 Impairment Effects of Combined Alcohol and Cannabis Use

Although there is indication that some alcohol and cannabis impairment effects may be similar, at least in the direction of effect (e.g., problems with steering variability and lane control [Lenné et al., 2010; Rakauskas et al., 2008; Ronen et al., 2008; Sexton et al., 2000, 2002; Weafer et al., 2008], increased reaction time [Kerr & Hindmarch, 1998; Liguori et al., 1998; Sexton et al., 2000]), there is also evidence of contrasting impairment effects by the two substances. Alcohol-impaired drivers tend to increase speed and reduce following distance (Gawron & Ranney, 1988; Weiler et al., 2000), whereas cannabis-impaired drivers appear to decrease speed and increase following distance (Lenné et al., 2010; Sexton et al., 2002). Given the frequency with which alcohol and cannabis are consumed together (Duhig et al., 2005; Pape et al., 2009), it is important to understand their combined effects on driver behavior.

Simulation and naturalistic studies examining the combined impact of alcohol and cannabis on driver performance are extremely few in number and differ based on analytical approach. In some cases, an interaction between alcohol and cannabis effects is assessed (e.g., Lenné et al., 2010; Stein et al., 1983), whereas in others, main effects are examined through direct comparisons between drivers impaired by alcohol or cannabis and drivers impaired by both substances. Generally, simulation and naturalistic studies have identified additive as opposed to synergistic effects of combined alcohol and cannabis use (Downey et al., 2013; Lamers & Ramaekers, 2001; Liguori et al., 2002; Ramaekers et al., 2000; Ronen et al., 2010; Sexton et al., 2002); however, there has been little consistency in findings for a number of behaviors, including speed (Downey et al., 2013; Ronen et al., 2010; Sexton et al., 2002; Smiley et al., 1975), stopping safety (Downey et al., 2013; Sexton et al., 2002), following distance (Attwood et al., 1981; Downey et al., 2013), and collision risk (Downey et al., 2013; Ronen et al., 2010; Sexton et al., 2002). General consistency has been found for lane control, indicating impairment relative to the use of either alcohol or cannabis alone (Attwood et al., 1981; Casswell, 1979; Downey et al., 2013; Hartman et al., in press; Ramaekers et al., 2001; Ronen et al., 2010; Sexton et al., 2002; Smiley et al., 1975).

The highly mixed set of results from both simulation and naturalistic studies is complicated by the fact that driving skills/behaviors (e.g., lane control) have been measured by a number of different variables (e.g., straddling barrier line, standard deviation of lateral position) both within and across studies, often producing differing results. Moreover, the use of varying dosages of alcohol and cannabis, along with the influence of other measured variables such as simulated time of day, level of driving experience, and frequency of cannabis use, makes comparison across studies difficult.

While no study has assessed the influence of age on driver impairment caused by combined alcohol and cannabis use, the majority of studies assessing impairment by combined use of these substances were based on drivers under 30 years of age. Thus, what little we do know about how driving skills and behaviors are affected by combined substance use is most applicable to the youngest generation of drivers. Even so, there may still be important age-related differences in the impact of the combined use of alcohol and cannabis within this segment of the driving population.

13.2.3 DUIA, DUIC, and Crash Risk among Young and Novice Drivers

A review of traffic crash data reveals two common trends: collisions are disproportionately experienced by young drivers and by those under the influence of alcohol and/or cannabis. In the United States, the rate of traffic fatalities per 100,000 registered vehicles in 1982 was 26.6, of which 15.8 (or approximately 60%) were alcohol-related (Yi et al., 2006). In the two decades that followed, the rate of fatal collisions dropped substantially to 17.6 per 100,000 drivers, yet the rate of alcohol-related fatalities dropped even further to 7 per 100,000 drivers—or 40% of total fatalities (Yi et al., 2006). In Canada, starting with the introduction of Breathalyzer legislation in 1969, which established a criminal code violation for drivers with a BAC greater than 0.08%, a nearly 20% decline in alcohol-related traffic fatalities was observed over the following 30 years (Asbridge et al., 2004).

In North America, the decline in alcohol-related fatal crashes has not been as substantial. A recent review of alcohol-related fatalities among adults in the United States finds little change since the turn of the century (Brady & Li, 2014; McCartt et al., 2013), with the proportion of fatal crashes involving alcohol remaining stable at around 43.5% for male drivers and between 24.7% and 27.7% for female drivers between the years 1999 and 2010 (Brady & Li, 2014). In Canada, alcohol-impaired fatalities have remained largely stable, with only modest reductions in impaired-driving fatalities (Vanlaar et al., 2012). Since 1997, the annual percentage of traffic fatalities where drivers have tested positive for alcohol has consistently remained in the range of 36%–39% (Vanlaar et al., 2012).

Outlining the prevalence of DUIA represents only part of our understanding of the public health and road safety impact of impaired driving. Equally, if not more important, is whether the consumption of alcohol (or cannabis) increases the risk of involvement in a traffic crash. In the alcohol realm, starting with the seminal Grand Rapids study (Borkenstein et al., 1974), a considerable body of research has examined this very question and, unequivocally, finds that driving while under the influence of alcohol poses a substantially increased crash risk. This research draws on data from a range of studies of drivers involved in fatal crashes, crashes resulting in injury, and property-damage crashes (Blomberg et al., 2005; Odero et al., 1997; Peck et al., 2008; Taylor et al., 2010; Zador et al., 2000). We also know that as the level of alcohol in the blood increases, crash risk follows suit. A BAC of 0.05 (a typical provincial administration sanction level) produces a 38% increase in the relative risk of a crash, whereas a BAC of 0.08 is associated with a 169% increase, and a BAC of 0.16 (double the criminal legal limit) leads to a greater than 2800% increase in relative crash risk (Blomberg et al., 2005).

We also note that alcohol-related fatalities have been disproportionately experienced by young and novice drivers. In 2013 in the United States, 18% of 16- to 20-year-olds and 33% of 21- to 24-year-olds had consumed alcohol prior to a fatal crash, higher than rates reported for drivers older than 45 years of age (National Center for Statistics and Analysis, 2014). In Canada, between 2000 and 2010, the proportion of traffic fatalities involving alcohol among young people 16–19 years of age has fluctuated between 33% and 47% (Traffic Injury Research Foundation [TIRF], 2013). These differential effects can be explained by the complicated relationship between alcohol consumption, driver age, and crash risk. Youth are more susceptible to the effects of alcohol consumption on driving performance, such that alcohol-impaired young and novice drivers have a much higher crash risk than adult drivers despite similar BAC levels; these effects cannot be explained by driver inexperience but are situated in the way young people experience alcohol impairment (Mayhew et al., 1986; Peck et al., 2008; Zador et al., 2000). Looking at US traffic fatality data, Zador et al. (2000) were able to determine that while fatal crash risk increased steadily with increasing BAC levels among all driver age groups, the relative risk of a fatal crash was found to decrease with increasing driver age at every BAC level. In other words, the effect of a particular BAC on crash risk was always higher for younger compared to older drivers (Zador et al., 2000).

In addition to age, other factors contextualize alcohol-related crashes. Driver sex plays an important role. Male drivers have historically been more heavily involved in crashes involving alcohol, in large part due to their higher rates of alcohol use and misuse and risk-taking propensity (Romano & Pollini, 2013; TIRF, 2013). This is also true among young and novice drivers; however, recent data suggest that fatal collision rates among young female drivers have risen at a faster rate than those of young males (TIRF, 2013; Tsai et al., 2010). The time of day and the presence of passengers, particularly other young people, also influence the likelihood of an alcohol-related crash (Romano & Pollini, 2013; Williams et al., 2012), as well as driver distraction (TIRF, 2013). Interestingly, when we consider the broader evidence around rates of DUIA among young and novice drivers and compare them to rates of alcohol-impaired fatalities, we are presented with a paradox. On the one hand, reported rates of DUIA, drawing from roadside and self-report surveys, show a clear decrease; yet, young and novice drivers continue to be heavily involved in fatal crashes—and disproportionately so (McCartt et al., 2013).

While we have a long history of studying the role of alcohol in traffic crashes, fewer studies have looked at cannabis-related traffic fatalities and the relationship between cannabis consumption and risk of crash. Data on trends in cannabis-related traffic fatalities are limited; however, those that exist consistently point to an ever-increasing prevalence of positive tests for cannabis in drivers injured or killed in a traffic crash (Bédard et al., 2007; Johnson et al., 2012; Sweedler et al., 2004). A detailed study of fatal crashes in six US states with high toxicology testing rates found that between 1999 and 2010, the presence of cannabis in drivers nearly tripled, increasing from 4.2% in 1999 to 12.2% in 2010, such that cannabis was the most prevalent drug other than alcohol detected in fatally injured drivers (Brady & Li, 2014). Studies from other jurisdictions and collected at different times point to an ever-increasing percentage of crash-involved drivers testing positive for cannabis (Asbridge et al., 2014; Biecheler et al., 2008; Drummer et al., 2004; Laumon et al., 2005; Mørland et al., 2011; Stoduto et al., 1993), including substantially higher rates in younger drivers (Mura et al., 2003, 2006; Office of National Drug Control Policy, 2011; Woodall et al., 2015). For instance, an analysis of Canadian fatality data by Beasley et al. (2011) revealed that of drivers who tested positive for drugs following their death in a car crash, 68.6% of those under the age of 19 and 54.2% of those aged 19–24 tested positive for cannabis, substantially higher rates than for older drivers (approximately 21% for drivers 35–55 years of age and 6% or less for drivers aged 55+). It is important to note that these rates will likely continue to rise given the emerging cannabis decriminalization and legalization in many jurisdictions, and the growing use and availability of medical marijuana (Anderson & Rees, 2014; Asbridge et al., 2014; Salomonsen-Sautel et al., 2014).

Beyond the observation that cannabis is increasingly appearing in the blood of drivers involved in traffic crashes, an area of particular concern is whether or not cannabis increases crash risk. Answering this question has posed many challenges due to the lack of available data drawn from high-quality studies. The difficulty in studying the role of cannabis in crashes is, in part, due to ethical issues associated with measuring cannabis. Unlike alcohol, where we can employ the Breathalyzer to obtain blood alcohol readings in drivers at the roadside, no similar universally accepted device exists for measuring cannabis (or other impairing drugs). As such, blood or urine samples are required to confirm the presence of drugs in drivers, which prove difficult to obtain from the general driving population (MacDonald et al., 2003). The legal status of cannabis generates further difficulty for researchers trying to recruit subjects from the roadside (Bates & Blakely, 1999; Ramaekers et al., 2004). Additionally, the relative risk of a traffic crash involving alcohol continues to be higher than that for cannabis and other drugs and, as such, may warrant less attention (Romano et al., 2014).

Studies of cannabis and collision risk have been slower to emerge than studies of alcohol. In the early 1970s, Smart (1974) studied self-reported cannabis use and the probability of traffic crash among college students. He found that cannabis users were involved in almost as many crashes under the influence of cannabis as they were when under the influence of alcohol. Based on a telephone survey of 6000 adolescents between the ages of 16 and 19 years of age, Hingson et al. (1982) found crash involvement to be related to the frequency of DUIC. The risk of collision increased substantially among young people who more frequently drove under the influence of cannabis compared to those who drove under the influence occasionally or not at all. More recently, a handful of self-report studies have found a significant association between cannabis use prior to driving and an increased risk of crash involvement (Asbridge et al., 2005; Fergusson & Horwood, 2001).

In order to obtain a comprehensive picture of DUIC and crash risk, we must look to well-designed studies that employ objective measures of both crash involvement and acute, precrash cannabis consumption. This necessitates a reliance on studies that employ epidemiologic designs (i.e., case–control, culpability, case crossover), in which the proportions of drivers testing positive for cannabis are compared between those involved in a crash (cases) and a sample of crash-free drivers (controls). We can then estimate the relative contribution of cannabis to crash risk, holding known confounders constant. Three recent systematic reviews and meta-analyses have considered this question (Asbridge et al., 2012; Elvik, 2013; Li et al., 2012) and draw similar conclusions; cannabis use prior to driving roughly doubles the likelihood of a subsequent motor vehicle crash. Findings were more robust when acute or "recent" cannabis consumption was measured in blood rather than in urine, where the presence of THC metabolites may be evident for days or weeks after consumption with no clear relationship to driver impairment, and the association was stronger for fatal crashes relative to crashes involving only injury (Asbridge et al., 2012).

As with alcohol, crash risk linked with DUIC appears to be elevated for younger drivers relative to older drivers (Mura et al., 2006), but the evidence is less definitive than for alcohol, given the lack of studies addressing this specific issue. This is, in part, due to differential use rates among younger drivers relative to older drivers. The small numbers of older drivers involved in crashes, and older "control" drivers not involved in crashes who have used cannabis, make estimates of age-specific crash risk more difficult (Asbridge et al., 2014). Most studies simply report on elevated crash risk for the entire population; as such, more research looking at the differential effects of cannabis on crash risk by driver age is warranted.

Finally, it would be remiss not to discuss the dangerous effects of the combined use of cannabis and alcohol on crash risk. A number of studies have demonstrated that drivers who use cannabis in combination with alcohol are at significantly greater risk of collision than those who use either substance alone (Drummer et al., 2004; Dubois et al., 2015; Longo et al., 2000). In fact, these effects are not only additive but synergistic (Dubois et al., 2015; Li et al., 2013), meaning that the effects on crash risk of using both substances together are greater than the sum of their effects if used alone. Although rates of driving after having used both substances concurrently are typically low, the implications for road safety are nonetheless important. Where this matters most is for drivers who consume each substance at levels below the legal threshold or the therapeutic level for impairment; when the substances are combined, drivers experience substantial impairment and an increased risk of being involved in a crash (Dubois et al., 2015).

13.2.4 Current Prevention Efforts

Current prevention efforts directed at DUIA and DUIC among young drivers in many ways parallel those directed at the general driving population, but there are targeted initiatives directed at younger or more inexperienced drivers as well. Those directed at DUIA have a longer history, with much more information on impact, while those directed at DUIC are more recent.

Efforts to educate young drivers about the hazards of DUIA and driving under the influence of other drugs have been seen by many as a cornerstone of effective prevention. For many years, classroom-based drug education and road safety education, including driver education programs, have included content directed to DUIA and, to a lesser extent, DUIC (Mann et al., 1988). The impact of classroom-based education on reducing DUIA and DUIC, and on improving driving skills in general, has been controversial. Early studies and reviews pointed to a lack of consistency in results, or even potential adverse effects, of these programs (Mann et al., 1986; Robertson & Zador, 1978). Subsequently, a meta-analysis of classroom-based driver education to prevent DUIA and riding with a drinking driver did conclude that education can prevent the latter, but no significant impact on DUIA was seen (Elder et al., 2005). More recent studies have assessed education in the context of graduated licensing programs, with indications that education may reduce subsequent collisions among new drivers (Shell et al., 2015; Vanlaar et al., 2009; Zhao et al., 2006).

Public education by the government and others has been a central part of efforts to control DUIA in the general population, and often, these efforts are targeted at young drivers. More recently, some public education campaigns have also targeted DUIC. While public education is considered an important component of programs to reduce DUIA and DUIC, many have questioned the value of these efforts (Babor et al., 2010; Vingilis & Coultes, 1990). A recent meta-analysis of the effect of road safety campaigns on collisions concluded that, overall, these campaigns were associated with a 9% reduction in collisions and that campaigns with a drinking–driving theme were associated with the strongest effects (Phillips et al., 2011). To date, no evaluations of programs directed at DUIC specifically have appeared in the literature.

Legal initiatives, including per se laws and administrative license suspension programs, are widely considered to have been central to the substantial reductions in drunk driving and resulting collisions, injuries, and fatalities that have occurred in many parts of the world in recent decades (Sweedler et al., 2004; Wickens et al., 2013). Unfortunately, it is difficult or impossible in most studies to determine how young drivers have been affected by these laws. However, it is reasonable to propose, in situations where a legal change (like the introduction of a per se law) has resulted in significant reductions in drunk-driving fatalities (e.g., Asbridge et al., 2004), that young drivers may be beneficially affected as well. While the impact of legal initiatives to address DUIA has been well documented, introduction of laws to address DUIC and their evaluation are at a much earlier stage (Huestis, 2015). Several jurisdictions have introduced per se laws directed at DUIC, either specifically or in the context of laws directed at a range of psychoactive drugs, but little is known about the impact of these laws (Huestis, 2015; Vindenes et al., 2012).

Additionally, some legal initiatives have been directed at young and novice drivers specifically. Graduated licensing systems (GLSs) are licensing systems for young or novice drivers that require drivers to successfully pass through one or more stages with varying driving restrictions before full licensing privileges are granted (Chapter 17; Foss et al., 2001; Mayhew, 2007; Waller, 2003). Among the restrictions typically applied in GLS systems are reduced- or zero-tolerance BAC requirements, requirements that an experienced driver be present, and restrictions on times of day and types of roadways that new drivers are allowed to use (Langley et al., 1996; Zhao et al., 2006). In general, GLS systems have been found to reduce collisions and resulting fatalities among young drivers (Chapter 17; Foss et al., 2001; Langley et al., 1996; Vanlaar et al., 2009). While it is difficult to determine the specific mechanism or licensing restriction that results in safety benefits, some evidence suggests that zero-tolerance BAC provisions and requirements for the presence of an experienced driver (often a parent) have resulted in reduced driving after drinking among young drivers (Cook et al., in press; Mann et al., 1997). Legal drinking age laws, which restrict the purchase and use of alcohol by individuals under specific ages, and zero-tolerance laws, which introduce zero- or reduced-BAC provisions for an extended period for young or new drivers, are in place in many jurisdictions (Voas et al., 2003). For example, in Ontario, Canada, the legal drinking age is 19, and drivers are required to have a BAC of zero until the age of 22 (Byrne et al., 2016). Evaluations of these laws have shown that they have been successful in reducing drunk driving, collisions, and fatalities among affected age groups (Byrne et al., 2016; Callaghan et al., 2014; Voas et al., 2003; Wagenaar and Toomey, 2002).

While some have recommended the introduction of laws equivalent to GLS and zero tolerance for DUIC, to date, very little evidence is available about how effective these laws are or might be. An interesting observation here is that cannabis prohibition in North America has not deterred use of this drug by a large proportion of young people, including driving under the influence of the drug (Adlaf et al., 2003). One study does suggest that new driver laws focused on DUIC could have beneficial effects. Cook et al. (in press) examined DUIA and DUIC among secondary school students with a driver's license in Ontario, Canada. In multivariate analyses controlling for age and other demographic measures, they found that the prevalence of both DUIA and DUIC increased significantly as drivers moved to graduated licensing stages with fewer restrictions. For example, as drivers moved from the first (G1) stage to the second (G2) stage, the proportions of drivers reporting DUIA increased from 4.8% to 16.7%, and the proportion reporting DUIC increased from 6.7% to 23.9%. Thus, some evidence suggests that GLS programs may exert similar beneficial effects on DUIC as on DUIA.

Rehabilitative or remedial measures for drivers apprehended for DUIA are part of comprehensive initiatives to address this problem in many jurisdictions. In general, these programs have shown success in reducing recidivism and improving health-related outcomes, including reductions in alcohol use (Ma et al., 2015; Mann et al., 1994; Robertson et al., 2009; Schulze et al., 2012; Stoduto et al., 2014; Wells-Parker et al., 2009). Again, researchers have typically not differentiated between age groups or between young drivers and others when assessing the effects of these programs, although research studies suggest that age may be an important moderator of program effects (Ma et al., 2015; Ouimet et al., 2013). Specific remedial programs for drivers apprehended for DUIC and other drugs have not been reported in the literature, but in some jurisdictions, individuals convicted of a drug-related driving offence may be required to complete a remedial program (Schulze et al., 2012). Although the impact of participation in these programs on driving measures for DUIC offenders has not been reported, some studies have shown that remedial program participation can reduce drug use as well as alcohol use (Stoduto et al., 2014).

13.2.5 Knowledge Gaps/Research Needs

The significant contributions of alcohol and cannabis to young driver collision risk means that understanding and addressing DUIA and DUIC among young drivers will remain a priority for research and prevention efforts. As seen in this chapter, much has been learned about how alcohol and cannabis affect driving behavior and collision risk, the epidemiology of DUIA and DUIC, and how collisions related to DUIA and DUIC can be prevented. Nevertheless, important gaps in knowledge remain.

One key issue is that much of the knowledge about these issues is derived from studies of the more general driving population, and thus, often, the applicability of this knowledge to younger drivers must be inferred. However, there are also important indications that younger drivers present special issues. For example, it is well known that an increasing BAC appears to increase collision risk at a much faster rate among young drivers (Zador et al., 2000). The reasons for this are unknown but have been linked to young drivers' inexperience with driving and with personal characteristics common to youth (e.g., increased risk-taking propensity) (Richer & Bergeron, 2009). Thus, studies to assess the effects of alcohol and cannabis on driving performance and on collision risk, specifically targeting young drivers, are needed to obtain a more complete understanding of how these drugs affect young drivers and how resulting collisions can be prevented.

Another gap relates to the relative lack of information on cannabis in comparison to alcohol. For many reasons, much more information is available on DUIA compared to DUIC, including effects on driving skills, collision risk, and also preventive measures. Since it now appears that DUIC may be as common as, or more common than, DUIA among young drivers in North America and other parts of the world (Adlaf et al., 2003; Dols et al., 2010; O'Malley & Johnston, 2013), this lack of information on the former must be seen as an important concern. This gap is particularly salient since several jurisdictions have recently legalized or decriminalized cannabis possession and others are considering such legislative shifts.

Related to the aforementioned needs is the current imbalance in prevention efforts directed to DUIA and DUIC, and to knowledge about the effectiveness of those efforts. While introduction of DUIA countermeasures and understanding of their impact has achieved a relatively mature status, prevention efforts directed at DUIC are far fewer, and evidence on their impact is almost nonexistent. Prevention of DUIA can help to guide DUIC prevention efforts, but it cannot be assumed that what we know about DUIA and its prevention can be directly and effectively transferred to the prevention of DUIC. Thus, an important priority for future research will be to establish the effects of efforts to prevent DUIC among young and new drivers.

Acknowledgments

The authors gratefully acknowledge the support of AUTO21, a member of the Networks of Centres of Excellence (NCE) program, which is administered and funded by the Natural Sciences and Engineering Research Council (NSERC), the Canadian Institutes of Health Research (CIHR), and the Social Sciences and Humanities Research Council (SSHRC), in partnership with Industry Canada, during the preparation of this chapter.

References

Adlaf, E.M., Mann, R.E., & Paglia, A. (2003). Drinking, cannabis use and driving among Ontario students. *Canadian Medical Association Journal, 168*(5), 565–566.

Anderson, B.M., Rizzo, M., Block, R.I., Pearlson, G.D., & O'Leary, D.S. (2010). Sex differences in the effects of marijuana on simulated driving performance. *Journal of Psychoactive Drugs, 42*(1), 19–30.

Anderson, D.M., & Rees, D.I. (2014). The legalization of recreational Marijuana: How likely is the worst-case scenario? *Journal of Policy Analysis & Management, 33*(1), 221–232.

Arnedt, J.T., Wilde, G.J.S., Munt, P.W., & MacLean, A.W. (2001). How do prolonged wakefulness and alcohol compare in the decrements they produce on a simulated driving task? *Accident Analysis & Prevention, 33*, 337–344.

Asbridge, M., & Langille, D. (2013). *Student Drug Use Survey in the Atlantic Provinces 2012: Technical Report, October, 2013*. Halifax, NS: Nova Scotia Department of Health and Wellness.

Asbridge, M., Mann, R.E., Flam-Zalcman, R., & Stoduto, G. (2004). The criminalization of impaired driving in Canada: Assessing the deterrent impact of Canada's first per se law. *Journal of Studies on Alcohol, 65*(4), 450–459.

Asbridge, M., Poulin, C., & Donato, A. (2005). Motor vehicle collision risk and driving under the influence of cannabis: Evidence from adolescents in Atlantic Canada. *Accident Analysis & Prevention, 37*(6), 1025–1034.

Asbridge, M., Hayden, J.A., & Cartwright, J.L. (2012). Acute cannabis consumption and motor vehicle collision risk: Systematic review of observational studies and meta-analysis. *British Medical Journal, 344*, e536.

Asbridge, M., Mann, R., Cusimano, M.D., Trayling, C., Roerecke, M., Tallon, J.M., Whipp, A., & Rehm, J. (2014). Cannabis and traffic collision risk: Findings from a case-crossover study of injured drivers presenting to emergency departments. *International Journal of Public Health, 59*(2), 395–404.

Attwood, D.A., Williams, R.D., Bowser, J.S., McBurney, L.J., & Frecker, R.C. (1981). The effects of moderate levels of alcohol and marihuana, alone and in combination on closed-course driving performance. Downsview, Ontario: Defence and Civil Institute of Environmental Medicine.

Babor, T., Caetano, R., Casswell, S., Edwards, G., Giesbrecht, N., Graham, K., Grube, J. et al. (2010). *Alcohol: No Ordinary Commodity: Research and Public Policy*. Oxford: Oxford University Press.

Bates, M.N., & Blakely, T.A. (1999). Role of cannabis in motor vehicle crashes. *Epidemiological Review, 21*(2), 222–232.

Beasley, E.E., & Beirness, D.J. (2012). *Alcohol and Drug Use among Drivers following the Introduction of Immediate Roadside Prohibitions in British Columbia: Findings from the 2012 Roadside Survey*. Ottawa, ON: Bierness & Associates.

Beasley, E.E., Beirness, D.J., & Porath-Waller, A. (2011). *A Comparison of Drug and Alcohol-Involved Motor Vehicle Driver Fatalities*. Ottawa, ON: Canadian Centre on Substance Abuse.

Beasley, E.E., Beirness, D.J., & Boase, P. (2013). Alcohol and drug use among drivers: British Columbia roadside surveys 2008–2012. In B. Watson & M. Sheehan (Eds.), *15th Proceedings of the 20th International Conference on Alcohol, Drugs and Traffic Safety*. Brisbane, Australia: ICADTS.

Bédard, M., Dubois, S., & Weaver, B. (2007). The impact of cannabis on driving. *Canadian Journal of Public Health, 98*(1), 6–11.

Begg, D., & Langley, J. (2004). Identifying predictors of persistent non-alcohol or drug-related risky driving behaviours among a cohort of young adults. *Accident Analysis & Prevention, 36*(6), 1067–1071.

Begg, D.J., Stephenson, S., Alsop, J., & Langley, J. (2001). Impact of graduated driver licensing restrictions on crashes involving young drivers in New Zealand. *Injury Prevention, 7*(4), 292–296.

Beirness, D.J., & Beasley, E.E. (2009). *Alcohol and Drug Use among Drivers. British Columbia Roadside Survey 2008*. Ottawa, ON: Canadian Centre on Substance Abuse.

Beirness, D.J., & Beasley, E.E. (2011). *Alcohol and Drug Use among Drivers. British Columbia Roadside Survey 2010*. Ottawa, ON: Canadian Centre on Substance Abuse.

Beirness, D.J., & Davis, C.G. (2006). *Driving under the Influence of Cannabis: Analysis Drawn from the 2004 Canadian Addiction Survey*. Ottawa, ON: Canadian Centre on Substance Abuse.

Beirness, D.J., Beasley, E.E., & McClafferty, K. (2015). *Alcohol and Drug Use among Drivers in Ontario: Findings from the 2014 Roadside Survey*. Toronto, ON: Ontario Ministry of Transportation.

Biecheler, M.B., Peytavin, J.F., SAM Group, Facy, F., & Martineau, H. (2008). SAM survey on "drugs and fatal accidents": Search of substances consumed and comparison between drivers involved under the influence of alcohol or cannabis. *Traffic Injury Prevention, 9*(1), 11–21.

Blomberg, R.D., Peck, R.C., Moskowitz, H., Burns, M., & Fiorentino, D. (2005). *Crash Risk of Alcohol Involved Driving: A Case-Control Study*. Stamford, CT: Dunlap and Associates.

Boak, A., Hamilton, H.A., Adlaf, E.M., & Mann, R.E. (2013). *Drug Use among Ontario Students, 1977–2013: Detailed OSDUHS Findings (CAMH Research Document Series No. 36)*. Toronto, ON: Centre for Addiction and Mental Health.

Borkenstein, R.F., Crowther, R.F., Shumate, R.P., Ziel, W.B., & Zylman, R. (1964). *The Role of the Drinking Driver in Traffic Accidents*. Bloomington, IN: Department of Police Administration, Indiana University.

Borkenstein, R.F., Crowther, R.F., & Shumate, R.P. (1974). The role of the drinking driver in traffic accidents (The Grand Rapids Study). *Blutalkohol, 11*(Suppl.), 1–131.

Brady, J.E., & Li, G. (2014). Trends in alcohol and other drugs detected in fatally injured drivers in the United States, 1999–2010. *American Journal of Epidemiology, 179*(6), 692–699.

Burns, P.C., Parkes, A., Burton, S., Smith, R.K., & Burch, D. (2002). *How Dangerous Is Driving with a Mobile Phone? Benchmarking the Impairment to Alcohol* (TRL Report TRL547). Crowthorne: Transport Research Laboratory.

Butters, J.E., Smart, R.G., Mann, R.E., & Asbridge, M. (2005). Illicit drug use, alcohol use and problem drinking among infrequent and frequent road ragers. *Drug & Alcohol Dependence, 80*(2), 169–175.

Byrne, P., Ma, T., Mann, R.E., & Elzohairy, Y. (2016). Evaluation of the general deterrence capacity of recently implemented (2009–2010) low and zero BAC requirements for drivers in Ontario. *Accident Analysis & Prevention, 88*, 56–67.

Callaghan, R.C., Gatley, J.M., Sanches, M., & Asbridge, M. (2014). Impacts of the minimum legal drinking age on motor vehicle collisions in Québec, 2000–2012. *American Journal of Preventive Medicine, 47*(6), 788–795.

Casswell, S. (1979). Cannabis and alcohol: Effects on closed-course driving behaviour. *Proceedings of the Seventh International Conference on Alcohol, Drugs and Traffic Safety*, 238–246.

Charlton, S.G., & Starkey, N.J. (2015). Driving while drinking: Performance impairments resulting from social drinking. *Accident Analysis & Prevention, 74,* 210–217.

Cook, S., Shank, D., Bruno, T., Turner, N.E., & Mann, R.E. (in press). Self-reported driving after drinking and after cannabis use among Ontario students: Associations with Graduated Licensing, risk taking and substance abuse. *Traffic Injury Prevention.*

Couper, F.J., & Peterson, B.L. (2014). The prevalence of marijuana in suspected impaired driving cases in Washington State. *Journal of Analytical Toxicology, 38,* 569–574.

Creaser, J.I., Ward, N.J., & Rakauskas, M.E. (2011). Acute alcohol impairment research in driving simulators. In D.L. Fisher, M. Rizzo, J.K. Caird, & J.D. Lee (Eds.) *Handbook of Driving Simulation for Engineering, Medicine and Psychology*. Boca Raton, FL: CRC Press.

Deshapriya, E.B.R., & Iwase, N. (1998). Impact of the 1970 legal BAC 0.05 mg% limit legislation on drunk-driver-involved traffic fatalities, accidents, and DWI in Japan. *Substance Use & Misuse, 33,* 2757–2788.

Dols, S.T., González, F.J.Á., Aleixandre, N.L., Vidal-Infer, A., Rodrigo, M.J.T., & Valderrama-Zurián, J.C. (2010). Predictors of driving after alcohol and drug use among adolescents in Valencia (Spain). *Accident Analysis & Prevention, 42*(6), 2024–2029.

Dott, A.B. (1972). *Effect of Marijuana on Risk Acceptance in a Simulated Passing Task.* Washington, DC: Public Health Service.

Downey, L.A., King, R., Papafotiou, K., Swann, P., Ogden, E., Boorman, M., & Stough, C. (2013). The effects of cannabis and alcohol on simulated driving: Influences of dose and experience. *Accident Analysis & Prevention, 50,* 879–886.

Drummer, O.H., Gerostamoulos, J., Batziris, H., Chu, M., Caplehorn, J., Robertson, M.D., & Swann, P. (2004). The involvement of drugs in drivers of motor vehicles killed in Australian road traffic crashes. *Accident Analysis & Prevention, 36*(2), 239–248.

Dubois, S., Mullen, N., Weaver, B., & Bédard, M. (2015). The combined effects of alcohol and cannabis on driving: Impact on crash risk. *Forensic Science International, 248,* 94–100.

Duhig, A.M., Cavallo, D.A., McKee, S.A., George, T.P., & Krishnan-Sarin, S. (2005). Daily patterns of alcohol, cigarette, and marijuana use in adolescent smokers and nonsmokers. *Addictive Behaviors, 30,* 271–283.

Eisenberg, D. (2003). Evaluating the effectiveness of policies related to drunk driving. *Journal of Policy Analysis & Management, 22,* 249–273.

Elder, R.W., Nichols, J.L., Shults, R.A., Sleet, D.A., Barrios, L.C., Compton, R., & Task Force on Community Preventive Services. (2005). Effectiveness of school-based programs for reducing drinking and driving and riding with drinking drivers: A systematic review. *American Journal of Preventive Medicine, 28*(5), 288–304.

Ellingstad, V.S., McFarling, L.H., & Struckman, D.L. (1973). *Alcohol, Marijuana and Risk Taking.* South Dakota University: Vermillion Human Factors Laboratory.

Elvik, R. (2013). Risk of road accident associated with the use of drugs: A systematic review and meta-analysis of evidence from epidemiological studies. *Accident Analysis & Prevention, 60,* 254–267.

Fergusson, D.M., & Horwood, L.J. (2001). Cannabis use and traffic accidents in a birth cohort of young adults. *Accident Analysis & Prevention, 33*(6), 703–711.

Fergusson, D.M., Horwood, L.J., & Boden, J.M. (2008). Is driving under the influence of cannabis becoming a greater risk to driver safety than drink driving? Findings from a longitudinal study. *Accident Analysis & Prevention, 40*(4), 1345–1350.

Foss, R.D., Stewart, J.R., & Reinfurt, D.W. (2001). Evaluation of the effects of North Carolina's 0.08% BAC law. *Accident Analysis & Prevention, 33,* 507–517.

Gawron, V.J., & Ranney, T.A. (1988). The effects of alcohol on driving performance on a closed-course and in a driving simulator. *Ergonomics, 31*(9), 1219–1244.

Hall, W., & Degenhardt, L. (2009). Adverse health effects of non-medical cannabis use. *Lancet, 374*(9698), 1383–1391.

Hartman, R.L., & Huestis, M.A. (2013). Cannabis effects on driving skills. *Clinical Chemistry, 59,* 478–492.

Hartman, R.L., Brown, T.L., Milavetz, G., Spurgin, A., Pierce, R.S., Gorelick, D.A., Gaffney, G., & Huestis, M.A. (2015). Cannabis effects on driving lateral control with and without alcohol. *Drug & Alcohol Dependence, 154,* 25–37.

Health Canada. (2013). *Canadian Alcohol and Drug Use Monitoring Survey (CADUMS).* Ottawa, ON: Health Canada.

Hingson, R., Heeren, T., Mangione, T., Morelock, S., & Mucatel, M. (1982). Teenage driving after using marijuana or drinking and traffic accident involvement. *Journal of Safety Research, 13*(1), 33–38.

Huestis, M.A. (2015). Cannabis-impaired driving: A public health and safety concern. *Clinical Chemistry, 61*(10), 1223–1225.

Hunault, C.C., Mensinga, T.T., Böcker, K.B.E., Schipper, M.A., Kruidenier, M., Leenders, M.E.C., de Vries, I., & Meulenbelt, J. (2009). Cognitive and psychomotor effects in males after smoking a combination of tobacco and cannabis containing up to 69 mg delta-9-tetrahydrocannabinol (THC). *Psychopharmacology, 204,* 85–94.

Johnson, M.B., Kelley-Baker, T., Voas, R.B., & Lacey, J.H. (2012). The prevalence of cannabis-involved driving in California. *Drug & Alcohol Dependence, 123*(1–3), 105–109.

Jones, A.W., & Neri, A. (1994). Age-related differences in the effects of ethanol on performance and behaviour in healthy men. *Alcohol & Alcoholism, 29,* 171–179.

Kann, L., Kinchen, S., Shanklin, S.L. et al. (2014). Youth risk behavior surveillance—United States, 2013. *Morbidity & Mortality Weekly Report: Surveillance Summaries*, *63*(Suppl 4), 1–168.

Kerr, J.S., & Hindmarch, I. (1998). The effects of alcohol alone or in combination with other drugs on information processing, task performance and subjective responses. *Human Psychopharmacology, 13,* 1–9.

Krueger, H.P., & Vollrath, M. (2000). Effects of cannabis and amphetamines on driving simulator performance of recreational drug users in the natural field. *Proceedings of the 15th International Conference on Alcohol, Drugs, and Traffic Safety.* Retrieved from http://www.icadts.org/proceedings/2000/icadts2000-150.pdf.

Lamers, C.T.J., & Ramaekers, J.G. (2001). Visual search and urban city driving under the influence of marijuana and alcohol. *Human Psychopharmacology—Clinical & Experimental, 16,* 393–401.

Langley, J.D., Wagenaar, A.C., & Begg, D.J. (1996). An evaluation of the New Zealand graduated driver licensing system. *Accident Analysis & Prevention, 28,* 139–146.

Laumon, B., Gadegbeku, B., Martin, J.L., & Biecheler, M.B. (2005). Cannabis intoxication and fatal road crashes in France: Population based case–control study. *British Medical Journal, 331*(7529), 1371.

Leadbeater, B.J., Foran, K., & Grove-White, A. (2008). How much can you drink before driving? The influence of riding with impaired adults and peers on the driving behaviors of urban and rural youth. *Addiction, 103*(4), 629–637.

Lenné, M.G., Dietze, P.M., Triggs, T.J., & Walmsley, S. (2010). The effects of cannabis and alcohol on simulated arterial driving: Influences of driving experience and task demand. *Accident Analysis & Prevention, 42,* 859–866.

Leung, S., & Starmer, G. (2005). Gap acceptance and risk-taking by young and mature drivers, both sober and alcohol-intoxicated, in a simulated driving task. *Accident Analysis & Prevention, 37,* 1056–1065.

Li, K., Simons-Morton, B.G., & Hingson, R. (2013). Impaired-driving prevalence among US high school students: Associations with substance use and risky driving behaviors. *American Journal of Public Health, 103*(11), e71–e77.

Li, M.C., Brady, J.E., DiMaggio, C.J., Lusardi, A.R., Tzong, K.Y., & Li, G. (2012). Marijuana use and motor vehicle crashes. *Epidemiologic Reviews, 34*(1), 65–72.

Liguori, A., Gatto, C.P., & Robinson, J.H. (1998). Effects of marijuana on equilibrium, psychomotor performance, and simulated driving. *Behavioural Pharmacology, 9*(7), 599–609.

Liguori, A., Gatto, C.P., & Jarrett, D.B. (2002). Separate and combined effects of marijuana and alcohol on mood, equilibrium and simulated driving. *Psychopharmacology, 163*, 399–405.

Longo, M.C., Hunter, C.E., Lokan, R.J., White, J.M., & White, M.A. (2000). The prevalence of alcohol, cannabinoids, benzodiazepines and stimulants amongst injured drivers and their role in driver culpability, Part II: The relationship between drug prevalence and drug concentration, and driver culpability. *Accident Analysis & Prevention, 32*(5), 623–632.

Ma, T., Byrne, P.A., Haya, M., & Elzohairy, Y. (2015). Working in tandem: The contribution of remedial programs and roadside licence suspensions to drinking and driving deterrence in Ontario. *Accident Analysis & Prevention, 85*, 248–256.

MacDonald, S., Anglin-Bodrug, K., Mann, R.E., Erickson, P., Hathaway, A., Chipman, C., & Rylett, M. (2003). Injury risk associated with cannabis and cocaine use. *Drug & Alcohol Dependence, 72*(2), 99–115.

Mann, R.E., Vingilis, E.R., Leigh, G., Anglin, L., & Blefgen, H. (1986). School-based programmes for the prevention of drinking and driving: Issues and results. *Accident Analysis & Prevention, 18*(4), 325–337.

Mann, R.E., Vingilis, E.R., & Stewart, K. (1988). Programmes to change individual behaviour: Education and rehabilitation in the prevention of drinking and driving. In M.D. Laurence, J.R. Snortum, & F.E. Zimring (Eds.), *The Social Control of Drinking and Driving*. Chicago, IL: University of Chicago Press, pp. 248–269.

Mann, R.E., Anglin, L., Wilkins, K., Vingilis, E.R., MacDonald, S., & Sheu, W.J. (1994). Rehabilitation for convicted drinking drivers (second offenders): Effects on mortality. *Journal of Studies on Alcohol, 55*(3), 372–374.

Mann, R.E., Stoduto, G., Anglin, L., Pavic, B., Fallon, F., Lauzon, R., & Amitay, O. (1997). Graduated Licensing in Ontario: Impact of the 0 BAL provision on adolescents' drinking-driving. In C. Mercier-Guyon (Ed.), *Alcohol, Drugs and Traffic Safety—T97*, Annecy, France: Centre d'Etudes et de Recherche en Medicin du Traffic, 1055–1060.

Mann, R.E., MacDonald, S., Stoduto, G., Bondy, S., Jonah, B., & Shaikh, A. (2001). The effect of introducing or lowering legal per se blood alcohol limits for driving: An international review. *Accident Analysis & Prevention, 33*, 569–583.

Mann, R.E., Brands, B., Macdonald, S., & Stoduto, G. (2003). *Impacts of cannabis on driving: An analysis of current evidence with an emphasis on Canadian data. A report prepared for Road Safety and Motor Vehicle Regulation Directorate, Publication No. TP 14179 E.* Ottawa, ON: Transport Canada.

Mayhew, D.R. (2007). Driver education and graduated licensing in North America: Past, present, and future. *Journal of Safety Research, 38*(2), 229–235.

Mayhew, D.R., Donelson, A.C., Beirness, D.J., & Simpson, H.M. (1986). Youth, alcohol and relative risk of crash involvement. *Accident Analysis & Prevention, 18*(4), 273–287.

McCartt, A.T., Farmer, C.M., & Eichelberger, A.H. (2013). US trends in late-night weekend alcohol-related fatal crashes and drinking and driving. In *International Conference on Alcohol, Drugs and Traffic Safety (T2013), 20th, 2013, Brisbane, Queensland, Australia*. Brisbane, Australia: ICADTS.

Mets, M.A.J., Kuipers, E., de Senerpont Domis, L.M., Leenders, M., Olivier, B., & Verster, J. C. (2011). Effects of alcohol on highway driving in the STISIM driving simulator. *Human Psychopharmacology, 26*, 434–439.

Miech, R.A., Johnston, L.D., O'Malley, P.M., Bachman, J.G., & Schulenberg, J.E. (2015). *Monitoring the Future National Survey Results on Drug Use, 1975–2014: Volume I, Secondary school students.* Ann Arbor, MI: Institute for Social Research, The University of Michigan.

Mørland, J., Steentoft, A., Simonsen, K.W., Ojanperä, I., Vuori, E., Magnusdottir, K., Kristinsson, J., Ceder, G., Kronstrand, R., & Christophersen, A. (2011). Drugs related to motor vehicle crashes in northern European countries: A study of fatally injured drivers. *Accident Analysis & Prevention, 43*(6), 1920–1926.

Moskowitz, H. (1985). Marihuana and driving. *Accident Analysis & Prevention, 17*(4), 323–345.

Moskowitz, H., Burns, M., Fiorentino, D., Smiley, A., & Zador, P. (2000). Driver characteristics and impairment at various BACs (DTNH-22-95-C-05000). Washington, DC: National Highway Traffic Safety Administration, US Department of Transportation.

Moskowitz, H., & Fiorentino, D. (2000). A review of the literature on the effects of low doses of alcohol on driving-related skills (No. HS-809 028).

Mura, P., Kintz, P., Ludes, B., Gaulier, J.M., Marquet, P., Martin-Dupont, S., Vincent, F. et al. (2003). Comparison of the prevalence of alcohol, cannabis and other drugs between 900 injured drivers and 900 control subjects: Results of a French collaborative study. *Forensic Science International, 133*(1), 79–85.

Mura, P., Chatelain, C., Dumestre, V., Gaulier, J.M., Ghysel, M.H., Lacroix, C., Kergueris, M.F. et al. (2006). Use of drugs of abuse in less than 30-year-old drivers killed in a road crash in France: A spectacular increase for cannabis, cocaine and amphetamines. *Forensic Science International, 160*(2), 168–172.

National Center for Statistics and Analysis. (2014). *Alcohol impaired driving: 2013 data. (Traffic Safety Facts. DOT HS 812 102).* Washington, DC: National Highway Traffic Safety Administration.

National Highway Traffic Safety Administration. (2013). *Traffic safety facts—2012 data: Alcohol-impaired driving.* DOT HS 811 870. Washington, D.C.: U.S. Department of Transportation.

O'Malley, P.M., & Johnston, L.D. (2013). Driving after drug or alcohol use by U.S. high school seniors, 2001–2011. *American Journal of Public Health, 103*(11), 2027–2034.

Odero, W., Garner, P., & Zwi, A. (1997). Road traffic injuries in developing countries: A comprehensive review of epidemiological studies. *Tropical Medicine & International Health, 2*(5), 445–460.

Office of National Drug Control Policy. (2011). *Teen drugged driving: A Community awareness activity toolkit.* Retrieved from http://www.whitehouse.gov/sites/default/files/ondcp/issues-content/drugged_driving_toolkit.pdf.

Organisation for Economic Co-operation and Development. (2006). Policy brief—Young drivers: The road to safety. Downloaded July 1, 2013, retrieved from http://internationaltransportforum.org/jtrc/safety/YDpolicyBrief.pdf.

Ouimet, M.C., Dongier, M., Di Leo, I., Legault, L., Tremblay, J., Chanut, F., & Brown, T.G. (2013). A randomized controlled trial of brief motivational interviewing in impaired driving recidivists: A 5-year follow-up of traffic offenses and crashes. *Alcoholism: Clinical and Experimental Research, 37*(11), 1979–1985.

Papafotiou, K., Carter, J.D., & Stough, C. (2005). The relationship between performance on the standardised field sobriety tests, driving performance and the level of Delta9-tetrahydrocannabinol (THC) in blood. *Forensic Science International, 155*(2–3), 172–178.

Pape, H., Rossow, I., & Storvoll, E.E. (2009). Under double influence: Assessment of simultaneous alcohol and cannabis use in general youth populations. *Drug & Alcohol Dependence, 101*, 69–73.

Peck, R.C., Gebers, M.A., Voas, R.B., & Romano, E. (2008). The relationship between blood alcohol concentration (BAC), age, and crash risk. *Journal of Safety Research, 39*(3), 311–319.

Phillips, R.O., Ulleberg, P., & Vaa, T. (2011). Meta-analysis of the effect of road safety campaigns on accidents. *Accident Analysis & Prevention, 43*(3), 1204–1218.

Poulin, C., Boudreau, B., & Asbridge, M. (2007). Adolescent passengers of drunk drivers: A multi-level exploration into the inequities of risk and safety. *Addiction, 102*(1), 51–61.

Quillian, W.C., Cox, D.J., Kovatchev, B.P., & Phillips, C. (1999). The effects of age and alcohol intoxication on simulated driving performance, awareness and self-restraint. *Age & Ageing, 28,* 59–66.

Rakauskas, M.E., Ward, N.J., Boer, E.R., Bernat, E.M., Cadwallader, M., & Patrick, C.J. (2008). Combined effects of alcohol and distraction on driving performance. *Accident Analysis & Prevention, 40,* 1742–1749.

Ramaekers, J.G., Robbe, H.W.J., & O'Hanlon, J.F. (2000). Marijuana, alcohol and actual driving performance. *Human Psychopharmacology—Clinical & Experimental, 15,* 551–558.

Ramaekers, J.G., Berghaus, G., van Laar, M., & Drummer, O.H. (2004). Dose related risk of motor vehicle crashes after cannabis use. *Drug & Alcohol Dependence, 73*(2), 109–119.

Richer, I., & Bergeron, J. (2009). Driving under the influence of cannabis: Links with dangerous driving, psychological predictors, and accident involvement. *Accident Analysis & Prevention, 41*(2), 299–307.

Robbe, H.W.J. (1998). Marijuana's impairing effects on driving are moderate when taken alone but severe when combined with alcohol. *Human Psychopharmacology—Clinical & Experimental, 13,* S70–S78.

Robertson, A.A., Gardner, S., Xu, X., & Costello, H. (2009). The impact of remedial intervention on 3-year recidivism among first-time DUI offenders in Mississippi. *Accident Analysis & Prevention, 41,* 1080–1086.

Robertson, L.S., & Zador, P.L. (1978). Driver education and fatal crash involvement of teenaged drivers. *American Journal of Public Health, 68*(10), 959–965.

Roehrs, T., Beare, D., Zorick, F., & Roth, T. (1994). Sleepiness and ethanol effects on simulated driving. *Alcoholism: Clinical & Experimental Research, 18,* 154–158.

Romano, E., & Pollini, R.A. (2013). Patterns of drug use in fatal crashes. *Addiction, 108*(8), 1428–1438.

Romano, E., Torres-Saavedra, P., Voas, R.B., & Lacey, J.H. (2014). Drugs and alcohol: Their relative crash risk. *Journal of Studies on Alcohol & Drugs, 75*(1), 56–64.

Ronen, A., Gershon, P., Drobiner, H., Rabinovich, A., Bar-Hamburger, R., Mechoulam, R., Cassuto, Y., & Shinar, D. (2008). Effects of THC on driving performance, physiological state and subjective feelings relative to alcohol. *Accident Analysis & Prevention, 40,* 926–934.

Ronen, A., Chassidim, H.S., Gershon, P., Parmet, Y., Rabinovich, A., Bar-Hamburger, R., Cassuto, Y., & Shinar, D. (2010). The effect of alcohol, THC and their combination on perceived effects, willingness to drive and performance of driving and non-driving tasks. *Accident Analysis & Prevention, 42,* 1855–1865.

Salomonsen-Sautel, S., Min, S.J., Sakai, J.T., Thurstone, C., & Hopfer, C. (2014). Trends in fatal motor vehicle crashes before and after marijuana commercialization in Colorado. *Drug & Alcohol Dependence, 140,* 137–144.

Schulze, H., Schumacher, M., Urmeew, R., Alvarez, J., Bernhoft, I.M., de Gier, H., Hagenzieker, M. et al. (2012). Driving under the influence of drugs, alcohol and medicines in Europe—Findings from the DRUID project. Lisbon: European Monitoring Centre for Drugs and Drug Addiction.

Sewell, R.A., Poling, J., & Sofuoglu, M. (2009). The effect of cannabis compared with alcohol on driving. *The American Journal on Addictions, 18,* 185–193.

Sexton, B.F., Tunbridge, R.J., Brook-Carter, N., Jackson, P.G., Wright, K., Stark, M.M., & Englehart, K. (2000). The influence of cannabis on driving (TRL Report TRL477). Crowthorne: TRL Limited.

Sexton, B.F., Tunbridge, R.J., Board, A., Jackson, P.G., Wright, K., Stark, M.M., & Englehart, K. (2002). The influence of cannabis and alcohol on driving (TRL Report TRL543). Crowthorne: TRL Limited.

Shell, D.F., Newman, I.M., Córdova-Cazar, A.L., & Heese, J.M. (2015). Driver education and teen crashes and traffic violations in the first two years of driving in a graduated licensing system. *Accident Analysis & Prevention, 82,* 45–52.

Smart, R.B. (1974). Marihuana and driving risk among college students. *Journal of Safety Research, 6*(2), 155–291.

Smiley, A. (1986). Marijuana: On-road and driving simulator studies. *Alcohol, Drugs, & Driving, 2,* 121–134.

Smiley, A. (1999). Marijuana: On-road and driving-simulator studies. In H. Kalant, W. Corrigall, W. Hall, & R.E. Smart (Eds.), *The Health Effects of Cannabis.* Toronto: Addiction Research Foundation, 173–191.

Smiley, A., LeBlanc, E., French, I., & Burford, R. (1975). The combined effects of alcohol and common psychoactive drugs. *Journal of the Canadian Society of Forensic Science, 8*, 57–64.

Smiley, A.M., Moskowitz, H., & Zeidman, K. (1981). Driving simulator studies of marijuana alone and in combination with alcohol. *25th Conference of the American Association for Automotive Medicine*, 107–116.

Society of Automotive Engineering. (2015). *Operational definitions of driving performance measures and statistics. Surface Vehicle Recommended Practice J2944*. Warrendale, PA: SAE.

Stein, A.C., Allen, R.W., Cook, M.S., & Karl, R.L. (1983). A simulator study of the combined effects of alcohol and marihuana on driving behavior—Phase II. Springfield, VA: National Technical Information Service.

Stoduto, G., Vingilis, E., Kapur, B.M., Sheu, W.J., McLellan, B.A., & Liban, C.B. (1993). Alcohol and drug use among motor vehicle collision victims admitted to a regional trauma unit: Semographic, injury, and crash characteristics. *Accident Analysis & Prevention, 25*(4), 411–420.

Stoduto, G., Mann, R.E., Flam-Zalcman, R., Sharpley, J., Brands, B., Butters, J., Smart, R.G., Wickens, C.M., Illie, G., & Thomas, R.K. (2014). Impact of Ontario's remedial program for drivers convicted of drinking and driving on substance use and problems. *Canadian Journal of Criminology & Criminal Justice, 56*(2), 201–217.

Sweedler, B.M., Biecheler, M.B., Laurell, H., Kroj, G., Lerner, M., Mathijssen, M.P.M., Mayhew, D., & Tunbridge, R.J. (2004). Worldwide trends in alcohol and drug impaired driving. *Traffic Injury Prevention, 5*(3), 175–184.

Taylor, B., Irving, H.M., Kanteres, F., Room, R., Borges, G., Cherpitel, C., Bond, J., Greenfield, T., & Rehm, J. (2010). The more you drink, the harder you fall: A systematic review and meta-analysis of how acute alcohol consumption and injury or collision risk increase together. *Drug & Alcohol Dependence, 110*(1), 108–116.

ter Bogt, T.F.M., de Looze, M., Molcho, M., Godeau, E., Hublet, A., Kokkevi, A., Kuntsche, E. et al. (2014). Do societal wealth, family affluence and gender account for trends in adolescent cannabis use? A 30 country cross-national study. *Addiction, 109*, 273–283.

Traffic Injury Research Foundation. (2013). Trends among fatally injured teen drivers, 2000–2010. Ottawa, ON: TIRF.

Tsai, V.W., Anderson, C.L., & Vaca, F.E. (2010). Alcohol involvement among young female drivers in US fatal crashes: Unfavourable trends. *Injury Prevention, 16*(1), 17–20.

Vanlaar, W., Mayhew, D., Marcoux, K., Wets, G., Brijs, T., & Shope, J. (2009). An evaluation of graduated driver licensing programs in North America using a meta-analytic approach. *Accident Analysis & Prevention, 41*(5), 1104–1111.

Vanlaar, W., Robertson, R., Marcoux, K., Mayhew, D., Brown, S., & Boase, P. (2012). Trends in alcohol-impaired driving in Canada. *Accident Analysis & Prevention, 48*, 297–302.

Vaughn, M.G., Define, R.S., DeLisi, M., Perron, B.E., Beaver, K.M., Fu, Q., & Howard, M.O. (2011). Sociodemographic, behavioral, and substance use correlates of reckless driving in the United States: Findings from a national sample. *Journal of Psychiatric Research, 45*, 347–353.

Vindenes, V., Jordbru, D., Knapskog, A.B., Kvan, E., Mathisrud, G., Slørdal, L., & Mørland, J. (2012). Impairment based legislative limits for driving under the influence of non-alcohol drugs in Norway. *Forensic Science International, 219*, 1–11.

Vingilis, E., & Coultes, B. (1990). Mass communications and drinking-driving: Theories, practices, and results. *Alcohol, Drugs & Driving, 6*(2), 61–81.

Voas, R.B., & Tippetts, A.S. (Eds.) (1999). *The Relationship of Alcohol Safety Laws to Drinking Drivers in Fatal Crashes*. Washington, D.C.: National Highway Traffic Safety Administration.

Voas, R.B., Tippetts, A.S., & Fell, J.C. (2003). Assessing the effectiveness of minimum legal drinking age and zero tolerance laws in the United States. *Accident Analysis & Prevention, 35*(4), 579–587.

Wagenaar, A.C., & Toomey, T.L. (2002). Effects of minimum drinking age laws: Review and analyses of the literature from 1960 to 2000. *Journal of Studies on Alcohol, 14, Supplement*, 206–225.

Waller, P.F. (2003). The genesis of GDL. *Journal of Safety Research, 34*(1), 17–23.

Weafer, J., Camarillo, D., Fillmore, M.T., Milich, R., & Marczinski, C.A. (2008). Simulated driving performance of adults with ADHD: Comparisons with alcohol intoxication. *Experimental & Clinical Psychopharmacology, 16*(3), 251–263.

Weiler, J.M., Bloomfield, J.R., Woodworth, G.G., Grant, A.R., Layton, T.A., Brown, T.L., McKenzie, D.R., Baker, T.W., & Watson, G.S. (2000). Effects of fexofenadine, diphenhydramine, and alcohol on driving performance: A randomized, placebo-controlled trial in the Iowa driving simulator. *Annals of Internal Medicine, 132,* 354–363.

Wells-Parker, E., Mann, R.E., Dill, P.L., Stoduto, G., Shuggi, R., & Cross, G.W. (2009). Negative affect and drinking drivers: A review and conceptual model linking dissonance, efficacy and negative affect to risk and motivation for change. *Current Drug Abuse Reviews, 2,* 115–126.

Whitehill, J.M., Rivara, F.P., & Moreno, M.A. (2014). Marijuana-using drivers, alcohol-using drivers, and their passengers: Prevalence and risk factors among underage college students. *JAMA Pediatrics, 168*(7), 618–624.

Wickens, C.M., Mann, R.E., Stoduto, G., Flam-Zalcman, R., & Butters, J. (2013). Alcohol control measures and traffic safety. *Alcohol Science Policy & Public Health,* 378–388.

Williams, A.F., West, B.A., & Shults, R.A. (2012). Fatal crashes of 16- to 17-year-old drivers involving alcohol, nighttime driving, and passengers. *Traffic Injury Prevention, 13*(1), 1–6.

Woodall, K.L., Chow, B.L., Lauwers, A., & Cass, D. (2015). Toxicological findings in fatal motor vehicle collisions in Ontario, Canada: A one-year study. *Journal of Forensic Sciences, 60*(3), 669–674.

World Health Organization. (2015). *Fact sheet #1—Road safety: Basic facts.* Downloaded May 12, 2015 from http://www.who.int/violence_injury_prevention/publications/road_traffic/Road_safety_media_brief_full_document.pdf?ua=1.

Yi, H.Y., Chen, C.M., & Williams, G.D. (2006). *Surveillance report# 76: Trends in alcohol-related fatal traffic crashes, United States, 1982–2004.* Bethesda, MD: National Institute on Alcohol Abuse and Alcoholism.

Young, M.M., Saewyc, E., Boak, A., Jahrig, J., Anderson, B., Doiron-Brun, Y., Taylor, S. et al. (2011). *The Cross-Canada Report on Student Drug Use.* Ottawa, ON: Canadian Centre on Substance Abuse.

Yu, J., Evans, P.C., & Perfetti, L. (2004). Road aggression among drinking drivers: Alcohol and non-alcohol effects on aggressive driving and road rage. *Journal of Criminal Justice, 32*(5), 421–430.

Zador, P.L., Krawchuk, S.A., & Voas, R.B. (2000). Alcohol-related relative risk of driver fatalities and driver involvement in fatal crashes in relation to driver age and gender: An update using 1996 data. *Journal of Studies on Alcohol, 61*(3), 387–395.

Zhao, J., Mann, R.E., Chipman, M., Adlaf, E., Stoduto, G., & Smart, R.G. (2006). The impact of driver education on self-reported collisions among young drivers with a graduated license. *Accident Analysis & Prevention, 38*(1), 35–42.

14

Attention Deficit Hyperactivity Disorder (ADHD)

Rosemary Tannock

Lana M. Trick

Abstract

Symptoms and Prevalence of ADHD. ADHD is a neurodevelopmental disorder characterized by persistent, developmentally inappropriate, and impairing levels of inattentive or hyperactive/impulsive behavior that affects 5%–7% of adolescents. ***ADHD and Driving.*** Marked individual differences exist, but compared to other drivers, as a group, drivers with ADHD report more adverse driving outcomes (e.g., collisions) and aberrant behaviors (e.g., angry/aggressive driving, speeding, errors with vehicle controls, more time with eyes off the road) and they may be more susceptible to the effects of fatigue and alcohol. ***Treatment.*** In teens and novice drivers with ADHD, behavioral interventions and pharmacological treatments (e.g., methylphenidate, atomoxetine) may be effective in improving driving performance. ***Implications for Teens, Parents, and Clinicians.*** The chapter concludes with practical advice for clinicians and parents of teens with ADHD. A dynamic, personalized, tiered approach to intervention is required to mitigate both stable risk factors associated with ADHD (e.g., inattention, impulsivity) as well as transient situational factors (e.g., sleepiness, distractions).

14.1 Introduction

The chapters up to this point have discussed factors that put teen and novice drivers at an elevated risk of collision. A number of these are relevant to the discussion of attention deficit hyperactivity disorder

(ADHD). For example, the problem of ADHD is best understood from a neurodevelopmental perspective, as it is associated with both delayed development of and dysregulation of the structure and function in major brain pathways (specifically, the frontal–subcortical–cerebellar pathways) that govern attention, inhibitory control, response to reward and emotion regulation, salience thresholds, and motor behavior. Thus, the discussion on developmental and social emotional factors in Chapters 9 and 10 may be of interest. Furthermore, teens with ADHD face some of the same challenges as other teens relating to peers (see Chapter 16). In fact, given that adolescents diagnosed with ADHD are more likely to associate with others who are also at risk, they may be especially likely to reinforce each other in dangerous driving practices, a tendency that may be particularly evident in girls (Cardoos et al., 2013; Laucht et al., 2007). However, there are issues that are unique to ADHD, ones that put teen drivers with ADHD at even higher risk than other drivers their age. For example, in an editorial, Winston et al. (2013) likened the combination of novice drivers with ADHD and the recent proliferation of distracting in-vehicle technologies to the perfect storm, suggesting the urgent need for additive interventions aimed at reducing the risk of young drivers with ADHD.

In this chapter, we begin with a synopsis of the key features of ADHD, including symptoms, prevalence, and comorbidities. We then present evidence on the effects of ADHD on driving. The chapter concludes with a discussion of treatments as they relate to a tiered, dynamic, and personalized approach to intervention, with a special view toward implications for clinicians, parents, and the teens themselves.

14.2 ADHD: Symptoms, Prevalence, Comorbidities

14.2.1 Symptoms

ADHD is one of the most common neurodevelopmental disorders of childhood, characterized by persistent, developmentally inappropriate, and impairing levels of inattentive or hyperactive/impulsive behavior or both (American Psychiatric Association, 2013: *DSM-V: Diagnostic and Statistical Manual of Mental Disorders*). A neurodevelopmental disorder is one where typical development of the brain is disturbed in some way before birth or in early infancy. These early disturbances result in a variety of impairments in attention, memory and learning, motor function, language, and emotional self-regulation. There is no known single risk factor that is necessary or sufficient to cause ADHD: most cases are believed to arise from several genetic and environmental risk factors that have small individual and combined effects that increase susceptibility to the disorder.

Inattentive behavior includes a variety of difficulties. For example, inattention includes problems concentrating on the task at hand until it is completed; being easily distracted by other things going on in the immediate context (texting, cell phone) or by one's own thoughts (mind wandering); procrastinating on starting a challenging task; losing things; forgetting appointments, deadlines, or other commitments; and having great difficulty organizing sequenced events, items, time, and thoughts. Symptoms of inattention tend to persist from childhood through adolescence and into adulthood, albeit with some changes in how they manifest themselves. For example, teens and adults with ADHD show more severe problems in concentration, organization, time management, and forgetfulness than do children with ADHD.

Hyperactive/impulsive behavior also includes a variety of difficulties. The nine possible symptoms of hyperactivity/impulsivity that are considered in the diagnosis of ADHD include being impatient and wanting immediate action; having difficulty waiting in line or taking turns in conversations, frequently making sudden decisions without thinking about the consequences; and constantly fidgeting, moving or feeling the need to move. Although high levels of overt hyperactivity are often thought to diminish in adulthood, growing evidence indicates that this is not the case: objectively measured hyperactivity persists into adulthood and is a more discriminative feature of ADHD than are computerized measures of inattention or impulsivity (e.g., Teicher et al., 2012).

In addition to these behavioral symptoms, ADHD is associated with deficits in a wide array of relatively independent cognitive domains, including executive functions (visual–spatial working memory, inhibitory control, vigilance and planning); reward regulation (preference for immediate smaller rewards over larger but delayed rewards, suboptimal decision making); emotional processing; arousal and activation; and motor control (e.g., Coghill et al., 2014; Willcutt et al., 2005). Although most, but not all, individuals with ADHD show deficits in one or two cognitive domains, very few show deficits across all. For example, many adults show no deficits in executive functions (Hervey et al., 2004; Moffitt et al., 2015). At this point, it is unclear if cognitive deficits drive the development of the disorder and give rise to the symptoms or if the deficits are simply correlates of ADHD.

There are no reliable "markers" of ADHD to date (e.g., no blood tests, brain scans, or cognitive tests that can rule in or rule out the diagnosis), so the diagnosis of ADHD remains a clinical one, based on procedures and measures specified in national clinical guidelines (e.g., Canadian ADHD Practice Guidelines). Diagnosis is based on clinical interviews, where the diagnostician asks about each ADHD symptom; the age of onset and resultant functional impairments; as well as the person's developmental, medical, and family history. The goal of the interview is to rule out other explanations of the symptoms as well as to determine whether the symptoms are more extreme, persistent, and impairing than expected for the developmental level of the person. Age is an important issue in diagnosis because the expression of ADHD symptoms and impairments changes throughout an individual's lifetime (e.g., for an overview of diagnostic issues, see Faraone et al., 2015).

Although ADHD is considered to be a categorical diagnosis in the field of medicine (meaning that clinicians determine that you do or don't "have" ADHD), most scientific evidence suggests that what is called "ADHD" constitutes the extreme and impairing tails of distributions of one or more heritable traits, so that the clinical cutoff in terms of the number of symptoms used to diagnose ADHD is somewhat arbitrary. This means that individuals who have subdiagnostic levels of ADHD symptoms may also manifest a similar range of impairments, including driving problems.

14.2.2 Prevalence

The worldwide prevalence of ADHD in children and adolescents is about 5%–7%, with no major differences between rates in Europe, North and South America, Asia, Australia, and Africa (Polanczyk et al., 2007, 2015; Thomas et al., 2015; Willcutt, 2012). The prevalence of ADHD in adults remains unclear but has been reported to be about 2.5% (Simon et al., 2009). At least two factors may explain the lower rates of ADHD in adulthood: genuine remission of symptoms or underdiagnosis due to the child-centric nature of the DSM diagnostic criteria that renders them difficult to apply to older adolescents and adults. Although there is a risk for both overdiagnosis and underdiagnosis of ADHD, its prevalence has not changed over the past three decades (Polanczyk et al., 2014).

The presumed multifactorial causation of ADHD is consistent with its marked heterogeneity in expression. For example, neuroimaging studies reveal a variety of structural and functional abnormalities in multiple cortical (dorsolateral and ventromedial prefrontal cortex) and subcortical areas (dorsal and ventral anterior cingulate, basal ganglia), extending into the amygdala and cerebellum, as well as delayed maturation of the cerebral cortex (see review by Faraone et al., 2015). Moreover, these brain differences change with age, with some normalizing and others appearing to remain fixed (Shaw et al., 2013). It is unclear how developmental changes in patterns of cortical thickness predict developmental changes in terms of ADHD symptoms or impairments.

14.2.3 Comorbidities

ADHD typically coexists with other disorders—a situation referred to as *comorbidity*. "Pure ADHD" is not common: comorbidity is the norm, not the exception. For example, in adults, ADHD is comorbid with one or more other disorders in 65%–89% of cases; for children, these comorbidities occur in 60%–100% of cases (see Sobanski, 2006, for a review). Of the comorbid disorders in adults with ADHD, depression, bipolar disorder, anxiety disorders, and substance abuse are the most commonly cited (for reviews, see Kooij et al., 2012; Sobanski, 2006; Wilens et al., 2002). Teens with ADHD are especially inclined to substance abuse problems related to alcohol and cannabis, which is of special concern given recent evidence that early use of cannabis is especially deleterious to the development of executive function, that is, working memory and selective attention (Tamm et al., 2013). In addition, teens with ADHD are also likely to be diagnosed with comorbid oppositional defiant disorder (ODD), which occurs in 45%–84% of cases, and some may also be diagnosed with conduct disorder (see Becker et al., 2015, for a review).

ADHD is also associated with sleep disorders, and children, adolescents, and adults may experience problems such as insomnia (late-onset sleep), excessive daytime sleepiness, and restless sleep (Becker et al., 2015; Cortese et al., 2009; Jan et al., 2011; Kooij et al., 2012). These sleep problems may be partly due to the stimulant medications used to control ADHD, but differences are found in unmedicated samples as well. Furthermore, up to 50% of those diagnosed with ADHD are also diagnosed with developmental coordination disorder, a disorder that affects the planning and timing of movements and is associated with clumsy or uncoordinated movements (Henderson & Henderson, 2002; Dewey et al., 2007; Langevin et al., 2015), though there are some who argue that ADHD per se may be related to timing deficits (Toplak & Tannock, 2005). Moreover, children, adolescents, and adults with ADHD manifest poor postural control (i.e., balance), especially when sensory signals are disrupted, as when closing their eyes or by blocking access to visual cues (e.g., Hove et al., 2015; Ren et al., 2014). In adults with ADHD, poor postural control was found to be associated with cerebellar anomalies (Hove et al., 2015). Notably, postural control is a key determinant of driving safety, at least among older drivers (Lacherez et al., 2014), and sleep deprivation can exacerbate postural control problems in young adults (Aguiar & Barela, 2015). However, the potential association between postural control in ADHD and driving outcomes (particularly under conditions of reduced visibility) has not yet been investigated.

From the perspective of the driving literature, what can be taken from this discussion is that there are almost too many reasons to expect that ADHD would have a negative impact on driving performance given all the associated disorders: substance abuse; abnormalities in sleep; problems with balance, motor coordination, and timing; and difficulties with emotional self-control. Given these comorbidities, there is sometimes uncertainly about the extent to which the problems are produced by the symptoms of ADHD *per se* as opposed to one of the comorbid disorders (e.g., Vaa, 2014). There are marked individual differences between drivers with ADHD depending on the comorbidity. Drivers with comorbid anxiety may drive very differently than ones with comorbid ODD, for example.

14.3 ADHD and Driving

In this section, we briefly discuss the literature on ADHD and driving, highlighting some of the most noteworthy findings. (Length restrictions for this chapter necessitate that this review be selective rather than comprehensive. For more comprehensive coverage, see Fuermaier et al., 2015.) Although there are a large number of studies on ADHD and driving, most do not focus solely on teens, and as a result, we will also present some of the literature on young adults. The section begins with discussions of the research based on reports (sometimes self-reports) of adverse driving outcomes (collisions, tickets, etc.) and questionnaires about specific driving behaviors. Then we discuss studies that involve direct measures of performance under controlled conditions, focusing on the impact of three factors that may have stronger effects on drivers with and ADHD than others: distraction, fatigue, and alcohol. The final section discusses the implications of in-vehicle automation for drivers with ADHD.

14.3.1 Driving Outcomes (e.g., Collisions, Tickets)

When looking at the impact of ADHD on driving, the most obvious place to look is differences in adverse driving outcomes, most notably crashes. It is important to consider these data in context, though. Individuals with ADHD are more likely to be involved in a variety of different types of mishaps—on the playground, in the home, at work, and on the road. A nationwide cohort study suggested that among children, adolescents, and young adults with ADHD, the mortality rate is 1.5 times higher than that of comparable others without disorders, mostly due to accidents (Dalsgaard et al., 2015). Comorbid ODD/ conduct disorder increases the risk to 2.0 times higher than others. Thus, individuals with ADHD are at higher risk of accidents in general, but vehicle collisions are one of the most common types of accident. At present, there is consensus that individuals with ADHD are at higher risk than other drivers. In fact, a large-scale review of over 20 studies based on self-reports, official driving records, and driving simulation concluded that drivers with ADHD had more collisions overall; had more highway, rear-end, fatal, and hit-and-run collisions; and are also more likely to be judged at fault in a collision (Fuermaier et al., 2015). Furthermore, this review reported that these studies indicated that drivers with ADHD had more citations for speeding and reckless driving, more convictions and arrests, more incidents of driving under the influence of alcohol or driving without a license, and more license suspensions. This general pattern is supported by two population-based studies on teens with ADHD (Chang et al., 2014; Redelmeier et al., 2010).

However, there is controversy about the exact extent of this difference between drivers with and without ADHD, particularly as it relates to confounding factors such as comorbid disorders and differences in driving exposure (the number of miles on the road). For example, Barkley et al. (1993) reported that drivers diagnosed with ADHD were three to four times more likely to have a collision than other drivers. In contrast, Vaa (2014) proposed that although drivers with ADHD have more adverse outcomes (are more likely to have speeding tickets, lose their driving license, or drive without a license), they are only 1.23 times more likely to have collisions than other drivers.

The discrepancies among studies reflect methodological differences. They involve different samples. Some studies do a better job controlling for the effects of comorbid disorders than others (Vaa, 2014). Conduct disorder is of particular concern because it is associated with delinquency and conflict with the law. It is a disorder diagnosed in childhood and adolescence, but for some, it develops into antisocial personality disorder in adulthood. Furthermore, individuals with conduct disorder are more likely to associate with peers who support delinquent behavior, and this may also contribute to the risks. Moreover, some studies do a better job than others of controlling for differences in driving exposure, i.e., the number of miles traveled (Vaa, 2014). This is a problem because the opportunities for collision are greater for those who drive farther and more often. Many studies do not even measure driving exposure, let alone control it, and thus, differences in collision risk between the ADHD and control groups may be partly due to differences in how frequently they drive or how far. However, even when self-report exposure data are included, they may not be very accurate. Because differences in collisions rates cannot be expected to emerge within a short period of time (collisions are relatively rare), studies of collisions often require that people report on activities that occur over an extended period of time (years). Many people do not have a very accurate idea of how far they drive over the course of a month, let alone over the course of years.

The solution to this problem would be to conduct a long-term naturalistic investigation, making use of instrumented vehicles to continuously monitor driving performance and exposure over a series of years. Aduen et al. (2015; see also Antin et al., 2011) conducted such a large-scale naturalistic study with instrumented vehicles that included a sample with ADHD, but the vehicle data are not yet available. Nonetheless, this study was better controlled than most in that it avoided the usual selection bias that occurs when individuals are chosen for participation based on specific clinical diagnoses. In this study, participants were only tested for ADHD after they had been selected for the study (a probability-based sampling procedure was used to select participants). Furthermore, although this study excluded individuals with antisocial personality disorder (related to conduct disorder), it did include another clinical group as a comparison: drivers diagnosed with depression. Results from this study reveal higher numbers of self-reported collisions and moving violations in drivers with ADHD than in either those diagnosed with clinical depression or a control group with no known disorder. Unlike drivers with depression, compared to drivers with no known disorder, drivers with ADHD were 2.2 times more likely to become involved in multiple automobile collisions, 2.1 times as likely to be at fault, and 2.3 times as likely to have multiple moving violations (Aduen et al., 2015).

14.3.2 Aberrant Driving Behaviors

Collisions and traffic citations do not happen every day, and consequently, the studies reported to this point rely on memory over a period of years. However, collisions are ultimately the result of specific driver behaviors that may emerge over the course of a day. Rosenbloom and Wurtz (2011) conducted a study where young adults with and without ADHD were required to keep a daily driving diary over 30 days. They recorded day-to-day behaviors such as "didn't use the seatbelt," "ran a red light," "ignored the speed limit," and "misread the sign and took the wrong exit at the roundabout," items taken from a modified version of the *Manchester Driving Behaviour Questionnaire* (DBQ) (Reason et al., 1990). Rosenbloom and Wurtz (2011) categorized the questions into two broad categories: faults (unintentional slips, lapses of attention, and errors with the vehicle controls) and violations (deliberate decisions to perform dangerous or illegal activities). The results indicated that compared to drivers without ADHD, drivers with ADHD reported more faults, though there was no significant difference in the number of violations. However, other studies using these measures indicate that there may also be differences in violations between samples with and without ADHD (e.g., Groom et al., 2015).

Drivers with ADHD may also deal with anger differently than others when driving. It is not uncommon for people to become frustrated with other drivers, but there may be differences in the way that people deal with frustration. Compared to other young drivers, young adults with ADHD are more likely to make negative comments about other drivers (Groom et al., 2015), and they also report doing reckless and aggressive things when angry, such as driving to "get even" (e.g., Oliver et al., 2015; see also Richards et al., 2002, 2006). Angry-aggressive driving is a factor in collision risk (see Wickens et al., 2013, for a review).

14.3.3 Direct Measures of Driving Performance

The studies so far involve reports and questionnaires. Although these kinds of investigation provide valuable information, they also have limitations. Many require participants to report on their own performance, but these reports may be biased one way or the other by the conditions of testing. People may overreport because the circumstance of testing may encourage them to focus on problems. In contrast, they may underreport in order to make a positive impression. Furthermore, there are always elements of subjectivity in self-reports (e.g., "What counts as a collision?"), and it is possible that drivers may not always be very aware of how they drive. (Knouse et al. [2005] report that teens with ADHD may have special difficulties in this regard.) Furthermore, it is impossible to control differences in the contextual factors surrounding driving. Drivers with ADHD may differ from other drivers in terms of the circumstances in which they drive. For example, they may drive different types of vehicles; they may be more likely to drive under extreme time pressures (e.g., late for appointments); and they may be more inclined to drive in the company of others who encourage and support reckless driving. There are a variety of contextual factors that contribute to collision risk, but these are generally not recorded or controlled in this type of study. Thus, it may not be the ADHD per se that is producing the problems but, rather, differences among groups in contextual factors surrounding driving.

One solution to this problem is to measure performance directly in such a way that all the drivers are using the same vehicle in the same circumstances. There have now begun to be a few driving studies that measure performance on the road this way (e.g., Cox et al., 2012), though because of the risk to participants and other road users, these studies are often conducted on test tracks, and they often have to be carried out with another driver (usually a driving instructor) to take over should conditions become dangerous. Even in experimental studies such as these, it is difficult to control all aspects of the driving environment, and safety considerations must be paramount.

However, it is important to understand how drivers with ADHD cope with the driving challenges associated with collisions, and for these types of study, driving simulators are necessary. Simulators permit performance to be measured directly and objectively (through instrumentation) under closely controlled conditions in the challenging situations associated with collisions without putting lives at risk. Simulation studies have their own limitations though. For one, there is no real consequence for poor driving in a simulator. No one is hurt; there is no property damage. This may encourage drivers to drive more recklessly, increasing the number of collisions (for a more detailed discussion, see Ranney, 2011). However, there are also reasons to expect that driving simulation studies would underestimate collisions when it comes to drivers with ADHD. Individuals with ADHD can perform very well for short time intervals, if sufficiently engaged. The novelty and excitement of being tested may be enough to help them perform well for short periods of time. Higher levels of physiological arousal facilitate better performance in drivers with ADHD, and indeed, most pharmacological treatments for ADHD involve stimulant medications that increase arousal. For drivers with ADHD, problematic behaviors are more likely to emerge on long boring drives, where they face the same things again and again at predictable intervals (e.g., Reimer et al., 2010). It is sometimes difficult to have simulation studies of adequate duration to ensure that these effects can be seen (see Chapter 25; see also Ranney, 2011). Furthermore, it impossible to completely replicate every aspect of real-world driving in a simulation. For example, simulation studies may underestimate the impact of emotions. Drivers with ADHD report more angry/aggressive behaviors in real-life driving, but it doesn't really make sense to get angry and "take revenge" on a simulated driver. In summary, it is important to acknowledge that simulation studies are not designed to estimate the prevalence of crashes in different groups. Instead, they are better equipped to compare groups in terms of how they respond to different kinds of driving challenges. In this section, we focus on three: distraction, fatigue, and alcohol.

14.3.3.1 The Impact of Distracting Activities Carried Out while Driving

Given that inattention is one of the primary symptoms of ADHD, there is good reason to expect that distracting tasks would be a key challenge for drivers with ADHD. Distraction may take several forms, but one common type occurs when drivers become involved in another activity (a secondary task) that interferes with their driving performance. For example, they may be eating, drinking, smoking, adjusting the radio, interacting with passengers, or using in-vehicle technologies (e.g., smartphones). If individuals with ADHD have special difficulties dealing with these types of distraction, this suggests that they will be more likely to be judged at fault in collisions where they were carrying out secondary tasks while driving. El Farouki et al. (2014) conducted a large-scale case responsibility study in which they recruited patients admitted to a hospital as a result of a vehicle collision. The patients answered questions about what they were doing and thinking immediately before the collision. They were also given tests to determine whether they had ADHD, depression, or anxiety disorders. The responsibility for the collision was assessed in an independent analysis, using a standard procedure (see Robertson & Drummer, 1994). Of those in the study, 54% were judged at fault in the collision, and 46% were not. The comparison involved the extent to which a variety of risk factors predicted which drivers would be judged at fault. The results of the study indicated that when considered in isolation, ADHD and external distraction were each a significant predictors of being judged at fault in the collision (odds ratios [ORs] = 2.18 and 1.47 more likely, respectively), but the combination of ADHD with external distraction resulted in the drivers being almost six times more likely to be judged at fault for the collision (adjusted OR = 5.79).

However, the results from El Farouki et al. (2014) are difficult to interpret, because they averaged across different types of external distraction. Some types of secondary tasks are more deleterious to driving than others. Texting is one of the more dangerous (for reviews, see Chapter 12; Caird et al., 2014; Simmons et al., 2016). For example, in an 18-month naturalistic study of 42 teen drivers that used a case–cohort approach, Klauer et al. (2014) found that teens were 3.9 times more at risk of collision or near collision while texting/browsing. Based on the work of El Farouki et al. (2014), it is difficult to ascertain whether secondary tasks in general have greater effects on drivers with ADHD (causing them to be more likely to drive in such a way that they are judged to be the driver at fault in a collision) or if drivers with ADHD are more likely to choose the types of secondary tasks that are especially dangerous. For example, although teens in general may be inclined to text while driving (e.g., Olsen et al., 2013), teens with ADHD may be especially at risk, perhaps because they are 2.85 times more likely to be diagnosed with Internet addiction than other teens (Ho et al., 2014; see also Chou et al., 2015). This may cause them to feel the need to be constantly online, on Twitter, Facebook, or Instagram or interacting with an online gaming community, while driving.

Thus, it is unclear whether individuals with ADHD have special difficulties dealing with secondary tasks or if they are just more likely to choose dangerous secondary tasks. The only way to find out is to test the drivers with and without ADHD under exactly the same conditions with exactly the same secondary tasks, and the safest way to accomplish this is using a driving simulator. A simulator study was carried out where 16- to 17-year-old drivers were tested in a variety of conditions, including one where they just drove and another where they drove while texting (Kingery et al., 2015; Narad et al., 2013). The performance of a group without ADHD was compared with that of a sample with ADHD. Even without a secondary task, the ADHD group had more variability in speed than the non-ADHD control, though there were no differences in driving speed, and overall texting had a negative effect on steering.

As expected, texting had the most pronounced negative effect on driving performance, especially steering. The group with ADHD had more variability in their steering (there were no significant differences between groups in driving speed). However, the secondary task did not have a significantly worse effect on the ADHD group than the non-ADHD group. On the contrary, the differences in steering performance between the ADHD and non-ADHD groups were largest when there was no secondary task. Eye movement analysis further supported this conclusion. When drivers glance away from the road for more than 2 seconds, it is particularly risky (e.g., Klauer et al., 2006), and drivers with ADHD made significantly more of these dangerously long off-road glances than non-ADHD drivers. However, the difference between the ADHD and non-ADHD groups in terms of the number of these dangerous off-road glances was largest in the conditions where there was no distracting secondary task. This result is consistent with another driving simulator study conducted with young adults (Reimer et al., 2010). Both ADHD and non-ADHD groups had more dangerously long off-road glances while texting.

14.3.3.2 Fatigue

Excessive daytime sleepiness is a well-known cause of accidents: up to 20% of car accidents are attributable to sleepiness at the wheel (e.g., Chapter 15; Connor et al., 2001; Sagaspe et al., 2010). Many with ADHD have comorbid sleep disorders. For example, they may have late-onset insomnia (they have trouble getting to sleep) and may experience a restless pattern of sleep that leaves them more tired than others in the morning, a pattern that may be even more pronounced in teens (Becker et al., 2015; Cortese et al., 2009; Jan et al., 2014; Kooij et al., 2012). Two recent studies demonstrated increased levels of daytime sleepiness in individuals with ADHD. One, a small-scale study of individuals with a confirmed clinical diagnosis of ADHD, showed that a significant proportion had a high level of daytime sleepiness as measured by an objective electrophysiological measure (Maintenance of Wakefulness Test). This sleepiness had a negative impact on their simulated driving performance (Bioulac et al., 2015). The second study, which used self-report data from a large-scale Internet-based epidemiological survey of registered drivers (N = 36,154), found that compared to drivers without ADHD, drivers with ADHD report more sleepiness behind the wheel and more sleep-related near misses (Philip et al., 2015).

Given that many with ADHD have a pattern of disturbed sleep that leaves them especially tired in the morning, it would be reasonable to expect that young drivers with ADHD might also be especially at risk at that time. This prediction was borne out in a driving simulation study (Reimer et al., 2007). Furthermore, this study also showed that compared to other drivers, those with ADHD had a bigger increase in the number of collisions between the second last and last sessions of the day (task-induced fatigue). The authors concluded that drivers with ADHD would be especially at risk on long monotonous sections of the highway when they were tired (early and late in the day).

14.3.3.3 The Effects of Alcohol on Drivers with ADHD

There is a great deal of literature that shows that driving under the influence of either cannabis or alcohol increases the risk of a collision (e.g., Asbridge et al., 2012; Elvik, 2012; Mann et al., 2010; Ogden & Moskowitz, 2004). Adolescents with ADHD are at higher risk than others their age of substance abuse disorder related to alcohol and cannabis (e.g., Charach et al., 2011), and some studies suggest that individuals with ADHD are more likely to drive under the influence (Fuermaier et al., 2015). Furthermore, comorbid sleep disorders cause many individuals with ADHD to struggle with fatigue (as discussed in the previous section), and alcohol and fatigue have interactive effects such that the deleterious effects of even small doses of alcohol are magnified when the individual is also fatigued (Howard et al., 2007). Finally, the recovery of motor performance after alcohol impairment takes longer in young adults with ADHD (Roberts et al., 2013). In fact, laboratory tasks suggest that for individuals with ADHD, alcohol may increase the bias that makes it difficult to ignore alcohol-related pictures when the task is simply to indicate whether a probe light is on the left or right side of the display (Roberts et al., 2012). Therefore, there are many reasons to expect that alcohol may have an even more deleterious effect on drivers with ADHD than other drivers.

Barkley et al. (2006) investigated the impact of two doses of alcohol in two groups of adults: a group with ADHD (unmedicated during testing) and a group without (Mean ages for the two groups = 29 and 33 years, respectively). This study involved a driving simulator, but there were also questionnaires where the participant and an external examiner blind to the manipulation evaluated the driver's performance. This study provided some support for the claim that alcohol has more deleterious effects on performance in drivers with ADHD in terms of the ratings of the external examiners (cf. Weafer et al., 2008).

14.3.4 ADHD Drivers and Automation: Is Technology the Solution?

In recent years, there has been a rapid increase in the number of in-vehicle technologies designed to assist or even take over driving for the driver: blind spot or lane departure warning systems, adaptive cruise control, automatic parking systems, automatic steering. Some of these are designed to compensate for problems of driver inattention, and consequently, there might be reasons to expect that drivers with ADHD would find them especially beneficial. However, at this point, these systems are designed for some situations and not others, and they are not perfect. Consequently, the driver still plays an important role in the vehicle—though perhaps it is a different role than in the past. Determining the impact of automation requires a consideration of how well drivers with ADHD could be expected to negotiate this new role.

Driving requires a number of activities: monitoring for hazards; controlling the lateral position of the vehicle (steering); controlling the longitudinal position (determining the speed, braking to avoid forward hazards); way finding; parking; etc. By reducing the demands of these tasks, these new technologies change what used to be an active task into a more passive and supervisory one: one where the driver has to sit and wait for extended periods of time, all the while staying vigilant for warnings, indications that the system is malfunctioning, or signals that the system needs to have the driver take over because the circumstances go beyond the ones for which it was manufactured. Many drivers struggle with this supervisory role because it entails long periods of inactivity punctuated by infrequent episodes that demand timely and sometimes immediate response. With full automation, it is difficult for drivers to stay focused on the road given that they are so rarely called upon to do anything. As a result, they may be even more tempted than usual to become immersed in some secondary activity while driving (e.g., Jamson et al., 2013), such as daydreaming, texting, working on the computer, or watching movies. Individuals with ADHD have special difficulties maintaining vigilance in situations such as these, where there are long periods of inactivity and boredom. In fact, there is evidence that their performance is at its worst in low-demand conditions. For example, in a driving simulation study, Cox et al. (2006) found that for a sample of adolescent males with ADHD, performance was worse in automatic than manual transmission vehicles (the latter makes more demands on the driver). The adolescents had an easier time paying attention when the demands of the drive were increased (cf. Forster et al., 2014, as it relates to cognitive load). Thus, there is reason to suspect that drivers with ADHD have special difficulties with the passive supervisory role required of them by automated or partially automated driving.

Furthermore, most of these automated systems are designed to help drivers who make inadvertent driving errors though lack of attention. However, for drivers with ADHD, lack of attention is only part of their problem. Sometimes, it is the deliberate activities that get drivers with ADHD into trouble, such as dangerous maneuvers performed to get even by angry/aggressive drivers (e.g., Oliver et al., 2015). Given that there are always methods to deactivate or circumvent automated systems, the problem of deliberate risk tasking/aggressive driving may still remain even with automation.

14.4 Treatments and Countermeasures

14.4.1 Pharmacological Treatments

There are two main types of medication for ADHD: psychostimulants and nonstimulants (Prince et al., 2015). Psychostimulants are either methylphenidate or amphetamine based (or a mixture), and extended-duration variants have been created. For example, short-acting methylphenidate (Ritalin) lasts only 3–4 hours, whereas OROS methylphenidate (Concerta) lasts up to 10–12 hours. The amphetamine-based medication lisdexamfetamine (Vyvanse) lasts up to 13–14 hours. Only the nonstimulant atomoxetine (Strattera) can last up to 24 hours (at least for some). Most individuals with ADHD take their medications on waking, early in the morning (e.g., 7 a.m.); consequently any beneficial effects may have worn off by the end of the day (e.g., 7 p.m.), though they may still need to drive. Although it would be possible for them to take another dose of stimulant later in the day, that could interfere with sleep—which would also exacerbate driving problems. This is an important consideration when evaluating the impact of medication on real-world driving—especially at night.

The gold standard for determining the effects of medication on performance are randomized control trials, where participants are randomly assigned to either a medication or placebo conditions (Gobbo & Louzã, 2014) identified 4 on-road driving studies; the remainder used either simulated driving (10 studies) or self-reports (1 study). Most of the studies evaluated the effects of psychostimulants (8 methylphenidate based, 4 amphetamine based); only three investigated atomoxetine. Converging evidence from these randomized controlled trials suggests that ADHD medication may improve but not normalize driving in teens and young adults with ADHD (see also Trick et al., 2013, with respect to age differences). However, most of the data came from driving simulator studies—and some might argue that these results would not necessarily generalize to real-life driving, where the consequences of poor driving performance are more severe (collisions, tickets, etc.).

Findings from a recent population-based study of driving also indicate beneficial effects of ADHD medication on reducing serious collisions, at least in men with ADHD (Chang et al., 2014). These researchers compared official reports of collisions resulting in serious injuries or deaths in individuals with ADHD when they were and were not taking medication (i.e., each person with ADHD served as his or her own control). In men with ADHD, the use of medication was associated with a 58% reduction in the risk of serious collisions.

Overall, there is some support for the idea that medications may improve performance in drivers with ADHD, particularly novice teen drivers with ADHD, though these medications do not fully normalize driving. However, to be effective, these medications must be taken regularly, and many teens periodically skip their medications (Molin et al., 2009). A study of novice teen drivers (Mean age = 17; 78% with only a learner's certificate) indicated that the effects of a single missed day of medication were most pronounced on the components of driving where the teens exhibited the least skill in baseline testing (Trick & Toxopeus, 2013). This suggests that consistent medication may be especially important for maintaining the controlled attentional processes necessary when developing a new skill (cf. Shiffrin & Schneider, 1977).

14.4.2 Behavioral (Nonpharmacological) Treatments

There is little rigorous research on the effects of behavioral interventions to improve driving in novice drivers with ADHD (for reviews, see Bruce et al., 2014; Classen & Monahan, 2013). Some of these interventions incorporate some of the best elements of training programs used with other drivers (see Chapters 18, 20, 21, and 28), but at this point, the sample sizes are too small to make firm conclusions about their effects on drivers with ADHD, and there are no long-term follow-ups. Nonetheless, one promising approach involves using office-based hazard perception training to facilitate the development of situational awareness in drivers with ADHD. Poulsen et al., (2010) used a brief (34 minutes) training program that involved commentary driving and hazard anticipation training (participants provided spontaneous and continuous commentary while watching a video shot from a driver's perspective under the supervision of a driving trainer providing feedback). This training improved hazard response time in 10 drivers with ADHD compared to a comparable ADHD control group who watched the same driving videos with a driver trainer but had no specific hazard training. Similarly, Fabiano et al. (2011) describe a comprehensive 8-week program for teens with ADHD and their parents called Supporting a Teen's Effective Entry to the Roadway (STEER). This program involves hazard perception training, monitored simulator training, instrumented vehicles to measure driving outside of training, parental feedback, and a graduated program where driving challenges are gradually introduced, promoting positive interactions between parents and their teens. However, this study only had seven participants: it is best viewed as a proof-of-concept study.

14.5 Conclusions and Recommendations

Converging evidence from studies using a variety of methods (official records, self-reports, simulated driving, monitored on-road driving) suggests that drivers with ADHD, particularly teen or novice drivers with ADHD, are at increased risk for poor driving outcomes. However, the research also shows that the increased risk is not universal among drivers with ADHD, and there are no robust predictors of who will be at risk. Moreover, our review of ADHD and the many factors contributing to risky driving indicates that universal intervention will not suffice. To reduce risky driving in individuals with ADHD, it will be necessary to consider a dynamic, personalized, tiered approach to intervention, which not only mitigates stable risk factors associated with ADHD (e.g., inattention, impulsivity) but also reduces driving risks associated with transient situational factors (e.g., sleepiness, distractions). Tiered intervention includes (1) *universal* interventions to meet the risks typically possessed by the target population, such as novice drivers (e.g., graduated driver licensing); (2) *selected* interventions of increased intensity to meet the needs of individuals with higher levels of risk than those typical of the population (e.g., extra support during initial driver training, optimized medication); and (3) *indicated* interventions that involve more resources and individual tailoring (e.g., technological monitoring, parent monitoring, restrictions on driving) to address the needs of those who have already experienced adverse outcomes (Winston et al., 2016).

14.5.1 Recommendations for Parents

14.5.1.1 Is My Teenager at Risk for Poor Driving Outcomes?

- In general, a diagnosis of ADHD increases the risk for poor driving outcomes: being a teen or novice driver with ADHD further increases the risk.
- It is not yet possible to predict with accuracy who will be at risk: not all individuals with ADHD have poor driving outcomes. The presence of neuropsychological deficits and additional diagnoses (e.g., ODD, conduct disorder, anxiety disorder, substance use/abuse) may increase the risk, as will certain conditions (e.g., long-distance drives on monotonous highways, fatigue, alcohol use, distraction).

14.5.1.2 What Can I as a Parent Do before Driving Begins?

- Be proactive and discuss when and whether your teen is ready to start to learn to drive. Discuss whether it might be helpful to delay starting to drive for a couple of years.
- Discuss the possibility that your teen may need extra help learning to drive (e.g., help studying driving theory, extra driving lessons, more than one on-road driving test).

14.5.1.3 Is There a Need to Disclose a Diagnosis of ADHD When Applying for a Driving License?

- There is no simple answer. The requirement to disclose a disability and use of medication varies from country to country and state to state. Many countries are in the process of developing specific recommendations for individuals with ADHD.
- In most countries, it is the licensing authority, not the physician, who makes the final determination of eligibility, even though physicians are required to conduct a medical fitness-to-drive examination. It is important to understand that "medical fitness to drive" is not based solely on a diagnosis. Rather, it includes assessment of the person's ability to accommodate to any medical problems and to function as a driver.
- Arguments against disclosing include the risk of stigma and/or penalization by insurance companies. It is important to remember that ADHD affects individuals differently. It is possible that drivers with ADHD may avoid risk with good clinical supervision and a keen understanding of their strengths and weaknesses as drivers.
- When your teenager is interested in learning to drive, discuss it with a physician. Your physician may recommend a functional assessment in addition to the office-based medication examination and will advise you about the use of medication while driving. However, in most countries, physicians have a statutory duty to report any concerns about a driver (including a driver with ADHD) who has demonstrated problems with driving or who is noncompliant with treatment to the licensing authorities.

14.5.1.4 Will My Teenager Need Extra Help Learning to Drive?

- Little is known about the challenges of driving training for teens and novice drivers with ADHD, but preliminary evidence suggests that teens/novice drivers with ADHD may find driving theory to be particularly challenging to master—in part because of the amount of reading required. They may experience problems remembering the content or understanding the questions in the theoretical training, and this may cause them to have to take the written test several times in order to pass. It is also possible that novice drivers with ADHD may need to take more on-road driving lessons to gain the same level of driving skill as peers without ADHD (Almberg et al., 2015).
- Teens and novice drivers with ADHD are advised to seek help and consider the use of accommodations during driver training (one strong reason to disclose diagnosis). These accommodations include the use of alternative methods to the traditional textbook/computer programs for learning driving theory (e.g., videos, picture-based material, use of comprehension strategies to assist with reading the lengthy text, note taking or a scribe to take notes).
- Seek driving instructors who have knowledge of and experience with ADHD and who are willing to support and train teens and novice drivers with ADHD.
- Parents/significant others may need to support teens and novice drivers with ADHD during the driver training (e.g., set contracts and rules, use positive reinforcement, support the mastery of driving theory as well as on the road driving).
- Additional specific training may be necessary. Office-based hazard training programs (e.g., Poulsen et al., 2010) or the STEER program (Fabiano et al., 2011) may be helpful options.

14.5.2 Key Points: Recommendations for Clinicians

- Be proactive and discuss driving risks with teens/novice drivers with ADHD as well as with their parents/significant others, well in advance of the typical age of application for a driving license.
- Conduct a medical fit-to-drive examination and determine whether there is a need to refer for a functional assessment (typically done by occupational therapists).
- Encourage parents/significant other to support teen and novice drivers with ADHD, especially in terms of studying driving theory and preparing for the written exam, and alert them of the potential need for teens to undertake more on-road driving instruction than typical.
- Monitor the driving experience and history of drivers with ADHD to ascertain typical mileage and timing of driving; determine the conditions that the driver finds particularly challenging.
- Keep abreast of the literature on the effects of medication on driving. Current evidence suggests that long-acting psychostimulants may improve (but not normalize) driving and reduce the risk of poor driving outcomes, particularly in teens or novice drivers with ADHD. Medication for comorbid disorders (e.g., antidepressants, anxiolytics, sleep enhancers) may cause drowsiness and therefore must be disclosed and monitored.
- Monitor medication effectiveness and compliance (especially coverage of driving times) and disclose concerns about the individual's driving performance.

Acknowledgments

Each author contributed equally in the preparation of this document (author names are listed alphabetically). This research was supported by grants from the Canadian Institutes of Health Research (Rosemary Tannock) and the Auto21 Network Centres of Excellence (Lana M. Trick).

References

Aduen, P.A. Kofler, M.J., Cox, D.J., Sarver, D.E., & Lunsford, E. (2015). Motor vehicle driving in high incidence psychiatric disability: Comparison of drivers with ADHD, depression, and no known psychopathology. *Journal of Psychiatric Research, 64*, 59–66.

Aguiar, S.A., & Barela, J.A. (2015). Adaptation of sensorimotor coupling in postural control is impaired by sleep deprivation. *PloS One, 10*(3), e0122340.

Almberg, M., Selander, H., Falkmer, M., Vaz, S., Ciccarelli, M., & Falkmer, T. (2015). Experiences of facilitators or barriers in driving education from learner and novice drivers with ADHD or ASD and their driving instructors. *Developmental Neurorehabilitation*, doi: 10.3109/17518423.2015.1058299.

American Psychiatric Association. (2013). *Diagnostic and Statistical Manual of Mental Disorders* (5th edition). Washington, DC: APA.

Antin, J., Lee, S. Hankey, J., & Dingus, T. (2011). The Second Strategic Highway Research Program: Design of the in vehicle driving behavior and crash risk study. Transportation Research Board, Report S2-S05-RR-1.

Asbridge, M., Hayden, J.A., & Cartwright, J.L. (2012). Acute cannabis consumption and motor vehicle collision risk: Systematic review of observational studies and meta-analysis. *British Medical Journal*, 344, e536.

Barkley, R.A., Guevremont, D.C., Anastopoulos, A.D., DuPaul, G.J., & Shelton, T.L. (1993). Driving-related risks and outcomes of attention deficit hyperactivity disorders in adolescents and young adults: A 3 to 5 year follow-up survey. *Pediatrics, 92*, 212–218.

Barkley, R.A., Murphy, K.R. O'Connell, T., Anderson, D., & Connor, D.F. (2006). The effects of two doses of alcohol on simulator driving performance in adults with attention-deficit/hyperactivity disorder, *Neuropsychology, 20*(1), 77–87.

Becker, S.P., Langberg, J.M., & Evans, S.W. (2015). Sleep problems predict comorbid externalizing behaviors and depression in young adolescents with attention deficit/hyperactivity disorder. *European Child and Adolescent Psychiatry, 24*, 897–907.

Bioulac, S., Chaufton, C., Taillard, J., Claret, A., Sagaspe, P., Fabrigoule, C., Bouvard, M.P., & Philip, P. (2015). Excessive daytime sleepiness in adult patients with ADHD as measured by the Maintenance of Wakefulness Test, an electrophysiologic measure. *The Journal of Clinical Psychiatry, 76*(7), 943–948.

Bruce, C., Unsworth, C., & Tay, R. (2014). A systematic review of the effectiveness of behavioural interventions for improving driving outcomes in novice drivers with attention deficit hyperactivity disorder (ADHD). *British Journal of Occupational Therapy, 77*(7), 348–357.

Caird, J.K., Johnston, K.A., Willness, C.R., Asbridge, M., & Steel, P. (2014). A meta-analysis of the effects of texting on driving. *Accident Analysis & Prevention, 71*, 311–318.

Cardoos, S.L., Loya, F., & Hinshaw, S.P. (2013). Adolescent Girls' ADHD symptoms and young adult driving: The role of perceived deviant peer affiliation. *Journal of Clinical Child & Adolescent Psychology, 42*(2), 232–242.

Chang, Z., Lichtenstein, P., D'Onofrio, B.M., Sjölander, A., & Larsson, H. (2014). Serious transport accidents in adults with attention-deficit/hyperactivity disorder and the effect of medication: A population-based study. *JAMA Psychiatry, 71*(3), 319–325.

Charach, A., Yeung, E., Climans, T., & Lillie, E., (2011). Childhood attention deficit/hyperactivity disorder and future substance use disorders: Comparative meta-analyses. *Journal of the American Academy of Child and Adolescent Psychiatry, 50*, 9–21.

Chou, W., Liu, T., Yang, P., Yen, C., & Hu, H. (2015). Multi-dimensional correlates of Internet addiction symptoms in adolescents with attention-deficit/hyperactivity disorder. *Psychiatry Research, 225*, 122–128.

Classen, S., & Monahan, M. (2013). Evidence-based review on interventions and determinants of driving performance in teens with attention deficit hyperactivity disorder or autism spectrum disorder. *Traffic Injury Prevention, 14*(2), 188–193.

Coghill, D.R., Seth, S., & Matthews, K. (2014). A comprehensive assessment of memory, delay aversion, timing, inhibition, decision making and variability in attention deficit hyperactivity disorder: Advancing beyond the three-pathway models. *Psychological Medicine, 44*(09), 1989–2001.

Connor, J., Whitlock, G., Norton, R., & Jackson, R. (2001). The role of driver sleepiness in car crashes: A systematic review of epidemiological studies. *Accident Analysis & Prevention, 33*(1), 31–41.

Cortese, S., Faraone, S.V., Konofal, E., & Lecendreux, M. (2009). Sleep in children with attention-deficit/hyperactivity disorder: Meta-analysis of subjective and objective studies. *Journal of the American Academy of Child and Adolescent Psychiatry, 48*, 894–908.

Cox, D.J., Punja, M., Powers, K., Merkel, R.L., Burket, R., Moore, M., Thorndike, F., & Kovatche, B. (2006). Manual transmission enhances attention and driving performance of ADHD adolescent males. *Journal of Attention Disorders, 10*(2), 212–216.

Cox, D.J., Davis, M., Mikami, A.Y., Singh, H., Merkel, R.L., & Burket, R. (2012). Long-acting methylphenidate reduces collision rates of young adult drivers with attention-deficit/hyperactivity disorder. *Journal of Clinical Psychopharmacology, 32*(2), 225–230.

Dalsgaard, S., Østergaard, S.D., Leckman, J.F., Mortensen, P.B., & Pedersen, M.G. (2015). Mortality in children, adolescents, and adults with attention deficit hyperactivity disorder: A nationwide cohort study. *Lancet, 385*, 2190–2196.

Dewey, D., Cantell M., & Crawford S.G. (2007). Motor and gestural performance in children with autism spectrum disorders, developmental coordination disorder, and/or attention deficit hyperactivity disorder. *Journal of International Neuropsychological Society, 13*, 246–256.

El Farouki, K., Lagarde, Orriols, L., Bouvard, M.P., Contrand, B., & Galera, C. (2014). The increased risk of road crashes in attention deficit hyperactivity disorder (ADHD) adult drivers: Driven by distraction? Results from a responsibility case–control study. *PLoS One 9*(12): e115002.

Elvik, R. (2012). Risk of road accident associated with the use of drugs: A systematic review and meta-analysis of evidence from epidemiological studies. *Accident Analysis & Prevention, 60,* 254–267.

Fabiano, G.A., Hulme, K., Linke, S., Nelson-Tuttle, C., Pariseau, M., Gangloff, B. et al. (2011). The Supporting a Teen's Effective Entry to the Roadway (STEER) Program: Feasibility and Preliminary Support for a Psychosocial Intervention for Teenage Drivers With ADHD. *Cognitive and Behavioral Practice, 18,* 267–280.

Faraone, S.V., Asherson, P., Banaschewski, T., Biederman, J., Buitelaar, J.K., Ramos-Quiroga, J.A., Rohde, L.A., Sonuga-Barke, E.J.S., Tannock, R., & Franke, B. (2015). Attention-deficit/hyperactivity disorder. *Nature Reviews Disease Primers,* Article number: 15020, doi:10.1038/nrdp.2015.20.

Forster, S., Robertson, D.J. Jennings, A., Asherson, P., & Lavie, N. (2014). Plugging the attention deficit: Perceptual load counters increased distraction in ADHD. *Neuropsychology, 28*(1), 91–97.

Fuermaier, A.B.M., Tucha, L. Evans, B.L., Koerts, J., de Waard, D., Brookhuis, K., Aschenbrenner, S., Johannes Thome, J., Lange, K.W., & Tuch, O. (2015). Driving and attention deficit hyperactivity disorder. *Journal of Neural Transmission,* 1–13. doi:10.1007/s00702-015-1465-6.

Gobbo, M.A., & Louzã, M.R. (2014). Influence of stimulant and non-stimulant drug treatment on driving performance in patients with attention deficit hyperactivity disorder: A systematic review. *European Neuropsychopharmacology, 24,* 1425–1443.

Groom, M.J., van Loon, E., Daley, D., Chapman, P., & Hollis, C. (2015). Driving behaviour in adults with attention deficit/hyperactivity disorder. *BMC Psychiatry, 15,* 175.

Henderson, S.E., & Henderson, L. (2003). Toward an understanding of developmental coordination disorder: Terminological and diagnostic issues. *Neural Plasticity, 10*(1–2), 1–13.

Hervey, A.S., Epstein, J.N., & Curry, J.F. (2004). Neuropsychology of adults with attention-deficit/hyperactivity disorder: A meta-analytic review. *Neuropsychology, 18*(3), 485.

Ho, R.C., Zhang, M.W.B., Tsan, T.Y., Toh, A.H., Pan, F., Lu, Y., Cheng, C. et al. (2014). The association between Internet addiction and psychiatric co-morbidity: A meta-analysis. *BMC Psychiatry, 14,* 183.

Hove, M.J., Zeffiro, T., Biederman, J., Li, Z., Schmahmann, J., & Valera, E.M. (2015). Postural sway and regional cerebellar volume in adults with attention-deficit/hyperactivity disorder. *NeuroImage: Clinical, 8,* 422–428.

Howard, M.E., Jackson, M.L., Kennedy, G.A., Swann, P., Barnes, M., & Pierce, R.J. (2007). The interactive effects of extended wakefulness and low-dose alcohol on simulated driving and vigilance. *Sleep, 30*(10), 1334–1340.

Jamson, A.H., Merat, N., Carsten, O.M.J., & Lai, F. (2013). Behavioural changes in drivers experiencing highly-automated vehicle control in varying traffic conditions. *Transportation Research Part C—Emerging Technologies, 30,* 116–125.

Jan, Y.W., Yang, C.M, & Huang, Y.S. (2011). Comorbidity and confounding factors in attention-deficit/hyperactivity disorder and sleep disorders in children. *Psychology Research and Behavior Management, 4,* 139–150.

Kingery, K.M., Narad, M., Garner, A.A., Antonini, T.N., Tamm, L., & Epstein, J.N. (2015). Extended visual glances away from the roadway are associated with ADHD- and texting-related driving performance deficits in adolescents. *Journal of Abnormal Child Psychology, 43,* 1175–1186.

Klauer, S.G., Dingus, T.A., Neale, V.L., Sudeweeks, J.D., & Ramsey, D. (2006). The impact of driver inattention on near-crash/crash risk: An analysis using the 100-car naturalistic driving study data. Report No. DOT HS 810594.

Klauer, S.G., Guo, F., Simons-Morton, B.G., Ouimet, M.C., Lee, S.E., & Dingus, T.A. (2014). Distracted driving and risk of road crashes among novice and experienced drivers. *New England Journal of Medicine, 370,* 54–59.

Knouse, L.E., Bagwell, C.L., Barkley, R.A., & Murphy, K.R. (2005). Accuracy of self-evaluation in adults with ADHD: Evidence from a driving study. *Journal of Attention Disorders, 8,* 221–234.

Kooij, J.J.S., Huss, M., Asherson, P., Akehurst, R., Beusterien, K., French, A., Sasané, R., & Hodgkins, P. (2012). Distinguishing comorbidity and successful management of adult ADHD. *Journal of Attention Disorders, 16*(5, Suppl.), 3S–19S.

Lacherez, P., Wood, J.M., Anstey, K.J., & Lord, S.R. (2014). Sensorimotor and postural control factors associated with driving safety in a community-dwelling older driver population. *The Journals of Gerontology Series A: Biological Sciences and Medical Sciences, 69*(2), 240–244.

Langevin, L.M., MacMaster, F.P., & Dewey, D. (2015). Distinct patterns of cortical thinning in concurrent motor and attention disorders. *Developmental Medicine & Child Neurology, 57*, 257–264.

Laucht, M., Hohm, E., Esser, G., Schmidt, M.H., & Becker, K. (2007). Association between ADHD and smoking in adolescence: Shared genetic, environmental and psychopathological factors. *Journal of Neural Transmission, 114*, 1097–1104.

Mann, R.E., Stoduto, G., Ialomiteanu, A., Asbridge, M., Smart, R.G., & Wickens, C.M. (2010). Self-reported collision risk associated with cannabis use and driving after cannabis use among Ontario adults. *Traffic Injury Prevention, 11*, 115–122.

Moffitt, T.E., Houts, R., Asherson, P., Belsky, D.W., Corcoran, D.L., Hammerle, M., Harrington, H. et al. (2015). Is adult ADHD a childhood-onset neurodevelopmental disorder? Evidence from a four-decade longitudinal cohort study. *American Journal of Psychiatry, 172*(10), 967–977.

Molina, B.S.G., Hinshaw, S.P., Swanson, J.M., Arnold, L.E., Vitiello, B., Jensen, P. S. et al. (2009). MTA at 8 years: Prospective follow up of children treated for combined-type ADHD in a multi-site study. *Journal of the American Academy of Child and Adolescent Psychiatry, 48*, 484–500.

Narad, M., Garner, A.A., Brassell, A.A., Saxby, D., Antonini, T.N., O'Brien, K.M., Tamm, L., Matthews, G., & Epstein, J.N (2013). Impact of distraction on the driving performance of adolescents with and without attention-deficit/hyperactivity disorder. *JAMA Pediatrics, 167*(10), 933–938.

Ogden, E.J.D., & Moskowitz, H. (2004). Effects of alcohol and other drugs on driver performance. *Traffic Injury Prevention, 5*, 185–198.

Oliver, M.L., Han, K., Bos, A.J., & Backs, R.W. (2015). The relationship between ADHD symptoms and driving behavior in college students: The mediating effects of negative emotions and emotion control. *Transportation Research Part F, 30*, 14–21.

Olsen, E.O., Shults, R.A., & Eaton, D.K. (2013). Texting while driving and other risky motor vehicle behaviors among US high school students. *Pediatrics, 131*, e1708–e1715.

Philip, P., Micoulaud-Franchi, J.A., Lagarde, E., Taillard, J., Canel, A., Sagaspe, P., & Bioulac, S. (2015). Attention deficit hyperactivity disorder symptoms, sleepiness and accidental risk in 36140 regularly registered highway drivers. *PLoS One, 10*(9), e0138004.

Polanczyk, G., de Lima, M.S., Horta, B.L., Biederman, J., & Rohde, L. A. (2007). The worldwide prevalence of ADHD: A systematic review and meta-regression analysis. *The American Journal of Psychiatry, 164*(6), 942-948.

Polanczyk, G.V., Willcutt, E.G., Salum, G.A., Kieling, C., & Rohde, L.A. (2014). ADHD prevalence estimates across three decades: An updated systematic review and meta-regression analysis. *International Journal of Epidemiology, 43*(2), 434–442.

Polanczyk, G.V., Salum, G.A., Sugaya, L.S., Arthur Caye, A., & Rohde, L.A. (2015). Annual Research Review: A meta-analysis of the worldwide prevalence of mental disorders in children and adolescents. *Journal of Child Psychology and Psychiatry, 56*(3), 345–365.

Poulsen, A.A., Horswill, M.S., Wetton, M.A. Hill, A., & Lim, S.M. (2010). A brief office-based hazard perception intervention for drivers with ADHD symptoms. *Australian and New Zealand Journal of Psychiatry, 44*, 528–534.

Prince, J.B., Wilens, T.E., Spencer, T.J., & Biederman, J. (2015). Stimulants and other medications for ADHD. In T. Stern, M. Fava, T. Wilens, and J. Rosenbaum (Eds.). *Massachusetts General Hospital Psychopharmacology and Neurotherapeutics.* (pp. 99–112). New York: Elsevier Press.

Ranney, T.A. (2011). Psychological fidelity: Perception of risk. In D.L. Fisher, M. Rizzo, J.K. Caird, and J. Lee (Eds.). *Handbook of Driving Simulation for Engineering, Medicine, and Psychology* (9-1–9-13). Boca Raton, FL: CRC Press.

Reason, J., Manstead, A., Stradling, S., Baxter, J., & Campbell, K. (1990). Errors and violations: A real distinction? *Ergonomics*, 33, 1315–1332.

Redelmeier, D.A., Chan, W.K., & Lu, H. (2010). Road trauma in teenage male youth with childhood disruptive behavior disorders: A population based analysis. *PLoS Medicine*, *7*(11), 1411.

Reimer, B., D'Ambrosio, L.A., Coughlin, J.F., Fried, R., & Biederman, J. (2007). Task-induced fatigue and collisions in adult drivers with attention deficit hyperactivity disorder. *Traffic Injury Prevention, 8*, 290–299.

Reimer, B., Mehler, B., D'Ambrosio, L.A., & Fried, R. (2010). The impact of distractions on young adult drivers with attention deficit hyperactivity disorder (ADHD). *Accident Analysis & Prevention, 42*, 842–851.

Ren, Y., Yu, L., Yang, L., Cheng, J., Feng, L., & Wang, Y. (2014). Postural control and sensory information integration abilities of boys with two subtypes of attention deficit hyperactivity disorder: A case–control study. *Chinese Medical Journal*, *127*(24), 4197–4203.

Richards, T.L., Deffenbacher, J.L., & Rosen, L.A. (2002). Driving anger and other driving-related behaviors in high and low ADHD symptom college students. *Journal of Attention Disorders, 6*(1), 25–38.

Richards, T.L., Deffenbacher, J.L., Rosén, L.A., Barkley, R.A., & Rodricks, T. (2006). Driving anger and driving behavior in adults with ADHD. *Journal of Attention Disorders, 10*(1), 54–64.

Roberts, W., Fillmore, M.T., & Milich, R. (2012). Drinking to distraction: Does alcohol increase attentional bias in adults with ADHD? *Experimental and Clinical Pharmacology, 20*(2), 107–117.

Roberts, W., Milich, R., & Fillmore, M.T. (2013). Reduced acute recovery from alcohol impairment in adults with ADHD. *Psychopharmacology, 228*, 65–74.

Robertson, M.D., & Drummer, O.H. (1994). Responsibility analysis: A methodology to study the effects of drugs in driving. *Accident Analysis & Prevention, 26*(2), 243–247.

Rosenbloom, T., & Wultz, B. (2011). Thirty-day self-reported risky driving behaviors of ADHD and non-ADHD drivers. *Accident Analysis & Prevention, 43*, 128–133.

Sagaspe, P., Taillard, J., Bayon, V., Lagarde, E., Moore, N., Boussuge, J., Chaumet, G., Bioulac, B., & Philip, P. (2010). Sleepiness, near misses and driving accidents among a representative population of French drivers. *Journal of Sleep Research*, *19*(4), 578–584.

Shaw, P., Malek, M., Watson, B., Greenstein, D., de Rossi, P., & Sharp, W. (2013). Trajectories of cerebral cortical development in childhood and adolescence and adult attention-deficit/hyperactivity disorder. *Biological Psychiatry*, *74*(8), 599–606.

Shiffrin, R.M., & Schneider, W. (1977). Controlled and automatic human information-processing. 2. Perceptual learning, automatic attending, and a general theory. *Psychological Review, 84*(2), 127–190.

Simon, V., Czobor, P., Balint, S., Meszaros, A., & Bitter, I. (2009). Prevalence and correlates of adult attention-deficit hyperactivity disorder: Meta-analysis. *British Journal of Psychiatry, 194*(3), 204–211.

Simmons, S., Hicks, A., & Caird, J.K. (2016). Safety-critical event risk associated with cell phone tasks as measured in naturalistic studies: A systematic review and meta-analysis. *Accident Analysis & Prevention, 87*, 161–169.

Sobanski, E. (2006). Psychiatric comorbidity in adults with attention-deficit/hyperactivity disorder (ADHD). *European Archives of Psychiatry Clinical Neuroscience, 256* [Suppl 1], I/26–I/31.

Tamm, L., Epstein, J.N., Lisdahl, K.M., Molinac, B., Tapert, S., Stephen, P., Hinshawe, S.P. et al. (2013). Impact of ADHD and cannabis use on executive functioning, in young adults. *Drug and Alcohol Dependence, 133*, 607–614.

Teicher, M.H., Polcari, A., Fourligas, N., Vitaliano, G., & Navalta, C.P. (2012). Hyperactivity persists in male and female adults with ADHD and remains a highly discriminative feature of the disorder: A case–control study. *BMC Psychiatry*, *12*(1), 190.

Thomas, R. Sanders, S., Doust, J., Beller, E., & Glasziou, P. (2015). Prevalence of attention-deficit/hyperactivity disorder: A systematic review and meta-analysis. *Pediatrics, 135*(4), e994–e1001.

Toplak, M., & Tannock, R. (2005). Time perception: Modality and duration effects in attention deficit/hyperactivity disorder (ADHD). *Journal of Abnormal Child Psychology, 33*(5), 639–654.

Trick, L.M., & Toxopeus, R. (2013). How missing a treatment of Mixed Amphetamine Salts Extended Release affects performance in teen drivers with ADHD. Poster at *7th International Symposium on Human Factors in Driving Assessment, Training, and Vehicle Design*. June 17–20, 2013. Bolton Landing, NY. (pp. 99–105). Iowa City, IA: University of Iowa Public Policy Centre.

Trick, L.M., Toxopeus, R., Jain, U., & Saliba, K. (2013). The effects of Multi-Layer Release Methylphenidate on driving in individuals with Attention-Deficit/Hyperactivity Disorder as a function of driver age. Poster presentation for the Annual Meeting of the *Vision Sciences Society*, May 10–15, 2013. Naples, Florida.

Vaa, T. (2014). ADHD and relative risk of accidents in road traffic: A meta-analysis. *Accident Analysis & Prevention, 62*, 415–425.

Weafer, J.M., Camarillo, D., Fillmore, M.T., Milich, R., & Marczinski, C.A. (2008). Simulated driving performance of adults with ADHD: Comparisons with alcohol intoxication. *Experimental and Clinical Psychopharmacology, 16*(3), 251–263.

Wickens, C.M., Mann, R.E., & Wiesenthal, D.L. (2013). Addressing driver aggression: Contributions from psychological science. *Current Directions in Psychological Science, 22*(5), 386–391.

Wilens, T.E., Biederman, J., & Spencer, T.J. (2002). Attention deficit/hyperactivity disorder across the lifespan. *Annual Review of Medicine, 53*, 113–131.

Willcutt, E.G. (2012). The prevalence of DSM-IV attention-deficit/hyperactivity disorder: A meta-analytic review. *Neurotherapeutics, 9*(3), 490–499.

Willcutt, E.G., Doyle, A.E., Nigg, J.T., Faraone, S.V., & Pennington, B.F. (2005). Validity of the executive function theory of attention-deficit/hyperactivity disorder: A meta-analytic review. *Biological Psychiatry, 57*(11), 1336–1346.

Winston, F.K., McDonald, C.C., & McGehee, D.V. (2013). Are we doing enough to prevent the perfect storm? Novice drivers, ADHD, and distracted driving. *JAMA Pediatrics, 167*(10), 892–894.

Winston, F.K., Puzino, K., & Romer, D. (2016). Precision prevention: Time to move beyond universal interventions. *Injury Prevention, 22*(2), 87–91.

15

Fatigue and Road Safety for Young and Novice Drivers

Jessica L. Paterson

Drew Dawson

Abstract

The Problem. Young drivers are overrepresented in motor vehicle accidents (MVAs), and injury from MVAs is a leading cause of death for youth. Fatigue is a significant contributor to MVAs for young drivers. ***Fatigue, Young Drivers, and Road Safety.*** Young drivers are vulnerable to fatigue because of biological predispositions, social factors, and inexperience. Fatigue manifests as reduced vigilance, impaired decision making, and poor hazard perception. Fatigue is also associated with dangerous driving behaviors such as aggressive driving, driver distraction, speeding, and drunk driving. Young drivers are not only more vulnerable to the negative effects of sleepiness and fatigue compared to older and more experienced drivers; they also have altered risk perception. ***Managing Fatigue Risk for Young Drivers.*** Principles of deterrence cannot be applied to fatigue as they are to other driving risks as there is no reliable, objective measure of fatigue. There is limited research addressing educational campaigns targeted at fatigued driving in young drivers. Finally, nighttime driving restrictions, enforced as part of graduated licensing systems and the primary method of control for this risk, are inadequate to manage the risks associated with fatigue. ***The Next Step in Managing Fatigue Risk for Young Drivers.*** Reducing fatigued driving behavior must rely on young drivers understanding the risk of driving while fatigued and self-regulating their behavior. However, research shows that young drivers may engage in fatigue driving despite being aware that they are impaired by sleepiness and fatigue. This failure of self-regulation could be addressed using the ego depletion model (EDM) to investigate how self-control is affected by fatigue and how perceived risk associated with fatigued driving influences self-regulatory behavior in young drivers.

15.1 Introduction

While fatigue can affect drivers of all ages, young adults are particularly vulnerable to fatigue due to biological predispositions, social factors, and inexperience. The consequences of fatigue for safety on the road have been researched in adult populations but less so in the case of young drivers. This is despite fatigue being "the most commonly identified physiological state found to impair drivers aged 16–24" (Wundersitz, 2012), involved in 20%–30% of all road deaths (Australian Transport Council, 2011), and that injury from motor vehicle accidents (MVAs) is a leading cause of death in young adults worldwide (World Health Organization, 2007).

In this chapter, we highlight fatigue as a serious risk factor that increases the likelihood of a young driver having a road crash. We begin with a general discussion of the determinants of fatigue and how these relate to young drivers in particular. Second, we discuss the consequences of fatigue for the driving task. Finally, we review existing approaches to managing fatigue-related risk for young drivers and make suggestions for future research.

15.2 Fatigue, Young Drivers, and Road Safety

15.2.1 Fatigue

Fatigue is "a biological drive for recuperative rest" (Williamson et al., 2011, p. 499) and is associated with a reduced capacity for mental and physical function (Brown, 1994). Fatigue can accumulate as a result of sleep loss, extended time awake, circadian disruption, and/or task-related factors (Brown, 1994; Williamson et al., 2011). Each of these factors is described briefly.

15.2.1.1 Determinants of Fatigue

Sleep loss, extended wake (i.e., extended time awake), circadian disruption, and task-related factors can, together or in isolation, result in fatigue (Williamson et al., 2011). Sleep loss can refer to any curtailing of sleep below habitual amounts. This can include a shortened sleep period (partial sleep loss) resulting from delayed sleep onset, advanced sleep offset, or a fragmented sleep period, and extend to a period of total sleep loss (≥24 hours). Most individuals will experience both partial and total sleep loss at some time in their lives, but partial sleep loss is the more common of the two forms.

Sleep loss and extended wake are linked constructs, to an extent, in that a reduction in sleep time necessarily results in more time awake. As time awake increases, so does the body's drive to sleep and, in turn, the appearance of symptoms of fatigue (discussed in Section 15.2.2). These symptoms can be relieved by through the use of temporary countermeasures such as caffeine but are most effectively reversed by sleep. If inadequate sleep is obtained following a period of extended wake, then symptoms of fatigue may become apparent again soon after waking. This is also true if the recovery sleep period occurs out of synchronicity with the body's circadian rhythm, a phenomenon that represents the role of circadian disruption in fatigue.

Circadian rhythms are oscillations in functioning that follow a 24-hour (approximately) rhythm. These rhythms are evident in body temperature, metabolism, and alertness, among other functions, as well as being critical in determining when sleep and wake are most likely to occur. Circadian rhythms align the body's sleep–wake cycle with the external environment, promoting wake during daylight hours and sleep during the nighttime. In line with this, wake that occurs during the biological night is likely to lead to fatigue, given that the propensity to sleep is increased. On the other hand, sleep that occurs during the biological day is likely to be of reduced restorative value, leading to fatigue during the next wake period.

Tasks completed during the wake period can also influence fatigue. Task-related factors can include the time spent on the task and the nature of the task itself. Driving is particularly sensitive to the task-related aspect of fatigue. There is a positive correlation between driving duration and fatigue, particularly for young drivers (Philip et al., 1999), and this has been associated with an increased rate of steering deviation (Philip et al., 2003). The monotony of the driving task has also been associated with fatigue and impaired performance. Under monotonous driving conditions, such as highway driving, fatigue can impair driving performance in as little as 20 minutes (Thiffault & Bergeron, 2003).

It is evident that fatigue can result from any combination of sleep loss, extended wake, circadian disruption, and/or task-related factors. There is significant evidence that youth are particularly vulnerable to experiencing fatigue. This evidence is presented here.

15.2.1.2 Fatigue and Young Drivers

Youth is defined as the period between 16 and 25 years of age. A young driver, then, is any driver falling within this age group (we recognize that this extends beyond the teen years, the focus of much of the material in the handbook). Because of legal restrictions on driving age, young drivers are often also novice drivers. The combination of youth and inexperience, combined with the dangerous task of driving, makes young drivers a group at increased risk of MVA. Concerningly, young drivers are a group also at increased risk of experiencing fatigue. The consequences of fatigued driving for road safety are discussed in Section 15.2.2. First, we must understand why young people are more vulnerable to experiencing fatigue.

Youth are vulnerable to fatigue for a number of reasons. As a result of ongoing cognitive and physical development, youth have an elevated sleep need (Dahl, 2008) and therefore need more time to achieve adequate sleep. However, youth may have less time available for sleep as a result of balancing multiple commitments, such as study, family and social commitments, and employment. The role of employment is a significant one. Young people in the workforce are more likely to work long days, irregular hours, and at night, resulting in circadian disruption and sleep curtailment (Martin et al., 2012). Evidence from a Canadian youth population indicates that work and study, combined with multiple other time pressures, result in sleep loss and negative health and safety outcomes for young workers (Laberge et al., 2011; Martin et al., 2012).

Youth also tend toward biologically delayed sleep schedules, which result in shorter sleep on weeknights and sleeping late on weekends to recover lost sleep. This pattern exacerbates delayed sleep schedules and results in cumulative increases in fatigue (Taylor et al., 2008). There is also evidence that youth are more vulnerable to fatigue (Lowden et al., 2009). Taken together, these factors contribute to increased fatigue risk for young drivers.

15.2.2 Fatigue and Road Safety

So far, we have discussed the definition of fatigue, the causes of fatigue, and why young drivers are a group at particular risk of experiencing fatigue. We have briefly mentioned that fatigue contributes to a substantial proportion of MVAs (Australian Transport Council, 2011) and that injury from MVAs is leading cause of death for young people worldwide (World Health Organization, 2007). The consequences of fatigue, for both safety on the road and general health and well-being, have been extensively researched in adult populations (for a review, see Connor et al., 2001) but less so in the case of young drivers. In this section, we describe the impairments associated with fatigue that can lead to MVAs for drivers of all ages, with a particular focus on the consequences for young drivers where possible.

Driving is a multifaceted task, drawing upon many different perceptual and cognitive constructs. Here, we highlight only a few of these, namely, decision making, vigilance, and hazard perception. Decision making, as a broad construct, is significantly impaired by sleep loss and fatigue. In their seminal review, Harrison and Horne (2000) highlight the multiple operational aspects of decision making that are affected by sleep deprivation and fatigue. In this case, effective decision making was defined as requiring "convergent, rule-based skills of logical, critical, and deductive reasoning but also can involve unique and unfamiliar circumstances necessitating a range of divergent skills" (Harrison & Horne, 2000, p. 238). Sleep deprivation has been shown to lead to impairments in decision making, including

- Avoiding distraction while handling large amounts of information
- Keeping track of critical tasks and updating approach strategies as situations develop
- Being able to think laterally and in innovative ways
- Assessing risk
- Maintaining motivation for tedious tasks
- Regulating mood and inhibiting behavior
- Demonstrating performance insight
- Remembering temporal information about events
- Effective communication

Evident in this list is that real-world decision-making skills typically involve dynamic and fluctuating environments. This is particularly true in the case of driving. These skills rely on the prefrontal cortex (PFC), which is highly sensitive to the effects of sleep deprivation and fatigue (Harrison & Horne, 2000). Indeed, PFC function has been demonstrated to be significantly impaired by sleep loss in young adults (Harrison et al., 2000; also see Chapter 9 on developmental perspectives).

Vigilance is another aspect of functioning that is diminished by fatigue. Vigilance refers to the ability to sustain attention for a task for an extended period of time (Thiffault & Bergeron, 2003). In a study of young (i.e., 20- to 26-year-old) male drivers, response time (a common measure of vigilance) increased with driving duration and was associated with worse driving performance (Ting et al., 2008). Aspects of driving performance measured included speed deviations, average headway, edge-line crossings, and lateral position. In another study of young (18- to 24-year-old) male drivers, sleep restriction was associated with increased sleepiness, increased response time, and inappropriate line crossings in a driving simulator (Philip et al., 2005). It is evident that driving duration and sleep history can influence fatigue and road safety for young drivers.

Hazard perception is another aspect of driving linked to crash risk and negatively affected by sleep loss and fatigue. Hazard perception is the ability to identify salient stimuli and interpret the likelihood and magnitude of associated risk (also see Chapters 6 and 7 in this handbook). Smith et al. (2009) found that novice drivers (17–24 years old, mean driving experience of 1.65 years) were slower to perceive hazards when sleepy, compared to when rested and compared to experienced drivers (28–36 years old, mean driving experience of 14.41 years). This decrease in speed of hazard perception was equivalent to 6.33 meters (~20 feet) of travel at 60 kph (the standard legal driving speed in Australia). Despite the strong link between hazard perception and crash risk (Horwsill & McKenna, 2004), and the demonstrated negative effects of sleep loss on hazard perception, there is limited research in this area. Additional aspects of hazard perception are addressed in depth in Chapters 6, 7, and 28.

So far, we have seen that fatigue has negative consequences for decision making, vigilance, and hazard perception, and that each of these functions is essential for road safety. In drivers of all ages, fatigue interacts with multiple risky driving behaviors to result in cumulative increases in risk. This has been demonstrated in the case of aggressive driving, driver distraction, speeding, and drunk driving (for a review, see Watling et al., 2013). Concerningly, young drivers are more likely to experience fatigue and are also more vulnerable to the negative effects of sleepiness and fatigue on driving performance compared to older and more experienced drivers (Watling et al., 2013). This is particularly troubling given that young drivers tend to have altered risk perception particularly when it comes to driving.

Young drivers are more likely to engage in risky driving behaviors (Scott-Parker et al., 2009), including fatigued driving (McGwin Jr. & Brown, 1999). In one study, younger male drivers reported a stronger liking for engaging in risky driving behavior, including driving while sleepy, compared to older drivers and young female drivers (Rhodes & Pivik, 2011). This is consistent with young driver research that shows that risky driving behavior, particularly drunk driving, is influenced by sensation seeking in young drivers (Fernandes et al., 2007; Scott-Parker et al., 2009). Low levels of perceived susceptibility to punishment or harm have also been associated with greater likelihood of driving when fatigued. This was particularly true in the case of young male drivers (Fernandes et al., 2010). Other factors increasing the likelihood of fatigued driving in young drivers include peer approval of the behavior and a general optimism bias regarding vulnerability to crash or punishment (Fernandes et al., 2007).

15.2.3 Managing Fatigue-Related Risk for Young Drivers

It is evident that young drivers are at increased risk for fatigue and that fatigue has significant negative effects on an individual's ability to drive safely. There are relatively few approaches currently used to manage this risk.

15.2.3.1 Deterrence Theory

Other risky driving behaviors such as speeding, drugs, and alcohol can be managed primarily using the principles of deterrence theory. Deterrence theory stipulates that individuals will engage in unlawful behavior when the perceived benefit is high. In turn, deterrence theory applies a significant *cost* when unlawful behavior is detected, for example, fines for exceeding the speed limit (Homel, 1988). Principles of deterrence are applied to objective measures of risky behaviors that increase the likelihood of an MVA and that can be measured (e.g., driving over the speed limit and driving while intoxicated). These principles are shown to be generally effective in controlling these behaviors and reducing MVAs (Fell et al., 2009; Tay, 2005). However, there is no known biomarker or reliable objective measure for fatigue, nor is there likely to be in the near future. Further, technologies to identify signs of fatigued driving are only in the early stages of development (Balkin et al., 2011; Eskandarian et al., 2007; McDonald et al., 2012, 2014). As a consequence, it is difficult to use the principles of deterrence theory to alter this behavior, as the ability for a third party to identify people driving while fatigued is limited.

15.2.3.2 Graduated Licensing Systems

Graduated licensing systems (GLSs) are another common way of managing driving risk for young drivers. GLSs limit exposure to risky driving situations by allowing driving privileges in a steplike way (Ferguson, 2003). There is evidence that certain GLS initiatives, particularly longer learner periods, passenger number restrictions, imposing a zero blood alcohol concentration, and mandating seat belt use, are associated with crash reduction (Senserrick & Whelan, 2003; also see Chapter 17 on graduated driver licensing [GDL]). In the case of managing fatigue risk, the primary approach has been to impose nighttime driving restrictions. As described in Section 15.2.1.1, sleep is biologically programmed to occur during the night. Thus, waking activity occurring during the biological night is likely to be characterized by significant feelings of fatigue and the associated impairments (see Section 15.2.2). Indeed, nighttime driving has been associated with a 5–10 times increase in crash risk for young drivers (Åkerstedt & Kecklund, 2001). Nighttime driving restrictions have been shown to reduce crash risk for young drivers; however, it is not clear what proportion of this reduction is related to fatigue risk or because of other risks associated with night driving, for example, aggressive and risky driving, driving with multiple passengers, and/or recreational driving (Senserrick & Whelan, 2003).

15.2.3.3 Educational Campaigns

Educational campaigns are often used to educate the public about the consequences of unsafe driving behaviors. There have been a limited number of public awareness campaigns addressing fatigued driving risk (for a review, see Fletcher et al., 2005), but few focusing on young drivers in particular and few that have been independently evaluated. In one evaluation study of the efficacy of fatigue-related road safety advertising, Tay and Watson (2002) found that individuals exposed to a fear-based antifatigue driving campaign were more likely to report intentions to avoid fatigued driving if they were also given coping strategies (these included power-napping, a coffee break, and swapping drivers), compared to a group shown the fear-based message alone. The group given strategies was also more likely to have avoided fatigued driving when follow-up questioning occurred 1–2 weeks later. Fear-based campaigns operate on the premise that the arousal of emotional tension will increase an individual's motivation to accept recommendations (Tay & Watson, 2002). It is currently unclear whether this approach would be effective in the case of young drivers.

15.2.4 Where Do We Go from Here?

As described previously, the primary approach to managing fatigued driving risk for young drivers is to impose restrictions limiting nighttime driving. However, the utility of driving restrictions is limited, as they may be broken and infractions may go undetected, resulting in at-risk young drivers being on the roads. Further, while fatigue risk is higher during the biological night, fatigue can occur at any time of day, dependent upon a number of factors. For example, a worker driving home from a night shift is likely to be significantly impaired by fatigue, and this time of day has been associated with significant fatigue-related driving performance impairment (Åkerstedt et al., 2005). In fact, reductions in hazard anticipation, hazard mitigation, and attention maintenance are observed in drivers compared during the morning and early evening, reductions that are presumably due in no small measure to fatigue (Hamid et al., 2013, 2014). Similarly, during a week of sleep reduced only slightly below habitual duration, fatigue can significantly impair vigilance during the day (Belenky et al., 2003). It is evident that reliance on nighttime driving restrictions is inadequate to address the risk posed by fatigue. Reliance on these restrictions may also take the focus off of education regarding the nature and consequences of fatigue-related risk for young drivers.

Reducing fatigued driving behavior and MVAs must therefore rely on young drivers understanding the risk of driving while fatigued and altering their own behavior based on the principles of perceived risk and self-regulation. As we have seen, there is a body of research addressing the factors that can influence a young driver to take the risk of driving when fatigued (i.e., Fernandes et al., 2007, 2010; Rhodes & Pivik, 2011; Scott-Parker et al., 2009). There is also a limited body of research addressing the factors that influence young drivers to *avoid* driving when fatigued. This research has shown that older drivers and drivers with negative attitudes toward sleepy driving were more likely to pull over and rest when experiencing sleepiness during a drive (Watling, 2014). In the case of younger drivers in particular, high emotional stability (low levels of anxiety, hostility, and impulsiveness) was associated with a lower likelihood of sleepy driving (Watling, 2014).

In a study investigating what young drivers (18–22 years old) would do should they feel impaired by fatigue while driving, 51.3% (n = 356) of the sample answered that they would “stop driving and do something in order to cope with the problem” (Lucidi et al., 2006, p. 305). An additional 48.4% (n = 336) answered that they would “continue driving but do something in order to cope with the problem” (Lucidi et al., 2006, p. 305). Strategies included opening the car window (18%), stopping to have a coffee (15%), increasing effort and attention (13%), and increasing the radio volume (12%). Only 0.3% (n = 2) participants responded that they would “ignore the problem and continue driving” (Lucidi et al., 2006, p. 305). These data suggest that young drivers may take steps to compensate for fatigue impairment when driving, or stop driving entirely. However, these data are based on a hypothetical incident, so it is not clear whether these behavioral intentions would be reflected in actual behavior.

In a driving simulation study, young drivers (20–28 years old) were exposed to moderate sleep restriction (woken at 5:00 a.m.) and then instructed to drive in a simulator (at either 9:00 a.m. or 2:00 p.m.) until they felt impaired by sleepiness. On average, young drivers stopped driving after approximately 40 minutes. Three participants experienced microsleeps during the driving test but waited an average of 12.3 minutes following these events to cease the driving task. Two of these participants were aware they had fallen asleep; the other was unsure (Watling & Smith, 2012). These data suggest that young drivers can become excessively sleepy following only moderate sleep restriction and during relatively brief stints of driving. Most importantly, these data show that young drivers may underestimate the dangers associated with driving while excessively sleepy. This is evidenced by the delay between microsleep occurrence and driving cessation. In line with this, it was revealed that young drivers frequently drove (>23% out of 2518 sampled driving episodes) despite perceiving themselves to be impaired by sleepiness (Smith et al., 2005). Indeed, previous research has shown that drivers often underestimate their own sleepiness (Horne & Reyner, 1999) and do not always stop despite the known increase in crash risk (Reyner & Horne, 1998). This is consistent with the results discussed previously showing driving continuing for up to 12 minutes after a microsleep (Watling & Smith, 2012).

It is evident not only that young drivers are particularly vulnerable to sleepiness and fatigue but also that they often fail to regulate their driving behavior when they are impaired by sleepiness and fatigue, despite being aware of the impairment. This suggests that young drivers may not fully appreciate the serious risks associated with driving while fatigued. The way that young drivers perceive fatigued driving risk is largely unknown, as is how their risk perception influences self-regulation of the behavior. The ego depletion model (EDM) may be one way of understanding the apparent failure of self-regulation evident in fatigued driving behavior for young drivers.

The EDM addresses failures of self-regulation, particularly in regard to health-related behaviors. The EDM stipulates that individuals have a limited capacity for self-control and once this is depleted, the ability to regulate behavior for other tasks will be impaired. This model proposes that it is more difficult to regulate a behavior in the early stages of uptake (Hagger et al., 2010) but that self-regulation skills can be strengthened over time (Muraven & Baumeister, 2000). In line with this, it may be the case that when young drivers become fatigued, their ability to self-regulate behavior is impaired (Hagger et al., 2010). This would then explain why young drivers, despite acknowledging their own sleepiness and fatigue, engage in fatigued driving even *after* experiencing a microsleep. There is evidence for improvements in self-regulation of behavior when individuals are made aware of the decision-making process and the likely outcomes of certain decisions (Chapman & Ogden, 2009; Oyserman et al., 2004; vanDellen & Hoyle, 2008). In the case of fatigued driving, this would involve connecting the decision to drive when fatigued with the potential consequences of having an MVA that results in harm to either themselves or other road users.

The next step then, in order to address the issue of fatigue risk for young drivers, is to investigate how risk is perceived when driving fatigued and how this perceived risk influences self-regulatory behavior (this is also related to the concept of skill and judgment calibration; see Horrey et al., 2015). Given that previous research has suggested potential differences in expectations of behavior (Lucidi et al., 2006) compared to actual behavior (Smith et al., 2005), a simulation study would provide the most realistic data feasible for a study of this high-risk behavior. This will provide a much-needed evidence base regarding the perception of risk associated with driving when fatigued for young drivers and may enable the development of interventions addressing fatigued driving for this high-risk group.

Perhaps one final possible way to reduce the effects of fatigue should be mentioned. As noted, it has been shown that among experienced drivers, fatigue impairs hazard anticipation, hazard perception, and attention maintenance skills (Hamid et al., 2013, 2014). It has also been shown that training programs can reduce the effect of fatigue on these three skills in more experienced drivers (Hamid, 2013) and that this same training program can improve these skills in novice drivers who are not fatigued (Yamani et al., 2014; see also Chapter 7 in this handbook). It does not seem too far a leap to imagine that the training program would also reduce the effects of fatigue on the hazard anticipation, hazard mitigation, and attention maintenance skills of teen drivers.

15.3 Conclusion

Fatigue is a significant contributor to road crashes for young drivers. Unlike other major driving risks like speed or drug and alcohol intoxication, fatigue is difficult to externally measure and regulate. As such, it is critical that young drivers are aware of the potentially serious consequences associated with driving when fatigued and self-regulate their own behavior to avoid these consequences. Enhancing self-regulation of fatigued driving for teen and novice drivers should be the focus of future research efforts. The EDM offers one theoretical approach on which to base a practical understanding of fatigued driving.

References

Åkerstedt, T., & Kecklund, G. (2001). Age, gender and early morning highway accidents. *Journal of Sleep Research, 10*(2), 105–110.

Åkerstedt, T., Peters, B., Anund, A., & Kecklund, G. (2005). Impaired alertness and performance driving home from the night shift: A driving simulator study. *Journal of Sleep Research, 14*, 17–20.

Australian Transport Council. (2011). *National Road Safety Strategy 2011–2020.*

Balkin, T. J., Horrey, W. J., Graeber, R. C., Czeisler, C. A., & Dinges, D. (2011). The challenges and opportunities of technological approaches to fatigue management. *Accident Analysis & Prevention, 43*(2), 565–572.

Belenky, G., Wesensten, N. J., Thorne, D., Thomas, M., Sing, H. C., Redmond, D. P., Russo, M. B., & Balkin, T. J. (2003). Patterns of performance degradation and restoration during sleep restriction and subsequent recovery: A sleep dose–response study. *Journal of Sleep Research, 12*, 1–12.

Brown, I. D. (1994). Driver fatigue. *Human Factors, 36*(2), 298–314.

Chapman, K., & Ogden, J. (2009). How do people change their diet? An exploration into mechanisms of dietary change. *Journal of Health Psychology, 14*(8), 1229–1242.

Connor, J., Whitlock, G., Norton, R., & Jackson, R. (2001). The role of driver sleepiness in car crashes: A systematic review of epidemiological studies. *Accident Analysis & Prevention, 33*(1), 31–41.

Dahl, R. E. (2008). Biological, developmental and neurobehavioural factors relevant to adolescent driving risks. *American Journal of Preventative Medicine, 35*(3S), S278–S284.

Eskandarian, A., Sayed, R., Delaigue, P., Blum, J., & Mortazavi, A. (2007). Advanced driver fatigue research. Washington, DC: U.S. Department of Transportation.

Fell, J. C., Fisher, D. A., Voas, R. B., Blackman, K., & Tippetts, A. S. (2009). The impact of underage drinking laws on alcohol-related fatal crashes of young drivers. *Alcoholism: Clinical and Experimental Research, 33*(7), 1208–1219.

Ferguson, S. A. (2003). Other high-risk factors for young drivers—How graduated licensing does, doesn't, or could address them. *Journal of Safety Research, 34*, 71–77.

Fernandes, R., Job, R. F. S., & Hatfield, J. (2007). A challenge to the assumed generalizability of prediction and countermeasure for risky-driving: Different factors predict different risk driving behaviors. *Journal of Safety Research, 38*, 59–70.

Fernandes, R., Hatfield, J., & Job, R. F. S. (2010). A systematic investigation of the differential predictors for speeding, drink-driving, driving while fatigued, and not wearing a seat belt, among young drivers. *Transportation Research Part F, 13*, 179–196.

Fletcher, A., McCulloch, K., Baulk, S. D., & Dawson, D. (2005). Countermeasures to driver fatigue: A review of public awareness campaigns and legal approaches. *Australian and New Zealand Journal of Public Health, 29*(5), 471–476.

Hagger, M. S., Wood, C., Stiff, C., & Chatzisarantis, N. L. D. (2010). Ego depletion and the strength model of self control: A meta-analysis. *Psychological Bulletin, 136*(4), 495–525.

Hamid, M. (2013). *Effect of Total Awake Time on Drivers' Performance and Evaluation of Training Interventions to Mitigate Effects of Total Awake Time on Drivers' Performance.* University of Massachusetts, Amherst, MA.

Hamid, M., Divekar, G., Borowsky, A., & Fisher, D. L. (2013). Using eye movements to evaluate the effect of total awake time on attention maintenance and hazard anticipation in a driving simulator. Paper presented at the Transportation Research Board 92nd Annual Meeting, Washington, DC.

Hamid, M., Samuel, S., Borowsky, A., & Fisher, D. L. (2014). Evaluation of effect of total awake time on driving performance skills—Hazard anticipation and hazard mitigation: A simulator study. Paper presented at the Transportation Research Board 93rd Annual Meeting, Washington, DC.

Harrison, Y., & Horne, J. A. (2000). The impact of sleep deprivation on decision making: A review. *Journal of Experimental Psychology: Applied, 6*(3), 236–249.

Harrison, Y., Horne, J. A., & Rothwell, A. (2000). Prefrontal neuropsychological effects of sleep deprivation in young adults—A model for healthy aging? *Sleep, 23*(8), 1067–1073.

Homel, R. (1988). *Policing and Punishing the Drinking Driver: A Study of General and Specific Deterrence.* New York: Springer-Verlag.

Horne, J. A., & Reyner, L. (1999). Vehicle accidents related to sleep: A review. *Occupational and Environmental Medicine, 56*(5), 289–294.

Horrey, W. J., Lesch, M. F., Mitsopoulos-Rubens, E., & Lee, J. D. (2015). Calibration of skill and judgment in driving: Development of a conceptual framework and the implications for road safety. *Accident Analysis & Prevention, 76*, 25–33.

Horwsill, M. S., & McKenna, F. P. (2004). Drivers' hazard perception ability: Situation awareness on the road. In S. Banbury & S. Tremblay (Eds.), *A Cognitive Approach to Situation Awareness: Theory and Application* (pp. 155–175). Farnham: Ashgate Publishing.

Laberge, L., Ledoux, E., Auclair, J., Thuilier, C., Gaudreault, M., Gaudreault, M., Veillette, S., & Perron, M. (2011). Risk factors for work-related fatigue in students with school-year employment. *Journal of Adolescent Health, 48*, 289–294.

Lowden, A., Anund, A., Kecklund, G., Peters, B., & Åkerstedt, T. (2009). Wakefulness in young and elderly subjects driving at night in a car simulator. *Accident Analysis & Prevention, 41*, 1001–1007.

Lucidi, F., Russo, P. M., Mallia, L., Devoto, A., Lauriola, M., & Violani, C. (2006). Sleep-related car crashes: Risk perception and decision-making processes in young drivers. *Accident Analysis & Prevention, 38*, 302–309.

Martin, J. S., Hébert, M., Ledoux, E., Gaudreault, M., & Laberge, L. (2012). Relationship of chronotype to sleep, light exposure, and work-related fatigue in student workers. *Chronobiology International, 29*(3), 295–304.

McDonald, A. D., Schwarz, C., Lee, J. D., & Brown, T. L. (2012). Real-time detection of drowsiness related lane departures using steering wheel angle. Paper presented at the Human Factors and Ergonomics Society Annual Meeting, Boston.

McDonald, A. D., Lee, J. D., Schwarz, C., & Brown, T. L. (2014). Steering in a random forest: Ensemble learning for detecting drowsiness-related lane departures. *Human Factors, 56*(5), 986–998.

McGwin Jr., G., & Brown, D. B. (1999). Characteristics of traffic crashes among young, middle-aged, and older drivers. *Accident Analysis & Prevention, 31*, 181–198.

Muraven, M., & Baumeister, R. F. (2000). Self-regulation and depletion of limited resources: Does self-control resemble a muscle? *Psychological Bulletin, 126*(2), 247–259.

Oyserman, D., Bybee, D., Terry, K., & Hart-Johnson, T. (2004). Possible selves as roadmaps. *Journal of Research in Personality, 38*, 130–149.

Philip, P., Taillard, J., Guilleminault, C., Salva, M. Q., Bioulac, B., & Ohayon, M. (1999). Long distance driving and self induced sleep deprivation among automobile drivers. *Sleep, 22*(4), 475–480.

Philip, P., Taillard, J., Klein, E., Sagaspe, P., Charles, A., Davies, W. L., Guilleminault, C., & Bioulac, B. (2003). Effect of fatigue on performance measured by a driving simulator in automobile drivers. *Journal of Psychosomatic Research, 55*, 197–200.

Philip, P., Sagaspe, P., Moore, N., Taillard, J., Charles, A., Guilleminault, C., & Bioulac, B. (2005). Fatigue, sleep restriction and driving performance. *Accident Analysis & Prevention, 37*, 473–478.

Reyner, L., & Horne, J. A. (1998). Falling asleep whilst driving: Are drivers aware of prior sleepiness? *International Journal of Legal Medicine, 111*(3), 120–123.

Rhodes, N., & Pivik, K. (2011). Age and gender differences in risky driving: The roles of positive affect and risk perception. *Accident Analysis & Prevention, 43*, 923–931.

Scott-Parker, B., Watson, B., & King, M. J. (2009). Understanding the psychosocial factors influencing the risky behaviour of young drivers. *Transportation Research Part F, 12*, 470–482.

Senserrick, T., & Whelan, M. (2003). Graduated driver licensing: Effectiveness of systems and individual components. Monash University Accident Research Centre.

Smith, S. S., Carrington, M., & Trinder, J. (2005). Subjective and predicted sleepiness while driving in young adults. *Accident Analysis & Prevention, 37*, 1066–1073.

Smith, S. S., Horwsill, M. S., Chambers, B., & Wetton, M. (2009). Hazard perception in novice and experienced drivers: The effects of sleepiness. *Accident Analysis & Prevention, 41*, 729–733.

Tay, R. (2005). The effectiveness of enforcement and publicity campaigns on serious crashes involving young male drivers: Are drink driving and speeding similar? *Accident Analysis & Prevention, 37*(5), 922–929.

Tay, R., & Watson, B. (2002). Changing drivers' intentions and behaviours using fear-based driver fatigue advertisements. *Health Marketing Quarterly, 19*(4), 55–68.

Taylor, A., Wright, H. R., & Lack, L. C. (2008). Sleeping-in on the weekend delays circadian phase and increases sleepiness the following week. *Sleep Biol Rhythms, 6*, 172–179.

Thiffault, P., & Bergeron, J. (2003). Monotony of road environment and driver fatigue: A simulator study. *Accident Analysis & Prevention, 35*, 381–391.

Ting, P.-H., Hwang, J.-R., Doong, J.-L., & Jeng, M.-C. (2008). Driver fatigue and highway driving: A simulator study. *Physiology and Behavior, 94*, 448–453.

vanDellen, M. R., & Hoyle, R. H. (2008). Possible selves as behavioral standards in self-regulation. *Self and Identity, 7*(3), 295–304.

Watling, C. N. (2014). Sleepy driving and pulling over for a rest: Investigating individual factors that contribute to these driving behaviours. *Personality and Individual Differences, 56*, 105–110.

Watling, C. N., & Smith, S. S. (2012). Too sleepy to drive: Self-perception and regulation of driving when sleepy. Paper presented at the Occupational Safety in Transport Conference, Gold Coast, QLD.

Watling, C. N., Armstrong, K. A., & Smith, S. S. (2013). Sleepiness: How a biological drive can influence other risky road user behaviours. Paper presented at the Australasian College of Road Safety (ACRS) Conference, National Wine Centre of Australia, Adelaide SA.

Williamson, A., Lombardi, D. A., Folkard, S., Stutts, J., Courtney, T. K., & Connor, J. L. (2011). The link between fatigue and safety. *Accident Analysis & Prevention, 43*(2), 498–515.

World Health Organization. (2007). *Youth and Road Safety*. Geneva: World Health Organization.

Wundersitz, L. N. (2012). An analysis of young drivers involved in crashes using in-depth crash investigation data. Centre for Automotive Safety Research.

Yamani, Y., Samuel, S., & Fisher, D. L. (2014). Evaluation of the effectiveness of a multi-skill program for training younger drivers on higher cognitive skills. Amherst, MA: Department of Mechanical and Industrial Engineering, University of Massachusetts.

16

Teen Driving Risk in the Presence of Passengers

Bruce Simons-Morton

Marie Claude Ouimet

Abstract

The Problem. Driving risk may be higher among teenage drivers in the presence of teenage passengers. However, there has been only one recent review of one aspect of the literature, and many questions remain, including the following: (1) What is the extent of the association between passengers and (a) teenage fatal and nonfatal crash risk and (b) risky driving or inattention? (2) How do these associations vary as a function of driving conditions and driver and passenger characteristics? ***Methods.*** Research was reviewed that examined the association of passenger presence and teenage driving outcomes using survey, crash database, observational, naturalistic, and simulation methods. ***Key Findings.*** While there is clear evidence that fatal crash risk among teenage drivers is higher in the presence of teenage passengers, the association between passenger presence and nonfatal crash risk is less clear. Risky driving is higher among teenage drivers in the presence of teenage passengers, particularly male passengers. In addition, teenage driver attention may sometimes be lower in the presence of teenage passengers, particularly when passengers talk with the driver. ***Conclusions.*** Risky driving and crash risk are often lower in the presence of adult passengers and higher in the presence of teenage passengers. The variability in associations with driving performance and outcomes may vary by driver and passenger characteristics and driving conditions.

16.1 Introduction

The purpose of this chapter is to examine current findings on passenger effects on teen driving outcomes. Before discussing the specific relationships between teen passengers and teenage risky driving behavior, it is important to understand the two general mechanisms by which passengers can increase risk: social influence and distraction.

16.1.1 Social Influence

With a passenger in the vehicle, driving becomes a social activity, with implications for driver behavior. In any environment, the presence of others stimulates people to try harder or otherwise modify their behavior to conform to group norms (Aronson et al., 2007; Cialdini, 2001). Notably, the classic research of Asch (1955) demonstrated that individuals' opinions are influenced by others. In one early experiment, participants were exposed to a group of seven experimental confederates, shown a series of objects, and asked to indicate aloud which matched. After two trials in which the confederates provided the correct responses, on the third trial, they unanimously provided the same incorrect response. Responding last, the respondent would experience the dilemma of either providing his actual perception of the correct response or conforming to the group's incorrect response. Compared to a control group, which provided the correct response in nearly all cases, about 50% of the time, those in the deception arm of the study agreed with the confederates in choosing the incorrect match. This was the first in a long line of studies demonstrating the powerful tendency to conform. While the nuances of conformity and the conditions under which it occurs are still active areas of research, conformity is the underlying principle for understanding social influence, in which one's behavior is influenced by perceptions of the expectations of others. Most people can be expected most of the time to prefer the rich social rewards of conformity to the isolation of independence (Cialdini, 2001).

Peer conformity is particularly common and powerful among adolescents (Brown & Larson, 2009; Steinberg, 2008). For example, teens who have friends who drink alcohol are very likely to begin drinking, and those who drink alcohol tend to have friends who drink (Simons-Morton, 2007). Peer similarity occurs when adolescents adopt the attitudes and behavior of their friends and select as friends others with similar attitudes and behaviors (Ennett & Bauman, 1994; Simons-Morton & Farhat, 2010). Substance use is not the only area of adolescent peer similarity and conformity.

Notably, the risky driving behavior of adolescents is similar within peer groups (Simons-Morton et al., 2011a). More importantly, the presence of a teenage passenger is significantly associated with increased fatal crash likelihood among young drivers (Chen et al., 2000; Ouimet et al., 2010). The effect of teen passengers on teenage crash risk could be due to distraction or to social influence but is not simply an effect of the presence of any passenger, because risk is greater in the presence of male than female teenage passengers and lower in the presence of adult passengers (Ouimet et al., 2010). Curiously, while the association between teen passengers and fatal crash risk is clear, the effect of teenage passengers on nonfatal crashes has not consistently been shown (Ouimet et al., 2015; Simons-Morton et al., 2011a). Hence, crash risk may vary according to the characteristics of the driver and passenger.

The mechanisms of peer influence are complicated, and an adequate discussion is beyond the limits of this chapter. Nevertheless, it can be said with certainty that conformity is at the core of social influence. The tendency to conform to prevailing social norms can be explained from an evolutionary biology perspective, given that we are essentially social creatures and to the extent that over the millennia, conformity would have increased acceptance by others, enhancing our chances for survival, while nonconformity would have decreased acceptance, survival, and procreation potential (Wilson, 2007). Therefore, the tendency to conform, embedded in our genetic heritage, may be particularly strong during adolescence. Neuroscience evidence indicates that the presence of peers heightens sensitivity of the adolescent brain's reward system to opportunities for risk taking (Chein et al., 2011). Hence, theories of adolescent development now emphasize the increased vulnerability to risk taking in the presence of peers and the enhanced reward salience for engaging in risk due to the imbalance between the relatively more rapid development of the affective processing than cognitive control systems (Chein et al., 2011; Lambert et al., 2014; Steinberg, 2008). The adolescent brain appears wired for increased risk taking in the presence of peers.

To some extent, conformity is the result of persuasion. A substantial body of research documents that people can be persuaded to think and behave in certain ways, although the elements of persuasion are complicated (Cialdini, 2001). Despite our predilection for independent thinking, each of us can be persuaded to do things we might not otherwise by our friends, relatives, and coworkers, not to mention our bosses and others with power over us. Persuasion works to the extent that it is linked to the consequences of nonconformity. Generally, we do what our parents, spouses, and friends want us to do to gain their approval and to avoid their disappointment in us, or worse, actual social exclusion. Similarly, we do what advertising suggests to the extent that it has persuaded us of the advantages of the product or concept, including the social advantages of what others will think of us in relation to it. With respect to driving, it is easy to imagine a teenage passenger persuading a reluctant teenager to drive faster, tailgate, pass the vehicle ahead, drive after drinking, and take other risks. It is not clear how often, when, or among whom this happens, but given the generally high-risk driving propensity of adolescents compared with adults (Simons-Morton et al., 2011b), it is easy to imagine this occurring. Shepherd et al. (2011) demonstrated that verbal persuasion effectively increased simulated risky driving, while informational influence to drive safely resulted in less risky driving.

Overt pressure and social norms are two, often related, forms of social influence. Usually, we are aware of being pressured overtly, but normative influence can be subtle. If we seek to conform to the norms of others, we need to determine what those norms and expectations might be. In some cases, information about expectations can be gained from communication or observation, but in many cases, it is based on perceptions from subtle clues. The norms of a person who drinks or smokes or verbalizes a preference for smoking or drinking are clear. But in many other cases, we assume social norms from subtle social cues that may not always be accurate. In any case, actual or perceived social norms, correctly deciphered or not, are strongly associated with behavior. For example, both peer and parent norms are associated with adolescent substance use (Simons-Morton & Chen, 2006; Simons-Morton & Farhat, 2010) and driving behavior (Simons-Morton et al., 2005, 2006, 2011a). Notably, teenage risky driving is associated with peer and parent norms. Teenagers who report that their parents expect them to drive safely (Simons-Morton et al. 2006) or would find out about their risky driving behavior and could do something about it (Simons-Morton et al., 2013) engage in less risky driving. Norms reflect the social acceptability of the behavior and are typically measured by assessing the prevalence (subjective norms) and peer acceptance (injunctive norms) of the behavior among relevant others (Fishbein, 1980). By conforming to the prevailing social norms, one increases the likelihood of social acceptance and approval.

While adolescents may be particularly susceptible to peer influence, not all are equally susceptible, and susceptibility might vary under certain conditions (Crone & Dahl, 2012; Romer, 2010). For example, social exclusion appears to increase conformity, at least among some people, particularly under conditions of uncertainty and particularly among adolescents who exhibit heightened neural sensitivity to rejection (Crone & Dahl, 2012). A recent study demonstrated that sensitivity to rejection was associated with risky driving in a simulator in the presence of a peer passenger (Falk et al., 2014).

Adolescents, even more than adults, may be more persuadable in certain contexts than others (Romer, 2010). Therefore, marketers use music, lighting, and design to affect moods conducive to persuasion. Possibly, drivers are particularly susceptible to social influence under certain contexts or moods (Taubman-Ben-Ari, 2012). Kahneman (2011), in his popular book, *Thinking, Fast and Slow*, notes that when we are relaxed, we tend not to be cognitively vigilant and therefore are more persuadable and suggestible than when we are alert and on guard. This might explain why teen passengers may increase teen driver crash risk particularly under certain conditions, such as late at night and when alcohol is involved. In addition to impairing judgment with respect to driving risk, alcohol and fatigue may reduce vigilance and increase reward sensitivity, thereby increasing the propensity to take risks in the presence of teen passengers. Hence, it seems likely that driving with passengers, like all social behavior, is influenced by passenger pressure and norms about driving risk, style, and safety.

16.1.2 Distraction

So far, we have emphasized social influence as the mechanism by which passengers can increase risk. Passengers can also increase risk through distraction. In simulation studies, distraction is generally measured by eye glance, sometimes by driving performance measures such as lane tracking. Increased cognitive load in the form of attention to things that are not driving related (such as passengers and cell phone messages) can reduce visual scanning and recognition of potential hazards (Caird et al., 2008; Horrey & Wickens, 2006). Passengers can also reduce driver attention to driving by physically touching the steering wheel, moving, talking loudly, and so on. As Kahneman (2011) argues, people have little capacity for engaging in more than one cognitively demanding task at a time. Because some activities are more interesting, compelling, attractive, and pleasant than other activities, people may prefer to pay attention to their phone or to a rowdy, engaging, interesting, familiar or unfamiliar passenger than to the forward roadway. Once one is engaged in a secondary task, it is difficult to fully reengage in the primary task of driving. This may help to explain why risky driving tends to be higher in the presence of teenage passengers and lower in the presence of adult passengers, who would be expected to exert overt and normative pressure on teenagers to drive carefully and pay attention to the driving task.

The topic of social influence and distraction is important and interesting because of the associations between passenger presence and crash outcomes, the complex ways that passengers might influence driving performance, and the potential for limiting the number of teenage passengers novice teenagers can carry via policies and parental limits. Indeed, all but five US states limit the number of teenage passengers that novice teenage drivers are allowed to carry during the first months of licensure (Insurance Institute for Highway Safety, 2014). Having established the theoretical basis for passenger influence, the purpose of this chapter is to examine current findings on passenger effects on teen driving outcomes. We focus on the following questions of interest with respect to teenage drivers:

1. Is passenger presence associated with higher crash and near-crash rates?
2. Is passenger presence associated with risky/dangerous driving?
3. Is passenger presence associated with reduced driver attention?
4. Do the associations between passengers and driver risk vary by the age, sex, or other characteristics of the driver and passenger?

16.2 Review of Literature by Method

A range of study types has been used to evaluate the relationship between passengers and teenage driving outcomes as defined by the aforementioned four questions. Principal among these are survey, crash database, observation, naturalistic, and simulation studies. Each of these approaches addresses only one or some of the questions of interest and has certain strengths and limitations in general and for research on this topic, which we note at the beginning of each section. We begin with survey research, which provides information about self-reported risk factors and outcomes. Then we review the evidence from crash database research. We then focus on the few observational and naturalistic studies on the topic, and finally, we review the available research that has used experimental methods. In each section, we focus on the findings relevant to the research questions of interest in this chapter. The discussion provides a summary of the state of the science as it relates to each research question across the study types as well as an estimate of the strength of the findings.

16.2.1 Survey Research

Survey research is convenient and often less expensive than other research methods, can accommodate large and diverse samples, and provides useful information about potential risk factors, although, of course, this information is subjective. Here, we have elected to review papers since 1990 that asked about actual driving performance and not simply intentions. Some researchers have combined objective measures of driving outcomes from naturalistic studies with self-report measures of potential risk factors. Here, we provide a brief review of findings from a sample of survey studies with or without other objective measures.

Some interesting information about the perceptions of adolescents about teenage passengers and drivers has been reported in cross-sectional surveys. Ulleberg (2004) surveyed a sample of 4397 Norwegian youth 16–23 years old, a large proportion of whom were accepting of riding with an unsafe driver. Females were more likely to tell the driver when they felt unsafe, while males perceived greater potential consequences of addressing unsafe driving. Meanwhile, in a telephone survey of 872 Australian respondents, including 224 young drivers (16–24 years old), Regan and Mitsopoulos (2003) found that drivers reported that 16- to 24-year-old passengers, compared to passengers aged 55 or older, were more likely to encourage teen drivers to engage in risky driving behavior by their presence or their comments.

In the US studies, Arnett et al. (1997) surveyed a sample of 17- to 18-year-old high school students ($N = 139$), of whom 59 also kept a 10-day log of their driving experiences. Participants reported higher speed while alone or with friends than with parents. Heck and Carlos (2008) surveyed a sample of more than 2000 seniors at 13 California high schools. Among the 1715 teenagers who reported driving, 38.4% indicated that teenage passengers distracted them while driving, mostly by talking or yelling, while 7.5% reported that teenage passengers deliberately distracted them, often by touching them or vehicle controls. In a large nationally representative survey of more than 5000 students from 68 high schools (Ginsberg et al., 2008), 10% reported that passengers play an important role in driving safety, making it the item with the lowest rating on a list of 25 items. Passenger behavior, such as acting wild (65%), trying to influence drivers' risky behavior (62%), smoking marijuana (48%), drinking alcohol (47%), and dancing or signing with drivers (21%), was estimated as playing a more important role. These perceptions were generally stronger in females than males. Mirman et al. (2012) surveyed a sample of 198 teen drivers and found, after controlling for several variables including gender, that those who drove more often with multiple peers had lower risk perceptions, less parental monitoring, and higher sensation-seeking propensity.

Some prospective associations between social norms and teenage risky driving have been reported. In a naturalistic study ($N = 42$) with objective measures of teen driving performance and surveys among these teens of potential risk factors measured at the start of the study, Simons-Morton and colleagues found that risky friends (measured with several questions about how many of one's five closest friends engaged in driving or nondriving risky behavior) was prospectively associated with speeding (Simons-Morton et al., 2012), elevated-g-force event rate, and crash/near-crash (CNC) rate (Simons-Morton et al., 2011a).

In summary, evidence from survey research supports the following contentions:

1. Passengers report increased driving risk associated with social influence to drive in a risky fashion and with passenger distraction, including talking and touching vehicle controls.
2. Risky driving and crash risk are higher among those whose social norms (in the form of having risky friends) favor risky driving.

16.2.2 Crash Database (Epidemiology) Analyses

Analysis of crash databases is the most prevalent approach to evaluating possible effects of teen passengers. Results from crash databases have been contradictory, however, sometimes showing increased risk with the presence of passengers and sometimes showing no significant results or even protective effects with passengers in the vehicle. While asking the same research questions, study design differences may explain these inconsistent findings. For example, some studies examined fatal crashes, while others focused on injury crashes. Another possibility is that some studies used national databases, while others used more regional databases. There is also a wide variation in terms of age for drivers and passengers included in the literature. Additionally, some studies controlled for exposure, such as kilometers driven, or they distinguished between culpability versus nonculpability, while others did not consider these factors. Finally, licensing age is lower in some countries (e.g., North America) compared to others (e.g., Europe).

Ouimet et al. (2015) conducted a systematic review of crash risk studies for young drivers (≤24 years old) with peer passengers. Studies needed to compare crashes with passengers and crashes while solo driving and include a measure of exposure, such as kilometers driven. Fifteen articles were included in the review. For fatal crashes, evidence from the literature consistently showed increased crash risk for young drivers with passengers; but for other types of crashes or a combination of crashes, the findings were inconsistent. Across studies, an increased fatal crash risk was observed in the following contexts: with at least one passenger versus solo driving; with two or more passengers versus solo driving; with male versus female passengers; and for younger versus older drivers. For other types of crashes (i.e., injury) or a combination of crashes (i.e., injury and fatal), increased risk was observed only for the following: with male versus female passengers and young drivers versus older drivers.

Curry et al. (2012) used the National Highway Traffic Safety Administration's (NHTSA) National Motor Vehicle Crash Causation Survey to examine precrash factors with on-scene crash investigations of 656 nationally representative serious crashes involving 677 teenage drivers. Compared to teenage males driving alone prior to the crashes, males accompanied by male or female peer passengers were more likely to be involved in risky behavior and be distracted by outside-vehicle factors. Female drivers were more likely to be distracted by inside-vehicle factors (such as looking at an object or a passenger, but not including talking with passengers); involvement in this type of distraction was higher for female drivers with male passengers.

In summary, evidence from crash database research supports the following contentions:

1. Increased fatal crash risk was observed in the following contexts: with at least one passenger versus solo driving; with two or more passengers versus solo driving; with male versus female passengers; and for younger versus older drivers. For other types of crashes (i.e., injury) or a combination of crashes (i.e., injury and fatal), increased risk was observed only for the following: with male versus female passengers and young drivers versus older drivers.
2. In pre-serious-crash periods in the presence of peer passengers, male drivers were more likely to be involved in risky behavior and be distracted by outside-vehicle factors, irrespective of passenger gender, while female drivers were more likely to be distracted by inside-vehicle factors, especially if passengers were males.

16.2.3 Observational Studies

Epidemiology studies provide useful information about the extent and variability in crash risk, while unobtrusive roadside observations can provide valid information about on-road driving behavior and driving conditions. Unfortunately, few observational studies have assessed the variability in driving performance and risk in relation to passenger presence. Also, one of the main difficulties in these studies is to correctly assess the age and sex of vehicle occupants, and some studies did not meet our criteria for inclusion. Some studies observed vehicles and classified drivers and passengers as ≤30 or >30 years old (not useful distinctions for this review) and examined risky behavior or distractions according to estimated driver (Baxter et al., 1990; Huisingh et al., 2015; Prat et al., 2015) and passenger age (Baxter et al., 1990). Here, we report the findings from observation studies conducted to evaluate the association of passengers by age and sex with driver performance.

McKenna et al. (1998) had observers record the sex and relative age of drivers and passengers (≤25 years old and >25 years old), and visually assess driver speed, following distance, and gap size between vehicles. When accompanied by a male peer passenger, compared to solo driving, male and female drivers drove faster and accepted smaller gaps. When accompanied by a female peer passenger, only males drove faster, and both male and female drivers accepted larger gaps. Overall, the study provided evidence of an association between the presence of young passengers and young driver risky driving behavior.

Simons-Morton et al. (2005) investigated two measures of risky driving, speed and close following, among teenagers (n = 471) and adult (n = 2251) drivers in the presence or absence of passengers in 10 public schools. A research assistant observed vehicles exiting high school parking lots a few meters from the exit. The assistant recorded the number and sex of the passengers and classified the age of the passengers as teens or adults. Another assistant videotaped the vehicles and recorded the license plate number for later identification. On nearby roadways where conditions allowed drivers to elect speed and following distance, a third research assistant used a video camera to measure headway and a laser gun (i.e., lidar) to calculate speed. The findings indicated that unaccompanied teenage drivers drove slightly faster and maintained a somewhat shorter average headway than usual traffic. However, relative to general traffic, both male and female teenage drivers on average drove significantly faster and closer to lead vehicles in the presence of a male teenage passenger. Conversely, in the presence of a female teenage passenger, male teenage driving was very similar to usual traffic with respect to average speed and headway. Overall, the observed rate of high-risk driving (defined as speed ≥15 mph above the posted speed limit and/or time headway of ≤1.0 second) was about more than twice higher for teenage male drivers with teenage male passengers than that of general traffic. This study provides perhaps the best evidence from an observation study of increased risky driving by teenagers carrying peer passengers.

Based on these few studies using unobtrusive observation, the evidence indicates the following:

1. Higher risky driving among young male and female drivers in the presence of young male passengers
2. Generally lower risky driving among male teenage drivers exposed to young female passengers

16.2.4 Naturalistic Driving Studies

Naturalistic driving research uses vehicle instrumentation with sensors that enable the collection of driving data over time. While naturalistic driving methods enable the collection of large amounts of data methods, the high cost limits sample sizes. In the Naturalistic Teenage Driving Study, 42 newly licensed teenage drivers were followed for 18 months after their vehicles were instrumented with accelerometers, GPS, cameras, and other equipment allowing the continuous assessment of numerous driving outcomes. Teenagers' CNC and elevated-gravitational-force events or kinematic risky driving (KRD) were much lower with adult passengers, compared with driving alone. Also, compared to solo driving, KRD was lower with teenage passengers, but there were no significant differences between CNC while driving with a teenager and alone (Simons-Morton et al., 2011a). Due to relatively small cells, it was not possible to examine results by driver and passenger gender. Other naturalistic studies also reported lower risky driving among novice teens with adult passengers (e.g., Prato et al., 2009).

Foss and Goodwin (2014) instrumented the vehicles of 52 high school students (38 beginners and 14 with experience) with devices that recorded video, audio, and vehicle information for 10 seconds before and 10 seconds after a KRD. The study showed that involvement in secondary tasks prior to the KRD was less frequent in the presence of teenage and adult passengers than while solo driving. When with passengers, loud conversation (12.6%) and horseplay (6.3%) were the interactions with passengers most present in recordings. Compared with one teenage passenger, these behaviors were significantly higher with two or more teenage passengers and significantly lower with siblings or parents.

With many other naturalistic driving studies underway in a number of countries, some with relatively large samples and many including young drivers, future research should provide detailed information about teen driving risk in relation to passenger presence. Based on the limited evidence, it cannot be concluded that teenage passengers increase the risk of nonfatal CNC and KRD. Naturalistic driving studies provide evidence of the following passenger effects among teenage drivers:

1. Adult passenger presence is negatively associated with CNC and KRD rates.
2. Teenage passenger presence is negatively associated with KRD rates and not significantly related to CNC rates.

16.2.5 Simulation Studies

Driving simulation provides substantial advantages for studying passenger effects on driving performance and the mechanisms by which passengers can increase risk, allowing experimentation in controlled and safe environments and measurement of a variety of driving outcomes, with reasonable generalization to on-road behavior (Fisher et al., 2007; Ouimet et al., 2011). Factors found to be associated with an outcome in survey, epidemiologic, observational, and naturalistic studies can be evaluated in well-controlled experimental studies to determine if the association holds when the exposure or potential risk factors are manipulated and other conditions are controlled. A small but growing number of recent simulation studies have examined the effects of passenger interactions on attention, generally measured by eye scan, and driving performance measures such as speed, intersection management, and gap. Here, we report first on experiments based on computerized driving games and then on simulation studies conducted in high-fidelity driving simulators, each providing different levels of realism based on the degree of participants' interaction with the device, the variety and complexity of scenarios participants encounter, and the measurement of outcomes. In games and driving simulators, participants can interact directly with the simulated environment to execute common driving maneuvers using the steering wheel, accelerator, and brake pedals, although simulators are more often connected to actual vehicles, while game consoles are more often desktop devices connected to a computer. With a driving simulator, researchers can program different scenarios and measure multiple driving outcomes depending on the research questions; in a game, scenarios and driving outcomes are limited and not programmable. Finally, a driving simulator can collect numerous driving outcomes, including speed, intersection management, and lane management; a desktop-based driving game might allow the collection of only a few outcomes that could be less sophisticated and realistic than driving simulation. Here, we review the small number of computerized driving game and driving simulation studies that have examined the effect of passengers on teenage simulated driving.

16.2.5.1 Computerized Driving Game Studies

Studies using computerized driving games have been conducted to examine and explain the effect of peer influence on simulated risky driving. The advantage of computer game driving consoles is their portability and low expense; however, most commercial games reward risky driving, crashes are common, and the consequences of crashing tend to be small, counter to the primary aim of actually driving to get to the destination with minimal risk. Also, sitting at a desk in front of a computer with a small wheel or joystick is not the same as being in an actual vehicle; hence, the realism for driving games may be low. Nevertheless, a previous review of the literature on randomized controlled trials and experiments using driving simulation did not show differences in the driver behaviors measured between computerized game and driving simulation devices (Ouimet et al., 2011). As such, the studies reviewed were designed in ways that provided useful information about the nature of teen passenger influence (Chein et al., 2011; Gardner & Steinberg, 2005; Shepherd et al., 2011).

Gardner and Steinberg (2005) randomly assigned male and female adolescents (aged 13–16 years old, $n = 106$), college students (18–22 years old, $n = 105$), and adults (24 years old and older, $n = 95$) to complete questionnaires and tasks alone or with two same-gender peers. One of the tasks was a computer game (Chicken) that allowed participants to engage in driving-related risk taking. Participants watched a moving vehicle on a computer screen, and they had to decide when to stop the vehicle to avoid a crash.

In each of the 15 trials, points could be gained if the vehicle was stopped before crashing into a brick wall. All three age groups took more risk (to the extent that they elected to stop the vehicle late) in the presence of peers than alone, but the two younger groups (13–16 and 18–22 years) took more risk than the oldest age group (24+ years). The study demonstrated an effect of peer presence on increased risk taking on the driving task compared with engaging in the task alone, consistent with the findings of observational studies (McKenna et al., 1998; Simons-Morton et al., 2005).

Shepherd et al. (2011) reported on two experiments ($N = 193$, study 1; $N = 89$, study 2) with college students randomized to play a driving video game (Euthusia Professional Racing for Sony PlayStation 2) alone or with two confederate peers who either applied pressure to take risks, were neutral, or were anti-risk. The basic findings were that verbal persuasion influenced risky simulated driving in the form of average speed, maximum speed, and number of crashes. In study 2, participants confirmed that informational persuasion (as in study 1) altered driving speed, as did informational persuasion plus subjective norms combined (manipulated by investing the confederate in the outcome and having him or her verbalize his or her expectations to the participants). In study 1, with only male drivers, participants who drove with passengers who encouraged them to take risks had a higher speed and number of crashes than those who drove with passengers who told them to drive slowly or who were neutral regarding risk taking; risk did not vary by passenger gender. In study 2, with both male and female participants, drivers verbally encouraged by female passengers to take more risk did so and had more crashes than those encouraged to drive safely; female drivers encouraged to drive riskily had more crashes than males so encouraged.

Chein et al. (2011) reported greater risky driving by male and female adolescents (14–18 years old, $n = 14$) in the Stoplight driving game when observed by same-gender peers than when unobserved by peers. The game consists of pressing a button to indicate if a vehicle progression on a computer screen should be stopped (or not) in four rounds of 20 intersections on a straight road. Participants played the game while their brains were being imaged using functional magnetic resonance imaging (fMRI). Participants demonstrated greater activation in reward-related brain regions and engaged in more risky driving decisions when observed by peers than when not observed by peers. There were no differences in the young adult group (19–22 years old, $n = 14$) and in a second adult group (24–29 years old, $n = 12$). This research suggests that peer passenger presence could promote riskier teenage driving performance by heightening sensitivity to the potential reward value of risky decisions, consistent with broader accounts of increased reward salience in teens (Albert et al., 2013).

16.2.5.2 Driving Simulator Studies

A number of driving simulation studies have shown no effect of passenger presence when passengers were silent but diminished driver attention when passengers talked to the drivers (Rivardo et al., 2008; Toxopeus et al., 2011). Notably, White and Caird (2010) examined the effect of passenger conversation and driver gender on the detection of potential road hazards in a sample of 40 young male and female drivers. Participants were randomly assigned to drive alone or with a confederate passenger from the opposite sex judged to be attractive and who was instructed to enter into a discussion with the driver. Results showed that there were more occasions in which the drivers "looked but failed to see" hazards with a passenger than while solo driving. The findings of these studies are consistent with an effect of inattention due to passengers talking, although it cannot be ruled out that driver behavior was affected by the drivers' perceptions of the passenger (young male drivers may have been distracted by the presence of the female confederate passenger or by cognitive load from the substance or amount of talking).

Other driving simulation studies were specifically designed to test effects on driver behavior of social norm manipulations. Ouimet et al. (2013) exposed 36 male participants to a male confederate passenger who dressed and acted in a manner consistent with either risk acceptance or risk aversion prior to driving the simulator but who did not converse with the driver or exert pressure during the drive. Compared to driving alone, teen drivers were less attentive to potential driving hazards (reduced scanning) but did not engage in more risky driving in the presence of a peer passenger (either risk accepting or risk averse). Contrary to expectations, some measures of risky driving were actually lower in the presence of the teenage passenger (i.e., longer gaps before initiating a left turn), particularly the risk-accepting passenger (i.e., longer headway). The results are consistent with an effect of teen passenger social norms on inattention but not on risky driving.

Simons-Morton et al. (2014) used a unique predrive priming activity to create participant perceptions of the confederate passenger's norms. A sample of 66 male teenagers were recruited to drive a simulator and randomly assigned to one of two groups: (1) driving alone and driving with a peer-aged confederate with risk-accepting social norms; or (2) driving alone and driving with a peer-aged confederate with risk-averse social norms. The order of presentation of driving alone or with the confederate passenger was counterbalanced. The confederate arrived late at the testing session and, in the risk-accepting condition, indicated that he would normally drive fast but hit every light; in the risk-averse condition, he explained that he tends to drive slowly anyway and hit every light. Then, while watching two videos of actual driving, the confederate indicated a preference for risky driving (in the risk-accepting condition) or for safer driving (in the risk-averse condition). Before the participant started to drive with the confederate as a passenger, the confederate also drove the simulator according to the assigned condition (e.g., faster if in the risk-accepting condition and slower if in the risk-averse condition), with the participant as a passenger. The driving task consisted of a series of intersections in which the light would turn yellow at different time intervals, some putting the drivers in the dilemma zone where they needed to stop quickly or pass through the intersection with a red light. Analyses showed a main effect of passenger presence on failing to stop at the yellow light and time spent in the intersection while the light was red. There was also a significant interaction by passenger type, with greater risky driving in the presence of the risk-accepting confederate compared to the risk-averse confederate. Exposure of teenage males to a risk-accepting confederate peer increased their risky simulated driving behavior compared with exposure to a risk-averse confederate peer. Also, drivers' visual scanning significantly narrowed horizontally and vertically in the presence of the confederate passenger (Pradhan et al., 2014), consistent with an effect of passenger presence on inattention, as reported by Ouimet et al. (2013). These experimental results are consistent with the hypotheses that (1) variability in teenage risky driving is due to perceived passenger social norms; and (2) inattention is greater in the presence of peer passengers with initial social priming.

Falk et al. (2014) examined the relationship between sensitivity to social exclusion, measured a week prior to driving simulation, and teen passenger effects on risky driving in a subsample of participants ($N = 36$) of the same study just described (Simons-Morton et al., 2014). In an fMRI scanner, participants participated in the computerized Cyberball game, where at first, the "ball" is shared equally among the participants, but eventually, the two confederate peers appeared to exchange it between themselves exclusively (actually, the ball passing by the peers was preprogrammed). Sensitivity to exclusion was measured by the neural response in the mentalizing and social pain brain networks during the neuroimaging task. Participants who expressed greater sensitivity to exclusion spent a greater percentage of time in the intersection while the light in was red, as measured in a driving simulation a week after fMRI testing. These results are consistent with increased risky driving behavior in the presence of a peer passenger. As in the main trial, there was a main effect of passenger presence (greater risky driving in the presence of a passenger than while solo driving) and an effect by passenger type, with greater risky driving in the presence of a risk-accepting than in the presence of a risk-averse passenger. The interaction between neural predictors and passenger type was not significant. However, the neural predictors were significantly associated with risk-taking behavior during the passenger drive, controlling for solo drive, passenger type, and susceptibility to peer pressure. The authors concluded that the findings are consistent with theory suggesting that sensitivity to exclusion leads to an increase in conformity (Crone & Dahl, 2008).

In another part of the study by Simons-Morton et al. (2014), novice teens' neural activity during a go/no-go response inhibition task was also measured during the fMRI session (Cascio et al., 2015). A week later, the 37 male participants drove in a simulator, as described previously. Increased activity in the response inhibition network, consistent with greater cognitive control, was associated with significantly safer driving in the presence of a risk-averse peer but no effect on risky driving when exposed to a risk-accepting confederate peer. Cascio and colleagues concluded that stronger inhibitory control may help reduce risky behavior in the presence of risk-averse peers. These findings also suggest that peer influences that discourage risk can play an important role in reducing risk in the presence of passengers.

The few experimental simulation studies conducted to date suggest that teenage passenger presence and passenger talking can reduce teenage driver attention, while teenage passenger social norms influence teenage risky driving and attention. Sensitivity to exclusion and cognitive control appear to serve as effect modifiers. Preliminary conclusions include the following:

1. Passenger presence was associated with simulated risky driving.
2. Passenger talking with and without prior social priming was found to reduce attention.
3. Passenger presence with social priming in the absence of talking was associated with reduced attention and differences in risky driving consistent with passenger norms, while passenger presence in the absence of talking without social priming was not associated with behavior change.
4. Verbal persuasion to increase risk or decrease risk did so.
5. The effect of passengers varied by the teenage driver characteristics of susceptibility to exclusion and cognitive control.

16.3 Discussion

A summary of findings by method is shown in Table 16.1, where the strength of the findings is coded according to the following: 0 = no evidence or not relevant; 1 = minimal evidence (one study); 2 = minimal evidence but not always consistent; 3 = moderate evidence but consistent (less than five studies); and 4 = strong, consistent evidence (five or more studies). Notably, each method provides information about different aspects of the relationship between passengers and teen driving performance. Here, we examine each question of interest across methods.

16.3.1 Is Passenger Presence Associated with Higher CNC Rates?

Strong evidence is obtained from analyses of crash databases that fatal crash risk is greater in the presence of teenage passengers. Analyses of nonfatal crash risk provide mixed evidence that it is higher in the presence of teenage passengers. Naturalistic research shows minimal evidence for lower CNC likelihood when teenage drivers carry adult passengers and no significant effects when they carry young passengers. There is no evidence from survey or simulation research.

16.3.2 Are Passengers Associated with Risky/Dangerous Driving?

Evidence of increased risk among teen drivers in the presence of teenage passengers is moderate in survey research, observational studies, and simulation studies, suggesting that driver behavior varies according to passenger norms and pressure. There is minimal evidence of increased risky driving from crash databases. In contrast, naturalistic research shows minimal evidence of a protective effect associated with passenger presence.

16.3.3 Are Passengers Associated with Lower Attention?

Surveys and simulation research provide moderate evidence that teenage passenger presence is associated with inattention among teenage drivers. The best evidence of passenger effects on inattention comes from simulation studies, where inattention to the forward roadway and scanning breadth have

TABLE 16.1 Evaluation of the Strength of the Evidence of the Association between Passenger Presence and Teenage Risky/Dangerous Driving and Crash Risk

Question	Survey	Crash Database	Observation	Naturalistic	Simulation
Is passenger presence associated with higher crash/near-crash rates?	(0)	(4) More fatal crash risk with teen passengers (2) More nonfatal or all-crash risk with teen passengers	(0)	(1) Less CNC with adults; negative association with teen passengers	(0)
Are passengers associated with risky/dangerous driving?	(3) More with teens	(1) More with teens	(3) More with teens	(1) Less KRD with adult and teen passengers	(3) More with teen passengers, social priming, and persuasion
Are passengers associated with lower attention?	(3) Less attention with teen passengers	(1) Less attention with teen passengers	(0)	(1) Less KRD associated with distraction with adult and teen passengers; more KRD associated with distraction with multiple teens compared to one	(3) Less attention with teen passengers talking
Does the association between passenger and driver risk vary by sex or other characteristics of the driver and passenger?	(3) More female passengers take action to reduce discomfort felt as passengers; more risky driving when friends are risky	(3) More fatal and nonfatal or all-crash risk with male versus female passengers	(3) More risky driving for male versus female drivers and with male versus female teen passengers (2) More risky driving among male teen drivers with female teen passengers	(0)	(1) More with higher sensitivity to exclusion (1) More with better cognitive control

Note: 0 = no evidence or not relevant; 1 = minimal evidence (one study); 2 = minimal evidence but not always consistent; 3 = moderate evidence but consistent (less than five studies); and 4 = strong, consistent evidence (five studies or more). CNC = crash/near crash; KRD = kinematic risky driving.

been shown to occur under a variety of passenger conditions, including when passengers talk to the teenage driver and when passengers exhibit particular norms regarding risk. There is minimal evidence of lower attention with passengers in crash databases and minimal evidence of a protective effect of passenger presence for involvement in a secondary task prior to KRD, though risk is higher with multiple passengers compared to one passenger. There is no evidence from observational studies.

16.3.4 Does the Association between Passenger and Driver Risk Vary by Sex or Other Characteristics of the Driver and Passenger?

Evidence from crash database studies provides moderate evidence for increased risk of fatal and nonfatal crashes with male passengers. Surveys and observation studies provide minimal to moderate evidence that male passengers are associated with teenage driving risk more than female teenage passengers, while naturalistic and simulation studies do not provide evidence of an effect of gender on teenage passengers. Experimental simulation research with male teenagers provides suggestive evidence about individual variability in risk, with social norms favoring risk, higher sensitivity to exclusion, and lower response inhibition linked to greater risky driving.

16.3.5 Limitations

The limitations of this review should be noted. A limited number of studies using any method have been conducted on passenger effects on teenage driving. Research is particularly needed about the individual variability in risk to answer the following questions: are certain drivers and certain passengers or combinations of drivers and passengers at particular risk, and is risk greater under certain driving conditions? Therefore, we can provide only tentative conclusions about the effects of passengers on teenage driving, pending further research.

16.4 Conclusions

The literature provides somewhat consistent evidence that teenage passengers increase teenage driving risk, particularly for fatal crashes. Simulation research provides evidence that social influence is one of the mechanisms by which passengers influence teenage risky driving. Observational studies and crash record research suggest that male teenage drivers may be more susceptible to peer influence than females, but not in all cases. Teenage passengers may also reduce driver attention. Of course, a great deal more needs to be learned about the nature of passenger effects on teenage driving. For example, some evidence suggests that male teenage drivers and passengers may pose the greatest risk, but little is known about the nature and conditions under which this risk occurs. Some drivers and some passengers may pose greater risk than others, through social influence or distraction, but it remains unclear if this is due to driver sex or individual characteristics. The picture that emerges is a complicated one suggesting that passenger influences may be somewhat conditional on the sex of the driver and passenger, and possibly on driving conditions. The growing body of thoughtful contemporary research on this topic promises to explain the remaining mysteries regarding the mechanism underlying teenage passenger effects on teenage driving risk (social influence and distraction) under certain driving conditions, such as the purpose of the trip, time of day, or even driver and passenger mood.

Acknowledgments

This work was supported in part by the intramural program of the Eunice Kennedy Shriver National Institutes of Child Health and Human Development (NICHD) of the National Institutes of Health (NIH); Marie Claude Ouimet was supported through a career award from the Quebec Health Research Fund (Fonds de recherche du Québec-Santé).

References

Albert, D., Chein, J., & Steinberg, L. (2013). The teenage brain: Peer influences on adolescent decision making. *Current Directions in Psychological Science, 22*(2), 114–120.

Arnett, J.J., Offer, D., & Fine, M.A. (1997). Reckless driving in adolescence: "State" and "trait" factors. *Accident Analysis & Prevention, 29*(1), 57–63.

Aronson, E., Wilson, T.D., & Akert, R.M. (2007). *Social Psychology* (6th ed.) Upper Saddle River, NJ: Pearson Education.

Asch, S.E. (1955). Opinions and social pressure. *Scientific American, 193,* 31–35.

Baxter, J.S., Manstead, A.S.R., Stradling, S.G., Campbell, K.A., Reason, J.T., & Parker, D. (1990). Social facilitation and driver behaviour. *British Journal of Psychology, 81,* 351–360.

Brown, B.B., & Larson, J. (2009). Peer relationships in adolescence. In R. Lerner & L. Steinberg (Eds.), *Handbook of Adolescent Psychology* (pp. 74–103). Hoboken, NJ: John Wiley & Sons.

Caird, J.K., Willness, C.R., Steel, Pl., & Scialfa, C. (2008). A meta-analysis of the effects of cell phones on driver performance. *Accident Analysis & Prevention, 40,* 1282–1293.

Cascio, C.N., Carp, J., O'Donnell, M.B., Tinney, Jr., F.J., Bingaham, R., Shope, J.T., Ouimet, M.C., Pradhan, A.K., Simons-Morton, B.G., & Falk, E.B. (2015). Buffering social influence: Neural correlates of response inhibition predict driving safety in the presence of a peer. *Journal of Cognitive Neursocience, 27*(1), 1–13.

Chein, J.M., Albert, D., O'Brien, L., Uckert, K., & Steinberg, L. (2011). Peers increase adolescent risk taking by enhancing activity in the brain's reward circuitry. *Developmental Science, 14*(2), F1–F10.

Chen, L.H., Baker, S.P., Braver, E.R., & Li, G. (2000). Carrying passengers as a risk factor for crashes fatal to 16- and 17-year-old drivers. *The Journal of the American Medical Association, 283*(12), 1578–1582.

Cialdini, R.B. (2001). *Influence: Science and Practice* (4th ed.). New York: HarperCollins.

Crone, E.A., & Dahl, R.E. (2012). Understanding adolescence as a period of social–affective engagement and goal flexibility. *National Review of Neuroscience, 13*(9), 636–650.

Curry, A.E., Mirman, J.H., Kallan, M.J., Winston, F.K., & Durbin, D.R. (2012). Peer passengers: How do they affect teen crashes? *Journal of Adolescent Health, 50*(6), 588–594.

Ennett, S.T., & Bauman, K.E. (1994). The contribution of influence and selection to adolescent peer group homogeneity: The case of adolescent cigarette smoking. *Journal of Personality and Social Psychology, 67,* 653–663.

Falk, E.B., Cascio, C., Carp, C., Tinney, F.J., Brook O'Donnell, M., Bingham, C.R., Shope, J.T., Ouimet, M.C., Pradhan, A.K., & Simons-Morton, B.G. (2014). Neural responses to exclusion predict susceptibility to social influence. *Journal of Adolescent Health, 54,* 522–531.

Fishbein, M. (1980). Attitudes and persuasion. *Annual Review of Psychology, 57,* 345–374.

Fisher, D.L., Pradhan, A.K., Pollatsek, A., & Knodler, M.A. (2007). Empirical evaluation of hazard anticipation behaviors in the field and on driving simulator using eye tracker. *Transportation Research Record, 2018,* 80–86.

Foss, R.D., & Goodwin, A.H. (2014). Distracted driver behaviors and distracting conditions among adolescent drivers: Findings from a naturalistic study. *Journal of Adolescent Health, 54,* 550–560.

Gardner, M., & Steinberg, L. (2005). Peer influence on risk taking, risk preference, and risky decision making in adolescence and adulthood: An experimental study. *Developmental Psychology, 41,* 625–635.

Ginsberg, K.R., Winston, F.K., Senserrick, T.M., Garcia-Espana, F., Kinsman, S., Quistberg, D.A., Ross, J.G., & Elliott, M.R. (2008). National young-driver survey: Teen perspective and experience with factors that affect driving safety. *Pediatrics, 121,* e1391–e1403.

Heck, K.E., & Carlos, R.M. (2008). Passenger distractions among adolescent drivers. *Journal of Safety Research, 39,* 437–443.

Horrey, W.J., & Wickens, C.D. (2006). Examining the impact of cell phone conversations on driving using meta-analytic techniques. *Human Factors, 48*(1), 196–205.

Huisingh, C., Griffin, R., & McGwin, G. (2015). The prevalence of distraction among passenger vehicle drivers: A roadside observational approach. *Traffic Injury Prevention, 16*(2), 140–146.

Insurance Institute for Highway Safety. (2014). *Effective Dates of Graduated Licensing Laws.* May 2014. Retrieved from http://www.iihs.org/iihs/topics/laws/graduatedlicenseintro?topicName=teenagers.

Kahneman, D. (2011). *Thinking, Fast and Slow.* New York: Farra, Straus, and Giroux.

Lambert, A.E., Simons-Morton, B.G., Cain, S.A., Weisz, S., & Cox, D.J. (2014). Considerations of a dual-systems model of cognitive development and risky driving. *Journal of Research on Adolescence, 24*(3), 541–550.

McKenna, F.P., Waylen, A.E., & Burkes, M.E. (1998). Male and female drivers: How different are they? Reading, UK: The University of Reading, AA Foundation for Road Safety Research.

Mirman, J., Albert, D., Jacobsohn, L.S., & Winston, F.K. (2012). Factors associated with adolescents' propensity to drive with multiple passengers and to engage in risky driving behaviors. *Journal of Adolescent Health, 50*(6), 634–640.

Ouimet, M.C., Simons-Morton, B.G., Zador, P.L., Lerner, N.D., Freedman, M., Duncan, G.D., & Wang, J. (2010). Using the U.S. National Household Travel Survey to estimate the impact of passenger characteristics on young drivers' relative risk of fatal crash involvement. *Accident Analysis & Prevention, 42*, 689–694.

Ouimet, M.C., Duffy, C.W., Simons-Morton, B.G., Brown, T.G., & Fisher, D.L. (2011). Understanding and changing the young driver problem: A systematic review of randomized controlled trials conducted with driving simulation. In D.L. Fisher, M. Rizzo, J.K. Caird, & J.D. Lee (Eds.), *Handbook of Driving Simulation for Engineering, Medicine and Psychology* (pp. 24.1–24.16). Boca Raton, FL: CRC Press.

Ouimet, M.C., Pradhan, A.K., Simons-Morton, B.G., Divekar, G., Mehranian, H., & Fisher, D.L. (2013). The effect of male teenage passengers on male teenage drivers: Findings from a driving simulator study. *Accident Analysis & Prevention, 58*, 132–139.

Ouimet, M.C., Pradhan, A.K., Brooks-Russell, A., Ehsani, J.P., Berbiche, D., & Simons-Morton, B.G. (2015). Young drivers and their passengers: A systematic review of epidemiological studies on crash risk. *Journal of Adolescent Health, 57*, S24–S35.

Pradhan, A.K., Li, K., Bingham, C.R., Simons-Morton, B.G., Ouimet, M.C., & Shope, J.T. (2014). Peer passenger influences on male adolescent drivers' visual scanning behavior during simulated driving. *Journal of Adolescent Health, 54*, S42–S49.

Prat, F., Planes, M., Gras, M.E., & Sullman, M.J. (2015). An observational study of driving distractions on urban roads in Spain. *Accident Analysis & Prevention, 74*, 8–16.

Prato, C.G., Toledo, T., Lotan, T., & Taubman-Ben-Ari, O. (2009). Modeling the behavior of novice young drivers during the first year after licensure. *Accident Analysis & Prevention, 42*(2), 480–486.

Regan, M.A., & Mitsopoulos, E. (2003). Understanding passenger influences on driver behaviour: Implications for road safety and recommendations for countermeasure development (Report #180). Victoria, Australia: Monash University Accident Research Centre.

Rivardo, M.G., Pacella, M.L., & Klein, B.A. (2008). Simulated driving performance is worse with a passenger than a simulated cellular telephone converser. *North American Journal of Psychology, 10*(2), 265–276.

Romer, D. (2010). Adolescent risk taking, impulsivity, and brain development: Implications for prevention. *Developmental Psychobiology, 52*, 263–276.

Shepherd, J.L., Lane, D.J., Tapscott, R.L., & Gentile, D.A. (2011). Susceptible to social influence: Risky "driving" in response to peer pressure. *Journal of Applied Social Psychology, 41*(4), 773–797.

Simons-Morton, B.G. (2007). Social influences on adolescent substance use. *American Journal of Health Behavior, 6*, 672–684.

Simons-Morton, B.G., & Chen, R.S. (2006). Over time relationships between early adolescent and peer substance use. *Addictive Behaviors, 31*(7), 1211–1223.

Simons-Morton, B.G., & Farhat, T. (2010). Recent findings on peer group influences on adolescent smoking. *Journal of Primary Prevention, 31*, 191–208.

Simons-Morton, B., Lerner, N., & Singer, J. (2005). The observed effects of teenage passengers on the risky driving behavior of teenage drivers. *Accident Analysis & Prevention, 37*, 973–982.

Simons-Morton, B.G., Hartos, J., Leaf, W., & Preusser, D. (2006). Increasing parent limits on novice young drivers: Cognitive mediation of the effect of persuasive messages. *Journal of Adolescent Research, 21*(1), 83–105.

Simons-Morton, B.G., Ouimet, M.C., Zhang, Z., Klauer, S.E., Lee, S.E., Wang, J., Chen, R., Albert, P., & Dingus, T.A. (2011a). The effect of passengers and risk-taking friends on risky driving and crashes/near crashes among novice teenagers. *Journal of Adolescent Health, 49*(6), 587–593.

Simons-Morton, B.G., Ouimet, M.C., Zhang, Z., Lee, S.L., Klauer, S.E., Wang, J., Chen, R., Albert, P.E., & Dingus, T.E. (2011b). Crash and risky driving involvement among novice adolescent drivers and their parents. *American Journal of Public Health, 101*(12), 2362–2367.

Simons-Morton, B.G., Ouimet, M.C., Chen, R., Klauer, S.G., Lee, S.E., Wang, J., & Dingus, T.A. (2012). Peer influence predicts speeding prevalence among teenage drivers. *Journal of Safety Research, 43*(5–6), 397–403.

Simons-Morton B.G., Bingham, C.R., Ouimet, M.C., Pradhan, A.K., Chen, R., Barretto, A., & Shope, J.T. (2013). The effect on teenage risky driving of feedback from a safety monitoring system: A randomized controlled trial. *Journal of Adolescent Health, 53*, 21–26.

Simons-Morton, B.G., Bingham, C.R., Falk, E.B., Li, K., Pradhan, A.K., Ouimet, M.C., Almani, F., & Shope, J.T. (2014). Experimental effects of injunctive norms on simulated risky driving among teenage males. *Health Psychology, 33*(7), 616–627.

Steinberg, L. (2008). A social neuroscience perspective on adolescent risk-taking. *Developmental Review, 28*(1), 78–106.

Taubman-Ben-Ari, O. (2012). The effects of positive emotion priming on self-reported reckless driving. *Accident Analysis & Prevention, 45*, 718–725.

Toxopeus, R., Ramkhalawansingh, R., & Trick, L. (2011). The influence of passenger–driver interaction on young drivers. In: Proceedings of the Sixth International Driving Symposium on Human Factors in Driver Assessment, Training, and Vehicle Design, Lake Tahoe, CA, pp. 66–72.

Ulleberg, P. (2004). Social influence from the back-seat: Factors related to adolescent passengers' willingness to address unsafe drivers. *Transportation Research Part F, 7*, 17–30.

White, C.B., & Caird, J.K. (2010). The blind date: The effects of change blindness, passenger conversation and gender on looked-but-failed-to-see (LBFTS) errors. *Accident Analysis & Prevention, 42*(6), 1822–1830.

Wilson, D.S. (2007). *Evolution for Everyone: How Darwin's Theory Can Change the Way We Think about Our Lives*. New York: Delacorte Press.

V

Licensing, Training, and Education

17

Addressing Young Driver Crash Risk through Licensing Policies

Robert D. Foss

Abstract

The Problem. Teenagers have higher crash rates than older, more experienced drivers. This primarily reflects the fact that learning to drive well takes time—years, not weeks or months—and teenagers are, by definition, less experienced. Despite efforts involving formal driver education, license testing, and traditional law enforcement, the high crash rate of young novice drivers has remained for decades. ***Graduated Driver Licensing***. In a new approach to dealing with the high teen crash rate, graduated driver licensing (GDL) has been adopted in the United States. Unlike previous efforts, GDL is based on several scientific principles as well as an understanding of the reasons that young novices crash so often. ***Key Results***. Numerous single- and multiple-state studies have found that GDL reduces crash rates among high school-age drivers. It does so largely by limiting the exposure of young drivers to high-risk conditions as they are learning, thereby offsetting the danger created by their minimally developed driving wisdom in the early months of driving.

17.1 Introduction

In 2015, teen drivers, the crashes they have, and programs meant to reduce these crashes were continually in the news headlines in the United States. Although young driver crash risks have been a concern for several years, they have not always received this amount of attention. Prior to the mid-1990s, teen driver crashes were not a major concern of the public, policy makers, or the research community. It was generally assumed that driver education programs dealt with the needs of young beginning drivers. This complacency quickly eroded in the 1990s as researchers and traffic safety practitioners began pointing to the high crash rate of young teen drivers and making the case for a different approach to licensing young drivers. This "new" approach—graduated driver licensing (GDL)—had been developed two decades earlier (Waller, 1976; Waller & Reinfurt, 1973). However, policy makers cited a variety of concerns and declined to embrace the new approach (Waller, 2003). The idea was resurrected in the early 1990s as researchers began looking more closely at young driver crash rates and sharing the troubling findings with road safety practitioners and policy makers. This chapter addresses young driver licensing in the United States during the past two decades, focusing heavily on GDL and its effects.

17.1.1 What Is a Teen Driver?

It's important to be clear what we mean by *teen driver.* Although any driver between the ages of 15 and 19 is technically a teen driver, we generally have more than merely an age range in mind when we use the term. The connotation of *teen driver* often involves assumptions about youthful impulsiveness, irresponsibility, or risk taking. However, as is explained, lack of driving experience is the more important issue than these assumed dispositions of adolescents in explaining their high crash rates. Accordingly, references to teen drivers here should be understood to mean drivers younger than 18—individuals who are both young and relatively inexperienced. Crash causation among more experienced drivers, even though they may still be younger than 20, is largely beyond the influence of licensing policy and is not addressed here.

17.2 Teenage Driver Crashes and Licensing

17.2.1 Teenage Driver Crash Counts and Rates

To understand the magnitude of the teenage driver crash problem, and how it changes over time, we need to examine rates rather than absolute counts. The reason for this is that exposure is a critically important factor in motor vehicle crashes. Regardless of their abilities, those who drive more and drive more often in high-risk conditions are more likely to crash. To get the clearest possible look at the crash risk resulting from individuals' driving behavior (both ability and style), we need to examine data with the effects of exposure removed (or at least minimized). We can do this by looking at crash rates—crash counts divided by some number that measures exposure. Figure 17.1 shows different ways of looking at teenage driver crashes. Both panels correctly portray information about crashes, but panel A provides a more useful look when—as is the case here—our purpose is to focus on how often individuals crash because of their own careless behavior or lack of skill. Panel A shows the likelihood of a crash, by age, per mile driven. This excludes those who don't drive at all and adjusts for the fact that driving more increases the chances of a crash, having little or nothing to do with one's ability. Panel B shows the proportion of the entire population of a given age who are involved in a crash as a driver. The patterns shown in the figure are quite different because the rates displayed reflect different sets of influences on crashes. Panel A shows that among teens who have begun driving, 16-year-olds are far more likely to crash than those who are even 1 year older. Here, it looks like the "teen driver problem" is mostly a matter of 16- and 17-year-olds. Panel B, on the other hand, seems to suggest that 18- and 19-year-olds are the larger part of

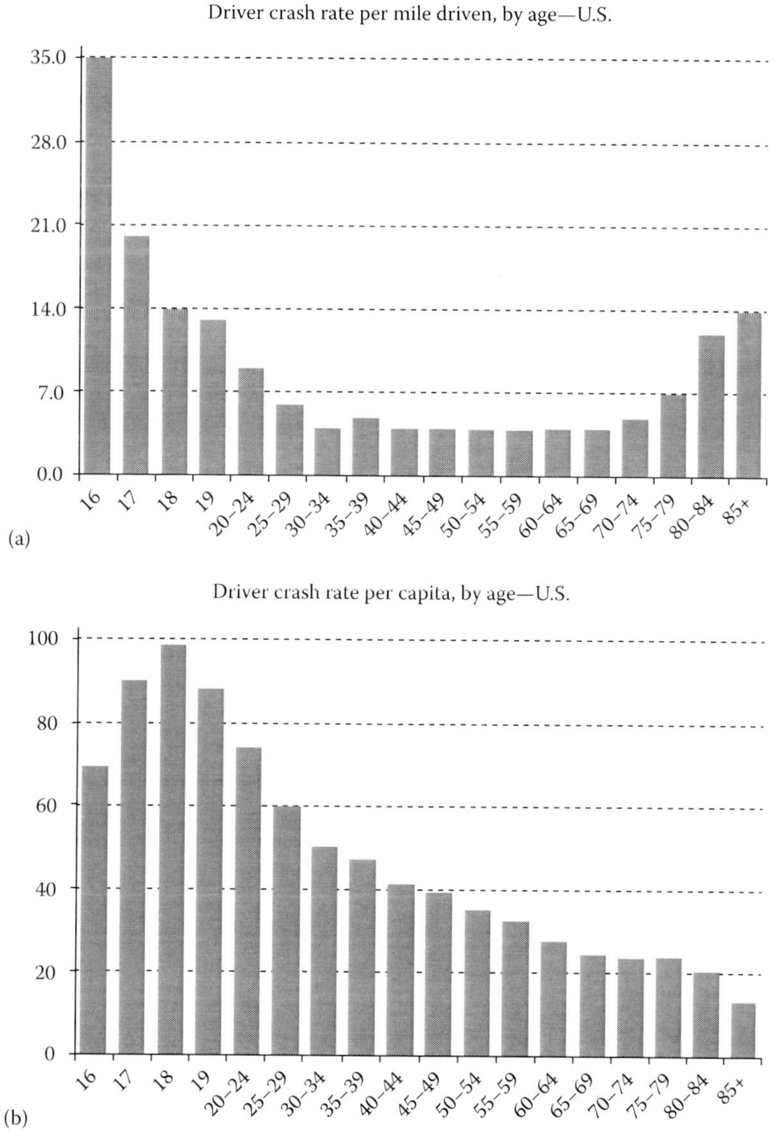

FIGURE 17.1 Crash involvement rates by driver age. (a) Rates per million miles driven. (b) Rates per 1000 population. (Adapted from Williams, A. F., *J. Safety Res.*, *34*(1), 5–15, 2003.)

the problem. These older teens crash more often because more of them are licensed and they drive substantially more than younger teenagers. Even though they are, on average, more experienced and mature, their resulting lower risk of crashing per mile driven is offset by the fact that they drive more, leading to higher per capita crash rates.

Although the data in Figure 17.1 are now more than a decade old (Williams, 2003), the age-related patterns they illustrate change slowly, if at all, over time. Portraying crash rates, rather than simply crash counts, highlights the two most central contributing factors in crashes—driver ability/demeanor and driving exposure. The illustration in panel A, showing the dramatic difference between 16- and 17-year-olds, signals the need to address teen driver crashes not as a matter of youthfulness (age), but as a matter of inexperience. Viewing the evidence this way clearly indicates the wrongheadedness of the common assumption that teenagers crash so often because they are careless risk-takers. It becomes clear that successfully reducing teen driver crashes requires efforts to address inexperience rather than focusing only on presumed misbehavior. GDL was designed to do just this (cf. Waller, 2003). The following section provides a brief discussion of the context within which GDL was designed to operate in the United States.

17.3 Driver Licensing in the United States

17.3.1 Driving Begins at an Early Age

The United States has traditionally allowed teenagers to begin driving at younger ages than is the case in most other countries. Currently, the minimum age to drive (without adult accompaniment) is 16 in most US states. Six mostly rural states allow driving at age 15½ or younger, and nine states require novices to be 16½ or older. In contrast, minimum licensing ages are higher throughout Europe (generally 18) and Australia (mostly 17, but 18 in Victoria) and in several Canadian provinces (16 in five provinces, at least 16 and 8 months in seven others). Moreover, both the licensing process—including training—and driving (gasoline costs in particular) are far less costly in the United States than in Europe. In combination, these factors have given rise to a substantial number of teenagers in the United States beginning the complex process of driving while they are still progressing through a series of personal, social, emotional, and biological developmental processes characteristic of adolescence (Keating, 2007; see also Chapters 9 and 10 in this volume).

Certain aspects of social, emotional, and neurocognitive development that occur during of adolescence can lead to behaviors that are inconsistent with the judicious caution needed to be the safest possible driver. For example, adolescents (ages 13–16) and youths (ages 18–22) are more inclined than adults to engage in risky behaviors (see Chapters 9–11 in this volume for elaboration on these matters). This tendency is exacerbated more for teenagers than adults by the mere presence of age peers (Gardner & Steinberg, 2005). Similarly, teenagers have long been known to be more impulsive or less deliberative and controlled than adults. Recent research suggests that this results from the differential rates at which various neurobiological systems develop. In particular, Steinberg (2007) has suggested that the tendency of adolescents to engage in impulsive, seemingly irrational behavior results not from their failure to know or care about the associated risks but, rather, from an underdeveloped ability to act in a way consistent with what they often know to be the better option.

Teenagers are frequently described by exasperated adults as "thinking they are invincible," to explain the many risky behaviors that naturally occur as an inescapable—and inescapably worrisome—feature of adolescence. However, there is virtually no evidence that teenagers actually believe they are invincible (Reyna & Farley, 2006). To the contrary, adolescents are more likely than adults to believe they are vulnerable to the vicissitudes of life. There is a critically important lesson here. Adolescents do engage in a range of risky behaviors, when driving and elsewhere, which give the appearance to adults of assumed invincibility. The flaw is in the assumption that adolescent behavior—in fact, the behavior of humans regardless of age—is controlled largely by deliberate, rational consideration. That is an outdated, demonstrably incorrect notion of how humans operate. Human behavior is far more complexly determined than is broadly assumed, and the role of rational consideration of possible costs and benefits of various actions is minimal in most behaviors (Kahneman, 2011).

17.3.2 The Potential and Need for Parent Involvement in Licensing

In addition to bringing the added complications of mixing beginning to drive with the turmoil of adolescence, the fact that driving begins relatively early in the United States raised the prospect that parents might be actively included in the licensing process (Keating, 2007; Simons-Morton & Ouimet, 2006;

Chapter 19). There appeared to be several potential benefits of extensive parental involvement as teenagers began driving. Parents almost always have greater access to, and influence over, teenagers than anyone outside the home. They are probably more motived to ensure that their children become competent and safe drivers as well. Parents can assist teens in learning the rudiments of driving by riding along and mentoring them as they practice. They can provide a second set of (wise, experienced) eyes that help avoid catastrophic consequences of the errors that novice drivers inevitably will make as they learn. Parents can also delay licensing or limit access to a vehicle until their children are sufficiently skilled and well-enough acquainted with the finer points to be good, safe drivers and not merely minimally competent.

17.4 Policy Approaches to Teenage Driver Safety

Long-standing approaches to addressing the crash risks of teenage drivers have focused on education about driving, testing knowledge, and punishment for transgressions. The high crash rates among teenage drivers suggest that these approaches have not worked particularly well. Consequently, they have given way during the past 20 years to the substantially different approach embodied in GDL, which is more in tune with principles of human learning, the nature of driving, and the reasons for the high crash rates seen among teenagers in the United States.

17.4.1 Education, Testing, and Punishment

17.4.1.1 Requiring Driver Education

Historically states have required individuals to pass a formal driver education class if they wish to be licensed before age 18. This is based on the reasonable supposition that individuals should have some formal preparation for a task as complex and potentially dangerous as driving before they are allowed to share the road with others. It turns out, however, that formal driver education as envisioned and delivered in the United States has never been shown to have any clear safety benefits (for summaries of this work, see Nichols, 2003; Williams et al., 2009; Chapter 18 in this volume). Following a large, costly, intensive study of a "supercharged" driver education program in the late 1970s, which failed to find evidence of notable benefits (Stock et al., 1983), many states began withdrawing financial support for driver education. For this reason, along with a move to emphasize core educational components, driver education was removed from the regular high school curriculum and came to be taught outside regular school hours. Nonetheless, driver education remains a requirement in most states for those who wish to obtain a license before age 18.

The general public continues to believe that completing a driver education class improves teenage driving abilities and behaviors, despite the large number of studies that have documented little or no safety effect. This belief reflects an outdated understanding of what safe driving entails and, especially, of how humans learn. Driver education in the United States has always been approached as a short course, involving about 30 hours of classroom time and 6 of actual on-road driving. Although this approach can provide a basic introduction to driving regulations, some facts, and general notions, it is unrealistic to think that an individual who knows nothing about driving could be transformed into a relatively competent driver with this minimal "dose" of passive learning and a mere 6 hours of actual driving practice.

A fundamental shortcoming of the traditional driver education model in the United States is that it rests on an outdated understanding of how humans learn and incorporate newly gained understanding into their behavioral repertoires. In brief, we learn how to do things largely by doing them, whether it's acquiring a second language, learning to paint, playing a musical instrument, engaging in a team sport, or even such a seemingly cerebral activity as understanding principles of physics. Currently, in formal driver education, individuals are largely *taught about* driving, rather than helped to *learn driving.*

Learning to drive, like learning a sport, requires being able to apply knowledge. Accordingly, if it is to succeed in helping novices learn, driver education should more closely resemble basketball, hockey, or soccer practice than a history classroom. Knowing rules of the road and what to do in various settings is important. But this alone does not make for a competent driver any more than it can produce a competent amateur athlete. Modern approaches to education—widely known as the "flipped classroom" (Bergmann & Sams, 2012)—regarding everything from writing to physics eschew the near-exclusive focus on factual learning. Instead, they concentrate on practicing principles that have been studied outside the classroom, recognizing that this is how the human brain consolidates knowledge (Morowitz & Singer, 1995). Until a similar transformation occurs in driver education, it will continue to produce individuals who know some things about driving but who are no more capable than those who have not taken a class.

17.4.1.2 Licensing Tests

As is the case with formal driver education, testing of drivers before issuing a license makes sense as a way to ensure that unqualified individuals are not allowed to drive. However, the amount and type of testing that are logistically feasible for licensing agencies in the United States to conduct are largely incapable of achieving the goal. This is the case for two reasons. First, testing can only focus on basic matters like knowing facts about driving, rules of the road, and being able to maneuver a vehicle with some degree of competence. Although these may be important, requiring drivers to pass tests of these abilities prevents only completely incompetent drivers from obtaining a license. Second, testing can only assess what drivers know, not what they will do. Virtually all drivers who take risks like speeding, texting, or driving after drinking know they should not do so, but they do it anyway. Testing cannot address such behavioral inclinations.

17.4.1.3 Law Enforcement and Threats of Punishment

Traffic law enforcement plays a central role in facilitating safe driving, but this role is often misunderstood. Traffic laws and their enforcement are meant to control extreme behaviors that are so dangerous that they have been declared illegal. These laws do not actually regulate the behavior of most drivers, who generally drive in a reasonably safe manner. Only a small subset of the driving public will drive after drinking, speed excessively, run red lights, or do other egregiously risky things. Law enforcement is meant to discourage such actions by drivers who might engage in them in the absence of well-enforced laws. Hence, there is a clear but limited role for law enforcement in controlling the excessive behaviors that some novice teenage drivers might consider. But such behaviors have relatively little to do with the higher crash risk of teenage drivers. Simply obeying traffic laws does not make for a competent, safe driver. Beyond violations, drivers do other things that are risky as well (Parker et al., 1995). As Reason et al. (1990) explain, numerous driving behaviors deviate from ideal. These range from minor slips, like taking a wrong turn; to mistakes such as misjudging the speed of oncoming traffic as one is about to make a left turn; to unintentional violations like exceeding the speed limit without realizing it. Similarly, the things that appear most important in the high crash rates of novice teenage drivers result more from what they don't yet know or understand than from deliberate misbehavior. In an effort to explain the high initial crash rate of novice teen drivers, McKnight and McKnight (2003) examined reports of 2000 nonfatal crashes involving teen drivers, looking at driver behaviors that led to the crashes. They found that the large majority of crashes "resulted from errors in attention, visual search, speed relative to conditions, hazard recognition, and emergency maneuvers, with high speeds and patently risky behavior accounting for but a small minority." In explaining the rationale for graduated licensing, Waller noted that it's unrealistic to believe that law enforcement can play a key role in reducing the young novice driver crash problem, because most young teen crashes result from unintentional mistakes, reflecting inexperience rather than misbehavior. There is nothing in the literature on learning to suggest that such mistakes can be influenced by threats of punishment.

17.4.2 Graduated Driver Licensing

Waller proposed adoption of a GDL system for new, young teenage drivers more than 40 years ago (Waller & Reinfurt, 1973). GDL differs notably from the approaches discussed previously. It is designed to directly address the primary reason for crashes among the youngest drivers (insufficient understanding resulting from inexperience). It does so based on the way humans learn to perform knowledge- and skill-based tasks (practice). Finally, it embodies the well-established principle that humans respond far more to policies than to information about risk, or admonitions to behave in a particular way (Etzioni, 1972).

GDL is structured to ensure that novices obtain extensive experience actually driving—the only way to learn—while simultaneously protecting them from the risks that go hand in hand with their limited understanding (Foss, 2007). As they gain experience, the conditions and limits designed to reduce their exposure to risk are progressively removed. Although the concept of GDL was laid out in the early 1970s, it was not adopted, with one limited exception (in Maryland in 1979 [see McKnight et al., 1983]), until 1987, when New Zealand implemented the first full GDL system. Evidence that this produced clear benefits (Frith & Perkins, 1992; Langley et al., 1996) ushered in the "modern" GDL era, beginning in the mid-1990s in the United States.

17.5 The Modern GDL Era

17.5.1 GDL Structure, Function, and Rationale

The essence of GDL is its structure as a multistage process that encourages novices to obtain substantial practical driving experience under conditions that minimize the riskiness of this practice as novices learn. The standard GDL system involves three license stages. In principle, progression through these stages is based on the accumulation of experience rather than merely reaching a certain age. However, all states have set a minimum age at which individuals can begin the licensing process, and some still apply the conditions of different stages based partly on age, rather than experience.

17.5.1.1 Stage 1—The Learner License

The first stage of GDL is an extended, mandatory learner period during which beginning drivers are allowed to drive only when accompanied by an experienced adult driver. In most US states, this stage lasts 6 months, even though we now know that that is not long enough for beginners to acquire the amount and variety of practice needed. Busy schedules of both teens and potential supervisors limit the amount of driving practice most beginners can realistically be expected to accomplish to less than 2 hours a week. Moreover, a 6-month learner stage allows individuals to become licensed to drive without ever driving during certain seasons and acquiring no experience, for example, with driving in the hazardous winter driving conditions found in many states. Currently, eight US states' GDL systems have a 12-month learner stage; three others require 9 months of supervised driving. Longer learner periods produce greater crash reductions among 16- and 17-year-old drivers, and some of this appears to be the direct result of improved learning (Masten & Foss, 2010).

The purpose of the mandatory learner stage is to provide ample opportunity for beginners to practice driving in real-world conditions so they can learn by doing. The purpose of the adult supervisor is to protect against risks resulting from the driver's inexperience and lack of deep understanding of driving. The adult can anticipate problems, see developing risks the novice won't, and provide warnings, or guidance if needed. The presence of an adult also discourages the kinds of ebullient, impulsive, and often risky actions that adolescents are prone to engage in (Steinberg, 2007). Supervisors cannot be amateur driving instructors and are not expected to be. However, they do have the opportunity to share their own well-developed wisdom about driving, serving as a mentor or coach for a new driver.

Incorporating experienced adult drivers—usually parents—into the licensing process is part of the genius of GDL. By comparison with professional driving instructors, adults are plentiful, are inexpensive (usually free), and generally know the teen better than anyone. Although they may not be the best of drivers, nearly all are good enough to avoid most collisions, and having driven for years, they certainly know far more about driving than the novices they accompany and mentor. Even though some of them engage in risky driving on occasion, they know they should not and are unlikely to encourage a novice teenager to drive in a similarly risky manner.

In efforts to ensure that learner drivers obtain enough driving practice, nearly all states mandate that beginners drive a certain number of hours while accompanied by an adult before progressing to the next licensing level. Some states attempt to ensure that practice actually takes place, requiring a driving log or certifying affidavit to be submitted to the licensing agency. Though these efforts to ensure that parents provide teens with a sufficient amount practice are far from perfect, there is some evidence that when done well, they are somewhat helpful (O'Brien et al., 2013).

The amount of driving practice required by learners varies widely across states, currently ranging from 0 to 70 hours, with a mode of 50. To encourage experience with driving in darkness, most states specify some nighttime driving (typically 10 hours) as a part of the requirement. Although these efforts to ensure adequate practice are well intentioned, the specified thresholds are based on no evidence about the amount of driving actually needed to reach some level of competence. Moreover, the required amounts of practice seem to be poorly publicized. A recent study found that many parents of recently licensed teen drivers didn't know, or remember, that there was an hourly requirement, and among those who did, a substantial proportion didn't know how many hours were required (O'Brien et al., 2013).

17.5.1.2 Stage 2—The Intermediate License

During the intermediate, or provisional, license stage, young novices are allowed to begin driving without adult accompaniment. However, driving in conditions known to be particularly risky (for teenagers) continues to require adult accompaniment. Most states (currently 44) limit drivers during this stage to carrying no more than one young passenger, unless there is also an adult in the vehicle. With one exception, all states also limit nighttime driving for intermediate licensees. Limits on both passengers and night driving are based on clear evidence that driving under those conditions is more risky for high school-age teens (Chen et al., 2000).

The logic for requiring an intermediate licensing stage, with limits on unaccompanied driving in the most risky conditions, is that although individuals who have driven for many months with an adult will have learned a lot, they won't yet be fully accomplished drivers. Consequently, they need continued protection from the full range of risks associated with driving as they continue to learn. In most cases, new drivers are able to acquire only several dozen hours of driving experience during the learner stage. Consequently, although they are no longer rank amateurs, they remain highly inexperienced when they move to the intermediate license stage. Up to this point, they have had no opportunity to learn how to handle being completely in charge of a vehicle, as they have only shared this responsibility with an adult "codriver." Driving with an adult or parent in the vehicle is qualitatively different from driving alone or with age peers. Not only do adolescents need experience being fully in charge—under relatively benign conditions—to round out learning to drive; they also need the opportunity to develop self-control as drivers. These can only be learned by driving without adult accompaniment.

As is the case for the learner stage, many states have not structured their GDL systems in a manner that ensures the maximum practice under the safest practical conditions. Most US GDL systems currently prohibit driving after midnight by intermediate drivers, reflecting a misunderstanding of night-driving risks for high school-age teens. It has long been known that fatal crash risks increase sharply well before midnight, nearly tripling by 10 p.m. (Chen et al., 2000). Moreover, the large majority of nighttime crashes involving high school-age teens occur well before midnight, because only a tiny fraction of driving by drivers this age takes place after midnight. Hence, restricting only late-night driving does little to protect novices from the risks of night driving.

17.5.1.3 Stage 3—The Full License

The third step in the GDL process is a full and essentially unrestricted license. This is the license that young beginners were able to obtain in the years before GDL was adopted merely by passing simple written and driving tests, with little or no driving practice. Some age-based restrictions remain in effect for drivers with a full license. The most notable and widespread of these is the prohibition on driving after drinking any alcohol until the age of 21, which applies in all 50 states and the District of Columbia.

17.5.1.4 Age-Based versus Experienced-Based Licensing

Unfortunately, in practice, the intent of GDL to make licensing levels contingent on increasing experience is not currently implemented as faithfully as one might hope. Too many jurisdictions continue to base progression through the system on attainment of a particular age rather than accumulation of experience. The most notable example of this is that in nearly all US states, GDL requirements no longer apply when drivers reach age 18. In a few states, GDL applies only to 15- or 16-year-olds. The wisdom of setting such arbitrary age cutoffs, with little or no scientific basis, has begun to draw the attention of researchers (Curry et al., 2014; Foss et al. 2014; Tefft et al., 2014) as well as policy makers.

17.5.2 Evaluations of GDL Programs

Dozens of studies have documented a decrease in 16-year-old driver crashes following introduction of a GDL system (summarized in Shope, 2007; Williams et al., 2012). Far fewer have found reductions among 17-year-old drivers. As noted earlier, there are two ways in which GDL might reduce crashes. The careful effort to build experience into the licensing process should make for better drivers, but the GDL process can also reduce crashes by limiting the total amount of driving, as well as the amount of driving in more risky conditions, by young beginners.

It turns out that the majority of the crash reductions produced by GDL result from limiting exposure of the youngest teens (Karaca-Mandic & Ridgeway, 2010). Drivers with learner licenses rarely crash, owing to the presence of an experienced adult driver in the vehicle. GDL reduces both the amount and the riskiness of driving by novice teenage drivers by virtue of the required lengthy learner period. With GDL in effect, fewer 16-year-olds are eligible to drive unaccompanied by an adult than was previously the case. The percentage reduction in crashes among teenage drivers is routinely found to be nearly equivalent to the reduction in the percentage licensed to drive solo.

Two studies have attempted to directly examine the extent to which GDL appears to have improved the driving abilities of novices (Foss et al., 2014; Masten & Foss, 2010). Both were conducted in North Carolina, which has one of the most comprehensive GDL systems in the United States (with a 12-month learner period for beginners, followed by limits on carrying more than one young passenger and driving after 9 p.m. for intermediate licensees). These studies each found evidence for a quite small (~5%) increase in the ability to avoid crashing among drivers licensed through the GDL system.

The differing findings about the effectiveness of GDL in reducing 16- versus 17-year-old crashes are explained by the fact that the primary mechanism by which it affects crashes is exposure reduction. In states that mandate only 6 months of supervised driving, a large proportion of 17-year-old drivers have progressed beyond the exposure-limiting elements of GDL and stand to benefit only from the apparently quite small improvements in their driving ability. The states where GDL has clearly been found to reduce 17-year-old driver crashes are those that mandate 12 months of supervised driving (Foss et al., 2001; Rios et al., 2006). The additional 6 months that must be spent in the learner license period means that many more 17-year-olds remain subject to the risk-reducing limits of GDL.

17.5.2.1 Effects of Individual GDL Elements

Although each of the many elements of a GDL system might possibly produce a safety benefit, GDL is meant to be an integrated system with mutually supportive elements, not simply an aggregation of independent pieces, with each having its own separate effect. Because GDL systems are enacted as packages of several elements, it has been largely impossible to determine how much the individual components have contributed to the overall effect. Nonetheless, the question of the relative influence of various pieces remains of interest to some, with the belief that this information might help to construct the ideal GDL system.

Some researchers have conducted sophisticated multivariable modeling analyses in an effort to estimate the independent effects of GDL elements (cf. Masten et al., 2013; McCartt et al., 2010). For reasons beyond the scope of this chapter, such analyses have proved incapable of providing sound estimates of the effects individual GDL elements produce. Nonetheless, they affirm the findings of other studies that a lengthy learner period (9–12 months), a limit of one teen passenger, and a night-driving limit beginning no later than 10 p.m. appear to reliably reduce fatal crash involvement. There is also suggestive evidence that increasing the minimum ages (by 6 months) for unaccompanied and unrestricted driving reduces fatal crash involvement by drivers under age 18.

Although the goal of identifying the most useful combination and calibration of GDL elements remains elusive, it is readily apparent that those elements that reduce either overall driving exposure or exposure to high-risk driving conditions are beneficial. This provides clear, principled (and not particularly surprising) guidance for structuring the most effective GDL system: those elements that effectively reduce exposure will reduce crashes among drivers to whom they apply.

17.6 The Future

Despite the adoption of GDL in most US states, teenage driver crashes remain all too common. As a result, there are questions about what can and should be done to further reduce this problem. A few promising options are briefly described here.

17.6.1 Taking Full Advantage of Current Evidence

Although most US states have embraced GDL in principle, nearly 80% still do not have adequately calibrated learner periods (they're too short) or night-driving limits (they begin too late in the evening). It is unquestionable that additional reductions in novice teenage driver crashes can be achieved by appropriately revising these states' licensing systems to reflect a current understanding of GDL, how it works, and what is needed for maximum effectiveness. No additional research is needed. Evidence-based recommendations simply need to be implemented by policy makers.

17.6.2 Engaging Parents More Effectively

It remains unclear whether parents are doing all they might to maximize their teens' learning during the initial GDL licensing stage. Recent research suggests that they are not doing what we would hope (Goodwin et al., 2014), but it is unclear how much can realistically be expected of them. This is currently a fertile area of research, with a good deal of effort examining ways to enlighten, assist, or motivate parents. Although early findings are somewhat discouraging (Curry et al., 2015), a good deal more will be known about this matter in another 5 years.

17.6.3 Extending the GDL Process to All Novice Drivers

In view of the fact that all novice drivers have higher crash rates early in their driving careers, regardless of the age at which they begin, it makes little sense to limit GDL to beginners under age 18. The United States is unique in having done so. Most jurisdictions outside the United States that have adopted GDL require all novices, or at least those up to about 25, to proceed through a graduated system. These differences reflect somewhat differing visions of GDL. In the United States, it has been thought of largely as a public health effort to ensure the physical well-being of adolescents, hence the heavy reliance on parents and the focus on limiting exposure to conditions that are particularly risky to high school-age teens. Outside the United States, GDL has been thought of and implemented more as a road safety matter, with greater reliance on more traditional law enforcement to ensure driver adherence to regulations. Although the US focus only on younger teens seems dubious, evidence on the effectiveness of GDL systems outside the United States is not sufficiently persuasive to suggest that the "international" approach is productive. Research on, and discussion of, the relative merits of these approaches has increased recently, and clearer answers are expected within the next few years.

References

Bergmann, J., & Sams, A. (2012). *Flip Your Classroom: Reach Every Student in Every Class Every Day.* Arlington, VA: International Society for Technology in Education.

Chen, L.-H., Baker, S. P., Braver E. R., & Li, G. (2000). Carrying passengers as a risk factor for crashes fatal to 16- and 17-year-old drivers. *JAMA*, 283, 1578–1582.

Curry, A. E., Pfeiffer, M. R., Durbin, D. R., Elliott, M. R., & Kim, K. H. (2014). Young Driver Crash Rates in New Jersey by Driving Experience, Age, and License Phase. AAA Foundation for Traffic Safety, Washington, DC.

Curry, A. E., Peek-Asa, C., Hamann, C. J., & Mirman, J. H. (2015). Effectiveness of parent-focused interventions to increase teen driver safety: A critical review. *Journal of Adolescent Health*, 57(1), S6–S14.

Etzioni, A. (1972). Human beings are not very easy to change after all. *Saturday Review*, 55(23), 45–47.

Foss, R. D. (2007). Improving graduated driver licensing systems: A conceptual approach and its implications. *Journal of Safety Research*, 38(2), 185–192.

Foss, R. D., Feaganes, J. R., & Rodgman, E. A. (2001). Initial effects of graduated driver licensing on 16-year-old driver crashes in North Carolina. *JAMA*, 286(13), 1588–1592.

Foss, R. D., Masten, S. V., & Martell, C. (2014). Examining the Safety Implications of Later Licensure: Crash Rates of Older vs. Younger Novice Drivers Before and After Graduated Driver Licensing. AAA Foundation for Traffic Safety, Washington, DC.

Frith, W. J., & Perkins, W. A. (1992). The New Zealand Graduated Licensing System. National Road Safety Seminar 2, Land Transport, Wellington, New Zealand, 256–278.

Gardner, M., & Steinberg, L. (2005). Peer influence on risk taking, risk preference, and risky decision making in adolescence and adulthood: An experimental study. *Developmental Psychology*, 41(4), 625–635.

Goodwin, A. H., Foss, R. D., Margolis, L. H., & Harrell, S. (2014). Parent comments and instruction during the first four months of supervised driving: An opportunity missed? *Accident Analysis & Prevention*, 69, 15–22.

Kahneman, D. (2011). *Thinking, Fast and Slow.* New York: Macmillan.

Karaca-Mandic, P., & Ridgeway, G. (2010). Behavioral impact of graduated driver licensing on teenage driving risk and exposure. *Journal of Health Economics*, 29(1), 48–61.

Keating, D. P. (2007). Understanding adolescent development: Implications for driving safety. *Journal of Safety Research*, 38(2), 147–157.

Langley, J. D., Wagenaar, A. C., & Begg, D. J. (1996). An evaluation of the New Zealand graduated driver licensing system. *Accident Analysis & Prevention*, 28(2), 139–146.

Masten, S. V., & Foss, R. D. (2010). Long-term effect of the North Carolina graduated driver licensing system on licensed driver crash incidence: A 5-year survival analysis. *Accident Analysis & Prevention*, 42(6), 1647–1652.

Masten, S. V., Foss, R. D., & Marshall, S. W. (2013). Graduated driver licensing program component calibrations and their association with fatal crash involvement. *Accident Analysis & Prevention*, 57, 105–113.

McCartt, A. T., Teoh, E. R., Fields, M., Braitman, K. A., & Hellinga, L. A. (2010). Graduated licensing laws and fatal crashes of teenage drivers: A national study. *Traffic Injury Prevention*, 11(3), 240–248.

McKnight, A. J., & McKnight, A. S. (2003). Young novice drivers: Careless or clueless? *Accident Analysis & Prevention*, 35(6), 921–925.

McKnight, A. J., Hyle, P., & Albrecht, L. (1983). Youth license control demonstration project. (DOT-HS-7-01765). National Technical Information Service, Springfield, VA.

Morowitz, H. J., Singer, J. L. (1995). *The Mind, the Brain, and Complex Adaptive Systems*. New York: Addison-Wesley.

Nichols, J. L. (2003). A review of the history and effectiveness of driver education and training as a traffic safety program. Presented at the NTSB Forum on Driver Education and Training, Prepared for the National Transportation Safety Board, Washington, DC.

O'Brien, N. P., Foss, R. D., Goodwin, A. H., & Masten, S. V. (2013). Supervised hours requirements in graduated driver licensing: Effectiveness and parental awareness. *Accident Analysis & Prevention*, 50, 330–335.

Parker, D., Reason, J. T., Manstead, A. S., & Stradling, S. G. (1995). Driving errors, driving violations and accident involvement. *Ergonomics*, 38(5), 1036–1048.

Reason, J., Manstead, A., Stradling, S., Baxter, J., & Campbell, K. (1990). Errors and violations on the roads: A real distinction? *Ergonomics*, 33(10–11), 1315–1332.

Reyna, V. F., & Farley, F. (2006). Is the teen brain too rational? *Scientific American Mind*, 17(6), 58–65.

Rios, A., Wald, M., Nelson, S. R., Dark, K. J., Price, M. E., & Kellermann, A. L. (2006). Impact of Georgia's teenage and adult driver responsibility act. *Annals of Emergency Medicine*, 47(4), 369.e1–369.e7.

Shope, J. T. (2007). Graduated driver licensing: Review of evaluation results since 2002. *Journal of Safety Research*, 38(2), 165–175.

Simons-Morton, B., & Ouimet, M. C. (2006). Parent involvement in novice teen driving: A review of the literature. *Injury Prevention*, 12(suppl 1), i30–i37.

Steinberg, L. (2007). Risk taking in adolescence new perspectives from brain and behavioral science. *Current Directions in Psychological Science*, 16(2), 55–59.

Stock, J. R., Weaver, J. K., Ray, H. W., Brink, J. R., & Sadoff, M. G. (1983). Evaluation of Safe Performance Secondary School Driver Education Curriculum Demonstration Project (DOT HS 806 568). Washington, DC: U.S. Department of Transportation.

Tefft, B. C., Williams, A. F., & Grabowski, J. G. (2014). Driver licensing and reasons for delaying licensure among young adults ages 18–20, United States, 2012. *Injury Epidemiology*, 1(1), 1–8.

Waller, P. F. (1976). Challenging the status quo in driver licensing. *Traffic Safety*, 76(10), 20–21.

Waller, P. F. (2003). The genesis of GDL. *Journal of Safety Research*, 34(1), 17–23.

Waller, P. F., & Reinfurt, D. R. (1973). *The Who and When of Accident Risk: Can Driver License Programs Provide Countermeasures*? Chapel Hill, NC: UNC Highway Safety Research Center.

Williams, A. F. (2003). Teenage drivers: Patterns of risk. *Journal of Safety Research*, 34(1), 5–15.

Williams, A. F., Preusser, D. F., & Ledingham, K. A. (2009). Feasibility study on evaluating driver education curriculum (DOT HS 811 108). Washington, DC: U.S. Department of Transportation.

Williams, A. F., Tefft, B. C., & Grabowski, J. G. (2012). Graduated driver licensing research, 2010–present. *Journal of Safety Research*, 43(3), 195–203.

18

The Training and Education of Novice Teen Drivers

Donald L. Fisher

Lisa Dorn

Abstract

The Chapter. Key advances have been made in recent years in the material that is taught in training programs to novice teen drivers, the method that is used to impart the training, and the medium that is used to deliver the training. ***Material.*** Unlike more general driver education programs with many objectives, training programs now exist that are focused specifically on reducing crashes by changing those behaviors known to be related to increases in risk. ***Methods.*** The new pedagogical techniques used to impart the training have in common the fact that they capture teen drivers making errors in judgment or vehicle control in both real and virtual environments, and then provide feedback to them (and sometimes parents) in real time or at some appropriate later time. ***Media.*** The new skills training programs have used platforms that are much more immersive, including in-vehicle recording systems on the open road as well as driving simulators, PCs, and nomadic devices. Training programs are also taking advantage of technological innovation in delivering interactive e-learning materials, often as part of a blended learning experience. Among those new programs that have been evaluated, many have proven effective at changing behaviors on driving simulators and in the field, some for periods of up to a year after little more than an hour of training. ***Limitations.*** (1) While most of the training programs have improved performance, performance is rarely at or near ceiling. (2) Only a handful of the studies, all with clear limitations, have examined directly the relationship between training and crashes and/or citations, making it difficult to determine just how big of an impact these new training programs will ultimately have on reducing new drivers' crash risks. (3) Very few studies have been carried out on a representative population of novice drivers, making it difficult to determine whether the great majority of training programs would work for the broad range of novice drivers in any one jurisdiction.

18.1 Introduction

Novice teen drivers are at a greatly inflated risk of crashing for the first 6–36 months of unsupervised driving. This is true worldwide, including countries such as the United States (McCartt et al., 2003), Canada (Mayhew et al., 2003), New Zealand (Lewis-Evans, 2010), and the Netherlands (Vlakveld, 2005). It has been estimated that during the first 6 months of unsupervised driving, novice teen drivers are some eight times more likely to be involved in serious crashes than are more experienced drivers (Preusser, 2003). Sadly, even after the introduction of graduated driver license programs, this is still the case (Foss et al., 2011). Not only are novice teen drivers at a greatly inflated risk of crashing, but motor vehicle crashes are the leading cause of serious morbidity and mortality for novice teens ages 15–20 (National Highway Traffic Safety Administration, 2014). The need for countermeasures to reduce this risk is undisputed as evidenced by national efforts to address this worldwide, including the United States (National Research Council, Institute of Medicine, and Transportation Research Board Program Committee for a Workshop on Contributions from the Behavioral and Social Sciences in Reducing and Preventing Crashes, 2007), Europe (Hatakka et al., 2003), and several of the Arab countries (e.g., Fisher, 2015), just to name a few.

Many different countermeasures have been used to decrease the crash rate among novice drivers. They include, among others, driver education, programs for peers (Chapter 16) and parents (Chapter 19), and laws designed to reduce the exposure of novice teen drivers to risky situations, such as graduated driver licensing (GDL) programs (Chapter 17). In this chapter, we will focus on only one countermeasure, driver education, and then on only one category of driver education program, in particular, the category of driver education program that is trying to address head on the factors that can decrease most of the crashes that lead to mortality and serious morbidity (as opposed to those more general programs that also address the skills necessary to be licensed).

Driver education for young people has had a difficult history. In a review of the impact of driver education (Roberts & Kwan, 2008), it was found that it both substantially reduced the time to licensing and had a negative impact on crash involvement (Kaestner, 1968; Nichols, 1970; Stock et al., 1983; Strang et al., 1982; Wynne-Jones & Hurst, 1985). Indeed, several reviews and evaluation studies of novice driver training programs have shown that crash rate improvements are rarely found and that in most cases, these programs have been associated with higher crash rates, perhaps due to earlier licensing and, therefore, an increased exposure to risk (Beanland et al., 2013; Christie, 2001; Elvik et al., 2009; Ker et al., 2005; Langford, 2006; Roberts & Kwan, 2008; Vernick et al., 1999). Only recently has there been any evidence that driver training can actually decrease crashes (e.g., Senserrick et al., 2009; Thomas et al., 2016).

Driver education is a much broader topic, and we need to acknowledge this. Driver education broadly defined covers the entire gamut of factors that need to be learned by the novice driver in order to obtain a license (Nichols, 2003). In fact, some have argued that driver education is primarily designed to do just this, i.e., that it was never intended solely or primarily as a method for giving novice drivers the skills to reduce serious crashes. Rather, its main purpose was to provide novice drivers with the knowledge and skills that they need in order to obtain their license, which includes passing a theory and practical driving test (Williams & Ferguson, 2004). This is an important goal, but it is not the goal of the driver education programs that we will discuss.

Virtually without exception, general driving training and education curriculums focus on teaching basic vehicle-maneuvering and observational skills and neglect a wide range of competencies required for safe driving. We believe that driver education should improve both skill *and* safety levels by focusing

Mercedes Benz Driving School, in which all driver coaches are trained in coaching methodology to address the higher levels of the GDE matrix using a self-assessment tool of personality-based emotional factors that are known to be associated with novice driver crash involvement (Mercedes-Benz, 2015). Client-centered learning is now beginning to be embraced by the UK regulatory authorities, and new driving instructor examinations focus on assessing these skills.

A different approach to developing self-evaluation competencies has been investigated in Norway (Dalland, 2012), in which learning to drive is considered a lifelong process. The learner driver curriculum is based on the GDE matrix, involving not just handling a vehicle in traffic but also how this is associated with a driver's motives. As such, the curriculum emphasizes self-knowledge, life values, and attitudes. Alongside practical work, including writing a log reflecting on events during practical driving lessons, a "reflecting team" discussion takes place to consider thoughts behind actions (Kjelsrud, 2010). Reflections about actions are encouraged in respect of other road users, in connection with the learner's own learning process, and internally, so that knowledge is absorbed and reflections can be independently formed in action. However, Dalland (2012) noted that while the curriculum emphasizes reflection and attitudes about actions within a constructivist paradigm, with the driver teacher being more of a coach than an instructor, the practical driving test remains rooted in the skills-based tradition.

While it can be seen that some European countries are implementing a new form of the practical driving test or client-centered learning approach to better incorporate self-assessment as one of the higher-level goals of the GDE matrix, no studies could be sourced that have attempted to evaluate the benefit of this approach on driver behavior or crash risk amongst young people, although one study has found that coaching can improve experienced drivers' driving skills and situational awareness according to assessments by trainers (Stanton et al., 2007).

18.3.2 Improving Impulse Control and Pressures to Engage in Risky Behaviors

Insight training aims to develop attitudinal and motivational competencies and to raise drivers' awareness of factors that contribute to crashes and of the potential risks when driving. An example of an intervention oriented toward developing self-awareness of risk-taking behavior is a program called Reduce Risk, Increase Student Knowledge (RRISK) in New South Wales (Australia), which targets 16-year-old students. RRISK has been running since the mid-1990s and seeks to reduce risky driving by developing or strengthening participants' resilience to pressures to engage in risky behaviors. It targets all high schools in a 1-day seminar, which includes a number of sessions using guest speakers, peer-facilitated group activities, a simulated crash scene, etc. Up-to-date information and statistics are provided regarding substance use and risk-taking behavior, including sessions on being a safer driver, risk factors, why young drivers are at a greater risk, and the most common causes of young driver crashes. A similar approach has been deployed widely in the United Kingdom since the late 1990s called Safe Drive Stay Alive (SDSA) (Fernandez-Medina et al., 2015).

Senserrick et al. (2009) compared the RRISK program to a 1-day road safety workshop that focused mainly on safe driving issues to address young driver crashes, fatalities, and injuries through the delivery of practical road safety education. The workshop involved community members, police, driving instructors, drug and alcohol educators, financial services advisors, and individuals recovering from a crash and aimed to develop positive attitudes toward speed, fatigue, alcohol, and seat belts. The numbers of police recorded infringements and crashes, as recorded by the NSW Police, over an average period of 2 years, were recorded. The two training groups were compared to a large cohort (n = 20,882) of newly licensed drivers who had been recruited specifically to act as a control group. Over the 2-year follow-up period, the two treatments groups did not differ from the overall cohort with regard to driving infringements (motoring offenses/moving violations). Further, there was no reduction in the crash involvement of those who had attended the road safety workshop. In contrast, the more general resilience-focused RRISK intervention was associated with a 44% reduced relative risk of crash involvement. In addition, research using self-report also found significant improvements in knowledge, attitudes, and behavior, compared to those that did not attend the program (Zask et al., 2006). The authors conclude that a program that focuses more generally on reducing risks and on building resilience has the potential to reduce crashes (Senserrick et al., 2009). However, the study used a self-selected sample, rather than a fully randomized control study, and thus the study, may suffer from self-selection bias.

Other training methods to address the higher levels of the GDE matrix have taken a different approach to tackling the young driver problem. It is well established that during the teen years, there is a tendency to seek out novel and stimulating experiences, and this can fail to be dampened by any parallel development of self-regulatory behaviors (Steinberg, 2004). In a study of learner drivers, cognitive behavior therapy was delivered to teach self-regulation and self-monitoring skills via personality constructs related to risk-taking behavior (Paaver et al., 2013). The training program consisted of a lecture and group work for 1.5 hours focusing on impulsivity and information processing style, including teaching novice drivers how to recognize impulsive tendencies in themselves and the situational factors that might trigger impulsive behavior. Participants were also asked to identify situations in which they were likely to behave in an impulsive manner or to take risks. In the group-work section of the program, participants were asked to complete a task that involved identifying the psychological factors involved in motor vehicle crashes, estimating their own risk for this kind of crash, and generating ways to decrease the risk. Crashes and traffic infringements as recorded by the police and traffic insurance databases over the 12 months following the intervention were collected. The trained group had almost half the number of traffic violations for speeding as the control group, and this difference was much larger for those with low and medium levels of impulsivity (those high in impulsivity were less affected by the training). There was no difference between the groups for crash involvement in a 12-month follow-up, but when the intervention group was compared with the control group plus participants assigned to the intervention group that did not complete the program, a decrease in the number of passive crashes (crashes where the driver was in the struck vehicle instead of the striking vehicle) was found.

In another group discussion-based method, Liu et al. (2009) reported findings from a pilot study aimed at improving communication skills between young drivers and their passengers given that young drivers have been found to be more likely to crash when carrying peer passengers (Chen et al., 2001). Training involved a 2-hour facilitated discussion workshop for young male drivers, and testing involved a simulator "drive." Compared with an untrained control group, trained participants' headway and speed choice were safer, while a higher percentage of the trained group made fewer mistakes and committed fewer violations. Trained passengers made more safe comments than did untrained passengers, confirming that young passengers can be trained to become a more positive influence using this kind of method.

These studies provide some early evidence of the potential benefits of addressing driver goals, motivation, and self-awareness on novice driver behavior, but more research is required to fully appreciate which training program components are particularly effective and how they can be improved.

18.4 Driver Education: Driving Skills

Next, we want to look at those driver training programs that address the lower levels of the GDE matrix. Since the mostly negative findings from a large-scale study of basic driver education undertaken in the 1970s in the United States that addressed the lower levels of the GDE matrix (Stock et al., 1983), the emphasis in North America until recently has been on GDL programs that are directed more toward enlightened enforcement and practice in the supervised driving period than basic driver education (Chapter 17). However, since the turn of the century, there has been an renewed emphasis in North America on driver training, in particular, on training the more complex driving skills such as hazard anticipation (e.g., Crundall et al., 2010), hazard mitigation (McKenna et al., 2006), and attention maintenance (Pradhan et al., 2011). This effort is likely to accelerate given the recent findings both that the introduction of an extended learner's permit phase still leaves novice drivers with a greatly inflated risk of crashing (Foss et al., 2011) and that training can actually decrease crashes, at least among teen males (Thomas et al., 2016).

The focus has been on training the skills of hazard anticipation, hazard mitigation, and attention maintenance because each such skill has been shown to be related to crashes, as discussed, separately, in Chapters 6 through 8. Moreover, the results of these studies are consistent with the conclusion from a seminal paper by McKnight and McKnight (2003) on teen drivers, which asks whether their crashes are primarily a function of their being careless or clueless. They argue that a large fraction of the crashes are due to the latter, not the former. The goals of the new training programs that address the lower-level goals of the GDE matrix are to make drivers less clueless and, in particular, more aware of the detailed ways in which risks can unfold so that drivers develop the driving skills that they need in order to avoid crashes.

18.4.1 Hazard Perception

For purposes of training, we can divide hazard anticipation into three different skills: prepotent, tactical, and strategic hazard anticipation skills (Engstrom et al., 2003; also see Chapters 6 and 28). Briefly, by *prepotent* hazard anticipation skills, we mean skills associated with the detection of hazards that are visible and clearly about to materialize (Sagberg & Bjornskau, 2006). By *tactical* hazard anticipation skills, we mean skills associated with the detection of hazards that are imminent but *latent* (Borowsky et al., 2007), either not visible (environmental prediction hazards [Crundall et al., 2012]) or, if visible, not immediately on a collision path with the driver (behavioral prediction hazards). Finally, by *strategic* hazard anticipation skills, we mean skills associated with the identification of latent hazards far upstream using information in cues (*precursors*) of where the latent hazards might materialize.

Because we refer throughout the review of training programs to latent hazards, we want to present at the very start an example of a latent hazard that requires both tactical skills (scanning toward the direction of the latent threat just before the latent threat might materialize) and strategic skills (scanning downstream for both precursors to latent hazards and the latent hazards themselves). Specifically, consider a scenario in which a vehicle is stopped in the parking lane in front of a marked midblock crosswalk (Figure 18.1). If the vehicle is a large one, it will block the driver's view approaching the crosswalk in the same direction of any potential pedestrians that might be entering the crosswalk. The pedestrian is the latent hazard because it may or may not materialize. There will often be precursors in such a scenario as well, either the crosswalk markings, a sign upstream of the crosswalk, or both.

18.4.1.1 Prepotent Hazard Detection Training

The European research project Training and Assessment using Interactive Evaluation tools and Reliable Methodologies (TRAINER), which took place between 2002 and 2004, led to the development of computer-based training (CBT) and simulator-based training programs as key elements. Interestingly, it

FIGURE 18.1 **(See color insert.)** Marked midblock crosswalk scenario. (Latent hazard is pedestrian that might emerge suddenly to the right in front of the truck.)

is the only training project whose effect on prepotent hazard detection has been evaluated. Following training on the CBT and simulator-based programs, participants' performance was evaluated on a driving simulator (Falkmer & Gregersen, 2003). Perhaps not surprisingly, there was no effect of training on participants' ability to detect prepotent threats, which could not be anticipated (experience could not help here).

18.4.1.2 Tactical Hazard Perception Training Programs

In the United States, the American Automobile Association (AAA) developed a risk awareness training program called DriverZED (Zero Errors Driving). In the first version of this program, participants sat in front of a PC and watched a total of 80 different scenarios filmed in city, town, and rural settings (Willis, 1998). The risky scenarios for training were selected specifically because of younger driver conflicts in those particular situations (Lonero et al., 1995). An evaluation of the original version of DriverZED was undertaken on a driving simulator using only vehicle variables and only scenarios that could be used to evaluate latent hazard anticipation (Fisher et al., 2002). Although differences in the hazard anticipation skills of untrained novice drivers, trained novice drivers, and experienced drivers were identified and in the predicted direction, it was not possible to determine whether the differences in performance were a function solely of differences in hazard anticipation skills or a combination of differences in both hazard anticipation and hazard mitigation skills.

In a more recent study of hazard anticipation training undertaken in the United States, as a part of a larger Driver Assessment and Training System (DATS), students navigated six routes on a simulator taking between 12 and 15 minutes (Park et al., 2006). The drivers were assigned to one of three simulator training modes: a single-monitor desktop simulator, a three-monitor desktop simulator, or a wide-screen vehicle cab simulator. Crash rates obtained from police accident reports of the drivers in the study were regressed on time since training. The linear increase in the *cumulative* crash rates of the participants in the three-screen vehicle cab simulator was clearly smaller than the increase in the cumulative crash rates of similar cohorts of drivers in California and Canada (which were indistinguishable from each other). Unfortunately, assignment of the participants to the experimental groups was not random, nor was assignment of the participants to experimental and control groups random. Thus, it is difficult to determine if the smaller increase in cumulative crash rates of the three-screen vehicle cab participants is a function of the differences in the participants or the difference between experimental and control groups.

One of the major problems confronting the development of training programs, for some time, was the assessment of when a driver does and does not recognize a latent hazard. Significant progress was possible when eye movements were used as an index of latent hazard perception. The assumption in these studies is that if a driver does not look toward a latent hazard, he or she cannot have determined that it is a location that requires monitoring (the converse does not hold necessarily, but it is much more likely that a driver who looks toward the location where a latent hazard will materialize will be able to respond quickly to the latent hazard than a driver who does not look toward the latent hazard).

In one of the early studies to use eye movements as a measure of the effects of tactical latent hazard anticipation training, large differences were obtained in trained and untrained novice drivers (Pollatsek et al., 2006; the fourth author of this study is an author of this chapter). The training program, Risk Awareness and Perception Training (RAPT), was PC based. The evaluation scenarios were deployed on a driving simulator. Effects of training were observed on latent hazard anticipation both on a driving simulator (Pollatsek et al., 2006) and in the field (Taylor et al., 2011), and they were evident both in scenarios that were similar to ones that were trained and ones that were quite dissimilar (Pollatsek et al., 2006). One study, which compared the training effects in the field and on the simulator, found large effects on eye fixations in both environments (Pradhan et al., 2009). For example, in the field, the trained group fixated on the critical region (area from which the latent hazard could materialize) 72.7% of the time, whereas the control group fixated on the critical region only 34.6% of the time. Finally, in a recent study in California, of 5000, 16-, 17-, and 18-year-old novice drivers (2500 in the RAPT cohort, 2500 in the control cohort) found a significant decrease in crashes among teen males a year after training (Thomas et al. 2016). The largest decreases in crashes, 43% and 35%, were observed in the 17- and 18-year-old males. There was no statistically significant change in the crashes of the female drivers. Of note, the training program took only 17 minutes on average to complete and was delivered right before the road test.

Elsewhere in the world, researchers at the Monash University Accident Research Center (MUARC) in Australia engaged in the development and evaluation of a novice driver training program (DriveSmart) that eventually combined CD-ROM and simulator training (Regan et al., 1999, 2000). This program focused on tactical hazard anticipation and attention maintenance. They first tried avoidance learning (near misses) as a way to teach hazard anticipation (Triggs & Regan, 1998). There was no effect of training. They then used a method that required the drivers to learn scanning, keeping ahead, and playing safe (SKAPS) (Regan et al., 1998), a method very similar to error training. This time, they found an effect of training on hazard anticipation in a posttraining evaluation.

The aforementioned studies have employed training programs that use as a dependent variable a measure of behavior linked directly to hazard anticipation, e.g., glancing toward a position where a latent hazard might appear or taking actions that might mitigate a latent hazard. Commentary driving is another way to gather information on what a driver understands about a hazard, either using videos of actual driving or putting teens inside a moving vehicle, either as a passenger or driver. Of note, commentary training began life within the armed and emergency services. Officers are trained to make predictions about what might happen next and then decide and report what action they would take to avoid developing hazards—without any pauses in the verbal report. It is not surprising, therefore, that compared with average drivers, police drivers demonstrated faster responses to hazards in a driving simulator (Dorn & Barker, 2005), given their advanced training and driving experience.

The question here is whether commentary driving might also show benefits even for those who have received relatively little training, such as learner drivers. The answer is yes. For example, having learner drivers listen to or produce a commentary while watching driving video clips has been shown to lead to improvements in subsequent hazard perception test scores (Isler et al., 2009; McKenna et al., 2006; Wetton et al., 2013) and reduced risk taking on a gap acceptance task (McKenna et al., 2006). Beneficial effects of commentary training have also been found for learner drivers who completed a driving simulator route containing several hazards (Crundall et al., 2010). Those trained in commentary had fewer crashes, slowed sooner on the approach to hazards, and applied pressure to the brakes sooner than the control group.

While commentary driver training is rather widespread as a method to improve hazard perception in the driver training industry amongst advanced drivers, recent evidence has emerged that implies that great care should be taken in how commentary training is used for young drivers. Previous studies showing a benefit of commentary training for average drivers have all required participants to produce commentaries in the training phase, but the subsequent assessment phase was always held in silence and not under conditions of live commentary. If young people are trained to produce a commentary while driving, there is the possibility that this could be viewed as a secondary task and, as such, could interfere with the primary task of vehicle control or responding to hazards. To test this, Young et al. (2014) conducted two experiments showing that producing a commentary slows responses to hazards on a concurrent hazard perception task.

18.4.1.3 Strategic Hazard Perception Training Programs

Researchers in the United Kingdom have focused on strategic hazard anticipation without the use of a driving simulator in the training and evaluation of the skill. Hazards were presented as video clips to novice drivers, and the training took approximately an hour, following which the drivers were evaluated using a set of video clips different from those used in the training phase (Chapman et al., 2002). The training proved effective at reducing the differences in the strategic scanning patterns of novice and experienced drivers. Chapman et al. noted that it is not sufficient to bring novice drivers to the point where their eye movement patterns resemble those of more experienced drivers, which is why their training focused on knowledge, scanning, and anticipation, rather than the modification of observable patterns of eye movements.

18.4.2 Novice Drivers and Hazard Mitigation

As with hazard anticipation, when we consider hazard mitigation, we need to discriminate between tactical and strategic skills. As we did previously, we define tactical hazard mitigation skills as those that are required to reduce the risk of crashing in a particular scenario. We define strategic hazard mitigation skills as largely driving at a speed and following at a distance appropriate for the environment in which one is traveling.

18.4.2.1 Tactical Hazard Mitigation Training

Three recent studies have been undertaken focusing on the effects of hazard mitigation training using similar, PC-based training programs (Hamid, 2013; Muttart, 2013; Yamani et al., 2016). All have shown the benefits of hazard mitigation training, although different populations of drivers have been used (Hamid studied nursing students; Muttart and Yamani as well as Samuel and Fisher focused on novice drivers), and different numbers of skills have been trained (Muttart concentrated on only one skill, hazard mitigation, while Hamid and Yamani as well as Samuel and Fisher targeted three skills, hazard mitigation, hazard anticipation, and attention maintenance). The focus here will be confined to the study by Muttart since it honed in on the target population and target skill of interest here, i.e., hazard mitigation among novice drivers.

As noted previously, prior to training a particular hazard mitigation skill, it is important to train the hazard anticipation skill upon which the mitigation depends. Muttart (2013) focused on training hazard mitigation in three different types of scenarios: intersection, run-off-the-road, and tangent (straight-section) scenarios. The training was conducted on a PC. The evaluation was done on a driving simulator. Hazard anticipation here, as previously, was measured by noting whether a driver glanced toward a latent hazard at a time when he or she could potentially mitigate the hazard.

An example of the hazard mitigation training for each of three types of scenarios in which teens most frequently crash will be described briefly here. First, in one of three intersection scenarios, the driver was traveling straight through an intersection. A large truck was in the adjacent left-turn lane blocking the driver's view of cars in the opposing lane across the intersection, cars that might be turning in front of the driver (because they too could not see the driver). The participant was taught first to anticipate the hazard using the error training method. Then the driver was taught, again using the error training method, to slow and therefore mitigate the hazard. Among the group of novice drivers (trained and untrained) who anticipated the hazard, the trained novice drivers mitigated the hazard better (slowed earlier and more in the scenario just described). Hazard mitigation training was shown to be effective among all three intersection scenarios.

Second, in the rear-end scenario to be discussed here, the driver approached a work zone. A worker downstream of a lead vehicle stepped precariously out of the work zone and into the travel lane. The lead vehicle slowed slightly. The trained drivers were taught to look for downstream events in a work zone that might cause a lead vehicle to slow suddenly (anticipation). And they were taught to keep a larger distance between their vehicle and the lead vehicle (mitigation). When the mitigation behaviors of the trained and untrained teen drivers who anticipated the hazard were compared, the trained drivers better mitigated the potentially threatening lead vehicle.

Third, in the curve scenario to be discussed here, the driver approached a curve. A vehicle could be seen stopped in the driver's travel lane at the apex of the curve. The teen driver was taught to anticipate the fact that a stopped vehicle could be present by looking ahead and around the curve. And the teen driver was taught to mitigate this situation by slowing appropriately. Among the teen drivers who anticipated the stopped vehicle, the trained teen drivers were more likely to slow than the untrained teen drivers.

Finally, not only have studies documented the relation between hazard mitigation training and safer behaviors, but also, one study documented a relation between tactical hazard mitigation training and a decrease in crash risk (Carstensen, 2002). Specifically, a variety of changes were introduced in Denmark's driver education system in the late 1980s. An evaluation study involving novice drivers was undertaken subsequently. The study showed that drivers trained using the new drivers' education program, which included the addition of components addressing perception, quick starts and stops, lane keeping, and velocity modulation, experienced a reduced crash risk compared to those who undertook the previous training modules. The comparison of crash risks was done before the new drivers' education program was put in place and after it had been in place for a while. Thus, other factors could explain the reduction in crash risk.

18.4.2.2 Strategic Hazard Mitigation Training

In an interesting twist on training strategic hazard mitigation skills, a determination was made of whether exposing teens to a video that included a commentary about threats that could become prepotent would not only allow the teens so exposed to respond more quickly to those threats (as discussed previously) but would also lead to improvements in the teens' strategic hazard mitigation skills (McKenna et al., 2006). Using a suite of different measures, it was found that there was a significant improvement among the trained teens in measures of strategic hazard mitigation (e.g., speed, gap acceptance) based on a program targeting tactical hazard anticipation skills. In subsequent experiments, the authors were able to verify that the difference in the observed performance of the trained and untrained teens was due to a specific improvement in hazard perception and not due to a nonspecific reduction in risk taking.

18.4.2.3 Combined Tactical and Strategic Hazard Mitigation Training

An altogether different study was performed to compare the effects of training higher-order skills versus vehicle-handling skills among novice drivers (Isler et al., 2011). In the higher-order skills training, tactical hazard anticipation and mitigation training were linked, as were strategic hazard anticipation and mitigation training. The authors reported that participants who received the higher-order skills training (commentary driving, PC-based videos, and on-road self-evaluation with a focus on both tactical and strategic hazard perception and hazard mitigation components) showed a statistically significant improvement in hazard perception (as indicated both on the road by visual search measures and in a PC-based test of hazard perception/mitigation) and hazard mitigation (as indicated in the PC-based test of hazard perception/mitigation and as indicated in safer attitudes toward following distances and overtaking behavior on a questionnaire). The second group, which only received vehicle-handling skills training, showed significant improvements in their on-road directional control and speed choice. However, this group showed no improvement in tactical hazard perception or hazard mitigation skills. This highlights the fact that hazard mitigation training programs without an anticipation component are unlikely to improve drivers' hazard mitigation skills.

18.4.3 Novice Drivers and Attention Maintenance

As with hazard anticipation, novice drivers differ from experienced drivers both in terms of tactical and strategic attention maintenance (Chapter 7). Attention maintenance has many meanings. We focus on the maintenance of attention when performing an in-vehicle task. At first glance, any training designed to improve drivers' performance attending to in-vehicle tasks may appear somewhat at odds with the effort to make younger drivers safer drivers. But while some in-vehicle tasks are unrelated to driving, there are many tasks that are related to driving and affect the safe performance of the vehicle. Examples include activating the defroster, adjusting the windshield wiper speed, turning on the emergency flashers, and glancing in rearview and side-view mirrors. For more experienced drivers familiar with the controls of a vehicle, the completion of these tasks may require multiple glances. But they are not apt to impose too heavy a cognitive load, nor result in many dangerously long glances inside the vehicle. However, as noted previously, novice drivers are still overwhelmed by the driving task. Performing tasks that require them to direct their gaze away from the forward roadway can be even more problematic for them than it can be for more experienced drivers. In fact, anyone who regularly rents cars knows just how difficult it can be to operate the controls of a vehicle with which one is unfamiliar. Thus, the training programs are targeting the performance of secondary in-vehicle tasks that are important to the safe operation of the vehicle.

18.4.3.1 Tactical Attention Maintenance Training

As noted previously and in Chapter 7, novice drivers are more likely to glance away from the roadway than experienced drivers for extended periods when attempting to perform a task inside the vehicle. In a recent field study, an evaluation was made of the efficacy of a PC-based training program, Focused Attention and Concentration Learning (FOCAL), that was designed to teach novice drivers not to glance away for these dangerously long periods of time (Pradhan et al., 2011). A FOCAL-trained group was compared to a placebo-trained group in an on-road test using an eye tracker to determine the duration of each glance inside the vehicle. The FOCAL-trained group made significantly fewer glances away from the roadway that were longer than 2 seconds than the placebo-trained group. Other measures indicated an advantage for the FOCAL-trained group as well. Another test of FOCAL was undertaken, only on a driving simulator (rather than on the road) and 3 months after training (rather than immediately). The logic for using a simulator rather than the road is that the road requires a driving instructor in the front passenger seat. Perhaps novice drivers would behave differently without a passenger, something not possible on the road but easily doable in a simulator. It was found that the effects of FOCAL extended for up to 3 months after training and that untrained novice drivers still took way too many dangerously long glances inside the vehicle (Divekar et al., 2013).

18.4.3.2 Strategic Attention Maintenance Training

There have been two approaches to strategic attention maintenance training. In the first approach, an attempt is made to decrease drivers' willingness to engage in secondary in-vehicle tasks in the presence of latent hazards (it is assumed that drivers would not do such in the presence of prepotent hazards and therefore, no training is needed for these latter hazards). Only one study has been undertaken to the best of our knowledge (Krishnan et al., 2015). In that study, a training program—Secondary Task Regulatory & Anticipatory Program (STRAP)—was developed that was designed to make drivers aware of latent hazards in the hope that they would then self-regulate their engagement in, postpone, or interrupt secondary tasks that they were performing at the time the latent hazard appeared. The secondary tasks included both tasks that required drivers to take their eyes off the road (e.g., operating the defroster) and those that did not (e.g., cell phone use). Participants were assigned to either STRAP or placebo training. After training, the groups navigated eight different scenarios on a driving simulator and were instructed to engage during the drive in as many secondary tasks as they wanted to as long as they felt safe doing so. The secondary tasks were fully user paced. The results indicated that STRAP-trained drivers were more likely to detect latent hazards and associated clues than placebo-trained drivers. This by itself is not surprising. But what is surprising, given that the STRAP-trained drivers were never explicitly told to limit their engagement in a secondary task in the area immediately preceding a latent hazard (the critical area), is the finding that the STRAP-trained drivers spent more than twice as long with their eyes on the forward roadway as the placebo-trained drivers and that the STRAP-trained drivers devoted less attention to the cell phone task in the critical area.

A second approach to strategic hazard anticipation training has been studied by Regan et al. (1998). They used Gopher's variable-priority (VP) training technique to increase participants' ability to multitask. The participants either received VP training or served as controls. Participants in the VP group were instructed over trials to systematically vary how much attention they gave to each task. VP subjects were found to perform significantly better than controls on measures of their driving performance when asked to perform three subtasks in a subsequent simulator evaluation. It is not clear how one balances the fact that drivers will be multitasking, and therefore, training that reduces the cognitive load while multitasking is good (will increase safety), with the fact that one wants to send a clear message that multitasking, except for safety-critical events, is bad.

18.5 Driver Education: Blended Programs

Recent advances in training programs have taken advantage of advances in the media that can be used to deliver the content. The following section will discuss the training programs using e-learning and, secondly, using in-vehicle data recorders (IVDRs). Each of these training programs addresses both the need to improve self-awareness and the need to improve skills.

18.5.1 Advances in Technology for Influencing Driver Motivations and Behavior: E-Learning

There is good reason to suppose that a new way of using media via e-learning might improve competencies at the upper three levels of the GDE matrix (Bixler, 2008; Means et al., 2010; Nelson, 2007; Saito & Miwa, 2007; Shen et al., 2007). Toward this end, in the United Kingdom, driver safety e-learning modules have been designed by a specialist driver education software company to encourage self-evaluation competencies by presenting scenario-based driving situations relevant to the kind of situations in which young drivers are particularly at risk (Buckley et al., 2014). Scenarios include being the driver on a night out with friends when one of them is playing around in the vehicle and distracting you, taking into account that the most effective GDL components appear to be restrictions on nighttime driving and limiting peer passenger numbers (McCarrt et al., 2009; Williams, 2007). Other examples include interactive scenarios in which you are driving at night and hazards unfold in a rural area, which carries greater risk of crash involvement for young people (Ayuso et al., 2014). The modules allow young drivers to report how they might typically respond in these situations, facilitating self-reflection and developing skills and knowledge in the use of effective coping strategies across many different high-risk scenarios (a form of error training). One study has evaluated the effect of this method on recorded traffic violations in a blended learning environment.

Specifically, the Young Driver Scheme (YDS) launched by the Thames Valley Police in the United Kingdom in 2008 invited drivers below the age of 25 who had committed a nonserious driving offense (mostly speeding) to take part in a driver education program instead of paying a fine and having penalty points added to their driving licence (see Conner & Lai, 2005, for a review). The YDS included an initial classroom-based workshop consisting of about 20–25 driving offenders. The workshop provided face-to-face interaction with a driver trainer who discussed various ways crashes can come about, attitudes, behaviors, distractions, and other factors. After completing the workshop, drivers were required to log onto a website hosting five e-learning modules (designed by a2om International Ltd). After completing each module, the participant was locked out for 4 days before the next one could be attempted. All five modules had to be passed within 28 days. Each module except the first one ended with a 25-question multiple-choice test on the module content. To pass the module, four-fifths of these questions needed to be answered correctly. If this score was not achieved, the participant had to go through the module again. Each module was visually engaging and interactive, taking about 20 minutes to complete. The participant could also go back and study any part of the completed module again at any time.

The content of each module included, first of all, an animated scenario in which a car is involved in a road traffic incident, and this is replayed several times from different perspectives, with risk factors pointed out from those involved; second, a passenger scenario in which an incident occurs when passengers want you to drive faster and interactive elements ask drivers to reflect on their decisions; third, a safety margin scenario in which interactive elements consider speed, stopping distances, and skidding; fourth, an overtaking scenario in which interactive elements point out the hazards involved and ask young drivers to reflect on their decisions; and finally, the same hazard perception training clips as evaluated by Isler et al. (2009), used to train visual search techniques for anticipating, spotting, and reacting to hazards, taking into account blind spots in the rearview and side mirrors, synchronized onto a single screen (a more complete discussion of hazard anticipation training programs occurred in the previous section).

The study to evaluate the effects of this training program monitored reoffending rates for YDS participants compared with a number of other driver groups, all below 25 years of age (af. Wåhlberg, 2011). One of the groups had been through a "speed awareness scheme" (SAS) for those that had been caught speeding, consisting of a 2.5-hour, classroom-based group session (20–24 participants), aimed specifically at the problem of speed and speeding. The fixed penalty notice group (FPN—fines only) consisted of three subgroups of drivers: those who had already taken an SAS course within a 3-year period before their current offense and had been caught committing another offense but were not eligible to be offered another course, those who drove excessively fast, and those who declined to take the SAS course. These drivers therefore had a somewhat different past, as compared to the YDS and SAS driver groups. Finally, a control group of young drivers was recruited via an e-mail campaign. Data on driving offenses for all driver groups were provided over a period of 6 months after each participant had completed the YDS and the SAS as well as offenses committed by those belonging to the control group and those who received an FPN. The findings showed a significant reduction in the number of offenses and penalty points for the YDS, while this was not the case for drivers who had been fined only (FPN) or had taken part in the SAS. The results seem to indicate a positive effect of the YDS for young driving offenders that resulted in a detectible behavioral change. Interestingly, the effect for offenses is stronger in the calculation where the time frame coincides with the start of the course, and not the offense. Again, we note that the YDS program addresses both self-awareness skills and other-awareness (hazard anticipation) skills.

18.5.2 Advances in Technology for Influencing Driver Motivations and Behavior: IVDRs

IVDRs are discussed at length in Chapter 20. Here we want to highlight their considerable impact on each of the upper three levels of the GDE matrix (Table 18.1). Specifically, consider their impact on awareness of one's own risky habits (e.g., speeding, level 1), awareness of one's own limitations (e.g., vis-à-vis maintaining attention, level 2), and awareness of one's skill levels (e.g., not overestimating one's own competencies and not underestimating the risks, level 3).

To illustrate this, we will discuss how app-based feedback via a smartphone after each journey, but not during, could be particularly beneficial for educating young drivers (this application is not discussed in Chapter 20). There has been a dramatic improvement in smartphone technology in which monitoring and mentoring drivers is particularly economical to implement, especially for insurance purposes. In recent studies, driver behavior improved when drivers were asked to look at a smartphone interface placed in a cradle on the vehicle windscreen presenting safety and fuel advice while driving at times considered safe (Birrell & Young, 2011; Birrell et al., 2014). The findings showed an almost 14% increase in mean headway to 2.3 seconds and an almost threefold reduction in time spent traveling closer than 1.5 seconds to the vehicle in front compared to a control condition with no feedback. There is also some evidence that students who accessed educational materials using a mobile phone were not significantly different on learning outcomes from students who did so using a traditional computer (Shih, 2007).

Via their smartphone app, drivers can be encouraged to drive smoothly and less aggressively and discouraged to drive for long periods, drive at excessive speeds, and use their mobile phone while driving. Feedback that is unambiguous, precise, and easily interpreted in terms of the discrepancy between observed and desired safe driving performance allows young drivers to link the feedback to the exact component of their at-risk driving for every journey (this is essentially real-time error learning, with mentoring self-administered and the possibility of mastery inherent in the delivery of the continuing feedback). In this way, app-based driver education could improve a driver's competency in particular skills (e.g., maintaining headway, level 3), in self-awareness of his or her own limitations (e.g., the difficulty of keeping under the speed limit, level 2), and in self-awareness of his or her risky habits (e.g., speeding, level 1). Some telematics service providers have even incorporated reflective questions, setting driving goals based on trends in the data and focusing on developing knowledge and skills across all levels of the GDE matrix (e.g., a UK-based company called The Floow). However, to date, there has been no academic research to evaluate whether postdrive app-based driver education of this nature is more beneficial for young driver safety compared with basic feedback on driver score and/or incentives.

18.6 Discussion

Unfortunately, little is known about the effectiveness of the many older-generation novice driver training programs currently delivered across the globe, and what is known suggests that they are not effective at reducing a number of crash- and violation-related measures (e.g., crash rates, cited violations, insurance claims, prosecutions, convictions for driving offenses, driver's license demerit points, etc.). However, there are a number of caveats that need to be introduced when considering the evaluation of the older generation of novice driver training programs.

For those studies that have evaluated the effect of training on violations and crashes, many are beset with major flaws that mean that the reliability and validity of their findings have been questioned (Beanland et al., 2013). For example, most evaluation studies have relied on self-report. Also, training and educational interventions are often too short in duration to have a long-term impact, but there are exceptions to this (af. Wåhlberg, 2011). There are many other factors influencing teen driver behavior that are likely to swamp the value of these interventions. For example, young people are regularly exposed to and are strongly influenced by the driving styles of their parents (Bianchi & Summala, 2004; Ferguson et al., 2001; Green & Dorn, 2008). Finally, the structure and content of driver training programs that have been evaluated generally do not include a comprehensive curriculum based on the GDE matrix to develop all the competencies a new driver needs to be safe. Most training programs focus on one level.

While some of the new generation of novice driver training programs show considerable promise, most transport and traffic authorities throughout the developed world are only at the stage of piloting a wide range of strategies and techniques to address the teen driver problem. Having said that, the results of evaluations of the effect of the new generation of training programs on behaviors with clear links to crashes (rather than of the effect of training programs on actual crashes and violations) are much more clear-cut. As discussed in Section 18.4 of this chapter, failures of hazard anticipation, hazard mitigation, and attention maintenance skills have all been clearly linked to increases in crashes. Novice drivers are clearly deficient in all three skills. And, training programs have been developed for all three skills that have been shown to transfer from one platform (typically a PC) to the open road; have been shown to generalize from the specific scenarios that were targeted in the training program to scenarios with conceptually similar, but perceptually different, characteristics; and have been shown to produce effects that last up to a year with only an hour's worth of training. In fact, in the case of hazard anticipation training, a program that lasts only 17 minutes has been found to reduce crashes among 17 and 18 year old males by 43% and 35% respectively (Thomas et al., 2016). However, whether the other broad suite of programs that change the behaviors that are known to increase crash risk and are known to be deficient in novice drivers actually lead to reductions in morbidity and mortality is not yet known. Studies are desperately needed to address the effectiveness of these programs on crash data.

Online learning also shows great promise. It can be enhanced by giving learners control of their interactions with media and prompting learner reflection. Studies indicate that manipulations that trigger learner activity or learner reflection and self-monitoring of understanding are effective when students pursue online learning as individuals. When used by itself, online learning appears to be as effective as conventional classroom instruction but not more so. However, given that e-learning is a low-cost method of educating thousands of drivers simultaneously, more research needs to evaluate its potential for improving the safety of young drivers.

Perhaps in the future, the insurance industry may take the driver education reins and widely disseminate telematics-based policies to millions of young drivers providing feedback on driving skills and facilitate the development of self-awareness of risky driving styles. Such feedback could also be supplemented with e-learning and error training to educate drivers even further. Early research on these schemes seems promising with the added bonus that monitoring and mentoring young drivers via pay-as-you-drive schemes are financially attractive for both the insurance company as well as the driver (pay-as-you-drive schemes index insurance rates to driving history as recorded by the IVDR). This means that participation in young driver education may circumnavigate the known difficulties in encouraging young people to develop their competencies across all levels of the GDE matrix given that many traffic and transport authorities are at a very early stage of development in this field.

In summary, young driver safety interventions appear to be rather fragmented and applied in a piecemeal fashion, leading some authors to call for a systems approach in which all stakeholders are involved rather than simply focusing on the errant driver (Scott-Parker et al., 2015). However, there is the beginning of an integration of approaches. Lower-level GDE matrix driving skills training programs are now being combined, no longer being offered as only separate units (Yamani et al., 2016). Parents are being integrated into the skills-based training programs. The same is true of the higher-level GDE self-awareness training programs as well. Perhaps in the not-too-distant future, a combination of a program targeting all levels of the GDE matrix that includes both parents and teens in the mix and supported by policy and regulatory systems will be a reality, not just a pipe dream.

Acknowledgments

Donald Fisher wishes to acknowledge the support of grants from the New England University Transportation Center to Michael Knodler as Principal Investigator (PI), from the Department of Transportation UTC program to Umit Ozguner as PI at Ohio State University, and from the Department of Transportation UTC program to Sue Chrysler as PI at the University of Iowa. No endorsement is intended of any of the training programs singled out in this chapter. Lisa Dorn is a consultant for The Floow.

References

af. Wåhlberg, A.E. (2011). Re-education of young driving offenders: Effects on recorded offences and self-reported collisions. *Transportation Research Part F: Traffic Psychology and Behaviour, 14*, 291–299.

Ayuso, M., Guillén, M., & Pérez-Mańin, A. (2014). Time and distance to first accident and driving patterns of young drivers with Pay-As-You-Drive insurance. *Accident Analysis & Prevention, 73*, 125–131.

Beanland, V., Goode, N., Salmon, P.M., & Lenne, M.G. (2013). Is there a case for driver training? A review of the efficacy of pre- and post-licence driver training. *Safety Sciences, 51*, 127–137.

Bianchi, A., & Summala, H. (2004). The "genetics" of driving behavior: Parents' driving style predicts their children's driving style. *Accident Analysis & Prevention, 35*, 655–659.

Birrell, S.A., & Young, M.S. (2011). The impact of smart driving aids on driving performance and driver distraction. *Transportation Research Part F, 14*(6), 489–493.

Birrell, S.A., Fowkes, M., & Jennings, P.A. (2014). Effect of using an in-vehicle smart driving aid on real-world driver performance. *IEEE Transactions on Intelligent Transportation Systems, 15*(4), 1801–1810.

Bixler, B.A. (2008). The effects of scaffolding student's problem-solving process via question prompts on problem solving and intrinsic motivation in an online learning environment. PhD dissertation, The Pennsylvania State University, State College, Pennsylvania.

Boccara, V., Delhomme, P., Vida-Gomel, C., & Rogalski, J. (2011). Development of student drivers' self-assessment accuracy during French driver training: Self-assessments compared to instructors' assessments in three risky driving situations. *Accident Analysis & Prevention, 43*, 1488–1496.

Borowsky, A., Shinar, D., & Oron-Gilad, T. (2007). Age, skill, and hazard perception in driving. *Proceedings of the Fourth International Driving Symposium on Human Factors in Driver Assessment, Training and Vehicle Design.*

Buckley, L., Chapman, R.L., & Sheehan, M. (2014). Young driver distraction: State of the evidence and directions for behavior change programs. *Journal of Adolescent Health, 54*, 516–521.

Carstensen, G. (2002). The effect on accident risk of a change in driver education in Denmark. *Accident Analysis & Prevention, 34*, 111–121.

Chapman, P., Underwood, G., & Roberts, K. (2002). Visual search patterns in trained and untrained novice drivers. *Transportation Research Part F: Traffic Psychology and Behaviour, 5*, 157–167.

Chen, L.-H., Braver, E.R., Baker, S.P., & Li, G. (2001). Potential benefits of restrictions on the transport of teenage passengers by 16 and 17 year old drivers. *Injury Prevention, 7*, 129–134.

Christie, R. (2001). *The Effectiveness of Driver Training as a Road Safety Measure: A Review of the Literature.* Melbourne: RACV.

Conner, M., & Lai, F. (2005). Evaluation of the effectiveness of the National Driver Improvement Scheme. Road Safety Report No. 64. Department for Transport, London.

Crundall, D., Andrews, B., van Loon, E., & Chapman, P. (2010). Commentary training improves responsiveness to hazards in a driving simulator. *Accident Analysis & Prevention, 42*, 2117–2124.

Crundall, D., Chapman, P., Trawley, S., Collins, L., van Loon, E., Andrews, B., & Underwood, G. (2012). Some hazards are more attractive than others: Drivers of varying experience. *Accident Analysis & Prevention, 45*, 600–609.

Dalland, E.B. (2012). The driving test in Norway: An intervention study. In L. Dorn (Ed.), *Driver Behaviour and Training,* (V) (pp. 81–99). Aldershot, UK: Ashgate.

Divekar, G., Pradhan, A., Masserang, K., Pollatsek, A., & Fisher, D. (2013). A simulator evaluation of attention maintenance training on glance distributions of younger novice drivers. *Transportation Research F: Traffic Psychology and Behaviour, 20*, 154–159.

Dorn, L. (2005). Driver coaching: Driving standards higher. In L. Dorn (Ed.), *Driver Behaviour and Training, Vol. II. Human Factors in Road and Rail Safety.* Aldershot: Ashgate.

Dorn, L., & Barker, D. (2005). The effect of driver training on simulated driving performance. *Accident Analysis & Prevention, 37*, 63–69.

Elvik, R., Høye, A., Vaa, T., & Sørensen, M. (2009). *The Handbook of Road Safety Measures*, 2nd ed. Bingley, UK: Emerald Books.

Engstrom, I., Gregerson, N., Hernetkoski, K., Keskinen, E., & Nyberg, A. (2003). *Young novice drivers, driver education and training: Literature review.* Linköping, Sweden: Swedish National Road and Transport Research Institute.

Falkmer, T., & Gregersen, N. (2003). The TRAINER project—The evaluation of a new simulator-based driver training methodology. In L. Dorn (Ed.), *Driver Behaviour and Training, Proceedings of the First International Conference on Driver Behaviour and Training* (pp. 317–330). Aldershot: Ashgate.

Ferguson, S., Williams, A., Chapline, J., & Reinfurt, D. (2001). Relationship of parent driving records to the driving records of their children. *Accident Analysis & Prevention, 22*, 229–234.

Fernandez-Medina, K., Wallbank, C., Helman, S., Hutchins, R., Palaganda, R., & Kinnear, N. (2015). An evaluation of Safe Drive Stay Alive in London. TRL CPR2043.

Fisher, D.L. (2015). Novice driver education: It's past and it's future. *Invited keynote address to be presented at the Third Forum on Traffic Safety: Youth and Traffic Safety*. Saudi Arabia: Damman

Fisher, D.L., Laurie, N.E., Glaser, R., Connerney, K., Pollatsek, A., Duffy, S.A., & Brock, J. (2002). The use of an advanced driving simulator to evaluate the effects of training and experience on drivers' behavior in risky traffic scenarios. *Human Factors, 44*, 287–302.

Foss, R., Martell, C., Goodwin, A., O'Brien, N., & UNC Highway Research Center. (2011). *Measuring Changes in Teenage Driver Characteristics During the Early Months of Driving.* Washington, DC: AAA Foundation for Traffic Safety.

Glendon, A.I., Dorn, L., Davies, D.R., Matthews, G., & Taylor, R.G. (1996). Age and gender differences in perceived accident likelihood and driver competencies. *Risk Analysis, 16*, 755–762.

Green, A., & Dorn, L. (2008). How do "Significant Others" influence young people's believes about driving? In L. Dorn (Ed.), *Driver Behaviour and Training Vol III. Human Factors in Road and Rail Safety.* Aldershot: Ashgate.

Gregersen, N.P., & Bjurulf, P. (1996). Young novice drivers: Towards a model of their accident involvement. *Accident Analysis & Prevention, 28*, 229–241.

Hamid, A. (2013). *Effect of total awake time on drivers' performance and evaluation of training interventions to mitigate effects of total awake time on drivers' performance (Dissertation).* Amherst, MA: Department of Mechanical and Industrial Engineering, University of Massachusetts.

Hatakka, M., Keskinen, E., Baughan, C., Goldenbeld, C., Gregersen, N.P., Groot, H., Siegrist, S., Willmes-Lenz, G., & Winkelbauer, M. (2002). From control of the vehicle to personal self-control: Groadening the perspectives of driver education. *Transportation Research Part F: Traffic Psychology and Behaviour, 5*, 201–215.

Hatakka, M., Keskinen, E., Buaghan, C., Goldenbeld, C., Gergersen, N., Groot, H., & Winkelbauer, M. (2003). *Basic driver training: New Models, EU-project (Final Report).* Turku, Finland: Department of Psychology, University of Turku.

Holyoak, K. (1984). Analogical thinking and human intelligence. In R. Sternberg (Ed.), *Advances in the Psychology of Human Intelligence* (pp. 199–230). Hillsdale, NJ: Erlbaum.

Isler, R., Starkey, N., & Williamson, A. (2009). Video-based road commentary training improves hazard perception of young drivers in dual task. *Accident Analysis & Prevention, 41*, 445–452.

Isler, R., Starkey, N., & Sheppard, P. (2011). Effects of higher-order driving skill training on young, inexperienced drivers' on-road driving performance. *Accident Analysis & Prevention, 43*, 1818–1827.

Ivancic, K., & Hesketh, B. (2000). Learning from errors in a driving simulation: Effects on driving skill and self-confidence. *Ergonomics, 43*, 1966–1984.

Kaestner, N. (1968). Research in driver improvement programs: The state of the art. *Traffic Quarterly, 22*, 497–520.

Keith, N., & Frese, M. (2008). Effectiveness of error management training: A meta-analysis. *Journal of Applied Psychology, 93*, 56–69.

Ker, K., Roberts, I., Collier, T., Beyer, F., Bunn, F., & Frost, C. (2005). Post-licence driver education for the prevention of road traffic crashes: A systematic review of randomised controlled trials. *Accident Analysis & Prevention, 37*, 305–313.

Kjelsrud, H. (2010). How can reflecting teams contribute to enhanced driving teacher learning? In L. Dorn (Ed.), *Driver Behaviour and Training (IV).* Aldershot, UK: Ashgate.

Krishnan, A., Samuel, S., Dundar, C., & Fisher, D. (2015). Evaluation of a training program (STRAP) designed to decrease young drivers secondary task engagement in high risk scenarios: A strategic approach to training hazard anticipation in young drivers. *Proceedings of the Transportation Research Board 94th Annual Meeting.* Washington, DC: National Academies Press.

Laapotti, S., Keskinen, E., Hatakka, M., & Katila, A. (2001). Novice drivers accidents and violations—A failure on higher or lower hierarchical levels of driving behavior. *Accident Analysis & Prevention, 33*, 759–769.

Langford, J. (2006). *Road Safety Implications of Further Training for Young Drivers.* Melbourne, Australia: Monash University Accident Research Centre.

Lewis-Evans, B. (2010). Crash involvement during the different phases of the New Zealand Graduate Driver Licensing System (GDLS). *Journal of Safety Research, 41*, 359–365.

Liu, C.C., Lenné, M.G., & Williamson, A.R. (2009). Development and evaluation of a young driver training program. Melbourne: Monash University Accident Research Centre.

Lonero, L., Clinton, K., Laurie, I., Black, D., Brock, J., & Wilde, G. (1995). *Novice Driver Education Curriculum Outline.* Washington, DC: AAA Foundation for Traffic Safety.

Matthews, M.L., & Moran, A.R. (1986). Age differences in male drivers' perception of accident risk: The role of perceived driving ability. *Accident Analysis & Prevention, 18*, 299–313.

Mayhew, D., Simpson, H., & Pak, A. (2003). Changes in collision rates among novice drivers during the first months of driving. *Accident Analysis & Prevention, 35*, 683–691.

McCarrt, A., Teoh, E., Fields, M., Braitman, K., & Hellinga, L. (2009). Graduated licensing laws and fatal crashes of teenage drivers: A national study. Arlington, VA: Insurance Institute for Highway Safety.

McCartt, A., Shabanova, V., & Leaf, W. (2003). Driving experience, crashes and traffic citations. *Accident Analysis & Prevention, 35*, 311–320.

McKenna, F., Horswill, M., & Alexander, J. (2006). Does anticipation training affect drivers' risk taking? *Journal of Experimental Psychology: Applied, 12*, 1–10.

McKnight, A., & McKnight, S. (2003). Young novice drivers: Careless or clueless? *Accident Analysis & Prevention, 35*, 921–925.

Means, B., Toyama, Y., Murphy, R., Bakia, M., & Jones, K. (2010). Evaluation of evidence-based practices in online learning: A meta-analysis and review of online learning. U.S. Department for Education. United States.

Mercedes-Benz. (2015). *Mercedes-Benz Driving Academy*. Accessed on August 29, 2015, retrieved from http://www.mb-drivingacademy.co.uk/uk/en/home.html.

Molina, J., Garcia-Ros, R., & Keskinen, E. (2014). Implementation of the driver training curriculum in Spain: An analysis based on the Goals of Driver Education (GDE) framework. *Transportation Research Part F: Traffic Psychology and Behaviour, 26*(Part A(0)), 28–37.

Muttart, J. (2013). *Identifying Hazard Mitigation Behaviors That Lead to Differences in the Crash Risk between Experienced and Novice Drivers (Dissertation)*. Amherst, MA: Department of Mechanical and Industrial Engineering, University of Massachusetts.

Mynttinen, S., Sundström, A., Koivukoski, M., Hakuli, K., Keskinen, E., & Henriksson, W. (2009). Are novice drivers overconfident? A comparison of self-assessed and examiner-assessed driver competences in a Finnish and Swedish sample. *Transportation Research Part F: Traffic Psychology and Behaviour, 12*(2), 120–130.

National Highway Traffic Safety Administration. (2014). *Traffic Safety Facts. 2012 Data. Young Drivers*. Accessed on August 3, 2015, retrieved from http://www-nrd.nhtsa.dot.gov/Pubs/812019.pdf.

National Research Council (US), Institute of Medicine (US) and Transportation Research Board (US) Program Committee for a Workshop on Contributions from the Behavioral and Social Sciences in Reducing and Preventing Crashes. (2007). *Preventing teen motor crashes: Contributions from the behavioral and social sciences: Workshop report*. Washington, DC: National Academies Press.

Nelson, B.C. (2007). Exploring the use of individualized, reflective guidance in an educational multi-user virtual environment. *Journal of Science Education and Technology, 16*(1), 83–97.

Nichols, J.L. (1970). Driver education and improvement programs. In N.W. Heimstra (Ed.), *Injury Control and Traffic Safety*. Springfield, IL: Charles C. Thomas.

Nichols, J. (2003). *A Review of the History and Effectiveness of Driver Education and Training as a Traffic Safety Program*. Washington, DC: National Transportation Safety Board.

Park, G., Cook, M., Allen, R., & Fiorentino, D. (2006). Automated assessment and training of novice drivers. *Advances in Transportation Studies: An International Journal, Special Issue*, 87–96.

Paaver, M., Eensoo, D., Kaasik, K., Vaht, M., Maestu, J., & Harro, J. (2013). Preventing risky driving: A novel and efficient brief intervention focusing on acknowledgement of personal risk factors. *Accident Analysis & Prevention, 50*, 430–437.

Pollatsek, A., Narayannan, V., Pradhan, A., & Fisher, D. (2006). The use of eye movements to evaluate the effect of PC-based risk awareness training on an advanced driving simulator. *Human Factors, 48*, 447–464.

Pradhan, A., Pollatsek, A., Knodler, M., & Fisher, D. (2009). Can younger drivers be trained to scan for information that will reduce their risk in roadway traffic scenarios that are hard to identify as hazardous? *Ergonomics, 52*, 657–673.

Pradhan, A., Divekar, G., Masserang, K., Romoser, M., Zafian, T., Blomberg, R., Thomas, F.D. et al. (2011). The effects of focused attention training (FOCAL) on the duration of novice drivers' glances inside the vehicle. *Ergonomics, 54*, 917–931.

Preusser, D. (2003). Young driver crash risk. *Annual Proceedings of the Association for Automotive Medicine* (pp. 527–532). Association for Advancement of Automotive Medicine.

Reeves, L., & Weisberg, R. (1994). The role of content and abstract information in analogical transfer. *Psychological Bulletin, 115*, 381–400.

Regan, M., Deery, H., & Triggs, T. (1998). Training for attentional control in novice car drivers. *Proceedings of the 42nd Annual Meeting of the Human Factors and Ergonomics Society* (pp. 1452–1456). Chicago, IL: Human Factors and Ergonomics Society.

Regan, M., Triggs, T., & Wallace, P. (1999). A CD ROM product for enhancing perceptual and cognitive skills in novice car drivers. *Proceedings of the 1st International Conference on Novice Driver Issues.*

Regan, M., Triggs, T., & Godley, S. (2000). Simulator-based evaluation of the DriveSmart novice driver CD-ROM training product. *Proceedings of the Road Safety Research, Policiing and Education Conference.* Brisbane, Australia.

Roberts, I.G., & Kwan, I. (2008). School-based driver education for the prevention of traffic crashes (Review). The Cochrane Collaboration, 4. Wiley.

Roelofs, E., Vissers, J., van Onna, M., & Kern, G. (2012a). Coaching young drivers in a second phase training programme. In L. Dorn (Ed.), *Driver Behaviour and Training* (V) (pp. 3–24). Aldershot, UK: Ashgate.

Roelofs, E., van Onna, M., Brookhuis, K., Marsman, M., & de Penning, L. (2012b). Designing developmentally tailored driving assessment tasks for formative purposes. In L. Dorn (Ed.), *Driver Behaviour and Training* (V) (pp. 61–80). Aldershot, UK: Ashgate.

Sagberg, F., & Bjornskau, T. (2006). Hazard perception and driving experience among novice drivers. *Accident Analysis & Prevention, 382*, 407–414.

Saito, H., & Miwa, K. (2007). Construction of a learning environment supporting learners' reflection: A case of information seeking on the Web. *Computers & Education, 49*(2), 214–29.

Scott-Parker, B., Goode, N., & Salmon, P. (2015). The driver, the road and the rules...and the rest? A systems based approach to young driver safety. *Accident Analysis & Prevention, 74*, 297–305.

Šeibokaitė, L., Endriulaitienė, A., Žardeckaitė-Matulaitienė, K., & Markšaitytė, R. (2011). Self-reported and experimental risky driving measurement: What to rely on in young drivers. Kaunas: Kaunas University of Technology, pp. 131–134.

Senserrick, T., Ivers, R., Boufous, S., Chen, H., Norton, R., Stenvenson, M., & Zask, A. (2009). Young driver education programs that build resilience have the potential to reduce road crashes. *Pediatrics, 124*, 1287–1292.

Shen, L.P., Leon, E., Callaghan, V., & Shen, R.M. (2007). Exploratory Research on an Affective e-Learning Model. In J. Fong & F.L. Wang (Eds.), *Proceedings of Workshop on Blended Learning' 2007*, pp. 267–278. Accessed on March 2, 2009 from http://www.cs.cityu.edu.hk/~wbl2007/WBL2007_Proceedings_HTML/WBL2007_PP267-278_Shen.pdf.

Shih, Y.E. (2007). Dynamic language learning: Comparing mobile language learning with online language learning. PhD diss., Minneapolis, MN: Capella University.

Stanton, N.A., Walker, G.H., Young, M.S., Kazi, T., & Salmon, P.M. (2007). Changing drivers' minds: The evaluation of an advanced driver coaching system. *Ergonomics, 50*, 1209–1234.

Steinberg, L. (2004). Risk-taking in adolescence: What changes, and why? *Annals of the New York Academy of Sciences, 1021*, 51–58.

Stock, J., Weaver, J., Ray, H., Brink, J., & Sadof, M. (1983). *Evaluation of Safe Performance Secondary School Driver Education Curriculum Demonstration Project.* Washington, DC: National Highway Traffic and Safety Administration.

Strang, P.M., Deutsch, K.B., James, R.S., & Manders, S.M. (1982). A comparison of on-road and off-road driver training. Canberra City, Australia: Department of Transport and Construction Office of Road Safety.

Sümer, N., Özkan, T., & Lajunen, T. (2006). Asymmetric relationship between driving and safety skills. *Accident Analysis & Prevention, 38*(4), 703–711.

Sundström, A. (2008). Self-assessment of driving skill—A review from a measurement perspective. *Transportation Research Part F Traffic Psychology and Behaviour, 11*(1), 1–9.

Taylor, T., Masserang, K., Divekar, G., Samuel, S., Muttart, J., Pollatsek, A., & Fisher, D. (2011). Long term effects of hazard anticipation training on novicd drivers measured on the road. *Proceedings of the 6th International Driving Symposium on Human Factors in Driver Assessment, Training, and Vehicle Design 2011.* Iowa City: Public Policy Center, University of Iowa.

Thomas, F.D., Rilea, S., Blomberg, R.D., Peck, R.C., & Korbelak, K.T. (2016). *Evaluation of the Safety Benefits of the Risk Awareness and Perception Training Program for Novice Teen Drivers* (DOT HS 812 235). Washington, DC: National Highway Traffic Safety Administration.

Triggs, T., & Regan, M. (1998). Development of a cognitive skills training product for novice drivers. *Proceedings of the 1998 Road Safety Research, Education and Enforcement Conference.* Wellington, New Zealand: Land Transport Authority.

Vernick, J.S., Li, G., Ogaitis, S., MacKenzie, E.J., Baker, S.P., & Geilen, A.C. (1999). Effects of high school driver education on motor vehicle crashes, violations, and licensure. *American Journal of Preventive Medicine, 16*(Suppl. I), 40–46.

Victoir, A., Eertmans, A., Van den Bergh, O., & Van den Broucke, S. (2005). Learning to drive safely: Social-cognitive responses are predictive of performance rated by novice drivers and their instructors. *Transportation Research Part F: Traffic Psychology and Behaviour, 8*, 59–71.

Vlakveld, W. (2005). *Young, Novice Motorists, Their Crash Rates, and Measures to Reduce Them: A Literature Study.* Leidschendam, The Netherlands: SWOV.

Wetton, M., Hill, A., & Horswill, M. (2013). Are what happens next exercises and self-generated commentaries useful additions to hazard perception trianing for novice drivers? *Accident Analysis & Prevention, 54*, 57–66.

Williams, A.F. (2007). Contribution of the components of graduated licensing to crash reductions. *Journal of Safety Research, 38*, 177–184.

Williams, A., & Ferguson, S. (2004). Driver education renaissance? *Injury Prevention, 10*, 4–7.

Williams, A., & Shults, R. (2010). Graduated licensing research 2007–present: A literature. *Journal of Safety Research, 41*, 77–84.

Willis, D. (1998). The impetus for the development of a new risk management training program for teen drivers. *Proceedings of the Human Factors and Ergonomics Society 42nd Annual Meeting* (pp. 1394–1395). Santa Monica, CA: Human Factors and Ergonomics Society.

Wynne-Jones, J.D., & Hurst, P.M. (1985). The AA Driver Training Evaluation. Traffic research. Report No. 33. Wellington, NZ: Ministry of Transport.

Yamani, Y., Samuel, S., & Fisher, D. (2016). Evaluation of the effectiveness of a multi-skill program for training younger drivers on higher cognitive skills. *Applied Ergonomics, 52*, 135–141.

Young, A., Chapman, P., & Crundall, D. (2014). Producing a commentary slows concurrent hazard perception responses. *Journal of Experimental Psychology: Applied, 20*, 285–294.

Zask, A., van Beurden, E., Brooks, L., & Dight, R. (2006). Is it worth the RRISK? Evaluation of the RRISK (Reduce Risk, Increase Student Knowledge) program for adolescents in rural Australia. *Journal of Adolescent Health, 38*, 495–503.

19

Parenting Teenaged Drivers: Training and Licensure

Jessica H. Mirman

Allison E. Curry

Abstract

In comparison to other domains of development and risk-taking during adolescence, there has been comparatively less research on how interactions with parents are associated with learning to drive. This is surprising given that motor vehicle crashes are a leading cause of morbidity and mortality among teenagers worldwide. However, in recent years, there has been a noticeable increase in research on parents of teenaged drivers as well as a proliferation of parent-directed interventions and programming to assist parents during teenagers' transition from passenger to drivers. In this chapter, we review the existing scientific literature on parental involvement in teenagers' path to licensure under the umbrella of the graduated driver licensing (GDL) system, thus providing a current overview of the state of the science.

19.1 Introduction

19.1.1 Chapter Overview

The overall goal of this chapter is to provide an overview of the existing scientific literature on parent–teen interactions across the learning-to-drive continuum. First, we provide a rationale for the importance of parents to actively facilitate and engage with their teen's learning-to-drive experience within the graduated driver licensing (GDL) policy environment. Next, we review parent–teen interactions in detail within the context of a three-step GDL system that includes: (1) a learner (permit) period, (2) an intermediate (probationary or restricted) period, and (3) full (unrestricted) licensure. Finally, we review interventions designed to optimize parent–teen interactions and conclude with developmental considerations as they relate to parental influence on licensure and training.

19.1.2 Parenting under the Auspice of GDL

Practically, parents guide teenagers through licensure, providing administrative and, often, financial support with respect to completing licensing paperwork, selecting and insuring a vehicle, determining when it is time to move to the next phase (within legal guidelines), providing driving practice supervision during the permit phase, and remaining engaged during the early independent driving period to ensure compliance with laws and home rules (Simons-Morton & Ouimet, 2006; Simons-Morton et al., 2002, 2008).

GDL provides parents with a structured support framework to help them navigate their teenagers' transition from child passenger to licensed driver. However, parents' knowledge and support of GDL varies. Some surveys of parents have found general support for the concept of GDL and overall parent engagement with learning to drive (Ferguson & Williams, 1996; Williams, 2011; Williams et al., 1998). However, this broad support does not mean that parents necessarily know the specific provisions, like the number of required practice hours (O'Brien et al., 2013). Nor does it mean that the broad support necessarily holds true when parents are asked about the support of specific provisions. Support for provisions might be influenced by local social norms and community practices; for example, some evidence suggests that rural communities are more supportive of lower licensing ages compared to more urban areas for practical reasons (Gill et al., 2013). However, other research has not found differences by urbanicity within state (*The North Carolina Graduated Driver Licensing System: Urban–Rural Differences*, 2001). Support for specific provisions has also been shown to vary according to affluence, gender, and whether one is a parent (Campbell et al., 2009).

Restrictions on cell phones, passengers, and nighttime driving are generally well supported, with teenagers being comparatively less supportive of passenger restrictions and older licensing ages than parents (Williams, 2011). In New Jersey, there is still high disapproval among parents and teenagers for using license plate decals to make the identification of teenaged drivers in the first two phases of GDL easier for law enforcement and other motorists, despite their use in other countries and evidence suggesting that decals reduce crashes (Curry et al., 2013, 2015a; Williams & McCartt, 2014). Finally, some research suggests that racial/ethnic disparities may exist, with GDL and other state-level traffic safety policies being less effective for these hard-to-reach groups (Curry et al., 2012; Garcia-Espana et al., 2012; Hartos et al., 2005; Jacobsohn et al., 2012). Community-based participatory action research may be needed to develop research studies that can identify reasons for these disparities and to guide the development of initiatives to promote compliance with state provisions. In the following three sections, we review parent–teen interactions in the context of GDL in more detail.

19.2 Learner's Permit

19.2.1 Goals of the Learner's Permit

This phase is intended to provide an opportunity for novice drivers to learn to drive under the guidance of an experienced and safe driver (Mayhew, 2003) and is the safest with respect to crash involvement (Gregersen et al., 2003; Mayhew et al., 2003; Williams et al., 1997). In most GDL systems, the teenaged applicant must first pass a knowledge test focusing on applicable laws and policies and basic vehicle maintenance to obtain a learner's permit. Jurisdictions vary with respect to the minimum age at entry, the minimum required holding period, the variety and quantity of supervised practice required (e.g., number of hours, types of driving conditions, and environments), the requirement that teens participate in a formal driver education program, and the processes by which these requirements will be satisfied (i.e., who will certify and be held accountable for these requirements being met).

19.2.2 Parent Role

For many teenagers, the role of practice supervisor—and furthermore the role of "certifier"—falls to parents. A 2006 nationally representative study of US public high school students found that 87% of licensed teenaged drivers reported that their parents were involved in teaching them to drive. Only 6.7% of teenagers reported teaching themselves (i.e., no involvement from a parent or instructor); these teenagers were more likely to be Hispanic/Latino, Black, and male. Thirty percent of US teenagers reported learning from both parents (mother and father) and a driver education instructor, 8% were taught only by an instructor, 8% reported learning from mothers only, 15.2% reported learning from fathers only, and 18% reported learning only from parents (no instructor involvement) (Ginsburg et al., 2011). Also in the United States, a two-parent supervisory team has been associated with a greater quantity of practice, potentially because of a greater number of practice opportunities, while no effect on overall practice hours was found for instructor involvement (Jacobsohn et al., 2012). Historically, most European countries have not emphasized accompanied driving by parents or the "lay instructor," but this may change as multistage GDL systems become more popular (Twisk & Stacey, 2007).

Given that the majority of teenagers have parents involved in the licensing process, it's important to understand what factors are associated with optimal supervision—and what optimal supervision might look like. Several studies have demonstrated heterogeneity in parents' perceptions of their competence as driving supervisors, desire to serve as practice supervisors, and engagement with the process (Goodwin et al., 2006; Mirman & Kay, 2011; Williams et al., 2006). Stronger parental intentions to be engaged supervisors have been found to be associated with stronger self-efficacy, perceived support, and normative pressure from teens, and beliefs that practice will be calm (Mirman et al., 2014b). Similarly, the family climate around practice driving and attitudes of the practice supervisor affect the teenager's perceptions about the utility of practice driving, attitudes about driving more generally, and potentially, future driving behavior (Taubman-Ben-Ari, 2010; Taubman-Ben-Ari & Katz-Ben-Ami, 2013). Parents of teenagers with permits have reported confidence that their teenagers were ready to begin to learn, and few anticipated that learning would be difficult (Sherman et al., 2004).

Parents can be successful in helping their teenagers obtain basic vehicle-maneuvering skills and follow the rules of the road; however, evidence from self-report and policy studies connecting the quantity of supervised practice to postlicense safety outcomes remains weak (Foss et al., 2012; O'Brien et al., 2013), and teenagers still demonstrate notable skill deficits at the end of the learner phase, including those associated with hazard perception and avoidance, difficulty with lane and speed management, and trouble with intersections, especially left turns (Braitman et al., 2008; Durbin et al., 2014b; Foss et al., 2011; McKnight & McKnight, 2003). The quantity and diversity of parent-supervised practice can increase the driving skill of prelicensed teenagers when measured using a validated observational assessment of driver skill (Durbin et al., 2014b; Mirman et al., 2014a,b), pointing both to the need to continue to find ways to help parents improve their supervision of learner teens and to the fact that methodological choices may influence study findings (Beanland et al., 2013).

Extant research suggests that in contrast to professional instructors, parents limit practice driving to low-risk environments, focus on basic mechanics of driving and vehicle maintenance as opposed to the situations and skill deficits most likely to cause a crash, lack the ability to effectively transfer their driving wisdom/experience to their teens, and often use routine trips to conduct practice drives, resulting in limited exposure to a variety of driving environments and conditions (Berg et al., 1999; Goodwin et al., 2010, 2012, 2013; Groeger & Banks, 2007; Mirman & Kay, 2011; Tronsmoen, 2010, 2011).

Little is understood regarding how parents view the role of formal driver education and how the instructor–teen–parent triad should interact to be most beneficial for the learner teen. Many parents report very minimal involvement with their teenager's instructor and mixed endorsement for a requirement compelling parents to be involved (Hartos & Huff, 2008). Parents have not been found to be successful as replacements for professional instructors; a Texas study found that when parents served as instructors, teenagers had more violations and crashes (Pezoldt et al., 2007). Finally, some studies suggest that parents might structure supervised practice differently depending on the gender of their child, and that these differences may impact driving attitudes, competencies, and safety (Gregersen et al., 2003; Taubman-Ben-Ari & Katz-Ben-Ami, 2013). For example, female teens perceived parents as more engaged in encouraging safe driving (e.g., stronger monitoring, more feedback on driving) compared to male teens (Taubman-Ben-Ari & Katz-Ben-Ami, 2013).

Parental involvement during the learner period of GDL can have important implications for skill acquisition, compliance with GDL provisions, and shaping how teenagers think about driving, as well as an important impact on teenagers' driving behaviors. Given the variable quality of parents' supervision of teenaged permit holders, additional research on how to best reach and support parents is warranted. Underrepresented teenagers and parents and teenagers who might be at higher risk for learning difficulties and/or risky driving represent specific subgroups that require more intensive focus from researchers.

19.3 Intermediate/Restricted License Holders

19.3.1 Goals of the Intermediate Phase

The intermediate phase of GDL is the most risky for new teenaged drivers (Mayhew et al., 2003). The overall goal of this phase is to keep the initial months of independent driving as low risk as possible while the newly licensed teenaged driver gains practical driving experience. Risk reduction is achieved by restricting drivers from situations associated with higher crash risk (e.g., driving with similarly aged passengers, driving at night). Some jurisdictions provide exemptions for teenagers to drive their younger siblings; however, risk of a fatality to a child passenger 8–12 years of age in a car driven by a teenaged driver less than 16 years of age is over seven times greater than the risk to a child passenger in a car driven by a driver at least 25 years old (Winston et al., 2008). Therefore, families should strongly consider safer alternatives to having a teenaged driver transport younger siblings.

19.3.2 Parent Role

Parents have several practical roles related to the transition to independent driver. They help to decide when the teenager is ready to apply for a license in terms of his or her driving skill and his or her maturity to handle driving-related responsibilities. Parents make choices regarding the type of vehicle their teenagers will drive, whether they will share the vehicle with other family members, and whether the teenagers will bear financial responsibilities for being a driver (e.g., paying for gas, insurance, licensing and registration, purchasing and maintaining the vehicle). Having shared access to a well-maintained vehicle with modern safety features, and avoiding driving pickups and SUVs, which have a higher risk of fatal rollovers compared to other vehicles, protects against crash and injury risk (Cammisa et al., 1999; Ferguson, 2003; Garcia-Espana et al., 2009; Rivara et al., 1998; Trowbridge et al., 2007).

In choosing a vehicle for their teenager, parents are motivated by safety, cost, and practical reasons (i.e., letting the teenager drive a car the family already owns, preference for low-maintenance vehicles). Additionally, geographic differences exist in parents' knowledge of specific safety features and willingness to permit sole access to vehicles (Hellinga et al., 2007). A national survey found that 83% of teenagers drive used vehicles, 22% drive SUVs, and 14% drive pickups (Eichelberger et al., 2014). Taking these all together, there is a clear need for continued education regarding safe vehicles for teenagers to drive, but these efforts will need to be delivered in light of real practical constraints parents face in choosing a vehicle.

Parents can also act as home-based GDL enforcers by reinforcing restrictions (e.g., passenger limits) through complementary house rules (Brookland et al., 2014; Hartos et al., 2000, 2001a). Research has shown that the type and strength of family-based restrictions on teenagers' early postlicense driving are modifiable via intervention (e.g., parent–teen agreements) and afford a protective benefit in terms of lower frequencies of risky driving (Beck et al., 2006; Hartos, 2002; Hartos et al., 2004, 2005).

In addition to limit setting, parents may have a positive influence on their teens' safe driving attitudes and behaviors by modeling safe driving behaviors (Schmidt et al., 2014), monitoring driving behaviors and free time (Farah et al., 2014; Laird, 2014; Mirman et al., 2012), and engaging in authoritative parenting practices (i.e., providing emotional support in the context of supportive rules) (Ginsburg et al., 2009). Several studies have demonstrated that parents and their teenagers have similar (i.e., correlations between the same constructs or similar "levels" of the same constructs) attitudes and appraisals of driving concepts (e.g., attitudes about safety, self-concept as a driver) (Lahatte & Le Pape, 2008; Taubman-Ben-Ari, 2010) and driving behaviors (e.g., citations and risky driving styles) (Ehsani et al., 2014; Ferguson et al., 2001; Taubman-Ben-Ari, 2010; Taubman-Ben-Ari & Katz-Ben-Ami, 2013; Taubman-Ben-Ari et al., 2005; Wilson et al., 2006). These associations likely result from the parenting practices described earlier, in addition to shared genetic and environmental factors. Parents and teenagers demonstrate less similarity in their perceptions of how involved parents should be in managing the teenagers' driving by setting limits or monitoring (Beck et al., 2005; Laird, 2014).

Parents can also have a safety-negative influence by modeling unsafe behaviors and by distracting teenagers directly (Schmidt et al., 2014). A survey of teenagers found that when they receive phone calls while driving, the caller is usually a parent; text messages are more often from friends. Teens reported that parents expect to be able to reach them and get mad if they don't answer their phone, and that they have to tell their parents where they are, creating a distracting and tense experience behind the wheel (Lavoie et al., 2014). Widely considered to be a critical public health problem, next steps in reducing distracted driving, or promoting engaged driving, include understanding how often and under what circumstances teenagers disengage from driving. Importantly, studies should not just focus on specific distractors, as rapid technological innovations will continually cycle new products in and out of markets faster than countermeasures can be developed (Durbin et al., 2014a).

19.4 Full License Holders

In many jurisdictions the restrictions of the intermediate stage automatically expire—either at the end of a designated length of time or when the driver reaches a certain age—and the driver transitions into an unrestricted or full license period; in other jurisdictions, the driver must present to a licensing office to transition to a full license. While in many international jurisdictions, GDL applies to older novice drivers, in the United States only a few states (Connecticut, New Jersey, Maine, Maryland) extend one or more GDL provisions to novice drivers older than 18 years of age. Given the safety benefit associated with GDL programs and that many drivers—and a disproportionate proportion of low-income drivers—do not get licensed until after age 18 (Curry et al., 2014; Tefft et al., 2014), extending GDL provisions to emerging adults raises the question of what kind of role parents might have, if any, toward facilitating their young adult's transition to licensed driver. For example, novice drivers who initiate learning to drive in their late teens and early 20s may obtain supervised practice from friends instead of parents or older family members, with unknown consequences. They may also be ill prepared for jobs that require driving. More research on this topic is needed.

19.4.1 Parent Role

The parental role for teenaged and young adult full license holders is not straightforward. About half of young adults aged 18–24 live at least some of the time with their parents, suggesting that it is certainly practically possible for parents to remain involved (Vespa et al., 2013). As young adults move out of the home and become more independent, it is more difficult and less developmentally appropriate for parents to enforce safe driving behaviors, and jurisdictions vary with respect to how old drivers are when they enter this stage. Developmentally, emerging adulthood has been proposed to be a relatively new stage between adolescence and adulthood that occurs among 18- to 25-year-old youth in communities supportive (politically, economically, and socially) of a prolonged period of identify exploration, flux, and delayed independence (Arnett, 2000). Young adults do need to increase their independent management of their own health behaviors, and the transition out of the home serves as landmark moment for many families in this regard (Lefkowitz, 2005). Of particular importance for emerging adults, especially given the emphasis on exploratory activities, is increased access to a variety of risky activities, decrease in parental supervision, and the impact that these changes have on their health (Fromme et al., 2009). This developmental period can be a time of great risk-taking in some domains, especially with respect to substance use (Arnett, 2005; Sussman & Arnett, 2014).

Understanding predictors of impaired driving is of particular importance given increased access to drugs and alcohol outside of the home, coupled with becoming legally allowed to purchase and consume alcohol at 21 years of age (in the United States). Driving impaired is strongly associated with other risk behaviors, including riding with an impaired driver, driving distracted, aggressive driving, and internalizing and externalizing psychiatric symptoms (e.g., depression, anxiety) (McDonald et al., 2014; Olsen et al., 2013; Whitehill et al., 2014). Although parents have a diminished role managing these behaviors from a supervisory perspective in late adolescence/early adulthood, parent practices (e.g., permissiveness) in earlier developmental periods can influence behavior later in life (Bingham & Shope, 2004; Laird, 2014). While not much is currently known about parent engagement with emerging adults on driving-related issues, this is an important area of future research given the interest in extending GDL. Other developmental research suggests that parents can continue to have an importance influence on young adults (Abar & Turrisi, 2008).

19.5 Interventions

Many interventions and resources have been developed for parents across the GDL continuum given the important role that parents have in training and licensure and the recognition that they could use more support. A variety of modalities have been successfully employed to deliver intervention content to parents, including in-vehicle data records (IVDRs), in-person training or orientation sessions, actual skills-based PC training programs such as Risk Awareness and Perception Training (RAPT), and web-based platforms to disseminate information and provide practical tools to facilitate parent supervision (e.g., electronic logging and rating tools, instructional videos, a framework for setting driving limits) (Farah et al., 2014; McGehee et al., 2007; Mirman et al., 2014c; Pollatsek et al., 2006; Simons-Morton et al., 2006, 2013; Taubman-Ben-Ari & Lotan, 2011; Toledo et al., 2012; Zakrajsek et al., 2013).

The most successful interventions, irrespective of modality, have a specified theoretical model that links program activities with short-, medium-, and long-term goals and actively engages parents (e.g., contact parents through phone calls or e-mails, or via in-person sessions), as opposed to passive content dissemination (Curry et al., 2015b). By definition, parent programs seek to alter the parent–teen relationship in some way by modifying parental psychological and/or behavioral variables: placement of restrictions on postlicense behaviors, increasing parent knowledge of learner skills and postlicense driving behaviors, and training parents to be more effective practice supervisors have shown the most promising results with respect to reductions in risky driving and increases in driving skill (Farah et al., 2014; Mirman et al., 2014c; Simons-Morton et al., 2006, 2013).

Formative and process evaluations of parent training interventions have shown parent and teen support for the concept of parent training and generally strong ratings of usability and acceptability of intervention content (Hartos et al., 2001b; Mirman et al., 2012b, 2014a; Ramirez et al., 2013; Winston et al., 2014). Research on the acceptability of IVDRs has shown that while parents and teenagers see value in the objectivity provided by such devices, concerns about cost, for parents, and about privacy invasion and trust violations, for parents and teens, are notable barriers to consistent use (Gesser-Edelsburg & Guttman, 2013; Guttman & Lotan, 2011).

Some states have developed parent orientation sessions as part of their GDL program (Fischer, 2013); however, these orientations have not yet been systematically evaluated for effectiveness. In parallel, there is a growing body of research that suggests that certain types of behaviors might be more difficult to cultivate due to waning engagement with the intervention over time, due to a perceived incompatibility with day-to-day life (e.g., planning drives ahead of time and increasing practice hours), and because less risky families tend to participate in evaluation research, making it difficult to ascertain the effect of the interventions on the most at-risk parents and teenagers (Guttman & Gesser-Edelsburg, 2011; Peek-Asa et al., 2014; Simons-Morton et al., 2006; Winston et al., 2014). Future gains can be made by developing strong collaborations between researchers and stakeholders. Making these connections will lead to research that establishes the types of sustainable platforms that can provide families with the necessary guidance and support to reduce the occurrence of motor vehicle crashes caused by teen drivers. Designing evidenced-based interventions that complement each phase of GDL was the first step. Testing their combined effectiveness in the real world and figuring out how to share these tools with various populations is next.

19.6 Developmental Considerations

The teenage years, and the developmental changes that come with them, can present unique, and often normative, challenges to the parent–teen relationship. These challenges include redefining the parent–youth relationship in light of adolescents' desire for autonomy, especially related to personal issues (Smetana & Asquith, 1994; Smetana et al., 2006); an increasing amount of time spent outside of the home (Dubas & Gerris, 2002; Larson et al., 1996); and willingness and ability to disagree with parents (Fuligni, 1998). For the most part, these changes are not harmful and are a healthy part of development, and while parents' roles may diminish in some areas, they can remain important providers of socioemotional and material support for youth (Steinberg, 2001; Steinberg et al., 1992).

Important to any discussion of parent–teen "interactions" or "parenting of teen drivers" is the recognition that these interactions are bidirectional, embedded in historical patterns of interactions, influenced by factors at other levels (e.g., biological, cultural, environmental), and potentially malleable (Lerner & Castellino, 2002). Further, as with earlier (and later) periods of development, teenagers have active roles in determining their developmental trajectories and outcomes. Teenagers vary in their preferences for activities and situations; how they interpret and learn from experiences, make inferences about themselves, and how the world works; how they regulate their internal states; how they elicit responses from other people; and how they engage with others. Collectively, these actions are thought of as the "active child" (or in this case, active teenager) in that people actively influence their development (Scarr & McCartney, 1983). Similarly, parents also vary along these dimensions. These concepts are critically important when it comes to characterizing (i.e., systematically and validly measuring) parent–teen interactions in the context of teen driving, determining what kinds of parent–teen interactions are most beneficial for teen drivers, and optimizing these interactions via intervention. Theoretical models seeking to explain teens' elevated crash likelihood compared to other populations have generally not considered bidirectionality or active teenagers (i.e., how teens might influence the kinds of supervision parents provide, which then influences teen behavior, and so on) (Beck et al., 2002; Runyan & Yonas, 2008; Shope & Bingham, 2008; Sommers & Ribak, 2008). Even when systems theories that are predicated on bidirectional interactions (e.g., bioecological model) have been invoked, there is a tendency to constrain interactions empirically as parents' unidirectional effect on teens (see Runyan & Yonas, 2008), and an examination of person-by-context effects is almost nonexistent (Carpentier et al., 2014).

Evidence does suggest that children and youth contribute to the ways they are parented (Fletcher et al., 2004; Granic & Patterson, 2006; Kerr et al., 2010). For example, some parents decrease support, control, and monitoring in response to teen problem behavior (Stice & Barrera, 1995). In a national study of parent–teen dyads during the learner period, stronger parental perceptions of perceived support from their teen related to practice driving (i.e., support provided by the teen to the parent) were associated with stronger parental self-efficacy, less parental stress, and greater intentions to remain engaged supervisors (Mirman et al., 2014b). Among teenagers, stronger perceptions about the risks of driving have been found to be associated with being more supportive of parental involvement and limit setting

(Laird, 2014). There is scant research in the parent–teen driving literature that examines how parents might modulate their engagement and change parenting practices as a result of their teenagers' behavior and as a result of prior behavioral patterns. None of this suggests that parents do not affect teen drivers' outcomes; rather, it suggests that the picture is just more complicated than what has generally been captured in existing theoretical models of teen crash involvement and empirical studies of parent–teen interactions in the context of licensure and training.

19.7 Overall Summary

The relationship between parents and teenagers as they progress through the GDL licensing system is complex. Undoubtedly, parents have a strong influence on their teenagers, and in turn, teenagers on their parents. GDL sets the stage for these interactions to unfold, but evidence-based psychoeducational programs are needed to complement strong GDL policy. A greater emphasis on reaching underserved populations globally is needed as is a more nuanced focus on reciprocal (parent–teen) interactions and how they predict stability and change in multiple domains related to learning to drive (skill, attitudes, behavior, self-regulation).

References

Abar, C., & Turrisi, R. (2008). How important are parents during the college years? A longitudinal perspective of indirect influences parents yield on their college teens' alcohol use. *Addictive Behaviors, 33*(10), 1360–1368. doi:10.1016/j.addbeh.2008.06.010.

Arnett, J. J. (2000). Emerging adulthood: A theory of development from the late teens through the twenties. *American Psychologist, 55*(5), 469–480.

Arnett, J. J. (2005). The developmental context of substance use in emerging adulthood. *Journal of Drug Issues, 35*(2), 235–254. doi:10.1177/002204260503500202.

Beanland, V., Goode, N., Salmon, P. M., & Lenné, M. G. (2013). Is there a case for driver training? A review of the efficacy of pre- and post-licence driver training. *Safety Science, 51*(1), 127–137. doi:10.1016/j.ssci.2012.06.021.

Beck, K. H., Hartos, J. L., & Simons-Morton, B. G. (2002). Teen driving risk: The promise of parental influence and public policy. *Health Education & Behavior, 29*(1), 73–84. doi:10.1177/109019810202900108.

Beck, K. H., Hartos, J. L., & Simons-Morton, B. G. (2005). Parent–teen disagreement of parent-imposed restrictions on teen driving after one month of licensure: Is discordance related to risky teen driving? *Prevention Science, 6*(3), 177–185. doi:10.1007/s11121-005-0001-6.

Beck, K. H., Hartos, J. L., & Simons-Morton, B. G. (2006). Relation of parent–teen agreement on restrictions to teen risky driving over 9 months. *American Journal of Health Behavior, 30*(5), 533–543. doi:10.5555/ajhb.2006.30.5.533.

Berg, H., Eliasson, K., Palmkvist, J., & Gregersen, N. P. (1999). Learner drivers and lay instruction—How socio-economic standing and lifestyle are reflected in driving practice from the age of 16. *Transportation Research Part F, 2*, 167–179.

Bingham, C. R., & Shope, J. T. (2004). Adolescent developmental antecedents of risky driving among young adults. *Journal of Studies on Alcohol, 65*(1), 84–94. Retrieved from http://www.ncbi.nlm.nih.gov/pubmed/15000507.

Braitman, K., Kirley, B. B., McCartt, A. T., & Chaudhary, N. K. (2008). Crashes of novice teenage drivers: Characteristics and contributing factors. *Journal of Safety Research, 39*(1), 47–54. doi:10.1016/j.jsr.2007.12.002.

Brookland, R., Begg, D., Langley, J., & Ameratunga, S. (2014). Parental influence on adolescent compliance with graduated driver licensing conditions and crashes as a restricted licensed driver: New Zealand Drivers Study. *Accident Analysis & Prevention, 69*, 30–9. doi:10.1016/j.aap.2013.06.034.

Cammisa, M. X., Williams, A. F., & Leaf, W. A. (1999). Vehicles driven by teenagers in four states. *Journal of Safety Research, 30*(1), 25–30. doi:10.1016/S0022-4375(98)00059-0.

Campbell, B. T., Chaudhary, N. K., Saleheen, H., Borrup, K., & Lapidus, G. (2009). Does knowledge of teen driving risks and awareness of current law translate into support for stronger GDL provisions? Lessons learned from one state. *Traffic Injury Prevention, 10*(4), 320–324. doi:10.1080/15389580903020527.

Carpentier, A., Brijs, K., Declercq, K., Brijs, T., Daniels, S., & Wets, G. (2014). The effect of family climate on risky driving of young novices: The moderating role of attitude and locus of control. *Accident Analysis & Prevention, 73*, 53–64. doi:10.1016/j.aap.2014.08.005.

Curry, A. E., Garcia-Espana, J. F., Winston, F. K., Ginsburg, K. R., & Durbin, D. R. (2012). Variation in teen driver education by state requirements and sociodemographics. *Pediatrics, 129*(3), 453–457. doi:10.1542/peds.2011-2303.

Curry, A. E., Pfeiffer, M. R., Localio, R., & Durbin, D. R. (2013). Graduated driver licensing decal law: Effect on young probationary drivers. *American Journal of Preventative Medicine, 44*(1), 1–7. doi:10.1016/j.amepre.2012.09.041.

Curry, A. E., Pfeiffer, M. R., Durbin, D. R., Elliott, M. R., & Kim, K. H. (2014). *Young Driver Licensing in New Jersey: Rates and Trends, 2006–2011* (pp. 1–22). Washington, DC: AAA Foundation for Traffic Safety.

Curry, A. E., Elliott, M. R., Pfeiffer, M. R., Kim, K. H., & Durbin, D. R. (2015a). Long-term changes in crash rates after introduction of a Graduated Driver Licensing decal provision. *American Journal of Preventative Medicine, 48*(2), 121–127. doi:10.1016/j.amepre.2014.08.024.

Curry, A. E., Peek-Asa, C., Hamann, C., & Mirman, J. H. (2015b). Effectiveness of parent-focused interventions to increase teen driver safety: A critical review. *Journal of Adolescent Health, 57*(1, Suppl.), S6–S14. doi:10.1016/j.jadohealth.2015.01.003.

Dubas, J. S., & Gerris, J. R. M. (2002). Longitudinal changes in the time parents spend in activities with their adolescent children as a function of child age, pubertal status, and gender. *Journal of Family Psychology, 16*(4), 415–27. Retrieved from http://www.ncbi.nlm.nih.gov/pubmed/12561286.

Durbin, D. R., McGehee, D. V, Fisher, D., & McCartt, A. (2014a). Special considerations in distracted driving with teens. *Annals of Advances in Automotive Medicine/Annual Scientific Conference, 58*, 69–83. Retrieved from http://www.pubmedcentral.nih.gov/articlerender.fcgi?artid=4001672&tool=pmcentrez&rendertype=abstract.

Durbin, D. R., Mirman, J. H., Curry, A. E., Wang, W., Fisher Thiel, M. C., Schultheis, M. T., & Winston, F. K. (2014b). Driving errors of learner teens: Frequency, nature and their association with practice. *Accident Analysis & Prevention, 72*, 433–439. doi:10.1016/j.aap.2014.07.033.

Ehsani, J. P., Simons-Morton, B. G., Xie, Y., Klauer, S. G., & Albert, P. S. (2014). The association between kinematic risky driving among parents and their teenage children: Moderation by shared personality characteristics. *Accident Analysis & Prevention, 69*, 56–61. doi:10.1016/j.aap.2014.03.015.

Eichelberger, A. H., Teoh, E. R., & Mccartt, A. T. (2014). *Vehicle Choices for Teenage Drivers: A National Survey of Parents July 2014*. Retrieved from http://www.iihs.org/frontend/iihs/documents/masterfiledocs.ashx?id=2071.

Farah, H., Musicant, O., Shimshoni, Y., Toledo, T., Grimberg, E., Omer, H., & Lotan, T. (2014). Can providing feedback on driving behavior and training on parental vigilant care affect male teen drivers and their parents? *Accident Analysis & Prevention, 69*, 62–70. doi:10.1016/j.aap.2013.11.005.

Ferguson, S. A. (2003). Other high-risk factors for young drivers—How graduated licensing does, doesn't, or could address them. *Journal of Safety Research, 34*(1), 71–77. doi:10.1016/S0022-4375(02)00082-8.

Ferguson, S. A., & Williams, A. F. (1996). Parents' views of driver licensing practices in the United States. *Journal of Safety Research, 27*(2), 73–81. doi:10.1016/0022-4375(96)00001-1.

Ferguson, S. A., Williams, A. F., Chapline, J. F., Reinfurt, D. W., & De Leonardis, D. M. (2001). Relationship of parent driving records to the driving records of their children. *Accident Analysis & Prevention, 33*(2), 229–34. doi:10.1016/S0001-4575(00)00036-1.

Fischer, P. (2013). *Promoting Parent Involvement in Teen Driving: An In-Depth Look at the Importance and the Initiatives*. Washington, DC: Governors Highway Safety Association.

Fletcher, A. C., Steinberg, L., & Williams-Wheeler, M. (2004). Parental influences on adolescent problem behavior: Revisiting Stattin and Kerr. *Child Development, 75*(3), 781–96. doi:10.1111/j.1467-8624.2004.00706.x.

Foss, R., Martell, C., Goodwin, A. H., & O'Brien, N. P. (2011). *Measuring Changes in Teenage Driver Crash Characteristics During the Early Months of Driving.* Washington, DC: AAA Foundation for Traffic Safety.

Foss, R., Masten, S. V, Goodwin, A. H., & O'Brien, N. P. (2012). *Role of Supervised Driving Requirements In a Graduated Driver Licensing Program (Report No. DOT HS 811 550.).* Washington, DC: NHTSA.

Fromme, K., Corbin, W. R., & Kruse, M. I. (2009). Behavioral risks during the transition from high school to college. *Developmental Psychology, 44*(5), 1497–1504. doi:10.1037/a0012614.

Fuligni, A. J. (1998). Authority, autonomy, and parent–adolescent conflict and cohesion: A study of adolescents from Mexican, Chinese, Filipino, and European backgrounds. *Developmental Psychology, 34*(4), 782–92. doi:10.1037/0012-1649.34.4.782.

Garcia-Espana, J. F., Ginsburg, K. R., Durbin, D. R., Elliott, M. R., & Winston, F. K. (2009). Primary access to vehicles increases risky teen driving behaviors and crashes: National perspective. *Pediatrics, 124*(4), 1069–1075. doi:10.1542/peds.2008-3443.

Garcia-Espana, J. F., Winston, F. K., & Durbin, D. R. (2012). Safety belt laws and disparities in safety belt use among US high-school drivers. *American Journal of Public Health, 102*(6), 1128–1134. doi:10.2105/AJPH.2011.300493.

Gesser-Edelsburg, A., & Guttman, N. (2013). "Virtual" versus "actual" parental accompaniment of teen drivers: A qualitative study of teens' views of in-vehicle driver monitoring technologies. *Transportation Research Part F: Traffic Psychology and Behaviour, 17*, 114–124. doi:10.1016/j.trf.2012.09.002.

Gill, S. K., Shults, R. a, Cope, J. R., Cunningham, T. J., & Freelon, B. (2013). Teen driving in rural North Dakota: A qualitative look at parental perceptions. *Accident Analysis & Prevention, 54*, 114–21. doi:10.1016/j.aap.2013.02.010.

Ginsburg, K. R., Durbin, D. R., Garcia-Espana, J. F., Kalicka, E. A., & Winston, F. K. (2009). Associations between parenting styles and teen driving, safety-related behaviors and attitudes. *Pediatrics, 124*(4), 1040–1051. doi:10.1542/peds.2008-3037.

Ginsburg, K. R., Winston, F. K., & Durbin, D. R. (2011). *Parents Teaching Teens to Drive: The Adolescent Perspective.* Philadelphia: The Children's Hospital of Philadelphia Research Institute and State Farm Mutual Insurance Company.

Goodwin, A. H., Waller, M. W., Foss, R., & Margolis, L. (2006). Parental supervision of teenage drivers in a graduated licensing system. *Traffic Injury Prevention, 7*(3), 224–231.

Goodwin, A. H., Foss, R., Margolis, L. H., & Waller, M. (2010). *Parents, Teens and the Learner Stage of Graduated Driver Licensing.* Washington, DC: AAA Foundation for Traffic Safety. doi:10.1037/e671962010-001.

Goodwin, A. H., O'Brien, N. P., & Foss, R. (2012). Effect of North Carolina's restriction on teenage driver cell phone use two years after implementation. *Accident Analysis & Prevention, 48*, 363–367. doi:10.1016/j.aap.2012.02.006.

Goodwin, A. H., Margolis, L. H., Foss, R., Harrell, S., O'Brien, N. P., & Kirley, B. B. (2013). *Improving Parental Supervision of Novice Drivers Using an Evidence-Based Approach.* Washington, DC: AAA Foundation for Traffic Safety.

Granic, I., & Patterson, G. R. (2006). Toward a comprehensive model of antisocial development: A dynamic systems approach. *Psychological Review, 113*(1), 101–31. doi:10.1037/0033-295X.113.1.101.

Gregersen, N. P., Nyberg, A., & Berg, H.-Y. (2003). Accident involvement among learner drivers—An analysis of the consequences of supervised practice. *Accident Analysis & Prevention, 35*(5), 725–730. doi:10.1016/S0001-4575(02)00051-9.

Groeger, J., & Banks, P. (2007). Anticipating the content and circumstances of skill transfer: Unrealistic expectations of driver training and graduated licensing? *Ergonomics, 50*(8), 1250–1263. doi:10.1080/00140130701318723.

Guttman, N., & Gesser-Edelsburg, A. (2011). "The little squealer" or "the virtual guardian angel"? Young drivers' and their parents' perspective on using a driver monitoring technology and its implications for parent–young driver communication. *Journal of Safety Research, 42*(1), 51–59. doi:10.1016/j.jsr.2010.11.001.

Guttman, N., & Lotan, T. (2011). Spying or steering? Views of parents of young novice drivers on the use and ethics of driver-monitoring technologies. *Accident Analysis & Prevention, 43*(1), 412–420. doi:10.1016/j.aap.2010.09.011.

Hartos, J. L. (2002). Parenting practices and adolescent risky driving: A three-month prospective study. *Health Education & Behavior, 29*(2), 194–206. doi:10.1177/109019810202900205.

Hartos, J. L., & Huff, D. C. (2008). Parent attitudes toward integrating parent involvement into teenage driver education courses. *Traffic Injury Prevention, 9*(3), 224–230. doi:10.1080/15389580801996521.

Hartos, J. L., Eitel, P., Haynie, D. L., & Simons-Morton, B. G. (2000). Can I take the car?: Relations among parenting practices and adolescent problem-driving practices. *Journal of Adolescent Research, 15*(3), 352–367. doi:10.1177/0743558400153003.

Hartos, J. L., Eitel, P., & Simons-Morton, B. G. (2001a). Do parent-imposed delayed licensure and restricted driving reduce risky driving behaviors among newly licensed teens? *Prevention Science: The Official Journal of the Society for Prevention Research, 2*(2), 113–122. doi:10.1023/A:1011595714636.

Hartos, J. L., Nissen, W. J., & Simons-Morton, B. G. (2001b). Acceptability of the checkpoints parent–teen driving agreement: Pilot test. *American Journal of Preventive Medicine, 21*(2), 138–141. doi:10.1016/S0749-3797(01)00330-0.

Hartos, J. L., Shattuck, T., Simons-Morton, B. G., & Beck, K. H. (2004). An in-depth look at parent-imposed driving rules: Their strengths and weaknesses. *Journal of Safety Research, 35*(5), 547–555. doi:10.1016/j.jsr.2004.09.001.

Hartos, J. L., Simons-Morton, B. G., Beck, K. H., & Leaf, W. A. (2005). Parent-imposed limits on high-risk adolescent driving: Are they stricter with graduated driver licensing? *Accident Analysis & Prevention, 37*, 557–562. doi:10.1016/j.aap.2005.01.008.

Hellinga, L. A., McCartt, A. T., & Haire, E. R. (2007). Choice of teenagers' vehicles and views on vehicle safety: Survey of parents of novice teenage drivers. *Journal of Safety Research, 38*(6), 707–713. doi:10.1016/j.jsr.2007.10.003.

Jacobsohn, L. S., Garcia-Espana, J. F., Durbin, D. R., Erkoboni, D., & Winston, F. K. (2012). Adult-supervised practice driving for adolescent learners: The current state and directions for interventions. *Journal of Safety Research, 43*(1), 21–28. doi:10.1016/j.jsr.2011.10.008.

Kerr, M., Stattin, H., & Burk, W. J. (2010). A reinterpretation of parental monitoring in longitudinal perspective. *Journal of Research on Adolescence, 20*(1), 39–64. doi:10.1111/j.1532-7795.2009.00623.x.

Lahatte, A., & Le Pape, M.-C. (2008). Is the way young people drive a reflection of the way their parents drive? An econometric study of the relation between parental risk and their children's risk. *Risk Analysis, 28*(3), 627–634. doi:10.1111/j.1539-6924.2008.01044.x.

Laird, R. D. (2014). Parenting adolescent drivers is both a continuation of parenting from earlier periods and an anticipation of a new challenge. *Accident Analysis & Prevention, 69*, 5–14. doi:10.1016/j.aap.2013.11.012.

Larson, R. W., Richards, M. H., Moneta, G., Holmbeck, G., & Duckett, E. (1996). Changes in adolescents' daily interactions with their families from ages 10 to 18: Disengagement and transformation. *Developmental Psychology, 32*(4), 744–754. doi:10.1037/0012-1649.32.4.744.

Lavoie, N., Lee, Y.-C., & Parker, J. (2014). Is that mom on the phone? Teen drivers and distraction. In *Applied Psychology Session*. Washington, DC: American Psychological Association.

Lefkowitz, E. S. (2005). "Things have gotten better": Developmental changes among emerging adults after the transition to university. *Journal of Adolescent Research, 20*(1), 40–63. doi:10.1177/0743558404271236.

Lerner, R. M., & Castellino, D. R. (2002). Contemporary developmental theory and adolescence: Developmental systems and applied developmental science. *The Journal of Adolescent Health, 31*(6 Suppl), 122–35. doi:10.1016/S1054-139X(02)00495-0.

Mayhew, D. R. (2003). The learner's permit. *Journal of Safety Research, 34*(1), 35–43. doi:10.1016/S0022-4375(02)00078-6.

Mayhew, D. R., Simpson, H. M., & Pak, A. (2003). Changes in collision rates among novice drivers during the first months of driving. *Accident Analysis & Prevention, 35*(5), 683–691. doi:10.1016/S0001-4575(02)00047-7.

McDonald, C. C., Sommers, M. S., & Fargo, J. D. (2014). Risky driving, mental health, and health-compromising behaviours: Risk clustering in late adolescents and adults. *Injury Prevention*, 1–8. doi:10.1136/injuryprev-2014-041150.

McGehee, D. V., Raby, M., Carney, C., Lee, J. D., & Reyes, M. L. (2007). Extending parental mentoring using an event-triggered video intervention in rural teen drivers. *Journal of Safety Research, 38*(2), 215–227. doi:10.1016/j.jsr.2007.02.009.

McKnight, J., & McKnight, S. (2003). Young novice drivers: Careless or clueless? *Accident Analysis & Prevention, 35*(6), 921–925. doi:10.1016/S0001-4575(02)00100-8.

Mirman, J. H., & Kay, J. (2011). From passengers to drivers: Parent perceptions about how adolescents learn to drive. *Journal of Adolescent Research, 27*(3), 401–424. doi:10.1177/0743558411409934.

Mirman, J. H., Albert, D., Jacobsohn, L. S., & Winston, F. K. (2012a). Factors associated with adolescents' propensity to drive with multiple passengers and to engage in risky driving behaviors. *The Journal of Adolescent Health, 50*(6), 634–640. doi:10.1016/j.jadohealth.2011.10.256.

Mirman, J. H., Lee, Y.-C., Kay, J., Durbin, D. R., & Winston, F. K. (2012b). Development of a web-based parent support program to improve quantity, quality, and diversity of teens' home-based practice driving. *Transportation Research Record: Journal of the Transportation Research Board, 2318*(-1), 107–115. doi:10.3141/2318-13.

Mirman, J. H., Albert, W. D., Curry, A. E., Winston, F. K., Fisher Thiel, M. C., & Durbin, D. R. (2014a). Teen driving plan effectiveness: The effect of quantity and diversity of supervised practice on teens' driving performance. *The Journal of Adolescent Health*, 1–7. doi:10.1016/j.jadohealth.2014.04.010.

Mirman, J. H., Curry, A. E., Wang, W., Fisher Thiel, M. C., & Durbin, D. R. (2014b). It takes two: A brief report examining mutual support between parents and teens learning to drive. *Accident Analysis & Prevention, 69*, 23–29. doi:10.1016/j.aap.2013.10.006.

Mirman, J. H., Curry, A. E., Winston, F. K., Wang, W., Elliott, M. R., Schultheis, M. T., Fisher Thiel, M. C., & Durbin, D. R. (2014c). Effect of the teen driving plan on the driving performance of teenagers before licensure: A randomized clinical trial. *JAMA Pediatrics*, 1–9. doi:10.1001/jamapediatrics.2014.252.

O'Brien, N. P., Foss, R., Goodwin, A. H., & Masten, S. V. (2013). Supervised hours requirements in graduated driver licensing: Effectiveness and parental awareness. *Accident Analysis & Prevention, 50*, 330–335. doi:10.1016/j.aap.2012.05.007.

Olsen, E. O., Shults, R. A., & Eaton, D. K. (2013). Texting while driving and other risky motor vehicle behaviors among US high school students. *Pediatrics, 131*(6), e1708–e1715. doi:10.1542/peds.2012-3462.

Peek-Asa, C., McGehee, D. V, & Ebel, B. E. (2014). Increasing safe teenaged driving time to integrate the growing evidence base. *JAMA Pediatrics, 168*(8), 703–704. doi:10.1542/peds.2011-2489.11.

Pezoldt, V., Womack, K. N., & Morris, D. E. (2007). *Parent-taught Driver Education in Texas: A Comparative Evaluation*. Washington, DC: National Highway Traffic Safety Administration.

Pollatsek, A., Narayanan, V., Pradhan, A., & Fisher, D. L. (2006). The use of eye movements to evaluate the effect of PC-based risk awareness training on an advanced driving simulator. *Human Factors*, (48), 447–464.

Ramirez, M., Yang, J., Young, T., Roth, L., Garinger, A., Snetselaar, L., & Peek-Asa, C. (2013). Implementation evaluation of steering teens safe: Engaging parents to deliver a new parent-based teen driving intervention to their teens. *Health Education & Behavior: The Official Publication of the Society for Public Health Education, 40*(4), 426–434. doi:10.1177/1090198112459517.

Rivara, F. P., Rivara, M. B., & Bartol, K. (1998). Dad, may I have the keys? Factors influencing which vehicles teenagers drive. *Pediatrics, 102*(5), E57.

Runyan, C. W., & Yonas, M. (2008). Conceptual frameworks for developing and comparing approaches to improve adolescent motor-vehicle safety. *American Journal of Preventative Medicine, 35*(3 Suppl), S336–S342. doi:10.1016/j.amepre.2008.06.019.

Scarr, S., & McCartney, K. (1983). How people make their own environments: A theory of genotype to environment effects. *Child Development, 54*(2), 424–435.

Schmidt, S., Morrongiello, B. A., & Colwell, S. R. (2014). Evaluating a model linking assessed parent factors to four domains of youth risky driving. *Accident Analysis & Prevention, 69*, 40–50. doi:10.1016/j.aap.2013.08.028.

Sherman, K., Lapidus, G., Gelven, E., & Banco, L. (2004). New teen drivers and their parents: What they know and what they expect. *American Journal of Health Behavior, 28*(5), 387–396. Retrieved from http://www.ncbi.nlm.nih.gov/pubmed/15482968.

Shope, J. T., & Bingham, C. R. (2008). Teen driving: Motor-vehicle crashes and factors that contribute. *American Journal of Preventative Medicine, 35*(3 Suppl), S261–S271. doi:10.1016/j.amepre.2008.06.022.

Simons-Morton, B. G., & Ouimet, M. C. (2006). Parent involvement in novice teen driving: A review of the literature. *Injury Prevention, 12 Suppl 1*, i30–i37. doi:10.1136/ip.2006.011569.

Simons-Morton, B. G., Hartos, J. L., & Leaf, W. A. (2002). Promoting parental management of teen driving. *Injury Prevention: Journal of the International Society for Child and Adolescent Injury Prevention, 8 Suppl 2*, ii24–ii30; discussion ii30–1. Retrieved from http://www.pubmedcentral.nih.gov/articlerender.fcgi?artid=1765491&tool=pmcentrez&rendertype=abstract.

Simons-Morton, B. G., L Hartos, J., Leaf, W. A., & Preusser, D. F. (2006). The effect on teen driving outcomes of the Checkpoints Program in a state-wide trial. *Accident Analysis & Prevention, 38*(5), 907–912. doi:10.1016/j.aap.2006.03.001.

Simons-Morton, B. G., Ouimet, M. C., & Catalano, R. F. (2008). Parenting and the young driver problem. *American Journal of Preventive Medicine, 35*(3 Suppl), S294–S303. doi:10.1016/j.amepre.2008.06.018.

Simons-Morton, B. G., Bingham, C. R., Ouimet, M. C., Pradhan, A., Chen, R., Barretto, A., & Shope, J. T. (2013). The effect on teenage risky driving of feedback from a safety monitoring system: A randomized controlled trial. *The Journal of Adolescent Health*, 1–6. doi:10.1016/j.jadohealth.2012.11.008.

Smetana, J. G., & Asquith, P. (1994). Adolescents' and parents' conceptions of parental authority and personal autonomy. *Child Development, 65*(4), 1147–1162.

Smetana, J. G., Metzger, A., Gettman, D. C., & Campione-Barr, N. (2006). Disclosure and secrecy in adolescent–parent relationships. *Child Development, 77*(1), 201–17. doi:10.1111/j.1467-8624.2006.00865.x.

Sommers, S. M., & Ribak, J. (2008). A model for preventing serious traffic injury in teens or "keep those teenagers out of our ICU!." *Dimensions of Critical Care Nursing, 27*(4), 143–151.

Steinberg, L. (2001). We know some things: Parent–adolescent relationships in retrospect and prospect. *Journal of Research on Adolescence, 11*(1), 1–19. doi:10.1111/1532-7795.00001.

Steinberg, L., Lamborn, S. D., Dornbusch, S. M., & Darling, N. (1992). Impact of parenting practices on adolescent achievement: Authoritative parenting, school involvement, and encouragement to succeed. *Child Development, 63*(5), 1266–1281.

Stice, E., & Barrera, M. (1995). A longitudinal examination of the reciprocal relations between perceived parenting and adolescents' substance use and externalizing behaviors. *Developmental Psychology, 31*(2), 322–334.

Sussman, S., & Arnett, J. J. (2014). Emerging adulthood: Developmental period facilitative of the addictions. *Evaluation & the Health Professions, 37*(2), 147–155. doi:10.1177/0163278714521812.

Taubman-Ben-Ari, O. (2010). Attitudes toward accompanied driving: The views of teens and their parents. *Transportation Research Part F: Traffic Psychology and Behaviour, 13*(4), 269–276. doi:10.1016/j.trf.2010.04.010.

Taubman-Ben-Ari, O., & Katz-Ben-Ami, L. (2013). Family climate for road safety: A new concept and measure. *Accident Analysis & Prevention, 54*, 1–14. doi:10.1016/j.aap.2013.02.001.

Taubman-Ben-Ari, O., & Lotan, T. (2011). The contribution of a novel intervention to enhance safe driving among young drivers in Israel. *Accident Analysis & Prevention, 43*(1), 352–359. doi:10.1016/j.aap.2010.09.003.

Taubman-Ben-Ari, O., Mikulincer, M., & Gillath, O. (2005). From parents to children—Similarity in parents and offspring driving styles. *Transportation Research Part F: Traffic Psychology and Behaviour, 8*(1), 19–29. doi:10.1016/j.trf.2004.11.001.

Tefft, B. C., Williams, A. F., & Grabowski, J. G. (2014). Driver licensing and reasons for delaying licensure among young adults ages 18–20, United States, 2012. *Injury Epidemiology, 1*(1), 4. doi:10.1186/2197-1714-1-4.

The North Carolina Graduated Driver Licensing System: Urban–Rural Differences. (2001). Chapel Hill, NC. Retrieved from http://www.hsrc.unc.edu/pdf/2001/Gdl_02_25.PDF.

Toledo, T., Lotan, T., Taubman-Ben-Ari, O., & Grimberg, E. (2012). Evaluation of a program to enhance young drivers' safety in Israel. *Accident Analysis & Prevention, 45*, 705–710. doi:10.1016/j.aap.2011.09.041.

Tronsmoen, T. (2010). Associations between driver training, determinants of risky driving behaviour and crash involvement. *Safety Science, 48*(1), 35–45. doi:10.1016/j.ssci.2009.05.001.

Tronsmoen, T. (2011). Differences between formal and informal practical driver training as experienced by the learners themselves. *Transportation Research Part F: Traffic Psychology and Behaviour, 14*(3), 176–188. doi:10.1016/j.trf.2010.11.009.

Trowbridge, M. J., McKay, M. P., & Maio, R. F. (2007). Comparison of teen driver fatality rates by vehicle type in the United States. *Academic Emergency Medicine, 14*(10), 850–855. doi:10.1197/j.aem.2007.06.038.

Twisk, D. a M., & Stacey, C. (2007). Trends in young driver risk and countermeasures in European countries. *Journal of Safety Research, 38*(2), 245–257. doi:10.1016/j.jsr.2007.03.006.

Vespa, B. J., Lewis, J. M., & Kreider, R. M. (2013). *America's Families and Living Arrangements: 2012 Population Characteristics.* Retrieved from https://www.census.gov/prod/2013pubs/p20-570.pdf.

Whitehill, J. M., Rivara, F. P., & Moreno, M. (2014). Marijuana-using drivers, alcohol-using drivers and their passengers: Prevalence and risk factors among underage college students. *JAMA Pediatrics, 168*(7), 618–624. doi:10.1001/jamapediatrics.2013.5300.Marijuana-Using.

Williams, A. F. (2011). Teenagers' licensing decisions and their views of licensing policies: A national survey. *Traffic Injury Prevention, 12*(4), 312–319. doi:10.1080/15389588.2011.572100.

Williams, A. F., & McCartt, A. T. (2014). Views of New Jersey teenagers about their state's policies for beginning drivers. *Journal of Safety Research, 48*, 1–6. doi:10.1016/j.jsr.2013.10.001.

Williams, A. F., Preusser, D. F., Ferguson, S. A., & Ulmer, R. G. (1997). Analysis of the fatal crash involvements of 15-year-old drivers. *Journal of Safety Research, 28*(1), 49–54. doi:10.1016/S0022-4375(96)00033-3.

Williams, A. F., Ferguson, S. A., Leaf, W. A., & Preusser, D. F. (1998). Views of parents of teenagers about graduated licensing systems. *Journal of Safety Research, 29*(1), 1–7. doi:10.1016/S0022-4375(97)00023-6.

Williams, A. F., Leaf, W. A., Simons-Morton, B. G., & Hartos, J. L. (2006). Parents' views of teen driving risks, the role of parents, and how they plan to manage the risks. *Journal of Safety Research, 37*(3), 221–226. doi:10.1016/j.jsr.2006.04.002.

Wilson, R. J., Meckle, W., Wiggins, S., & Cooper, P. J. (2006). Young driver risk in relation to parents' retrospective driving record. *Journal of Safety Research, 37*(4), 325–332. doi:10.1016/j.jsr.2006.05.002.

Winston, F. K., Kallan, M. J., Senserrick, T. M., & Elliott, M. R. (2008). Risk factors for death among older child and teenaged motor vehicle passengers. *Archives of Pediatric & Adolescent Medicine, 162*(3), 253–260. doi:10.1001/archpediatrics.2007.52.

Winston, F. K., Mirman, J. H., Curry, A. E., Pfeiffer, M. R., Elliott, M. R., & Durbin, D. R. (2014). Engagement with the TeenDrivingPlan and diversity of teens' supervised practice driving: Lessons for Internet-based learner driver interventions. *Injury Prevention*, 1–6. doi:10.1136/injuryprev-2014-041212.

Zakrajsek, J. S., Shope, J. T., Greenspan, A. I., Wang, J., Bingham, C. R., & Simons-Morton, B. G. (2013). Effectiveness of a brief parent-directed teen driver safety intervention (checkpoints) delivered by driver education instructors. *The Journal of Adolescent Health, 53*(1), 27–33. doi:10.1016/j.jadohealth.2012.12.010.

20

Feedback Technologies to Young Drivers

Tomer Toledo

Tsippy Lotan

Abstract

Newly licensed drivers are overrepresented in road crashes, especially during their first months of solo driving. Changes in licensing and development of training programs have addressed this inflated risk with a focus on gradual exposure of young drivers to the various risky driving conditions and circumstances, and training toward improvement of hazard anticipation, hazard mitigation, and attention maintenance.

Recent advancements in advanced driver assistance systems (ADAS) can help young drivers reduce their risks. In this chapter, we focus on feedback technologies that monitor and provide feedback to young drivers and their parents. These systems have been in the market, mostly in research studies and insurance programs, for over a decade. Most studies showed large reductions (over 50%) in safety-related event rates when feedback was provided. However, only one study, yet, managed to show a positive correlation with crash rates of teen drivers.

Despite these positive indications regarding safety benefits, feedback systems to young drivers are still far from being widely acknowledged, accepted, and used.

Barriers to acceptance and usage from the young drivers' point of view include invasion of privacy and restriction of independence, lack of trust, and fear that the feedback will become an instrument for punishment and limitation of car use. Similarly, parents express concerns related to privacy, parent–young driver relationship, and erosion of trust.

This chapter suggests directions to overcome the barriers through improved operation, increased motivation, and incentives. These directions include improving ease of installment, the pay-how-you-drive insurance model, increased parent involvement, incentives, positive positioning of the feedback systems, privacy protection, and legislation. The importance of positive media discourse, as well as the recommendation to install the system prior to the onset of the solo driving, can also contribute to wide acceptance and effective usage.

Finally, it is important to note that effective operation of feedback systems requires a sustainable process that incorporates feedback indicating deviations from safe driving behavior of young drivers over time.

20.1 What Are Feedback Technologies?

With current licensure practices, the formal training and monitoring of young drivers end immediately after passing the driving test, or at the end of a provisional licensing period within graduated driver licensing (GDL) programs. However, the empirical evidence shows that at this point, young drivers still lack the skills, experience, and understanding to perform well as drivers. They also often lack the maturity to drive responsibly, in particular in the presence of peer pressure. Furthermore, parents' presence, which played a significant role during the learner and provisional phases, disappears, leaving the young drivers on their own. Thus, continuous training, monitoring, and feedback on driving to the young, already licensed, drivers could play a useful role in the intermediate period until they gain experience, skills, and maturity.

Driving is a continuance of actions and interactions between the driver and the environment in which the driver operates. According to traditional classifications, feedback is defined by provision of information about a system or process that may affect a change in the process. In the context of driving, feedback generally pertains to information about the environment in which the driving occurs. This definition is clearly very broad, as information is everywhere, pertains to numerous aspects that can have an effect on driving, and is continuously generated and accumulated.

In the context of driving, feedback systems filter the huge amounts of data available into information that pertains to specific behaviors, conditions, predicted risks, and interactions. Dynamic feedback systems are often user specific, updated in real time, and can be continuously accumulated and provided. These can be further subdivided into alerting and monitoring systems. Alerting systems focus on specific behaviors (e.g., distance keeping, speed management, distraction, and fatigue) and provide real-time alerts to drivers for the purpose of avoiding specific actual risks, mitigating undesired behavior, or preventing dangerous behaviors while driving. Their focus is on preventing various situations before they occur or before they escalate. Monitoring systems document and accumulate actual (driving) behaviors, analyze them according to specific measures, determine their potential safety implications, and provide feedback to either the driver or other parties (e.g., parents, insurers).

In this chapter, we address monitoring systems. In particular, the focus is on systems that are tailored for young drivers. The feedback can be provided in near real time about an event that has already occurred (as opposed to an alerting system, which provides information in real time about an event that is about to occur) or off-line. The feedback is user specific, meaning that it pertains to actual behavior of a specific driver. In some cases, it may also be situation specific, meaning that it pertains to specific behavior occurring in specific circumstances. We will not address alerting systems that do not store data on behavior.

The reason for this focus on monitoring systems is threefold. First, there are (yet) hardly any studies on young drivers' reaction to and interactions with alerting systems (such as forward collision or lane departure warning systems), the circumstances that triggered the provision of the alert, the actual response (if any) to alerts, and their impact on crash risks. Second, alerting technologies can be (wrongly) interpreted as taking away the responsibility from the driver. For example, a forward collision warning system that alerts the driver about the high probability of a rear-end crash into the lead vehicle might cause the driver to (over)trust its reliability and be less attentive to distance keeping or attention maintenance. Hence, drivers can be tempted to shift responsibility from themselves to the alerting technology and therefore increase their risk. Monitoring systems, on the other hand, do not take responsibility away from the driver but, rather, collect and provide information that can eventually (if handled correctly) help drivers modify their behavior to be safer. Third, a major key success factor for the effectiveness of monitoring feedback systems is the presence and involvement of a relevant entity that cares about the safety of the specific driver. For young drivers, this entity is often the parents. Indeed, monitoring feedback systems provide parents with actual tools to be actively involved in the driving behavior of their children. Hence, the relevance of such systems to young drivers and their parents is direct and measurable and has already proven to be effective.

20.2 How Does It Work?

Various feedback applications have been developed and tested. Some examples include those reported by McGehee et al. (2007), Toledo et al. (2008), and Lerner et al. (2010). While they vary in specific features and capabilities, they commonly incorporate several generic tasks, as described in Figure 20.1.

20.2.1 Measurement

The core technology may include one type or several types of sensors that provide the raw information about the state of the vehicle and its surroundings. The technology may use sensors that already exist in the vehicle or in another device that the technology is attached to (e.g., a smartphone or navigation

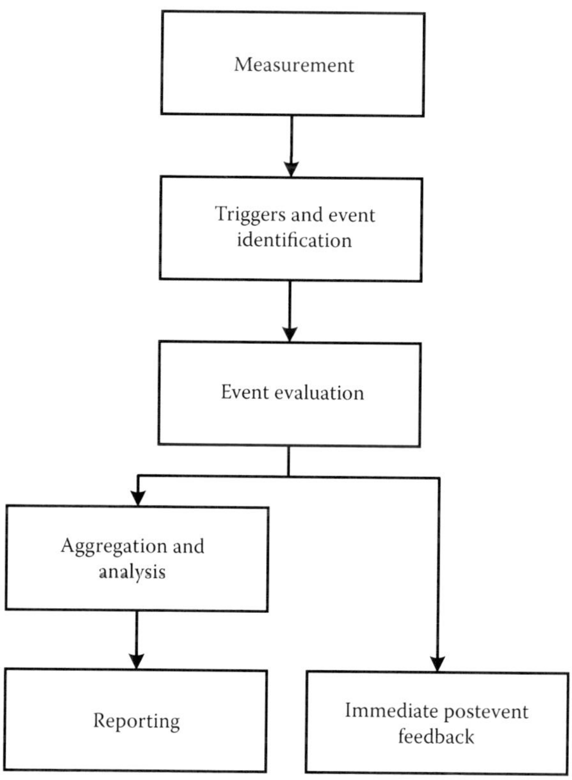

FIGURE 20.1 Generic framework of a feedback system.

device). It may also incorporate its own integral sensors, which make the technology stand-alone. Typically, several information items are collected, with the corresponding types of sensors, for example,

1. G-forces that operate on the vehicle. These are typically measured by accelerometers in two or three dimensions.
2. Speed of the vehicle measured by a vehicle speed sensor or derived from a global positioning system (GPS).
3. Position of the vehicle, typically measured using a GPS receiver.
4. Use of seat belts. This and the state of various other systems in the vehicle may be obtained from the on-board diagnostics (OBD) or using a camera inside the vehicle.
5. Identification of the driver and presence of passengers may be detected by a camera or by the use of magnetic keys, Bluetooth readers, or similar devices.
6. Distance and headway from the vehicle in front (and potentially other vehicles and obstacles). This is measured by range sensors or cameras.
7. Lane keeping and lane departure measured by range sensors or cameras according to lane markings.
8. Time stamp on measurements.

20.2.2 Triggers and Event Identification

The collected measurements are processed to detect various types of events. This step is necessary in order to reduce the large amount of raw data to meaningful information. In some cases, event identification may be straightforward and performed within the sensor system. For example, braking or speeding events may be defined when deceleration rates or speed values, respectively, exceed certain thresholds. Identification of other events, such as cornering or lane departures, typically involves complex image-processing or pattern recognition algorithms.

20.2.3 Event Evaluation

The identified events are evaluated in terms of their severity and predicted associated risk. The evaluation takes into account parameters of the events, such as the values of the variables of interest that were measured (e.g., maximum deceleration value in a braking event), other relevant variables (e.g., the speed at which the event took place), and the event duration.

Some applications use different types of information to trigger events and, subsequently, to identify and evaluate them. This is relevant in particular applications that use video cameras in order to avoid recording information continuously. They define specific conditions on other types of data that, when satisfied, events are identified. For example, when acceleration values measured by accelerometers exceed a certain threshold, a video camera is triggered, and a potential event is generated. The video data themselves are used to determine whether or not a relevant event has actually occurred, and if so, what its severity is.

Feedback from the applications may be provided in two ways:

1. Off-line, aggregated feedback on the overall driving
2. Immediate feedback related to a specific event that has just occurred

20.2.4 Immediate Postevent Feedback

Feedback is provided shortly after the event has ended. It is typically delivered to the driver in the vehicle and involves an auditory and/or minimal visual signal. A distinction is made here between alerts that are aimed to warn the driver of a potential risk and give feedback on a past event. The latter aims to help the driver identify and avoid risky situations in the future through increased awareness and a sense of monitoring. It should be noted that some current applications provide both warning and feedback capabilities. For the purpose of monitoring, immediate postevent feedback may also be provided outside the vehicle. For example, an automated text message detailing the event that has occurred may be delivered to the parents of a novice driver. This can serve two purposes: First, it allows providing immediate postevent feedback to another person who can act, in real time, to prevent the driver from continuing to undertake undesired driving behaviors. Simply knowing that this threat exists might have a moderating effect on driving behavior (Ernest-Jones et al. 2011; Van Bommel et al. 2014). Second, provision of immediate feedback in the vehicle may be distracting to the driver. We are not aware of any studies evaluating this effect. But, it is clear that the risk of distraction has led to implementation of very limited and minimal forms of immediate feedback in current feedback systems.

20.2.5 Aggregation and Analysis

Summary statistics on events are accumulated and analyzed for the purpose of providing feedback on driving performance. These typically involve calculation of the rates of events (normalized by driving time or mileage) or other similar measures. These statistics are supposed to indicate the risk of crash involvement. However, only few studies have evaluated the association between these statistics and crash risk. Wahlberg (2006, 2007) and Bagdadi and Várhelyi (2011) found significant correlations between g-force measures and self-reported crash rates. The former used celerations (changes in speed), while the latter used jerk (changes in acceleration) as the predicting variable. Klauer et al. (2009) showed a significant association between the classifications of drivers based on their crash records and the rates of lateral and longitudinal acceleration events and yaw rates. Toledo et al. (2009) found a positive connection between the overall rates of events defined by complex g-force patterns and recorded crash rates. Simons-Morton et al. (2012) developed a composite measure that takes into account the rates of several types of events. The composite measure was positively associated with crash rates. This study is the only one that addressed teen drivers.

In order to simplify the presentation of the results to drivers, some applications classify the event rates into categories based on predefined ranges of values.

20.2.6 Reporting

Periodic reports are presented to the driver, and in the context of novice drivers, often to their parents. These reports summarize the exposure and aggregate event statistics (or classifications) for the driver over the reporting period (e.g., week or month). In order to make the presented data more meaningful, they are presented sometimes in comparison to the mean of a reference group of drivers. Trends in the events statistics over time are also shown. Reports are provided through dedicated webpages or e-mails. Figure 20.2 shows an example of a web-based monthly driver report that was used by Toledo et al. (2008). The chart at the top of the figure shows the various trips the driver made in the month. Each square represents a trip. The x-axis indicates the day of the month, and the y-axis indicates the accumulation of trips within each day. The color-coding of trips expresses classification in terms of event rates. Detailed information on each trip is presented as shown at the bottom of the figure.

20.3 How Effective Are They?

Relatively few studies evaluated the effect of feedback technologies on driving behavior and risk, in particular in the context of novice drivers. Table 20.1 summarizes the setup and results of these studies.

Individually, the studies summarized in the table suffer from limitations of the experimental design and small sample sizes. In fact, the study of Farah et al. (2014) is the only one with more than a hundred participants. Most of the studies did not include control groups and so relied only on within-subject

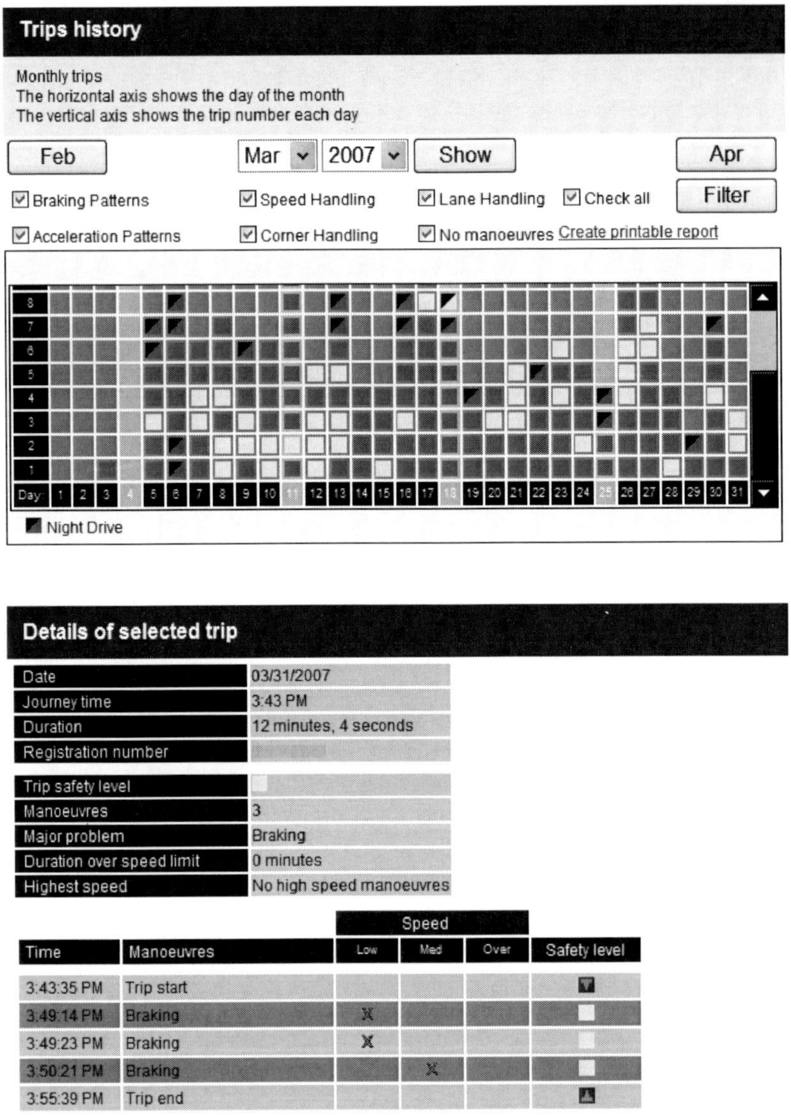

FIGURE 20.2 **(See color insert.)** An example of a web-based driver report. The selected trip is the third trip taken on March 31, a yellow square outlined in red.

analysis. The two exceptions are the studies by Carney et al. (2010) and Farah et al. (2014). Interestingly, the feedback effects in these two studies are among the lowest reported.

Most of the studies showed large reductions (over 50%) in safety-related event rates when feedback was provided. Farmer et al. (2010) found 43% reductions. Farah et al. (2014) and Prato et al. (2010) found lower reductions, less than 20%. The studies with lower reductions were all based on g-force events. This approach may generate more false events that are not affected by the feedback compared to video-based events. McGehee et al. (2007) and Farmer et al. (2010) also found reductions in seat belt nonuse. Feedback provided both in the vehicle and off-line yields lower event rates compared to providing only one of the feedback types (Farmer et al. 2010; Simons-Morton et al. 2013). This seems to indicate that both types are useful and complementary. Off-line feedback provided to both parents and novice drivers is the most useful to reduce event rates, followed by feedback to the parents only, and then to the novice drivers only

TABLE 20.1 Studies That Evaluated the Effect of Driving Technologies

Reference	Description
McGehee et al. (2007)	**Sample size and composition:** N = 25, 16- to 17-year-old drivers in Iowa with up to 1 year driving experience. **Measurement system:** Video-based recording of events that are triggered by g-force conditions, seat belt use. **Feedback:** *Real time*—in-vehicle LED lights. *Off-line*—weekly mail summaries of events and video clips to both parents and teens. **Experiment phases and duration:** Baseline—9 weeks, no feedback. Intervention—40 weeks with feedback. **Study groups:** Treatment only (no control). **Results:** *Impact*—Reduction of 58% and 76% in event rates in first and second 8-week feedback periods. 15% decrease in seat belt nonuse. *Group effects*—Large reductions up to 89% in event rates within intervention for drivers with high event rates in baseline. No significant effect on drivers with low event rates in baseline.
Carney et al. (2010)	**Sample size and composition:** N = 18, 16- to 17-year-old drivers in Minnesota with up to 5 months of unsupervised driving experience. **Measurement system:** Video-based recording of events that are triggered by g-force conditions. **Feedback:** *Real time*—in-vehicle LED lights. *Off-line*—weekly mail summaries of events and video clips to parents, who were asked to discuss them with the teens. **Experiment phases and duration:** Baseline—6 weeks, no feedback. Intervention—40 weeks with feedback. Second baseline—6 weeks immediately after intervention, no feedback. **Study groups:** Treatment only (no control). **Results:** *Impact*—Reduction of 61% in event rates during intervention. Almost no effect on drivers with low event rates in baseline. 64% reduction for drivers with high event rates. *After intervention*—No significant change from intervention to second baseline. *Feedback access*—67% of novice drivers talked with parents about feedback at least once per month.
Farmer et al. (2010)	**Sample size and composition:** N = 84, 16- to 18-year-old drivers in Washington, DC, with 0–15 (median: 3) months of unsupervised driving experience. **Measurement system:** G-force events, seat belt use, speeding. **Feedback:** *Real time*—in-vehicle sounds. *Off-line*—web-based reports to parents only. **Experiment phases and duration:** Baseline—2 weeks, no feedback. Intervention—20 weeks with feedback. Second baseline—2 weeks immediately after intervention, no feedback. **Study groups:** (a) Feedback off-line only to parents, (b) in vehicle to teens and off-line to parents, (c) in vehicle in all cases and off-line only if no immediate improvement in behavior, (d) control. **Results:** *Impact*—Reduction of up to 43% in events, up to 98% in seat belt nonuse, up 37% in speeding during interventions (a, b, and c) compared to control. Most are not statistically significant. Within intervention, initially large reduction in events, and then gradual increase. *After intervention*—No large change from intervention to second baseline. *Group effects*—Lower impact for group with feedback off-line only within intervention and second baseline. *Feedback access*—Average of one log-in per week. Added e-mail notifications to participants during experiment. Log-in values reduced to 0.5 log-ins per week.
Musicant and Lampel (2010)	**Sample size and composition:** N = 32, 17- to 24-year-old drivers in the United Kingdom. **Measurement system:** G-force events, speeding. **Feedback:** *Real time*—in-vehicle LED lights. *Off-line*—web-based reports and weekly e-mails to parents only. **Experiment phases and duration:** Baseline—14 weeks, no feedback. Intervention—up to 22 weeks with feedback. **Study groups:** Treatment only (no control). **Results:** *Impact*—Reduction of 59% in event rates during intervention compared to baseline. Reduction is higher for drivers with high scores in baseline.

(*Continued*)

TABLE 20.1 (CONTINUED) Studies That Evaluated the Effect of Driving Technologies

Reference	Description
Prato et al. (2010)	**Sample size and composition:** N = 62, 17-year-old drivers in Israel within first year of driving. **Measurement system:** G-force events, speeding **Feedback:** *Real time*—in-vehicle LED lights. *Off-line*—web-based reports to parents and teens. **Experiment phases and duration:** Baseline—18 weeks, no feedback. Intervention—35 weeks with feedback. **Study groups:** Treatment only (no control). **Results:** *Impact*—Event rates lower by 15% if both parents and novice drivers access feedback, and 21% higher if only novice drivers access feedback compared to when only the parents access feedback. *Feedback access*—49% only parents, 22% only novice driver, 28% both.
McGehee et al. (2013)	**Sample size and composition:** N = 90, 14.5- to 15.5-year-old school license drivers, 16- to 17-year-old drivers with and without prior school license in Iowa. **Measurement system:** Video-based recording of events that are triggered by g-force conditions, seat belt and cellphone use, passengers in the vehicle. **Feedback:** *Real time*—in-vehicle LED lights. *Off-line*—weekly mail summaries of events and video clips to parents, who were asked to discuss them with the teens. **Experiment phases and duration:** Baseline—4 weeks, no feedback. Intervention—16 weeks with feedback. Second baseline—4 weeks immediately after intervention, no feedback. **Study groups:** (a) School license drivers, (b) regular license drivers with and (c) without previous school license driving experience. There is no control group with respect to the feedback. School licenses allow younger teens to drive only to and from school. **Results:** *Impact*—Reduction of 84%, 58%, and 68% in event rates for school license drivers, regular license drivers with and without prior experience compared to the first baseline, respectively. *After intervention*—increase in event rates for school drivers in second baseline compared to intervention (still 61% lower compared to first baseline). No significant change with regular license holders.
Simons-Morton et al. (2013)	**Sample size and composition:** N = 88, novice drivers in Michigan with under 4 weeks of unsupervised driving experience. **Measurement system:** Video-based recording of events that are triggered by g-force conditions. **Feedback:** *Real time*—in-vehicle LED lights. *Off-line*—weekly e-mails to parents and teens, website access with video clips. **Experiment phases and duration:** Baseline—2 weeks, no feedback. Intervention—13 weeks with feedback. **Study groups:** (a) Feedback in vehicle only, (b) both in vehicle and off-line. No no-feedback control group. **Results:** *Impact*—Reduction of 54% in event rates with both in-vehicle and off-line feedback compared to in vehicle only. *Feedback access*—91% of parents accessed feedback at least once. Average of 0.78 log-ins per week. Log-ins on last week about one-fourth of first week.
Farah et al. (2014)	**Sample size and composition:** N = 217, 17- to 22-year-old (mean, 17.5) drivers in Israel with less than 1.5 month of driving experience. **Measurement system:** G-force events, speeding. **Feedback:** *Real time*—in-vehicle LED lights. *Off-line*—web-based reports to parents and novice drivers. **Experiment phases and duration:** 52 weeks with feedback. Covers 3 months of supervised and 9 months unsupervised driving within Israeli GDL program. **Study groups:** (a) Off-line feedback to novice drivers only, (b) off-line feedback to both novice drivers and parents, (c) training to parents in addition to feedback, (d) control. **Results:** *Impact*—Reduction of 10% in event rates for feedback groups and 29% for group with additional training over the control during unsupervised driving. No differences in supervised driving period. *Group effects*—Differences between training and no-training groups are only and feedback-only groups are only for drivers with higher initial event rates (Musicant et al. 2013). *Feedback access*—Initially high access rates. Decline by 60% within 6 months, and 90% within 12 months.

(Farah et al. 2014). Observations of feedback access (through web log-ins or self-reports) indicate that most participants access the feedback and use it for intrafamilial communications (Carney et al. 2010; Farah et al. 2014; Farmer et al. 2010; Simons-Morton et al. 2013). However, the rates of feedback access decline drastically from the beginning to the end of the experiments (Farah et al. 2014; Simons-Morton et al. 2013). Within the intervention periods, the reductions in event rates are maintained over the entire period (Carney et al. 2010; McGehee et al. 2007, 2013). However, when a second baseline was used at the end of the intervention, event rates did not increase substantially in this period (Carney et al. 2010; Farmer et al. 2010; McGehee et al. 2013). An exception to this is a group of 14.5- to 15.5-year-old drivers (McGehee et al. 2013). Among the participants, McGehee et al. (2007), Carney et al. (2010), and Musicant and Lampel (2010) found that the reductions in event rates are proportionally larger for drivers who initially had high event rates than for drivers who had lower event rates. This may be partially a result of regression to the mean. But, drivers who initially had low event rates did not exhibit an increase in event rates. They generally had small or negligible reductions in event rates.

20.4 How Far Are We from Wide Dissemination?

Feedback technologies for young drivers have been around for almost a decade. Their effectiveness has been demonstrated in a number of studies, showing reductions in event rates and (although in a limited number of cases) a connection to crash rates. Nevertheless, we are still far away from wide dissemination.

There are two major target audiences to be considered when investigating acceptance and usage of feedback technologies: the young drivers' population and their parents. Clearly, those two groups have very different views on acceptance and usage, especially in nonsymmetric situations, when parents are exposed to the driving feedback of their teen driving and not vice versa. In this section, we discuss barriers to acceptance as well as mechanisms to overcome them through incentives and positive motivation.

20.4.1 Barriers to Acceptance and Usage

Views of teen drivers regarding in-vehicle feedback technologies were qualitatively studied in Gesser-Edelsburg and Guttman (2013). Personal individual interviews were conducted with 26 teens who had used the technology. Eighteen focus group interviews were conducted with teens who had not. The main findings were that the teens' views of feedback technologies were analogous to their negative views on actual parental accompaniment, namely, invasion of privacy and restriction of independence. Other issues that were raised by teens referred to lack of trust and fear that the feedback would become an instrument for punishment and limitation of car use. However, teens also attributed positive functions to the technology as an "objective" indicator that would be acceptable to both parents and teen drivers when they performed well or needed to improve in particular driving maneuvers. Yet, the overriding perceptions of the technology were negative, and they were viewed as an extension of parental supervision that could introduce tensions and frictions into the parent–young driver relationship.

Another finding was that teens had mixed and contradictory views of their parents as role models. This has implications for their acceptance of parental authority regarding safer driving advice. The feedback from a more objective technological device could thus be viewed as more credible. These findings indicate that parents' involvement through the use of feedback technologies requires guidance. It is hence recommended that parents' "accompaniment" both directly and through use of feedback technology must be branded not as a means for parents to employ sanctions but, rather, as means to enhance driving skills.

Guttman and Lotan (2011) report on parents' attitudes and views regarding the use and ethics of driver-monitoring technologies. The study is based on phone interviews of 906 parents of young drivers. Most believed that parents should feel morally obligated to install it. When cost was not a consideration, most said they would, and believed other parents would also be willing to install the technology. About half of the respondents expressed willingness to install the technology after being told about its estimated cost. The rest saw the monetary cost as a barrier to installation. Considerations of the family relations were viewed as an incentive. Parents who supported the installation believed that it would serve as a trigger for parent–young driver communication. Those who did not support installation thought that it would erode trust in the parent–young driver relationship. Parents who expressed great concern about their teen driving were also more likely to state that they would be willing to install the technology regardless of its monetary cost. Most said parents should have access to the monitoring data.

Guttman and Gesser-Edelsburg (2011) conducted interviews with 79 young drivers and their parents who had experienced driving with a feedback system. The main concerns raised were related to privacy, parent–young driver relationship, self-esteem and confidence, constructive use of the feedback data, and the limitations of the documentation that can be provided by the technology.

It is important to mention that although parents seem to be very positive when told about the technology and state that they will definitely use it for their teens, once they were offered to do so, even at no cost and with monetary incentive, in a research setting, they often chose not to.

20.4.2 Overcoming the Barriers: Incentives and Motivations

It is evident that there are many barriers to the acceptance and usage of feedback systems by its two major target audiences: parents and teens. In order to reach wide dissemination, these barriers have to be dealt with. Some of the barriers, mentioned in the previous section, have been addressed with varying degrees of success; others still need to be recognized and addressed:

Technology improvement. This is probably the easiest barrier to tackle, as it requires dealing with equipment rather than with human perceptions and behavior. Indeed, feedback systems are becoming friendlier in several aspects, specifically regarding their ease of installation and deinstallation. Some systems include a direct connection to the car's OBD, which can be self-installed, requires only simple calibration, and provides smartphone-based feedback and even stand-alone smartphone apps.

Monetary incentives. Probably the most promising direction in this category is usage-based insurance, also known as pay as you drive or pay how you drive. In these schemes, car insurance costs also depend on dynamic driving parameters, such as actual extent of driving, driving locations, time of driving, and the telematics-based measures of the quality/safety of driving. Several insurance companies, mostly in the United States, are already implementing telematics usage-based insurance, with specific programs for young drivers (Bolderdijk et al. 2011). Understanding the effect of the various premium structures offered on different drivers (e.g., those with base high- and low-risk driving behavior) enrolling in these programs and on their behavior is still an open research question.

Parents remain the key agents for the introduction, effective usage, and long-term maintenance of safe driving with the help of feedback systems. However, research findings and market behavior show that typically, parents do not adopt this role voluntary and need guidance and motivation to do so. Several programs have been suggested to involve parents in their teen's driving using contracts, agreements, and structured processes. Some of the programs include use of feedback systems (e.g., Farah et al. 2013, 2014; McGehee et al. 2007). Research suggests that useful parental involvement is crucial to enhance their teen's safe driving (Farah et al. 2014; Prato et al. 2010; Shimshoni et al. 2015). Feedback systems provide a validated tool to help parents perform their role. However, their involvement needs to be legitimized (mostly by their teens), clarified, and structured, as well as sustained over time.

Incentive-based systems. Ultimately, teens will choose whether or not to install a feedback system and use the feedback it provides. However, their acceptance requires clear motivation or incentives. Recently, there have been some attempts to generate such feedback systems. A good example is the stand-alone smartphone app RefuelMe. When activated, it accumulates safe mileage (mileage driven with no unsafe events) and reports to a sponsor (typically the parent), who in turn provides the teen with a prespecified reward. In this way, teens are driving to earn desired rewards. Two pilot studies (Lotan et al. 2014; Musicant et al., 2015) used this system with group rewards. These rewards are decided by the group members, who accumulate the required mileage together. In the experiments, this scheme led to high levels of use of the system by the group members and even others in their social networks. But, once the rewards were achieved, the use of the app practically stopped. Hence, the question of how to move from external rewards to intrinsic motivation remains and cannot be answered by incentives only.

Positive positioning. Feedback systems should be positively positioned as steering young drivers, not spying on them. In order to avoid one of the major barriers to acceptance, namely, privacy intrusion, the systems should be presented as helping young drivers and not as spying on them. Hence, they should be informative, with a focus on recommended tips to achieve safe behavior rather than on emphasis of unsafe behaviors (Guttman and Lotan 2011). This is in agreement with findings from other public health domains (e.g., illness prevention, smoking cessation, and physical activity), according to which gain-framed messages appear to be more effective than loss-framed messages (Gallagher and Updegraff 2012).

User-friendliness. Feedback systems compete in today's highly digital and overstimulated world; hence, they have to be attractive. Successful design of effective feedback needs to build on a comprehensive understanding of the driver, the feedback, and their interactions (Lee 2007; Feng & Dommez 2013). In order to make the systems attractive to teens and encourage their effective usage even without tangible rewards, special attention should be devoted to the development of advanced user interfaces, learning capabilities, user-specific adjustments, dynamic scores and comparisons, competitions, gaming, and social network implementations.

Privacy protection. A major concern regarding the use of feedback systems is intrusion of privacy. Several guidelines have been suggested to address this issue (Guttman 2012). First, use of the system has to be transparent and clear between the parent and teen, meaning that teens are fully aware of the information that is being collected. Second, data collection has to be minimal and limited to safety-related driving. For example, location data were available in the study of Shimshoni et al. (2015), but they were not reported to parents since teens expressed concerns about their location being exposed to their parents. Hence, the message from parents to teens was, "I don't care where you drive, I care how you drive," which contributed to acceptance by teens. Third, exposure to the data collected can be dynamic and conditioned on actual driving behavior; as long as there is no indication of unsafe driving behavior, data are not transferred to parents, but once certain thresholds are reached, parents are notified. Fourth, even in the extreme case, in which privacy is fully protected and only the young drivers are exposed to their driving information, feedback can be effective. In the study by Farah et al. (2014), an experimental group with this type of feedback performed better than the control group.

Legislation can naturally contribute to wide use of feedback systems. These systems can, for example, be incorporated into the GDL process as an intermediate stage between accompanied and independent driving. It may also be used to verify adherence to other restrictions or as means to allow controlled driving bypassing the restrictions. For example, the system can monitor night driving, or night driving may be allowed when the system is functional. To the best of our knowledge, such measures have not been implemented. Legislation would need to address obstacles regarding standardization and validation of the systems as well as enforcement mechanisms and guarantee of availability to all.

Change of media discourse. A major obstacle to wide dissemination of feedback systems lies in their positioning and marketing. Most news items and media coverage target the parents and position these technologies as "Big Brother," tracking, monitoring, and even spying. They are rarely presented as driving assistance tools, teaching and learning aids, or safety-enhancing systems. Clearly, this framing does not encourage young drivers to install feedback technologies with monitoring capabilities.

Timing of installation. Young drivers are eager to drive. Hence, it is recommended that the feedback system be installed before they start their independent driving. It is expected that the objection of young drivers to the feedback systems would be lower if it is preinstalled before they start independent driving. Furthermore, if installed early, they can be used as a monitoring and learning tool during the learner phase.

20.5 Summary

It is expected that advanced technological systems will become more widely used and accepted. Feedback monitoring systems can lead to significant decreases in the frequencies of undesired behaviors that they monitor and provide feedback on. This is, in particular, important for young drivers, who exhibit high crash rates during their first year of driving.

Research clearly shows that correct implementation of feedback systems to young drivers can bring large reductions (over 50%) in safety-related event rates. Although the evidence is still limited, there seems to be a connection between these behaviors and crash risks. Thus, feedback systems have a potential to improve the safety of young drivers if accepted and implemented correctly. Currently, feedback systems are still far away from being widely accepted and used. A number of barriers to acceptance by young drivers exist, including invasion of privacy, restriction of independence, lack of trust, and limitations of the feedback systems. In order to reach wide dissemination, the technological aspects of the systems have to be optimized and simplified, but more importantly, clear motivation and incentives (monetary or other) must be put in place in order to overcome the basic objection of young drivers to being monitored. Additionally, the feedback systems need to be positively positioned and marketed in order to help young drivers use them as an acceptable tool to increase their safety.

Finally, it should be kept in mind that as with many advanced new technologies, there might be deterioration in proper and effective usage of the system over time. Hence, processes to encourage sustainable effective use must be put in place to ensure the safety of young drivers and their environment in the long run.

References

Bagdadi, O. and Várhelyi, A., 2011. Jerky driving—An indicator of accident proneness? *Accident Analysis & Prevention* 43(4), 1359–1363.

Bolderdijk, J.W., Knockaert, J., Sleg, E.M. and Verhoef, E.T., 2011. Effects of Pay-As-You-Drive vehicle insurance on young drivers' speed choice: Results of a Dutch field experiment. *Accident Analysis & Prevention* 43(3), 1181–1186.

Carney, C., McGehee, D.V., Lee, J.D., Reyes, M.L. and Raby, M., 2010. Using an event triggered video intervention system to expand the supervised learning of newly licensed adolescent drivers. *American Journal of Public Health* 100(6), 1101–1106.

Ernest-Jones, M., Nettle, D. and Bateson, M., 2011. Effects of eye images on everyday cooperative behavior: A field experiment. *Evolution and Human Behavior* 32, 172–178.

Farah, H., Musicant, O., Shimshoni, Y., Toledo, T., Grimberg, E., Omer, H. and Lotan, T., 2013. The first year of driving: Can an in-vehicle data recorder and parental involvement make it safer? *Transportation Research Record* 2327, 26–33.

Farah, H., Musicant, O., Shimshoni, Y., Toledo, T., Grimberg, E., Omer, H. and Lotan, T., 2014. Can providing feedback on driving behavior and training on parental vigilant care affect male teen drivers and their parents? *Accident Analysis & Prevention* 69, 62–70.

Farmer, C.M., Kirley, B.B. and McCartt, A.T., 2010. Effects of in-vehicle monitoring on the driving behavior of teenagers. *Journal of Safety Research* 41(1), 39–45.

Feng, J. and Dommez, B., 2013. Design of effective feedback: Understanding driver, feedback and their interaction. 7th International Driving Symposium on Human Factors in Driver Assessment, Training and Vehicle Design. No 62.

Gallagher, K.M. and Updegraff, J.A., 2012. Health message framing effects on attitudes, intentions and behavior: A meta-analysis review. *Annals of Behavioral Medicine* 43(1), 101–116.

Gesser-Edelsburg, A. and Guttman, N., 2013. "Virtual" versus "actual" parental accompaniment of teen drivers: A qualitative study of teens' views of in-vehicle driver monitoring technologies. *Transportation Research Part F* 17, 114–124.

Guttman, N., 2012. 'My son is reliable': Young drivers' parents' optimism and views on the norms of parental involvement in youth driving. *Journal of Adolescent Research* 28(2), 241–268.

Guttman, N. and Gesser-Edelsburg, A., 2011. "The little squealer" or "The virtual guardian angel"? Young drivers' and their parents' perspective on using a driver monitoring technology and its implications for parent–young driver communication. *Journal of Safety Research* 42(1), 51–59.

Guttman, N. and Lotan, T., 2011. Spying or steering? Views of parents of young novice drivers on the use and ethics of driver-monitoring technologies. *Accident Analysis & Prevention* 43(1), 412–420.

Klauer, S.G., Dingus, T.A., Neale, V.L., Sudweeks, J.D. and Ramsey, D.J., 2009. Comparing real-world behaviors of drivers with high versus low rates of crashes and near-crashes, National Highway Traffic Safety Administration Report DOT HS 811 091.

Lee, J.D., 2007. Technology and teen drivers. *Journal of Safety Research* 38(2), 203–213.

Lerner, N., Jenness, J., Singer, J., Klauer, S., Lee, S., Donath, M., Manser, M. and Ward, N., 2010. An exploration of vehicle-based monitoring of novice teen drivers: Final report, National Highway Traffic Safety Administration Report DOT HS 811 333.

Lotan, T., Musicant, O. and Grimberg, E., 2014. Can young drivers be motivated to use smartphone based driving feedback?—A pilot study. Preprints of the Transportation Research Board 93rd Annual Meeting, Washington DC.

McGehee, D.V., Raby, M., Carney, C., Lee, J.D. and Reyes, M.L., 2007. Extending parental mentoring using an event-triggered video intervention in rural teen drivers. *Journal of Safety Research* 38(2), 215–227.

McGehee, D.V., Reyes, M.L. and Carney, C., 2013. Age vs. experience: Evaluation of a video feedback intervention for newly licensed teen drivers. Iowa Department of Transportation Report No. TPF-5(207).

Musicant, O. and Lampel, L., 2010. When technology tells novice drivers how to drive. *Transportation Research Record: Journal of the Transportation Research Board* 2182, 8–15.

Musicant, O., Lotan, T. and Grimberg, E., 2015. The potential of group incentive schemes to encourage use of driving safety apps. *Transportation Research Record: Journal of the Transportation Research Board* 2516, 1–7.

Musicant, O., Farah, H., Toledo, T., Shimshoni, Y., Omer, H. and Lotan, T., 2013. The potential of IVDR feedback and parental guidance to improve novice male young drivers' behavior. *Proceedings of the Seventh International Driving Symposium on Human Factors in Driver Assessment, Training, and Vehicle Design.*

Prato, C.G., Toledo, T., Lotan, T. and Taubman-Ben-Ari, O., 2010. Modeling the behavior of novice young drivers during the first year after licensure. *Accident Analysis & Prevention* 42(2), 480–486.

Shimshoni, Y., Farah, H., Lotan, T., Grimberg, E., Dritter, O., Musicant, O., Toledo, T. and Omer, H., 2015. Effects of parental vigilant care and feedback on novice driver risk. *Journal of Adolescence* 38, 69–80.

Simons-Morton, B.G., Zhang, Z., Jackson, J.C. and Albert, P.S., 2012. Do elevated gravitational-force events while driving predict crashes and near crashes? *American Journal of Epidemiology* 175, 1075–1079.

Simons-Morton, B.G., Bingham, C.R., Ouimet, M.C., Pradhan, A.K., Chen, R., Barretto, A. and Shope, J.T., 2013. The effect on teenage risky driving of feedback from a safety monitoring system: A randomized controlled trial. *Journal of Adolescent Health* 53(1), 21–26.

Toledo, T., Musicant, O. and Lotan, T., 2008. In-vehicle data recorder for monitoring and feedback on drivers' behavior. *Transportation Research Part C* 16(3), 320–331.

Van Bommel, M., van Prooijen, J.W., Elffers, H. and van Lange, P.A.M., 2014. Intervene to be seen: The power of a camera in attenuating the bystander effect. *Social Psychology and Personality Science* 5(4), 459–466.

Wahlberg, A.E., 2006. Speed choice versus celeration behavior as traffic accident predictor. *Journal of Safety Research* 37, 43–51.

Wahlberg, A.E., 2007. Aggregation of driver acceleration behavior data: Effects on stability and accident prediction. *Safety Science* 45(4), 487–500.

21

Simulation-Based Training for Novice Car Drivers and Motorcycle Riders: Critical Knowledge Gaps and Opportunities

Michael G. Lenné

Eve Mitsopoulos-Rubens

Christine Mulvihill

Abstract

Simulation is a widely accepted form of training in domains such as aviation, and the potential for using simulation-based training for surface transportation modes has long been recognized. Despite this potential, there has been relatively little research conducted to test this potential formally. The central question addressed within this chapter considers the research evidence that simulation is a mechanism or tool through which to impart effectively the intended competencies in the training of novice car drivers and novice motorcycle riders and, critically, that its use does not translate to negative outcomes. In the current chapter, evidence of effectiveness is assessed against three core criteria: learning on the simulation task itself, transfer of training, and retention of competencies. The majority of simulator-based research to date has used vehicle and motorcycle setups and desktop-based applications to support the training of higher-order competencies such as hazard perception and control of attention in novice car drivers and, to a much lesser extent, novice motorcycle riders. The focus on higher-order competencies is quite appropriate given their role in safe driving and riding in the long term. While some progress has been made in demonstrating that certain examples of simulation-based training can be a possible means through which to impart the critical competencies, further research is required to better round out the evidence base and to address the many unresolved issues that remain. At the core of these issues is an appreciation that simulation is just a training tool and that, in the case of novice driver and novice rider learning, at least, simulation should never be used as a substitute for real-world practice, serving instead potentially to support and augment real-world practice as part of a staged learning approach.

21.1 Introduction

Simulation, in theory, has much to offer as a training tool for imparting the critical competencies that are needed for safe driving and riding in the long term. In particular, relative to training in an actual vehicle or motorcycle, simulation can provide trainees with the opportunity to practice a range of driving maneuvers in a safe, risk-free environment and, in so doing, help support novice drivers and riders in the learning process.

The supporting technology is evolving at a rapid rate. Simulators of increasingly higher physical fidelity, once considered to be out of reach for the beginning driver or rider, are likely to present themselves as an option in the not-too-distant future and are likely to be more affordable to operators. This brings with it the need to understand what role simulation-based training can play in the learning of novice car drivers and novice motorcycle riders. Central to this issue, and the focus of this chapter, is an examination of the recent research evidence regarding the effectiveness of simulation-based training for drivers and riders. In addressing this central issue, we consider effectiveness along the following dimensions: (1) learning on the target simulation task, (2) transfer of training, and (3) retention of those competencies over time.

Simulators vary, for example, on the basis of the extent to which they look and feel like the real system. In this chapter, we consider the full spectrum of simulators in their application to novice driver and rider training, ranging from full car and motorbike setups through to lower-cost desktop-based simulators. The focus is on publications that have used simulation as a training delivery method and not solely those that have used simulation to evaluate training. Taking this approach reduces the number of relevant publications considerably. In defining training simulators, here, we adopt a definition of human-in-the-loop simulation that requires human participation and where an operator's inputs to the simulation influence the outcomes. This definition includes cases where the driver or rider can make direct inputs to the simulation, for example, that impact speed or lateral control, and these inputs produce subsequent changes in the scenario. At a practical level, simulation according to this definition requires, at a minimum, the ability for the participant to interact dynamically with the driving or riding environment through the use of a steering wheel or other mechanism to make adjustments to lateral position or navigation more broadly; an accelerator and/or brake pedal or other mechanism to make adjustments to speed; or other inputs that change the way the participant experiences the scenario. Human-in-the-loop simulation is also defined as interaction simulation (Lilienthal & Moroney, 2009), which covers the essence of our definition here.

There are, of course, many other approaches to training that could be incorporated within a broader definition of simulation (see McDonald et al., 2015, for a review). For example, some training programs typically do not require any equipment other than the training software and the standard components of a "home" PC. These are typically focused on hazard perception where participants view various still or moving images and respond to events as prompted with mouse clicks on areas of interest (e.g., Isler et al., 2009; Pradhan et al., 2009). While the efficacy of these programs is considered elsewhere in book, these types of programs are not included in our definition. By way of linking with other chapters in the handbook, the reader may wish to refer to Chapter 25 for more information on the methods one needs in order to use simulators in research and training and to Chapters 5 through 8 to learn more about the areas of hazard anticipation, hazard mitigation, and attention maintenance.

Further, in this chapter we focus on *novice* operators as opposed to *teens*. Whether a novice operator is also a teen is dependent on a given jurisdiction's minimum licensing age. Thus, it is important to acknowledge that not all novice operators are teens, and this is particularly true for novice motorcycle riders. For example, in Victoria, Australia, the minimum age at which an individual can apply for a car driver learner permit is 16 years, and for a car driver license, 18 years. Moreover, there is evidence that individuals are delaying licensure—and that this is an international trend (Tefft et al., 2014). Aspiring motorcycle riders in Victoria must wait until they are at least 18 years of age before they can apply for a rider learner permit. In practice, most riders do not begin their riding career until their 30s or 40s, with most also experienced car drivers.

It should be noted at the outset that the significant majority of the recent literature has addressed training for novice car drivers, and for higher-order competencies, typically operationalized as hazard perception and control of attention. This focus of this class of competencies is entirely appropriate given the role these skills play in safe driving (Deery, 1999; Gregersen & Bjurulf, 1996), and forms the focus area within this current chapter.

It is our intention not only to provide a critique of the relevant studies that have used simulation as a training tool but also to highlight some of the theoretical and practical issues that remain outstanding. Simulation has, for many years, held much promise as a training delivery method, and we raise issues here that still need to be resolved if this potential is to be realized.

21.2 Important Issues in Training Effectiveness

Realizing the potential benefits associated with simulation-based training is not a trivial undertaking. As a general rule, simulators that are used for driver and rider training tend to be of lower fidelity than driving simulators that are used for research purposes. This is largely a function of the need for training simulators to be low cost, transportable, and easy to use with minimal or no specialist training. However, it is important to note that low physical fidelity does not necessarily imply low effectiveness (Park et al., 2005), and while there are many types of simulators available, a system with a high level of physical realism is not necessarily better when it comes to training. The level of realism that is appropriate, and indeed, whether simulation itself is appropriate and likely to be effective, will depend on numerous considerations, including the overall purpose of the training, training content, timing of training delivery (in terms of stage of learning), amount of training, the trainees, logistical/pragmatic issues such as accessibility and affordability, and the capability and motivation of the trainer. Moreover, for best effect, simulators need to be integrated into the total learning framework, which involves several training methods, strategies, and tools, for addressing the overall training requirements (Salas et al., 2006).

Discussions on simulator effectiveness in training in a range of road safety settings typically center on the issue of fidelity, and the importance of matching fidelity to training requirements. At least two types of fidelity can be distinguished: physical (or task) fidelity and functional (or instructional) fidelity.

Physical fidelity is the degree to which the simulator looks and feels like the real-world operational system. The two main factors considered to influence physical fidelity are the levels of visual detail and motion (Thompson et al., 2009). At one extreme, lower-fidelity simulators typically comprise a single PC monitor; simplified vehicle controls (e.g., mouse, keyboard, steering wheel, and pedals built for gaming); and a fixed base. At the other extreme, simulators with very high physical fidelity include a large projection screen, giving a wide field of view and a highly immersive experience; actual vehicle controls; and a full motion base. Training simulators are typically of lower physical fidelity than those used for research programs; however, this alone should not be used as an indicator of their efficacy for training.

Functional fidelity is the extent to which the simulator acts like the real-world operational system. The consensus is that if there is good functional fidelity, then there is likely to be a high positive transfer of learned skills to the operational environment. Thus, in selecting a simulator for use in operator training, at least functional fidelity should be high. However, while high functional fidelity is necessary, a low level of physical fidelity may be sufficient. These decisions will ultimately depend on the purpose of the training, the nature of the tasks to be trained, and the role that simulation ultimately can play in the broader educative framework to support and augment the development of novice drivers and riders.

21.3 Evidence of Simulation-Based Training Effectiveness

It is worth revisiting the dimensions being considered here with respect to effectiveness: (1) learning on the target simulation task, (2) transfer of training to real-life situations in a real vehicle, and (3) retention of those competencies over time.

21.3.1 Novice Car Drivers

The following section is structured to consider those studies that use vehicle-based (i.e., relatively higher physical fidelity) versus desktop-based methods (i.e., relatively lower physical fidelity) to *deliver* training, noting that a variety of methods for evaluation can be used across both approaches to training delivery.

21.3.1.1 Vehicle-Based Simulators

Table 21.1 presents a summary of the recent research, including a description of each study, methods, outcomes, and limitations. In light of recent advances in simulation capabilities and use, the review was focused on those publications in the previous 10 years (2005 onward) with selected older publications of a seminal nature. Brief inspection of Table 21.1 highlights the range of measures that have been used to measure training performance such as speed and braking patterns (Wang et al., 2010b) and number of hazards and collisions successfully avoided (Allen et al., 2011).

In addressing the first element of effectiveness, learning on the target task, several studies have shown modest effects. One of the earlier studies completed in 2003, the European Union TRAINER project, aimed to develop new, improved, and cost-effective driver training methodology for use in European countries. The intention was that this training methodology supplement current driver training practice in each country, rather than replace it (Nalmpantis et al., 2005). The specific purpose of the training was to provide beginning drivers with the opportunity to gain experience in driving and handling the vehicle, and also to enhance their skill in risk awareness (Falkmer & Gregersen, 2003). As described by Nalmpantis et al. (2005), the project involved the development of a PC-based multimedia tool, simulator software, and a modular driving simulator, of which there were two versions: a "low-cost" simulator and a "mean-cost" simulator. The low-cost simulator is a static driving simulator with one monitor giving a horizontal viewing angle of 40°. The mean-cost simulator is a semidynamic driving simulator with three screens providing a horizontal viewing angle of 120°. This simulator is able to simulate lateral motion as well as vehicle vibrations.

An evaluation study conducted in Sweden using research simulator at the Swedish National Road and Transport Research Institute (VTI) revealed that the training was associated with safer driving performance in some driving situations (Falkmer & Gregersen, 2003). The research involved a group of learner drivers who were randomly divided into three groups: (1) the mean-cost simulator group (MC), (2) the low-cost simulator group (LC), and (3) the control group (CG). Each group of drivers completed the current driver education and training provided in Sweden. In turn, participants in the MC group received training in the multimedia tool and the mean-cost simulator, while participants in the LC group completed training in the multimedia tool and the low-cost simulator. Participants in the control group received no additional training. Participants were then tested in the VTI research simulator and performance-tested in six different scenarios.

The general expectation of the study was that, regardless of scenario, the participants in the MC group would perform more safely than those in the LC group and, in turn, those participants in the control group. Differences between the groups were observed for three of the six scenarios. Falkmer and Gregersen (2003) concluded that while there were some benefits associated with the TRAINER program, there was still potential for improvement and a requirement to analyze additional driving performance measures. It was argued that the absence of a significant difference between the trained groups and the control group for certain scenarios could be due to the absence of such scenarios in the training program. That is, the results may be indicative of a lack of transfer from the training to the test.

Finally, it is interesting to note that Falkmer and Gregersen (2003) did not measure (or at least report to have measured) participants' driving performance prior to training. Thus, it is not known whether participants across the three groups were performing similarly prior to undertaking the training or and whether the posttraining performance of the MC and LC groups represented a significant improvement over their pretraining performance.

TABLE 21.1 Summary of Research on the Efficacy of Simulation-Based Training: Vehicle-Based and Desktop Applications

Source	Participants/Conditions	Simulation Characteristics (Fidelity and Technical Specifications)	Simulation Content and Target Skills/Abilities Measured	Type of Study and Assessment Method	Findings
Allen et al., 2011	Novice drivers ($n = 67$). Simulation training (n unspecified). No training (n unspecified).	Fixed-base, desktop PC with single monitor and gamelike controls. Displays view from front roadway.	8 different traffic scenarios comprising 4 key hazardous events designed for training hazard perception. Performance feedback with summary of errors given at end of each training session.	Experimental. Pretest and posttest driving performance on training simulator using nonidentical scenarios for each test. Objective measures of performance were percentage of accidents avoided, intersection violations, speed exceedances, and lane-keeping errors.	*Learning*: Trained participants provided better collision avoidance accuracy and fewer speed limit violations than untrained participants.
Carpentier et al., 2013	Novice drivers. Training ($n = 15$). Control ($n = 14$).	Fixed-base driving simulator with force-feedback controls and 135° field of view, 3 screens.	Commonly occurring hazardous scenarios in different traffic environments. Driving instructor provided information and feedback for 10 hazardous situations in terms of what to look out for and how to respond.	Experimental. Performance was assessed pretraining and posttraining; and retention and transfer were assessed 2–4 weeks posttraining. Glance behavior was measured to assess hazard detection (collisions, mirror use, detection time, and correct hazard detection). Training scenario hazards were not identical to pretraining and posttraining scenario hazards.	*Learning and retention*: Training group had a significant reduction in mean detection time from pretraining to posttest and a further reduction during retention test. The reduction was significantly larger in the training group than in the control group. Training group had a higher mean percentage of correctly detected hazards than the control group at posttraining, better retention, and better use of mirrors. The effects of training on collisions were inconclusive. *Transfer*: The trained group performed better on the detection tasks for near compared to far-transfer scenarios. The results for correctly identified hazards, collisions, and mirror use were not conclusive.

(*Continued*)

TABLE 21.1 (CONTINUED) Summary of Research on the Efficacy of Simulation-Based Training: Vehicle-Based and Desktop Applications

Source	Participants/Conditions	Simulation Characteristics (Fidelity and Technical Specifications)	Simulation Content and Target Skills/Abilities Measured	Type of Study and Assessment Method	Findings
Cox et al., 2009	Novice drivers. Simulation ($n = 9$). Control ($n = 9$).	Virtual reality, realistic controls, high-fidelity simulator, 180° field of view, rearview and side-view mirror images, fixed base.	Five simulated driving scenarios incorporating low-demand traffic conditions for practicing basic driving skills (e.g., using indicator) and high-demand traffic conditions for practicing higher-order cognitive skills (e.g., decision making). Instructor gave feedback after each scenario.	Experimental. Training and pretraining and posttraining on-road driving performance (steering, turning, speed control, braking, vision, attention, attitude) assessed by an instructor using 3-point subjective rating scales and blind to condition.	Participants who received simulator training performed significantly better on the road 2–4 days after training on all measures.
Panou et al., 2010	25 novice drivers—simulator training in the use of advanced driver assistance systems (ADAS).	Not specified.	Participants drove six different ADAS scenarios across different driving environments and traffic density levels.	Quasi-experimental. Participants provided user acceptance and workload ratings in response for a collision avoidance warning system (CAS) and lane departure warning (LDW) system before and after driving each simulated traffic scenario. Driving instructors provided an assessment of the participants' workload and driving performance for the two systems.	Participants and instructors rated both systems as being "average," with the number of errors performed being less than the number perceived. Efficiency, effectiveness, learnability, and potential impact were estimated at the same levels.

(Continued)

TABLE 21.1 (CONTINUED) Summary of Research on the Efficacy of Simulation-Based Training: Vehicle-Based and Desktop Applications

Source	Participants/Conditions	Simulation Characteristics (Fidelity and Technical Specifications)	Simulation Content and Target Skills/Abilities Measured	Type of Study and Assessment Method	Findings
Pradhan et al., 2011	Novice drivers. Simulation training ($n = 19$). Control training ($n = 18$).	Interactive desktop PC with a single monitor, controlled with mouse. Displays traffic from inside the vehicle.	Video clips of ordinary driving scenarios. Forward Concentration and Attention Learning (FOCAL) training to teach novices how to reduce their glance durations to less than 2 seconds while performing an in-vehicle (map-reading) task accurately.	Experimental. Objective pretraining and posttraining assessments on the simulation used in training. Eye movements used for on-road posttraining assessment to measure number of on- and off-road glances in different scenarios and for different in-vehicle tasks (e.g., operating windshield wipers).	*Learning*: Simulation-trained participants showed a significantly greater reduction in the proportion of glances away from the forward roadway than controls when tested in the simulator. *Transfer*: When tested on the road, simulation-trained participants had significantly lower proportions of tasks with glances that exceeded various thresholds and a lower proportion of glances over those thresholds compared to the control group.
Vidotto et al., 2011	Young novice motorcyclists. Simulator training ($n = 207$). Control (road safety lesson) ($n = 203$).	Fixed-base HRT motorcycle simulator with single 19-inch LCD computer monitor. Simple handlebar and foot pedal motorcycle controls. Displays view from front roadway.	12 different traffic courses (10 minutes), each containing 8 hazard scenarios of potential accidents based on real-world events. Software replays events and provides feedback each time the rider has an accident, and overall advice on riding performance is given at the end of each course.	Experimental. Performance assessed by number of hazards avoided and an instructor's (unspecified) subjective assessment of overall safety.	The mean proportion of avoided hazards increased as a function of the number of tracks performed, with the size of improvement being greater in the experimental group.

(Continued)

TABLE 21.1 (CONTINUED) Summary of Research on the Efficacy of Simulation-Based Training: Vehicle-Based and Desktop Applications

Source	Participants/Conditions	Simulation Characteristics (Fidelity and Technical Specifications)	Simulation Content and Target Skills/Abilities Measured	Type of Study and Assessment Method	Findings
Vlakveld et al., 2011	Novice drivers. Simulator-based Risk Awareness and Perception Program (Sim RAPT) training ($n = 18$). Control ($n = 18$).	Fixed-base interactive driving simulator with full-cab sedan, 135° field of view, and 3 screens.	10 scenarios: 7 hazard anticipation scenarios with common latent hazards and 3 scenarios with no high-priority hazards. 3 different drives per hazard anticipation scenario: (1) no hazards materialize, (2) same as 1 but hazards materialize aggressively, (3) same as 2 but hazards materialize less aggressively. Feedback given after drive 2. Feedback comprised an instruction video showing actual participant movements followed by recommended movements for optimal hazard detection.	Experimental. Performance measures: number of correctly and timely anticipated latent hazards on the test drives based on eye movements and independent assessments by two experimenters blinded to condition.	*Near-transfer learning*: Trained drivers fixated correctly significantly more often than untrained drivers (84% versus 57% of the time respectively). *Far-transfer learning*: Trained drivers fixated correctly significantly more often than untrained drivers (71% versus 53% of the time respectively).
Wang et al., 2010b	Novice drivers. Training group (exposed to hazardous situations in simulator; feedback on performance by watching own driving and an experienced driver's performance in the same scenarios; $n = 16$). No-training control group ($n = 16$).	Fixed-base interactive driving simulator, realistic controls, 180° field of view.	Participants were required to negotiate their way through 8 different traffic scenarios containing hazards/hazardous events.	Experimental. Hazard handling performance in terms of crashes/near crashes assessed by two independent researchers blind to condition. Workload and hazard anticipation measured by NASA Task Load Index (TLX). Speed control. Comparisons before and after training.	*Learning*: compared to control group, trained drivers demonstrated better hazard anticipation and handling performance; earlier hazard detection and responding; lower overall workload; and better speed control prior to hazards.

(*Continued*)

TABLE 21.1 (CONTINUED) Summary of Research on the Efficacy of Simulation-Based Training: Vehicle-Based and Desktop Applications

Source	Participants/Conditions	Simulation Characteristics (Fidelity and Technical Specifications)	Simulation Content and Target Skills/Abilities Measured	Type of Study and Assessment Method	Findings
Wang et al., 2010a	Novice drivers. Simulation-based error training (SET) (n = 16). Video-based guided error training (VGET) (n = 16).	SET training = desktop PC with single monitor and video-game steering wheel. Displays view from front roadway. VGET training = as per SET but not interactive.	8 risky urban traffic scenarios as used by Wang et al. (2000b). SET participants were required to detect and respond to hazards, whereas VGET watched videos of error and errorless driving performances.	Experimental. Pretraining and posttraining test of driving performance on a realistic fixed-base simulator. Number of errors and response distance assessed objectively by simulator, hazard handling performance (different levels of accident avoidance) assessed by two independent researchers blind to condition; mental workload, confidence, metacognition, and intrinsic motivation assessed by self-report surveys.	*Transfer*: Compared to the VGET group, the SET group made significantly fewer errors and displayed shorter response distances 1 week posttraining and demonstrated better hazard handling performance. SET demonstrated higher levels of intrinsic motivation and metacognition than VGET.

Vlakveld et al. (2011) hypothesized that novice drivers' overrepresentation in crashes could be the result of poor visual scanning for hazards that are hidden from view. Using an advanced fixed-base driving simulator, they examined the effectiveness of a program designed to address this potential failure. The performance of 18 novice drivers who received the training program of approximately 60 minutes' duration was compared to that of a control group of 18 novice drivers who did not receive the training. Training comprised 10 scenarios: 7 hazard anticipation scenarios with common latent hazards and 3 scenarios with no high-priority hazards. There were three versions of each hazard anticipation scenario, and participants drove through each one in succession: (1) a hazard detection drive in which the possible hazards did not materialize; (2) an error drive, which was the same as the hazard detection drive except that the hazards materialized aggressively, and (3) an improvement drive, which was the same as the error drive except that hidden hazards materialized less aggressively.

At the conclusion of each error drive, interactive feedback via an instruction video and voice-over was given to show participants how they had performed and how they could have performed in order to detect hazards as early as possible. Following training, participants completed an evaluation drive on a different simulator to that used in training so that learning could be assessed. Learning was assessed in two ways: first, by measuring the extent to which performance during the training transferred to performance on *similar* scenarios during the evaluation (this is known as *near transfer*) and, second, by measuring the extent to which performance during training transferred to performance on *different* scenarios during the evaluation (this is known as *far transfer*). Performance was measured in terms of the number of correctly identified latent hazards in the test drives and was assessed independently by two experimenters who were blind to condition. Hazard identification was correct if the participants' gaze directions allowed for timely detection of the hazard such that a crash could be averted. The results showed that for near-transfer learning, the trained drivers fixated correctly significantly more often than untrained drivers (84% versus 57% of the time, respectively). For far-transfer learning, trained drivers fixated correctly significantly more often than untrained drivers (71% versus 53% of the time, respectively).

Other recent studies have also compared performance for a treatment (training) group to that of a control group, an exception being that of Allen et al. (2011). Hazard perception training improved simulator performance compared to a control group on a number of measures, including mirror use (Carpentier et al., 2013); hazard anticipation, braking patterns, and number of collisions or near collisions avoided (Wang et al., 2010a,b); and instructor-based ratings of hazard handling performance (Wang et al., 2010a,b). No studies have measured transfer of training to the on-road environment. With regard to the third element, retention of competencies over time, learning is typically assessed only a short time after training, ranging from immediately after training (Cox et al., 2009; Wang et al., 2010a) to 7 days after training (Allen et al., 2011).

21.3.1.2 Desktop Applications

While the majority of desktop applications involve methods including video-based training that do not meet the definition of simulation adopted in this chapter, there are a small number that do involve a level of participant control and interaction with the scenario. Desktop simulation has been used to train higher-order cognitive skills in novice drivers other than hazard perception (Pradhan et al., 2011; Regan et al., 2000a). DriveSmart focuses on the development of attentional control skills using a single-monitor PC setup. While a large component of the program involves viewing video clips, there is a second component to the program that does sit within the definition of simulation adopted in this chapter, namely, a virtual road environment in which the user controls a car using the computer mouse. This component of DriveSmart is called the "Concentration" module. Through the use of variable priority training, this module is designed to enhance attentional control skills by giving trainees extensive practice at simultaneously completing two competing tasks and learning to prioritize their attention across the two tasks. One of the two tasks in this module requires the user to maintain a constant distance between his or her own car and another car approximately 15 meters ahead. During this exercise, the speed of the lead vehicle varies between 60 and 80 kph. The trainee controls the speed of his or her own car by pressing the left mouse button (to accelerate) and the right mouse button (to decelerate) in response to the changes in speed of the leading car. The second task requires the trainee to scan the computer monitor for a series of one-digit numbers that appear in random locations on the screen, and then to perform a numeric calculation on these numbers. Trainees are required to prioritize their attention to each of the two tasks in ratios that vary between exercises. Performance feedback is provided at the end of each trial.

The instructional effectiveness of DriveSmart was evaluated through an experiment using an advanced driving simulator (Regan et al., 2000b). Learner drivers aged between 16 years 11 months and 17 years 10 months and with between 40 and 110 hours of driving experience participated in the study. Performance on the baseline drive (i.e., mean speed) did not differ significantly between control and training groups.

One week after training, and again 4 weeks later, all participants performed several drives in the simulator. Evidence of heightened risk perception skill was observed for the treatment group relative to the control group in approximately half of the traffic scenarios analyzed. In the remaining scenarios, critically, no negative effects of the training (i.e., better risk perception skill among the controls relative to the treatment group) were observed. The positive training effects generalized to risky traffic situations that were not encountered during training and persisted for at least 4 weeks after training. Further, comparison of mean speed on the baseline drive with that on an exit drive completed 4 weeks after training revealed significantly lower speeds among participants who completed the DriveSmart training than among those who did not (Regan et al., 2000b). Since July 2000, DriveSmart has been distributed to all learner drivers in Victoria. In 2014, the Transport Accident Commission made some minor updates to the look and feel of DriveSmart, and moved it from CD-ROM format to being available online. Users register to complete the program online at the following website: http://www.drivesmart.vic.gov.au.

Building upon earlier research exploring the effects of imparting risk perception and risk awareness training to novice drivers (Pollatsek et al., 2006; Pradhan et al., 2005, 2006, 2009), a similar approach is employed in the Forward Concentration and Attention Learning (FOCAL) program (Pradhan et al., 2011). FOCAL teaches novice drivers to increase their attention to the forward roadway by reducing the amount of time spent on a range of necessary but potentially distracting in-vehicle tasks, including checking the speedometer and operating windshield wipers. A video plays in real time. The participants decide themselves when to view either the forward roadway or the in-vehicle task, but not both. They control how long each view is available to them. Compared to controls, Pradhan et al. (2011) found that participants receiving focused-attention training on a single-monitor desktop PC demonstrated a significantly greater reduction in the proportion of glances away from the forward roadway as measured by an eye-tracking device. Pretraining and posttraining performance was assessed on the same day. Transfer of training to the on-road environment was positive, with the proportion of glances exceeding various thresholds being significantly lower in the simulation-trained group compared to the control group (e.g., in 59.5% of the scenarios, the maximum glance was over 2 seconds for the FOCAL-trained group, whereas in a full 75.9% of the scenarios, the maximum glance was over 2 seconds for the control group). The evaluation was conducted immediately after the training (D. Fisher, personal communication, March 20, 2015).

More recently, Zafian et al. (2014) examined the efficacy of an interactive desktop program, Road Aware, on trainees' subsequent hazard detection performance in a driving simulator. While the vehicle's speed and lane position were controlled by the Road Aware training program, 24 novice drivers used keyboard controls to pan visually left and right of the forward roadway view, to look over their left or right shoulders, and to scan the mirrors. Participants also identified objects and objects and areas that could present a potential hazard via mouse click. A further 24 participated in a placebo training program that involved viewing videos. The simulator evaluation of both programs occurred with 2 days of training for almost all of the participants and focused on assessments of glance behaviors as a measure of hazard identification. Similar to the trend in findings reported by Vlakveld et al. (2011), superior levels of hazard identification, in terms of correct glance behavior, were reported in the simulator evaluation for both scenarios that were related and unrelated to the training program context, as implemented through the use of near and far-transfer scenarios.

These desktop training approaches do demonstrate benefits to varying degrees in evaluations that occur within 4 weeks of training, but typically within hours or days of training. The contribution of this form of training, compared to less interactive video-based programs and more interactive driving simulator-based programs, in discussed later.

21.3.2 Novice Motorcycle Riders

While the vast majority of recent research into hazard perception and attentional control has concentrated on novice car drivers, there is now a small and growing body of work assessing these issues in a motorcycle context, much of it facilitated by the development of the Honda Rider Trainer (HRT). As described by Vidotto et al. (2011), the HRT is a simulator that is powered by PC technology. Visual images are displayed on a screen (e.g., 19-inch LCD monitor), which is positioned in front of the user. Interaction is achieved through typical motorcycle controls: handlebars fitted with active throttle, front brake and clutch, and foot pegs with rear brake and shift lever. The HRT preprogrammed courses utilize several scenarios to cover a range of driving settings (e.g., city, urban, rural). Several features of the simulator are programmable, including engine size (small, medium, large), transmission type (automatic, manual), and lighting conditions (day, night, fog). Each course comprises either seven or eight hazardous events. Voice instructions guide users along the preprogrammed courses. In the case of a collision, the course is paused, and the events that led up to the crash are replayed. Each course is followed with a replay of the entire drive. This is often coupled with commentary and feedback on the individual user's performance.

In partnership with Honda Motor Europe Ltd., researchers at the University of Padua in Italy carried out a program of research with the overall aim of evaluating the HRT to train effectively the hazard perception and responding skills of novice motorcycle riders (e.g., Alberti et al., 2012; Spagnolli et al., 2009; Vidotto et al., 2008, 2011). Also known as the Safe Motorcyclist Awareness and Recognition Trainer (SMART), the HRT has been "specifically designed to give riders a safe bridge between a typical beginning riding course (which often takes place in a parking lot) and the real-world scenario of riding in traffic and on public roads" (http://www.motorcycle.com/how-to/honda-smartrainer-86756.html).

Vidotto et al. (2011) report on a study involving 410 participants, aged 14–15 years. It is not stated whether participants had any prior driving or riding experience. The simulator was set up with a small engine size, automatic transmission, and daylight conditions. Participants were assigned to either a control group (CG) or an experimental group (EG). Over three sessions, EG participants completed 12 training courses of 10 minutes each (including replay). Three courses were completed at each session, with the first two courses of the first session and the last two courses of the third session providing measures of pretraining and posttraining performance, respectively. CG participants were divided into four groups. All four groups completed two courses on their third session to obtain a measure of "post-training" performance. Two of these groups (CG1 and CG2) also completed two "pretraining" courses on their first session. On the second session, CG1 took part in a classroom lesson ("passive training"), which covered the topics of driving code, hazard perception, and hazard awareness. CG3 also took part in the passive training. Prior to their first session, all participants completed two practice courses to familiarize themselves with the simulator.

The results were largely positive, supporting the use of the HRT as an effective means for imparting training in hazard perception to novice riders. For Vidotto et al. (2011), the dependent variable of interest was the proportion of avoided hazards. Two findings are particularly noteworthy. First, EG participants showed an increase in the number of avoided hazards as a function of the number of courses completed. From the 1st to the 12th course, participants in the EG experienced a proportional improvement of approximately 16% in hazard avoidance. The mean proportion of avoided hazards for the 12th course approximated 93%, leading the authors to speculate that this level of performance is "a good approximation of the maximum attainable performance level" for an HRT trainee. The second noteworthy finding is that, posttraining, EG participants had a significantly higher proportion of avoided hazards than each of the control groups. Further, the control groups that undertook the passive training (CG1 and CG3) demonstrated, posttraining, a higher proportion of avoided hazards than each of the two control groups that did not undertake the passive training. Thus, while the full simulator training (12 courses) provided the most benefit, passive training coupled with a subset of courses (two to four) in the simulator led to greater improvement in hazard avoidance than completion of two to four simulator training courses alone. The strong possibility remains, therefore, that the heightened posttraining performance of the EG was simply a reflection of more simulator practice rather than any specific benefits associated with training.

Tagliabue et al. (2013) reported similar results using a similar experimental setup and participant pool comprising 207 14- to 15-year-old high school students. In their first study, those authors confirmed earlier findings by demonstrating similar improvements in hazard avoidance conferred through 12 HRT-based training sessions spread 1 week apart over 3 weeks. An interesting addition to this study was the classification of scenes used in training into those that required an attention shift or not. Hazard avoidance performance was superior in those scenes requiring no attention shift. In their second study, with 60 participants, Tagliabue et al. (2013) showed that the HRT-based training led also to improvements in a visual detection task, using stimuli not context relevant for driving, in parallel with improvements in hazard avoidance. Those authors propose that the training program enhanced overall training efficiency and useful field of view. The absence of a control group does bring into question the real contribution of the training program to the results observed.

Alberti et al. (2012) provide complementary evidence in support of the HRT. In their study, 14 participants aged 20–25 years, with no riding experience, completed a single practice course for familiarization purposes followed by four training courses in an urban environment and under daylight conditions while having their eye movements recorded. The HRT was programmed to simulate a motorcycle with a small engine size and automatic transmission. The critical dependent measure was first-fixation latency—that is, elapsed time from when the hazard first appears to when the participant first fixates on the area containing the hazard. From the first to the fourth course, there was a significant decrease in first-fixation latency, implying that, as a result of the training, hazards are fixated earlier. This result was taken to support the hypothesis that riders trained on the HRT have heightened hazard anticipatory abilities and that this is observed after as little as 16–24 minutes of training. The findings must be considered in light of the fact that no control group was used, and so it is not possible to separate out the roles of the training program and simulator experience broadly in driving the observed results. This concern is particularly relevant given that the participants had previously reported no riding experience. It is not yet known whether the skills augmented as part of training using the HRT transfer to real-world riding and remain resistant to degradation without practice following training.

21.4 Discussion

The chapter aimed to focus on the recent evidence bearing on the effectiveness of the use of driving simulation for training novice car drivers and motorcycle riders. An evaluation framework is used to guide these assessments that focuses on evidence of target competencies being learnt during training, transfer of training, and retention of competencies over time.

The balance of the research that was examined focused on novice car drivers, although in the last 5 years, attention has been directed to novice motorcycle riders. In the current review, higher-order cognitive skills training was demonstrated to be beneficial using relatively simple simulations representing the driver's field of view and using low-fidelity controls. Evidence for transfer of nontechnical skills training has not been established, due to challenges in identifying and measuring the link between performance and crash risk.

Some of the research should be interpreted with caution as a result of important limitations in the study design and evaluation methods. These limitations include failure to include a control group; use of only subjective measures of performance; failure to measure pretraining performance; use of noninferential statistical analyses; poor or no explanation of the training approach; nonrandom assignment to control or treatment group; and use of identical training and testing scenarios. Despite an overall improvement in the study design and evaluation methods over the last 5 years, many of the limitations identified here were evident in a review of earlier research in this field (Goode et al., 2013).

Most studies fail to measure retention adequately, if at all, and the very few studies that have measured it have done so only a short time after training (in most cases immediately after, but in some cases up to 4 weeks after). Next to retention, transfer of training is perhaps the most important measure of a program's success. In order to demonstrate a real benefit of training, it is critical to examine on-road or in-field performance, particularly crash or near-crash events, since the aim of training is to improve safety in the real world. If the real benefits of training are to be established, it is imperative that research efforts focus on transfer of training.

Time and resource constraints are largely responsible for many of the methodological limitations associated with the training evaluations. This is particularly so in transfer-of-training evaluations, where a large sample size is often necessary to demonstrate an effect, the studies are long in duration, and participant numbers tend to fall between initial retention and follow-up testing. These challenges often restrict the level of scientific rigor with which studies can be designed, conducted, and evaluated.

It is interesting to consider how the use of simulation for novice driver and rider training sits in the context of its use within other domains, notably with regard to evaluation. In particular, demonstrating effective transfer of training to the real-world environment is widely acknowledged as a critical phase in any training assessments. This is one area where there are more efforts directed in domains such as aviation. Thorough and rigorous evaluation of training effectiveness and cost–benefit analysis requires the conduct of labor-intensive transfer studies in which simulator training is conducted and performance on the simulator is compared with subsequent performance in the aircraft (e.g., Pfeiffer et al., 1991; Taylor et al., 1999). In their simplest form, transfer studies involve comparing performance in the actual aircraft of a group of trainees who receive simulator training with a control group who receive training in the actual aircraft (Rolfe & Caro, 1982). Transfer studies of this nature are very labor intensive and costly and are therefore rarely conducted (Bell & Waag, 1998). In the aviation domain, several transfer studies have been conducted to explore the impact of simulator fidelity on training effectiveness. This research challenges the belief that simulators of the highest visual fidelity offer the most effective training of procedural tasks (Johnson, 1981; Lintern, 1991; Lintern et al., 1990). The majority of aviation transfer studies are referred to as quasi-transfer studies, as defined by Lintern and colleagues. This refers to designs that examine transfer from several simulator training sessions to a test simulator configuration rather than a design that examines transfer from simulator to the actual aircraft. While novice driver and rider training can benefit from these approaches to training evaluation, the development of on-road approaches to the assessment of driver behavior provides strong opportunities to conduct full-transfer studies where the transfer of simulation-based training is assessed on the road (e.g., Underwood, 2013).

Does simulation have more to offer than digital training tools that are not simulator based? We think so, but the evidence is not yet there when we apply a strict definition of what constitutes simulation. At a theoretical level, there are reasons to believe that it will have benefits for particular capabilities, such as attentional control, where the trainee might need to manage competing tasks and demands. In these situations, providing a more realistic and embedded representation of those demands in training is likely to provide benefits in performance and subsequent transfer over and above desktop-based training tools. However, for other competencies, such as hazard perception, this may not be the case, and digital training tools that are not simulation based might be suitable. Clear guidance based on the research evidence is required on how best to integrate simulation-based training into practice. The research evidence and guidance is currently lacking and remains a significant knowledge gap. In closing, our approach to evaluating recent research shows that there is a clearer picture emerging that demonstrates that the learning of competencies during simulation-based training in novice drivers and, to a lesser extent, riders, can be achieved using relatively low-level simulation. Progress against the other two dimensions of evaluation, transfer and retention, has been more modest. There remains little guidance on the level and type of simulation; type of training; duration of training; and combined influence of each of these variables on retention and transfer of training in the real world.

While the technology is rapidly evolving, it is important that researchers and practitioners work through how best to realize the potential for simulation within the learning context. Many important questions and issues remain. Central to these is an appreciation that simulation is just a training tool and that, in the case of novice driver and novice rider learning, at least, simulation should not be used as a substitute for real-world practice, serving instead to support and augment real-world practice through a staged learning approach. Further, as is the case with all road safety programs, it is critical that sufficient resources are invested into the evaluations of simulation-based training programs and that these evaluations use the appropriate methodologies.

Acknowledgments

The material presented here is in part drawn from research conducted by the Monash University Accident Research Centre for VicRoads that used funds made available through the Motorcycle Safety Levy, and for the Defence Science and Technology Organisation. We are also appreciative of the comments provided by the editors on an earlier version of this chapter.

Finally, we acknowledge the major contribution of the late Emeritus Professor Tom Triggs, whose great work in this area has guided and shaped our thinking in the simulation and training space amongst many others.

References

Alberti, C., Gamberini, L., Spagnolli, A., Varatto, D., & Semezato, L. (2012). Using an eye-tracker to assess the effectiveness of a three-dimensional riding simulator in increasing hazard perception. *Cyberpsychology, Behavior and Social Networking, 15* (5), 274–276.

Allen, W., Park, G., Terrace, S., & Grant, J. (2011). Detecting transfer of training through simulator scenario design: A novice driver training study. *Proceedings of the Sixth International Driving Symposium on Human Factors in Driver Assessment, Training and Vehicle Design.*

Bell, H.H., & Waag, W.L. (1998). Evaluating the effectiveness of flight simulators for training combat skills: A review. *International Journal of Aviation Psychology, 8*, 223–242.

Carpentier, A., Wang, W., Jongen, E., Hermans, E., & Brijs, T. (2013). Training hazard perception of young novice drivers: A driving simulator study. *92nd TRB Annual Meeting Compendium of Papers DVD*, pp. 1–15.

Cox, C.V., Moncrief, R., Wharam, R., Mourant, R., & Cox, D.J. (2009). Does virtual reality driving simulation training transfer to on-road driving in novice drivers? A pilot study. *Chronicle of the American Driver and Traffic Safety Education Association, 57*, 9–25.

Deery, H. (1999). Hazard and risk perception among young novice drivers. *Journal of Safety Research, 30*, 225–236.

Falkmer, T., & Gregersen, N.P. (2003). The TRAINER project—The evaluation of a new simulator-based driver training methodology. In L. Dorn (Ed.), *Driver Behaviour and Training* (pp. 317–330). Aldershot: Ashgate.

Goode, N., Salmon, P.M., & Lenné, M. (2013). Simulation-based driver and vehicle crew training: Applications, efficacy and future directions. *Applied Ergonomics, 44*, 435–444.

Gregersen, N.P., & Bjurulf, P. (1996). Young novice drivers: Towards a model of their accident involvement. *Accident Analysis & Prevention, 28*, 229–241.

Isler, R., Starkey, N., & Williamson, A. (2009). Video-based commentary training improves hazard perception of young drivers in a dual task. *Accident Analysis & Prevention, 41*, 445–452.

Johnson, S.L. (1981). Effect of training device on retention and transfer of a procedural task. *Human Factors, 23,* 257–272.

Lilienthal, M.G., & Moroney, W.F. (2009). Appendix A: Glossary of Modeling Terms. In D. Vincenzi, J. Wise, M. Mouloua, & P. Hancock (Eds.), *Human Factors in Simulation and Training* (pp. 417–422). Boca Raton, FL: CRC Press.

Lintern, G. (1991). An informational perspective on skill transfer in human–machine systems. *Human Factors, 33,* 251–266.

Lintern, G., Roscoe, S.N., & Sivier, J.E. (1990). Display principles, control dynamics, and environmental factors in pilot training and transfer. *Human Factors, 32,* 299–317.

McDonald, C.C., Goodwin, A.H., Pradhan, A.K., Romoser, M.R.E., & Williams, A.F. (2015). A review of hazard anticipation training programs for young drivers. *Journal of Adolescent Health, 57,* S15–S23.

Nalmpantis, D., Naniopoulos, A., Bekiaris, E., Panou, M., Gregersen, N.P., Falkmer, T., Baten, G., & Dols, J.F. (2005). "TRAINER" project: Pilot applications for the evaluation of new driver training technologies. In G. Underwood (Ed.), *Traffic and Transport Psychology* (pp. 141–156). Amsterdam: Elsevier.

Panou, M.C., Bekiaris, E.D., & Touliou, A.A. (2010). ADAS module in driving simulation for training young drivers. *13th International IEEE Annual Conference on Intelligent Transportation Systems* (pp. 1582–1587). Madeira Island, Portugal.

Park, G.D., Allen, R.W., Rosenthal, T.J., & Fiorentino, D. (2005). Training effectiveness: How does driving simulator fidelity influence driver performance? *Proceedings of the Human Factors and Ergonomics Society 49th Annual Meeting* (pp. 2201–2205). Orlando, FL.

Pfeiffer, M.G., Horey, J.D., & Butrimas, S.K. (1991). Transfer of simulated instrument training to instrument and contact flight. *International Journal of Aviation Psychology, 1,* 219–229.

Pollatsek, A., Narayanaan, V., Pradhan, A., & Fisher, D.L. (2006). Using eye movements to evaluate a PC-based risk awareness and perception training program on a driving simulator. *Human Factors, 48,* 447–464.

Pradhan, A.K., Fisher, D.L., & Pollatsek, A. (2005). The effects of PC-based training on novice drivers' risk awareness in a driving simulator. *Proceedings of the Third International Driving Symposium on Human Factors in Driver Assessment, Training and Vehicle Design* (pp. 81–87). Rockport, ME.

Pradhan, A.K., Fisher, D.L., & Pollatsek, A. (2006). Risk perception training for novice drivers. Evaluating duration of effects of training on a driving simulator. *Transportation Research Record, 1969,* 58–64.

Pradhan, A.K., Pollatsek, A., Knodler, M., & Fisher, D.L. (2009). The effects of focused attention training on the duration of novice drivers' glances inside the vehicle. *Ergonomics, 54,* 917–931.

Pradhan, A.K., Pollatsek, A., Knodler, M., & Fisher, D.L. (2011). Can younger drivers be trained to scan for information that will reduce their risk in roadway traffic scenarios that are hard to identify as hazardous? *Ergonomics, 52,* 657–673.

Regan, M.A., Triggs, T.J., & Godley, S.T. (2000a). Evaluation of a novice driver CD-ROM based training program: A simulator study. *Proceedings of the IEA 2000/HFES 2000 Congress* (pp. 2-334–2-337). San Diego, CA.

Regan, M.A., Triggs, T.J., & Godley, S.T. (2000b). Simulator-based evaluation of the DriveSmart novice driver CD-ROM training product. *Proceedings of the 2000 Road Safety Research, Policing and Education Conference* (pp. 315–320). Brisbane.

Rolfe, J.M., & Caro, P.W. (1982). Determining the training effectiveness of flight simulators: Some basic issues and practical developments. *Applied Ergonomics, 13,* 243–250.

Salas, E., Wilson, K.A., Priest, H.A., & Guthrie, J.W. (2006). Design, delivery, and evaluation of training systems. In Salvendy, G. (Ed.), *Handbook of Human Factors and Ergonomics*, 3rd edition (pp. 472–512). Hoboken, NJ: John Wiley & Sons.

Spagnolli, A., Gamberini, L., Furtan, S., Bertoli, L., Chalambalakis, A., Scottini, R., & Turra, P. (2009). Users' performance with a rider trainer: The role of social setting. *Proceedings of CHI Italy 2009.* Rome.

Tagliabue, M., Da Pos, O., Spoto, A., & Vidotto, G. (2013). The contribution of attention in virtual moped riding training of teenagers. *Accident Analysis & Prevention, 57*, 10–16.

Taylor, H.L., Lintern, G., Hulin, C.L., Talleur, D.A., Emanuel, T.W.J., & Phillips, S.I. (1999). Transfer of training effectiveness of a personal computer aviation training device. *International Journal of Aviation Psychology, 9*, 319–335.

Tefft, B.C., Williams, A.F., & Grabowski, J.G. (2014). Driver licensing and reasons for delaying licensure among young adults ages 18–20, United States, 2012. *Injury Epidemiology, 1*, 4.

Thompson, T.N., Carroll, M.B., & Deaton, J.E. (2009). Justification for use of simulation. In Vincenzi, D.A., Wise, J.A., Mouloua, M., & Hancock, P.A. (Eds.), *Human Factors in Simulation and Training* (pp. 39–48). Boca Raton, FL: CRC Press.

Underwood, G. (2013). On-road behaviour of younger and older novices during the first six months of driving, *Accident Analysis & Prevention, 58*, 235–243.

Vidotto, G., Bastianelli, A., Spoto, A., Torre, E., & Sergeys, F. (2008). Using a riding trainer as a tool to improve hazard perception and awareness in teenagers. *Advances in Transportation Studies, 16*, 51–60.

Vidotto, G., Bastianelli, A., Spoto, A., & Sergeys, F. (2011). Enhancing hazard avoidance in teen-novice drivers. *Accident Analysis & Prevention, 43*, 247–252.

Vincenzi, D.A., Wise, J.A., Mouloua, M., & Hancock, P.A. (Eds.). (2008). *Human Factors in Simulation and Training*. Boca Raton, FL: CRC Press.

Vlakveld, W., Romoser, M.R.E., Mehranian, H., Diete, F., & Fisher, D.L. (2011). Does the experience of crashes and near crashes in a simulator-based training program enhance novice driver's visual search for latent hazards? *Transportation Research Record, 2265*, 153–160.

Wang, Y.B., Zhang, W., & Salvendy, G. (2010a). A comparative study of two hazard handling training methods for novice drivers. *Traffic Injury Prevention, 11*, 483–491.

Wang, Y.B., Zhang, W., & Salvendy, G. (2010b). Effects of a simulation based training intervention on novice drivers' hazard handling performance. *Traffic Injury Prevention, 11*, 16–24.

Zafian, T.M., Samuel, S., Borowsky, A., & Fisher, D.L. (2014). Can young drivers be trained to better anticipate hazards in complex driving scenarios? A driving simulator study. *Presentation at the 93rd Transportation Research Board Annual Meeting*, TRB, National Research Council, Washington, DC.

Recommended Reading

Fisher, D.L., Rizzo, M., Caird, J., & Lee, J.D. (Eds.). (2011). *Handbook of Driving Simulation for Engineering, Medicine and Psychology*. Boca Raton, FL: CRC Press.

Hancock, P.A., Vincenzi, D.A., Wise, J.A., & Mouloua, M. (Eds.). (2008). *Human Factors in Simulation and Training*. Boca Raton, FL: CRC Press.

Triggs, T.J. (1994). Human performance and driving: The role of simulation in improving young driver safety. *Proceedings of the 12th Triennial Congress of the International Ergonomics Association* (Vol. 1, pp. 23–26). Toronto, Canada.

22

Teen and Novice Drivers: The Road from Research to Practice

John Nepomuceno

Donald L. Fisher

Tsippy Lotan

Abstract

The Chapter. As the content of this handbook indicates, there is much that is known about the behaviors of teen and novice drivers and their response to various interventions designed to change and improve those behaviors. But nowhere has it been made clear exactly how the research gets translated into practice, into programs that are intended to reduce novice driver crashes. This chapter is an attempt to do just that. We begin with a discussion of the general role of various entities in the implementation of interventions designed either to provide teens and novice drivers with the skills, motivation, and technologies to drive more safely or to screen those who do not have such skills. Then, four major interventions are used to illustrate how these entities are instrumental in completing the trip along the road from research to practice. It is clear from these interventions that the road is by no means a straight one and depends on political, social, individual, and scientific contributions coming together synergistically at a particular moment in time, either to launch a portion of a program (e.g., graduated driver licensing [GDL] was a state-by-state effort) or to launch an entire program (e.g., license screening). ***Limitations.*** There are several obvious limitations. Perhaps the most significant one is that the approach taken in this chapter is largely based on case studies. These cases are by no means exhaustive, and therefore, the lessons learned are limited.

22.1 Introduction

There is a near-unanimous agreement among researchers in the field of teen and novice driving that much can be done to reduce the crash risk of our most vulnerable road users by changing their behaviors. For many of us in the research community, it is a time to act, but we are not quite sure how to do so without access to the substantial resources or political levers needed to make the translation from research to practice a reality. Nevertheless, successful programs to reduce crashes have been introduced at all levels, and we chronicle examples of such programs. We hope that this chapter offers some insights into how the road from research to practice has been traveled in the past, and how it might be traveled in the future.

To begin, federal and state governments around the world, along with various national and international foundations, institutes, and organizations, fund research to develop and evaluate interventions that will reduce teen crashes (and more generally all automobile crashes) by whatever means, by changing either the behavior of the vehicle or the behavior of the driver or both. We are interested in this chapter almost exclusively in those evidence-based interventions designed either to change novice driver behaviors or to introduce better screens of novice drivers for behaviors that are linked to crashes.

The incentives for federal, state, and local jurisdictions to develop and evaluate such interventions are clear. The resulting morbidity and mortality from automobile crashes is simply unacceptable on both societal and financial levels. Automobile crashes are the ninth leading cause of death worldwide and are projected to be the seventh leading cause of death by the year 2030 (World Health Organization, 2013). They are the leading cause of death among our younger drivers (World Health Organization, 2013). However, the aforementioned jurisdictions face multiple pressures on tight budgets and so are forced to make decisions based not just on costs. However, for-profit companies, institutes, foundations, and organizations function under a related, but perhaps less complex, set of incentives. For-profit companies need to make a business case, not a political case (though within the company, certainly, political priorities will govern what is chosen as well as business priorities). And institutes, foundations, and organizations can often leave aside the business case and serve directly the public good as they so define it. Understanding the complex web of incentives is key to unraveling just which interventions designed to reduce teen and novice driver crashes developed in the lab will find the support that they need to become embedded in practice once they leave the lab. As such, regarding translating research into practice, this chapter is as much a discussion of the necessary incentives as it is of the actual translational activities that have been undertaken.

It should be mentioned at the outset that the discussion of the road from research to practice necessarily complements the general discussions in Chapters 23 and 24, respectively, on the history of research in developed countries and in low- and middle-income countries, focusing for the most part on the developed countries. And the discussion of the road from research to practice insofar as it focuses on four cases studies of interventions that have traveled this road—novice driver training programs, in-vehicle monitoring devices, graduated driver licensing (GDL) laws and hazard anticipation tests at licensure—also complements, respectively, the material in Chapters 18, 20, 17, and 28. We refer the reader to these latter chapters for other details.

22.2 The Various Entities at the End of the Road

As noted previously, there are a number of entities that get involved at the end of the road in the translation of research into practice. These include individual insurance companies, insurance (-related) foundations and institutes, international organizations, private foundations, automobile companies, legislatures, and regulatory agencies. Other entities such as third-party vendors, tech companies, and telecommunications giants also have a role to play, but we do not specifically speak to their contributions.

22.2.1 Individual Insurance Companies

Most insurance companies compete for business with one another. Therefore, any intervention that benefits all insurance companies will be hard to justify from a business standpoint for any single insurance company. There are two ways that insurance companies can benefit from funding the development and evaluation of an intervention and/or translating that research into an intervention that can be used by the public. First, benefits can come from whatever long-term gains that result from increases in the numbers of teens purchasing insurance that arise through incentives, such as reductions in insurance premiums to teens and novice drivers that participate in the intervention (e.g., a novice driver training program) or that introduce the intervention into their vehicle (e.g., in-vehicle monitoring and feedback systems). Second, insurance companies can also benefit from marketing efforts that leverage research results; for example, showcasing to parents that their company has gone the extra mile to keep their children safe. This can provide the company favorable exposure at both the local and national level.

Sometimes, automobile insurance is the responsibility of a particular jurisdiction, not the responsibility of a particular company. Australia and New Zealand are two such countries. As an example, consider the Transport Accident Commission (TAC) in the state of Victoria in Australia. The TAC, unlike private insurers throughout the world, is a state-owned insurance corporation that provides the compulsory portion of the auto insurance in Victoria (TAC, 2015). In state-run insurance companies like the TAC, there is a clear business case for reducing crashes among all drivers, especially the most vulnerable. TAC's mission statement includes promoting road safety, improving Victoria's trauma system, and supporting those who have been injured on its roads: "Reducing the frequency and severity of transport accidents not only saves lives and avoids serious injuries, it also reduces claims. This provides savings to the Victorian community and ensures the long-term financial viability of the transport accident scheme" (TAC, 2015).

22.2.2 Foundations and Institutes

The majority of foundations and institutes that are active at the end of the road from research to practice are associated with insurance companies in one way or another. The insurance company foundations, unlike the individual insurance companies, have a broader mandate and can fund research that could be translated into interventions that benefit all teen and novice drivers (and all drivers for that matter). Examples in the United States include the American Automobile Association Foundation for Traffic Safety (AAA FTS) and the Arbella Insurance Foundation, with obvious relations to particular insurance companies. There are also nonprofit groups such as Advocates for Highway and Auto Safety. The latter is a consortium of consumer, medical, public health, and safety groups along with insurance companies, which undertakes activities that lead to the adoption of federal and state laws, policies, and programs that can prevent motor vehicle crashes. Similarly, in Israel, there is the Israel Insurance Association Research Fund. The motivation that foundations and international organizations have for implementing the more broadly based interventions flows both out of the safety mandates of these entities and from the clear visibility they provide these foundations and organizations among teens and their parents.

Insurance institutes, like insurance company foundations, also have a broad mandate. However, they are more likely to conduct the research in-house. In the United States, these include the Insurance Institute for Highway Safety and the Liberty Mutual Research Institute for Safety. Along with undertaking research that supports the implementation of behavioral interventions that benefit both teen and novice drivers and reduce loss claims for insurance companies (like the evaluation of GDL programs), insurance company foundations and institutes will often educate lawmakers when invited, or legislative and regulatory bodies in areas where they think these bodies could be instrumental in enacting effective legislation to reduce crash risks. Thus, they too are a key player in the road from research to practice.

Perhaps one of the most prominent foundations not associated with an insurance company is the Or Yarok Association for Safer Driving in Israel (Or Yarok, 2015). Its activities include applied projects, research, public advocacy, and public motivation. While there are many foundations like Or Yarok that advocate for road safety for teens, what makes Or Yarok unique among such foundations is that it initiates advocacy and interventions only when they are clearly based on research: "Research provides the basis to all of Or Yarok's activities and accompanies all aspects of its operation. Or Yarok promotes an Evidence-Based-Safety approach both in its own activities and in the overall road safety community in Israel" (Or Yarok, 2015).

22.2.3 International Organizations

Internationally, there are the global efforts such as those undertaken by the World Health Organization (Decade of Action for Road Safety 2011–2020) and the United Nations (Road Safety Collaboration). These efforts were motivated by the fact that it became increasingly clear at the turn of the century that road traffic injuries were a growing international public health issue, especially among low- and middle-income countries, with fatalities expected to increase by 80%. Moreover, it was estimated that road traffic injuries cost these countries somewhere between 1% and 2% of their gross national product, which is more than they receive in total development aid (World Health Organization, 2004). And it is teen and novice drivers who are often most at risk. Although these efforts are one step removed from practice, they often are critical to taking that final step. For example, manuals are now available that explain how to implement proven programs that reduce pedestrian casualties, increase seatbelt and helmet compliance, and reduce impaired driving and speeding, among others. Clearly, several of these can have a direct impact on teen and novice driver crashes.

22.2.4 Automobile Companies

Historically, automobile manufacturers are most often known for their contributions to improvements in vehicle behaviors, not for their efforts to develop and evaluate programs that impact novice driver behaviors. In addition to funding efforts to improve the safety of their vehicles, automobile manufacturers have, over the past decade, developed and funded the evaluation of all manner of collision warning systems, which are designed to provide drivers with the information they need to drive more safely (NHTSA, 2015). The automobile manufacturers' incentives for making both of the aforementioned improvements are well known and not of direct relevance here (IIHS, 2015a). But, these improvements are largely responsible for the large decrease in crashes over the past several decades and so are worth pointing out (IIHS, 2015b).

More recently, automobile manufacturers have developed and deployed novice driver training programs. The incentives for automobile manufacturers to develop such programs are relatively simple. Development serves as a marketing tool. For example, Mercedes-Benz actually offers on road lessons to novice drivers (Mercedes-Benz, 2015). Toyota has produced a web-based program, TeenDrive365: In School, which provides material for both educators and novice drivers (Toyota, 2015). And Ford has the MyKey® option, which directly targets young drivers, by helping parents control seatbelt use, top vehicle speeds, and sound system volume (Ford Media Center, 2013). However, there is relatively little incentive to evaluate the effectiveness of these programs, and so, perhaps not surprisingly, there has been no analysis of the effectiveness of either of these programs, at least to the best of our knowledge. As a final comment, it is of interest to note that automobile manufacturers have funded efforts in the area of distraction and teen driving (Toyota CSRC, 2012). As is true of any basic research, many additional steps must be taken before the road from research to practice is successfully traversed.

22.2.5 Legislatures

In many countries, government legislation is beleaguered by all sorts of competing interests at different levels of government. Yet legislators are the ones that must eventually pass the laws that make a program a required one. For example, consider the GDL programs in the United States. Getting a law passed in any one state is a complex process. In any given state, this requires not only legislators willing to sponsor a given law but also usually a combination of the support of the governor, other elected officials, and public opinion leaders; the pressure of media attention as evidenced by continuing press events and editorials; and the advocacy of interest groups, in this case, victims/survivors of teen driving (Gillan, 2006). Many of the processes are required at the federal level, only on a much larger scale. However, unlike the state legislatures, the federal government does not pass laws that states must necessarily follow with regard to driving safety. Rather, as with impaired driving legislation, at least in the United States, the receipt of federal monies may be subject to compliance.

22.2.6 Regulatory Agencies

Regulatory agencies have a different role to play than legislatures. For example, in the United States, at the state level, the Departments of Motor Vehicles are responsible for the actual content of the driver licensing exams. The composition of those exams, both the theory portion (the written portion) and the road test, can potentially have a big impact on who actually receives a license to drive. As the heads of the Department of Motor Vehicles are usually political appointees, a particular agency could not realistically take any actions outside of tacit approval from the governor and other legislative bodies. The incentives for change therefore typically need not only be internal to a given regulatory agency but can be external to that agency as well.

22.3 Novice Driver Training Programs

In this section, we discuss four case studies where research has made it into practice. In these four cases, the complex interweaving of incentives described previously has played out synergistically. Perhaps there is something to be learned from this. To begin, consider novice driver training programs. Many such programs have been developed. But only recently have researchers begun to evaluate programs that have targeted behaviors known to be linked to crashes. The research has shown a significant reduction in these behaviors (Chapter 18), and in one case an actual reduction in crashes (Thomas et al., 2016). Here, we focus on those evidence-based training programs in the United States and Canada, Australia and New Zealand, and Europe that have been made widely available to teen and novice drivers in those regions and how this came to pass.

22.3.1 United States and Canada

To begin, consider the novice driver training programs in the United States and Canada that are now available to the public and that are based on programs that have been evaluated, either on a driving simulator or in the field. Most such evidence-based training programs are now offered by insurance companies. The question is why now rather than, say, 10 or 20 years ago.

As noted throughout the handbook, standard driver education programs have done little to reduce the crash risk of teen and novice drivers (Chapter 18; Nichols, 2003). Therefore, it was difficult for any single insurance company to make a business case for undertaking the research and development needed to determine whether a training program could be designed that would be effective in reducing crashes. In fact, after the DeKalb study in the United States, which showed that the best driver education programs provided little reduction in risk (Stock et al., 1983), the federal government itself largely turned its back on the funding of novice driver education for the next 30 years, and advisedly so (Nichols, 2003), becoming seriously engaged again only after the turn of the century.

However, the development and evaluation of novice driver training programs did not entirely languish over the 30-year period. For example, in the late 1990s, the AAA FTS developed a training program, DriverZED (Zero Errors Driving), which was one of the first to target training for driver behaviors that were likely precursors of crashes, including, specifically, hazard anticipation (Willis, 1998). This program was evaluated and proved, at least on a driving simulator, to lead to safer behaviors (better hazard anticipation and hazard mitigation) among teen and novice drivers who had been trained than among those who had received no training (Fisher et al., 2002). The program has since been through several revisions and, as of 2015, is available for a minimal charge on the AAA FTS website (AAA Foundation for Traffic Safety, 2015b). The AAA does not offer a reduction in the premiums for teens completing the DriverZED training program. Rather, the training program was developed by the foundation under its general mission statement: "To identify traffic safety problems, foster research that seeks solutions, and disseminate information and educational materials" (AAA Foundation for Traffic Safety, 2015a).

Perhaps one of the major factors motivating the AAA FTS to take a leap during the late 1990s was the advance in the way that driver education training could be delivered. The digital age had arrived. Videos, real or virtual (simulated), could be manipulated, and drivers could be trained in an immersive environment, which never before had been possible. Slowly but steadily, the federal government has stepped up its funding of research on novice driver education, including agencies such as the National Insitutes of Health, the Centers for Disease Control, and the National Highway Traffic Safety Administation. And as more and more programs were developed and evaluated, the evidence began to mount that novice drivers could be trained on skills known to reduce crash risk, such as hazard anticipation, hazard mitigation, and attention maintenance (Chapter 18). Most recently, a hazard anticipation skills training program has been shown not only to improve hazard anticipation skills among teens, but to reduce crashes among male teen drivers (Thomas et al., 2016).

Individual insurance companies now saw an opportunity not only to reduce the dispproportionate number of crashes among teen and novice drivers but also to increase their visibility among teens and their parents or to provide reductions in premiums to those who completed the training programs successfully. State Farm developed a program called Road Aware (State Farm, 2015), which was evaluated and proved to reduce novice drivers' risky driving behaviors (Zafian et al., 2014). The Arbella Insurance Foundation funded the development of a related training program, which was evaluated and also proved to reduce novice drivers' risky driving behaviors (Vlakveld et al., 2011). The initial indications are that the latter training program reduces crashes and citations by up to 20% (Zhang et al., in press). In both cases, travel at the end of the road from research to practice was made possible by individual insurance companies who, in turn, benefited from research sponsored by both insurance foundations and various federal agencies.

22.3.2 Australia and New Zealand

As noted previously, in Australia and New Zealand, motor vehicle personal liability insurance is compulsory. In most states in Australia, it is the state itself that underwrites the insurance (Australian Insurance Law Association, 2010). Given the incentive that the state has for reducing crashes when it underwrites that insurance, it is perhaps not surprising that in one of those states where such is the case, Victoria, the underwriter (the TAC) was the among the first to sponsor the development and evaluation of a novice driver training program that clearly demonstrated an effect on the hazard mitigation, hazard anticipation, and attention maintenance behaviors of novice drivers (Regan et al., 2000). This research has had a huge impact on the field, reigniting the push for new programs of driver skills training that have continued apace for the past 15 years, programs that have now been implemented, as noted previously, by a number of different individual insurance companies. However, curiously, to the best of our knowledge, there are no skills-based training requirements in any of the jurisdictions in Australia. However, as we will discuss in a bit, there are skills-based screening requirements that followed from the research funded by the TAC.

As described in Chapter 18 and elsewhere in the handbook, whereas some training programs focus on complex driver skills, others focus on driver attitudes. In the state of New South Wales (Australia), programs have been implemented throughout the school systems that focus on attitudes and, in particular, on developing the knowledge, understanding, and decision-making skills needed to deal with teen and novice driver issues such as alcohol and drug use, risk taking, speeding, fatigue, peer influence, occupant restraints, driver and passenger distraction, and the requirements of the Graduated Licensing Scheme (NSW Department of Education and Communities, 2015). It appears available information that the program was developed by the New South Wales Department of Education and Communities in cooperation with the New South Wales Centre for Road Safety and is based on an earlier study that showed potential reductions in crashes when such a program was in place (Senserrick et al., 2009). Interestingly, in this particular state, compulsory insurance is acquired from private-sector companies.

In New Zealand, the Accident Compensation Corporation insures all teen drivers. A free online training program that targets situation awareness, hazard perception, and risk management for novice drivers has been introduced (Isler & Isler, 2015). The development of the online training program was sponsored by the Accident Compensation Corporation and the New Zealand Transport Agency. It is arguably the case that the development of the online program occurred both because the Accident Compensation Commission was a government entity and also because the intervention was developed by researchers who had themselves undertaken studies that suggested that the approach succeeded (Isler et al., 2009). In both Australia and New Zealand, for the most part, the insurance companies were involved in the road from research to practice along the entire journey because of their much broader mandate.

22.3.3 Europe

The situation is very different in Europe than it is in the United States and Canada or Australia and New Zealand (Chapter 18). Entire countries in Europe have adopted new novice driver training programs. For example, in Norway, the novice driver curriculum is based on a program developed by various members of the European Union known as the Goals for Driver Education (GDE) matrix (Hatakka et al., 2003). The curriculum includes an emphasis on all levels of driving competency, from handling a vehicle to understanding the motives for driving (Dalland, 2012). In Finland, training of a sort has been required at the very time of licensure, where a novice driver must provide an assessment of his or her driving skills, which the instructor then compares with his or her assessment of the novice driver's skills, providing structured feedback where there are differences (Mynttinen et al., 2009). The shift from a focus on teaching crash-imminent driving to a focus on the training of self-assessment was occasioned by studies that showed that the training of crash-imminent driving skills (e.g., learning how to control a skid) actually increased the likelihood of a crash (Christie, 2001). It became clear that what was needed was both to improve novice drivers' anticipation of risk and to make sure that their self-assessment of their skills did not become overinflated or underinflated (e.g., Horrey et al., 2015). These programs have proven generally beneficial, although the jury is still out (Washington et al., 2011).

Regardless of whether the postlicense training programs now in place for novice drivers are effective, the question is why such programs have made it into practice nationwide in so many European countries (Sweden, Finland, Austria, Luxembourg, and Norway). Although our answer is largely based on speculation, we cannot help but believe that in no small measure, the safety climate in many European countries has much more influence on the public and elected officials than is true, say, in a country such as the United States. Vision Zero, which is now a multinational attempt to reduce road traffic fatalities to zero, had its birth in Sweden and was approved by its parliament in October 1997 (Goodyear, 2014). As of today, in the United States, the push toward zero traffic fatalities has a national footprint, and was embraced most recently at a Leadership Summit held by the Department of Transportation in Washington, DC in April of 2016.

22.4 In-Vehicle Feedback Systems

Consider next in-vehicle feedback systems. As described in Chapter 20, we can now monitor many activities of teen drivers with in-vehicle devices. Importantly, these systems can deliver feedback to teen drivers on their behaviors in real time (e.g., speed, lateral acceleration) that exceed critical thresholds, and the systems can transmit this behavior to parents, either in real time or at some later time. These programs have been shown to reduce especially risky behaviors of teen and novice drivers. And they are now offered worldwide to teen and novice drivers, showing that they can lead to reductions in risky behaviors such as speeding (Bolderdijk et al., 2011). Here, we look briefly at how this came about both in the United States and in Israel.

Note that we do not speak to recent advances in the aforementioned systems, which include cameras that can process information in real time, because they have not been evaluated to date. Such systems provide forward collision warnings, lane change warnings, and pedestrian and bicycle warnings, to name just a few. However, they are deserving of further monitoring (e.g., Mobileye, 2012).

22.4.1 United States

In the United States, the ability to monitor behavior inside the vehicle through the on-board diagnostics system has provided insurance companies and private companies with a motive to become involved in the development and evaluation of these systems (Simons-Morton et al., 2011a). Thus, potentially, insurance companies could improve teen and novice driver safety by linking premiums to safe driving practices, assuming that the safe driving practices could be measured and linked to crashes. American Family Insurance and Ltyx/DriveCam were among the first to partner together to provide funding for the potential benefits of in-vehicle video feedback systems as a way to measure behaviors that were assumed to be risky, using a population of rural drivers (McGehee et al., 2007). In this case, it was the tail wagging the dog, the initial studies being funded by the private sector. Only later did federal agencies decide to follow up on the results and enter the picture (Carney et al., 2010), showing eventually that not only was there a decrease in the behaviors known to be risky (Simons-Morton et al., 2014), but there was also a relation between these behaviors and crash risk (Simons-Morton et al., 2011b). There is now a large push for what is often referred to as usage-based insurance, where premiums are tied to driving habits as monitored by event data recorded (in-vehicle monitoring devices, sometimes referred to as "black boxes"), regardless of the age of the driver (Lieber, 2014).

22.4.2 Israel

In Israel, much of the research on in-vehicle feedback technology systems has been funded by the Israel Insurance Association Research Fund. It has been shown that in-vehicle feedback systems during the first 9 months of solo driving can reduce risky behaviors by up to 29% if both teens and parents are involved and the parents are given training on how to increase attention to the roadway by their children who are getting their license (17–22 years old) (Farah et al., 2014).

Given the evidence on effectiveness, and in an effort both to formalize the requirements for monitoring systems used by young drivers (and professional drivers) and to regulate their implementation, the National Road Safety Agency initiated a call for a standard for such systems. Indeed, in 2010, after an intense 18 months of expert deliberations, a standard for these monitoring systems (Standard 5905) was approved by the Standards Institution of Israel (2010). The publication of the standard, together with wide media coverage and strong lobbying of Or Yarok (the largest NGO dealing with road safety in Israel), were among the triggers to insurance companies to offer reduced premiums to young drivers who install and use monitoring systems. It should be noted that compulsory insurance rates for young drivers in Israel are about double the rates of experienced drivers. The AIG Safe Drive initiative, for example, offers a 10% reduction on insurance premiums for new young drivers joining the program and a reduction of up to 30% for renewal depending on a composite safety score of the drivers actually driving the vehicle.

Another initiative to advance dissemination of monitoring systems among young drivers was established through adoption of the system by the army. The Israeli Defense Forces (IDF) have equipped more than 7000 vehicles with G-based monitoring systems and plan to enlarge the dissemination to 10,000 vehicles. The population of soldiers driving these vehicles is composed of mostly young drivers during their mandatory military service. The military implementation is based on feedback provided directly to the commanders. Needless to say, monitoring by commanders (instead of by parents) provides a natural and easy implementation channel. Commanders are given weekly reports and are notified in real time of extreme behaviors. The three main triggers for real-time alerts are (1) crash indication—an extreme g-force measurement; (2) excessive speed compared to the allowed speed; and (3) unidentified trip—soldiers are expected to identify themselves through their *soldier ID* at the beginning of each trip. If they fail to do so, the vehicle transmits a note on unidentified trip to the commander. The IDF implementation is evaluated internally, and based on positive results regarding safety and fuel consumption (not publicly available), the project continues and is being expanded.

Despite the success of such feedback systems in the army, their adoption outside of the army has been stalled because of a number of concerns such as the invasion of privacy and erosion of the parent–young driver relationship (Guttman & Lotan, 2011). This is a situation where it would appear that the insurance industry would like to see a broader introduction of what is a proven safety technology, but the public concerns about its adoption have slowed the progress considerably (Weinstock, 2015).

22.5 Graduated Driver Licensing

A number of individuals who are key players in the travel along the road from research to practice, including those most intimately involved in its eventual success, have written about this journey as it applies to graduated licensing programs (GDL) in the United States and around the world (e.g., Waller, 2003; Williams, 2005; Williams et al., 2015; also see Chapter 17). Here, we talk only about the development of the program in the United States and the Netherlands, where it has proved particularly difficult. Its adoption in countries such as New Zealand has been much simpler, and readers are encouraged to delve into their experience (Langley et al., 1996).

22.5.1 United States

The adoption of GDL laws is an amazing success story, but why it became a success story is not entirely clear (Williams, 2005). The concept of a GDL program was first introduced by Waller in the early 1970s (Waller & Reinfurt, 1973). The National Highway Traffic Safety Administration (NHTSA) quickly followed suit, and congress offered financial incentives for states to adopt it (Teknekron, Inc., 1977), providing a model law in the process. But full adoption took over 20 years, with not much happening between the early 1970s and 1980s. It was not until Florida adopted a three-stage GDL program (learner phase, restricted license phase, and unrestricted license phase) in 1996 that the tide turned.

The question that Williams asked in one of his several histories of the adoption of GDL programs is why it took so long (Williams, 2005). He provided three possible answers. First, part of the answer is that the evidence began to accumulate through research studies from around the world that GDL programs either could be or were a success. This included studies completed in North America before the GDL programs were widely adopted that strongly pointed to the need for just such programs (Williams et al., 1995), studies completed after the adoption of the first GDL programs in the United States that showed an impact of GDL programs on crashes (Foss et al., 2001; Shope et al., 2001), and studies completed later in New Zealand, which had already adopted a GDL program that clearly showed that such programs were saving lives (Begg & Stephenson, 2003). But by itself, this was probably not enough.

Rather, Williams goes on to discuss a second answer to the question of why it took so long to adopt what was, by all accounts, a program of licensing rooted in science and for which considerable evidence already existed (Williams, 2005). In particular, Williams goes on to describe a possible change in the safety culture or the social climate as the new millenium began to creep over the horizon. Baby boomers were now parents of children who were just coming to be of driving age. Some have speculated that this generation was more aware of the risks entailed by driving and also actively more concerned (Waller, 2003). Such a view is consistent with the formation of influential organizations like Mothers Against Drunk Driving in 1980. Baby boomers, perhaps because of Vietnam, felt empowered to take action in general, and traffic safety was just one of those arenas.

Williams goes on to consider a third answer to his question. In particular, public attention is generally focused in any one arena on only one or two questions. As the public attention to one issue waxes, it wanes to another issue. The fluctuation of public attention to any one issue has been the subject of a number of research articles (e.g., Hilgartner & Bosk, 1988). In the driving arena, alcohol engaged the public's attention in the 1980s and 1990s, attention first increasing and then decreasing (Ross, 1997). Also, competing for attention during this period of time were seatbelt laws. But, as interest in these issues waned, other issues had a chance to rise to the surface. And for the reasons cited previously, the growing body of research and the change in safety culture, the time was ripe for attention to the novice driver and ways to prevent crashes that were evidence based. GDL rose to the top of heap, and it stands as a pillar of continuing success.

As Williams says in his concluding remarks, the take-away message for researchers is that "good research indicating that a measure can reduce losses on the highway can be a starting point, but it can take much more to get such a measure adopted—and 'much more' may involve things interested parties have no control over and may not even understand." The road from research to practice is one definitely less traveled and often leading nowhere. But, the intergral is much more than the sum of its parts, and today, it is estimated that among 16-year olds, fatal crashes have been reduced by up to 41% (McCartt et al., 2010). There are few interventions championed by researchers in the field of traffic safety that have had as large an impact on lives saved as GDL programs. Even though the road from research to practice often leads nowhere, the effort pays off over time, and quite remarkably so.

22.5.2 The Netherlands

The introduction of the GDL in the Netherlands was less cumbersome because it did not involve 50 different jurisdictions. But the process appeared equally politically complex (the material that follows is based on personal communication with Willem Vlakveld). Until very recently, the Dutch driving license system remained unchanged, although multiple attempts to introduce a GDL system were initiated. Briefly, the emphasis of the existing system was on the driving test, both skills and theory. From their 18th birthday on, teens could start to take driving lessons from a certified driving instructor. There was not a minimum number of hours of behind-the-wheel training required before a learner could register for the driving skills test. (However, as a rule, on average, 40 hours of behind-the-wheel training from a certified driving instructor was required to pass the test.) Neither was there a minimum number of hours of theory instruction required to register for the test. When the driving instructor considered a learner to be ready for the test, the learner could take the test. When the learner failed the test, he or she had to take more formal driving lessons for a second attempt until the test was passed. A graduated driving license system did not exist, and learners were not allowed to drive with a layperson (e.g., a parent) in order to gain experience before the driving test was passed.

In the 1990s, research institutes started to raise concern about the driving license system in the Netherlands and proposed the introduction of a GDL system and a driver education system that incorporated higher-order skills training such as hazard anticipation training and risk awareness training. In 2003, this resulted in a very ambitious plan by the Ministry of Transport. However, soon after the presentation of this plan, the government collapsed, and there was no longer a majority in parliament to support this plan. The driving schools argued that the introduction of GDL would result in fewer formal driving lessons. They had a strong lobby and successfully made the case to enough members of parliament that a GDL system was not safe by itself. In 2005, the Dutch driving license authority wrote a somewhat less ambitious plan in which they related the introduction of a GDL program to a driving education system they had developed. The then-minister of transport rejected some elements of GDL in this plan because he thought it would make driver education too expensive, requiring testing at multiple points in time that was not required in the current licensing exam, including a separate testing after the learner's permit phase of hazard anticipation skills.

It was not until 2011 that a simplified GDL system was introduced in the Netherlands. This was in no small measure a reaction to the fact that an evaluation had just been completed in Germany of their GDL system and it proved to reduce crash rates significantly among novice drivers in the first 2 years after their 18th birthday (Willmes-Lenz et al., 2010). Teens now could begin their driver training and take a theory test any time after they had reached the age of 16 years and 6 months. After having passed the theory test, they could take the skills test any time after they had reached the age of 17. When they had passed the driving skills test and obtained their license, they were only allowed on the road when accompanied by an experienced driver, their coach, until they were 18 years old. This is a kind of permit phase that all GDL systems have. The driving school lobby had no objections, because the driving test remained the same, and learners could only start accompanied driving after they have passed the driving test.

22.6 Hazard Perception Testing at Licensure

Hazard perception testing is now a part of the licensing requirements around the world, including Australia, the United Kingdom, and the Netherlands. Here we focus on how it came to be that hazard perception testing was introduced into driver license exams in each of these three countries.

To begin, the discovery of an association between the ability to perceive hazards and the likelihood of a crash goes back more than half a century now (Spicer, 1964, as cited by Pelz and Krupat, 1974). The material in Chapter 28 goes into great detail about the efforts since to further cement the relation between hazard perception skills and crashes, including research that shows a strong correlation between hazard perception test scores and crash risk, measures of on-road driving behavior, key factors associated with crash risk, and experience. Thus, there has been a long history of research supporting the potential use of hazard perception testing at licensure as a way to screen novice drivers.

22.6.1 United Kingdom

The United Kingdom was actually the first to introduce hazard perception testing into its licensing exam for novice drivers. Some of the original research on hazard perception testing was done in the United Kingdom, showing a clear difference between the hazard perception skills of novice and experienced drivers (McKenna and Crick, 1991a,b). There then followed an intensive period of public education by several of the principal researchers and key officials at the Department for Transport who thought that the test was important to introduce (Directgov, 2010). For example, the interested principals gave presentations to key stakeholders such as the Department for Transport, the Parliamentary Advisory Committee on Transport Safety, the Transport Select Committee (Westminster and Scotland), members of parliament, and parties who would be affected by any implementation, including the Driving Standards Agency, driving instructor associations, and the Automobile Association (AA) (Research Excellence Framework, 2014). This resulted in the team's hazard perception test being adopted by the Department for Transport as part of all new driver assessments in November 2002 (Directgov, 2010). The test used by the Department for Transport uses the same methods as the one that was originally developed by McKenna and Crick, and this remains a core component of the driving test today (Research Excellence Framework, 2014). Over a million hazard perception tests are taken as part of the statutory driving test in the United Kingdom every year (e.g., 1.3 million hazard perception tests were taken in 2012) (Driver and Vehicle Standards Agency, 2015). This is one of the few cases where an individual from academia has collaborated with individuals from federal regulatory agencies to push successfully for an intervention that is evidence based.

22.6.2 The Netherlands

The history of the introduction of hazard perception testing into the theory portion of the driving license exam is intertwined with the history of the attempt to introduce a GDL program into the Netherlands (see Section 22.5.2). (The material in this section is based on personal communications with Willem Vlakveld.) Briefly, as noted previously in the discussion of the introduction of the GDL program into the Netherlands, in 2003, a plan was put forth by the Ministry of Transport, which included not only the elements of a GDL program but also higher-order skills training, such as hazard anticipation. Moreover, this proposed plan incorporated a hazard perception test into the theory test. The introduction of this hazard perception test was proposed after consultation with the British driving license authority. However, this plan was not instituted at that time, because there was a change of government (there was no longer a majority in the parliament to support the plan). A hazard perception test was again proposed in the 2005 plan put forward by the Dutch driving license authority as part of an expanded GDL system. Again, as noted previously, the then-minister of transport rejected the plan and, along with it, a separate hazard perception test after the learner's permit phase. However, it was decided to incorporate a hazard perception test into the traditional theory test. After this decision, Stichting Wetenschappelijk Onderzoek Verkeersveiligheid (SWOV) (the Institute for Road Safety Research in the Netherlands) and the Dutch driving license authority Centraal Bureau Rijvaardigheidsbewijzen (CBR) started to develop a hazard perception test. Their cooperation was enough to lead to the introduction of the hazard perception test into the theory portion of the driving test in 2009 (Vlakveld, 2011).

22.6.3 Australia

Finally, we have already talked about the clear incentives for state-run insurance entities to reduce crashes. This incentive applies across the board to other state jurisdictions. In this context, it is perhaps not surprising that in Victoria, it was VicRoads (the state government Department of Transport) that pushed for the inclusion of a test for hazard perception skills in its licensing exam. The test was developed in the early 1990s (Hull & Christie, 1992). A separate test was designed for the state of Queensland by some of the individuals most centrally involved in research on hazard perception (Wetton et al., 2011). Although the state of Queensland does not have the same incentive as the state of Victoria (the state of Queensland does not underwrite compulsory insurance), the spread of evidence-based interventions like GDL can take on a life of its own, which may be the case here.

22.7 Discussion

We started off by noting that the road from research to practice has been traversed and traversed successfully in a number of different cases. As this handbook makes clear, there is a great deal that is known around the world about programs that can reduce crashes and the behaviors that lead to crashes. However, despite the overwhelming evidence that such programs work as intended, globally, very little that could be accomplished has actually been put into practice by either developed or low- and middle-income countries. The review in this chapter suggests that although there is no magic bullet, without research, there will never be the progress we all hope will one day be made. Research is a necessary ingredient in all of the major successes (Williams, 2005). But it is by no means sufficient.

In jurisdictions where there is a federal entity with a clear financial incentive to undertake both research and practice (such as some states in Australia), this additional ingredient is all that is needed. However, in jurisdictions where such is not the case, at least four additional ingredients seem to be necessary for success above and beyond scientific evidence, two of those four discussed previously (also see Williams, 2005). First, there often needs to be a spark that ignites an advance in practice that is widely recognized as effective and for which the country is prepared to move forward. This was the case with GDL programs in the United States, where the first state to introduce a multicomponent GDL program was Florida, when a representative's personal experience highlighted the importance of such programs. Second, the culture needs to be one that supports safety. This is true in Europe (with Vision Zero) and in the United States (with Mothers Against Drunk Driving).

Third, the attention of the nation to a particular issue in a given area seems to be limited to one or, at most, two areas of concern. So while various concerns dominated the traffic safety focus of the United States when GDL programs were blossoming in other countries, once those concerns diminished, there was ample capacity to change focus to something like GDL. Finally, it would appear in some cases that not only do all of the aforementioned need to come together, but also, the efforts of a small set of individuals over a long period of time are needed to complete the trip along the road from research to practice. This seems to be the case in the core group in the United Kingdom that eventually was able to make hazard perception testing a part of the license test: more power to these generous and committed souls.

Acknowledgments

A number of individuals read over sections of this chapter, and some even contributed to portions of this chapter, for which the authors are very grateful. They include Allan Williams, Mark Horswill, Willem Vlakveld, and Dan McGehee. Of course, the authors are responsible for any errors in content.

References

AAA Foundation for Traffic Safety. (2015a). *About the AAA Foundation for Traffic Safety*. Accessed September 4, 2015, retrieved from AAA Foundation for Traffic Safety: https://www.aaafoundation.org/about-aaa-foundation-traffic-safety#mission.

AAA Foundation for Traffic Safety. (2015b). *Driver-ZED*. Accessed September 4, 2015, retrieved from Driver-ZED 3.0: Driver Education Interactive DVD for Teen Driving Safety: http://www.driverzed.org/home/?CFID=360019&CFTOKEN=60595501.

Australian Insurance Law Association. (2010). Mandatory Insurance: Legal and Economic Myths and Realities. *AIDA World Congress*. PAIRS. Accessed October 24, 2015, retrieved from http://www.aida.org.uk/docs/Australia%20.doc.

Begg, D., & Stephenson, S. (2003). Graduated driver licensing: The New Zealand experience. *Journal of Safety Research, 24*, 99–106.

Bolderdijk, J., Knockaert, J., Sleg, E., & Verhoef, E. (2011). Effects of Pay-As-You-Drive vehicle insurance on young drivers' speed choice: Results of a Dutch field experiment. *Accident Analysis & Prevention, 43*, 1181–1186.

Carney, C., McGehee, D., Lee, J., Reyes, M., & Raby, M. (2010). Using an event triggered video intervention system to expand the supervised learning of newly licensed adolescent drivers. *American Journal of Public Helath, 100*, 1101–1106.

Christie, R. (2001). *The Effectiveness of Driver Training as a Road Safety Measure: A Review of the Literature*. Noble Park, Victoria, Australia: Royal Automobile Club of Victoria Ltd. Report 1 of 3. Accessed October 25, 2015, retrieved from http://www.education.vic.gov.au/Documents/school/teachers/health/effectdriver.pdf.

Dalland, E. (2012). The driving test in Norway: An intervention study. In L. Dorn (Ed.), *Driver Behaviour and Training*. Aldershot: Ashgate.

Directgov. (2010). *The hazard perception test (HPT) explained*. Accessed September 26, 2015, retrieved from The UK Government Web Archive: http://webarchive.nationalarchives.gov.uk/+/www.direct.gov.uk/en/motoring/learnerandnewdrivers/theorytest/dg_4022535.

Driver and Vehicle Standards Agency. (2015). *Car driving theory test: operational statistics*. Accessed September 26, 2015, retrieved from https://www.gov.uk/government/statistics/car-driving-theory-test-operational-statistics.

Farah, H., Musicant, O., Shimshoni, Y., Toledo, T., Grimberg, E.O., & Lotan, T. (2014). Can providing feedback on driving behavior and training on parental vigilant care affect male teen drivers and their parents? *Accident Analysis & Prevention, 69*, 62–70.

Fisher, D.L., Laurie, N.E., Glaser, R., Connerney, K., Pollatsek, A., Duffy, S.A., & Brock, J. (2002). The use of an advanced driving simulator to evaluate the effects of training and experience on drivers' behavior in risky traffic scenarios. *Human Factors, 44*, 287–302.

Ford Media Center. (2013). *News: MyKey*, June 11. Accessed January 9, 2015, retrieved from Ford Motor Company MEDIACENTER: https://media.ford.com/content/fordmedia/fna/us/en/news/2013/06/11/ford-mykey—now-on-6-million-vehicles—helps-parents-keep-teens-.html.

Foss, R., Feaganes, J., & Rodgman, E. (2001). Initial effects of graduated driver licensing on 16-year-old driver crashes in North Carolina. *Journal of the American Medical Association, 286*, 1588–1592.

Gillan, J. (2006). Legislative advocacy is key to address teen driving deaths. *Injury Prevention, 12* (Suppl 1) i44–i48.

Goodyear, S. (2014). *The Swedish approach to road safety: The accident is not the major program.* Accessed October 25, 2015, retrieved from CityLab: The Atlantic Monthly Group: http://www.citylab.com/commute/2014/11/the-swedish-approach-to-road-safety-the-accident-is-not-the-major-problem/382995/.

Guttman, N., & Lotan, T. (2011). "The little squealer" or "The virtual guardian angel"? Young drivers' and their parents' perspective on using a driver monitoring technology and its implications for parent–young driver communication. *Accident Analysis & Prevention, 43*, 412–420.

Hatakka, M., Keskinen, E., Baughan, C., Goldenbeld, C., Gregersen, N.P., Groot, H., Siegrist, S., Willmes-Lenz, G., & Winkelbauer, M. (2003). *Basic Driver Training: New Models, EU-project (Final Report).* Turku, Finland: Department of Psychology, University of Turku.

Hilgartner, S., & Bosk, C. (1988). The rise and fall of social problems: A public arenas model. *American Journal of Sociology, 94*, 52–78.

Horrey, W.J., Lesch, M.F., Mitsopoulos-Rubens, E., & Lee, J.D. (2015). Calibration of skill and judgment in driving: Development of a conceptual framework and the implications for road safety. *Accident Analysis & Prevention*, 76, 25–33.

Hull, M., & Christie, R. (1992). Hazard perception test: The Geelong trial and future development. Paper presented at the National Road Safety Seminar, Wellington, New Zealand.

Insurance Institute for Highway Safety: Highway Loss Data Institute (IIHS). (2015a). *Search test results.* Accessed September 5, 2015, retrieved from Safey ratings: http://www.iihs.org/iihs/ratings.

Insurance Institute for Highway Safety: Highway Loss Data Institute (IIHS). (2015b). *Saving lives: Improved vehicle designs bring down death rates (Status Report, Vol. 50, No. 1).* Insurance Institute for Highway Safety: Highway Loss Data Institute. Accessed September 20, 2015, retrieved from http://www.iihs.org/iihs/sr/statusreport/article/50/1/1.

Isler, R., & Isler, N. (2015). Free Online Training in Situation Awareness, Hazard Perception. *ResearchGate*, 1–8. Accessed August 5, 2015, retrieved from http://www.researchgate.net/publication/265569485_Free_Online_Training_in_Situation_Awareness_Hazard_Perception_and_Risk_Management_for_Learner_Drivers_in_New_Zealand.

Isler, R., Starkey, N., & Williamson, A. (2009). Video-based road commentary training improves hazard perception of young drivers in dual task. *Accident Analysis & Prevention, 41*, 445–452.

Langley, J., Wagenaar, A., & Begg, D. (1996). An evaluation of the New Zealand graduated driver licensing system. *Accident Analysis & Prevention, 28*, 139–146.

Lieber, R. (2014). Lower your car insurance bill, at the price of some privacy. *New York Times*, August 15. Accessed September 13, 2015, retrieved from http://www.nytimes.com/2014/08/16/your-money/auto-insurance/tracking-gadgets-could-lower-your-car-insurance-at-the-price-of-some-privacy.html?_r=0.

McCartt, A., Teoh, E., Fields, M., Braitman, K., & Hellinga, L. (2010). Graduated licensing laws and fatal crashes of teenage drivers: A national study. *Traffic Injury Prevention, 11*, 240–248.

McGehee, D., Raby, M., Carney, C., Lee, J., & Reyes, M. (2007). Extending parental mentoring using an event-triggered video intervention in rural teen drivers. *Journal of Safety Research, 38*, 215–227.

McKenna, F., & Crick, J. (1991a). Experience and expertise in hazard perception. In G. Grayson, & J. Lester (Eds.), *Behavioural Research in Road Safety 1990* (pp. 39–46). Crowthorne: Transport Research Laboratory.

McKenna, F., & Crick, J. (1991b). *Hazard Perception in Drivers: A Methodology for Testing and Training (Final Report).* Crowthorne: Transport Research Laboratory.

Mercedes-Benz. (2015). *Mercedes-Benz Driving Academy.* Accessed August 29, 2015, retrieved from http://www.mb-drivingacademy.co.uk/uk/en/home.html.

Mobileye. (2012). *Newsletter—Q1 2015.* Accessed January 9, 2015, retrieved from Mobileye: http://www.mobileye.com/press-room/retail-partners/mobileye-wants-to-increase-the-awareness-of-driving-safety-of-young-traffic-participants-in-training/.

Mynttinen, S., Sundström, A., Koivukoski, M., Hakuli, K., Keskinen, E., & Henriksson, W. (2009). Are novice drivers overconfident? A comparison of self-assessed and examiner-assessed driver competencies in Finish and Swedish sample. *Transportation Research Part F: Traffic Psychology and Behavior, 12*, 120–130.

National Highway Traffic Safety Administration (NHTSA). (2015). *Estimating safety benefits of new technologies.* Accessed September 20, 2015, retrieved from NHTSA: http://www.nhtsa.gov/Research/Crash+Avoidance/ci.Estimating+Safety+Benefits+of+New+Technologies.print.

National Safety Council. (2015). *Traffic deaths have increased six consecutive months.* Accessed October 10, 2015, retrieved from http://www.nsc.org/learn/about/Pages/Traffic-deaths-have-increased-six-consecutive-months.aspx.

Nichols, J. (2003). *A Review of the History and Effectiveness of Driver Education and Training as a Traffic Safety Program.* Washington, DC: National Transportation Safety Board.

NSW Department of Education and Communities. (2015). *Driver education and driver training in NSW public schools.* Accessed September 7, 2015, retrieved from NSW Department of Education and Communities: http://www.curriculumsupport.education.nsw.gov.au/policies/road/driver/index.htm.

Or Yarok. (2015). *Or Yarok Association for Safety Driving.* Accessed September 13, 2015, retrieved from https://www.oryarok.org.il/webfiles/fck/about%20or%20yarok(2).pdf.

Pelz, D., & Krupat, E. (1974). Caution profile and driving record of undergraduate males. *Accident Analysis & Prevention, 6*, 45–58.

Regan, M., Triggs, T., & Godley, S. (2000). Simulator-based evaluation of the DriveSmart novice driver CD-ROM training product. *Proceedings of the Road Safety Research, Policing and Education Conference.* Brisbane, Australia.

Research Excellence Framework. (2014). *Improving road safety by developing a hazard perception test for drivers (Impact case study REF3b).* Accessed September 26, 2015, retrieved from http://www.reading.ac.uk/web/FILES/reas/Improving_road_safety_by_developing_a_hazard_perception_test_for_drivers.pdf.

Ross, H. (1997). The rise and fall of drunk driving as a social problem in the USA. In C. Mercier-Guyon (Ed.), *Proceedings of the 14th International Conference on Alcohol, Drugs, and Traffic Safety Vol. 1.* Accessed October 26, 2015, retrieved from http://www.icadtsinternational.com/files/documents/1997_002.pdf.

Senserrick, T., Ivers, R., Boufous, S., Chen, H., Norton, R., Stenvenson, M., & Zask, A. (2009). Young driver education programs that build resilience have the potential to reduce road crashes. *Pediatrics, 124*, 1287–1292.

Shope, J., Molnar, L., Elliott, M., & Waller, P. (2001). Graduated driver licensing in Michigan: Early impact on motor vehicle among 16-year-old drivers. *Journal of the American Medical Association, 286*, 1598–1599.

Simons-Morton, B.G., Ouimet, M.C., Zhang, Z., Klauer, S.E., Albert, P.S., & Dingus, T.A. (2011a). Crash and risky driving involvement among novice teenagers. *American Journal of Public Health, 101*, 2362–2367.

Simons-Morton, B.G., Ouimet, M.C., Zhang, Z., Klauer, S.E., Lee, S.E., Wang, J., Albert, P.S., & Dingus, T.A. (2011b). Crash and risky driving involvement among novice adolescent drivers and their parents. *American Journal of Public Health*, 101, 2362–2367.

Simons-Morton, B.G., Bingham, C., Ouimet, M.C., Pradhan, A., Chen, R., Barretto, A., & Shope, J. (2014). The effect of teenage risky driving of feedback from a safety monitoring system: A randomized controlled trial. *Journal of Adolescent Health, 53*, 21–26.

Spicer, R. (1964). *Human Factors in Traffic Accidents.* Honolulu, HI: Department of Health.

State Farm. (2015). *Teen Driver Safety Home.* Accessed September 5, 2015, retrieved from Welcome to State Farm Teen Driver Safety Website: http://teendriving.statefarm.com/road-aware.

Stock, J., Weaver, J., Ray, H., Brink, J., & Sadof, M. (1983). *Evaluation of Safe Performance Secondary School Driver Education Curriculum Demonstration Project.* Washington, DC: National Highway Traffic and Safety Administration.

Teknekron, Inc. (1977). *Model for Provisional (Graduated) Licensing of Young Novice Drivers (DOT Report HS-802-313).* Washington, DC: US Department of Transportation.

Thomas, F., Rilea, S., Blomberg, R., Peck, R., & Korbelak, K. (2016). *Evaluation of the safety benefits of the Risk Awareness and Perception Training program for novice drivers (DOT HS 812 235).* Washington, DC: National Highway Traffic Safety Administration.

Toyota. (2015). *TeenDrive365.* Accessed September 7, 2015, retrieved from Toyota TeenDrive365: http://www.toyota.com/teendrive365//?srchid=sem:google:Brand_365:Brand:TeenDrive365:td365:11.21.2013&gclid={SI:gclid}&gclid=CjwKEAjw67SvBRC1m5zPv4GboAUSJAB6MJlkwWByymlGMaqRXL8VfR_5hfSbaNew2VpL3GsElzPsFBoC-zjw_wcB.

Toyota CSRC. (2012). *Toyota CSRC News and Events: Teen Driver Distraction Study Release,* November 27. Accessed September 7, 2015, retrieved from Teen Driver Distraction Study Release: http://www.toyota.com/csrc/teen-driver-distraction-study-release.html.

Transport Accident Commission. (2015). *What we do.* Accessed September 5, 2015, retrieved from What we do—TAC—Transport Accident Commission: http://www.tac.vic.gov.au/about-the-tac/our-organisation/what-we-do.

Vlakveld, W.P. (2011). Hazard anticipation of young novice drivers (Ph.D. dissertation). Groningen, NL: University of Groningen.

Vlakveld, W.P., Romoser, M., Mehranian, H., Diete, F., & Fisher, D. (2011). Does the experience of crashes and near crashes in a simulator-based training program enhance novice drivers' visual search for latent hazards? *Transportation Research Record, 2265*, 153–160.

Waller, P. (2003). The genesis of GDL. *Journal of Safety Research, 34*, 17–23.

Waller, P., & Reinfurt, D. (1973). *The Who and When of Accident Risk: Can Driver License Programs Provide Countermeasures?.* Chapel Hill, NC: University of North Carolina Highway Safety Research Center.

Washington, S., Cole, R., & Herbel, S. (2011). European advanced driver training programs: Reasons for optimism. *IATSS Research, 34*, 72–79.

Weinstock, J. (2015). *PAYD—A new era in car insurance pricing.* Accessed September 20, 2015, retrieved from http://www.igudbit.org.il/eng/Index.asp?ArticleID=444&CategoryID=430&Page=2.

Wetton, M., Hill, A., & Horswill, M. (2011). The development and validation of a hazard perception test for use in driver licensing. *Accident Analysis & Prevention, 49*, 1759–1770.

Williams, A. (2005). The fall and rise of graduate driver licensing in North America. *Transportation Research Circular E-C0721* (pp. 153–160). Washington, DC: Transportation Research Board. Accessed October 26, 2015, retrieved from http://onlinepubs.trb.org/onlinepubs/circulars/ec072.pdf.

Williams, A., Preusser, D., Ulmer, R., & Weinstein, H. (1995). Characteristics of fatal crashes of 16-year-old drivers: Implications for licensure. *Journal of Public Health Policy, 16*, 347–360.

Williams, A., McCartt, A., & Sims, L. (2015). *History and current status of state Graduated Driver Licensing (GDL) laws in the United States.* Arlington, VA: Insurance Institute for Highway Safety. Accessed October 26, 2015, retrieved from http://www.iihs.org/frontend/iihs/documents/masterfiledocs.ashx?id=2091.

Willis, D. (1998). The impetus for the development of a new risk management training program for teen drivers. *Proceedings of the Human Factors and Ergonomics Society 42nd Annual Meeting* (pp. 1394–1395). Santa Monica, CA: Human Factors and Ergonomics Society.

Willmes-Lenz, G., Prücher, F., & Groβmann, H. (2010). Evaluation of the novice driver training models "Accompanied driving from 17" and "Voluntary further training seminards for holders of probationary driving license." Results up to November 2009. Bergisch Gladbach, Germany: BAST, Federal Highway Research Institute.

World Health Organization. (2004). *World Report on Road Traffic Injury Prevention.* Geneva: World Health Organization. Accessed October 25, 2015, retrieved from http://apps.who.int/iris/bitstream/10665/42871/1/9241562609.pdf.

World Health Organization. (2013). *Global Status Report on Road Safety: Time for Action.* Geneva: World Health Organization.

Zafian, T., Samuel, S., Borowsky, A., & Fisher, D. (2014). Can young drivers be trained to better anticipate hazards in complex driving scenarios? A driving simulator study. *Proceedings of the Transportation Research Board 93rd Annual Meeting.* Washington, DC: National Academy of Sciences.

Zhang, T., Li, J., Thai, H., Zafian, T., Samuel, S., & Fisher, D.L. (in press). Evaluation of the effect of a novice driver training program on crashes & citations. *Proceedings of the Human Factors and Ergonomics Society Annual Meeting*, Washington, DC.

VI

International Perspective

23

Developed Nations

Willem Vlakveld

Abstract

The Problem. When in a country, mass motorization is accomplished, teens usually start to drive as soon as the legal system allows them to do so. What all countries with mass motorization have in common is an overrepresentation of teen drivers in car crashes. Countries differ widely in the moment mass motorization was realized. They also differ in culture, road infrastructure, traffic composition, licensing system, and overall traffic safety level. These differences between countries have an effect on countermeasures and on research programs regarding the young novice driver problem. An overview of differences in countermeasures and research programs of developed countries is needed. ***Types of Countermeasures.*** At the highest level, countermeasures can be divided into measures that adapt the teen driver to the traffic and road system and measures that adapt the traffic and road system to the teen driver. In the first type of countermeasure, programs have been developed that try to enhance the skills, knowledge, attitudes, and motives that are essential for safe driving, most notably standard classroom and behind-the-wheel driver education programs. In the second type of countermeasure, task demands are reduced in such a way that they do not exceed the capabilities of the teen driver, such as a restriction on driving with peers. These measures are mostly measures that reduce exposure to potentially dangerous traffic situations. ***Research Programs.*** There are studies about the types of crashes in which young novice drivers are overrepresented, studies about the causes of these crashes, and studies about development and evaluation of countermeasures to these crashes. These studies about the causes can be about biological aspects such as brain development, gender, and mental disorders (e.g., ADHD). They can be on lifestyle and peer pressure, and they can be on factors that temporarily reduces one's driving capabilities, such as alcohol, illicit drugs, fatigue, distraction, and emotions. Problems with the execution of the driving task such as visual search, hazard perception, and decision making can also be the subject of young novice driver studies. Finally, studies about the causes can be on factors related to exposure such as driving at night, driving with passengers, driving in older cars, and driving with speeds that are too high for the circumstances. Studies can also be on the development and the evaluation of countermeasures. ***Key Results.*** When the young novice driver problem starts to arise in a country, the emphasis is on the improvement of basic driving skills and on enforcement. Later on, the emphasis is on the improvement of so-called higher-order skills (hazard perception, risk awareness, risk acceptance, self-evaluation) and on the restriction of early driving to relatively safe circumstances so that teen drivers can gain experience in low-risk environments, restrictions such as those imposed by graduated driver licensing (GDL). In countries or other jurisdictions (e.g., states and provinces) with a low license age, the emphasis is on GDL and parental involvement. In countries with a late license age, the emphasis is on the improvement of higher-order skills and on postlicense driver education.

23.1 Introduction

The fact that young novice drivers are overrepresented in car crashes is not new, and this overrepresentation is true for all developed countries, despite differences in culture, history, and overall traffic safety levels between these countries (OECD, 2006). Although the overall traffic safety levels have improved considerably throughout the developed world, the overrepresentation of young drivers between 16 and 24 years of age is persistent (Elvik, 2010). The ratio of the fatality rate of young drivers to the fatality rate of middle-aged drivers in developed countries appears to be more or less constant over time (OECD, 2006). Although the problem is persistent, this does not mean that no research has been carried out and no measures have been taken to reduce this problem. This chapter is about differences in in research programs and countermeasures among countries and continents in which mass motorization has been accomplished and most teens start to drive as soon as the legal system allows them to do so. Chapter 24, a related chapter, discusses much the same issues in low- and middle-income countries.

This chapter starts with a brief history of the young novice driver research. After this historic overview, the causes of the overrepresentation of teen drivers are discussed. The next two sections are about countermeasures. The first section on countermeasures is about improvement of the skills, attitudes, and motivations to drive safely. The second section on countermeasures is about reducing the exposure to situations that are risky for teen drivers.

23.2 History and Research Trends in Different Parts of the World

What are the research trends with regard to the young novice driver problem? How have these trends developed over time? And, do trends differ per continent? In order to provide an answer to these questions, use was made of a database that was created by Hagenzieker et al. (2014). These researchers investigated if there are developments over time in road safety research. In order to do this, they created a database with all the studies they could find in peer-reviewed journals and conference papers about road safety. They selected studies in the Scopus database of Elsevier (http://www.scopus.com) and the Web of Knowledge (WoK) of Thomson Reuters (http://www.wokinfo.com) that had terms such as "road safety," "traffic safety," "driver," "driving," and various combinations with "accident" or "crash," "car," "automobile," "auto," "vehicle," "road user," "pedestrian," "bicycle," "motorcycle," and "speed" in the title, the abstracts, or the keywords. After this selection, they removed duplicates and irrelevant papers (e.g., papers on the effect of cycling on one's health). The final set included 26,536 different papers on road safety research. This work was carried out in 2011, and only studies from 2010 and before were included.

Hagenzieker et al. (2014) did not explore their database on studies regarding the young driver problem. However, use could be made of their database. A selection of papers in their data set was made that had the words "teen driver(s)," "teenage driver(s)," "young driver(s)," "learner driver(s)," or "novice driver(s)" in either the title, the abstract, or the keywords. In their database of 26,536 articles about road safety research, 534 articles were about young novice drivers. This is 2% of all the articles. Figure 23.1 shows the annual number of journal articles about young drivers in the period 1966–2010.

The first article appeared in 1966. The title of this article was "Traffic Safety Starts with Youngsters." In the 2010, the last year in the database, 56 articles on young novice drivers were published. The increase in articles over the past decades is not very informative, because the total number of scientific articles has

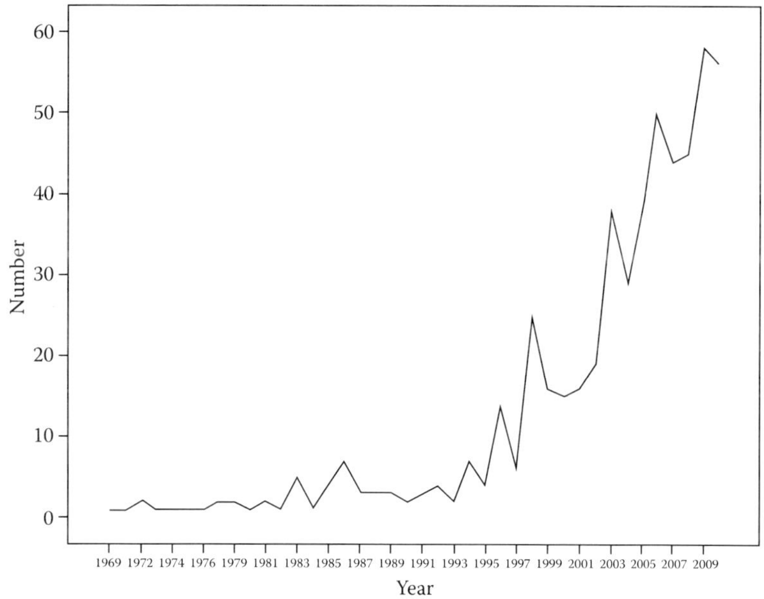

FIGURE 23.1 Annual number of journal articles on young drivers and road safety in the period 1998–2010.

increased enormously. Figure 23.2 shows the percentage of articles on young drivers among all articles on road safety in each 5-year period from 1996 to 2010.

Figure 23.2 shows that scientific interest for the young novice driver problem increased considerably in the second half of the 90s. In the period from 2005 to 2010, the last period in the database, almost 3% of all the scientific publications about road safety were about young novice drivers.

Where do the publications about young novice drivers and road safety come from? In Figure 23.3, the countries are presented that had at least 10 publications in the data set.

Almost half of all the publications stem from the United States. When scientists started to publish on young novice drivers, these studies were mostly about crash rate. This means that they were about the fact that young novice drivers were overrepresented in crashes and about the crash types that were

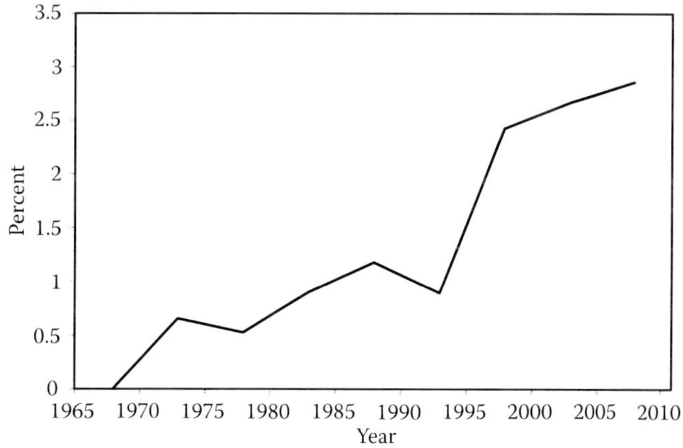

FIGURE 23.2 Percentage of peer-reviewed articles on young drivers among all peer-reviewed articles on road safety within each 5-year period from 1996 to 2010.

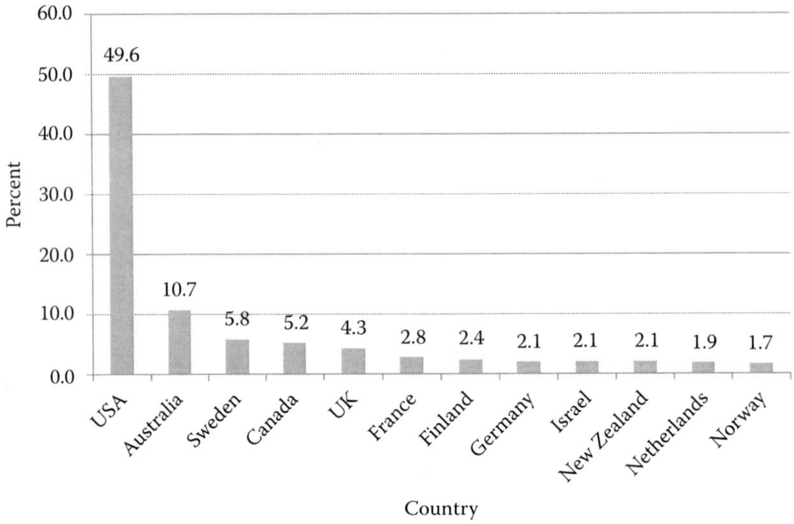

FIGURE 23.3 Percentage of all publications about young novice drivers per country. Only countries with at least 10 publications in the database are included.

typical for young novice drivers. In the 1990s, graduated driver licensing (GDL) became the main topic, especially in the USA. Recent topics in the young novice driver literature are distraction, the effect of passengers on driving behavior, and parental involvement. Figure 23.4 shows the topics of the papers about young novice drivers in North America (the United States and Canada), where only topics comprising more than 2% of all the publications on young novice drivers in North America are included.

Overall, most studies in North America have been about GDL. Driver education and training has received less attention in North America, although when the young novice driver problem started to arise in the United States during the late 1960s and the 1970s, driver education was one of the

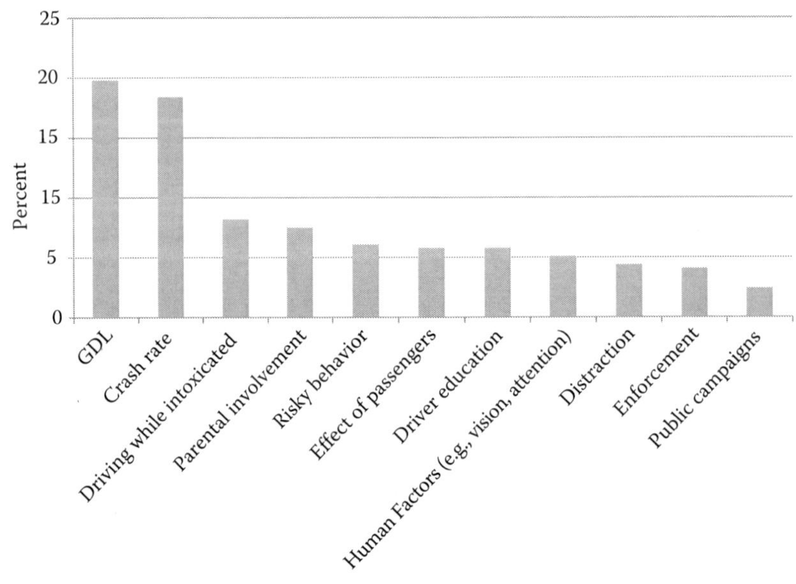

FIGURE 23.4 Subjects comprising more than 2% of all the publications of the young novice driver studies in North America (United States and Canada).

main topics. Figure 23.5 shows the topics of the papers about young novice drivers in Australia and New Zealand, where again, only topics comprising more than 2% of all the publications in Australia and New Zealand were included.

New Zealand was the first country in the world that introduced GDL (Begg & Langley, 2009). This was in 1987. Australia was also an early adapter of GDL. Despite this fact, research articles about GDL are less abundant in Australia and New Zealand than in North America. Driver education has received relatively more attention in Australia and New Zealand than in North America (14% in the former two countries versus 6% in North America). Note that cohort studies have quite often been conducted in Australia and New Zealand but not in North America. Cohort studies are, for instance, longitudinal studies in which background factors, behavioral factors, and personality factors are obtained at an early age (mostly before young people start to drive), and later on, it is determined whether these factors are good predictors for crash involvement (e.g., Begg et al., 1999). Figure 23.6 shows the frequency of paper topics about young novice drivers in Europe, including Israel.

Driver education is the dominant subject in Europe, and there are only a few European studies on GDL (19%). In continental Europe (the countries of the European Union, except the United Kingdom and Ireland), studies on driver education are not so much on acquisition of higher-order cognitive skills such as hazard perception skills but on the risk awareness, self-awareness, attitudes, and motivation to drive safely of novice drivers (e.g., Molina et al., 2014). Compared to North America, Australia, and New Zealand (3%), relatively more studies in Europe have been conducted about the hazard perception skills of novice drivers (6%). Europe is the only continent in which over 2% of all the studies were about in-vehicle driving style monitoring devices. These monitoring devices are electronic devices in the car that provide immediate feedback to the young driver in the car and mostly also feedback to the parents at home when the young driver shows reckless driving behavior (e.g., is speeding, is making sharp turns, is accelerating fast, and is braking abruptly). The fact that in Europe, more than 2% of all the studies about young novice drivers and road safety were about this subject is mainly due to studies from Israel (Farah et al., 2014; Guttman & Lotan, 2011; Prato et al., 2010; Taubman et al., 2014).

The results in this section are based on a literature search that had some limitations. Only a limited number of keywords for road safety and young novice drivers were applied to search important, but not

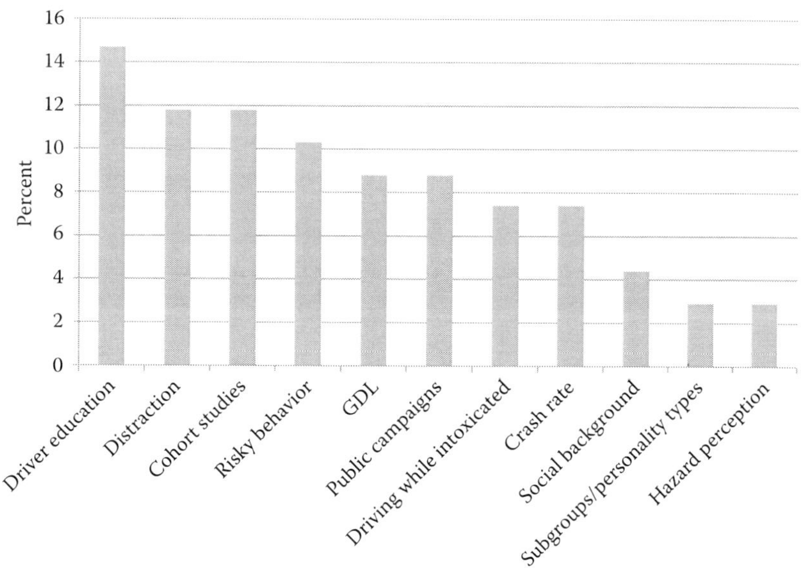

FIGURE 23.5 Subjects comprising more than 2% of all the publications of the young novice driver studies in Australia and New Zealand.

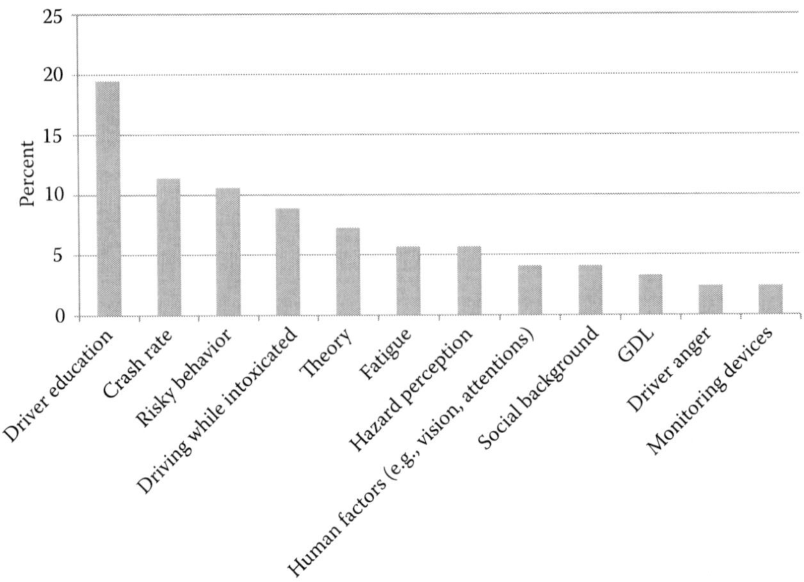

FIGURE 23.6 Subjects comprising more than 2% of all the publications of the young novice driver studies in Europe, including Israel.

all, electronic scientific databases. Only studies in English were included. Nowadays, scientists from countries with a different mother tongue also publish in English journals, but this trend is relatively new. The oldest studies, most of the time, have no abstract, and because of this, they will be less often included. It is sometimes difficult to conclude from the title and the abstract what a study is about, and a study can be on more than one subject (e.g., scanning [human factors] and driver training). When an article covered more than one subject, a decision was made about what the main subject was.

23.3 Research on Causes

I will first describe briefly the various subjects that have been the focus of research programs in developed countries and then, at the end of this section, discuss the different emphases among the developed countries. Figure 23.7 presents the overview of the subjects of studies about why teen drivers are overrepresented in crashes.

At the top are the biological aspects, in particular, those that are characteristic for adolescence (also see Chapter 9). Below this category are the social and cultural factors (Chapters 10 and 16). Both the biological factors and the social and cultural factors are constituent not only for traffic behavior but also for what teens do with regard to other aspects such as drug use and risky sexual behaviors. The third category from the top is about the transient factors that reduce task capabilities (Chapters 12 through 15). The fourth and the fifth rows from the top are about hazard anticipation in traffic and task execution (Chapters 6 through 8). The sixth row is different from the others and is about exposure to risks (also Chapter 8). The four factors in columns to the right of each category in Figure 23.7 are supposed to be representative of that category. There is no direct relationship between a factor in one row and the factor directly above or underneath that factor in another row.

Studies about biological aspects, especially on brain development, come from the United States (e.g., Cascio et al., 2015; Chein et al., 2011; Falk et al., 2014; Telzer et al., 2015). Studies about the sociological and psychological aspects regarding the overrepresentation of teen drivers come from all developed countries. Sweden has quite often conducted studies about the socioeconomic background of

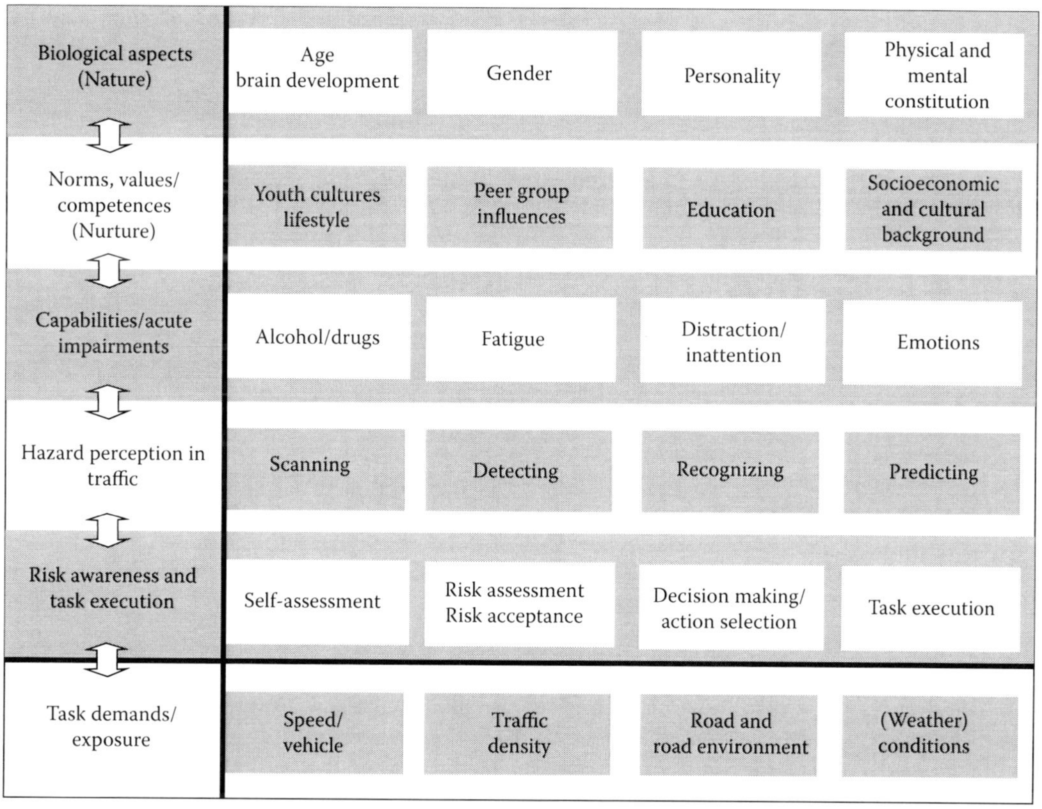

FIGURE 23.7 Taxonomy of subjects about studies on why teen drivers are overrepresented in crashes.

young drivers and crash involvement (e.g., Hasselberg & Laflamme, 2008, 2009; Hasselberg et al., 2005; Laflamme et al., 2005; Murray, 1998). Studies on peer pressure stem mainly from North America (e.g., Chen et al., 2000; Ouimet et al., 2010; Pradhan et al., 2014; Simons-Morton et al., 2005, 2011; White & Caird, 2010). All developed countries conduct studies about the things that reduce the driving capabilities of young drivers, such as alcohol, illicit drugs, fatigue, distraction, and emotions. Hazard perception has rarely been studied in Europe, except in the United Kingdom. The countries that have published the most studies about the hazard perception skills of young drivers are the United Kingdom, Israel, and Australia (e.g., Borowsky et al., 2009, 2010; Crundall & Underwood, 1998; Horswill et al., 2015; McKenna & Horswill, 1999; Smith et al., 2009; Underwood & Crundall, 1998; Underwood et al., 2003; Wetton et al., 2011). Risk awareness and risk acceptance have relatively often been the subject of study in European countries (e.g., Gregersen, 1996; Hatakka et al., 2002; Katila et al., 2004; Laapotti et al., 2001; Møller & Gregersen, 2008). The fact that young drivers more often drive in circumstances that are riskier for all drivers (older vehicles, at night, with speeds that are too high for the circumstances, etc.), shown in the bottom row of Figure 23.7, has been the subject of study in all developed countries.

23.4 Research on Countermeasures

In principle, two types of countermeasures are possible: measures that limit the exposure to circumstances in which young novice drivers are in particular at risk (see Chapters 17, 19, and 22) and measures that are intended to make young novice drivers better drivers by improving their skills, attitudes, and motivations to drive safely (Chapters 18 through 22). Driver training is a method to improve skills and attitudes. When young novice drivers drive safer, because they have better skills, do not over estimate their own competences, and do not underestimate the risks, and are also motivated to drive safely, the crash rate will decrease. Exposure measures are, for instance, raising the age limit for solo driving, restrictions on nighttime driving, and restrictions on driving with peers.

In the past decades in North America (United States and Canada) and in Australia and New Zealand, more emphasis has been put on exposure measures through the introduction of GDL systems, whereas in Europe, more emphasis has been put on reduction of crash risk through measures that lower crash risk, such as driver education. This difference in preferences probably has two causes: (1) the age limit has always been lower in North America and in Australia and New Zealand than in Europe, and (2) evaluation studies in the past showed high school driver education in North America and Australia to have no effect on crash rate (see, for an overview, Engström et al., 2003). These two causes are elaborated upon as follows:

- *Age limit.* Due to the high age limit in most European countries, the intermediate phase of GDL, the phase in which solo driving is allowed but with restrictions, is difficult to realize. In order to have effective restrictions in this phase such as no nighttime driving and no driving with peers, the aid of parents is required because enforcement by the police is difficult to realize. Parents have more influence on their children when they are 16 years of age than when they are 18 years of age.
- *No effect of professional driver education.* Especially, the DeKalb County study (Stock et al., 1983) has been very influential. To date, it is by far the most rigorous evaluation study of driver education that has ever been conducted as it probably is the only study on crash rate with a large sample size and random assignment. The results of this study, however, have been disputed and have been reanalyzed (Lund et al., 1986; Peck, 2011). In the latest reanalysis, Peck (2011) concludes that the extensive driver training program of the DeKalb County study did in fact result in a small short-term reduction of crashes and violations after licensing, but this small positive effect of the extensive training program was overshadowed by the fact that this training program also led to early licensing. Early licensing means more exposure and more crashes.

23.4.1 Improving Skills, Attitudes, and Motivation

The measures that try to improve the skills, attitudes, and the motivation to drive safely are predriver education, basic driver training, postlicense training for novice drivers, rehabilitation courses for young offenders, and stand-alone training programs in higher-order skills such as hazard anticipation.

23.4.1.1 Predriver Education

Although predriver education programs are not part of the legislative licensing system in a country, they are often present. The aim of these interventions is not teaching teens to drive but to raise awareness of the road environment and the complexity of driving among young people before they learn to drive (Senserrick, 2007). Waylen and McKenna (2008) found that risky attitudes toward driving can be observed in adolescents long before they learn to drive. It is the aim of predriver education programs to change these risky attitudes into safety-oriented attitudes. An example of such a program is Crash Magnets (http://www.crashmagnets.com/) in Scotland (Mann & Lansdown, 2009). Predriver training programs are used in Europe and Australia. Mercedes Benz has developed a predriver training program that is called Road Sense. In Germany, the United Kingdom, and the Netherlands, youngsters can attend this program before they start learning to drive. There is no clear evidence that these predriver education programs improve attitudes regarding safe driving (Glendon et al., 2014; Poulter & McKenna, 2010).

23.4.1.2 Professional Basic Driver Education

In some countries (e.g., Denmark, Germany, and the Netherlands), young people can only take driving lessons from qualified driving instructors in order to prepare themselves for the driving test. They are not allowed to train with an older, more experienced driver sitting next to them (i.e., someone who is not qualified as a driving instructor) before they have passed the driving test. In contrast to what many people tend to think, there is no evidence that professional basic driver training provided by qualified instructors is more effective in reducing crash rates than informal driver training provided by lay persons (mostly parents) (Beanland et al., 2013; Christie, 2001; Elvik et al., 2009; Engström et al., 2003; Lonero & Mayhew, 2010; Mayhew & Simpson, 2002; Senserrick & Haworth, 2005; Wells et al., 2008). However, this result should be interpreted with some caution. Driver training programs are difficult to evaluate with crash rate as the dependent variable, because crashes are rare and random assignment of participants is difficult to realize. Equally important to recognize is that the research community has not provided driving instructors (or parents, for that matter) with tools for training driving skills that have a proven effect on crash rate. It is only very recently that training programs have been developed that have been shown to have an effect on those behaviors that are known to be linked to crashes (see Section 23.3 in this chapter).

Within the project Guarding Automobile Drivers through Guidance Education and Training (GADGET) of the European Union, the so-called Goals for Driver Education (GDE) matrix was developed (Hatakka et al., 2002). The GDE matrix is based on the assumption that the driving task can be described as a hierarchy (see Table 23.1). The idea of the hierarchical approach is that abilities and preconditions on a higher level influence the demands, decisions, and behavior on a lower level (Keskinen et al., 1999; Michon, 1985). At the lowest level is *vehicle control* (steering, braking, etc.). This level is called the *operational level* by Michon. A level higher is the *mastering traffic situations* level. Michon called this the *tactical level*, and it deals with decision making in traffic ("Can I now safely turn right?", "Can I now overtake that car?", etc.). A level above the level of mastering traffic situations is the *trip-related context and considerations* level. Michon called this the *strategic level*. This level is about decisions regarding the trip and vehicle use ("Can I still take the car now that I have consumed alcohol?", "Can I drive safely in these adverse weather conditions?", "Do these passengers distract me?", etc.). The highest level is about lifestyle and values and norms. This level is called *personal characteristics, ambitions, and competences* and was invented by Keskinen. Questions at this level are, "What does driving mean for me?", "How important is road safety for me?", and "How good am I in suppressing intentions to commit unsafe behavior (e.g., speeding)?" For each of these, four levels were indicated: the knowledge, the skills, the factors that can jeopardize performance at a particular level, and the abilities of the drivers to consider their own limitations and reflect upon their own behavior and attitudes at a particular level (Table 23.1).

TABLE 23.1 Goals for Driver Education (GDE) Matrix

	GDE Matrix		
	Knowledge and Skills	Risk-Increasing Factors	Self-Evaluation
1. Goals for life and skills for living	Lifestyle, group norms, motives, personal values	Risk acceptance, sensation seeking, group norms, peer pressure	Impulse control, awareness of safety negative motives, awareness of own risky habits
2. Goals and context of driving (trip related)	Modal choice, route choice, awareness of peer pressure in the car	Alcohol, fatigue, distraction, extra motives (e.g., impress peer passengers)	Self-awareness of own limitations and awareness of risks of alcohol, fatigue, peer pressure, et cetera
3. Mastery of traffic situations	Applying rules of the road, hazard anticipation	Disobeying rules, close following, information overload, no attention for vulnerable road users	Calibration skills (not overestimating one's own competences and not underestimating the risks)
4. Vehicle maneuvering	Car control, knowledge of protection systems	Not fully automated vehicle-handling skills, no seatbelts, poor vehicle maintenance	Calibration of car control skills

Source: Adapted from Hatakka, M. et al., *Transp. Res. Part F Traffic Psychol. Behav.*, 5(3), 201–215, 2002.

The GDE matrix has influenced basic driver programs in some European countries (Norway, Sweden, Finland, Germany, Austria, Switzerland, and Spain) (e.g., Molina et al., 2014). In these countries, more emphasis is put on the two higher levels of driving and on self-assessment (the third column of the GDE matrix). However, it is not known whether changes in basic driver training programs under the influence of the GDE matrix have resulted in a decline in crash rates after licensing. There are some indications that basic driver education programs that also address higher-order skills such as hazard perception and risk awareness lower the crash rate of novice drivers (Carstensen, 2002; Hirsch et al., 2006).

23.4.1.3 Postlicense Training for Novice Drivers

Postlicense training courses, also known as second-phase training programs, are a typical European phenomenon, probably because European countries have no full-fledged GDL systems. In some European countries, learners can practice before the driving test under the supervision of an experienced driver, but not one European country has put into force an intermediate stage with restrictions. Obligatory postlicense training programs exist in Finland, Luxembourg, Austria, Switzerland, and Estonia. Almost all other European countries have voluntary postlicense training programs (e.g., Pass Plus in the United Kingdom) (see, for an overview of European postlicense training programs, Washington et al., 2011). Newly licensed drivers attend these training courses in the first year or first 2 years after licensing. The idea behind the postlicense training for novices is that lessons in traffic insight and self-assessment have little effect during initial driver training, as learners have no experience yet to reflect upon.

The Finnish postlicense training program consists of a self-evaluation; a so-called feedback drive in real traffic with a driving instructor in order to evaluate the driving style of the novice driver; on-track training, not intended to learn skills but to raise risk awareness and self-awareness; and a group discussion. A similar and even more elaborate compulsory postlicense program was introduced in Austria in 2003. The Finish postlicense training did not seem to have reduced the self-reported crash rate considerably, whereas in Austria, the introduction of the training seemed to have reduced the self-reported crash rate (Mynttinen et al., 2010).

Besides the fact that evaluation studies of postlicensing training programs are inconclusive, they start at a moment in one's driving career when the crash risk is already rapidly declining. The crash rate is the highest directly after teens start to drive unaccompanied (either without restrictions or with restrictions such as not driving with peers and not driving in hours of darkness), and the crash rate declines rapidly in the first months after the start of solo driving (Foss et al., 2011; McCartt et al., 2003; Sagberg, 1998). It takes, however, years of driving experience before the crash rate no longer declines (Maycock et al., 1991; McCartt et al., 2009; Vlakveld, 2011). This implies that although some experience is required to make postlicense training effective, postlicense training should start as early after the start of solo driving as possible. When postlicense training has an effect on crash rate, the longer after the start of solo driving postlicense training takes place, the lower its contribution to a decline of crash rate will be.

23.4.1.4 Rehabilitation Courses for Novice Drivers

Rehabilitation courses are also known as driver improvement programs or remedial driver education programs. Some European countries have a stricter demerit point system for newly licensed drivers (e.g., the United Kingdom, Germany, and France), and when novice drivers exceed a certain amount of points in their first year or first years after licensing—sometimes one offense is enough (e.g., in the case of driving under the influence of alcohol or drugs)—they have to attend a special training program. When they do not attend this training program, their driving license is suspended. Ker et al. (2005) conducted a meta-analysis that included 18 rehabilitation courses (for both novice drivers and experienced drivers). No effect on the crash rate was found. However, in another meta-analysis (Masten & Peck, 2004), a small but significant drop in crash rate was found for "group meetings" and "individual meetings." In these group meetings and individual meetings, no distinction was made between teen drivers and older drivers. In Germany, the introduction of a stricter demerit point system and compulsory training programs for young drivers resulted in a small drop in the crash rate of young male drivers but not of young female drivers (Meewes & Weissbrodt, 1992). In the United Kingdom, the introduction of a stricter demerit point system for newly licensed drivers resulted in a small but statistically significant decrease in the rate of offenses in the second year of driving but not in the first year after licensing (Simpson et al., 2002).

23.4.1.5 Stand-Alone Higher-Order Training Programs

Several short training programs for novice drivers have been developed to improve their hazard perception skills and their awareness of risks in traffic situations. These training programs have been developed in the United Kingdom, Australia, New Zealand, and the United States. They are interactive PC-based training programs or training programs in which a simple driving simulator is used (Chapman et al., 2002; Crundall et al., 2010; Fisher et al., 2002, 2006; Isler et al., 2009; McKenna & Crick, 1997; McKenna et al., 2006; Pollatsek et al., 2006; Pradhan et al., 2009; Regan et al., 2000; Vlakveld et al., 2011; Wang et al., 2010; Wetton et al., 2013). Although all these training programs improve the hazard perception skills of novice drivers, their effect on the crash rate is unknown. However, hazard perception training was not chosen without forethought. In fact, it is only one of several skills that has been shown to be linked to crashes (Horswill et al., 2015).

Another higher-order cognitive skill that has recently been the focus of more and more research is attention maintenance, which is defined as the ability to keep safety-critical glances inside the vehicle to less than 2 seconds. Especially long glances are known to increase the crash rate of all drivers (Horrey & Wickens, 2007; Klauer et al., 2014). Training programs now exist that have been shown to decrease the frequency of the especially long glances of novice drivers in field settings (Pradhan et al., 2011). Again, while reductions in long glances are known to be associated with decreases in crashes, it is not known whether the training, which leads to a reduction in especially long glances, also leads to a reduction in crashes.

Hazard perception and attention maintenance training programs try to improve the more cognitive higher-order skills. There are also higher-order training programs for young drivers that try to improve attitudes and the motivation to drive safely (e.g., Gregersen, 1996). In Sweden, these training programs are called *insight training programs*. Nyberg and Engström (1999) found that the effect on self-reported behavior of these insight training programs was limited. It is interesting to note that Senserrick et al. (2009) found that an Australian insight training program for young drivers on road safety issues only did not lower the crash rate, whereas what was called a *resilience training program* that was not only about road safety but also about drugs and unsafe sex did lower the crash rate.

23.4.1.6 Training Devices

Traditionally, young people learn to drive in a car. Professional driving instructors use a car with a dual-brake system. For learning the rules of the road and how to apply them, traditionally, textbooks are used. These things are changing. For learning to drive, in some countries, driving lessons can partly be replaced by simulator lessons (related material is discussed in Chapter 25, focusing on simulation). In the Netherlands, for example, approximately 150 rather simple driving simulators (simulators without a moving base that do not simulate the feeling that one is driving) are in use to teach learner drivers the most basic driving skills before they start to drive with a driving instructor on the open road. After an initial increase, the number of training simulators in the Netherlands has remained the same since 2010. Although simulators have existed since the 1960s, for a long time, they remained too expensive to be used widely for training purposes. However, PCs have become ever more powerful, and cheaper. This means that the price of simple simulators (that is, simulators that do not imitate the feeling that one is driving) has gone down considerably. Because driving lessons are rather expensive in the Netherlands, training devices that can save money on driving lessons are very attractive, certainly for the larger driving schools. De Winter et al. (2009) found that the chance to pass the driving test in the Netherlands was 4%–5% higher for learners who first took simulator lessons before they started with lessons on the open road than for learners who immediately started with driving lessons on the open road.

Driving simulators are used not only to train the most basic driving skills but also to train higher-order skills of novice drivers, in particular, hazard perception skills (Vlakveld et al., 2011; Wang et al., 2010) and attention maintenance skills (Pradhan et al., 2011). Vlakveld et al. (2011) found that novice drivers detected potential hazards considerably more often after a simulator-based training in hazard perception. In this training program, use was made of the didactic principle of *error management training* (Keith & Frese, 2008). Error management training is a training method that involves active exploration as well as explicit opportunities for learners to make errors during training and to learn from them. In China, Wang et al. (2010) developed simulator-based training for novice drivers on hazard perception that was also based on error management training, and that was also successful. In this training, the actions of the learners in the simulator were recorded while they missed potential hazards or detected hazards too late. These video clips were played back to them. Learners also watched videos in which experienced drivers anticipated the potential hazards correctly. Whether simulator-based hazard perception training also reduces crash rate has not been investigated yet. However, not all simulator-based training programs for young novice drivers improve higher-order skills. In Israel, a more general simulator-based training program for young novice drivers that was not exclusively about hazard anticipation had no effects on the intention to drive safely and did not improve traffic safety knowledge (Rosenbloom & Eldror, 2014).

Although driving simulators without a moving base have become cheaper, they are still expensive. Much cheaper are interactive computer-based multimedia training programs. These, training programs can be about hazard perception and risk awareness (e.g., Pradhan et al., 2009) but they can also be about the dangers of distraction (Pradhan et al., 2011). There are also more general computer-based interactive training programs for young novice drivers, such as DriverZED (Fisher et al., 2002) and Road Aware (Samuel et al., 2013) in the United States, DriveSmart in Australia (Regan et al., 2000), and the online training program eDrive in New Zealand (Isler & Starkey, 2012). All the mentioned interactive computer-based training programs improved behavior (better hazard anticipation, shorter durations of eyes off the road when engaged in a secondary task). In Germany, Weiss et al. (2013) developed a paper version of a hazard perception and risk awareness training program and an interactive multimedia computer version of the same training program. Learners who completed the computer version detected hazards sooner and were less overconfident about their own skills than learners who had completed the pen-and-paper training. Although computer-based interactive multimedia training programs seem to be effective, their effect on crash rate has not been investigated yet.

23.4.2 Reducing Exposure to Situations in Which Young Drivers Are at Risk

The simplest way to reduce exposure to risks is to postpone the age at which teens are allowed to drive without supervision. The driving test can also be seen as a way to limit exposure to those minimally qualified. The aim of this test is to allow access to the open road only for those teen drivers who have proved they possess the competences to drive safely. GDL is also a way to limit exposure because learners are only allowed to practice in safe circumstances (Chapter 17). GDL is implemented in North America, Australia, and New Zealand but is only partly implemented in Europe. No European country has a phase in which solo driving is allowed with restrictions (e.g., not in the dark and not with peers). Because GDL is the subject of study in Chapters 17 and 19, it is not discussed in this chapter.

23.4.2.1 Higher Age Limits

Not only because they are inexperienced but also because they are young and not yet fully matured, age contributes to the elevated crash rate of teen drivers. The older teenagers are when they start solo driving, the lower the crash rate in the beginning of their driving career will be. Elvik et al. (2009) conducted a meta-analysis on studies about licensure age. Ten studies were included. Table 23.2 shows the estimates of this analysis on the decline in percentages of crashes in the first year of full licensing.

TABLE 23.2 Effects on Accidents of Increasing the Driver's Age at the Time of Obtaining the (Full) Driving License by 1 Year During the First Year after Obtaining the (Full) License

Increase of Driver's Age at the Time of Obtaining the (Full) Driving License	Percentage Change in the Number of Accidents		
	Types of Accidents Affected	Best Estimate	95% Confidence Interval
From 16 to 17 years	All accidents	−10	(−20, +5)
From 17 to 18 years	All accidents	−7	(−15, +1)
From 18 to 19 years	All accidents	−6	(−17, +4)
From 19 to 20 years	All accidents	−6	(−22, +13)
From 20 to 21 years	All accidents	−5	(−29, +27)

Source: Elvik, R. et al., *The Handbook of Road Safety Measures* (2nd ed.), Emerald Group, Bingley, UK, p. 767, 2009.

Table 23.2, which is mainly based on studies from the United Kingdom, shows that it is likely, but not certain, that it is beneficial for road safety to raise the age limit.

23.4.2.2 The Driving Test

Driving tests, both the knowledge test and the practical test, are intended to exclude from public roads drivers who do not possess the competences to drive safely. Driving tests can only function as a good selection tool when they are both reliable and valid. Studies about the reliability of the practical driving test are scarce. Baughan and Simpson (1999) examined the reliability of the British practical driving test in 1999. Candidates who had never done the practical driving test before but had prepared themselves for the road test and thought they were ready underwent the test. Thirty-four percent passed, and 66% failed. However, the candidates did not receive the results of this test. Three days later, they did the practical driving test again with another examiner who did not know they had undergone the test before. Of the candidates who passed the first test, 53% failed to pass the test for the second time, and 47% passed the test again. Of those who failed to pass the test the first time, 76% failed again, but 24% passed this second test. These results indicate that, at least in 1999, the British practical driving test was not very reliable. However, the results are not unexpected. A reliable road test is difficult to achieve because there can be differences in consensus between examiners whether a certain action should be interpreted as safe or as unsafe. The assessment of the skills of a candidate is always a matter of interpretation. Assessment is, in particular, difficult to achieve because the traffic situations differ per candidate. A candidate can be so unfortunate that he or she encounters many different hazardous situations during the test, whereas another candidate encounters no hazardous situation during the test.

The validity of both the practical test and the theory test are also difficult to measure. In almost all developed countries, the vast majority of learners will ultimately pass the test, although many will need more than one attempt. Those who do not pass the test, even after many attempts, are not allowed to drive and thus cannot be exposed to the risk of crashes as car drivers. It is therefore not possible to compare the crash rate for drivers who have passed the test with drivers who have not passed this test, in order to determine whether drivers who fail the driving test have a higher crash rate than drivers who pass it. Despite this problem, some attempts have been made to test validity of the driving test. Baughan and Sexton (2002) examined whether there was an association between the number of faults during the practical driving test in the United Kingdom and self-reported crash involvement in the first 6 months after having passed the driving test. They found that there was no association between the number of driving faults during the test and crash rate. In a cohort study from the United Kingdom, no association was found between the number of attempts that were required to pass the practical test and the crash rate in the first years after having obtained the license. However, for young female drivers, there was an indication that the more practical tests were needed to pass this test, the higher the crash rate was after licensure (Maycock & Forsyth, 1997). In Finland, Hatakka et al. (2002) found that the better male drivers performed in the practical test, the more they were involved in crashes, and the more traffic offenses they committed. The mentioned indications of the poor validity of the practical test are not as surprising as they may seem. On-road practical tests generally require demonstration of adequate skill in car control and adequate application of the rules of the road in standard traffic situations. Analyses of the underlying causes of crashes in which young novice drivers are involved have revealed that a lack of these skills is rarely the cause of their crash involvement (Clarke et al., 2006; Curry et al., 2011; McKnight & McKnight, 2003).

Almost all jurisdictions require that applicants first have to pass a theory test before they can start with the learner phase in countries or states with a GDL system and before they can do the practical test. The theory test covers topics in addition to knowledge of the rules of the road and how to apply these rules in traffic situations (e.g., a picture of a traffic situation with the question, "Is the red car in this situation allowed to turn right?"). Topics may also be about vehicle safety and maintenance and about knowledge of risky driving behavior (e.g., the risks of driving while intoxicated). Some countries (e.g., the United Kingdom, most states of Australia, and the Netherlands) have incorporated a hazard perception test in their knowledge test. The United Kingdom introduced the theory test in 1996. The introduction of this theory test in the driving test had a small positive effect on attitudes regarding road safety and had a small positive effect on behavior. However, no statistically significant effect was found on crash rate (Simpson et al., 2002). Elvik et al. (2009) conducted a meta-analysis on studies about the effect of the knowledge test and found that the effect on crash rate was exactly zero. Although theory tests do not seem to have an effect on crash rate, results on a hazard perception test that is incorporated in the licensing system seem to predict crash risk (Horswill et al., 2015).

23.4.3 Conclusions with Regard to Training and Exposure

In North America during the past decades, the emphasis has been on the implementation of GDL systems and not on driver education, but recently, a renewed interest in driver education has been developing. In Europe, on the other hand, the focus has been on driver education and not on GDL, but now, a tendency toward GDL systems is beginning to emerge. In Australia and New Zealand, it has always been both: driver education and GDL. There are indications in North America that at least some people are starting to think about the incorporation of higher-order skills training such as hazard perception training and risk awareness training in a GDL system (Blomberg & Fisher, 2012; Lonero & Mayhew, 2010). It is important to make a distinction between the incorporation of basic driver training and the incorporation of higher-order skills training in a GDL system. Incorporation of basic driver training in a GDL system is probably not beneficial for road safety. Mayhew et al. (2003) found that in Nova Scotia, when learners could reduce the phase of accompanied driving (learner phase) from 6 to 3 months when they completed a recognized driver education program, the crash rates did not decrease.

In Europe, more and more countries are incorporating elements of GDL in their driving license system. For instance, in Germany, Austria, and the Netherlands learners, were not allowed to practice while supervised by an older, more experienced driver and could only train with a professional driving instructor. In the past years, these three countries have incorporated a phase with supervised driving (the learner phase of a GDL system) in their licensing system for young drivers. In Germany and the Netherlands, young drivers now have to drive under supervision of an older and experienced driver (mostly a parent) after they have passed the national driving test (both the knowledge test and the practical test) if they are younger than 18 years of age. It is estimated that the introduction of the learner phase for young drivers in Germany has reduced the number of crashes in the first 18 months of solo driving (after their 18th birthday) by 17%–36% (Willmes-Lenz et al., 2010).

In Australia, the function of driver education within a GDL system is being reconsidered as well. The emphasis is on driver education that promotes self-awareness and calibration skills (Bailey, 2009). Calibration is the balancing of perceived task demands and perceived capabilities (Horrey et al., 2015; Mitsopoulos et al., 2006). Young drivers not only have to learn to "see" the possible hazards but also not to overestimate their capabilities and not to underestimate the risks (e.g., Weiss et al., 2013).

In Europe, risk awareness and self-awareness of young novice drivers has been a dominant subject of study in young novice driver research since the 90s. This is probably caused by the influence of the GDE matrix (see Table 23.1). The very fact that young novice drivers also have to learn to see the hazards has been neglected in Europe, except in the United Kingdom. However, training and testing of hazard perception has recently also become subject of study of young novice drivers in continental Europe (e.g., Huestegge et al., 2010; Malone & Brünken, 2013; Vidotto et al., 2011; Vlakveld, 2014).

References

Bailey, T. J. (2009). Self-awareness and self-monitoring–Important components of best educational practice for novice drivers. *Journal of the Australasian College of Road Safety, February*, 45–51.

Baughan, C., & Sexton, B. F. (2002). Do driving test errors predict accidents?: Yes and no. In G. B. Grayson (Ed.), *Behavioural Research in Road Safety XI: Proceedings of the 11th Seminar on Behavioural Research in Road Safety* (pp. 252–268). Crowthorne: Transportation Research Laboratory.

Baughan, C., & Simpson, H. (1999). Consistency of driving performance at the time of the L-test, and implications for driver testing. In G. B. Grayson (Ed.), *Behavioural Research in Road Safety* (Vol. IX, pp. 206–214). Crowthorne: Transport Research Laboratory.

Beanland, V., Goode, N., Salmon, P. M., & Lenné, M. G. (2013). Is there a case for driver training? A review of the efficacy of pre- and post-licence driver training. *Safety Science, 51*(1), 127–137. doi: 10.1016/j.ssci.2012.06.021.

Begg, D., & Langley, J. (2009). A critical examination of the arguments against raising the car driver licensing age in New Zealand. *Traffic Injury Prevention, 10*(1), 1–8.

Begg, D., Langley, J. D., & Williams, S. M. (1999). A longitudinal study of lifestyle factors as predictors of injuries and crashes among young adults. *Accident Analysis & Prevention, 31*(1–2), 1–11.

Blomberg, T. I. F. D., & Fisher, D. L. (2012). *A Fresh Look at Driver Education in America*. Washington, DC: National Highway Traffic Safety Administration.

Borowsky, A., Oron-Gilad, T., & Parmet, Y. (2009). Age and skill differences in classifying hazardous traffic scenes. *Transportation Research Part F, 12*, 277–287.

Borowsky, A., Shinar, D., & Oron-Gilad, T. (2010). Age, skill, and hazard perception in driving. *Accident Analysis & Prevention, 42*(4), 1240–1249.

Carstensen, G. (2002). The effect on accident risk of a change in driver education in Denmark. *Accident Analysis & Prevention, 34*(1), 111–121. doi: 10.1016/s0001-4575(01)00005-7.

Cascio, C. N., Carp, J., O'Donnell, M. B., Tinney, F. J., Bingham, C. R., Shope, J. T., Ouimet, M. C., Pradhan, A. K., Simons-Morton, B. G., & Falk, E. B. (2015). Buffering social influence: Neural correlates of response inhibition predict driving safety in the presence of a peer. *Journal of Cognitive Neuroscience, 27*(1), 83–95. doi: 10.1162/jocn_a_00693.

Chapman, P., Underwood, G., & Roberts, K. (2002). Visual search patterns in trained and untrained novice drivers. *Transportation Research Part F, 5*, 157–167.

Chein, J., Albert, D., O'Brien, L., Uckert, K., & Steinberg, L. (2011). Peers increase adolescent risk taking by enhancing activity in the brain's reward circuitry. *Developmental Science, 14*(2), F1–F10. doi: 10.1111/j.1467-7687.2010.01035.x.

Chen, L. H., Baker, S. P., Braver, E. R., & Li, G. (2000). Carrying passengers as a risk factor for crashes fatal to 16- and 17-year-old drivers. *The Journal of the American Medical Association, 283*(12), 1578–1582. doi: 10.1001/jama.283.12.1578.

Christie, R. (2001). *The Effectiveness of Driver Training as a Road Safety Measure: A Review of the Literature*. Noble Park, Victoria, Australia: Royal Automobile Club of Victoria (RACV) Ltd.

Clarke, D. D., Ward, P., Bartle, C., & Truman, W. (2006). Young driver accidents in the UK: The influence of age, experience, and time of day. *Accident Analysis & Prevention, 38*(5), 871–878.

Crundall, D., & Underwood, G. (1998). Effects of experience and processing demands on visual information acquisition in drivers. *Ergonomics, 41*(4), 448–458.

Crundall, D., Andrews, B., van Loon, E., & Chapman, P. (2010). Commentary training improves responsiveness to hazards in a driving simulator. *Accident Analysis & Prevention, 42*(6), 2117–2124. doi: 10.1016/j.aap.2010.07.001.

Curry, A. E., Hafetz, J., Kallan, M. J., Winston, F. K., & Durbin, D. R. (2011). Prevalence of teen driver errors leading to serious motor vehicle crashes. *Accident Analysis & Prevention, 43*(4), 1285–1290.

De Winter, J. C. F., De Groot, S., Mulder, M., Wieringa, P. A., Dankelman, J., & Mulder, J. A. (2009). Relationships between driving simulator performance and driving test results. *Ergonomics, 52*(2), 137–153.

Elvik, R. (2010). Why some road safety problems are more difficult to solve than others. *Accident Analysis & Prevention, 42*(4), 1089–1096.

Elvik, R., Høye, A., Vaa, T., & Sørensen, M. (2009). *The Handbook of Road Safety Measures* (2nd ed.). Bingley: Emerald Group.

Engström, I., Gregersen, N. P., Hernetkoski, K., Keskinen, E., & Nyberg, A. (2003). *Young Novice Drivers, Driver Education and Training: Literature Review.* Linköping, Sweden: Swedish National Road and Transport Research Institute.

Falk, E. B., Cascio, C. N., Brook O'Donnell, M., Carp, J., Tinney Jr, F. J., Bingham, C. R., Shope, J. T., Ouimet, M. C., Pradhan, A. K., & Simons-Morton, B. G. (2014). Neural responses to exclusion predict susceptibility to social influence. *Journal of Adolescent Health, 54*(5, Supplement), S22–S31. doi: 10.1016/j.jadohealth.2013.12.035.

Farah, H., Musicant, O., Shimshoni, Y., Toledo, T., Grimberg, E., Omer, H., & Lotan, T. (2014). Can providing feedback on driving behavior and training on parental vigilant care affect male teen drivers and their parents? *Accident Analysis & Prevention, 69*, 62–70. doi: 10.1016/j.aap.2013.11.005.

Fisher, D. L., Laurie, N. E., Glaser, R., Connerney, K., Pollatsek, A., & Duffy, S. A. (2002). Use of a fixed-base driving simulator to evaluate the effects of experience and PC-based risk awareness training on drivers' decisions. *Human Factors, 44*(2), 287–302.

Fisher, D. L., Pollatsek, A. P., & Pradhan, A. K. (2006). Can novice drivers be trained to scan for information that will reduce their likelihood of a crash? *Injury Prevention, 12, Suppl I*, i25–i29.

Foss, R. D., Martell, C. A., Goodwin, A. H., & O'Brien, N. P. (2011). Measuring changes in teenage driver crash characteristics during the early months of driving. Washington, DC: AAA Foundation for Traffic Safety.

Glendon, A. I., McNally, B., Jarvis, A., Chalmers, S. L., & Salisbury, R. L. (2014). Evaluating a novice driver and pre-driver road safety intervention. *Accident Analysis & Prevention, 64*, 100–110. doi: 10.1016/j.aap.2013.11.017.

Gregersen, N. P. (1996). Young drivers' overestimation of their own skill—An experiment on the relation between training strategy and skill. *Accident Analysis & Prevention, 28*(2), 243–250.

Guttman, N., & Lotan, T. (2011). Spying or steering? Views of parents of young novice drivers on the use and ethics of driver-monitoring technologies. *Accident Analysis & Prevention, 43*(1), 412–420.

Hagenzieker, M. P., Commandeur, J. J. F., & Bijleveld, F. D. (2014). The history of road safety research: A quantitative approach. *Transportation Research Part F: Traffic Psychology and Behaviour, 25, Part B*, 150–162. doi: 10.1016/j.trf.2013.10.004.

Hasselberg, M., & Laflamme, L. (2008). Road traffic injuries among young car drivers by country of origin and socioeconomic position. *International Journal of Public Health, 53*(1), 40–45.

Hasselberg, M., & Laflamme, L. (2009). How do car crashes happen among young drivers aged 18–20 years? Typical circumstances in relation to license status, alcohol impairment and injury consequences. *Accident Analysis & Prevention, 41*(4), 734–738.

Hasselberg, M., Vaez, M., & Lucie, L. (2005). Socioeconomic aspects of the circumstances and consequences of car crashes among young adults. *Social Science & Medicine, 60*(2), 287–295.

Hatakka, M., Keskinen, E., Gregersen, N. P., Glad, A., & Hernetkoski, K. (2002). From control of the vehicle to personal self-control; broadening the perspectives to driver education. *Transportation Research Part F: Traffic Psychology and Behaviour, 5*(3), 201–215. doi: 10.1016/S1369-8478(02)00018-9.

Hirsch, P., Maag, U., & Laberge-Nadeau, C. (2006). The role of driver education in the licensing process in Quebec. *Traffic Injury Prevention, 7*(2), 130–142. doi: 10.1080/15389580500517644.

Horrey, W. J., & Wickens, C. D. (2007). In-vehicle glance duration distributions, tails, and model of crash risk. *Transportation Research Record, 2018*, 22–28.

Horrey, W. J., Lesch, M. F., Mitsopoulos-Rubens, E., & Lee, J. D. (2015). Calibration of skill and judgment in driving: Development of a conceptual framework and the implications for road safety. *Accident Analysis & Prevention, 76*, 25–33. doi: 10.1016/j.aap.2014.12.017.

Horswill, M. S., Hill, A., & Wetton, M. (2015). Can a video-based hazard perception test used for driver licensing predict crash involvement? *Accident Analysis & Prevention, 82*, 213–219. doi: 10.1016/j.aap.2015.05.019.

Huestegge, L., Skottke, E. M., Anders, S., Müsseler, J., & Debus, G. (2010). The development of hazard perception: Dissociation of visual orientation and hazard processing. *Transportation Research Part F, 13*, 1–8.

Isler, R. B., & Starkey, N. J. (2012). *Driver Education and Training as Evidence-Based Road Safety Interventions.* Paper presented at the Australian Road safety Research, Policing and Education Conference, Wellington, New Zealand. Retrieved from http://acrs.org.au/files/arsrpe/Isler%20and%20Starkey%20-%20Driver%20Education%20and%20Training%20as%20evidence%20based%20road%20safety%20interventions.pdf.

Isler, R. B., Starkey, N. J., & Williamson, A. R. (2009). Video-based road commentary training improves hazard perception of young drivers in a dual task. *Accident Analysis & Prevention, 41*(3), 445–452. doi: 10.1016/j.aap.2008.12.016.

Katila, A., Keskinen, E., Hatakka, M., & Laapotti, S. (2004). Does increased confidence among novice drivers imply a decrease in safety?: The effects of skid training on slippery road accidents. *Accident Analysis & Prevention, 36*(4), 543–550. doi: 10.1016/S0001-4575(03)00060-5.

Keith, N., & Frese, M. (2008). Effectiveness of error management training: A meta-analysis. *Journal of Applied Psychology, 93*(1), 59–69. doi: 10.1037/0021-9010.93.1.59.

Ker, K., Roberts, I., Collier, T., Beyer, F., Bunn, F., & Frost, C. (2005). Post-licence driver education for the prevention of road traffic crashes: A systematic review of randomised controlled trials. *Accident Analysis & Prevention, 37*(2), 305–313. doi: 10.1016/j.aap.2004.09.004.

Keskinen, E., Hatakka, M., Katila, A., Laapotti, S., & Peräaho, M. (1999). Driver training in Finland. *IATSS Research, 23*(1), 78–84.

Klauer, S. G., Guo, F., Simons-Morton, B. G., Ouimet, M. C., Lee, S. E., & Dingus, T. A. (2014). Distracted driving and risk of road crashes among novice and experienced drivers. *New England Journal of Medicine, 370*(1), 54–59. doi: 10.1056/NEJMsa1204142.

Laapotti, S., Keskinen, E., Hatakka, M., & Katila, A. (2001). Novice drivers' accidents and violations—A failure on higher or lower hierarchical levels of driving behaviour. *Accident Analysis & Prevention, 33*(6), 759–769. doi: 10.1016/S0001-4575(00)00090-7.

Laflamme, L., Vaez, M., Hasselberg, M., & Kullgren, A. (2005). Car safety and social differences in traffic injuries among young adult drivers: A study of two-car injury-generating crashes in Sweden. *Safety Science, 43*(1), 1–10.

Lonero, L., & Mayhew, D. R. (2010). *Teen Driver Safety: Large-Scale Evaluation of Driver Education Review of the Literature on Driver Education Evaluation.* Washington, DC: AAA Foundation for Traffic Safety.

Lund, A. K., Williams, A. F., & Zador, P. (1986). High school driver education: Further evaluation of the Dekalb County study. *Accident Analysis & Prevention, 18*(4), 349–357. doi: 10.1016/0001-4575(86)90048-5.

Malone, S., & Brünken, R. (2013). Assessment of driving expertise using multiple choice questions including static vs. animated presentation of driving scenarios. *Accident Analysis & Prevention, 51*, 112–119. doi: 10.1016/j.aap.2012.11.003.

Mann, H. N., & Lansdown, T. (2009). Pre-driving adolescent attitudes: Can they change? *Transportation Research Part F: Traffic Psychology and Behaviour, 12*(5), 395–403. doi: 10.1016/j.trf.2009.05.003.

Masten, S. V., & Peck, R. C. (2004). Problem driver remediation: A meta-analysis of the driver improvement literature. *Journal of Safety Research, 35*(4), 403–425. doi: 10.1016/j.jsr.2004.06.002.

Maycock, G., & Forsyth, E. (1997). *Cohort Study of Learner and Novice Drivers Part 4: Novice Driver Accidents in Relation to Methods of Learning to Drive, Performance in the Driving Test and Self Assessed Driving Ability Behaviour*. Crowthorne: Transport Research Laboratory.

Maycock, G., Lockwood, C. R., & Lester, J. F. (1991). *The Accident Liability of Car Drivers*. Crowthorne: Transport and Road Research Laboratory.

Mayhew, D. R., & Simpson, H. M. (2002). The safety value of driver education an training. *Injury Prevention, 8*(suppl 2), ii3–ii8. doi: 10.1136/ip.8.suppl_2.ii3.

Mayhew, D. R., Simpson, H. M., Desmond, K., & Williams, A. F. (2003). Specific and long-term effects of Nova Scotia's graduated licensing program. *Traffic Injury Prevention, 4*(2), 91–97. doi: 10.1080/15389580309866.

McCartt, A. T., Shabanova, V. I., & Leaf, W. A. (2003). Driving experience, crashes and traffic citations of teenage beginning drivers. *Accident Analysis & Prevention, 35*(3), 311–320.

McCartt, A. T., Mayhew, D. R., Braitman, K. A., Ferguson, S. A., & Simpson, H. M. (2009). Effects of age and experience on young driver crashes: Review of recent literature. *Traffic Injury Prevention, 10*(3), 209–219.

McKenna, F. P., & Crick, J. (1997). *Developments in Hazard Perception; Prepared for Road Safety Division of DETR* (Vol. TRL Report 297). Crowthorne: Transport Research Laboratory.

McKenna, F. P., & Horswill, M. S. (1999). Hazard perception and its relevance for driver licensing. *IATSS Research, 23*(1), 36–41.

McKenna, F. P., Horswill, M. S., & Alexander, J. L. (2006). Does anticipation training affect drivers' risk taking? *Journal of Experimental Psychology: Applied, 12*(1), 1–10.

McKnight, A. J., & McKnight, A. S. (2003). Young novice drivers: Careless or clueless? *Accident Analysis & Prevention, 35*(6), 921–925. doi: S0001457502001008.

Meewes, V., & Weissbrodt, G. (1992). Führerschein auf Probe-Auswirkung auf die Verkehrssicherheit. *Schriftenreihe der Bundesanstalt für Straßenwesen, Unfall- und Sicherheitsforschung Straßenverkehr*. Bergisch Gladbach: Bast.

Michon, J. A. (1985). A critical review of driver behaviour models: what do we know, what should we do? In L. Evans & R. C. Schwing (Eds.), *Human Behavior and Traffic Safety* (pp. 487–525). New York: Plenum Press.

Mitsopoulos, E., Triggs, T., & Regan, M. (2006, 29 May–1 June). *Examining novice driver calibration through novel use of a driving simulator*. Paper presented at the SimTecT 2006 Simulation conference, Melbourne, Australia.

Molina, J. G., García-Ros, R., & Keskinen, E. (2014). Implementation of the driver training curriculum in Spain: An analysis based on the Goals for Driver Education (GDE) framework. *Transportation Research Part F: Traffic Psychology and Behaviour, 26, Part A*, 28–37. doi: 10.1016/j.trf.2014.06.005.

Møller, M., & Gregersen, N. P. (2008). Psychosocial function of driving as predictor of risk-taking behaviour. *Accident Analysis & Prevention, 40*(1), 209–215.

Murray, Å. (1998). The home and school background of young drivers involved in traffic accidents. *Accident Analysis & Prevention, 30*(2), 169–182.

Mynttinen, S., Gatscha, M., Koivukoski, M., Hakuli, K., & Keskinen, E. (2010). Two-phase driver education models applied in Finland and in Austria—Do we have evidence to support the two phase models? *Transportation Research Part F: Traffic Psychology and Behaviour, 13*(1), 63–70. doi: 10.1016/j.trf.2009.11.002.

Nyberg, A., & Engström, I. (1999). "Insight": An evaluation: An interview survey into driving test pupils' perception of the "Insight" training concept at the Stora Holm Driver Training Centre. Linköping: VTI.

OECD. (2006). Young drivers; the road to safety (T. R. Centre, Trans.). Paris: Organisation for Economic Co-operation and Development.

Ouimet, M. C., Simons-Morton, B. G., Zador, P. L., Lerner, N. D., Freedman, M., Duncan, G. D., & Wang, J. (2010). Using the U.S. national household travel survey to estimate the impact of passenger characteristics on young drivers' relative risk of fatal crash involvement. *Accident Analysis & Prevention, 42*(2), 689–694.

Peck, R. C. (2011). Do driver training programs reduce crashes and traffic violations?—A critical examination of the literature. *IATSS Research, 34*(2), 63–71. doi: 10.1016/j.iatssr.2011.01.001.

Pollatsek, A., Narayanaan, V., Pradhan, A. K., & Fisher, D. L. (2006). Using eye movements to evaluate a PC-based risk awareness and perception training program on a driving simulator. *Human Factors, 48*(3), 447–464.

Poulter, D. R., & McKenna, F. P. (2010). Evaluating the effectiveness of a road safety education intervention for pre-drivers: An application of the theory of planned behaviour. *British Journal of Educational Psychology, 80*(2), 163–181. doi: 10.1348/014466509x468421.

Pradhan, A. K., Pollatsek, A., Knodler, M., & Fisher, D. L. (2009). Can younger drivers be trained to scan for information that will reduce their risk in roadway traffic scenarios that are hard to identify as hazardous? *Ergonomics, 52*(6), 657–673.

Pradhan, A. K., Divekar, G., Masserang, K., Romoser, M., Zafian, T., Blomberg, R. D., Thomas, F. D. et al. (2011). The effects of focused attention training on the duration of novice drivers' glances inside the vehicle. *Ergonomics, 54*(10), 917–931. doi: 10.1080/00140139.2011.607245.

Pradhan, A. K., Li, K., Bingham, C. R., Simons-Morton, B. G., Ouimet, M. C., & Shope, J. T. (2014). Peer passenger influences on male adolescent drivers' visual scanning behavior during simulated driving. *Journal of Adolescent Health, 54*(5, Supplement), S42–S49. doi: 10.1016/j.jadohealth.2014.01.004.

Prato, C. G., Toledo, T., Lotan, T., & Taubman-Ben-Ari, O. (2010). Modeling the behavior of novice young drivers during the first year after licensure. *Accident Analysis & Prevention, 42*(2), 480–486.

Regan, M. A., Triggs, J. T., & Godley, S. T. (2000). *Simulator-based evaluation of the DriveSmart novice driver CD-Rom training product*. Paper presented at the Road Safety Research, Policing and Education Conference, Brisbane, Queensland, Australia.

Rosenbloom, T., & Eldror, E. (2014). Effectiveness evaluation of simulative workshops for newly licensed drivers. *Accident Analysis & Prevention, 63*, 30–36. doi: 10.1016/j.aap.2013.09.018.

Sagberg, F. (1998). *Month-by-month changes in accident risk among novice drivers.* Paper presented at the 24th International Conference of Applied Psychology, San Francisco, August 9–14.

Samuel, S., Zafian, T., Borowsky, A., Romoser, M. R. E., & Fisher, D. L. (2013). *Can young drivers learn to anticipate hidden hazards: A driving simulator study?* Paper presented at the 7th International Driving Symposium on Human Factors in Driver Assessment, Training, and Vehicle Design, Lake George, NY.

Senserrick, T. (2007). Recent developments in young driver education, training and licensing in Australia. *Journal of Safety Research, 38*(2), 237–244. doi: 10.1016/j.jsr.2007.03.002.

Senserrick, T., & Haworth, N. (2005). Review of literature regarding national and international young driver training, licensing and regulatory systems. Clayton, Victoria, Australia: Monash University Accident Research Centre (MUARC).

Senserrick, T., Ivers, R. Q., Boufous, S., Chen, H. Y., Norton, R., Stevenson, M. R., van Beurden, E., & Zask, A. (2009). Young driver education programs that build resilience have potential to reduce road crashes. *Pediatrics, 124*(5), 1287–1292. doi: 10.1542/pes.2009-0659.

Simons-Morton, B., Lerner, N., & Singer, J. (2005). The observed effects of teenage passengers on the risky driving behavior of teenage drivers. *Accident Analysis & Prevention, 37*(6), 973–982.

Simons-Morton, B. G., Ouimet, M. C., Zhang, Z., Klauer, S. E., Lee, S. E., Wang, J., Chen, R., Albert, P., & Dingus, T. A. (2011). The effect of passengers and risk-taking friends on risky driving and crashes/near crashes among novice teenagers. *Journal of Adolescent Health, 49*(6), 587–593. doi: 10.1016/j.jadohealth.2011.02.009.

Simpson, H. M., Chinn, L., Stone, J., Elliot, M. A., & Knowles, J. (2002). *Monitoring and Evaluation of Safety Measures for New Drivers*. Wokingham, UK: Transport Research Laboratory.

Smith, S. S., Horswill, M. S., Chambers, B., & Wetton, M. (2009). Hazard perception in novice and experienced drivers: The effects of sleepiness. *Accident Analysis & Prevention, 41*(4), 729–733.

Stock, J. R., Weaver, J. K., Ray, H. W., Brink, T. R., & Sadof, M. G. (1983). Evaluation of safe performance secondary school driver education curriculum project, Final Report. Washington, DC: National Highway Safety Administration.

Taubman-Ben-Ari, O., Musicant, O., Lotan, T., & Farah, H. (2014). The contribution of parents' driving behavior, family climate for road safety, and parent-targeted intervention to young male driving behavior. *Accident Analysis & Prevention, 72*, 296–301. doi: 10.1016/j.aap.2014.07.010.

Telzer, E. H., Ichien, N. T., & Qu, Y. (2015). Mothers know best: Redirecting adolescent reward sensitivity toward safe behavior during risk taking. *Social Cognitive and Affective Neuroscience, 10*(10), 1383–1391. doi: 10.1093/scan/nsv026.

Underwood, G., & Crundall, D. (1998). Effects of experience and processing demands on visual information acquisition in drivers. *Ergonomics, 41*(4), 448–458.

Underwood, G., Chapman, P., Berger, Z., & Crundall, D. (2003). Driving experience, attentional focusing, and the recall of recently inspected events. *Transportation Research Part F, 6*(4), 289–304.

Vidotto, G., Bastianelli, A., Spoto, A., & Sergeys, F. (2011). Enhancing hazard avoidance in teen-novice riders. *Accident Analysis & Prevention, 43*(1), 247–252.

Vlakveld, W. P. (2011). *Hazard Anticipation of Young Novice Drivers*. Leidschendam, the Netherlands: SWOV Institute of Road Safety Reseach.

Vlakveld, W. P. (2014). A comparative study of two desktop hazard perception tasks suitable for mass testing in which scores are not based on response latencies. *Transportation Research Part F: Traffic Psychology and Behaviour, 22*, 218–231. doi: 10.1016/j.trf.2013.12.013.

Vlakveld, W. P., Romoser, M., Mehranian, H., Diete, F., Pollatsek, A., & Fisher, D. L. (2011). Do crashes and near crashes in simulator-based training enhance novice drivers' visual search for latent hazards? *Transportation Research Record, 2265*, 154–160.

Wang, Y., Zhang, W., & Salvendy, G. (2010). Effects of a simulation-based training intervention on novice drivers' hazard handling performance. *Traffic Injury Prevention, 11*(1), 16–24.

Washington, S., Cole, R. J., & Herbel, S. B. (2011). European advanced driver training programs: Reasons for optimism. *IATSS Research, 34*(2), 72–79. doi: 10.1016/j.iatssr.2011.01.002.

Waylen, A. E., & McKenna, F. P. (2008). Risky attitudes towards road use in pre-drivers. *Accident Analysis & Prevention, 40*, 905–911.

Weiss, T., Petzoldt, T., Bannert, M., & Krems, J. (2013). Calibration as side effect? Computer-based learning in driver education and the adequacy of driving-task-related self-assessments. *Transportation Research Part F: Traffic Psychology and Behaviour, 17*, 63–74. doi: 10.1016/j.trf.2012.10.001.

Wells, P., Tong, S., Sexton, B. F., Grayson, G. B., & Jones, E. (2008). *Cohort II: A Study of Learner and New Drivers. (Vol. 1—Main Report)*. London: Department for Transport.

Wetton, M. A., Hill, A., & Horswill, M. S. (2011). The development and validation of a hazard perception test for use in driver licensing. *Accident Analysis & Prevention, 43*(5), 1759–1770. doi: 10.1016/j.aap.2011.04.007.

Wetton, M. A., Hill, A., & Horswill, M. S. (2013). Are what happens next exercises and self-generated commentaries useful additions to hazard perception training for novice drivers? *Accident Analysis & Prevention, 54*, 57–66. doi: 10.1016/j.aap.2013.02.013.

White, C. B., & Caird, J. K. (2010). The blind date: The effects of change blindness, passenger conversation and gender on looked-but-failed-to-see (LBFTS) errors. *Accident Analysis & Prevention, 42*(6), 1822–1830. doi: 10.1016/j.aap.2010.05.003.

Willmes-Lenz, G., Prücher, F., & Groβmann, H. (2010). Evaluation of the novice driver training models "Accompanied driving from 17" and "Voluntary further training seminars holders of probationary driving license." Results up to November 2009. Bergisch Gladbach, Germany: BAST, Federal Highway Research Institute.

24

Road Safety and Teen and Novice Drivers in Low- and Middle-Income Countries

Adnan A. Hyder

Jeffrey C. Lunnen

Abstract

Currently the ninth leading cause of death worldwide, road traffic injuries are projected to be the seventh leading cause by 2030 and thus represent a major, continued, and often neglected threat to public health. Road traffic injuries are the leading cause of death for people aged 15–29 years worldwide. Research from high-income countries (HICs) indicates that teen and novice drivers have a higher risk of crash and death than other drivers, but there are few studies in low- and middle-income countries (LMICs), where the burden is highest. Addressing traditional risk factors for road traffic injury such as seat belt and motorcycle helmet use, speed, and alcohol, in combination with emerging risk factors like distracted driving, will reduce fatal and nonfatal road traffic injuries among teens and novice drivers worldwide. Graduated driver licensing has proven effective for reducing crashes and related deaths among young drivers in HICs; similar programs should also be implemented in LMICs as well. While there are challenges to the implementation of evidence-based programs in LMICs, the United Nations Global Decade of Action for Road Safety 2011–2020 represents an opportunity for governments, nongovernmental organizations, the private sector, academics, and youth themselves to come together to reduce the large number of crashes, injuries, deaths, and disability among teen and novice drivers worldwide.

24.1 Introduction

Currently the ninth leading cause of death worldwide, road traffic injuries (RTIs) are projected to be the seventh leading cause by 2030 and thus represent a major, continued, and often neglected threat to public health (WHO 2013). According to the most recent global health estimates from the World Health Organization (WHO), road traffic crashes were responsible for more than 1.25 million deaths in 2012 (WHO 2014). Traffic crashes also result in as many as 50 million nonfatal injuries each year, and many nonfatal RTIs result in millions of people with short- and long-term disability (WHO 2009). RTIs, both fatal and nonfatal, are responsible for economic losses of between 1% and 3% of annual gross domestic product, especially in low- and middle-income countries (LMICs) (Peden & Hyder 2002). In addition, the personal and social impact of death and disability from RTIs, or "intangible costs," are staggering (Pérez-Núñez et al. 2012).

More than 90% of fatal RTIs occur in LMICs, although these countries have just 48% of the world's total registered vehicles (WHO 2009, 2014). The highest absolute numbers of fatal RTIs are observed in the WHO region of Southeast Asia, whereas rates per 100,000 population are highest in the African region. Although the majority of fatal RTIs are observed among people aged 30–49 years, RTIs are actually the leading cause of death for young people aged 15–29 years (WHO 2009). In 2012, there were 325,736 fatal RTIs among this age group, which translates to an unadjusted rate of approximately 18 deaths per 100,000 population (WHO 2014).

So-called vulnerable road users, i.e., motorcycle users, bicyclists, and pedestrians, account for approximately 50% of all fatal RTIs according to the 2013 *Global Status Report on Road Safety* (WHO 2013). Car occupants and unspecified road users account for the remaining 31% and 19%, respectively (WHO 2013). If the proportions of fatalities comprised by different road user categories at the global level hold for all age groups, approximately 100,978 fatal RTIs observed among 15- to 29-year olds could be car occupants, whether drivers or passengers (WHO 2014). Motorcycle use is a major mode of transportation for young people in many LMICs, especially in Southeast Asia and in parts of Latin America; in many contexts, teens and novice drivers use cars and motorcycles interchangeably depending upon socioeconomic status and need.

In general, studies of teen and novice drivers in the scientific literature posit inexperience, overconfidence, or a combination to explain the higher risk of road traffic crashes and, likewise, injury and death among this population (McKnight & McKnight 2003; also see Chapters 5, 7, and 8). Learning how to drive a car or a motorcycle requires a person to master visual, mechanical, and cognitive skills, but this process may be more difficult for teens since their brains are still developing (Chapter 9). It does not help matters that adolescents may also be more susceptible to peer influence or peer pressure than adults (Simons-Morton et al. 2011; also see Chapter 16). Indeed, when other risk factors are added, such as driving at night and driving with other teenage male passengers, the risk of road traffic crashes among teen drivers further increases (Simons-Morton et al. 2005, 2011). This is why such drivers are considered especially vulnerable, and ought to be considered vulnerable road users.

In this chapter, we address road safety and youth, especially teen and novice drivers, from an international perspective, with a particular focus on LMICs. We define the global burden of RTIs among 15- to 29-year olds including deaths, years of life lost (YLLs), years of life lost to disability (YLDs), and disability-adjusted life years (DALYs), and associated risk factors. We also introduce intervention strategies that have proven effective in reducing crashes and fatal and nonfatal RTIs among drivers, and acknowledge challenges to their application in LMIC settings. We conclude by introducing the United Nations Decade of Action for Road Safety 2011–2020 as an opportunity for the global community to come together to address RTIs among teen and novice drivers.

24.2 Global Burden of RTIs among People Aged 15–29 Years

The WHO estimates that although the majority of fatal RTIs (N = 364,462) are observed among people aged 30–49 years (29%), RTIs represent the leading cause of death for people aged 15–29 years worldwide. In 2012, the absolute number of fatal RTIs among this age group was 325,736, which translates to a rate of 18.1 deaths per 100,000 population (WHO 2014) (Figure 24.1).

The highest absolute number of fatal RTIs among people aged 15–29 years is observed in Southeast Asia (80,883 deaths), but rates of fatal RTIs are highest in Africa (24.4 per 100,000 population) (Figure 24.2). Consistent with global trends, the overwhelming majority (~80%) of fatal RTIs among people aged 15–29 years are observed among males; the absolute number of deaths among males aged 15–29 years

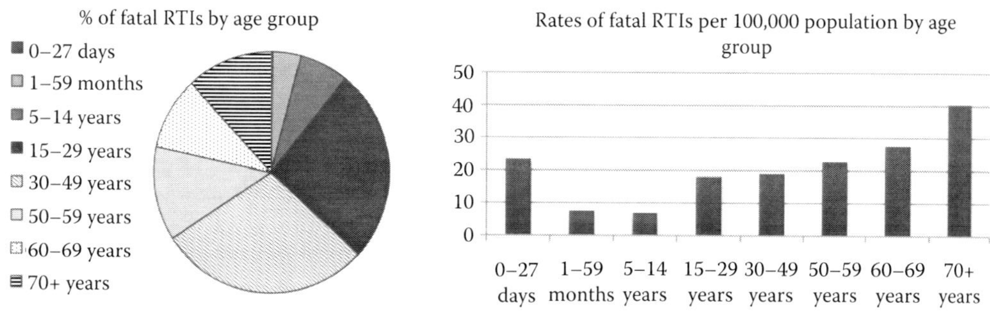

FIGURE 24.1 Global burden of fatal road traffic injuries by age group. Note: RTIs = road traffic injuries. (From World Health Organization, *Global Health Estimates 2014 Summary Tables: Deaths by Cause, Age and Sex, by WHO region, 2000-2012*, World Health Organization, Geneva, 2014.)

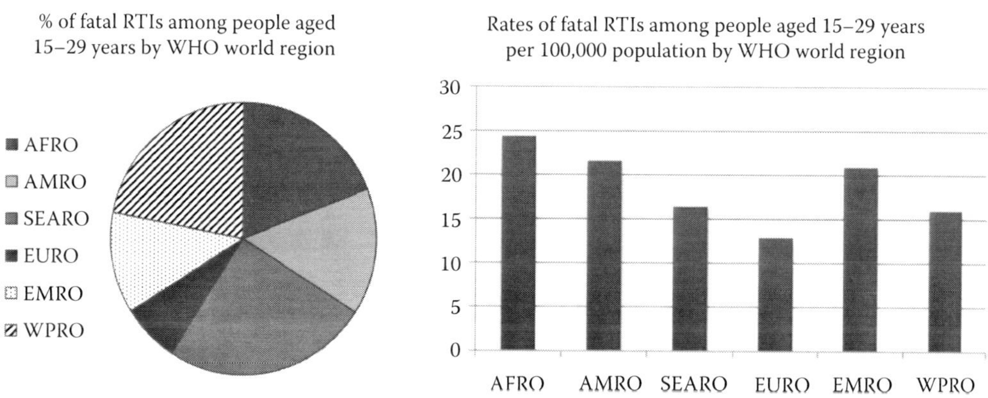

FIGURE 24.2 Regional trends in fatal RTIs among people aged 15–29 years. Notes: AFRO = WHO African Region; AMRO = WHO Region of the Americas; EMRO = WHO Eastern-Mediterranean Region; EURO = WHO European Region; RTIs = road traffic injuries; SEARO = WHO Southeast Asia Region; WPRO = WHO Western Pacific Region. (From World Health Organization, *Global Health Estimates 2014 Summary Tables: Deaths by Cause, Age and Sex, by WHO region, 2000–2012*, World Health Organization, Geneva, 2014.)

is more than 3.5 times higher than their female counterparts (Figure 24.3). The greatest differential between the sexes is observed in the Eastern Mediterranean region, where males contribute to ~85% of fatal RTIs.

RTIs impact, on population health can also be measured in other ways; for example, in YLL and YLD. A YLL is calculated by multiplying the absolute number of deaths by the standard life expectancy at the age of death in years (Hyder et al. 2012). A YLD is calculated by multiplying the number of incident cases by a disability weight and the average duration of the case until remission or death. These metrics can be viewed independently or together to determine the burden of disease within a population. Summing YLL and YLD creates a population health metric: the DALY. One DALY can be thought of as one lost year of "healthy" or disease-free life. The sum of these DALYs across a population (or the burden of disease) can be thought of as a measurement of the gap between current and ideal health status—where the entire population lives to an advanced age, without disease and disability (Hyder et al. 2012).

In 2012, RTIs contributed to a total of 64 million YLLs, of which 22.5 million YLLs (35%) were among people aged 15–29 years, or roughly 1256 per 100,000 population. YLLs were more than 3.5 times higher among males than among females aged 15–29 years. As is true of the general population, absolute numbers of YLLs were highest in Southeast Asia, whereas rates were highest in Africa (Figure 24.4) (WHO 2014). These indicators reflect the impact of RTIs not on the absolute level of mortality but on the YLL as a result of young deaths and time lost from ill health and morbidity—all outcomes that negatively impact both national health and economic status.

In 2012, RTIs contributed to 13 million YLDs, or approximately 197 per 100,000 population (WHO 2014). At the global level, the absolute number of YLDs was highest in the Western Pacific region, and the highest rate per 100,000 population was observed in the European region. RTIs were responsible for 3.5 million YLDs or ~193 per 100,000 population among people aged 15–20 years—and again, rates were highest in the European region.

RTIs contributed to 78.7 million DALYs, and were the eighth leading cause of DALYs worldwide in 2012 (WHO 2014). Among people aged 15–29 years, RTIs resulted in 25.9 million DALYs, which translates to approximately 1445 per 100,000 population. The highest absolute number of DALYs from RTIs, for both the general population and people aged 15–29 years, was observed in the Southeast Asia region; the highest rates were found in the African and Eastern Mediterranean Region (Figure 24.5).

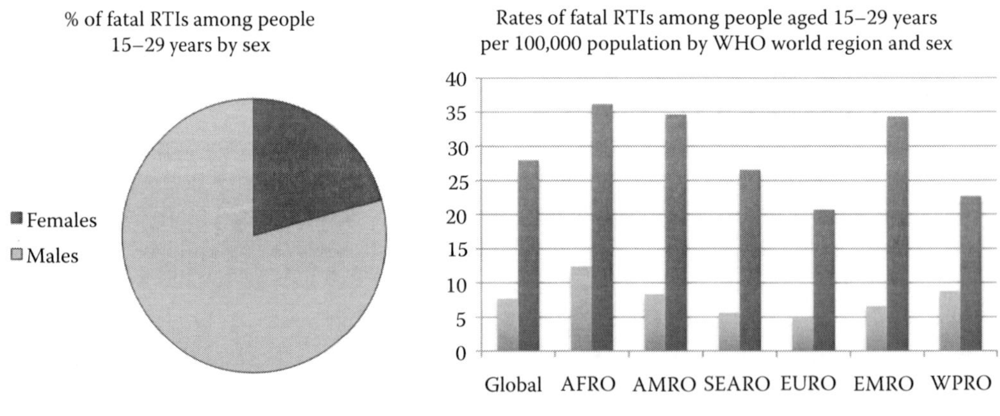

FIGURE 24.3 Sex differences in fatal road traffic injuries among people aged 15–29 years. Notes: AFRO = WHO African Region; AMRO = WHO Region of the Americas; EMRO = WHO Eastern-Mediterranean Region; EURO = WHO European Region; RTIs = road traffic injuries; SEARO = WHO Southeast Asia Region; WPRO = WHO Western Pacific Region. (From World Health Organization, *Global Health Estimates 2014 Summary Tables: Deaths by Cause, Age and Sex, by WHO region, 2000–2012*, World Health Organization, Geneva, 2014.)

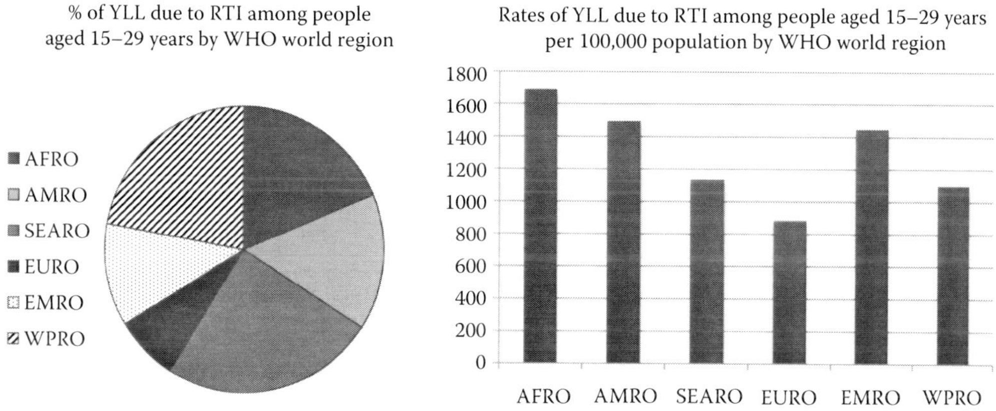

FIGURE 24.4 Regional trends in years of life lost due to RTI among people aged 15–29 years. Notes: AFRO = WHO African Region; AMRO = WHO Region of the Americas; EMRO = WHO Eastern-Mediterranean Region; EURO = WHO European Region; RTIs = road traffic injuries; SEARO = WHO Southeast Asia Region; WPRO = WHO Western Pacific Region; YLLs = years of life lost. (From World Health Organization, *Global Health Estimates 2014 Summary Tables: Deaths by Cause, Age and Sex, by WHO region, 2000–2012*, World Health Organization, Geneva, 2014.)

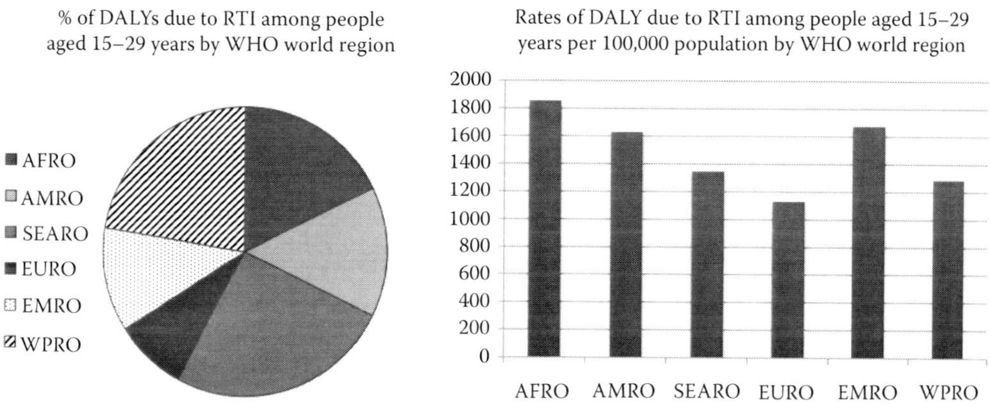

FIGURE 24.5 Regional trends in disability-adjusted life years due to RTI among people aged 15–29 years. Notes: AFRO = WHO African Region; AMRO = WHO Region of the Americas; DALYs = disability-adjusted life years; EMRO = WHO Eastern-Mediterranean Region; EURO = WHO European Region; RTIs = road traffic injuries; SEARO = WHO Southeast Asia Region; WPRO = WHO Western Pacific Region. (From World Health Organization, *Global Health Estimates 2014 Summary Tables: Deaths by Cause, Age and Sex, by WHO region, 2000–2012*, World Health Organization, Geneva, 2014.)

24.3 Current Perspectives on Teen and Novice Driving

Despite the global burden of RTIs being highest in LMICs, especially among people aged 15–29 years, most of our understanding of teen and novice driver behavior, and related outcomes, comes from studies conducted in HICs (Chapter 23). As stated earlier, the scientific literature from these contexts emphasizes some combination of inexperience and overconfidence (or even "carelessness" or "cluelessness") being responsible for the increased risk of crashes among teens and novice drivers (McKnight & McKnight 2003). Learning how to drive requires a person to master new visual, mechanical, and cognitive skills.

For adolescent drivers, crash risk is exacerbated by their physiology, that is, the incomplete maturation of cognitive and motor skills, including working memory, visual–spatial attention, and speed of processing related to their brain development (Romer et al. 2014; also see Chapter 9).

Driver age is an important risk factor; for example, a study by Chapman et al. (2014) found that among participants of a graduated driver licensing (GDL) program, crash rates among teens remain elevated relative to older novice drivers for years subsequent to licensure. Even with the same level of experience and driving exposure, younger novices tend to have higher crash rates than do older novices, with even just a 1-year increase in age at licensure being associated with having fewer crashes (Chapman et al. 2014; Vlakveld 2005; Chapter 17). A study from Australia (Scott-Parker et al. 2013b) found that there was a smaller subset of the young driver population that was especially at high risk for crashes and RTIs, in which higher percentages of anxiety, depression, and reward and sensation seeking were reported (also see Chapters 10 and 11). An Italian study identified three separate subgroups ("risky drivers, worried drivers and careful drivers") and found that drivers in the first group had more tickets, a greater number of crashes with vehicle damage or injury, and a negative attitude regarding road safety, and thought their risk of crash involvement was lower than their peers (Lucidi et al. 2010). All of these above findings suggest that the interaction between inexperience, overconfidence, and personality type, not one or the other, predisposes teen and novice drivers to engage in certain risky behaviors.

Key risk factors for road traffic crashes among teen and novice drivers include the lack of seat belts and motorcycle helmets, mobile phone use, alcohol use, and speeding, all of which are global risks and critical to address (also see Chapters 11 and 12). Also, driving at night and driving in the presence of teenage male passengers are risks (Simons-Morton et al. 2005; Chapter 16). Since adolescents may be more susceptible to peer pressure than their adult counterparts, certain risky behaviors may be normalized. A study among newly licensed teenage drivers in the United States found that crashes and near crashes were lower in the presence of an adult passenger and higher among teenage drivers with more risk-taking friends. Those with risky friends were more likely to engage in risky driving and experience crashes or near crashes (Simons-Morton et al. 2011).

Although the physical environment can have a protective effect, in many LMICs roads, and cities were simply not planned or designed to accommodate the current mix and flow of road users. And they lack the resources to engage in expensive redesign projects to separate road users that have proven effective in several HICs. Although preference has been given to users of four- or more-wheeled vehicles in the design of many LMIC road systems, these vehicles have never enjoyed complete dominion. LMIC roads are a mix of motorized and nonmotorized transportation, as well as the pedestrian population. Teen and novice drivers in LMICs must contend with uneven or unpaved road surfaces; dangerous potholes; the absence of sidewalks; pedestrians and vendors on the road; the lack of traffic signs, signals, and lights; unmarked pedestrian crossings; etc. Therefore the physical environment itself can also be a risk factor to teen and novice drivers around the world. Additionally, vehicles available to teen and novice drivers in LMICs may be older and lacking protective equipment—for example, LMICs often import or sell older or discontinued vehicles that would not meet safety standards in many HICs; thus, a teen or novice driver may be at risk of crash or injury because of the physical environment or his or her own vehicle—in LMICs.

24.4 Interventions for Reducing Crashes and Injuries among Teens and Novice Drivers

The 2013 *Global Status Report on Road Safety* indicates that RTIs worldwide require persistent attention; Table 24.1 describes some common interventions. Many countries have implemented and enforced effective legislation to increase seat belt/child restraint and motorcycle helmet use, and reduce speeding and drunk driving. High levels of enforcement and maintaining a high perception thereof among the general public, and especially teens and novice drivers, are essential to the success of such measures. Unfortunately, just 7% of the world's population is covered by comprehensive legislation that

TABLE 24.1 Introduction to Interventions for Road Traffic Injury Prevention and Control

Best Practices	Promising Practices	Ineffective Practices
• Comprehensive road safety legislation and proper enforcement • Mandatory seat belt use • Compulsory motorcycle helmet usage • Motorcycle helmet standards • Drunk-driving restriction • Speed limits • Graduated driver licensing programs • Speed and red-light cameras • Traffic calming measures • Roadway lighting • Guardrails • Emergency response services • Continuing education for trauma care	• Bans on mobile phone use and texting • Alcohol ignition locks • Police patrols • Incentive programs	• Driver education programs

Note: Table based on a table developed by Hyder et al. 2013.

addresses all of the aforementioned risk factors (WHO 2013). Additionally, in many LMICs, enforcing laws remains a challenge—although this is a very cost-effective way to reduce RTIs (Table 24.2). In this section, we explain briefly why addressing traditional risk factors for RTIs is vital before addressing emerging risk behaviors such as mobile phone use while driving.

Seat belt use has been shown to reduce a person's risk of death by 40–50% if a front-seat passenger and 25–75% if a rear-seat passenger (Elvik & Vaa 2004; Zhu et al. 2007). When a motor vehicle crashes, a car occupant without a seat belt will continue to move forward at the same speed at which the vehicle was traveling before the crash and will be catapulted forward into the structure of the vehicle—most likely the steering wheel column if driving, the dashboard if a front-seat passenger, or the front seat if a rear-seat passenger. A passenger can also be ejected if not belted. While 111 countries, covering 69% of the world's population, have comprehensive seat belt laws covering all vehicle occupants, less than half of all countries have data on seat belt wearing rates, with this number disproportionately lower in LMICs (6% in low-income countries and 43% in middle-income countries) compared to HICs (80%) (WHO 2013). For example, a cross-country analysis of seat belt wearing rates in Egypt, Mexico, Russia, and Turkey found that use ranged from 4% to 75% among drivers (Vecino-Ortiz et al. 2014).

In the event of a road traffic crash, wearing a standard motorcycle helmet can reduce a motorcycle user's risk of death by 40% and the risk of serious head injury by 70% (Liu et al. 2005). Helmet legislation and solid enforcement thereof can increase motorcycle helmet use and reduce head injuries (WHO 2009). Currently, 90 countries, which amounts to approximately 77% of the world's population, have comprehensive helmet laws—requiring that both drivers and passengers use helmets (WHO 2014). Motorcycle helmets must meet international standards and should also be affordable; otherwise, we should expect to see nonstandard helmets with limited efficacy. Helmet laws should be enforced around the clock as well; for example, a study (Bachani et al. 2013a) from Cambodia found that although the overall helmet-wearing rate was 63.8% among motorcycle drivers, just 48.8% wore helmets at night.

TABLE 24.2 Cost-Effectiveness of RTI Interventions in Low- and Middle-Income Countries

Intervention	Cost per DALY (USD)
Improved enforcement (LMIC average)	5.25
Speed bumps (or humps) at top-25-percentile dangerous junctions (LMIC average)	8.89
Motorcycle helmets (Thailand)	467

Note: Table based on table developed by Bishai et al. 2008. DALY = disability adjusted life year; LMICs = low- and middle-income countries.

Speeding is a major risk factor for RTIs in nearly all countries. Traveling at faster speeds increases the likelihood of a crash, and the severity of both the crash and the resulting injuries (WHO 2013). Speeding is of particular concern for young novice drivers due to their inexperience in detecting and responding appropriately to driving hazards (Scott-Parker et al. 2013a; Chapter 11). Additionally, if a young novice driver holds a favorable socially shared image—a prototype—of a "typical" risky young driver, the more likely he or she is to speed if the circumstance were to arise since the behavior is thought to be normative (Ouellette et al. 1999). Often, a difference of 10 kph can mean the difference between life and death in a crash, and so it is vital that a teen or novice driver adopt model safe behaviors. Only 59 countries, covering just 39% of the world's population, have implemented an urban speed limit of 50 kph or less and allow local authorities to reduce these limits. Although most countries have speed limit laws, their enforcement is often lacking: only 26 countries rate enforcement of their national speed limits as "good" (WHO 2013). For example, a study from Kenya found that than 40% of vehicles were speeding (Bachani et al. 2013b).

Traffic calming measures that reduce speed can be used to address the safety of young road users, especially teen and novice drivers who are prone to speeding. Traffic calming can include installing speed bumps or humps, mini-roundabouts, and designated pedestrian crossings; introducing visual changes to the road surface such as zebra crossings; improving the surface of the road itself or installing rumble strips; changing road lighting; redistributing traffic blocking roads; and creating one-way streets. Evidence from HICs indicates that traffic calming can reduce injuries and deaths and is applicable in LMICs (Toroyan & Peden 2007).

Drinking and driving increases the risk of being involved in a crash and the severity of resulting injuries. Driving impairment begins with very low levels of alcohol consumption. The vast majority of adult drivers are affected or impaired with a BAC of 0.05 g/dL while at a BAC of 0.1 g/dL, crash risk is approximately five times higher than that of someone with a BAC level of zero (Compton et al. 2002; Hurst et al. 1994; Chapter 13). A study measuring the prevalence of drinking and driving in two Vietnamese provinces found that between 8% and 33% of drivers were found to be driving under the influence of alcohol (Bachani et al. 2013c). Young and novice drivers who drink and drive have a greatly increased risk of a crash compared to more experienced drivers (Peden et al. 2004; Zador 1991). Setting and enforcing legislation on BAC limits of 0.05 g/dL for all drivers and 0.02 g/dL for young (teen) and novice drivers can lead to significant reductions in alcohol-related crashes. According to the WHO, just 23% of countries apply a BAC of 0.02 g/dL or less for young and novice drivers; such legislation is more common in HICs (46%) as compared to middle-income countries (16%) and low-income countries (9%) (WHO 2013).

Distracted driving is a relatively new but established risk factor for crashes and RTIs. Distractions can be visual, manual, or cognitive, and these increase one's chances of crashing a vehicle. Specific distracted behaviors can include eating, reading, talking, smoking, applying makeup, and also the use of cellular phones and portable electronic devices (Chapter 12). Brodsky and Slor (2013) found that novice teen drivers who played music inside the car that they preferred exhibited elevated positive moods but also frequent and severe driver miscalculations and inaccuracies, violations, and aggressive driving. Drivers who text, i.e., send text messages, are 23 times more likely to be in a crash (National Highway Traffic Safety Administration 2006). The proliferation of mobile phones in LMICs makes mobile phone use—talking or texting—while driving of particular concern. In a US study, Carter et al. (2014) found that 92% of teen drivers reported regularly engaging in distracted driving behaviors. In Mexico, a study of mobile phone use among car drivers found a prevalence of 10.8% (Vera-López et al. 2013). A recent study from the United States found that texting bans were associated with a 7% reduction in crash-related hospitalizations among all age groups—only marginal reductions were seen among teens ("adolescents") (Ferdinand et al. 2015).

GDL programs explicitly address the unique risks faced by teen and novice drivers by mandating a multistep planned and supervised practice. Where GDL systems exist, a driver must obtain a learner's permit and a provisional license before a full license. Learner's permits and provisional licenses have restrictions, which are lifted over the course of the defined period. GDL programs are effective in reducing crash rates among young drivers, and those with more restrictions result in greater fatality reduction (Russell et al. 2011; Chapter 17). Since teen and novice drivers' elevated risk of experiencing a road traffic crash is universal, GDL programs should be implemented in LMIC settings. However, there is no evidence of any LMICs implementing GDL programs nationwide.

Even when the aforementioned risk factors are addressed, crashes may still happen. Therefore, timely, high-quality, prehospital emergency medical services (EMS) rendered at the scene of a road traffic crash are vital. In places where formal EMS exists, usually with ambulances, they are most effective if their equipment, training, infrastructure, and operations are standardized. To prevent death among teens, staff must have equipment, supplies, and training appropriate for this population. And indeed, pediatric emergency medical care (<18 years) ought to be part of a well-developed EMS system. Improving the organization and planning of trauma care services is an affordable and sustainable way to increase both the quality of care and outcomes (Peden et al. 2004). Managing RTIs in the acute phase can prevent survivors from developing permanent disabilities. And as we know from HICs, permanent disability can be avoided with improved rehabilitation services. Rehabilitation helps to minimize the impact of disabilities and return injured people to active life.

24.5 Challenges to Implementation of Effective Measures in LMICs

Simply put, many LMICs do not prioritize RTIs, despite the enormous health, economic, and social burden they place on people, their health systems, and their economies. This is due to a host of reasons, which include but are not limited to not understanding injury as a health problem, a history of weak governance, poor legislation, finite resources to address health problems, and low technical capacity. This stands in comparison to HICs such as Sweden, which has a Vision Zero policy, i.e., no deaths from automobile crashes.

The dominant orientation of public health systems in LMICs is toward infectious diseases; part of this is related to the biomedical sector not fully understanding "energy" as the cause of injury (rather than a microbiologic agent). Since nonmedical approaches are often employed, intersectoral collaboration is minimal, and trust across divisions in LMIC governments is low; therefore, injury prevention becomes less important, and resource allocation is disproportionately low. Often in LMICs, a single person (or "focal point") is the only person working on all types of injuries. For example, in the absence of national road safety councils, or boards, with operational budgets, RTI prevention loses out to other health problems. The *2004 World Report on Road Traffic Injury Prevention* offered a set of concrete suggestions for countries to improve road safety, and chief among them was that governments should identify a lead agency to guide national road safety strategies (Peden et al. 2004). According to the 2013 *Global Status Report on Road Safety*, approximately 89% (n = 182) of the countries surveyed had a lead agency for road safety (WHO 2013). However, only 77% of 159 countries reported that their lead agencies were actually funded for effective action.

As mentioned in the previous section, only 7% of the world's population is covered by comprehensive laws that address traditional risk factors for road safety. In many LMICs, there is space for tougher laws and better enforcement, especially as regards teen and novice drivers. However, since the majority of studies on the effectiveness of road safety interventions were conducted in HICs, there is often reluctance to apply lessons learned there in LMIC contexts. The scarcity of publicized successes in injury prevention in LMICs can also be attributed to the "know–do" gap, which is understood as a slow translation of knowledge into programs and delayed diffusion of evidence demonstrating potentially effective interventions. Furthermore, there is a lack of human technical capacity for injury prevention in many LMICs, which contributes to the implementation of programs without an evidence base. Capacity development across functions like research, police development, implementation efficiency, and management expertise is also necessary. Recent efforts such as the Bloomberg Philanthropies–funded Global Road Safety Program (2010–2015)—a $125 million investment—have aimed to close the know–do gap in at

TABLE 24.3 Examples of Activities That Help Teen and Novice Drivers by Road Safety Pillar

	United Nations Decade of Action for Road Safety Pillars				
Road Safety Actor	1 Road Safety Management	2 Safer Roads and Mobility	3 Safer Vehicles	4 Safer Road Users	5 Post-Crash Response
Governments	– Establish and fund a lead agency for road safety – Develop a strategy with a focus on teen and novice drivers – Set short- and long-term targets for teen and novice drivers	– Promote safe operation, maintenance, and improvement of existing roads – Develop new roads in accordance with best practices	– Incentivize import and export of motor vehicles with high levels of road user protection	– Set and seek compliance with road safety legislation – Require GDL for teen and novice drivers	– Develop or strengthen prehospital and hospital systems – Provide early rehabilitation – Establish road user insurance schemes
Nongovernmental organizations (NGOs)	– Advocate on behalf of teen and novice drivers – Apply pressure to governments to create a lead agency for road safety and ensure it is funded – Insist on preparation of high-quality data on teen and novice drivers	– Demand safe roads for all road users – Bring attention to poor road design – Share research on safe road infrastructure with governments – Advocate for greater capacity building in safe infrastructure	– Push government and the private sector to purchase, operate, and maintain vehicles with advanced technology and occupant protection	– Require governments to implement GDL	– Promote strong EMS
Private industry	– Invest in company and social road safety efforts	– Invest in safe road design	– Produce new vehicles with seat belts, anchorages, airbags, electronic stability control, and antilock braking systems	– Produce campaigns to discourage key risk factors for RTI among teens	– Invest in public–private partnerships for EMS
Academia	– Conduct studies showing benefit of lead agencies for road safety – Offer technical assistance to lead agencies	– Research benefits of safe road infrastructure – Demonstrate impact of traffic calming measures	– Evaluate impact of safe design	– Demonstrate benefit of GDL in LMICs – Measure exposure and risk factors in new settings – Evaluate GDL in LMICs	– Demonstrate benefit of EMS on health outcomes related to RTI in teen and novice drivers in LMICs
Youth	– Reach out to lead agency for road safety on causes of teen and novice driver crashes	– Demand safe roads for teen drivers	– Promote vehicles with safety features	– Challenge teens to adhere to proper licensing practices	– Advocate for EMS

least nine participating LMICs. Countries who participated (Brazil, Cambodia, China, India, Kenya, Mexico, Russia, Turkey, and Vietnam) received support from a consortium of international partners with expertise in domains such as implementation, social marketing, capacity development for police, monitoring and evaluation, and sustainable transportation. A formal evaluation will determine which components worked best, and where; however, the fact of the program's very existence is a success—providing visibility and, along with it, some pressure to deliver on the promise of road safety.

24.6 Recommendations in Light of the Global Decade of Action for Road Safety

RTIs need not be inevitable and can be prevented through the implementation of a systems approach to road safety. Measures include better land-use, urban, and transport planning; designing safer roads and requiring audits for new projects; improving vehicle safety; promoting public transport; managing speed more effectively via enforcement and traffic calming; setting and enforcing comprehensive legislation on seat belt, motorcycle helmet, and mobile phone use, speeding, and the consumption of alcohol and other drugs; and improving post-crash care. In March 2010, the United Nations General Assembly resolution 64/255 proclaimed a Decade of Action for Road Safety 2011–2020 with the goal of stabilizing and then reducing the forecasted level of fatal RTIs around the world by increasing activities at national, regional, and global levels.

Actors such as governments, nongovernmental organizations (NGOs), private industry, academics, and youth each have their role to play. The Decade proposes activities that should be implemented in accordance with five pillars: road safety management; safer roads and mobility; safer vehicles; safer road users; and post-crash response. In Table 24.3, we outline different activities that each of these actors could undertake in order to improve road safety for teen and novice drivers, with special relevance to LMIC settings. Several countries have invested in road safety interventions, and the international road safety community has worked hard to successfully get road safety included in the newly agreed-upon global Sustainable Development Goals (SDG). Other Decade of Action for Road Safety successes include the establishment of the World Day of Remembrance for Road Traffic Victims (the third Sunday in November each year), the launch of the 2013 *Global Status Report on Road Safety* (referenced throughout this chapter), the organization of United Nations Road Safety Week, and the launch of the Global Alliance for Care of the Injured. More work must be done.

24.7 Conclusions

RTIs are the leading cause of death among people aged 15–29 years worldwide, and without concerted global action, this will likely continue. RTIs are preventable, however, and governments, NGOs, the private sector, academics, and youth must work together to get teen and novice drivers to adopt safer road safety practices, especially in LMICs. Activities should focus on normalizing seat belt and motorcycle helmet use, obeying posted speed limits, discouraging consumption of alcohol and other drugs, and avoiding mobile phones while driving. GDL programs should be taken up across the globe. Additionally, these same actors must also address road design and maintenance, vehicle safety, and post-crash care. A systems approach to road safety represents an opportunity to reduce a large number of crashes, injuries, deaths, and disability among teen and novice drivers, and other road users, worldwide.

Acknowledgments

The authors would like to acknowledge the necessary and often thankless work of road safety professionals and organizations the world over—and especially in LMICs. Now, more than ever, we need you.

References

Bachani, A. M., Branchini, C., Ear, C., Roehler, D. R., Parker, E. M., Tum, S., Ballesteros, M. F., & Hyder, A. A. (2013a). Trends in prevalence, knowledge, attitudes and practices of helmet use in Cambodia: Results from a two-year study. *Injury*, 44(4 Suppl), S24–S30. doi: 10.1016/ S0020-1383(13)70210-9.

Bachani, A. M., Hung, Y. W., Mogere, S., Akungah, D., Nyamari, J., Stevens, K. A., & Hyder, A. A. (2013b). Prevalence, knowledge, attitude and practice of speeding in two districts in Kenya: Thika and Naivasha. *Injury*, 44(4 Suppl), S24–S30. doi: 10.1016/S0020-1383(13)70209-2.

Bachani, A. M., Jessani, N. S., Pham, C. V., Quang, L. N., Nguyen, P. N., Passmore, J., & Hyder A. A. (2013c). Drinking & driving in Viet Nam: Prevalence, knowledge, attitudes, and practices in two provinces. *Injury*, 44(4 Suppl), S38–S44. doi: 10.1016/S0020-1383(13)70211-0.

Bishai, D., Asilmwe, B., Abbas, S., Hyder, A. A., & Bazeyo, W. (2008). Cost-effectiveness of traffic enforcement: Case study from Uganda. *Injury Prevention*, 14, 223–227. doi: 10.1135/ip.2008.018341.

Brodsky, W., & Slor, Z. (2013). Background music as a risk factor for distraction among young-novice drivers. *Accident Analysis & Prevention*, 59, 382–393. doi: 10.1016/j.aap.2014.06.022.

Carter, P. M., Bingham, C. R., Zakrajsek, J. S., Shope, J. T., & Sayer, T. B. (2014). Social norms and risk perception: Predictors of distracted driving behavior among novice adolescent drivers. *Journal of Adolescent Health*, 54(5 Suppl), S32–S41. doi: 10.1016/j.jadohealth.2014.01.008.

Chapman, E. A., Masten, S. V., & Browning, K. K. (2014). Crash and traffic violation rates before and after licensure for novice California drivers subject to different driver licensing requirements. *Journal of Safety Research*, 50, 125–139. doi: 10.1016/j.jsr.2014.05.005.

Compton, R. P., Blomberg, R. D., Moskowitz, H., Burns, M., Peck, R. C., & Fiorentino, D. D. (2002). Crash risk of alcohol impaired driving. In: D. R. Mayhew & C. Dussault (Eds.), *Proceedings of the 16th International Conference on Alcohol, Drugs and Traffic Safety*, Montreal, 4–9 August 2002. Montreal, Société de l'assurance automobile du Québec, 2002: 39–44.

Elvik, R., & Vaa, T. (2004). *The Handbook of Road Safety Measures*. Amsterdam: Elsevier Science.

Ferdinand, A. O., Menchemi, N., Blackburn, J. L., Sen, B., Nelson, L., & Morrisey, M. (2015). The impact of texting bans on motor vehicle crash-related hospitalizations. *American Journal of Public Health*, 105(5), 859–865. doi: 10.2105/AJPH.2014.302537.

Hurst, P. M., Harte, D., & Frith, W. J. (1994). The Grand Rapids dip revisited. *Accident Analysis & Prevention*, 26, 657–654.

Hyder, A. A., Puvanachandra, P., & Morrow, R. H. (2012). Measuring the health of populations: Explaining composite indicators. *Journal of Public Health Research*, 1, 3, 222–228. doi: 10.4081/jphr.2012.e35.

Hyder, A. A., Puvanachandra, P., & Allen, K. A. (2013). *Road traffic injury and trauma care: Innovations for policy (road trip)*. World Innovation Summit for Health Forum Report. Available from http://wish-qatar.org/summit/inaugural-summit/reports/forum-reports.

Liu, B., Ivers, R., Norton, R., Blows, S., & Lo, S. K. (2005). Helmets for preventing injury in motorcycle riders. *The Cochrane Database of Systematic Reviews*, 5.

Lucidi, F., Giannini, A. M., Sgalla, R., Mallia, L., Devoto, A., & Reichmann, S. (2010). Young novice driver subtypes: Relationship to driving violations, errors, and lapses. *Accident Analysis & Prevention*, 42(6), 1689–1696. doi: 10.1016/j.aap.2010.04.008.

McKnight, A. J., & McKnight, A. S. (2003). Young novice drivers: Careless or clueless? *Accident Analysis & Prevention*, 35(6), 921–925.

National Highway Traffic Safety Administration. (2006). *The Impact of Driver Inattention on Near-Crash/Crash Risk: An Analysis Using the 100-Car Naturalistic Driving Study Data*. Washington, DC: US Department of Transportation.

Ouellete, J. A., Gerrard, M., Gibbons, F. X., & Reis-Bergan, M. (1999). Parents, peers, & proto-types: Antecedents of adolescent alcohol expectancies, alcohol consumption, and alcohol-related life problems and rural youth. *Psychology of Addictive Behaviors*, 13, 183–197.

Peden, M., & Hyder, A. A. (2002). Road-traffic injuries are a global public health problem. *BMJ*, 324(7346), 1153.

Peden, M., Scurfield, R., Sleet, D., Mohan, D., Hyder, A. A., Jarawan. E., & Mathers, C. (Eds.). (2004). *World Report on Road Traffic Injury Prevention*. Geneva: World Health Organization.

Pérez-Núñez, R., Pelcastre-Villafuerte, B., Híjar, M., Avila-Burgos, L., & Celis, A. (2012). A qualitative approach to the intangible cost of road traffic injuries. *International Journal of Injury Control and Safety Promotion*, 19(1), 69–79. doi: 10.1080/17457300.2011.603155.

Romer, D., Lee, Y. C., McDonald, C. C., & Winston, F. K. (2014). Adolescence, attention allocation, and driving safety. *Journal of Adolescent Health*, 54(5 Suppl), S6–S15. doi: 10.1016/j.jadohealth.2014.10.202.

Russell, K. F., Vandermeer, B., & Hartling, L. (2011). Graduated driver licensing for reducing motor vehicle crashes among young drivers. *Cochrane Database Systematic Review*, 10:CD003300. doi: 10.1002/14651858.

Scott-Parker, B., Hyde, M. K., Watson, B., & King, M. J. (2013a). Speeding by young novice drivers. What can personal characteristics and psychosocial theory add to our understanding? *Accident Analysis & Prevention*, 50, 242–250. doi: 10.1016./j.aap.2012.04.010.

Scott-Parker, B., Watson, B., King, M. J., & Hyde, M. K. (2013b). Revisiting the concept of the "problem young driver" within the context of the "young driver problem": Who are they? *Accident Analysis & Prevention*, 59, 114–152. doi: 10.1016/j.aap.2013.05.009.

Simons-Morton, B., Lerner, N., & Singer, J. (2005). The observed effects of teenage passengers on the risky driving behaviors of teenage drivers. *Accident Analysis & Prevention*, 37(5), 973–982.

Simons-Morton, B. G., Ouimet, M. C., Zhang, Z., Klauer, S. E., Lee, S. E., Wang, J., Chen, R., Albert, P., & Dingus, T. A. (2011). The effect of passengers and risk-taking friends on risky driving and crashes/near crashes among novice teenagers. *Journal of Adolescent Health*, 49(6), 587–593. doi: 10.1016/j.jadohealth.2011.02.009.

Toroyan, T., & Peden, M. (Eds.). (2007). *Youth and Road Safety*. Geneva: World Health Organization.

Vecino-Ortiz, A., Bishai, D., Chandran, A., Bhalla, K., Bachani, A. M., Gupta, S., & Hyder, A. A. (2014). Seatbelt wearing rates in middle income countries: A cross-country analysis. *Accident Analysis & Prevention*, 71, 115–119. doi: 10.1016/j.aap.2014.04.020.

Vera-López, J. D., Pérez-Núñez, R., Híjar, M., Hidalgo-Solórzano, E., Lunnen, J. C., Chandran, A., & Hyder, A. A. (2013). Distracted driving: Mobile phone use while driving in three Mexican cities. *Injury Prevention*, 19(4), 276–279. doi: 10.1136/injuryprev-2012-040496.

Vlakveld, W. (2005). *Young, Novice Motorists, Their Crash Rates, and Measures to Reduce Them: A Literature Study (SWOV-rapport R-2005-3)*. Leidschendam, The Netherlands: SWOV (Institute for Road Safety Research).

World Health Organization. (2009). *Global Status Report on Road Safety: Time for Action*. Geneva: World Health Organization.

World Health Organization. (2011). *Mobile Phone Use: A Growing Problem of Driver Distraction*. Geneva: World Health Organization.

World Health Organization. (2013). *Global Status Report on Road Safety 2013: Supporting a Decade of Action*. Geneva: World Health Organization.

World Health Organization. (2014). *Global Health Estimates 2014 Summary Tables: Deaths by Cause, Age and Sex, by WHO Region, 2000–2012*. Geneva: World Health Organization.

Zador, P. L. (1991). Alcohol-associated relative risk of fatal driver injuries in relation to driver age and sex. *Journal of Studies of Alcohol*, 52, 302–310.

Zhu, M., Cummings, P., Chu, H., & Cook, L. J. (2007). Association of rear seat safety belt use with death in a traffic crash: A matched cohort study. *Injury Prevention*, 13, 183–185.

VII

Methods

25

Driving Simulators and the Teen Driver: Models, Methods, and Measurement

Donald L. Fisher

Michael G. Lenné

Lana M. Trick

William J. Horrey

Abstract

The Chapter. Driving simulators have come down radically in cost and ease of operation, some requiring nothing more than a stand-alone tablet or head-mounted display. Correspondingly, their use has expanded greatly, in terms of research, testing, and training. Researchers wanting to study teen and novice drivers using simulators and practitioners wanting to apply them can sometimes feel overwhelmed in deciding which simulator to purchase for a particular function. We hope to make this decision easier by giving readers interested in teen and novice driver behaviors, and those largely unfamiliar with driving simulators as a way of studying such behaviors, a sense of the following: the range of driving simulators now available; some of the most critical issues to consider when deciding what type of driving simulator to purchase; the different dependent variables that can be measured, including both vehicle and driver behaviors; the problems created by simulator sickness and how to mitigate these problems; and a range of special considerations. ***Limitations and Recommendations.*** (1) We do not cover here the important question of just how well training drivers on a simulator transfers to the real world. Those issues are covered in Chapter 21. (2) The scope of this chapter does not afford ample opportunity to delve deeply into some of the issues that are raised. As such, we aim to cover those topics in just enough depth to give the reader some appreciation for the range of design issues and other considerations. (3) There are many other ways of testing and training teen and novice drivers in addition to driving simulators. Some of these methods are covered in this handbook (e.g., hazard anticipation tests, Chapter 28; naturalistic studies, Chapter 26; in-vehicle feedback, Chapter 20); others are not and should be considered as well.

25.1 Introduction

If you are reading this chapter, there is a good chance that you already have a particular use for a driving simulator in mind, either as way to evaluate some aspect of teen and novice driver behavior or as a way to train some driving skill. In some cases, a driving simulator is used only to create a virtual world; in other cases, the simulator affords a fully interactive experience where the driver controls all or most aspects of the vehicle. As a tool, driving simulators can be readily applied in the evaluation of drivers as well as in the training of drivers. Simulators are especially important in research on teen and novice drivers because these drivers are already at a higher risk of collision than other drivers. Simulators permit their performance to be assessed in the challenging situations associated with collisions without putting young drivers or other road users at risk. There are a variety of other ways that simulators can be used with teen and novice drivers though.

First, consider the use of driving simulators in training. Many training programs now exist that use simulations of the virtual world to present drivers with scenarios that target the teaching of higher-order driving skills such as hazard anticipation, hazard mitigation, and attention maintenance (Chapters 6 through 8 and 18). The simulations are used because they can capture scenarios that, in the real world, are either too dangerous to drive or too expensive to stage. Further, simulation also affords repeatable presentation of environments designed to develop specific skills in novice drivers while also supporting more objective trainee assessment—often a limitation in many training programs. As driving simulators become less expensive, they are being deployed much more widely with the full set of controls that allow participants to choose their own path through the virtual world. For example, over 150 simulators are now in use in the Netherlands for driver training (SWOV, 2010).

Second, consider the use of driving simulators in evaluation. In terms of practical application, this could soon include their use at licensure for both the theory test (Chapter 28) and the field drive (Mayhew et al., 2011). On the theoretical side, this includes evaluations of the differences in the behaviors of novice and experienced drivers (Chapters 5 through 8), along with the differences in various subpopulations of teen and novice drivers, such as those diagnosed with attention deficit hyperactivity disorder (ADHD) and those who are neurotypical (Chapter 14). It also includes evaluations of the effects of fatigue (Chapter 15), distraction (Chapter 12), and alcohol and THC (Chapter 13) on the performance of teen and novice drivers, just to list a few of the factors that have been studied.

The discussion so far provides a very short list of the ways that driving simulators can be used. More comprehensive treatments are available elsewhere (e.g., Fisher et al., 2011; Caird & Horrey, 2011). The goal here is not to describe these uses exhaustively but, rather, to underscore the importance of experimental control when devising the scenes and scenarios that are used in training and evaluation. In this chapter, we briefly describe some issues that are generic to simulation research, generally including what it is developers can and cannot control when it comes to devising scenes and scenarios, noting of course that these requirements will differ markedly between training and evaluation purposes. In fact, scene and scenario control is one of the most critical factors when it comes to simulator selection. We then go on to discuss several of the different simulators and how they vary in terms of the developers' ability to control what the driver sees in the scenes and scenarios, along with some of the dependent measures that can be employed. We conclude the chapter with a discussion of several important considerations that come hand in hand with driving simulator applications, including scenario design considerations (especially related to hazards and crashes), the problems caused by simulator sickness, and other issues such as distributed simulators and participant recruitment.

We should note at the outset that we will only be discussing simulators that are used in research, because for these, the virtual worlds can be modified more easily. In contrast, in some simulators, the worlds cannot be easily modified. These simulators are typically more straightforward to operate and are typically used for training; they allow novice drivers to get a feel for what is safe and what will result in a crash, and they provide drivers with specific set scenarios. Such simulators also have been used in driving-under-the-influence/driving-while-intoxicated prevention programs as well as in texting-and-driving prevention programs. Examples include the simulators sold by Drive Square (2015) and Mechanical Simulation (2015). Throughout this document, we mention names only for the readers' convenience; no endorsement of the products is implied or intended.

25.2 Types of Simulators

There are potentially almost as many different levels of research simulators as there are simulators themselves. In many cases, cost will be an important factor in the decision, with other factors such as system requirements for the research objectives often a secondary consideration. The costs are ever changing, but as a rough estimate, full-scale simulators can cost as little as $100,000. Desktop simulators can cost as little as $10,000. Portable simulators now include smartphones, tablets, and head-mounted simulators (HMSs). They can cost as little as $1,000. Finally, augmented-reality (AR) simulators are still under development for the most part, and the cost is hard to determine; nonetheless, they warrant brief description here.

25.2.1 Full Scale

Full-scale simulators are defined as ones that give the driver an immersive experience and let the driver maintain the illusion that he or she is sitting in an actual vehicle. Drivers in such simulators have access to nearly the same amount of total visual information as in the real world. The driver navigates through that world either in an actual car in which the participant sits or in a built-up cab (BUC), which may contain as little as a driver's seat, foot pedals for the brake and accelerator, a steering wheel, and a full or partial instrumentation cluster (e.g., speedometer). Full-scale simulators can cost tens of millions of dollars if the car is an actual one, the car employs 6 degrees of freedom (DOF) of motion (x, y, z, pitch, roll, and yaw), and the virtual world is projected on floor-to-ceiling screens that surround the driver's vehicle. Globally, there are many leading examples of such simulators, including the National Advanced Driving Simulator (NADS [The University of Iowa, 2015]); Sim IV, located at VTI (Swedish National Road and Transport Research Institute) in Gothenburg and TNO's (Netherlands Organisation for Applied Scientific Research) driving simulator in Soesterberg, the Netherlands; and Toyota Motor Corporation's driving simulator at the Higashifuji Technical Center located in Susono City (McNamara, 2009). Given current prices, it is possible now to buy or construct 6-DOF simulators for much less than $1 million and motion-based simulators that use gaming bases with racing seats and wheel and pedal controls for less than $100,000. These developments might particularly appeal to a novice participant group and help with psychological immersion in the scenario.

Alternatively, much more affordable full-scale simulators are available if one wants the participants to have an immersive experience but is willing to sacrifice motion and an actual car for a driving cockpit (also called a built up cab or BUC). It increases the sense of realism considerably when the BUC vibrates in ways similar to an actual car moving over a roadway (Plouzeau et al., 2013). Motion is not necessary for the study of a great many different behaviors (Greenberg, 2011; although for a counterexample, see Salmon et al., 2011, with touchscreen interfaces). If it is important for a researcher to have complete control over the virtual worlds one builds, i.e., over the scenes and scenarios that comprise the virtual world, then the driving simulator is likely to cost at least $100,000. Examples include the simulators manufactured by Drive Safety (2014), Oktal (2015), and Realtime Technologies, Inc. (RTI) (2015). (In fact, some simulators, the most recent versions allow users to upload virtual worlds built with a standard computer-aided design [CAD] package like Civil 3D, though this can be a time-consuming process.) There are a variety of different types of simulator, though, and for a broader survey, see the work of Slob (2008).

However, if one is willing to give up some of the control one has over scenes and scenarios, this can lower the cost to half of the above mentioned $100 thousand, and even less sometimes, and the ease of programming a given scenario often increases by an order of magnitude. The "cost" here is the potential reduction in simulator functionality and the reduced scope of research questions that can be addressed. Examples include simulators manufactured by Systems Technology, Inc. (STI) (2013). Building a road geometry that is not in the database requires going to the manufacturer, and even if one does that, the control of other vehicles in the scenario is typically limited. For example, in some simulators, it is difficult to have another vehicle in an opposing lane turn across the front of the participant's vehicle at the precise time that the participant's vehicle is approaching the intersection. Roundabouts are a problem with some simulators. Further, there are often no intersections where the crossroads have horizontal curvatures embedded in them. Nonetheless, the advantage of these types of simulator is that one can construct complex scenes and scenarios in a fraction of the time that it takes to do so using more sophisticated software, which gives the researcher good, though not complete, control of design.

25.2.2 Desktop

Desktop simulators involve computer monitors (often three) that sit on a desk and a steering wheel mounted to the desk. Some also have foot pedals under the desk or gaming controls for braking and speed control. Interestingly, control over the virtual environment is not necessarily sacrificed when moving to a desktop simulator. Both RTI and NADS manufacture simulators that provide researchers full control over the creation of scenes and scenarios that comprise a virtual world. The advantage of using the NADS desktop driving simulator (the MiniSim) is that code used to create the virtual world can be uploaded for later use on the advanced simulator at the University of Iowa. However, there are desktop simulators without full scene and scenario functionality that have most other core functions. Primary among these is the aforementioned STI.

25.2.3 Portable

Although some vehicle cockpit (BUC) configurations are arguably portable, we consider here three types of truly portable simulators that have recently become available: tablets, smartphones, and HMSs. Perhaps the flashiest are the recently introduced HMSs, including the Oculus Rift (Strater, 2014) and FOVE (de Looper, 2015), which allow a true 3-D experience. HMSs have been available for some time (Romoser et al., 2005) but were associated with high rates of simulator sickness (Stoner et al., 2011) and thus not suitable for research applications. This is not true of the current HMSs, at least the ones that we have tested. The virtual reality experience delivered by the new HMSs is three dimensional because slightly different images are delivered to the right and left eyes.

Tablets and smartphones are not as flashy, but they are certainly more widely available. In some training applications, participants can themselves control steering (by gently tilting the tablet from 0° to 90°), acceleration (by toggling a button upward), braking (by toggling the same button downward), and scanning (by toggling a button toward the left and/or right) the roadway through 180° (Krishnan et al., 2015). The virtual environment for such nomadic devices is often created using Shockwave 3-D (Director), a language used for many games. Suitable virtual environments can also now be created with Flash 3-D, which does not require high-end graphics (gaming) cards and can be published on iOS and Android devices (note that Director 12 also publishes to iOS).

25.2.4 Augmented Reality

Head-mounted displays (HMDs) not only can be used while sitting still to project a virtual world in three dimensions (as in Section 27.2.3), but they can also be worn while driving an actual vehicle to overlay computer-generated information on the view of the roadway ahead or enhancements of existing objects (Davies, 2015). One of the keys is to ensure that the virtual content that is displayed aligns with and is anchored to elements in the real world as your head moves freely within the cabin of the automobile. The AR optics can also be used to display directions on the road, making it unnecessary to look at your GPS and see whether this is the exact left turn that must be made in order to get where you are going or it is the one 10 yards downstream—something that can be confusing when relying on only verbal directions. Potential collisions can also easily be made apparent. For example, it could be made clear to a driver when a light turned green that a car was running the red light and might pose a threat. It is unclear when this technology will be available to the research community, but it will revolutionize the world of simulation in terms of its uses in training and evaluation, albeit with a necessary consideration of unintended consequences, such as attention capture and tunneling.

25.2.5 What Type of Simulator Is the Right One for You?

The decision regarding what type of simulator to choose will likely be largely determined by space and budgetary concerns. For larger or more sophisticated setups, additional constraints and considerations might factor into the decision as well. For example, building infrastructure can play a role, including reinforced floors, ceiling clearances, vibration and noise pollution, available circuits, air-conditioning needs, as well as many others. It is important to note that hardware and software costs make up only one part of the expenditure for a simulator facility: experienced staff or students are a necessary part of a successful and productive laboratory.

More pertinently for the current chapter, what types of things should one consider for novice and teen driver applications? Obviously, the specific areas of interest (e.g., intersection safety, training) will factor heavily in the decision-making process. That is, it is important to match system specifications with research or training needs. For example, if one were interested in studying teen driver behavior at intersections, a wide field of view might be necessary to more fully assess visual scanning behaviors on approach to the intersection. Two seats would be required to study passenger influences in a simulator, and so on. In addition to considering research objectives, the needs and expectation of the participant groups are also worthy of consideration. For example, younger generations are more tech savvy, and consequently, full cabs, motion, and great visuals are likely more important to achieve psychological immersion with this group.

25.2.6 Performance Measurement

Although the physical layout and fidelity of a prospective simulator are important, it is also important to consider what types of measures can be derived from the system. It is important to consider not only variables collected by the simulator (vehicle behaviors) but other variables that can be collected by eye tracking (Chapter 27) and physiological monitors as well. The dependent variables one can collect shape and constrain the types of studies that can be conducted. Thus, it is important to consider the range of measures needed to test a study hypothesis before making a decision on which simulator to use.

Readers are presumably familiar with the many vehicle behaviors one can gather, including speed, acceleration, position with respect to lane center and the edge line, time and distance headway, and brake pressure, just to name a few. These are typically sampled 30 times a second in most driving simulators but typically aggregated over time or driving distance for analysis. Data are either recorded automatically as part of a participant's drive or selectively recorded depending on the researcher's preferences. Some of the most commonly used measures in research on teen drivers include driving speed and speed variability, braking response time to hazards (hazard RT), and the standard deviation of lateral position (SDLP), which measures the standard deviation of the distance between the driver's vehicle and the center of the lane (a good measure of steering or lane-keeping performance).

Driver behaviors can also be captured with cameras and include the foot position (e.g., on the accelerator, hovering somewhere between the accelerator and brake), hand position (e.g., no, one, or two hands on the steering wheel; holding a smartphone), and head position (which can be captured either directly or indirectly by the forward-facing camera on a head-mounted eye tracker, by remote eye-tracking or camera systems, or through integrated motion capture systems). As is clear from this brief list, some driver behaviors do not require anything in the way of exotic equipment to capture. The trade-off, in some cases, is in the amount of postprocessing of data or coding that is required to bring the data into a useable form. Head- or body-mounted cameras cost typically on the order of tens of dollars to several hundred dollars.

Also, the simulator environment oftentimes presents an ideal environment for the integration of supplementary measurement systems, such as eye trackers (head mounted or remote, integrated with motion trackers or not), electroencephalography (EEG), electrocardiography (ECG), galvanic skin conductance, near-infrared spectroscopy (NIRS), etc. However, the integration and synchronization of multiple systems is not always trivial.

25.3 The Virtual World: Scenes and Scenarios

We begin this section by making a distinction that we think clarifies an important part of the discussion around the functionality of simulators: the distinction between scenes and scenarios. A scene is generally defined as the world without any of the moving elements incorporated into it—essentially, the road and road infrastructure and the built and natural landscapes. This encompasses objects that change systematically (a priori) over time as a drive plays out (e.g., traffic signals, variable message signs), but it does not include any contingent changes, which are dependent on the location or timing of some other event (e.g., a traffic signal that changes to red as the driver approaches within 100 yards of it). In contrast, scenarios are defined as the complete set of static and moving objects along with the various contingencies that change the movement or appearance of those objects in the virtual world—these are the elements that contribute to a dynamic and interactive user experience. So, for example, sometimes a researcher will want a constant time headway or distance headway between a participant's vehicle and a lead vehicle. The contingency here is between the lead vehicle and participant's vehicle's speed and resulting headway distance.

25.3.1 Scenes

There are three things about the development of scenes that need to be understood. First, many simulator programs come with premade tiles. Tiles are typically square (e.g., 1000 feet by 1000 feet) and are relatively bare except for the roadway geometry. So, for example, one tile might contain a T-intersection with a stop sign at the top of the *T*. Another tile might be a four-way signalized intersection, again with a very specific set of lanes. It often seems like the choices are limitless when one first looks at the tiles that come with a particular simulator. Road geometry notwithstanding, many simulator platforms offer the capability of adding environmental elements to the scenes, such as vegetation, buildings, and other static elements. This affordance can expand the diversity of the environments—even when using the same or similar base tiles. This being said, unless one can design a wide array of road geometries and pavement markings (e.g., number of lanes, placement and angle of cross streets, location of Jersey barriers, layout of roundabouts, insertion of bicycle lanes, and so on), one can easily reach the limitations of preexisting libraries of tiles. This problem can generally be avoided if the simulator software can read CAD programs such as Civil 3D (Autodesk, 2015), which support design without constraints and different roadway geometries. This problem is especially likely to arise when studying novice drivers' behavior at crash hot spots (areas of high risk), particularly intersections. These differences in details about the exact geometry of intersections cannot necessarily be captured using generic tiles. The ability to model the geometry of an intersection precisely is a tremendous advantage for our understanding of why novice drivers are at risk in a given intersection environment.

Second, there is a limitation that is so important even though it is only encountered occasionally: some types of research question require that the simulation developer construct objects (e.g., vehicles, pedestrians, and the built environment) and surfaces (e.g., pavement markings, sign text, and figures) and locate them at specific locations on the road or in the environment. For example, in a study of the effect of advanced yield markings on novice driver's ability to anticipate a latent hazard (pedestrian) (e.g., Fisher & Garay-Vega, 2012), there would need to be a way to create in the scene a truck stopped in a parking lane in front of a marked midblock crosswalk, obscuring a pedestrian who might be crossing from the right. The developer would also need to be able to design the advance yield markings and to place them at the correct distance upstream of the crosswalk in order to study their effect. Depending upon the flexibility of the simulator software, this could potentially be achieved by an experienced simulator programmer.

Finally, one needs to be concerned about the quality of the visual image that is presented. A sign that subtends a given visual angle is usually legible at a much farther distance in the real world than it is in the simulator (Andersen, 2011). Glare is almost impossible to render (Chrysler & Nelson, 2011), and nighttime conditions are equally difficult to duplicate realistically. The latter could be especially important if one were interested in the study of fatigue (Chapter 15). If signs in the simulator do not have the retroreflectivity or legibility that they have in the real world, then the differences in behavior one finds between teen drivers who are fatigued and those who are not fatigued might be entirely artifactual. The quality of the visual image might also influence the complexity of the surrounding road and roadside environments that can be used.

25.3.2 Scenarios

Most users may find, even after having read the aforementioned, that their particular use of a driving simulator to study teen and novice drivers does not require much in addition to what the manufacturer offers when it comes to scene generation capabilities. However, in more sophisticated laboratories or for more complex research questions, users often have a very profound need for flexible scenario development capabilities. There are three primary capabilities that someone purchasing a new driving simulator for research uses should consider when it comes to the creation of scenarios (for a more detailed discussion, see Kearney & Grechkin, 2011). An assumption here is that researchers will seek to carefully create a repeatable participant experience, which necessitates that dynamic entities in the simulated environment can be carefully controlled.

First, it is absolutely essential in most studies to be able to control the trajectory and timing of the movement of other vehicles, road users, and entities around the participant's vehicle, typically triggered by the speed, location, and even timing of the participant's vehicle. So, for example, a line of parked cars may appear ahead of the lead vehicle. A researcher may wonder whether a novice driver notices the fact that one of the parked vehicles backs up and then starts pulling out into the driver's travel lane (not enough to cause a crash but enough to encourage slowing down and possibly steering to the lane center). Thus, a researcher must be able to determine where and when (in relation to the participant's vehicle) the lead vehicle starts pulling out, what path it takes, where it starts, and also the corresponding responses made by the driver. More elaborate scenarios, in terms of dynamics and constraints, can rapidly explode the amount of care and caution needed in planning and programming the scenarios (not to mention the need for a lot of testing).

Second, you will want to make sure that your scenario development tools allow you to control tightly when and how objects in the environment change states as a function of one's own location and speed as well as those of other vehicles. For example, consider a signalized intersection. You may want the signal to change from red to green as the novice driver gets close to the intersection, hoping to determine whether the novice driver glances to the sides for vehicles that might be running a red light. This requires that you be able to maintain a green signal for as long as it takes the participant driver to approach it and then have it change to red to then observe the hypothesized glance behavior. Often, this can require creation of special traffic signals where the signal phasing can be controlled. The standard signals may not allow you to adjust the timing of the signal in the fashion needed. And this, in turn, requires you to be able to have full control over the creation of the objects in your scenes.

Third, you will want to be able to control a range of environmental and vehicle factors such as the visibility (fog, rain), time of day, and level of traffic (the ambient vehicles), along with the headlights and turn signals of other vehicles—all of which might have some bearing on the underlying research questions or which might relate to the desired training application.

25.4 Design of Driving Simulator Studies: Issues and Considerations

The number of peer-reviewed articles on young drivers has grown from an average of three annually back in the 1960s and 1970s to an average of more than 50 annually in the last 5 years (Chapter 23). The growth in driving simulators suggests that the percentage of these studies that use a driving simulator is also steadily increasing. Certainly, enough is known to provide researchers interested in using driving simulators to learn more about teen and novice drivers with some generally accepted good practices.

Not surprisingly, the treatment of special populations in a simulator study, in terms of recruitment and screening strategies, handling, and selection of measures, bears some consideration (e.g., ADHD, nonneurotypical drivers, fatigued drivers, those driving under the influence of alcohol or drugs, distracted drivers, drivers in company of passengers). In some cases, the fidelity as well as the physical setup of a particular simulator can impact how feasible certain studies are (e.g., use of a full cab for studies of passenger effects versus a steering mock-up). These provisions notwithstanding, the use of simulators in such studies can provide a very valuable resource for expanding our knowledge and understanding of the risks faced by novice and teen drivers.

There are a number of special considerations that bear keeping in mind when conducting experiments on a driving simulator with teens and novice drivers. We address what we think are the most important in the next sections. These include discussions of simulator sickness and methods to reduce it; of how best to study crashes and near crashes; how to run distributed, asynchronous simulation; and of various different recruitment methods.

25.4.1 Simulator Adaptation Syndrome

Some people experience simulator adaptation syndrome (SAS) when they are in a driving simulator. SAS is also known as simulator sickness: symptoms of this syndrome include disorientation, dizziness, eye strain, stomach awareness, and in severe cases, nausea. Participants experience symptoms to varying degrees, from a minor feeling of unease to full-blown nausea, but if a participant is feeling uncomfortable, it is important that the simulation be stopped and the participant be cared for. At this point, it is not clear why, but teens do not experience the symptoms of SAS as strongly as other groups; fewer have to be dropped due to SAS. In contrast, in older drivers, particularly ones over the age of 65, up to 40% of the participants may become so uncomfortable that the simulation has to be halted (Allen & Reimer, 2006; Romoser & Fisher, 2009). Teens may get SAS too, and though the rates are typically lower—depending on the conditions in the drive—they can be up to 30% (Taylor et al., 2014).

Researchers using simulators should be trained to recognize the early symptoms of SAS so that they can take the appropriate measures to make the symptoms abate (Kennedy et al., 1987). Continuous monitoring is necessary to ensure that a simulation is halted before the participant becomes physically ill. However, there are measures that can be taken to prevent SAS or at least reduce its likelihood. For example, some people are more likely to experience SAS than others, and there are tests that can be used to identify and screen out the ones at risk. These tests are not perfect, but they at least identify those at the most risk. Furthermore, there are things to consider about the design of the simulation. SAS is more likely to occur if the driver is required to make large numbers of abrupt turns and stops within a short period of time. As well, participants should be given a warm-up drive so that challenges such as these are introduced gradually, and the drives should be kept to a relatively short period, at least at first. Lowering the ambient temperature of a simulator laboratory can also decrease symptoms of SAS. When using a driving simulator, it is important to monitor the degree to which participants feel symptoms, and this can be done in runtime as well as with posttrial questionnaires. Thus, it is important to consider SAS when designing a study. A more thorough discussion of this issue can found elsewhere (e.g., Stoner et al., 2011).

25.4.2 Driver Sensitization to the Conditions Associated with Crashes

Many investigators will want to study crashes in a driving simulator. After all, that is what often attracts researchers to simulators as platforms for studying driving behavior: they present no (or minimal) risk while offering precise control over the scenarios. But there is an obvious problem when it comes to studying the effect of some factor on crash risk. As soon as a driver crashes, he or she may become hypersensitive and behave in ways in the driving simulator that he or she would not have in the real world (this can manifest as an adverse reaction to the crash event or in terms of the participant's realization of the determination of the demand characteristics of a particular study). It is important to avoid sensitizing the driver to hazards because they encounter too many in a short period of time; a second caution is to make sure that the participant does not become oversensitized to the appearance of hazards (Ranney, 2011).

There are several ways to reduce the problems that come about through hypersensitivity. First, a study could have one and only one crash scenario placed at the very end of an experiment (or sprinkle suitably few crash scenarios throughout the experiment). However, the data one can gather here are typically limited to one observation per participant, often way too few to achieve the power that one needs. A related possibility, used by several investigators, is to allow crashes to occur but to give no indication to the participant that such has occurred or to make their occurrence somewhat ambiguous (e.g., allow the objects to pass through one another with no disruption to the control dynamics or motion, have a colliding vehicle immediately disappear). The specifics of such an approach would need to be tailored according to the needs of the particular application.

Second, in order to reduce hypersensitivity to crashes, it is possible to study near crashes. This has been the tactic used in naturalistic studies. Specifically, near crashes are 10–15 times more likely in naturalistic studies than are crashes (Klauer et al., 2006). Given the relatively small number of crashes in such studies, it became of interest to determine whether the factors that led to near crashes were the same as the factors that led to crashes. The conclusion is that much can be learned about crashes by studying near crashes (Guo et al., 2010).

Third, one could use other surrogate measures of crash likelihood, measures that have been linked to crashes. These include the standard deviation of lane position (Ghosh et al., 2012) and acceleration noise (Boonspiripant, 2009).

Fourth, perhaps the most often-used indices of the likelihood of a crash are behaviors that are indicative of a failure to anticipate a hazard, mitigate a hazard, or maintain attention. These are all covert methods because an actual hazard is never materialized, although these behaviors have been associated with crashes (e.g., Chapter 6). It is important to note that studies of hazard anticipation call for careful thought and planning as well, in terms of coordinating the different aspects of the scenario.

25.4.3 Distributed, Asynchronous Simulation

Researchers will often want to study the effects of interventions on similar populations of drivers as tested at different locations (e.g., in different states or countries) at different points in time. In today's environment, this requires not only that several locations have the same model of driving simulator but also often that special precautions be taken in the development of the scenes and scenarios. Typical of the problems that can occur are ones with the representation of different objects that can move throughout the environment. One site might have a particular need to focus on commercial vehicles. Thus, the pointer to vehicle #52 might bring up a Volvo tractor trailer. However, another installation might have a particular need to target SUVs. Thus, the pointer to vehicle #52 might be a particular model of Subaru Forester. The computing skill required to execute successfully distributed simulation can be quite extensive for some driving simulators. For other models of driving simulators, there are no problems whatsoever. Researchers with an interest in distributed, asynchronous simulation should speak at length with a company's technical staff and others using the company's simulators to make sure that what seems very simple on the surface is actually doable. A related feature afforded by these simulations is the ability to study interactions between road users, and between road user groups, in more naturalistic ways. That is, rather than interacting with the computer-generated behavior of another vehicle, the interaction contains more realistic behaviors.

25.4.4 Recruitment

Recruitment of teen and novice drivers can be difficult. We wish we could provide our readers with a simple, inexpensive way to do such. For studies requiring relatively few participants (100 or fewer), we have used driving schools extensively. They can be incredibly helpful because, at least in some states, the teens will use the driving school for the on-the-road exam. Thus, the driving school will be able to provide participants who have just received their restricted license, participants who are still in the learner's permit stage, and participants who have had no on-the-road experience. For larger studies, it is almost a requirement that one be able to involve some licensing agency. So, for example, for one study of the effects of a training program on novice drivers, the training program was made available at the registry of motor vehicles when teens came to be licensed (Thomas et al., 2016). Generally, appropriate cash payments and prize draws can enhance the appeal of such studies for the teen driver audience.

25.4.5 Community of Users

We believe that it is important to have as broad a community of users as possible who are familiar with the driving simulator. This makes possible the sharing of information, either individually or through manufacturer-based user groups. Several simulator developers have had active user groups at various points in time. A broad community of users is especially important in the competitive world of driving simulation, where companies can disappear overnight. The larger is the installed user base, the more likely it is that an individual can find someone who knows someone who can fix hardware or software problems. Regardless, it is critical that one remain always on good terms with the technical staff. In fact, soon after one of the coauthors purchased a half-million-dollar driving simulator, the manufacturer went out of business. But because relations with the technical staff were nurtured throughout the first several years that the manufacturer was in business, the driving simulator itself lived a happy life for another 15 years.

25.5 Conclusions

- Simulation is a very effective tool for studying many aspects of novice teen driver safety. These include a range of training and skill development programs, which can also be used for evaluating an even wider range of factors and countermeasures in this group. Key drivers for this advantage stem from the repeatability with which environments can be presented to teen drivers, along with the greater depth in measurement precision that use of a simulator affords.
- Many lessons have been learned about how to configure a driver simulator for use with teen drivers. These lessons can be reframed into guidelines for use. For example, these range from the specification of simulator requirements to match research questions—a focus on scanning behavior at

intersections will benefit from a wider field of view, while a focus on driver–passenger interactions will require a two-seat simulator.

- In selecting a simulator, there are decisions to be made in trading ease of use for flexibility in scene and scenario design, along with issues around performance measurement and the use of vehicle-based measures in combination with other ocular and physiological measures. With respect to vehicle measures, speed, hazard RT, and SDLP are the most commonly used measures of behavior, and sensitive to a range of manipulations in teen driver research, including inexperience itself, alcohol and drug use, and distraction.
- Prospective users of driving simulators need to be aware of potential issues and challenges, including simulator adaptation sickness, sensitivity to crash scenarios, recruitment, and distributed systems, among others.

Acknowledgments

This research was supported by a grant from the AUTO21 Network Centres of Excellence (Lana M. Trick).

References

Allen, R., & Reimer, B. (2006). New approaches to simulation and the older operator. *Advances in Transportation Studies: An International Journal, Special Issue*, 3–8.

Andersen, G. (2011). Sensory and perceptual factors. In D. C. Fisher, M. Rizzo, & J. Lee (Eds.), *Handbook of Driving Simulation for Engineering, Medicine and Psychology* (pp. 8-1–8-11). Boca Raton, FL: CRC Press.

Autodesk. (2015). *Autocad Civil 3D*. Accessed August 13, 2015, retrieved from http://www.autodesk.com/products/autocad-civil-3d/overview.

Boonspiripant, S. (2009). *Speed Profile Variation as a Surrogate Measure of Road Safety Based on GPS-Equipped Vehicle Data (Dissertation)*. Atlanta: Georgia Tech.

Caird, J.K., & Horrey, W.J. (2011). Twelve practical and useful questions about driving simulation. In D. Fisher, M. Rizzo, J. Caird, & J. Lee (Eds.), *Handbook of Driving Simulation for Engineering, Medicine, and Psychology*. Boca Raton, FL: CRC Press.

Chrysler, S., & Nelson, A. (2011). Design and evaluation of signs and pavement markings using driving simulators. In D. Fisher, J. Caird, M. Rizzo, & J. Lee (Eds.), *Handbook of Driving Simulation for Medicine, Engineering and Psychology*. Boca Raton, FL: CRC Press.

Davies, A. (2015). Mini's working on augmented reality driving goggles. *Wired*, April 19. Accessed August 20, 2015, retrieved from: http://www.wired.com/2015/04/minis-working-augmented-reality-driving-goggles/.

de Looper, C. (2015). Samsung invests in FOVE, an eye-tracking virtual reality headset. *Tech Times*, June 25. Accessed August 14, 2015, retrieved from: http://www.techtimes.com/articles/63701/20150625/samsung-invests-fove-eye-tracking-virtual-reality-headset.htm.

Drive Safety, Inc. (2014). Home. Accessed August 14, 2015, retrieved from Drive Safety: http://www.drivesafety.com/.

Drive Square. (2015). *Driving Simulators*. Accessed August 13, 2015, retrieved from Drive Square: http://www.drivesquare.com/drivingsimulator/.

Fisher, D., & Garay-Vega, L. (2012). Advance yield markings and drivers' performance in response to multiple-threat scenarios at mid-block crosswalks. *Accident Analysis & Prevention, 44*, 35–41.

Fisher, D., Rizzo, M., Caird, J., & Lee, J. (Eds.). (2011). *Handbook of Driving Simulation for Engineering, Medicine and Psychology*. Boca Raton, FL: CRC Press.

Ghosh, D., Jamson, S., Baxter, P., & Elliott, M. (2012). Continuous measures of driving performance on an advanced office-based driving simulator can be used to predict simulator task failure in patients with obstructive sleep apnoea syndrome. *Thorax, 67*, 815–821.

Greenberg, J. (2011). Physical fidelity of driving simulators. In D. Fisher, M. Rizzo, J. Caird, & J. Lee (Eds.), *Handbook of Driving Simulation for Engineering, Medicine and Psychology.* Boca Raton, FL: CRC Press.

Guo, F., Klauer, S., McGill, M., & Dingus, T. (2010). *Task 3—Evaluating the Relationship Between Near-Crashes and Crashes: Can Near-Crashes Serve as a Surrogate Safety Metric for Crashes?* Washington, DC: National Highway Traffic Safety Administration.

Kearney, J., & Grechkin, T. (2011). Scenario authoring. In D. Fisher, J. Caird, M. Rizzo, & J. Lee (Eds.), *Handbook of Driving Simulation for Engineering, Medicine and Psychology.* Boca Raton, FL: CRC Press.

Kennedy, R., Berbaum, K., Lilienthal, M., Dunlap, W., & Mulligan, B. (1987). *Guidelines for Alleviation of Simulator Sickness Symptomatology.* Washington, DC: Naval Air Systems Command.

Klauer, S., Dingus, T., Neale, V., Sudweeks, J., & Ramsey, D. (2006). *The Impact of Driver Inattention on Near-Crash/Crash Risk: An Analysis Using the 100-Car Naturalistic Driving Study Data.* Washington, DC: National Highway Traffic Safety Administration.

Krishnan, A., Samuel, S., Romoser, M., & Fisher, D. (2015). A study to evaluate the effectiveness of a tablet-based training program for younger drivers addressing distraction head on. *Proceedings of the 59th Human Factors and Ergonomics Society Annual Meeting* (pp. 1651–1655). Thousand Oaks, CA: SAGE Publications.

Mayhew, D., Simpson, H., Wood, K., Lonero, L., Clinton, K., & Johnson, A. (2011). On-road and simulated driving: Concurrent and discriminant validation. *Journal of Safety Research, 42*, 267–275.

McNamara, P. (2009). Inside Toyota's $15M driving simulator. *CAR Magazine*, November 24. Accessed August 13, 2015, retrieved from: http://www.carmagazine.co.uk/features/opinion/phil-mcnamara/inside-toyotas-15m-driving-simulator/.

Mechanical Simulation. (2015). *Driving Simulators.* Accessed August 13, 2015, retrieved from Mechanical Simulation: https://www.carsim.com/products/ds/.

Oktal: Simulation in motion. (2015). Home. Accessed December 2015, retrieved from Oktal automotive simulators website: http://www.oktal.fr/en/automotive/range-of-simulators/software.

Plouzeau, J., Paillot, D., Akykent, B., & Merienne, F. (2013). Vibrations in dynamic driving simulator: Study and implementation. *CONFERE 2013* (pp. 1–8). HAL Archives-Ouvertes. Accessed August 14, 2015, retrieved from https://hal.archives-ouvertes.fr/hal-00844571/document.

Ranney, T. (2011). Psychological fidelilty: The perception of risk. In D. Fisher, M. Rizzo, J. Caird, & J. Lee (Eds.), *Handbook of Driving Simulation for Engineering, Medicine and Psychology.* Boca Raton, FL: CRC Press.

Realtime Technologies, Inc. (2015). Home. Accessed August 14, 2015, retrieved from Realtime Technologies, Inc: http://www.simcreator.com/index.htm.

Romoser, M., & Fisher, D.L. (2009). The effect of active and passive strategies on improving older drivers' scanning for intersections. *Human Factors, 51*, 652–668.

Romoser, M., Fisher, D., Mourant, R., Wachtel, J., & Sizov, K. (2005). The use of a driving simulator to assess senior driver performance: Increasing situational awareness through post-drive one-on-one advisement. *Proceedings of the Third International Driving Symposium on Human Factors in Driver Assessment, Training and Vehicle Design.* Iowa City: Public Policy Center, University of Iowa.

Salmon, P.M., Lenné, M.G., Triggs, T., Goode, N., Cornelissen, M., & Demczuk, V. (2011). The effects of motion on in-vehicle touch screen system operation: A battle management system case study. *Transportation Research Part F: Traffic Psychology and Behaviour, 14*(6), 494–503.

Slob, J.J. (2008). State-of-the-Art driving simulators, a literature survey. *DCT Report*, 107.

Stoner, H., Fisher, D., & Mollenhauer, M. (2011). Simulator and scenario factors influencing simulator sickness. In D. Fisher, M. Rizzo, J. Caird, & J. Lee (Eds.), *Handbook of Driving Simulation for Engineering, Medicine and Psychology.* Boca Raton, FL: CRC Press.

Strater, L. (2014). *Newsletter for the Human Factors and Ergonomics Society Virtual Environments Technical Group*. Accessed August 5, 2015, retrieved from http://tg.hfes.org/vetg/newsletters/VETG-Newsletter-2014-10.pdf.

SWOV. (2010). *Simulators in driver training*. Accessed August 13, 2015, retrieved from SWOV Fact Sheet, Institute for Road Safety, Leidschendam, the Netherlands: https://www.swov.nl/rapport/Factsheets/UK/FS_Simulators_in_driver_training.pdf.

Systems Technology, Inc. (2013). *STISIM Drive*. Accessed August 14, 2015, retrieved from Systems Technology, Inc.: http://www.stisimdrive.com/.

Taylor, T., Roman, L., McFeaters, K., Romoser, M., Borowsky, A., Merritt, D.J., Pollatsek, A., Lee, J.D., & Fisher, D.L. (2014). *Cell phone conversations impede latent hazard anticipation while driving, with partial compensation by self-regulation in more complex driving scenarios* (Technical Report). Amherst, MA: Arbella Insurance Human Performance Laboratory, University of Massachusetts.

The University of Iowa. (2015). *The National Advanced Driving Simulator*. Accessed August 14, 2015, retrieved from Overview: http://www.nads-sc.uiowa.edu/overview.php.

Thomas, F.D., Rilea, S., Blomberg, R.D., Peck, R.C., & Korbelak, K.T. (2016). *Evaluation of the Safety Benefits of the Risk Awareness and Perception Training Program for Novice Teen Drivers* (DOT HS 812 235). Washington, DC: National Highway Traffic Safety Administration.

26

Using Naturalistic Driving Methods to Study Novice Drivers

Sheila G. Klauer

Johnathon Ehsani

Bruce Simons-Morton

Abstract

The Problem. Teenage drivers are overrepresented in fatal and injury crashes on our nation's roadways. While crash databases have allowed for a basic understanding of some of the contributing factors of crashes, these databases do not allow for deeper understanding of driver behavior. And while simulators and test tracks allow driver behavior to be tested, they are not real-world environments where drivers are exposed to their normal daily pressures. ***Role of Naturalistic Driving Studies.*** Naturalistic driving studies allow researchers to study novice driver behavior on public roadways in the seconds leading up to actual crashes and near crashes as well as a broad spectrum of traffic and roadway experiences. ***Naturalistic Driving Method.*** An overview of the naturalistic driving method will be provided, including a clear description and definition of naturalistic driving, a discussion of how this method fits into the traffic safety researcher's toolbox, and an examination of the benefits and limitations of the method. ***How to Design a Naturalistic Driving Study.*** This chapter will show researchers how to design a naturalistic driving study step by step, covering research protocol development, data collection and quality control, data coding, data coding quality control, and potential analyses that are well suited to naturalistic data. ***Key Analyses and Results from the Naturalistic Teenage Driving Study.*** Key analyses conducted with previous naturalistic driving studies will be presented, along with findings from these analyses. This section is first meant to inform researchers about the types of analyses that have been done. Secondly, because broad spectrums of analyses have not yet been performed, it is meant to entice future researchers with analyses that can and should be conducted going forward.

26.1 Introduction

Motor vehicle crashes are the leading cause of death and injury for teenagers. In the United States, the fatal crash rate for 16- and 17-year-old drivers is three times higher than for drivers aged 20 years, and over two times higher than for drivers aged 18 through 20 years (Centers for Disease Control and Prevention, 2012). This higher crash rate for novice drivers, even when compared to near-age peer groups, indicates that they are a vulnerable population of road users who are likely committing driver errors that lead to increased risk on the road. Understanding exactly what those errors may be and subsequently improving crash countermeasures requires additional research.

To better understand contributing factors associated with crashes for populations of all ages, traffic safety researchers have previously used epidemiological research methods (e.g., Chapter 5). These methods rely on data from actual crashes that were collected retrospectively. Conversely, to better understand driver behavior, traffic safety researchers have used controlled experimentation to precisely measure driving performance (e.g., simulation studies; Chapter 25). These data collection methods are prospective but conducted in contrived scenarios that may or may not generalize to actual driving on public roadways. While both methods have provided valuable insights into traffic safety issues, recent technological improvements have enabled a third methodology: naturalistic driving studies (NDSs). The naturalistic method allows traffic safety researchers to study driver behavior in situ, or in real-world traffic environments, with precise, objective measurements. Most uniquely, this method allows for driver behavior to be observed in the seconds leading up to actual crashes; however, it also allows for the observation of driver behavior in a wide variety of different traffic and roadway geometries.

NDSs span the gap between epidemiological studies and empirical studies by assessing data on actual crashes (as with epidemiological databases) and using high-resolution, detailed driving performance data (as with empirical, controlled experiments). Given the unique nature of the data they supply, naturalistic studies have become an important tool in the study of traffic safety that researchers should familiarize themselves with and consider as part of their efforts to reduce fatalities and injuries on our roadways. Several NDSs have been conducted in the United States, but there are also NDSs currently ongoing in many countries worldwide including Canada, Australia, European Union member countries, and China, to name a few.

Table 26.1 presents the benefits and limitations of NDSs. Broadly speaking, naturalistic studies provide access to data such as detailed kinematic data and video of driver behavior. The limitations of this data collection include the fact that it requires more resources and that the data collection process is logistically complex. Researchers interested in embarking on naturalistic driving projects should review the pros and cons and carefully evaluate them in the light of their key research questions to ensure accurate and useful results for their study.

TABLE 26.1 Benefits and Limitations of Naturalistic Driving Studies

Benefits	Limitations
Detailed driving performance (kinematic) in the seconds leading up to safety events.	Naturalistic studies tend to be expensive and require long periods of data collection.
Video data of driver behavior for baseline driving and in the seconds leading up to crashes and near crashes.	Video data need to be coded manually, which can become resource intensive.
Driver behavior is collected under real driving circumstances and daily pressures (high external validity).	Crashes are typically of lower severity than those collected in crash databases (typically no fatal crashes and a large proportion of non-police-reported crashes).
Typically supplies a wide variety of questionnaire/survey data for participants.	Sample of drivers may be biased toward those who are not opposed to being captured on video for long periods of time.
Can utilize data from crashes, near crashes, and/or other safety surrogates from the same data set.	No experimental control.

26.1.1 Results from NDSs

Several NDSs have been conducted to date. The 100-Car Naturalistic Driving Study, which was conducted in 2003 and 2004, was the first large-scale NDS that primarily examined driver behavior associated with crashes and near crashes (Dingus et al., 2006). Additional NDSs have been conducted on novice teenage drivers, older drivers, motorcycle drivers, and commercial-vehicle drivers (Farmer et al., 2010; Goodwin et al., 2012; Hickman et al., 2010; Simons-Morton et al., 2013, 2015).

Technologically, NDSs have contributed to improvements in the design of in-vehicle controls (heating, ventilation, and air-conditioning [HVAC]; infotainment systems), driver monitoring and feedback systems, and collision mitigation systems. Results from NDSs have also informed new public policy initiatives, particularly for commercial-vehicle drivers in terms of hours-of-service regulations and bans on cell phone texting (Hickman et al., 2010).

Although most NDSs have been strictly observational in nature, some studies of teen drivers have also incorporated a form of intervention, such as monitoring and feedback, to potentially affect a change in normal driving behavior. Both approaches have provided important data that have moved the needle on our understanding of the risks and dangers of teenage driver behavior.

This chapter will focus on the design of NDSs that are solely observational in nature. While monitoring and feedback studies are useful, these studies are primarily focused on determining whether driver behavior can be changed in some manner. This chapter will be devoted to the design of NDSs that strictly observe driving behavior under normal and uncontrolled roadway conditions.

26.2 Design of Teenage NDSs

Naturalistic studies are powerful tools for assessing driver behavior in real-world environments. While naturalistic studies offer researchers many advantages, they also have their limitations. This section will highlight the pros and cons of naturalistic studies with the goal of helping researchers design successful studies that benefit the field of traffic safety. This section provides a how-to for researchers interested in conducting an NDS with novice drivers by discussing the key steps involved, common pitfalls to avoid, and the importance of maintaining critical project goals.

The primary goal of NDSs is to capture driver behavior on public roadways under typical driving conditions. It is important to keep reminders that participants are in a driving study to a minimum. Therefore, the data collection systems should be as unobtrusive as possible in the cabin of the vehicle, and any required participant interaction with sensors should be eliminated. Researcher interaction with participants should also be restricted, including the collection of questionnaire/survey data and/or collection of driving performance data that requires interaction with the participant. The resulting data set will therefore be as representative of normal driving as possible.

26.2.1 Protocol Development

26.2.1.1 Research Questions

The development of research questions is the first step in the design of any successful study. If those questions require an NDS in order to be answered, then the entire study must be designed with the primary goal of obtaining data to answer these research questions. The second goal of all NDSs is the creation of a rich database of driving data that can be used to answer a diverse range of research questions, some not yet imagined! The NDS must always be constrained by the collection of data that are necessary to answer the primary research questions.

In NDSs, research questions can be broad in scope, precisely focused, or a combination of the two. Common research questions for NDSs might include measures of driving exposure. For example, "What is the variability in vehicle miles traveled among a population of novice drivers?" This question could be answered with continuous mileage data collected per trip per novice driver.

An example of a more focused question would be, "At what point in the trip do teenage drivers typically fasten their seat belts?" This question could be answered by identifying seat belt information in the vehicle network and/or the development/purchase of a seat belt sensor and calculation of the time between ignition on and the fastening of the driver's seat belt. This will provide a data set in which variability of timing of seat belt fastened, speed of vehicle, and even road type being traversed at time of seat belt activation could be assessed for every trip. Research questions that require driving performance comparisons under precise driving conditions (e.g., braking behavior toward a lead vehicle at signalized intersections) are far more difficult to analyze using NDSs and are best answered using controlled experiments.

26.2.1.2 Selection of Measures

NDSs typically collect different types of data during the course of the driving study. Survey/questionnaire data are typically collected for all of the participants at either one or multiple milestones during data collection. Vehicle data are collected continuously (from ignition on to ignition off) using vehicle sensor data and on-board diagnostic data (OBD2). Video data are also collected continuously using either one or multiple cameras that could include the forward view, driver's face, rear view, over-the-driver's-shoulder view, driver foot/pedal view, and/or views of other locations. Some level of audio data may be collected along with the video data as well, though these data have historically been more difficult to obtain in naturalistic driving environments for a number of ethical reasons. The primary research questions should guide the selection of surveys, vehicle sensors, and video streams so that project resources are efficiently expended.

Data collection sensors in the vehicle typically include accelerometers, GPS, lane position, and vehicle speed. Table 26.2 provides a list of possible sensors and driving performance measures that have been used in previous NDSs. This list is not exhaustive but, rather, should be used as a guide to identify the most important driving measures and provide examples of potential sensors/dependent measures that can be calculated to evaluate driving performance. If a commercial system is being used, there may be less flexibility in choosing the types of sensors used to measure driving performance; however, there are many useful metrics that can be calculated using accelerometers and speed alone.

Multiple video data channels are typically collected, ranging from studies with only two channels (forward view and interior of the vehicle cabin) to those with up to eight channels of video that include rear view, over-the-shoulder views, driver foot/pedal camera, etc. One naturalistic study of child passenger safety seats, for example, had a camera mounted to observe the child passenger in the rear seat (Charleton et al., 2010). Camera angles and viewing angles should be configured carefully and with adequate testing to ensure that the correct information is being captured.

TABLE 26.2 List of Driving Performance Sensors and Their Subsequent Dependent Measures

Sensor	Dependent Measure	Metric
Longitudinal acceleration (y acceleration)	Longitudinal deceleration (hard braking) Longitudinal acceleration (fast start)	G-force
Lateral acceleration (x acceleration)	Hard cornering	G-force (negative value is hard left turn; positive value is hard right turn)
Vehicle speed	Traveling above the speed limit or too fast for conditions	Miles per hour Kilometers per hour
Lane position	Distance across lane line Distance to lane line	Feet, inches, centimeters, etc.
Global positioning system (GPS)	Location of vehicle	Latitude, longitude, and elevation
Forward radar	Range to a lead vehicle, time to collision with lead vehicle	Meters to lead vehicle, meters per second to lead vehicle

In certain cases, and depending on the aims of the particular study, interviews have also been used either prior to data collection, at the conclusion of data collection, or after a safety-critical event of interest. These interviews have been used to better understand drivers' attitudes and perspectives on the instrumentation in the vehicle and/or other components of data collection, to find out if the drivers were involved in any crashes/near crashes that may not yet have been identified, and/or to gauge how often the drivers drove the instrumented vehicle versus other vehicles. A self-report of time that a teenager drove the instrumented vehicle versus another family vehicle (not instrumented) is particularly important when teen driving exposure/experience is a principal metric to obtain.

Data obtained from interviews conducted after a crash have been very informative with regard to sleep patterns in the days preceding the crash, any medications taken, potential impairment, and/or other important driver behavior characteristics that may not be obtainable via video only. These data may reflect transitory medication/sleep habits that would not be available from questionnaire/survey data collected at the beginning or end of the study. These interviews are also effective at obtaining drivers' perspectives, or at least their memory, of the event, which could provide insight into additional contributing factors not easily discernible from the video.

26.2.1.3 Obtaining Institutional Review Board Approval

Conducting NDSs presents numerous ethical issues for researchers to consider. Participants' privacy is the primary risk of participating in an NDS. These types of studies collect a wide variety of personally identifying information (PII) that researchers must diligently protect. PII may include not only face camera views, but also forward camera views that capture participants, participant family members, or pedestrians. Some additional camera views with the potential to capture passengers and/or the driver (over the shoulder) need to be protected as well. Also, given that drivers move around within a vehicle, even the best-designed camera angles will capture PII for brief periods of time, and so care needs to be taken to ensure that PII is removed from videos after the fact.

PII also extends to the collection of continuous GPS data. GPS data collected at the beginning and end of every trip theoretically allow another individual to determine a participant's location of residence, employment, and other frequently traveled routes within a certain degree of accuracy. Therefore, these data must also be protected to ensure that they are not used in ways that could cause harm to NDS participants.

In the United States, the National Institutes of Health issues a certificate of confidentiality (CoC), which, under a congressional mandate, allows data collection with the provision that if these data were ever subpoenaed for any reason, they would be considered protected. Researchers conducting studies in other countries should inquire whether a similar certificate is available in their respective country.

All of the data to be collected as part of the NDS must be described as clearly and concisely as possible in any applications to institutional review boards as well as in the informed consent forms that participants will sign. The cameras, camera angles, and vehicle sensors must all be presented to the participants. Participants must also be presented with CoCs if obtained and in place to protect their personal data. For this reason, the informed consent forms (and institutional review board/research ethics board applications) tend to be lengthier than those for the average driving research study.

26.2.1.4 Recruitment of Participants

Participant recruiting methods used for NDSs are similar to those used for conducting any other research study where a volunteer participant is followed for an extensive period. In addition to newspaper advertisements, random-selection phone-calling is carried out, and fliers are distributed in public locations and on vehicles in parking lots. Options for recruiting novice drivers include recruiting through the public school systems, the department of motor vehicle offices, and/or private driving schools.

Novice drivers are typically younger than 18 years of age and thus can only provide informed *assent*. Their parent or legal guardian must provide informed consent for participation in any research study. Given that mutual agreement is required, it is important to obtain informed consent and assent separately. Thus, both parents and teenage participants should be asked for consent/assent in separate locations to ensure that both are fully in agreement and that neither of the parties is being coerced to participate in any way.

26.2.1.5 Installation and Uninstallation of Data Collection Systems

Depending on the complexity of the data collection instruments, an appropriate amount of time must be scheduled for every vehicle installation, and participants should be made aware of the approximate amount of time that they will need to leave their vehicle with the research team.

As part of recruitment, researchers must also have clearly thought through which types of vehicles can effectively be instrumented. The design of the sensor suite, driven by the research questions for the study, will dictate vehicle characteristics required for installation. For example, if vehicle network speed (vehicle speed that is obtained from the OBD2 port) is required, then all participants must possess a vehicle that has an OBD2 port or is newer than a 1996 model, and sometimes even newer, depending on the vehicle make/model.

It is difficult to foresee possible issues that could occur during installation. Therefore, it is essential that researchers perform practice installation and uninstallation on a variety of vehicle makes/models/years to assess prior to data collection what types of vehicles work well and what types should be excluded from participation. The installation and uninstallation of data collection systems should be timed to appropriately inform participants and effectively budget resources for these stages of the study. This step should not be underestimated, as problems with installations can result in participants developing negative attitudes and becoming frustrated with poor installation procedures (e.g., taking too long) and/or data loss. Both scenarios can reduce the usefulness of the resulting data set and should be minimized as much as possible.

Uninstallation is an ideal time for researchers to administer any exit interviews and/or collect any additional questionnaires, and potentially provide final payment to participants. Researchers may consider withholding final payment until equipment has been returned and final survey/questionnaire data are collected.

26.2.2 Data Collection

This section covers the key phases of data collection, which encompass data collection protocols, data quality assurance/quality control (QA/QC), and completion of data collection. While these steps are somewhat similar to the steps required for an instrumented-vehicle study, given that the instrumented vehicles in a naturalistic study will not be under the researcher's constant control, there are additional factors that must be considered and planned for to ensure high quality in the resulting naturalistic driving data set.

26.2.2.1 Data Collection

Naturalistic data collection can be performed using a range of methods. As technology continues to improve, new methods will be developed as well. This section will describe current approaches to data collection; however, this should not be considered an exhaustive discussion of all possible methods.

Data collection can be continuous or based on triggers that exceed a certain threshold. Continuous data collection records data from ignition on to ignition off. Triggered data collection occurs continuously and is buffered, but data are only saved and stored if a performance threshold is exceeded (e.g., if a driver brakes at greater than −0.45 *g*). Depending on the research question, a combination of both triggered and continuous data collection can be used.

Continuous data collection will result in large data sets. As a rule of thumb, continuous data collection for tens of vehicles for one month would result in 1 terabyte; 100 vehicles for 12 months would result in 10–20 terabytes; 1000 vehicles for 24 months would result in several petabytes. To date, neither cell phone nor wireless communication can efficiently transfer data sets this large. As a result, research staff members are required to visit Data Acquisition System (DAS)-equipped vehicles to swap and replace storage devices on a routine basis. Cell phone chips onboard these vehicles can be designed to transmit bytes of data to allow researchers to monitor the quality of the data. This design feature is very helpful for ensuring the highest-quality resulting naturalistic driving data set, as data collection in the field can be problematic.

26.2.2.2 Quality Assurance/Quality Control

As the data are collected, having a QA/QC plan in place is also critical. Data collection in a high-vibration environment such as a vehicle can cause frequent problems and errors. Thus, it is critical that the data are reviewed as they are collected and that all cameras and sensors are checked to make certain that they are functioning appropriately and accurately. Any issues with cameras or vehicle sensor data should be corrected and modified to ensure the highest possible data quality. This QA/QC process must begin as soon as data collection begins and must continue throughout the data collection process. As researchers plan a naturalistic study, this step must be included in the budget and integrated into the process, or the risk of data loss or poor-quality naturalistic data will increase substantially.

26.2.2.3 Data Collection—Lessons Learned

Interested readers should review previous NDSs' lessons-learned sections that are available in several different publications (Dingus et al., 2006, 2015). These chapters contain a wealth of information for researchers who have little or no experience conducting an NDS. Some of the key lessons learned from the Second Strategic Highway Research Program (SHRP2) NDS are as follows:

- Recruiting a diverse participant pool requires a diverse range of recruiting strategies. The amount of work that each of these strategies requires should not be underestimated.
- Participants do not show up for appointments. Develop backup plans/strategies.
- Data sensors/cameras fail. Successful NDS have QC processes in place at every step in the process, and sometimes, they have redundant processes in place.
- Emergencies happen. Have a set protocol for research staff to follow when an emergency happens. Examples include a severe crash, an instrumented vehicle being stolen, an instrumented vehicle catching fire in a public parking garage, and an instrumented vehicle being driven into a secured government facility and detained.

26.2.2.4 Postprocessing Naturalistic Data

The vehicle parametric data, collected as part of NDSs, may require postprocessing prior to data analysis. Postprocessing typically consists of using programs to identify anomalies in the data and then automatically correcting the inconsistencies found therein. Common examples of these anomalies include a malfunctioning or poorly calibrated sensor, such as an accelerometer that is not zeroed, or vehicle speed that is reporting a single, inaccurate value (not zero). Interested readers should refer to the lessons-learned sections from other naturalistic studies for a comprehensive list of common naturalistic data postprocessing steps.

The goal of data coding is to define a complete set of data elements that will make the database more directly useful for search and analysis. Data coding using naturalistic driving data involves training multiple data coders to review the video and driving performance data and to record driver behaviors. To glean the rich driver behavior data from the video, some form of data coding will likely be required to aid in the data's usability. Examples of coding protocols include the sequence of actions in the seconds prior to crashes and near crashes, and recording environmental and roadway variables that are not automatically recorded by the DAS.

26.2.2.4.1 Event Selection

As it is not generally feasible to review and record all the video data that have been collected, a strategy is required to identify relevant periods of driving where further data coding would be beneficial (i.e., automated trigger). Examples include using hard braking/cornering events to identify safety-critical events, using GPS data to identify specific roadway geometries (e.g., multilane intersection), or using random selection to examine the prevalence of certain behaviors, such as secondary task engagement.

After the sample of events has been selected and identified, trained data coders review the events. Coders watch the appropriate video and then record the event of interest and all other factors that may be relevant, such as driver behavior, environmental and roadway variables, and scenario-specific variables of interest. These elements might include the following:

- **Event variables:** Variables used to establish the scenario and sequence of events prior to and throughout the critical event. These variables include event severity, event nature (e.g., conflict with lead versus crossing vehicle), preincident maneuver, precipitating event, driver reaction, postmaneuver control, information about other drivers/vehicles/objects involved (e.g., type, position, maneuvers, impairments), and fault assignment.
- **Driver variables:** Variables used to systematically describe the driver's state prior to and during the critical event. These include driver ID, driver behavior (e.g., speeding, aggressive driving), driver impairments (e.g., drowsiness, anger, substance abuse), secondary task engagement and duration (e.g., cell phone use), placement of hands on the wheel, visual obstructions, and seat belt use.
- **Environmental variables:** Variables used to describe environmental and/or roadway conditions consistent with the National Highway Traffic Safety Administration's General Estimates System (GES) and other crash databases. These include roadway surface condition, traffic flow, number of travel lanes, traffic density, traffic control device at event onset, relation to junction of vehicle at event onset, roadway alignment (e.g., curve, grade), locality type (e.g., residential, interstate), ambient lighting, weather, and windshield wiper status.

The research objectives and research questions will dictate the types of data triggering and coding that are required for each project. The researcher will then develop the triggers and the coding protocols for the trained data coders to follow to ensure that the appropriate information is recorded. Data coding QA/QC is a critical element and is the topic of the next section of this chapter.

26.2.2.4.2 Data coding QA/QC

A suggested data coding QA/QC workflow has been developed, tested, and implemented by the Virginia Tech Transportation Institute (Klauer et al., 2012). Three phases of this workflow are discussed briefly:

- **Protocol development:** The focus of the protocol and coding will depend largely on the research questions. The research protocol is the document that trained data coders will use/refer to when coding these data. This data dictionary will provide all operational definitions for each variable of interest. The coding protocol will provide a set of precise step-by-step procedures and how-tos for the entire coding process. Researchers will typically lead the development of a preliminary protocol and data dictionary based on the research questions. The protocol is tested against a sample of video segments by an experienced coder. The protocol is refined and retested until the research team is satisfied that the protocol captures the variables of interest.
- **Coder training:** The protocol is used to train a small team of coders (typically three to four at a time to keep the initial QC manageable). Once the formal training session has been completed, coders begin coding under the supervision of a senior coder. Ideally, a more experienced coder should review the majority of each newly trained coder's work. Corrections are made, and detailed feedback is provided to coders for review. Once coders are able to meet reliability standards of 90% accuracy in this initial review, they may be assigned clips to code for the study.
- **Data coding QC:** During ongoing data coding, QC can be maintained using spot checks, interrater tests, and intrarater tests. A spot check is a second person's review of data coding work where corrections can be made and comments provided to the original coder to confirm and/or improve

accuracy. Interrater and intraraters tests are used periodically during a coding project to ensure consistency both between coders at a given point in time (inter) and within individual coders over time (intra). The frequency of these tests may be determined by the duration and complexity of the coding.

26.2.3 Data Analysis

Data analysis techniques will depend largely on the research questions and the type of data used. The following sections highlight a variety of different analytic techniques that have been used to answer some common research questions.

26.2.3.1 Driving Exposure

Using continuously collected data, the number of trips, miles driven, and minutes driven can be calculated for each participant in a study. These measures can be combined with meteorological data and sunset/sunrise data to determine the weather and light conditions during these drives, and estimate the amount of driving occurring under different conditions. GPS data can also be used to examine the different types of trips completed by participants (e.g., from home to work, or to a new location).

26.2.3.2 Prevalence of Risky Behaviors

The prevalence of certain driving behaviors can be assessed using a random selection of trip segments. These segments can be coded to assess the frequency of behaviors under typical driving conditions.

26.2.3.3 Assessing Crash Risk

Crash risk refers to the amount of risk a driver assumes (above the normal level of risk of driving/riding in a motor vehicle) when engaging in a particular behavior. Crash risk can be assessed by comparing the frequency of the behavior during a safety-critical event (e.g., secondary task engagement) to the frequency of the behavior in the randomly selected trip segments.

Key results from teenage NDSs have pointed to the role played by secondary tasks in safety-critical events. Klauer et al. (2014) found that secondary tasks, such as dialing cell phones, reaching for objects, texting, and looking at roadside objects, increase the risk of crash and near-crash occurrence for novice drivers. Notably, these are all tasks that require both visual engagement and manual manipulation on the part of the driver. The increased risk for visual–manual tasks was especially true for teenage drivers compared with experienced drivers. Other secondary tasks, such as talking on cell phones, were shown to not increase crash and near-crash risk. It is hypothesized that during these tasks, drivers generally have their eyes on the forward roadway and are thus more able to respond to traffic appropriately (Fitch et al., 2013; Klauer et al., 2006, 2014).

Studies evaluating eye-glance data have found that regardless of the eye-glance metric, the longer novice drivers' eyes are off the forward roadway (using either a sum of sequential glances and/or single-glance durations), the greater the risk of crash and near crash (Klauer et al., 2010; Simons-Morton et al., 2014). Simons-Morton et al. (2014) also found that single-glance durations, specifically glance durations when using wireless devices, resulted in higher risk estimates as glance duration increased compared to the same analysis that incorporated any single glance off the forward roadway.

Other studies using risky driving maneuvers, also called *kinematic risky driving behaviors*, have shown that risky driving maneuvers occur four times more frequently for novice drivers than for experienced drivers (Simons-Morton et al., 2011). Further analyses of kinematic risky driving behaviors have suggested that the rate of kinematic risky driving events for novice drivers is nearly equivalent to that of experienced drivers when their parents are passengers in the vehicle. This result suggests that novice drivers are capable of driving like experienced drivers but that such behavior is not transferred to unsupervised situations.

26.2.4 Existing Naturalistic Driving Databases for Qualified Researchers

Large-scale NDSs produce powerful databases that contain a plethora of driving performance and driver behavior information. SHRP2, the largest NDS conducted to date, has made significant amounts of data available to qualified human-subjects researchers online at https://insight.shrp2nds.us/. The data dictionaries and data queries are available in addition to the data sets, which can be downloaded. Traffic safety researchers can use these data to answer a wide variety of driving safety research questions. The National Academies, the sponsor of the study, is hopeful that many researchers will use these data for that purpose.

Smaller-scale naturalistic driving databases for both light-vehicle and commercial-truck data are also available online at http://forums.vtti.vt.edu/index.php?/index?. At the time of this publication, there are several efforts underway by other groups to allow qualified researchers access to other data sets, and it is widely believed that the sharing of data will become the modus operandi for future studies.

Regardless of the specific NDS, transportation safety researchers have only skimmed the surface of the power of NDS data. New and unique analyses are being conceived that will provide insights into driver behavior never before obtainable. Some novel new analyses include using speed and acceleration data to model fuel usage; analyses evaluating driver behavior in different roadway and traffic environments continue to expand through the use of video imaging techniques. The power of NDS has also been coupled with other research methods to figure out ways to use NDS video and data to improve driving simulation studies as well as instrumented-vehicle studies. This is a great time for transportation researchers as the new tool of NDS is publically available and waiting for researchers to analyze and learn how to improve transportation safety worldwide.

Acknowledgments

The authors would like to thank Michael Buckley for his skilled technical editing as well as the editors of this book for their thoughtful comments that greatly improved this chapter.

References

Centers for Disease Control and Prevention. (2012). Web-based Injury Statistics Query and Reporting System (WISQARS) [Online]. National Center for Injury Prevention and Control, Centers for Disease Control and Prevention (producer). Accessed February 25, retrieved from http://www.cdc.gov/injury/wisqars/.

Charleton, J., Koppel, S., Kopinathan, C., & Taranto, D. (2010). How do children really behave in restraint systems while traveling in cars? Paper presented at the 54th Annual Scientific Conference of the Association for the Advancement of Automotive Medicine. Vol. 54. Las Vegas, NV.

Dingus, T.A., Hankey, J.M., Antin, J.F., Lee, S.E., Eichelberger, E.G., Stulce, K.E., McGraw, D., Perez, M.A., & Stowe, L. (2015). *Naturalistic Driving Study: Technical Coordination and Quality Control.* (Report No. S2-SO6-RW-1). Washington, DC: Transportation Research Board.

Dingus, T.A., Klauer, S.G., Neale, V.L., Petersen, A., Lee, S.E., Sudweeks, J., Perez, M.A., Hankey, J., Ramsey, D., Gupta, S., Bucher, C., Doerzaph, Z.R., Jermeland, J., & Knipling, R.R. (2006). *The 100-Car Naturalistic Driving Study: Phase II—Results of the 100-Car Field Experiment.* (Interim Project Report for DTNH22-00-C-07007, Task Order 6; Report No. DOT HS 810 593.) Washington, DC: National Highway Traffic Safety Administration.

Farmer, C.M., Kirley, B.B., & McCartt, A.T. (2010). Effects of in-vehicle monitoring in the driver behavior of teenagers. *Journal of Safety Research, 41*, 39–45.

Fitch, G.M., Soccolich, S.A., Guo, F., McClafferty, J., Fang, Y., Olson, R.L., Perez, M.A., Hanowski, R.J., Hankey, J.M., & Dingus, T.A. (2013). The Impact of Hand-Held and Hands-Free Cell Phone Use on Driving Performance and Safety-Critical Event Risk. Technical Report No DOT HS 811 757. Washington, DC: National Highway Traffic Safety Administration.

Goodwin, A.H., Foss, R.D., Harrell, S.S., & O'Brien, N.P. (2012). *Distracted Driving among Newly Licensed Teen Drivers. Report to the Automobile Association of America Foundation for Traffic Safety.* Washington, DC: AAA Foundation for Traffic Safety.

Hickman, J.S., Hanowski, R.J., & Bocanegra, J. (2010). Distraction in commercial trucks and buses: Assessing prevalence and risk in conjunction with crashes and near-crashes. Report No. FMCSA -RRR-10-049. Retrieved from http://www.fmcsa.dot.gov/facts-research/research-technology/report /Distraction-in-Commercial-Trucks-and-Buses-report.pdf. Washington, DC: Federal Motor Carrier Safety Administration.

Klauer, S.G., Guo, F., Sudweeks, J.D., & Dingus, T.A. (2010). An Analysis of Driver Inattention Using a Case-crossover Approach on 100-car Data. (Report No. DOT HS 811 334). Washington, DC: National Highway Traffic Safety Administration.

Klauer, S.G., Dingus, T.A., Neale, V.L., Sudweeks, J.D., & Ramsey, D.J. (2006). The Impact on Driver Inattention on Near Crash/Crash Risk: An Analysis Using the 100 Car Naturalistic Driving Study Data (Report No. DOT HS 810 594). Washington, DC: National Highway Traffic Safety Administration.

Klauer, S.G., Perez, M., & McClafferty, J. (2012). Naturalistic Driving Studies and Data Coding and Analysis Techniques. Book chapter in B. Porter (Ed.) *Handbook of Traffic Psychology*. Elsevier, New York.

Klauer, S.G., Guo, F., Simons-Morton, B.G., Ouimet, M.-C., Lee, S.E., & Dingus, T.A. (2014). Distracted driving and risk of road crashes among novice and experienced drivers. *The New England Journal of Medicine, 370*, 54–59.

Simons-Morton, B.G., Ouimet, M.C., Zhiwei, Z., Lee, S.L., Klauer, S.E., Wang, J., Chen, R., Albert, P.E., & Dingus, T.E. (2011). Risky driving among novice teenagers and their parents. *American Journal of Public Health, 101*(12), 2362–2367.

Simons-Morton, B.G., Bingham, C.R., Ouimet, M.C., Pradhan, A.K., Chen, R., Baretto, A., & Shope, J. (2013). The effect of teenage risky driving of feedback from a safety monitoring system: A randomized control trial. *Journal of Adolescent Health, 53*(1), 21–26.

Simons-Morton, B.G., Guo, F., Klauer, S.G., Ehsani, J.P., & Pradhan, A.K. (2014). Keep your eyes on the road: Young driver crash risk increases according to duration of distraction. *Journal of Adolescent Health, 54*, S61–S67.

Simons-Morton, B.G., Klauer, S.G., Ouimet, M.C., Guo, F., Albert, P.S., Lee, S.E., Ehsani, J.P., Pradhan, A.K., & Dingus T.A. (2015). Naturalistic teenage driving study: Findings and lessons learned. *Journal of Safety Research, 54*, 29–44.

27

Eye Movements—Utility, Method, and Measurements

Donald L. Fisher

William J. Horrey

Jeff K. Caird

Bryan Reimer

Abstract

The Chapter. The measurement of eye movements while driving has an almost 50-year history. Gains in fundamental knowledge about driving, specifically regarding teen and novice drivers, are briefly reviewed. Basic tenets of eye movements as well as common measures are described as well as eye-tracking systems that have been applied in many research settings. We also present several issues and challenges that are inherent in the study of eye movements in an attempt to offer newcomers to eye tracking some added insight. ***Limitations and Recommendations.*** (1) We do not exhaustively review the existing literature on eye movements and driving. Some of this material is covered elsewhere (Chapters 6, 7, and 12). (2) Space constraints preclude an in-depth coverage of some of the inherent issues. We hope that a brief introduction to some of these issues will spur interested readers into digging deeper.

27.1 Introduction

The study of eye movements has led to tremendous insights into human attention and cognition, not to mention how we actively seek and process information in the world around (or, conversely, what information we fail to process). Eye movements align the foveal region of the retina where acuity is highest with objects that require processing and recognition (i.e., fixations [Land, 2006]). In the beginning of eye movement measurement, scientists identified the anatomy of the eye and created the technical means to determine the muscular activity of the eyes while looking at stationary and moving objects (see Wade, 2010, for the history of eye tracking). For example, in one of the earliest examples of eye tracking, Yarbus (1967) put trackers on contact lenses that stayed in place through suction to determine where people looked. It has since been discovered that where people look depends greatly on the content of the material that is being shown to them and the goals of the person (see also Chapter 7). For example, when observers look at paintings, objects with high contrast or detail and biological significance such as humans and animals are more often fixated (Livingstone, 2002). In dynamic settings such as driving, properties such as contrast, size, luminance change, color, and motion (so-called bottom-up properties) naturally draw the eyes to certain locations in the visual field.

Another fundamental discovery has been that the task requirements imposed on a viewer affect where and how images are scanned (Yarbus, 1967). For example, in the context of driving, if drivers are told to look for signs, this goal affects how they scan the traffic environment (e.g., Ho et al., 2001). If they are told to drive as they ordinarily would, a different pattern of saccades and fixations emerges. When considering the task requirements, it is also important to know whether the task requires selective (focus on a particular object) or divided (focus on multiple objects simultaneously) attention. The nature of these requirements also imposes a variety of constraints on eye movement patterns such as the duration and frequency of fixations and the interleaving of attention to multiple tasks (e.g., Horrey et al., 2006).

In the driving context, the information gathered from fixations and pursuit movements (i.e., the tracking of moving targets) is used for vehicle control, contextual and hazard anticipation, and navigation. From a vast array of potential elements in the visual field, drivers process relevant information as it supports these overlapping tasks. Finding and extracting relevant information is essential to safe driving. However, not all of these tasks require foveal vision. For example, parafoveal (the area just outside the fovea) and peripheral vision are often sufficient for vehicle control (e.g., Summala et al., 1996).

The breadth of visual search may change as driving demands increase, such as at higher speeds and when preparing to turn at an intersection, changing lanes in traffic, or maneuvering through a curve. Other contexts (e.g., car following, sign processing) require concentrated glances to extract relevant control and navigation information. Through these processes, the novice driver perceptually learns where to look through many hours of practice and exposure to patterns of traffic and environment.

A number of literature reviews have surveyed eye movements and driving. Some address driving as a simple extension of everyday tasks (e.g., Land, 2006), whereas others focus on a more comprehensive review of driving research (e.g., Green, 2007). For example, Serafin (1994) provides an interesting synthesis of eye movements and driving literature for straight and curved driving.

The purpose of this chapter is to describe (1) how the measurement and study of eye movements have advanced our understanding of novice and teen drivers, (2) what measurements can be gleaned from the study of eye movements, (3) what types of eye-tracking systems are available, and (4) what are some of the prevailing issues and challenges associated with the study of eye movements. In some cases, technical and procedural issues are briefly introduced, although an in-depth treatment is covered within the citations that are provided. Material in this chapter overlaps with a number of the other chapters (in particular, Chapters 6 through 8, 12, and 28).

27.2 Novice and Teen Eye Movements

Novice and teen drivers (aged 15–19) have the highest rates of fatalities and injuries per mile or kilometer driven of any other age group (WHO, 2015; Chapter 5). Failures in looking altogether (e.g., glances inside the vehicle when a hazard materializes) and failures in looking at the correct location at the right time (e.g., the location of a potential hazard before it materializes) are major contributors to teen and novice driver injuries and fatalities. For example, McKnight and McKnight (2003) identified distractions and failing to search ahead, to the side, and to the rear of the vehicle as contributors to crashes in teen drivers. Curry et al. (2011) also identified inadequate surveillance of the traffic environment for hazards as a contributor to crashes, which accounted for 21.3% of teen precrash recognition errors. The evolution of knowledge about where, when, and why scanning and attention errors occur is synthesized in this review (see also Chapter 7).

27.2.1 Scanning Patterns

In what is regarded as a classic study, Mourant and Rockwell (1972) compared novice and experienced drivers' eye movements on the road. In their study, novice driver eye movements were recorded before, during, and after the completion of a driver education course. Experienced drivers scanned more widely horizontally during neighborhood and freeway encounters with turns, traffic, lane changes, and stops. Concentrations of fixations for novice drivers were closer to the hood and to the right of their vehicles than was the pattern for experienced drivers. Novice drivers looked at their speedometers more frequently to gauge their speed and rarely used their rearview and side-view mirrors compared to experienced drivers.

These differences are also consistent with research reported by Chapman and Underwood (1998). They found that experienced drivers adapt their visual scanning patterns to different road situations, while novice drivers tend to use the same scanning patterns for all road types (also see Borowsky et al., 2009). Results indicating a smaller range of horizontal scans have also been reported by Deery (1999) and Underwood (2007), suggesting difficulty integrating eye movements and controlling the vehicle on the part of novice drivers, which presumably relates to the cognitive load required before the automation of driving skill occurs.

Differences in novice and experienced drivers' scanning patterns are thought to reflect differences in visual search skill and processing. One explanation for the result that the focus of the attention of novice drivers (as indexed by the location of their eyes) is close to the front of vehicles and to the right is that novices use foveal search rather than peripheral vision for lane keeping (Mourant & Rockwell, 1972). This may represent a difficulty that novice drivers have processing information in the periphery unrelated to control of the vehicle and then using this information to control the vehicle. For experienced drivers, this demand on cognitive and motor resources is a reasonable one; for novice drivers, it may be too high as younger drivers are learning. However, Milloy et al. (2010) questioned if novices are indeed overloaded or simply unaware of the benefit of increased checks around them and to their mirrors. Studies have indicated that when workload is manipulated by controlling route familiarity, drivers in the low-workload conditions demonstrated increased visual searches to the left, indicating the possibility that a reduced workload allows drivers to focus on areas where hazards are more likely to emerge (i.e., oncoming traffic [Mourant et al., 1969]).

27.2.2 Hazard Anticipation and Mitigation

As noted, drivers who fail to anticipate latent hazards (hazards that are hidden or visible but not yet materialized) are much more likely to be in a crash (Horswill & McKenna, 2004). The study of eye movements has been integral to our understanding of hazard anticipation in novice and teen drivers (Chapter 6). Large differences have been found in novice and experienced drivers' ability to anticipate traffic hazards, whether these be other vehicles, bicyclists, or pedestrians (Borowsky et al., 2010; Pradhan et al., 2005; Underwood, 2007). Moreover, hazard anticipation has also been used as a dependent variable in studies that evaluate the effect of different treatments on novice drivers, such as training (Chapter 18; Pollatsek et al., 2006; Samuel et al., 2014).

Eye movements are also essential to understanding why it is that so many novice drivers fail to mitigate scenarios in which latent hazards can appear (Chapter 8). For example, studies can shed insight into whether novice drivers sometimes anticipate hazards but fail to effectively mitigate them or whether they fail to anticipate them altogether. The answer turns out to be a little bit of both. Even when one controls for hazard anticipation, one finds that novice drivers are still more likely to fail to mitigate the hazard (Muttart et al., 2013, 2014).

27.2.3 Attention Maintenance

Finally, consider the use of eye movements to differentiate between novice and experienced drivers' ability to maintain attention on the driving task (also see Chapters 7 and 12). There are typically two aspects to attention maintenance that have been linked to crashes on the open road and in a driving simulator: (1) drivers must keep whatever glances they make inside the vehicle to a relatively short duration (Horrey & Wickens, 2007; Simons-Morton et al., 2014), and (2) drivers must keep whatever glances they make to the sides of the road to a relatively short duration (Divekar et al., 2012).

Eye movements have been used to identify the differences in novice drivers in each of these two areas. For example, novice drivers are much more likely to take especially long in-vehicle (off-road) glances than are experienced drivers (Chan et al., 2010). See Chapters 7 and 12 for a broader discussion of visual attention allocation and distraction.

27.2.4 Summary

This section is not intended to provide an exhaustive review or synthesis of the vast number of studies of eye movements in novice and teen drivers (though see Appendix 27.1 for a summary of some of this work); rather, it is intended to underscore the utility and importance of this form of measurement. Novice and teen drivers' approach to the driving task is quite different from that of older, more experienced drivers. Eye movements not only reflect some of these differences in a compelling manner but also offer insight into areas where training or other areas of targeted remediation can improve the overall safety of this group of drivers. In the following sections, we elaborate on more of the technical aspects of eye monitoring and measurement, including a discussion of key measurements, eye-tracking systems, and some of the inherent challenges. We focus these remarks on novice drivers whenever it makes most sense to do so.

27.3 Measures, Systems, and Technical, Procedural, and Data Challenges

Here, we describe in order (1) the mechanics of eye movements; (2) the different ways in which eye movements have been summarized; (3) how often and how precise the measurements of eye location must be in order to compute the summary measure; (4) how best to compile the summary measures of eye movements; (5) how to measure information acquired in the periphery; (6) how to measure eye scanning patterns; (7) how the different systems that are available for measuring eye movements work; and (8) the types of challenges that are faced whenever one collects data about eye behaviors and then attempts to analyze them.

27.3.1 A Primer on Eye Movements

A number of different types of eye movements have been identified (Holmqvist et al., 2011, pp. 21–23; Land, 2006; Palmer, 1999; Rayner, 1998). The major classes of eye movements relevant to our discussion involve the change of locus of attention. Saccades are ballistic shifts in the orientation of gaze that largely couple with movements in the locus of attentional processes. Interspersed between saccadic eye movements are fixations and smooth pursuit movements that largely govern the foveal retrieval of information (i.e., the acquisition of visual information). The fovea is the central portion of the retina with the highest sensitivity to visual information. This represents approximately 2° of visual angle, beyond which one's sensitivity to detailed (high-resolution) visual information diminishes. Fixations are defined as periods of minimal to no movement of the eye, allowing for detailed information to be gathered from the foveated object. Smooth pursuit movements predictively track an object in motion, rather than remaining fixed on a given point in space. In the context of driving, smooth pursuit movements have been largely ignored in the literature, or they are, for simplified interpretation, treated as fixations (Reimer & Sodhi, 2006). Although the vision science community has developed a strong theoretical understanding of the visual acquisition of foveal and peripheral information, the analysis of eye movement data in the driving context is often reduced to an analysis of the periods of continuous information retrieval of a given area of interest (AOI) or target (also called glances, see below), with certain historically developed definitions that greatly simplify the way such data are interpreted.

27.3.2 What Types of Summary Measures of Eye Movements Are There?

The International Organization for Standards (ISO) (2001, 2002) and the Society of Automotive Engineers (SAE) (2000) have developed standards for the interpretation of eye movements in a driving context. The use of measurement and analysis approaches consistent with those defined by ISO and SAE facilitate more effective comparisons across studies and experimental setups. Even with these definitions, however, many studies do not sufficiently define deviations in interpretation of eye movement measures (e.g., data validation procedures, sampling characteristics, measurement limitations, etc.), making it difficult for readers and reviewers to cohesively interpret or replicate work.

ISO (2002, p. 2) defines glance duration as the "time from the moment at which the direction of gaze moves towards a target (e.g., the interior mirror) to the moment it moves away from it." In essence, the definition of a glance includes a preceding saccadic movement and the subsequent one or more fixation points bounded by a target (predetermined AOI). As such, several shifts in attention that may occur between the fixation points are bound together within the construct of a glance.

While the construct of a glance greatly simplifies recording and analytical considerations, the vision science literature would suggest that saccades and fixations need to be considered separately, as perception is suppressed during the saccade (Zuber & Stark, 1966). As such, accurate time course modeling of information retrieval can only be considered in a more fine-grained analysis of fixations, with or without smooth pursuit movements, which is not provided for under the ISO definition of a glance. Regardless, the ISO definition of a glance is broadly used and leads to several popular metrics widely considered in the study of driver behavior. These definitions include the following:

- *Glance frequency* is the number of glances to a target within a predefined time period, or during a predefined task, where each glance is separated by at least one glance to a different target (ISO 2002, p. 2).
- *Total glance time (TGT)* is broadly considered as the total time that an individual looks at a target, summed over all target glances. This metric is often considered in relation to a functionally relevant period of time, such tuning the radio, making a phone call, etc. (e.g., Driver Focus–Telematics Working Group, 2006).
- *Total eyes-off-the-road time (TEORT)* is broadly defined as the total time during which a driver is glancing away from the forward roadway, regardless of locus, over a region of interest, such as a driver performing a particular task. This is a foundational metric in the National Highway Traffic Safety Administration (NHTSA) visual–manual driver distraction guidelines for in-vehicle electronic devices (NHTSA, 2013).
- *Percent road center (PRC)* is defined as the proportion of a predefined interval during which the driver looks at a specified region within the forward roadway (Victor et al., 2005; Wang et al., 2014).

The mean duration of individual (single) glances and the percentage of glances exceeding a specific threshold such as 1.8 or 2.0 seconds (e.g., long-duration glances; Driver Focus–Telematics Working Group, 2006; NHTSA, 2013) are often considered within the context of either TGT or TEORT. NHTSA (2013), in its recommended guidelines for evaluating the visual distraction associated with in-vehicle visual–manual electronic devices, takes a somewhat idiosyncratic approach to what is considered TEORT. NHTSA's definition of "away from the forward road scene" means that glances to other driving-relevant locations (e.g., instrument cluster, mirrors, etc.) are included in TEORT calculations. However, one should note that the use of NHTSA's definition for HMI evaluation purposes is limited to a simple driving scenario that would be expected to minimize driving relevant TEORT. Earlier efforts (e.g., Driver Focus–Telematics Working Group, 2006) focused on assessing glances to locations relevant to a task under evaluation (e.g., radio display, controls, buttons, etc.). More recent efforts (Muñoz et al., 2015) have developed methods that quantify differences in visual scanning transition patterns as a foundation for modeling glance allocation strategies in the context of interface evaluation, environmental sensing, etc. By looking at glance allocation strategies, in addition to more traditional clustering-based methods of assessing attention allocation, the approaches provide a foundation for considering how drivers move their eyes spatially around the vehicle and its surroundings as they navigate through time.

The same summary measures of eye behaviors have been used for novice drivers as have been used for experienced drivers. However, some more complex summary measures predominate. For example, with hazard anticipation, the question one wants to ask is whether a driver looks at a particular interval in time (often called the launch zone) toward a particular area of the scene (often called the target zone) (Pradhan et al., 2005). Thus, it is important not only that the driver glanced toward a particular target but also that the driver glanced at a particular time toward that target. Otherwise, the glance is ineffectual (cannot be used to mitigate a potential hazard).

27.3.3 How Often and Precise Must Information Be about the Location of the Eye in Order to Compute the Summary Measures?

Sampling frequency is an important consideration with respect to the temporal accuracy with which eye movements (glances, fixations, or saccades) can be identified. While ISO (2001, 2002) does not define a minimum sampling frequency (frames per second [fps]) for eye tracking in an automotive context, SAE (2000) does provide an illustrative example of data reduction using the context of 30 fps video. In contrast, the eye-tracking literature (Holmqvist et al., 2011, p. 30) would suggest that a minimum sampling frequency of 250 Hz is needed to reasonably categorize the initiation and conclusion of a saccadic movement. We need this information in order to know when a glance (as defined by ISO 2001) begins and ends. Eye-tracking systems often report metrics based upon a running average of raw measurements without also reporting the algorithms used to compute the averages, thus further complicating the degree to which one can accurately classify the time course of eye movements (Reimer & Sodhi, 2006).

While laboratory studies of eye movements in vision science prize spatial precision and high sampling rates, analyses of eye movements in the context of driving, especially with novice drivers, do not always require high-precision collection. In fact, the large saccades to and from the road or to and from a target beside the road that determine crash likelihoods can be captured with far less specificity. Signals on the order of 15 fps can capture most relevant behaviors, as fixation times and fixation locations are generally greater than 150 milliseconds (Carpenter, 1988), often much longer. What is critical in the interpretation of eye movements is an understanding of just how much accuracy one needs in terms of the sampling rate to categorize behaviors of interest and then working within those bounds.

27.3.4 How Can Eye Movements Best Be Reduced to Summary Measures?

ISO (2002) provides a reasonably simplified, well-documented analytical approach that reduces the perceptual process to a series of glances that can be readily recorded through commercially available eye-tracking technology (see Section 27.3.7) or through experienced coders' visual inspection of a recording of the driver's eyes. Human-coded eye movement data, based on video of the driver's face, are adequate to support many analytical approaches (Mehler et al., 2016; Smith et al., 2005) and are a critical method with which to validate the output of an eye tracker (Reimer et al., 2013). Manual coding is not without limitations. For example, discriminating the true target of an eye movement becomes harder for human observers as a function of the target's eccentricity in the visual field (cf., Schieber et al., 1997). However, despite its limitations, manual coding can be highly useful in applied research with novice drivers.

27.3.5 What about Measures of Information in the Periphery?

As discussed to this point, eye movements are used to foveate (focus on) relevant parts of the scene. The visual periphery, nonetheless, can be a source of important (perhaps best considered as blurry) information (Anstis, 1998). While visual information in the periphery is less accessible than in the fovea, it represents more than 99% of the entire visual field and is essential in a driving context. This is especially true when vehicles, pedestrians, or bicyclists are unexpectedly approaching from the side. While the literature does not provide a well-developed theory on the relative contribution of the foveal and peripheral visual field to driving, several studies have considered the role of peripheral vision in safe driving (e.g., Horrey et al., 2006; Owsley & McGwin, 1999).

Bearing in mind the importance and utility of peripheral vision and the overall movement of the eyes, researchers have examined metrics of attentional allocation using different approaches than those focused on specific objects in the environment. These metrics include the useful field of view, among others (Ball et al., 2002). Here, it is determined just how far out into the periphery information can be gathered when the driver is and is not performing a foveally positioned secondary task and when the information in the periphery stands out or is surrounded by distractors. During the measurement of the useful field of view, the participant is given only a very brief glance at the stimulus array, and there is no time for movement of the eyes or the head. Interestingly, this is one of the few measures where novice drivers may be better than experienced drivers given that the size of the useful field of view generally decreases with age. However, no one has compared the useful field of view of novice and experienced drivers where the targets would be ones that were not random, but rather, were ones related to traffic safety. Given that older drivers know better what to expect, it may well be that these expectations actually benefit them and that they have a larger useful field of view than do novice drivers.

27.3.6 What about Measures of Scanning Patterns?

It is useful to summarize the differences in scanning patterns. The movement of the eyes across a broad region of interest can be studied using variability metrics or the percentage of time centered on an AOI such as the road center or across the coordinate system of the eye tracker (Harbluk et al., 2007; Reimer et al., 2012; Sodhi et al., 2002; Wang et al., 2014). Such analytic approaches can average across fixations, smooth pursuit movements, and saccades and provide metrics that appear sensitive to changes in experience and cognitive workload, just to name a few. In fact, earlier in this chapter, it was noted that novice drivers scanned less far side to side than did experienced drivers.

27.3.7 What Types of Eye-Tracking Systems Are There?

Inasmuch as there are a variety of measures that can be derived from the monitoring of eye movements, there are many different vendors and types of eye-tracking systems. In general, eye-tracking systems can be broadly classified as being either head mounted or remotely mounted, although there remain a wide range of features and variations within each class. In some cases, the underlying optics or algorithms are similar across class. In this section, we describe some of the characteristics of these different types of systems, while highlighting some of the practical considerations for each. And we point out, where relevant, what features are particularly useful when studying novice drivers.

27.3.7.1 Head-Mounted Eye-Tracking Systems

This type of system has been widely used in many basic and applied settings. The name derives from the fact that the cameras and optics that monitor the eye are mounted onto a helmet or hat worn by the participant. Recently, more and more manufacturers are building the optics into lightweight glasses—a move that can make the equipment less cumbersome and also, depending on the nature of the study, can free up the participants' heads for other types of physiological measurement (e.g., electroencephalography [EEG]). The system optics often include two cameras, one directed toward the scene (i.e., the general direction where the eyes/face is pointed) and one directed at the eye. Additionally, small infrared (IR) illuminators are sometimes used to brighten the view of the eye camera; for other systems, IR illumination is a critical part of the determination of point of regard. For these latter systems, IR is used to bounce a small point of light off of the cornea (referred to as the corneal reflection [CR]). Some more advanced systems consider multiple IR sources and CRs. In conjunction with a calibration procedure, where the participant is asked to look at a series of fixed and known locations (usually a grid or array of targets), the system (through matrix mathematical transforms) can determine the point of regard based on the relative location of the CR and the pupil. Systems based on CR have long been a cornerstone for eye-tracking systems; however, systems that rely on computer vision algorithms to determine the direction of the pupil are now common as well (Al-Rahayfeh & Faezipour, 2013).

Head-mounted systems have many advantages. One of the advantages for this type of system is that they tend to be less expensive than many remote tracking systems. Additionally, they tend to be easy to deploy and fairly robust in terms of reliable data collection, not to mention that the angular error (i.e., the accuracy of the calculated point of regard) of these systems can be better than many remote systems, especially multicamera ones. Also, many of these systems operate from a laptop or tablet, making them more portable than fixed, remote systems.

However, one drawback relates to the fact that the optics are mounted on the head. While the system can reliably provide x and y coordinates for the momentary point of regard, the frame of reference (the forward-looking camera) is always moving, and so mapping these coordinates onto features and locations in the world can require a large amount of postprocessing of data. Many vendors provide software that supports this activity, but it can still be a time-consuming process. Other innovations, such as the integration of eye movements with motion capture suites or head-tracking equipment, can offer some shortcuts to linking eye data to a global reference system, as can other approaches, such as the use of markers mounted in the world that can be identified by computer vision algorithms (e.g., identifying and logging data according to dynamic AOIs).

Other drawbacks include the following (1) the head-tracking equipment is cumbersome, even though the material can be lightweight; (2) wearing it for extended periods of time can be uncomfortable and fatiguing; and (3) in some cases, the equipment itself might constrain natural scanning behavior (e.g., a driver might be less inclined to turn his or her head to the sides when wearing a wired helmet). Newer systems with wireless transmission of data make this much less of a problem now. Finally, because the optics are proximally located, they are subject to being knocked about by participants (e.g., scratching head, rubbing eyes)—which can disrupt the calibration and quickly render a session's data unsalvageable (although recalibration is possible in real time and, sometimes, even after the fact, assuming one has fixed targets that one knows that the participant is looking toward at a particular moment in time, e.g., a stop sign). To ensure that a data stream is robust to optical movement, validation of measurements across a recording (drive, task, etc.) is critical.

27.3.7.2 Remote Eye-Tracking Systems

As the name denotes, the cameras and optics for remote eye-tracking systems are located near the participant or driver but are nonintrusive or contact-free. The least sophisticated remote eye-tracking system—and perhaps *eye-tracking system* is a generous term—is a video camera aimed at the participant's face or eyes. High-resolution cameras are clearly a benefit in these types of uses. The setup and administration of this type of system can be elegant in its simplicity; however, there are obvious constraints. For starters, because this approach cannot calibrate the direction of the eyes to precise or known locations in space, the experimenter is generally limited to a rather coarse assessment of where the individual is looking. For example, with a properly mounted camera, one can distinguish—with reasonable confidence—between drivers' glances to the road and glances inside the vehicle (or away from the road) (see Smith et al., 2005, for a more in-depth discussion on the reliability of the approach). Thus, many important experimental questions can be addressed using such a technique—particularly in situations where knowledge of precise x and y coordinates is not critical. Probably the major downside is that video does not output data in a useable format; it must be coded. Frame-by-frame coding of video can be enormously time-consuming. For some—those without wide access to ready and willing undergraduate students—these time costs can be prohibitive. Methods of automating the arduous task of glance identification in an automobile are being explored with significant success through machine learning approaches (Fridman, Langhans et al., in press; Fridman, Lee et al., 2016). The approaches leverage large manually labeled data sets of driver glances' locations (i.e., frame-by-frame coded video) to make predictions of the glance orientation of new data frames with accuracies of over 90%.

More sophisticated remote eye-tracking systems run the gamut as well and can include a single or multiple cameras. Desktop single-camera systems are utilized in many basic lines of research that can be administered from a simple workstation (i.e., single computer monitor). Many such systems operate on the same principles of CR as described previously for head-mounted systems. In general, these approaches favor precision in angular accuracy, and so they often employ chin and forehead rests to immobilize the head, or even bite bars. That being said, the precision is limited to the calibration space (e.g., the computer monitor); any glances outside of this range tend to be lost or inaccurate.

It follows that multicamera systems are more appropriate in situations where the participant is expected to make larger and/or more frequent head movements as they perform a task. Such systems tend to use computer vision algorithms along with a calibration process that tracks facial features (versus traditional CR). Although the technology and algorithms in support of multicamera systems have grown in the past decade, the introduction of head movements can often present some challenges for the robustness of data collection (described in Section 27.3.8).

27.3.8 What Are Some of the Challenges Concerning Data Acquisition, Handling, Processing, and Interpreting?

27.3.8.1 What System Is Right for You?

Deciding what eye tracker meets your needs requires a consideration of the trade-offs associated with different types of systems as well as the specific situational needs or constraints. For the latter, some experiments or environments simply cannot accommodate a particular system. For example, because there is no experimenter present, a field operational test or naturalistic driving study generally cannot rely on a system that must be set up or calibrated for each use. Currently, this means that only camera systems are possible, which, as noted previously, have real disadvantages. However, it is likely that technological advances will soon offer increasing options.

When trade-offs are possible, one consideration involves the desired or required degree of angular accuracy of the system (degrees of visual angle). Were one concerned only with whether a teen driver was looking at the road ahead or at a friend in the passenger seat, the angular accuracy of the system might be less important, as a coarse appraisal of eye or head movements would offer sufficient insight. However, if one needs to know whether drivers fixated on a precise target or location in the world (e.g., a potential hazard located down the road), more system precision is necessary. This issue is also related to the quality of the recorded images, especially for systems that rely not only on the x and y coordinates of the point of regard but also on the recorded video. Using the last example, in determining whether the driver looked at a target, an experimenter also needs to be able to resolve the target object in the video.

Other considerations include the ease of calibration, which depends on factors such as the eye shape, presence of contacts or glasses, and light levels, as well as the time to learn and achieve calibration reliably. Although important in all contexts, timely and reliable calibration might be critical in some operational settings (e.g., in field environments or with special populations of novice drivers). Ideally, some systems can alert users when calibration has drifted and recalibration is needed. Some postprocessing software now allows users modest capability to make *post hoc* corrections for calibration drift.

Not completely unrelated to the ease and maintenance of calibration, the operational environment can impact the performance of systems (i.e., the percentage of good, useable data). As noted previously, a user might be interested in tracking eye movements across a wide operational field of view, and so a tracker with an effective range of only 30° of visual field would not be practical (generally, however, head-mounted eye trackers are fine in this case because an individual will typically move his or her head when large eye movements are required). Moreover, some environments are very hostile for eye trackers, including areas of high glare or sharp fluctuations in lighting conditions, high vibration, and other noise. Discovering the constraints of different types of systems in a given context can sometimes be a frustrating process of trial and error. Unfortunately, successful implementation in one setting does not necessarily mean that the same system will work as well in a similar context. Such limitations would be particularly important to consider if, for example, one wanted to evaluate a novice driver's eye behaviors during the nighttime, an especially high-risk period for these drivers.

27.3.8.2 Data Synchronization

In some cases, it is important to synchronize data from eye trackers with data from other systems, such as driving simulators and physiological monitors. Often, inserting manual markers in the separate data streams can be an effective yet simple approach (i.e., flagging independent data streams and later using these flags to synchronize the separate time stamps). Where more precision is necessary or where it is too cumbersome to manually flag multiple data streams, one can sometimes digitally synchronize data streams by sending and logging time-stamp information, sent from one system to another, or by using specialized software that is intended to coordinate and log data from multiple channels into one data stream. More novel approaches using vibration and steering data for automated synchronization have also been proposed (Fridman, Brown et al., 2016). Some vendors offer or bundle such software with their system. Other systems offer provisions to send or receive data from other sources for the purposes of integration (e.g., software development kit [SDK]). Other systems are strictly stand-alone, with no means for digital synchronization with external systems. Knowing the capabilities of the system under consideration and, better still, that other users have successfully achieved a similar integration can greatly impact one's level of confidence in the prospects for doing so. With novice drivers, the integration of eye data with vehicle and driver data can be especially important in studying hazard mitigation where, as described previously, hazard mitigation depends on successful hazard anticipation. In developing a data synchronization plan, it is critical to understand the time course of the activity of interest (e.g., a novice driver changing lanes over several seconds) and the synchronized time course in which various data streams (lateral control of the vehicle, eye movements, etc.) can support the assessment of questions of interest. While subsecond synchronization is critical for predictive modeling, it not always needed for general observation of behaviors.

27.3.8.3 Sampling Rates

The frequency at which a system collects data (e.g., 50 Hz, 100 Hz) might be a concern for some applications. Higher rates translate to more data per unit time and, hence, larger data files—a consideration for data management. Yet, higher rates can also translate to more precise data in the time dimension, especially for saccadic measures. In some cases, users want to employ gaze-contingent event triggers, such that the information or environments perform some function based on where the eyes are currently directed. For example, one might wish to have a lead vehicle brake sharply, but only at the moment when the driver diverts his or her eyes toward the vehicle console. Generally, these applications call for higher rates (e.g., 240 Hz). Such a capability would be critical when trying to understand why novice drivers are overinvolved in rear-end collisions, being the striking vehicle in most cases.

27.3.8.4 Postprocessing

Data management also calls for some consideration, especially for systems that log eye parameters as well as video files. Depending on the number of cameras, resolution, and video compression algorithms, some systems can only record data continuously for a couple of hours. In most experimental situations involving multiple blocks or trials, these limits can readily be accommodated; however, users looking to examine eye movements for extended periods of time should plan accordingly. This would be true, for example, if one wanted to study the effect of a training program on the eye behaviors of a novice driver over tens or hundreds of hours of naturalistic driving.

The more common issue associated with postprocessing involves the transformation of the raw data into the more meaningful metrics described in Section 27.3.2. The raw data themselves are uninformative: a row of data amounts to a snapshot (e.g., 1 of 30 or 100 per second) that captures whether the pupil was detected and the x and y coordinates, among others. Only through the application of algorithms or coding rules across multiple rows of data can one begin to glean information regarding fixations and glances, blinks, and saccades. This process can be very time-consuming. Moreover, this usually calls for effort from multiple individuals; this, in turn, requires that a careful and clear coding scheme is developed such that each individual is distilling the same data elements (see Smith et al., 2005, for additional details). Reporting measures of interrater reliability (e.g., kappa) are encouraged and, in some cases, necessary. This would be true with data on an experiment with novice drivers and hazard anticipation where the question was whether the driver looked at a particular region at a particular point in time. This information needs to be coded by individual raters since the AOI is changing in time and space. Fortunately, software can greatly facilitate this process when the AOI remains more or less stable (e.g., the interior of the car).

In addition to facilitating the distillation of various metrics (e.g., Section 27.3.2), postprocessing software can also offer many different features that can aid in the analysis and interpretation of eye data. For example, data visualization tools can allow users to examine scan paths and heat map overlays, perform link analyses, and build probability matrices for transitions between AOIs. While not strict requirements for many applications, potential users might find them worthy when considering different eye-tracking solutions for understanding novice driver behavior. For example, building probability matrices between different AOIs could be very useful when trying to develop a real-time algorithm that monitors driver state.

27.4 Conclusions

As described in this chapter, eye movements have long been studied in basic and applied settings, and the many diverse measures that can be derived from eye data have greatly expanded our understanding of attention, cognition, and information processing (Rayner, 1998). They are particularly important and relevant for our understanding of drivers, including novice and teen drivers (Taylor et al., 2013). Arguably, without knowledge of eye movements, we would still not be clear about the critical differences between novice and experienced drivers' hazard anticipation (Pradhan et al., 2005), hazard mitigation (Muttart et al., 2013), attention maintenance (Chan et al., 2010), and vehicle control skills (Mourant & Rockwell, 1972), exactly those skills that are so critical to the safety of the driver.

One can ask whether these differences are due to novice drivers being careless or simply inexperienced or uninformed. McKnight and McKnight (2003), using police crash records, argue that these differences suggest that novice drivers, for the most part, are clueless, not careless. In terms relevant to this chapter, these differences point to an underdeveloped understanding or mental model on the part of novice drivers regarding the appropriate prioritization of the most pertinent areas for information acquisition, not to mention an imbalance in the relative demands of various driving subtasks (e.g., more attention required for basic vehicle control in inexperienced drivers: see also Chapter 7). To take this one step further, it is experiments using eye behaviors to understand novice drivers that have added greatly to the evidence that novice drivers are clueless rather than careless (Pollatsek et al., 2006).

Measures of eye movements can be gleaned from many different types of systems, including head-mounted and remote solutions. Each type of system carries with it benefits and drawbacks, and the ultimate utility of a given system is highly contingent upon the user's needs and preferences. For example, if a researcher were only interested in the degree to which a particular in-dash infotainment unit diverted a novice driver's eyes away from the road, a simple video camera system might be sufficient. However, if one were more concerned with the study of a novice driver's glances to specific targets in the environment (e.g., road signs), a spatially accurate system, whether head mounted or remote, would be necessary. While space constraints do not permit a full and elaborated discussion of all of the issues inherent to eye tracking, we hope that the chapter provided a useful introduction and overview.

There is much that remains to be known, and we believe that eye movements will prove to be a key to expanding our understanding of why it is that novice drivers crash more frequently than experienced drivers (though admittedly only one of several questions needing answer). For example, we now know that the learner's permit period does not change the sharp reduction in crashes that occurs over the first year after a teen driver receives a restricted (solo) license (Foss et al., 2011). We cannot hope to know what the novice driver is learning unless we know the information to which he or she is attending. Eye movement records will give us this information. As another example, consider special populations. We know that teens with attention deficit hyperactivity disorder (ADHD) make especially long glances away from the forward roadway (Kingery et al., 2015). We cannot hope to determine whether training programs designed to reduce the frequency of these especially long glances will work unless we record the eye behaviors of the ADHD teens before and after training. The unsafe driving behaviors of other special populations seem equally likely to yield to the insights provided by eye trackers. Finally, as a third example, consider the increasingly automated vehicles of the future. These vehicles will require that the system know the driver state moment by moment whenever control needs to be transferred from the automated driving suite to the driver. Teens, as has been discussed, are easily distracted, more so than adults. Real-time knowledge of where the driver is focusing will be especially important to know. Eye trackers are now inexpensive enough to be considered for this function. It remains to be seen whether technology can advance far and fast enough to make autocalibration a reality in the near future.

Acknowledgments

Grants from the AUTO21 Network of Centres of Excellence (AUTO21 NCE) and the Canadian Foundation for Innovation (CFI) to JKC funded a number of projects and capital equipment that aided this review.

Appendix 27.1

Research has delineated a number of basic strategies or patterns of eye movements that occur in certain circumstances while driving. This table provides a summary of these results. Inclusion of studies in the table is not intended to be comprehensive but merely representative of a progression of knowledge.

Context, Group, or Condition	Historical and Representative Studies
Speed	• Miura (1979): Speed reduces the breadth of search and fixation durations and increases the frequency of fixations in a complex, demanding road environment. • Spijkers (1992): With greater speed, drivers become more efficient at focusing on relevant information sources within and along the roadway.

(*Continued*)

Context, Group, or Condition	Historical and Representative Studies
Curves	• Shinar et al. (1977): Drivers fixate on curves in advance of turning and lateral cues while turning. • Land and Horwood (1995): Drivers look into the curve for information cues and away from the direction of travel. • Lehtonen et al. (2013): Look-ahead fixations in rural curves aid the driver in planning and oncoming vehicle anticipation. Higher workload resulted in shorter look-ahead fixations.
Signs	• Mourant et al. (1969): Fixation durations and distributions to various information sources while driving. • Luoma (1988): Signs are almost always foveated. • Zwahlen and Schnell (1998): Empirically derived model of minimum required legibility distance for symbolic warning signs. • Caird et al. (2007): In-vehicle advanced warning signs were looked at longer and more frequently by younger drivers.
Straight	• Serafin (1994): Synthesis of glance distributions to roadway, traffic, and signs across studies.
Lane change	• Robinson et al. (1972): Number of fixations and sequence (mirrors, direction, of movement) depend on traffic present. • Olson et al. (2005): Before changing lanes to the left, the frequency of glances to the rearview mirror, left mirror, and blind spot.
Intersections	• Miura (1986): Drivers process deeper at each fixation point when scanning an intersection. • Land (1992): A description and model of eye–head coordination scanning at an intersection. • Theeuwes (1996): Top-down search of intersections for expected information such as signs and pedestrian locations predominates search even when unexpected information is available. • Milloy et al. (2010): Novice drivers were less likely to sample their rearview mirror when approaching a late yellow light.
Car following	• Mourant et al. (1969): Route familiarity reduces fixation durations on average. • Sivak et al. (1986): Concentration of glances were in the center of the rear window, arguing for high-mounted brake lights. • Crundall et al. (2004): Car following reduces the spread of visual search, especially at night. • Tijerena et al. (2004): Drivers look away briefly from lead vehicles if optical expansion is near zero.
Steering control	• Senders et al. (1967): Frequency and duration of attention to the traffic environment is related to uncertainty of visual information about self position and other vehicles. • Land and Horwood (1996): Distal information is used to anticipate road curvature, and proximal information is used to fine-tune lane position.
Novice drivers (first 6 months)	• Mourant and Rockwell (1972): Novice drivers concentrated their gaze in a small area and looked closer to the vehicle than experienced drivers. • Chapman et al. (2002): A training intervention increased novice drivers' search for hazards. • Underwood et al. (2002): Constrained visual search of a motorway for hazards in novice drivers appears to be due to limited mental models. • Falkmer and Gregersen (2001, 2005): Novice drivers with cerebral palsy have less flexible search strategies. Novice drivers did not fixate closer to the vehicle. • Fisher et al. (2006): Training of novice drivers to look for hazards was effective. • Olson et al. (2007): Glances of novice drivers to mirrors improved after 6 months of driving on-road (also see Underwood, 2013). • Milloy et al. (2010): Novice driver eye movements when encountering a late yellow light at intersections.
Older drivers	• Owsley and McGwin (1999): Disease effects on vision and driving. • Ho et al. (2001): Older drivers require longer and more frequent fixations to search for information from traffic signs than younger drivers.
Motorcycle riders	• Hosking et al. (2010): Inexperienced riders were less flexible in their visual search for hazards than experienced riders. • Ohlhauser et al. (2011): Car drivers who were also motorcycle riders searched and responded to motorcycles more flexibly.

(*Continued*)

Context, Group, or Condition	Historical and Representative Studies
In-vehicle distractions	• Dingus et al. (1988): An array of in-vehicle activities, including navigation and dashboard tasks, were examined on public roads. Total glance time, lane exceedences, and single glances longer than 2 seconds were used to stratify tasks. • Recarte and Nunes (2003): Cognitive tasks result in a reduction in the breadth of scanning. • Green and Shah (2004): A review of eye movements to telematics is presented. • Chisholm et al. (2008): Glance frequency and duration to MP3 players affect driving performance. • Caird et al. (2014): Meta-analysis of eye movements in texting indicates large effect sizes for eyes off road. Reading and typing texts increases the frequency, and duration of glances away from the road affects vehicle control and detection of hazards. • Tivisten and Dozza (2014): Sampling of distractions is frequently timed relative to the demand of the roadway.
External distractions	• Eye movements to external distractions are reviewed by Milloy and Caird (2011).
Alcohol	• Eye-steering control becomes progressively less coordinated as blood alcohol content (BAC) increases (Marple-Horvat et al., 2007).
Night driving	• Mortimer and Jorgeson (1974): The breadth of fixations is constrained by low beams, and fixations are concentrated in the lane ahead.

References

Al-Rahayfeh, A.M.E.R., & Faezipour, M.I.A.D. (2013). Eye tracking and head movement detection: A state-of-art survey. *Translational Engineering in Health and Medicine*, 2100212.

Anstis, S. (1998). Picturing peripheral acuity. *Perception, 27*, 817–826.

Ball, K., Wadley, V.G., & Edwards, J.D. (2002). Advances in technology used to assess and retrain older drivers. *Gerontechnology, 1*(4), 251–261.

Borowsky, A., Oron-Gilad, T., & Parmet, Y. (2009). Age and skill differences in classifying hazardous traffic scenes. *Transportation Research Part F: Traffic Psychology and Behaviour, 12*(4), 277–287.

Borowsky, A., Shinar, D., & Oron-Gilad, T. (2010). Age, skill and hazard perception in driving. *Accident Analysis & Prevention, 42*, 1240–1249.

Caird, J.K., Chisholm, S.L., & Lockhart, J. (2007). Do in-vehicle advanced signs enhance older and younger drivers' intersection performance? Driving simulation and eye movement results. *International Journal of Human-Computer Studies, 66*, 132–144.

Caird, J.K., Johnston, K., Willness, C., Asbridge, M., & Steel, P. (2014). A meta-analysis of the effects of texting on driving. *Accident Analysis & Prevention, 71*, 311–318.

Carpenter, R. (1988). *Movements of the Eyes* (2nd edn). London: Pion Limited.

Chan, E., Pradhan, A.K., Pollatsek, A., Knodler, M.A., & Fisher, D.L. (2010). Are driving simulators effective tools for evaluating novice drivers' hazard anticipation, speed management, and attention maintenance skills? *Transportation Research: Part F, 13*, 343–353.

Chapman, P.R., & Underwood, G. (1998). Visual search of driving situations: Danger and experience. *Perception 27*, 951–964.

Chapman, P., Underwood, G., & Roberts, K. (2002). Visual search patterns in trained and untrained novice drivers. *Transportation Research: Part F, 5*, 157–167.

Chisholm, S., Caird, J.K., & Lockhart, J. (2008). The effects of practice with MP3 players on driving performance. *Accident Analysis & Prevention, 40*, 704–713.

Crundall, D.E., Shenton, C., & Underwood, G. (2004). Eye movements during intentional car following. *Perception, 33*, 975–986.

Curry, A.E., Hafetz, J., Kallan, M.J., Winston, F.K., & Durbin, D.R. (2011). Prevalence of teen driver errors leading to serious motor vehicle crashes. *Accident, Analysis & Prevention, 43*, 1285–1290.

Deery, H.A. (1999). Hazard and risk perception among young novice drivers. *Journal of Safety Research, 30*(4), 225–236.

Dingus, T.A., Antin, J.F., Hulse, M., & Wierwille, W.W. (1988). Attentional demand requirements of an automobile moving-map navigation system. *Transportation Research: Part A, 23*(4), 301–315.

Divekar, G., Pradhan, A.K., Pollastek, A., & Fisher, D.L. (2012). External distractions: Evaluation of their effect on younger novice and experienced drivers behavior and vehicle control. *Transportation Research Record, 2321*, 15–22.

Driver Focus–Telematics Working Group. (2006). *Statement of Principles, Criteria, and Verification Procedures on Driver-Interactions with Advanced In-Vehicle Information and Communication Systems*. Washington, DC: Alliance of Automobile Manufacturers.

Falkmer, T., & Gregersen, N.P. (2001). Fixation patterns of learner drivers with and without cerebral palsy (CP) when driving in real environments. *Transportation Research: Part F, 4*, 171–185.

Falkmer, T., & Gregersen, N.P. (2005). A comparison of eye movement behavior of inexperienced and experienced drivers in real traffic environments. *Optometry and Vision Science, 82*(8), 732–739.

Fisher, D.L., Pollatsek, A., & Pradhan, A. (2006). Can novice drivers be trained to scan for information that will reduce the likelihood of a crash? *Injury Prevention, 12*(Suppl. 1), i25–i29.

Foss, R., Martell, C., Goodwin, A., O'Brien, N., & UNC Highway Research Center. (2011). *Measuring Changes in Teenage Driver Characteristics During the Early Months of Driving*. Washington, DC: AAA Foundation for Traffic Safety.

Fridman, L., Brown, D.E., Angell, W., Abdic, I., Reimer, B., & Young Noh, H. (2016). Automated synchronization of driving data using vibration and steering event. *Pattern Recognition Letters, 75*(1), 9–15.

Fridman, L., Lee, J., Reimer, B., & Victor, T. (2016). "Owl" and "Lizard": Patterns of head pose and eye pose in driver gaze classification. *IET Computer Vision*, doi: 10.1049/iet-cvi.2015.0296.

Fridman, L., Langhans, P., Lee, J., & Reimer, B. (in press). Driver gaze region estimation without using eye movement. *IEEE Intelligent Systems*. http://arxiv.org/pdf/1507.04760.pdf.

Green, P. (2007). Where do drivers look while driving (and for how long)? In R. Dewar & P. Olson (Eds.), *Human Factors in Traffic Safety* (2nd edn) (pp. 57–82). Tucson, AZ: Lawyers and Judges Publishing.

Green, P., & Shah, R. (2004). *Safety Vehicles Using Adaptive Interface Technology (Task 6). Task Time and Glance Measures of the Use of Telematics: A Tabular Summary of the Literature*. Washington, DC: NHTSA.

Harbluk, J.L., Noy, Y.I., Trbovich, P.L., & Eizenman, M. (2007). An on-road assessment of cognitive distraction: Impacts on drivers' visual behavior and braking performance. *Accident Analysis & Prevention, 39*(2), 372–379.

Ho, G., Scialfa, C.T., Caird, J.K., & Graw, T. (2001). Visual search for traffic signs: The effects of clutter, luminance, and aging. *Human Factors, 43*(2), 194–207.

Holmqvist, K., Nyström, M., Andersson, R., Dewhurst, R., Jarodzka, H., & Van de Weijer, J. (2011). *Eye Tracking: A Comprehensive Guide to Methods and Measures*. Oxford: Oxford University Press.

Horrey, W.J., & Wickens, C.D. (2007). In-vehicle glance duration: Distributions, tails, and a model of crash risk. *Transportation Research Record, 2018*, 22–28.

Horrey, W.J., Wickens, C.D., & Consalus, K.P. (2006). Modeling drivers' visual attention allocation while interacting with in-vehicle technologies. *Journal of Experimental Psychology: Applied, 12*(2), 67–78.

Horswill, M.S., & McKenna, F.P. (2004). Drivers' hazard perception ability: Situation awareness on the road. In S. Banbury & S. Tremblay (eds.) *A Cognitive Approach to Situation Awareness: Theory and Application*. Aldershot, UK: Ashgate, pp. 155–175.

Hosking, S.G., Liu, C.C., & Bayly, M. (2010). The visual search patterns and hazard responses of experienced and inexperienced motorcycle riders. *Accident Analysis & Prevention, 42*, 196–202.

ISO 15007-2. (2001). *Road vehicles—Measurement of Driver Visual Behaviour with Respect to Transport Information and Control Systems—Part 2: Equipment and Procedures*. Geneva: International Standards Organization.

ISO 15007-1. (2002). *Road Vehicles—Measurement of Driver Visual Behavior with Respect to Transport Information and Control Systems—Part 1: Definitions and Parameters (ISO Committee Standard 15007-1)*. Geneva: International Standards Organization.

Kingery, K.M., Narad, M., Garner, A.A., Antonini, T.N., Tamm, L., & Epstein, J.N. (2015). Extended visual glances away from the roadway are associated with ADHD- and texting-related driving performance deficits in adolescents. *Journal of Abnormal Child Psychology, 43*(6), 1175–1186.

Land, M.F. (1992). Predictable eye–head coordination while driving. *Nature, 359*, 318–320.

Land, M.F. (2006). Eye movements and the control of actions in everyday life. *Progress in Retinal and Eye Research, 25*, 296–324.

Land, M.F., & Horwood, J. (1995). Which parts of the road guide steering? *Nature, 377*, 339–340.

Land, M., & Horwood, J. (1996). The relations between head and eye movements during driving. *Vision in Vehicles, 5*, 153–160.

Lehtonen, E., Lappi, O., Kotkanen, H., & Summala, H. (2013). Look-ahead fixations in curve driving. *Ergonomics, 56*(1), 34–44.

Livingstone, M. (2002). *Vision and Art: The Biology of Seeing*. New York: Harry N. Abrams.

Luoma, J. (1988). Drivers' eye fixations and perceptions. In A.G. Gale, M.H. Freeman, C.M. Haslegrave, P. Smith, & S.P. Taylor (Eds.), *Vision in Vehicles II* (pp. 231–237). Amsterdam: Elsevier Science.

Marple-Horvat, D.E., Cooper, H.L., Glibey, S.L., Watson, J.C., Mehta, N., Kaur-Mann, D., Wilson, M., & Keil, D. (2007). Alcohol badly affects eye movements linked to steering, providing for automatic in-car detection of drink driving. *Neuropsycholpharmacology, 33*, 849–858.

McKnight, J.A., & McKnight, S.A. (2003). Young novice drivers: Careless or clueless. *Accident, Analysis & Prevention, 35*, 921–925.

Mehler, B., Kidd, D., Reimer, B., Reagan, I., Dobres, J., & McCartt, A. (2016). Multi-modal assessment of on-road demand of voice and manual phone calling and voice navigation entry across two embedded vehicle systems. *Ergonomics, 59*(3), 344–367.

Milloy, S., & Caird, J.K. (2011). External driver distractions: The effects of video billboards and wind farms on driving performance. In D.L. Fisher, M. Rizzo, J.K. Caird, & J.D. Lee (Eds.), *Handbook of Driving Simulation for Engineering, Medicine, and Psychology* (pp. 16-1–16-14). Boca Raton, FL: CRC Press.

Milloy, S., Caird, J.K., Ohlhauser, A., & Pearson, A. (2010). Do responses differ between novice and experienced driver when a late yellow light is encountered? *Proceedings of the 54th Annual Conference of the Human Factors and Ergonomics Society* (pp. 2081–2085). San Francisco.

Miura, T. (1979). Visual behavior in driving: An eye movement study. *Departmental Bulletin Paper: Faculty of Human Sciences*, Osaka University, *5*, 253–289.

Miura, T. (1986). Visual search in intersections: An underlying mechanism. *IATSS Research, 16*(1), 42–49.

Mortimer, R.G., & Jorgeson, C.M. (1974). *Eye Fixations of Drivers in Night Driving with Three Headlight Beams (Rep. No. UM-HSRI-HF-74-17)*. Ann Arbor, MI: Highway Safety Research Institute.

Mourant, R.R., & Rockwell, T.H. (1972). Strategies of visual search by novice and experienced drivers. *Human Factors, 14*(4), 325–335.

Mourant, R.R., Rockwell, T.H., & Rackoff, N.J. (1969). Drivers' eye movements and visual workload. *Highway Research Record, 292*, 1–10.

Muñoz, M., Reimer, B., & Mehler, B. (2015). Exploring new qualitative methods to support a quantitative analysis of glance behavior. *Proceedings of the 7th International Conference on Automotive User Interfaces and Interactive Vehicle Applications (AutomotiveUI '15)*, Nottingham, UK.

Muttart, J., Fisher, D., Pollatsek, A., & Marquard, J. (2013). Comparison of anticipatory glancing and risk mitigation of novice drivers and exemplary drivers when approaching curves. *Proceedings of the Seventh International Driving Symposium on Human Factors and Driver Assessment, Training, and Vehicle Design*. Iowa City: Public Policy Center, University of Iowa.

Muttart, J., Fisher, D., & Pollatsek, A. (2014). Comparison of anticipatory glancing and risk mitigation of novice drivers and exemplary drivers when approaching intersections in a driving simulator. *Proceedings of the Transportation Research Board 93rd Annual Meeting.* Washington, DC: National Academy of Sciences.

NHTSA (National Highway Traffic Safety Administration). (2013). *Visual-manual NHTSA Driver Distraction Guidelines for In-vehicle Electronic Devices (Docket No. NHTSA-2010-0053).* Washington, DC: U.S. Department of Transportation National Highway Traffic Safety Administration (NHTSA).

Ohlhauser, A., Milloy, S., & Caird, J.K. (2011). Driver responses to motorcycle and lead vehicle braking events: The effects of motorcycling experience and novice versus experienced drivers. *Transportation Research: Part F, 14,* 472–483.

Olson, E.C.B., Lee, S.E., & Wierwille, W.W. (2005). Eye glance behaviour during lane changes and straight-ahead driving? *Transportation Research Record, 1937,* 44–50.

Olson, E.C.B., Lee, S.E., & Simons-Morton, B.G. (2007). Eye movement patterns for novice drivers: Does 6 months of driving experience make a difference? *Transportation Research Record, 2009,* 8–14.

Owsley, C., & McGwin, G. (1999). Vision impairment and driving. *Survey of Ophthalmology, 43*(6), 535–550.

Palmer, S.E. (1999). *Vision Science: Photons to Phenomenology.* Cambridge, MA: MIT Press.

Pollatsek, A., Fisher, D.L., & Pradhan, A. (2006). Identifying and remedying failures of selective attention in younger drivers. *Current Directions in Psychological Science, 15*(5), 255–259.

Pradhan, A.K., Hammel, K.R., DeRamus, R., Pollatsek, A., Noyce, D.A., & Fisher, D.L. (2005). The use of eye movements to evaluate the effects of driver age on risk perception in an advanced driving simulator. *Human Factors, 47,* 840–852.

Rayner, K. (1998). Eye movements in reading and information processing: 20 years of research. *Psychological Bulletin, 124,* 372–422.

Recarte, M.A., & Nunes, L.M. (2003). Effects of verbal and spatial-imagery tasks on eye movements while driving. *Journal of Experimental Psychology: Applied, 6*(1), 31–43.

Reimer, B., & Sodhi, M.S. (2006). Detecting eye movements in dynamic environments. *Behavior Research Methods, 38*(4), 667–682.

Reimer, B., Mehler, B., Wang, Y., & Coughlin, J.F. (2012). A field study on the impact of variations in short-term memory demands on drivers' visual attention and driving performance across three age groups. *Human Factors, 54*(3), 454–468.

Reimer, B., Mehler, B., Dobres, J., & Coughlin, J.F. (2013). The effects of a production level "voice-command" interface on driver behavior: Reported workload, physiology, visual attention, and driving performance (MIT AgeLab Tech. Rep. No. 2013-17A). Massachusetts Institute of Technology, Cambridge, MA.

Robinson, G.H., Erickson, D.J., Thurston, G.L., & Clark, R.L. (1972). Visual search by automobile drivers. *Human Factors, 14*(4), 315–323.

Samuel, S., Borowsky, A., & Fisher, D.L. (2014). Text messaging while driving: User experience and interface type. *Proceedings of the Transportation Research Board 93rd Annual Meeting 2013,* Washington, DC.

Schieber, F., Harms, M.L., Berkhout, J., & Spangler, D.S. (1997). Precision and accuracy of video-based measurements of driver's gaze location. *Proceedings of the Human Factors and Ergonomics Society 41st Annual Meeting* (pp. 929–933). Santa Monica, CA: HFES.

Senders, J.W., Kristofferson, A.B., Levison, W.H., Dietrich, C.W., & Ward, J.L. (1967). The attentional demand of automobile driving. *Highway Research Record, 195,* 15–33.

Serafin, C. (1994). *Driver Eye Fixations on Rural Roads: Insight into Safe Driving Behavior (UMTRI Rep. No. 94-21).* Ann Arbor, MI: The University of Michigan Transportation Research Institute.

Shinar, D., McDowell, E.D., & Rockwell, T.H. (1977). Eye movements and curve negotiation. *Human Factors, 19*(1), 63–71.

Simons-Morton, B.G., Guo, F., Klauer, S.G., Ehsani, J.P., & Pradhan, A.K. (2014). Keep your eyes on the road: Young driver crash risk increases according to the duration of distraction. *Journal of Adolescent Health, 54*, S61–S67.

Sivak, M., Conn, L.S., & Olson, P.E. (1986). Driver eye-fixations and the optimal location for automobile brake lights. *Journal of Safety Research, 17*, 13–22.

Smith, D.L., Chang, J., Glassco, R., Foley, J., & Cohen, D. (2005). Methodology for capturing driver eye glance behavior during in-vehicle secondary tasks. *Transportation Research Record, 1937*(1), 61–65.

Society of Automotive Engineers. (2000). *Definition and Measures Related to the Specification of Driver Behavior Using Video-Based Techniques (SAE Recommended Practice J2364).* Warrendale, PA: Society of Automotive Engineering.

Sodhi, M., Reimer, B., & Llamazares, I. (2002). Glance analysis of driver eye movements to evaluate distraction. *Behavior Research Methods, Instruments, & Computers, 34*(4), 529–538.

Spijkers, W. (1992). Distribution of eye-fixations driving—Effects of road characteristics and driving speed as assessed by two eye-movements registration devices. *IATSS Research, 16*(1), 27–34.

Summala, H., Nieminen, T., & Punto, M. (1996). Maintaining lane position with peripheral vision during in-vehicle tasks. *Human Factors, 38*(3), 442–451.

Taylor, T., Pradhan, A.K., Divekar, G., Romoser, M., Muttart, J., Gómez, R., Pollatsek, A., & Fisher, D.L. (2013). The view from the road: The contribution of on-road glance-monitoring technologies to understanding driver behavior. *Accident Analysis & Prevention, 58*, 175–186.

Theeuwes, J. (1996). Visual search at intersections: An eye movement analysis. In A.G. Gale, I.D. Brown, C.M. Haslegrave, & S.P. Taylor (Eds.), *Vision in Vehicles V* (pp. 125–134). Amsterdam: Elsevier Science.

Tijerena, L., Barickman, F.S., & Mazzae, E.N. (2004). *Driver Eye Glance Behaviour during Car following (Rep. No. DOT HS 809 723).* Washington, DC: NHTSA.

Tivisten, E., & Dozza, M. (2014). Driving context and visual–manual phone tasks influence glance behavior in naturalistic driving. *Transportation Research: Part F, 26*, 258–272.

Underwood, G. (2007). Visual attention and the transition from novice to advanced driver. *Ergonomics, 50*(8), 1235–1249.

Underwood, G. (2013). On-road behaviour of younger and older novices during the first six months of driving. *Accident Analysis & Prevention, 58*, 235–243.

Underwood, G., Chapman, P., Bowden, K., & Crundall, D. (2002). Visual search while driving: Skill and awareness during inspection of a scene. *Transportation Research: Part F, 5*, 87–97.

Victor, T.W., Harbluk, J.L., & Engström, J.A. (2005). Sensitivity of eye-movement measures to in-vehicle task difficulty. *Transportation Research Part F: Traffic Psychology and Behaviour, 8*(2), 167–190.

Wade, N.J. (2010). Pioneers of eye movement research. *i-Perception, 1*(2), 33–68.

Wang, Y., Reimer, B., Dobres, J., & Mehler, B. (2014). The sensitivity of different methodologies for characterizing drivers' gaze concentration under increased cognitive demand. *Transportation Research Part F: Traffic Psychology and Behaviour, 26*, 227–237.

World Health Organization (WHO). (2015). *Global Status Report on Road Safety 2015.* Geneva: World Health Organization.

Yarbus, A. (1967). *Eye Movements and Vision.* New York: Plenum Press.

Zuber, B.L., & Stark, L. (1966). Saccadic suppression: elevation of visual threshold associated with saccadic eye movements. *Experimental Neurology, 16*(1), 65–79.

Zwahlen, H.T., & Schnell, T. (1998). Driver eye scanning behavior when reading symbolic warning signs. In A.G. Gale, I.D. Brown, C.M. Haslegrave, & S.P. Taylor (Eds.), *Vision in Vehicles VI* (pp. 3–11). Amsterdam: Elsevier.

28

Hazard Perception Tests

Mark S. Horswill

Abstract

It is now 50 years since the publication of the first known study involving a hazard perception test for drivers (Spicer, 1964), and hazard perception tests are being increasingly used as part of driver licensing procedures around the world. They typically involve drivers watching images of traffic situations and indicating their awareness of potentially dangerous events, on the assumption that a greater awareness of such events has implications for an individual's crash liability while driving. Consistent with this assumption, test scores have been found to be associated with crash risk, on-road driver ratings, and key factors linked with crash risk (such as distraction, fatigue, alcohol, and speed choice), as well as differentiating between high- and low-risk driver groups. This chapter will discuss our current understanding of hazard perception tests, including their validity, issues surrounding test content, alternative test implementations, and what high scorers might be doing that low scorers are not.

28.1 What Are Hazard Perception Tests?

Fifty years ago, Spicer (1964, cited by Pelz & Krupat, 1974), working for the Department of Health in Honolulu, reported a test in which drivers viewed 11 films of traffic situations and were asked to identify the essential features of each situation. Young crash-involved drivers were less able to perceive these features than crash-free drivers. This raised the possibility of a causal link between drivers' ability to perceive potentially dangerous traffic events and their crash risk. If such a causal link exists, then we may have a means of not only predicting a driver's crash risk but also manipulating it. This is the premise behind hazard perception tests.

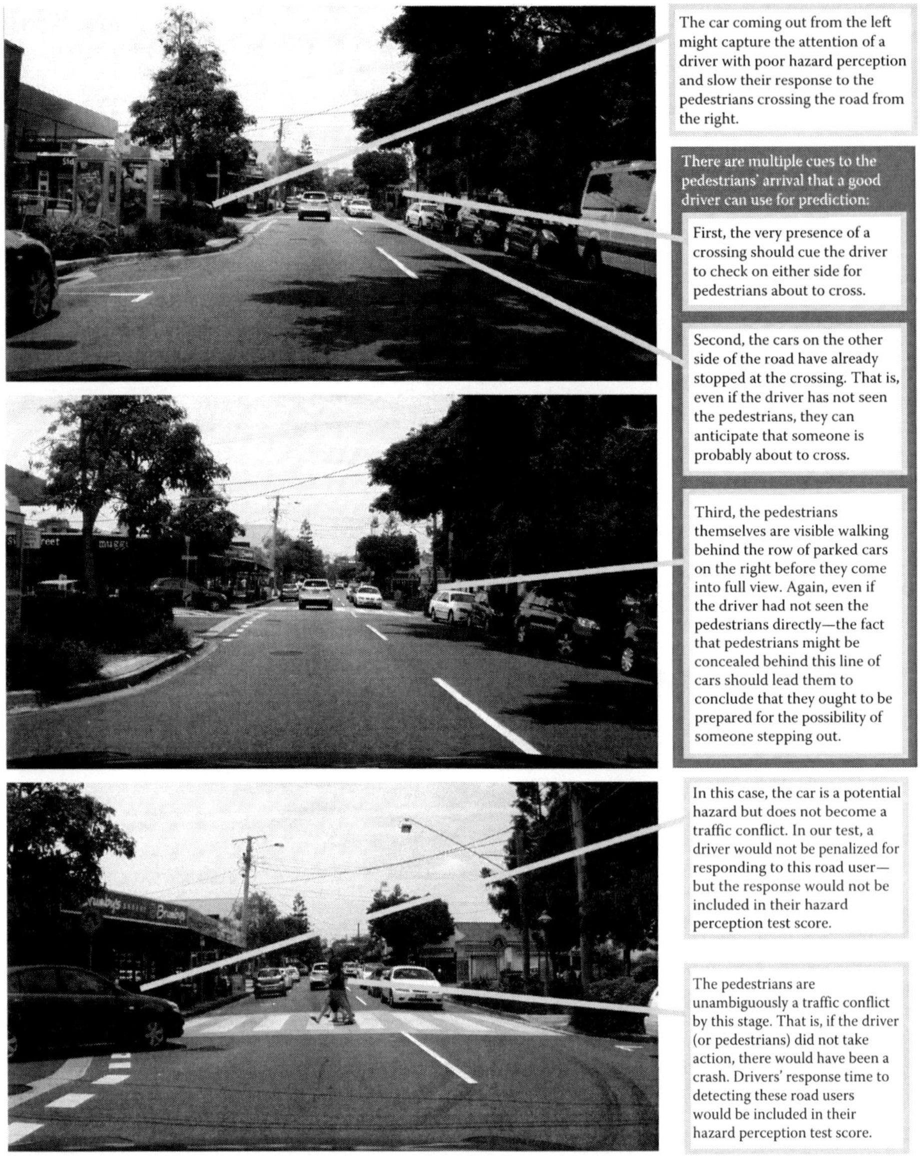

FIGURE 28.1 These images are frames from one of the video clips used in our latest hazard perception test developed for research purposes. The incident illustrates a traffic conflict that can be predicted from multiple precursors. This situation might be challenging for a novice driver because there is more than one potential source of danger competing for attention.

Since Spicer's report, a number of different hazard perception tests for drivers have been developed (e.g., Borowsky et al., 2010; Crundall et al., 2003; Grayson & Sexton, 2002; McKenna & Crick, 1991; Pelz & Krupat, 1974; Scialfa et al., 2011; Watts & Quimby, 1979; Wetton et al., 2011). In the typical modern test, drivers have to anticipate potential hazards in videos of traffic situations filmed from the driver's perspective. For example, in the test that my team devised for driver licensing in the state of Queensland in Australia (Wetton et al., 2011), drivers are instructed to predict likely traffic conflicts in 15 video clips of genuine traffic situations, filmed from the driver's perspective and viewed on a computer. A traffic conflict is defined as any situation in which the driver is required to take evasive action to avoid a collision with another road user (other road users can include other moving vehicles, pedestrians, bicycles, parked cars, and so on). Drivers use the computer mouse to click, as early as possible, on any road user they anticipate is likely to be involved in a traffic conflict with the vehicle in which the camera is mounted (see Figure 28.1). Their hazard perception score is the mean response time to the traffic conflicts.

While hazard perception tests have generally been developed for research purposes, they are also employed as screening measures in some driver licensing procedures, where drivers need to pass a hazard perception test as part of the process of earning an unrestricted driving license. The rationale is that this will force drivers to achieve a certain minimum competency in hazard perception. Other uses for hazard perception tests include evaluating fitness to drive more generally, for example, for individuals recovering from head trauma (Preece et al., 2010, 2011) and older drivers (Anstey et al., 2012; Horswill et al., 2008, 2013).

28.2 Do Hazard Perception Tests Succeed in Measuring What They're Attempting to Measure?

The most fundamental question we need to ask about hazard perception tests is whether they do the job that they are intended to. That is, do they effectively measure drivers' competence at perceiving hazardous situations to the extent needed to allow us to predict their crash risk? There are plenty of reasons why this might not be the case. Hazard perception tests are not measuring driver behavior during actual driving, the response mode (e.g., clicking a computer mouse) can be unrealistic, and most tests do not measure how drivers respond once they have detected a hazard (where a driver may detect a hazard early but then fails to deal with it appropriately). This means that the onus is on researchers to provide compelling evidence for test validity. So—what evidence exists?

We can test the proposition that a hazard perception test measures drivers' hazard perception competence effectively by generating hypotheses relating to how we could expect the test to perform if it is valid. First, we might expect test scores to be associated with crash risk. Second, we might expect test scores to be associated with measures of on-road driving behavior. Third, we might expect associations between test scores and some of the key factors associated with crash risk, such as distraction/inattention, fatigue, speed, and alcohol. Fourth, we might expect test scores to be able to differentiate high-crash-risk driver groups (e.g., young novice drivers) from lower-crash-risk groups (e.g., mid-age experienced drivers). In the following sections, I examine the evidence relating to each of these validity hypotheses (for discussions of the psychometric reliability of hazard perception tests, I refer the reader to Horswill and McKenna [2004], Scialfa et al. [2014], and Wetton et al. [2011]).

28.2.1 Hazard Perception Test Scores Are Associated with Crash Involvement and On-Road Measures

The original retrospective crash relationship shown by Spicer (1964, cited in Pelz & Krupat, 1974) has been replicated using a number of different hazard perception tests (Boufous et al., 2011; Cheng et al., 2011; Darby et al., 2009; Horswill et al., 2010; McKenna & Horswill, 1999; Pelz & Krupat, 1974; Quimby et al., 1986; Rosenbloom et al., 2011; Transport and Road Research Laboratory, 1979; Horswill et al., 2015). Prospective links between hazard perception test scores and crash involvement have also been reported by Congdon (1999), specifically for fatal and serious crashes, and Wells et al. (2008), specifically for non-low-speed public-road crashes for which drivers accepted some blame. Also, Horswill et al. (2015) found hazard perception test scores predicted active crashes (defined as non-parking crashes in which the driver's vehicle was moving) in the year following the test. It is worth noting that the first version of the hazard perception test developed for licensing in the United Kingdom failed to predict crash involvement, but a second version could (Grayson & Sexton, 2002). In the light of all of the methodological problems associated with using crash involvement as a validity criterion (Horswill & McKenna 2004; Horswill et al., 2010), obtaining any statistically reliable relationship between a single computer-based behavioral test and crash involvement could be regarded as an achievement.

In addition, significant associations have been reported between hazard perception test response times and observer ratings of driving performance (on dimensions such as anticipation, skill, attentiveness, and safety) while the driver travels a set route on public roads (Grayson et al., 2003; Mills et al., 1998; Ross et al., 2013; Wood et al., 2013). This also supports the proposal that hazard perception test scores reflect real driving to some degree.

28.2.2 Hazard Perception Test Scores Are Associated with Crash Predictors

A colleague once asked me why I was researching hazard perception when much of the variance in drivers' crash involvement could be accounted for by factors such as distraction/inattention, fatigue, alcohol, and speeding. In a retrospective rebuttal to this comment, I present evidence that links hazard perception test scores to all of these factors.

First, a range of distraction types have been found to disrupt hazard perception performance (Borowsky et al., 2014, 2015; Horswill & McKenna, 1999; Savage et al., 2013; Taylor et al., 2013). Second, with respect to fatigue effects, Smith et al. (2009) found that the level of sleepiness significantly impacted hazard perception test scores for younger novice drivers (though not for mid-age experienced drivers). Similarly, Hamid et al. (2014) reported that total time awake affected drivers' hazard anticipation skill, as assessed in a test using gaze direction as the outcome measure. Third, Grayson et al. (2003) found that better hazard perception test scores were associated with lower speeds during an on-road drive. Also, McKenna et al. (2006) found that hazard perception training not only improved hazard perception test scores; it also resulted in reduced speeding intentions in potentially hazardous situations. Finally, hazard perception test scores have been shown to be compromised by alcohol consumption, consistent with its effect on real driving (Deery & Love, 1996; West et al., 1993).

This evidence raises the possibility that hazard perception competence, as measured by hazard perception tests, could be causally implicated in many crashes. That is, when a driver crashes as a result of distraction/inattention, fatigue, speeding, or alcohol consumption, it could be argued that these incidents reflect a failure of hazard perception on some level.

28.2.3 Hazard Perception Test Scores Can Differentiate High-Crash-Risk and Low-Crash-Risk Driver Groups

Given the high sample sizes required for crash studies (Horswill et al., 2010) and the practical difficulties with running on-road studies, many researchers have chosen to validate their hazard perception tests by determining whether they can differentiate a high-risk group of drivers (typically young novices) from a lower-risk group of drivers (typically midage experienced drivers). Many studies have reported significant novice/experienced driver differences in hazard perception test scores (Borowsky et al., 2009, 2010; Horswill et al., 2008; McKenna & Crick, 1991; Quimby & Watts, 1981; Scialfa et al., 2011, 2012; Smith et al., 2009; Wallis & Horswill, 2007; Wetton et al., 2010, 2011). However, a few studies have failed to find such differences for some tests, namely, early versions of the UK licensing hazard perception test (Chapman & Underwood, 1998; Crundall et al., 2003) and a Norwegian hazard perception test (Sagberg & Bjornskau, 2006). Some researchers (Crundall et al., 2012) have speculated that these failures to find novice/experienced driver differences are due to differences in test content, with the implication that not all hazard perception tests are equally valid and that it is not trivial to create a valid test.

28.3 What Are High Scorers Doing in the Hazard Perception Test That Low Scorers Are Not Doing?

Hazard perception competence is usually conceptualized in the literature in terms of situation awareness (Fisher & Strayer, 2014; Horswill & McKenna, 2004; McGowan & Banbury, 2004). That is, those who achieve higher scores on hazard perception tests are thought to have a sophisticated representation of the traffic around them, and they can use this mental model to predict likely outcomes from any given situation. Endsley (1995) defines situation awareness as comprising (1) perception (e.g., noticing all the relevant elements in the traffic environment), (2) comprehension (e.g., understanding how those traffic elements act and interact), and (3) projection (e.g., being able to make predictions about what is likely to happen next in any given traffic situation on the basis of perception and comprehension). This maps onto advanced police driver training techniques for hazard perception (Coyne et al., 2007), in which drivers are told to "read the road" in order to extract predictive cues to allow them to anticipate potentially dangerous situations as early as possible. That is, the expert driver is seen as actively seeking out information rather than simply reacting to unfolding events.

Groeger (2000) suggests an alternative mechanism that might explain differences between high and low scorers on hazard perception tests, where the high scorers' superior hazard perception ability is a result of a more passive, implicit pattern-matching process. Expert drivers are regarded as having an extensive library of hazardous events attained from a long driving career, and when they encounter a potential hazard, they can identify it as such relatively effortlessly because it reminds of them of previous similar events they have encountered (that is, hazard perception can be regarded as cued recall from memory). In contrast, the novice driver does not have such an extensive library of experience and hence is slower in identifying the hazard. Another account, along the same lines, is that expert drivers have high familiarity with nonhazardous situations, and hence, when a novel hazardous situation occurs, it is more likely to capture expert drivers' attention, because it stands out as being abnormal (Groeger, 2000). While active and passive mechanisms behind hazard perception competence need not be mutually exclusive, some research has found that the hazard perception response times of experienced drivers were slowed to a proportionally greater extent than those of novices by a demanding secondary task (McKenna & Farrand, 1999). This has been argued to be more consistent with active accounts of hazard perception competence (Horswill & McKenna, 2004) as it implies that the mechanism by which experienced drivers gain their hazard perception advantage over novices is more likely to be an effortful, cognitive-demanding process that is vulnerable to disruption.

What other mechanisms might account for individual differences in hazard perception test scores? One possibility is that some of the variance in scores reflects response style rather than skill. For instance, it might be possible for drivers to obtain faster hazard perception response times in tests by adopting a lower threshold for what constitutes a hazard (that is, they are willing to classify a broader range of events as being hazards), without changing their perception of the road environment (Wallis & Horswill, 2007). If response style could be shown to account for all variability in test scores, then it would raise serious concerns about the validity of hazard perception tests—where individual differences in test scores may be reflecting a test-specific strategy rather than something that reflects real driving. However, this seems unlikely given the evidence cited earlier that hazard perception test scores do map onto real driving to some degree. Also, if experienced drivers were extracting the same information from a traffic environment as novices (and only gaining a better score as a result of a different response style), then we would not expect to find the differences in eye-scanning patterns that emerge across different levels of driving expertise in both hazard perception tests and real driving (e.g., Underwood et al., 2011), where experienced drivers' search strategies appear to reflect a greater level of skill.

In addition, in a response-style account, where Rowe (1997) asked both experienced and novice drivers to complete a hazard perception test and then gave them an unexpected memory test for both hazard in the test and nonhazard elements. He found that experienced drivers were more likely to remember the hazards than novices but not the nonhazards. This is not consistent with a response-style account, in which we would not expect high scorers to remember the hazards more than the low scorers. It is, however, consistent with the idea that the experienced drivers engaged in more elaborate processing of hazards (and that this memory advantage was specific to hazards) than novices, and this is difficult to explain using solely a response-style mechanism.

A related consideration is whether hazard perception test scores just reflect differences in risk-taking propensity, where mid-age experienced drivers are simply more risk averse than young novice drivers, and this leads to them setting a lower threshold at which they would consider responding to a developing hazard (resulting in a faster response time to that hazard), without the need for group differences in any skill. However, if this were true, then one would predict an association between risk ratings of hazards and associated hazard perception test scores. Research has found close to zero correlations between drivers' risk ratings both of driving in general and of individual hazards, and hazard perception test scores, both at the level of the overall test and at the level of the individual hazard (Farrand & McKenna, 2001; Wallis & Horswill, 2007).

28.4 What Sort of Hazards Ought to Be in a Hazard Perception Test?

One of the lessons from the validity evidence presented earlier is that it is not trivial to produce a valid hazard perception test. Notably, the initial version of the UK hazard perception test was not associated with crash involvement and did not yield novice/experienced differences in some studies (Grayson & Sexton, 2002). The question is, what are the ingredients required for a valid test?

A number of researchers have suggested that differences in hazard perception test validity can be explained by differences in the nature of the hazards used (Crundall et al., 2012; Grayson & Sexton, 2002; Sagberg & Bjornskau, 2006; Wetton et al., 2011). There seems to be some consensus that an important element of a successful hazard is that there are cues that allow the hazard to be predicted and hence yield response-time differences between those able to interpret those cues and those who cannot. One possible problem with some low-validity tests is that they may include hazards that are inherently unpredictable, where drivers are likely to have similar response times regardless of their level of hazard perception ability (Crundall et al., 2012; Grayson & Sexton, 2002; Jackson et al., 2009; Wetton et al., 2011). There is some evidence to back up this view. Sagberg and Bjørnskau (2006), in a post hoc analysis of their hazard perception test, concluded that the potential for anticipation appeared to be a key element involved in hazards that could discriminate between novice and experienced drivers. Consistent with this, Pradhan et al. (2005) found that novice drivers were less likely to fixate hazard precursors than more experienced drivers. Refining this further, Crundall et al. (2012) classified hazard precursors as either behavioral (anticipatory cues to the hazard that were directly associated with the road user[s] involved in the hazard, such as a pedestrian about to walk onto the road) or environmental (features of the roadway that could signal the potential for a hazard, such as a blind bend or junction). They found that novices were less likely to fixate the behavioral precursors to hazards and less likely to fixate the environmentally cued hazards than more experienced drivers.

Beyond consideration of item content, one common technique to maximize test validity is to use empirical criterion keying to select hazards. This involves choosing hazards on the basis of empirical validity evidence. Relationships between individual hazards and either crash involvement or on-road driving performance could, in principle, be used for this purpose, but for practical reasons, the usual approach is to rank hazards on their ability to discriminate between young novice and older experienced driver groups (Scialfa et al., 2011; Wetton et al., 2010, 2011). One issue with this approach is that it is possible to end up with a hazard perception test containing a pool of hazards that is not representative of every possible hazard type, because some types of hazard may have inherently better validity than others. I would argue that this should not be regarded as a problem (whereas a test that lacks empirical validity would undoubtedly be a problem), because the goal is not to capture drivers' reactions to anything that might happen on the road. Instead, the test should be viewed as attempting to measure the level of sophistication in drivers' mental model of the traffic environment (McKenna & Crick, 1991), in line with the situation awareness conception of hazard perception discussed previously, and some types of hazards might be more effective in achieving this goal than others.

Incidentally, in case the reader is wondering whether the novice/experienced differences reported as validity evidence earlier might be a direct result of selecting hazards on the basis of their ability to discriminate between novices and experienced drivers, it is worth noting that we did find overall novice/experienced driver group differences for our hazard pool before we selected out any items (Wetton et al., 2010, 2011).

28.5 Variations on a Theme: Alternative Approaches to Measuring Hazard Perception Ability

In previous sections, the focus has been on video-based hazard perception tests in which participants identify hazards in real time, as this is the model typically used in driver licensing. However, there are alternatives that address some of the limitations of traditional tests.

A number of researchers have used gaze direction as a measure of hazard perception ability, where an eye tracker is used to determine whether a driver fixates areas of interest relevant to detecting hazards. A key attraction of this approach is that it does not require drivers to do something that they would not normally do while driving (such as click on hazards with a computer mouse). This means we can avoid asking individuals explicitly to respond to hazards (or, in our case, traffic conflicts), given that this may affect how they respond. However, eye-tracking measures may not yet be practical for licensing purposes given the time and expense involved in collecting and processing the data. Also, the usual caveat to interpreting eye-tracking data applies, where just because someone is looking at something, it does not necessarily follow that they are also paying attention to it.

Another way of assessing hazard perception competence is to measure how individuals choose to drive in the presence of a developing hazard. Crundall et al. (2012) used a simulator in which individuals drove a virtual car through various computer-generated hazardous traffic situations. One of the measures of hazard perception was drivers' chosen approach speed leading up to each hazard (they also measured gaze direction). The use of driver-controlled simulators to measure hazard perception has the advantage of capturing how drivers will act if and when they detect a hazard (Underwood et al., 2011). This addresses a couple of limitations of traditional video-based tests, namely that (1) video-based tests do not take account of what people will do once they have detected the hazard and (2) watching a traffic scene unfold may elicit different visual search behavior compared with actively controlling a car in the same situation. On the other hand, one drawback of giving test takers control of the vehicle is that it means that hazards are no longer equivalent across participants (e.g., variations in approach speed to a hazard changes the nature of the hazard, so drivers will not be experiencing the same hazard as each other). It is also not clear whether using footage that is simulated affects drivers' behavior—where the typical standard of computer-generated imagery used in driving simulators still does not result in stimuli that look as realistic as a filmed traffic scene. As an alternative to a driving simulator approach to measuring speed, McKenna et al. (2006) developed a video-based test of speed choice, in which drivers had to indicate their speed preferences in various filmed traffic scenes—where some scenes included embedded hazards and others did not. They found that drivers trained in hazard perception chose lower speeds when hazards were present.

Another approach to measuring hazard perception is to evaluate the prediction element of situation awareness more explicitly. Jackson et al. (2009) found that experienced drivers were better than novices at accurately predicting what happened next in video clips of traffic that were stopped just before a hazard occurred. Similar measures include situation awareness probes (McGowan & Banbury, 2004), where the video is paused and test takers are asked questions about what is happening. However, it should be noted that this type of approach has been found to also act as an intervention to improve situation awareness (McGowan & Banbury, 2004), which may undermine its assessment properties.

Other methods used to assess hazard perception ability include retrospective judgments on traffic clips (Renge et al., 2005; Vlakveld, 2014), the use of static images of traffic (e.g., Pradhan et al., 2009; Scialfa et al., 2012; Wetton et al., 2010), plan-view schematic representations of traffic (e.g., Pollatsek et al.,

2006), real driving on a closed track with staged hazards (Lee et al., 2008; Wood, 2002), and real driving on public roads (Wood et al., 2013). Focusing on the latter example, there are advantages and disadvantages to measuring hazard perception during real driving. For example, on one hand, it is a more direct measure of the real task. On the other hand, it is possible that, during a real-world drive that typically lasts less than an hour, drivers may not encounter enough suitable traffic conflicts to allow reliable assessment. It typically takes my team 3 hours of driving to capture one traffic conflict that is suitable for our hazard perception test. However, some researchers (Taylor et al., 2011) have dealt with this issue by measuring fixations on potential hazards instead (i.e., elements that a good driver ought to monitor but do not necessary develop into traffic conflicts), as these are more frequent.

28.6 The Future of Hazard Perception Testing

In sum, we have a method for assessing an important component of drivers' competency that appears to have value as a crash predictor. Where can we go from here? While some of the options for measuring hazard perception described here might be currently impractical for mass testing for novice drivers (e.g., the use of eye tracking and controllable driving simulators), one would presume that it is only a matter of time before technology renders these possibilities more feasible. One possibility is that the hazard perception test of the future could be given more crash prediction power by combining a number of different ways of assessing hazard perception skill, using convergent evidence to gain the benefits of each technique while offsetting their individual limitations (the testable assumption being that a battery of hazard perception measures might account for more variation in crash risk than any single measure). For mass testing, this might require automation of techniques like eye tracking, which may not be trivial—but might not be impossible either. Another possibility is that, as controllable driving simulators become more realistic and affordable, the driving test could consist of exposing drivers to hundreds of reconstructed scenarios based on real crashes in their locale, where drivers would not receive their license unless they could demonstrate that they could reliably avoid crashing in all of these situations (some researchers have already been developing simulator scenarios based on real crash data, e.g., McDonald et al., 2012). A different approach might be to take advantage of the recent availability of low-cost driving cameras, which continually record video of the forward view from the vehicle and can be triggered to capture footage when unusual g-forces are detected (potentially indicating, amongst other things, the failure to deal with a hazard). There is some promising evidence that such systems may be valuable for novice driver training (McGehee et al., 2007), but in principle, they could also be used as part of formal driver licensing. In practice, this might require advances in the ability of such devices to detect hazardous situations, but it could be that, instead of the typical 40-minute on-road driving test, learner drivers' entire on-road driving experience becomes their driving test. This would mean that the assessment would be able to capture how the driver dealt with rare hazardous encounters that are unlikely to occur in the typical on-road driving test, in addition to addressing the issue that test takers may not drive as they normally do during a traditional driving test.

So, what has been achieved in the 50 years between the first publication on hazard perception in 1964 and the present? On November 7, 2014, the UK hazard perception test received the Prince Michael International Road Safety Award (Driver and Vehicle Standards Agency, 2014); the 1.5 million tests per year that have been conducted in the United Kingdom since 2002 have been credited with reducing crashes by an estimated 11.3% (Wells et al., 2008). This has been equated to an estimated annual saving of £89.5 million (US$142 million) via the prevention of 8535 damage-only and 1076 injury accidents per year (Dr. S. Helman, head of transport psychology, Transport Research Laboratory, UK, November 13, 2014, personal communication). As a result, the hazard perception test was described as being responsible for a considerable improvement to road safety (Driver and Vehicle Standards Agency, 2014).

References

Anstey, K.J., Horswill, M.S., Wood, J.M., & Hatherly, C. (2012). The role of cognitive and visual abilities as predictors in the multifactorial model of driving safety. *Accident Analysis & Prevention, 45*, 766–774. doi: 10.1016/j.aap.2011.10.006.

Borowsky, A., Oron-Gilad, T., & Parmet, Y. (2009). Age and skill differences in classifying hazardous traffic scenes. *Transportation Research Part F Traffic Psychology and Behaviour, 12*(4), 277–287. doi: 10.1016/j.trf.2009.02.001.

Borowsky, A., Shinar, D., & Oron-Gilad, T. (2010). Age, skill, and hazard perception in driving. *Accident Analysis & Prevention, 42*(4), 1240–1249. doi: 10.1016/j.aap.2010.02.001.

Borowsky, A., Horrey, W.J., Liang, Y., Simmons, L., Garabet, A., & Fisher, D.L. (2014). Memory for a hazard is interrupted by performance of a secondary in-vehicle task. *Proceedings of the Human Factors and Ergonomics Society Annual Meeting.*

Borowsky, A., Horrey, W.J., Liang, Y., Garabet, A., Simmons, L., & Fisher, D.L. (2015). The effects of momentary visual disruption on hazard anticipation and awareness in driving. *Traffic Injury Prevention, 16*(2), 133–139.

Boufous, S., Ivers, R., Senserrick, T., & Stevenson, M. (2011). Attempts at the practical on-road driving test and the hazard perception test and the risk of traffic crashes in young drivers. *Traffic Injury Prevention, 12*(5), 475–482. doi: 10.1080/15389588.2011.591856.

Chapman, P., & Underwood, G. (1998). Visual search of driving situations: Danger and experience. *Perception, 27*, 951–964.

Cheng, A.S.K., Ng, T.C.K., & Lee, H.C. (2011). A comparison of the hazard perception ability of accident-involved and accident-free motorcycle riders. *Accident Analysis & Prevention, 43*(4), 1464–1471. doi: 10.1016/j.aap.2011.02.024.

Congdon, P. (1999). *VicRoads Hazard Perception Test, Can It Predict Accidents?* Camberwell, Victoria, Australia: Australian Council for Educational Research.

Coyne, P., Mares, P., & MacDonald, B. (2007). *Roadcraft: The Police Driver's Handbook.* London: The Stationery Office.

Crundall, D., Chapman, P., Phelps, N., & Underwood, G. (2003). Eye movements and hazard perception in police pursuit and emergency response driving. *Journal of Experiment Psychology: Applied, 9*(3), 163–174. doi: 10.1037/1076-898X.9.3.163.

Crundall, D., Chapman, P., Trawley, S., Collins, L., van Loon, E., Andrews, B., & Underwood, G. (2012). Some hazards are more attractive than others: Drivers of varying experience respond differently to different types of hazard. *Accident Analysis & Prevention, 45*, 600–609. doi: 10.1016/j.aap.2011.09.049.

Darby, P., Murray, W., & Raeside, R. (2009). Applying online fleet driver assessment to help identify, target and reduce occupational road safety risks. *Safety Science, 47*(3), 436–442. doi: 10.1016/j.ssci.2008.05.004.

Deery, H.A., & Love, A.W. (1996). The effect of a moderate dose of alcohol on the hazard perception profile of young drink-drivers. *Addiction, 91*, 815–827.

Driver and Vehicle Standards Agency. (2014). Hazard perception test wins road safety awards. Accessed November 12, retrieved from https://www.gov.uk/government/news/hazard-perception-test-wins-road-safety-award.

Endsley, M.R. (1995). Towards a theory of situation awareness in dynamic systems. *Human Factors, 37*(1), 32–64.

Farrand, P., & McKenna, F.P. (2001). Risk perception in novice drivers: The relationship between questionnaire measures and response latency. *Transportation Research Part F, 4*, 201–212. doi: 10.1016/S1369-8478(01)00024-9.

Fisher, D.L., & Strayer, D.L. (2014). Modeling situation awareness and crash risk. *Annals of Advances in Automotive Medicine, 58*, 33–39.

Grayson, G., & Sexton, B. (2002). *The Development of Hazard Perception Testing*. Crowthorne: Transport Research Laboratory.

Grayson, G.B., Maycock, G., Groeger, J.A., Hammond, S.M., & Field, D.T. (2003). *Risk, Hazard Perception and Perceived Control*. Crowthorne: Transport Research Laboratory.

Groeger, J.A. (2000). *Understanding Driving*. Hove: Psychology Press.

Hamid, M., Samuel, S., Borowsky, A., & Fisher, D.L. (2014). Evaluation of Effect of Total Awake Time on Driving Performance Skills–Hazard Anticipation and Hazard Mitigation: A Simulator Study. Transportation Research Board 93rd Annual Meeting.

Horswill, M.S., & McKenna, F.P. (1999). The effect of interference on dynamic risk-taking judgments. *British Journal of Psychology, 90*(5), 189–199. doi: 10.1348/000712699161341.

Horswill, M.S., & McKenna, F.P. (2004). Drivers' hazard perception ability: Situation awareness on the road. In S. Banbury & S. Tremblay (Eds.), *A Cognitive Approach to Situation Awareness: Theory and Application*, pp. 155–175. Aldershot: Ashgate.

Horswill, M.S., Hill, A., & Wetton, M. (2015). Can a video-based hazard perception test used for driver licensing predict crash involvement? *Accident Analysis & Prevention, 82*, 213–219. doi: 10.1016/j.aap.2015.05.019.

Horswill, M.S., Marrington, S.A., McCullough, C.M., Wood, J., Pachana, N.A., McWilliam, J., & Raikos, M.K. (2008). The hazard perception ability of older drivers. *The Journals of Gerontology Series B: Psychological Sciences and Social Sciences, 63*(4), 212–218. doi: 10.1093/geronb/63.4.P212.

Horswill, M.S., Anstey, K.J., Hatherly, C.G., & Wood, J. (2010). The crash involvement of older drivers is associated with their hazard perception latencies. *Journal of the International Neuropsychological Society, 16*(5), 939–944. doi: 10.1017/S135561771000055X.

Horswill, M.S., Sullivan, K., Lurie-Beck, J.K., & Smith, S. (2013). How realistic are older drivers' ratings of their driving ability? *Accident Analysis & Prevention, 50*, 130–137. doi: 10.1016/j.aap.2012.04.001.

Jackson, L., Chapman, P., & Crundall, D. (2009). What happens next? Predicting other road users' behaviour as a function of driving experience and processing time. *Ergonomics, 52*(2), 154–164. doi: 10.1080/00140130802030714.

Lee, S.E., Klauer, S.G., Olsen, E.C.B., Simons-Morton, B.G., Dingus, T.A., Ramsey, D.J., & Ouimet, M.C. (2008). Detection of road hazards by novice teen and experienced adult drivers. *Transportation Research Record*, (2078), 26–32. doi: 10.3141/2078-04.

McDonald, C.C., Tanenbaum, J.B., Lee, Y.C., Fisher, D.L., Mayhew, D.R., & Winston, F.K. (2012). Using crash data to develop simulator scenarios for assessing novice driver performance. *Transportation Research Record*, (2321), 73–78. doi: 10.3141/2321-10.

McGehee, D.V., Raby, M., Carney, C., Lee, J.D., & Reyes, M.L. (2007). Extending parental mentoring using an event-triggered video intervention in rural teen drivers. *Journal of Safety Research, 38*, 215–227. doi: 10.1016/j.jsr.2007.02.009.

McGowan, A.M., & Banbury, S.P. (2004). Evaluating interruption-based techniques using embedded measures of driver anticipation. In S. Banbury & S. Tremblay (Eds.), *A Cognitive Approach to Situation Awareness*, pp. 176–192. Aldershot: Ashgate.

McKenna, F.P., & Crick, J.L. (1991). *Hazard Perception in Drivers: A Methodology for Testing and Training*. Crowthorne: Transport and Road Research Laboratory.

McKenna, F.P., & Farrand, P. (1999). The role of automaticity in driving. In G.B. Grayson (Ed.), *Behavioural Research in Road Safety IX*. Crowthorne: Transport Research Laboratory.

McKenna, F.P., & Horswill, M.S. (1999). Hazard perception and its relevance for driver licensing. *Journal of the International Association of Traffic and Safety Sciences, 23*(1), 26–41.

McKenna, F.P., Horswill, M.S., & Alexander, J.L. (2006). Does anticipation training affect drivers' risk taking? *Journal of Experimental Psychology: Applied, 12*(1), 1–10. doi: 10.1037/1076-898X.12.1.1.

Mills, K.L., Hall, R.D., McDonald, M., & Rolls, G.W.P. (1998). The effects of hazard perception training on the development of novice driver skills. Department for Transport (UK). Accessed November 8, retrieved from http://www.dft.gov.uk.

Pelz, D.C., & Krupat, E. (1974). Caution profile and driving record of undergraduate males. *Accident Analysis & Prevention, 6*, 45–58.

Pollatsek, A., Narayanaan, V., Pradhan, A., & Fisher, D.L. (2006). Using eye movements to evaluate a PC-based risk awareness and perception training program on a driving simulator. *Human Factors, 48*(3), 447–464.

Pradhan, A.K., Hammel, K.R., DeRamus, R., Pollatsek, A., Noyce, D.A., & Fisher, D.L. (2005). Using eye movements to evaluate effects of driver age on risk perception in a driving simulator. *Human Factors, 47*(4), 840–852.

Pradhan, A.K., Pollatsek, A., Knodler, M., & Fisher, D.L. (2009). Can younger drivers be trained to scan for information that will reduce their risk in roadway traffic scenarios that are hard to identify as hazardous? *Ergonomics, 52*(6), 657–673. doi: 10.1080/00140130802550232.

Preece, M.H.W., Horswill, M.S., & Geffen, G.M. (2010). Driving after concussion: The acute effect of mild traumatic brain injury on drivers' hazard perception. *Neuropsychology, 24*(4), 493–503. doi: 10.1037/a0018903.

Preece, M.H.W., Horswill, M.S., & Geffen, G.M. (2011). Assessment of drivers' ability to anticipate traffic hazards after traumatic brain injury. *Journal of Neurology Neurosurgery and Psychiatry, 82*(4), 447–451. doi: 10.1136/jnnp.2010.215228.

Quimby, A.R., & Watts, G.R. (1981). *Human Factors and Driving Performance.* Crowthorne: Transport and Road Research Laboratory.

Quimby, A.R., Maycock, G., Carter, I.D., Dixon, R., & Wall, J.G. (1986). *Perceptual Abilities of Accident Involved Drivers.* Crowthorne: Transport and Road Research Laboratory.

Renge, K., Ishibashi, T., Oiri, M., Ota, H., Tsunenari, S., & Mukai, M. (2005). Elderly drivers' hazard perception and driving performance. In G. Underwood (Ed.), *Traffic and Transport Psychology: Theory and Application*, pp. 91–99. London: Elsevier.

Rosenbloom, T., Perlman, A., & Pereg, A. (2011). Hazard perception of motorcyclists and car drivers. *Accident Analysis & Prevention, 43*(3), 601–604. doi: 10.1016/j.aap.2010.08.005.

Ross, R.W., Scialfa, C., Cordazzo, S., & Bubric, K. (2013). Predicting Older Adults' On-Road Driving Performance. 7th International Driving Symposium on Human Factors in Driver Assessment, Training, and Vehicle Design.

Rowe, R.M. (1997). Anticipation in skilled performance. Department of Psychology, University of Reading.

Sagberg, F., & Bjornskau, T. (2006). Hazard perception and driving experience among novice drivers. *Accident Analysis & Prevention, 38*, 407–414. doi: 10.1016/j.aap.2005.10.014.

Savage, S.W., Potter, D.D., & Tatler, B.W. (2013). Does preoccupation impair hazard perception? A simultaneous EEG and Eye Tracking study. *Transportation Research Part F: Traffic Psychology and Behaviour, 17*(0), 52–62. doi: 10.1016/j.trf.2012.10.002.

Scialfa, C.T., Deschenes, M.C., Ference, J., Boone, J., Horswill, M.S., & Wetton, M. (2011). A hazard perception test for novice drivers. *Accident Analysis & Prevention, 43*(1), 204–208. doi: 10.1016/j.aap.2010.08.010.

Scialfa, C.T., Borkenhagen, D., Lyon, J., Deschenes, M., Horswill, M., & Wetton, M. (2012). The effects of driving experience on responses to a static hazard perception test. *Accident Analysis & Prevention, 45*, 547–553. doi: 10.1016/j.aap.2011.09.005.

Scialfa, C.T., Pereverseff, R.S., & Borkenhagen, D. (2014). Short-term reliability of a brief hazard perception test. *Accident Analysis & Prevention, 73*, 41–46. doi: 10.1016/j.aap.2014.08.007.

Smith, S.S., Horswill, M.S., Chambers, B., & Wetton, M. (2009). Hazard perception in novice and experienced drivers: The effects of sleepiness. *Accident Analysis & Prevention, 41*(4), 729–733. doi: 10.1016/j.aap.2009.03.016.

Spicer, R.A. (1964). *Human Factors in Traffic Accidents.* Honolulu, HI: Department of Health.

Taylor, T.G.G., Masserang, K.M., Pradhan, A.K., Divekar, G., Samuel, S., Muttart, J.W., Pollatsek, A., & Fisher, D.L. (2011). Long term effects of hazard anticipation training on novice drivers measured on the open road. *Proceedings of the Sixth International Driving Symposium on Human Factors in Driver Assessment, Training, and Vehicle Design.*

Taylor, T., Pradhan, A.K., Divekar, G., Romoser, M., Muttart, J., Gomez, R., Pollatsek, A., & Fisher, D.L. (2013). The view from the road: The contribution of on-road glance-monitoring technologies to understanding driver behavior. *Accident Analysis & Prevention, 58*, 175–186. doi: 10.1016/j.aap.2013.02.008.

Transport and Road Research Laboratory. (1979). *A Hazard Perception Test for Drivers*. Crowthorne: Transport and Road Research Laboratory.

Underwood, G., Crundall, D., & Chapman, P. (2011). Driving simulator validation with hazard perception. *Transportation Research Part F Traffic Psychology and Behaviour, 14*(6), 435–446. doi: 10.1016/j.trf.2011.04.008.

Vlakveld, W.P. (2014). A comparative study of two desktop hazard perception tasks suitable for mass testing in which scores are not based on response latencies. *Transportation Research Part F: Traffic Psychology and Behaviour, 22*, 218–231.

Wallis, T.S.A., & Horswill, M.S. (2007). Using fuzzy signal detection theory to determine why experienced and trained drivers respond faster than novices in a hazard perception test. *Accident Analysis & Prevention, 39*(6), 1177–1185. doi: 10.1016/j.aap.2007.03.003.

Watts, G.R., & Quimby, A.R. (1979). *Design and Validation of a Driving Simulator*. Crowthorne: Transport and Road Research Laboratory.

Wells, P., Tong, S., Sexton, B., Grayson, G., & Jones, E. (2008). *Cohort II: A Study of Learner and New Drivers*. London: Department for Transport.

West, R., Wilding, J., French, D., Kemp, R., & Irving, A. (1993). Effect of low and moderate doses of alcohol on driving hazard perception latency and driving speed. *Addiction, 88*, 527–532. doi: 10.1111/j.1360-0443.1993.tb02059.x.

Wetton, M.A., Horswill, M.S., Hatherly, C., Wood, J.M., Pachana, N.A., & Anstey, K.J. (2010). The development and validation of two complementary measures of drivers' hazard perception ability. *Accident Analysis & Prevention, 42*(4), 1232–1239. doi: 10.1016/j.aap.2010.01.017.

Wetton, M.A., Hill, A., & Horswill, M.S. (2011). The development and validation of a hazard perception test for use in driver licensing. *Accident Analysis & Prevention, 43*(5), 1759–1770. doi: 10.1016/j.aap.2011.04.007.

Wood, J.M. (2002). Age and visual impairment decrease driving performance as measured on a closed-road circuit. *Human Factors, 44*(3), 482–494.

Wood, J.M., Anstey, K.J., Horswill, M.S., & Lacherez, P.F. (2013). Evaluation of screening tests for predicting older driver performance and safety assessed by an on-road test. *Accident Analysis & Prevention, 50*, 1161–1168. doi: 10.1016/j.aap.2012.09.009.

VIII

The Way Forward

29

What Is Learned in Becoming a Competent Driver?

Allan F. Williams

Jean T. Shope

Robert D. Foss

Abstract

The Chapter. Regardless of the age at which driving begins, there is a rapid decrease in crashes during the first several months of independent (unsupervised) driving. Understanding the learning processes involved could lead to ways to accelerate this decrease. Existing research to examine what is learned is limited, focuses almost exclusively on hazard anticipation, and often involves inadequate research designs. ***Discussion, Limitations, and Recommendations.*** A broader range of approaches, guided by theoretical models of driving, is needed. Longitudinal studies with multiple measurements during the early months of driving are essential to adequately address learning in the early months of driving.

29.1 Introduction

One of the most striking and consistent observations in the highway safety field is the rapid decrease in crash rates that occurs over the first several months of novices' independent driving. This pattern exists whether driving begins at 16, 17, 18, or older (Chapman et al., 2014; Lewis-Evans, 2010; Masten & Foss, 2010; Twisk & Stacey, 2007). But what accounts for this rapid decrease is not well understood. A better understanding of why and how this decrease occurs may lead to ways to accelerate the rate at which drivers improve. In 2008, the Transportation Research Board (TRB) Subcommittee on Young Drivers noted the lack of research addressing this topic, and the question of how driving competency develops was identified as one research priority among several (TRB, 2013).

Since 2008, there has been considerable research on young drivers. A 2015 special issue of the *Journal of Adolescent Health* covered progress in dealing with the other research priority areas identified in 2008.

However, there has been scant research on learning to drive, and a paper on this topic was not included in the special issue. Noting this continuing deficiency, the TRB subcommittee held a 1½-day workshop in the summer of 2015 to review what is known and not known about how driving competency develops, and to identify what research is needed to begin filling this knowledge gap. The specific question posed was, "What changes to reduce crash rates so quickly during the early months of unsupervised driving?"

29.2 The Challenge

With just a few practice lessons, rudimentary driving skills (steering, braking, turning, etc.) can be gained if not mastered. But piloting a car is only a small part of the driving competence necessary to negotiate roads and traffic safely. Driving competently is a complex task requiring the application of a wide range of psychomotor, cognitive, social, and emotional skills. We know that developing competence in performing a complex task such as driving requires rehearsal of the skills being learned. Experience gained over an extended period of time results in focused attention and appropriate responses becoming automatic (Ericsson, 2002, 2005; Keating, 2007). Presumably, beginning drivers go through a trial-and-error period on the road as they acquire experience dealing with various driving situations. The question is whether the learning processes that are operating as experience is gained, and that contribute to the rapid decrease in crashes, can be identified and influenced, thus lessening the error rate.

There is presently limited information on how driving competence develops and how this process might be influenced. Discussions during the 2015 workshop provided some information about what is known and not known about becoming a competent driver, and considered some of the risk factors for beginning drivers. However, these discussions made relatively little headway in identifying what types of research are needed.

29.3 Approaches to the Question

There are two sources of information that might help us understand how and why the rapid decrease in crashes takes place over the first few months of independent driving. One source is research that directly addresses behaviors presumably related to driving competence, for example, the ability to identify potential threats in the driving environment. The other source is research on driver behavior and crash characteristics. This area includes *naturalistic driving* studies that can measure changes in driver behavior during the first few months of driving, using cameras and vehicle instrumentation, as well as studies that examine changes in crash types and crash-contributing factors during early driving. These studies can provide clues about what is being learned. For example, if some driving skills are learned more quickly than others, this finding may be reflected in differential changes in crash patterns.

Existing information from these sources, as revealed in the workshop, is limited. Research on driver learning processes has been narrowly focused and often based on inadequate research methods. Analyses of changes in driver behavior have provided few clues that would help explain the rapid decrease in crashes in the first few months of driving. This commentary does not attempt a review of the literature on learning to drive. Rather, it discusses the limited scope and inadequate methods of the existing studies, and offers suggestions for future research.

29.4 Becoming a Competent Driver

Studies of driving competence have focused almost exclusively on how novices deal with hazards. It is obvious that being able to identify dangerous situations (hazard anticipation) and then evaluate and respond to them appropriately (hazard mitigation) is important in avoiding crashes. Paying attention to the forward roadway (attention maintenance) is a necessary condition for hazard anticipation to occur. All three of these abilities are important, and each would be expected to improve as driving experience is gained. Most of the research to date has dealt only with hazard anticipation. Studies routinely find that "novice" drivers are inferior to "experienced" drivers in hazard anticipation (Deery, 1999; Grayson & Sexton, 2002; Mourant & Rockwell, 1972; Pradhan et al., 2005; Underwood et al., 2005). However, there are problems with many of these studies. There is no accepted standard for measuring hazard anticipation, and the diversity of measures used hinders interpretation of this literature. Interpretation is also complicated by the tendency of researchers to equate youthfulness with inexperience, for example, identifying 16- to 19-year-olds as novices, although they may vary widely from having considerable driving experience and quite well-developed hazard perception skills to being complete novices. This latter point highlights the importance of research that measures novice drivers' changes in hazard perception capabilities over time (preferably month by month). Without such a longitudinal focus on the initial months of driving, it is not possible to understand what changes in those first few months of driving that plays a central role in the decrease in crashes, or to adequately assess opportunities for training interventions. There has been virtually no longitudinal research on the development of hazard anticipation skills. Precise definitions of driving experience, and fine-grained analyses of changes over the first months of driving, will be essential to advance our understanding of how novice drivers improve so quickly.

There are some encouraging findings from recent research. Although new license holders have been found to look away from the road longer while engaged in secondary tasks than those licensed for a longer period of time, there is some evidence that attention maintenance can be improved through training (Divekar et al., 2013; Pradhan et al., 2011). Hazard anticipation skills more broadly can be trained to an extent (McDonald et al., 2015), and there is some indication that this training transfers to on-road driving, rather than being evidenced only in simulated driving (Pradhan et al., 2009; Taylor et al., 2011). One recent study found encouraging, albeit weak, evidence that hazard perception training may reduce crash involvement of some novice drivers (Thomas et al., 2015).

29.5 Driver Behavior Changes

One study of crash characteristics indicated that some common mistakes made in the first months of driving, such as failure to yield and inattention, decline even more sharply than the rapid decrease in crashes overall (Foss et al., 2011). Some types of crashes, particularly those involving running off the road, also decline quickly (Foss et al., 2011; Mayhew et al., 2003; Sagberg, 1998). Other types of crashes show different patterns in the early months of independent driving (Foss et al., 2011), suggesting that either learning is not involved or it is less important than other issues, such as increased exposure.

Naturalistic driving studies can examine changes in driver behavior and errors over time. This approach is particularly valuable. A naturalistic study in rural Virginia found slight increases in distance driven during the initial months of driving (Lee et al., 2011). A similar study of more urbanized driving in Israel (Musicant & Benjamini, 2012) found an initial sharp increase followed by a decrease in exposure during the initial months of unsupervised driving. These early findings suggest that changing exposure is unlikely to explain the rapid decrease in crashes among novices, although the Israeli study raises the intriguing possibility that changes in exposure to novel driving settings—which declined markedly in the early months—might be a critical factor in the declining crash risk among novices. Additional naturalistic driving studies are needed to examine this possibility and to incorporate a greater variety of driving conditions, licensing regulations, and cultural contexts than are presently covered in the few existing naturalistic studies of beginning drivers.

One naturalistic study reported that "kinematic risky driving" (events like hard braking) did not change over the early months of driving but that novice drivers increasingly sped and engaged in distracting activities during this period (Klauer et al., 2014; Simons-Morton et al., 2012). Moreover, there is some evidence that novices become less cautious as they gain experience (Chapman et al., 2014; Underwood, 2013). Most of these changes would be expected to increase rather than decrease crash risk. Thus, based on the handful of relevant studies available, we have so far gained little understanding of what is actually being learned, or otherwise changes, that accounts for the rapid decrease in crashes among novice drivers.

29.6 Future Directions

Several paths of inquiry are needed to reveal what accounts for the rapid decrease in novice driver crashes. Additional studies of driving exposure, crash characteristics, individual differences, and naturalistic measures of driving during the first months of driving are needed to elaborate and clarify what is presently known about driver behavior during this period. Longitudinal studies of changes in hazard perception capabilities during early driving may help to pinpoint their importance in contributing to the rapid decrease in crashes. Findings from such studies might then inform efforts to accelerate the development of competence through training programs.

Future progress will be enhanced by a greater focus on issues that are suggested by theoretical considerations. Current thinking in cognitive psychology conceptualizes behavior as driven by the interplay between conscious, relatively slow deliberative processes and near-instantaneous, nonconscious *intuitive* processing (Bechara et al., 1997; Damasio, 1994; Kahneman, 2011). The extent to which conscious effort plays a role in action depends on the behavior in question. Driving is almost entirely intuitive, with little time available for deliberation before acting (Kahneman, 2011). Recent theoretical models of driving reflect this (Charlton & Starkey, 2011; Fuller, 2005; Lewis-Evans & Rothengatter, 2009; Vaa, 2013). Charlton and Starkey's tandem process model (2011) is an effort to incorporate key elements of earlier conceptualizations in a single model to explain the ways in which people process and respond to information while engaged in continuous actions like driving.

These theories provide guidance for future inquiry into what develops that underlies the rapid decrease in crashes in the early months of driving. Recent experimental research using skin conductance response to measure differences between novice and experienced drivers in nonconscious ("feeling") responses to risky driving situations, before any behavior is observed, signals that measures of important nonconscious or preconscious processes proposed in recent theoretical models are possible (Kinnear et al., 2013).

In sum, much of novice driver learning appears to be implicit, with the knowledge then nonconsciously deployed. Studies that recognize this, and that are designed to carefully measure these phenomena, are necessary to advance our understanding of novice drivers.

29.7 Conclusion

The critical question of why or how novice driver crash rates decline so rapidly within just a few months has rarely been considered. Existing research that might shed light on the matter is sparse and is generally not designed to address the question. Understanding this dramatic phenomenon will require longitudinal studies, with multiple measurements during the first months of driving to document changes in several possible indicators of what novice drivers are learning, or what else is changing, that produces the rapid decrease in crash rates.

Acknowledgments

This chapter is based on a report to the Eunice Kennedy Shriver National Institute of Child Health and Human Development (NICHD) following the TRB Subcommittee on Young Drivers midyear meeting in August 2015.

References

Bechara, A., Damasio, H., Tranel, D., & Damasio, A.R. (1997). Deciding advantageously before knowing the advantageous strategy. *Science, 275*, 1293–1294.

Chapman, E.A., Masten, S.V., & Browning, K.K. (2014). Crash and traffic violation rates before and after licensure for novice California drivers subject to different licensing requirements. *Journal of Safety Research, 50*, 125–138.

Charlton, S.G., & Starkey, N.J. (2011). Driving without awareness: The effects of practice and automaticity on attention and driving. *Transportation Research Part F: Traffic Psychology and Behaviour, 14*, 456–471.

Damasio, A.R. (1994). *Descartes' Error: Emotion, Rationality and the Human Brain.* New York: Grossett/Putnam.

Deery, H.A. (1999). Hazard and risk perceptions among novice drivers. *Journal of Safety Research, 30*, 225–236.

Divekar, G., Pradhan, A.K., Masserang, K.M., Reagan, I., Pollatsek, A., & Fisher, D.L. (2013). A simulator evaluation of the effects of attention maintenance training on glance distributions of younger novice drivers inside and outside the vehicle. *Transportation Research Part F: Traffic Psychology and Behaviour, 20*, 154–169.

Ericsson, K.A. (2002). Attaining excellence through deliberate practice: Insights from the study of expert performance. In M. Ferrar (Ed.) *The Pursuit of Excellence through Education*, pp. 21–55. Mahwah, NJ: Erlbaum.

Ericsson, K.A. (2005). Recent advances in expertise research: A commentary on the contributions to the special issue. *Applied Cognitive Psychology, 192*, 223–241.

Foss, R.D., Martell, C.A., Goodwin, A.H., & O'Brien, N.P. (2011). *Measuring Changes in Teenage Driver Crash Characteristics during the Early Months of Driving.* Washington DC: AAA Foundation for Traffic Safety.

Fuller, R. (2005). Towards a general theory of driver behaviour. *Accident Analysis & Prevention, 37*, 461–472.

Grayson, G.B., & Sexton, B.F. (2002). The development of hazard perception testing. Report TRL558, Crowthorne: Transportation Research Laboratories.

Kahneman, D. (2011). *Thinking, Fast and Slow.* New York: Farrar, Straus and Giroux.

Keating, D.P. (2007). Understanding adolescent development: Implications for driving safety. *Journal of Safety Research, 38*, 147–157.

Kinnear, N., Kelly, S.W., Stradling, S., & Thomson, J. (2013). Understanding how drivers learn to anticipate risk on the road: A laboratory experiment of affective anticipation of road hazards. *Accident Analysis & Prevention, 50*, 1025–1033.

Klauer, S.G., Guo, F., Simons-Morton, B.G., Ouimet, M.-C., Lee, S.E., & Dingus, T.A. (2014). Distracted driving and risk of road crashes among novice and experienced drivers. *The New England Journal of Medicine, 370*, 54–59.

Lee, S.E., Simons-Morton, B.G., Klauer, S.G., Ouimet, M.C., & Dingus, T.A. (2011). Naturalistic assessment of novice teenage crash experience. *Accident Analysis & Prevention, 43*, 1472–1479.

Lewis-Evans, B. (2010). Crash involvement during the different phases of the New Zealand graduated driver licensing system (GDLS). *Journal of Safety Research, 41*, 359–365.

Lewis-Evans, B., & Rothengatter, T. (2009). Task difficulty, risk, effort and comfort in a simulated driving task—Implications for Risk Allostasis Theory. *Accident Analysis & Prevention, 41*, 1053–1063.

Masten, S.V., & Foss, R.D. (2010). Long-term effect of the North Carolina graduated driver licensing system on licensed driver crash incidence: A 5-year survival analysis. *Accident Analysis & Prevention, 42*, 1647–1652.

Mayhew, D.R., Simpson, H.M., & Pak, A. (2003). Changes in collision rates among novice drivers during the first months of driving. *Accident Analysis & Prevention, 35*, 683–691.

McDonald, C.C., Goodwin, A.F., Pradhan, A.K., Romoser, M.R., & Williams, A.F. (2015). A review of hazard anticipation training programs. *Journal of Adolescent Health, 57*, 515–523.

Mourant, R.R., & Rockwell, T.H. (1972). Strategies of visual search by novice and experienced drivers. *Human Factors, 14*, 325–335.

Musicant, O., & Benjamini, Y. (2012). Driving patterns of novice drivers—A temporal spatial perspective. No. 12-1388. Washington, DC: Transportation Research Board 91st Annual Meeting.

Pradhan, A.K,. Hammel, K.R., DeRamus, R., Pollatsek, A., Noyce, D.A., & Fisher, D.L. (2005). Using eye movements to evaluate effects of driver age on risk perception in a driving simulator. *Human Factors, 47*, 840–852.

Pradhan, A.K., Pollatsek, A., Knodler, M., & Fisher, D.L. (2009). Can younger drivers be trained to scan for information that will reduce their risk in roadway traffic scenarios that are hard to identify as hazardous? *Ergonomics, 52*, 657–673.

Pradhan, A.K., Divekar, G., Masserang, K., Romoser, M., Zafian, T., Blomberg, R.D., Thomas, F.D. et al. (2011). The effects of focused attention training on the duration of novice drivers' glances inside the vehicle. *Ergonomics, 54*, 917–931.

Sagberg, F. (1998). Month-by-month changes in accident risk among novice drivers. In *Proceedings of the 24th International Congress of Applied Psychology*, San Francisco.

Simons-Morton, B.G., Ouimet, M.C., Wang, J., Chen, R., Klauer, S.G., Lee, S.E., & Dingus, T. (2012). Peer influence predicts speeding prevalence among teenage drivers. *Journal of Safety Research, 43*, 397–403.

Taylor, T.G., Masserang, K.M., Pradhan, A.K., Divekar, G., Samuel, S., Muttart, J.W., Pollatsek, A., & Fisher, D.L. (2011). Long term effects of hazard anticipation training on novice drivers measured on the open road. In *Proceedings of the International Driving Symposium on Human Factors in Driver Assessment, Training, and Vehicle Design* (p. 187).

Thomas, F.D., Rilea, S.L., Blomberg, R.D., Peck, R.C., & Korbelak, K.T. (2015). *Evaluation of the Safety Benefits of the Risk Awareness and Perception Training Program for Novice Teenage Drivers*. Draft report. Washington, DC: National Highway Traffic Safety Administration.

TRB. (2013). *Motor Vehicle Crashes and Injuries Involving Teenage Drivers: Future Directions for Research.* Transportation Research Circular, E-C180, Washington DC. Retrieved from http://onlinepubs.trb.org/onlinepubs/circulars/ec180.pdf.

Twisk, D.A., & Stacey, C. (2007). Trends in young driver risk and countermeasures in European countries. *Journal of Safety Research, 38*, 245–257.

Underwood, G. (2013). On-road behaviour of younger and older novices during the first six months of driving. *Accident Analysis & Prevention, 58*, 235–243.

Underwood, G., Phelps, N., Wright, C., Van Loon, E., & Galpin, A. (2005). Eye fixation scanpaths of younger and older drivers in a hazard perception task. *Ophthalmic and Physiological Optics, 25*, 346–356.

Vaa, T. (2013). Proposing a risk monitor model based on emotions and feelings: Exploring the boundaries of perception and learning. In M.A. Regan, J.D. Lee, & T.W. Victor (Eds.), *Driver Distraction & Inattention: Advances in Research and Countermeasures* (Vol. 1). Burlington, VT: Ashgate Publishing.

30

The Handbook of Teen and Novice Drivers: The Way Forward

Donald L. Fisher

Jeff K. Caird

William J. Horrey

Lana M. Trick

Abstract

Heraclitus famously said that the only constant is change. Although around the world, novice teen driver crashes have, for over half a century, been stuck at unacceptably high numbers, change is afoot. Many of the chapters in this handbook have articulated the important changes and progress that have been made in our understanding of and approaches to the safety of novice and teen drivers over the last 25 years. We are convinced that change will continue to prevail such that, in the not-too-distant future—say 10–20 years—we can dramatically reduce novice and teen fatalities, in both developed countries (Chapter 23) and low- and middle-income countries (Chapter 24).

Here we describe the way forward in research, first over the next 10 years and then further out into the future, drawing from the lessons we have learned from the many authors who have graciously contributed to this handbook. We believed before we started this handbook, and believe even more strongly after reading and rereading the chapters, that the study of novice teen drivers represents the unique opportunity to study the interaction of research issues central to so many different areas, including issues related to multitasking, visual search and attention, parental and teen interactions, peer pressure, emotional and psychosocial development, training, and the role of technology in training, to name just a few.

30.1 What We Understand

As we hope has been made clear, there has been a revolution not only in what we understand about novice and teen drivers but also in what can be done to reduce fatalities and injuries within this most vulnerable cohort. We know the types of crashes in which novice drivers are overinvolved: intersection, run-off-road, and rear-end crashes (Chapter 5). We know the complex set of skills, both cognitive (hazard anticipation, Chapters 6; attention maintenance, Chapter 7; and hazard mitigation, Chapter 8) and social (Chapter 10), that novice teen drivers need in order to avoid crashes. We know the factors that increase crash risk, both external factors, such as speeding (Chapter 11), distraction (Chapter 12), alcohol and THC (Chapter 13), and teen passengers (Chapter 16), and internal factors, either biological (fatigue, Chapter 15) or developmental (Chapter 9; attention deficit hyperactivity disorder [ADHD], Chapter 14).

Moreover, we now have effective strategies to reduce novice teen driver crashes (Chapter 22), the most important of which are the graduated driver licensing (GDL) programs (Chapter 17) and the licensing exams, which incorporate an evaluation of teens' hazard anticipation skills (Chapter 28). In fact, there are a variety of strategies that are now showing real promise, new strategies involving in-vehicle feedback technologies (Chapter 20) and simulator training (Chapter 21), strategies that might be augmented with parental involvement (Chapter 19). Furthermore, with modification, driver education and training (Chapter 18), a strategy that initially seemed hopeless, may prove effective under the right circumstances.

This progress has been possible in no small measure because of the advances in the technologies we have to study novice teen driver behaviors. We have seen dramatic increases in our ability to gather data in the field with the advent of naturalistic studies (Chapter 26). We have seen equally dramatic increases in our ability to gather data in controlled settings, through driving simulation (Chapter 25) and eye tracking (Chapter 27).

Yet, despite the aforementioned progress, we are still far short of our goals. Road traffic injuries are the leading cause of death among individuals between 15 and 29 years old (World Health Organization, 2015). There were an estimated 325,736 road traffic injuries that resulted in fatalities in 2012 among this age group (WHO, 2015). Of these deaths, it is estimated that over 100,000 fatal road traffic injuries occurred to occupants of an automobile, either as drivers or as passengers (Chapter 29). The other injuries occurred to vulnerable road users such as pedestrians, bicyclists, and motorcyclists. If those between 15 and 29 represent the immediate future of countries around the world, more needs to be done to reduce their deaths and injuries. Clearly, there is much still to be done.

30.2 The Next 10 Years of Research, Practice, and Policy

As we describe here, advances in technology may ultimately reduce novice teen driver fatalities and injuries to the point where they no longer cast the pall over their health and well-being that they do today. But, at least for the next 10 years, we expect that there is much to be done before the advances in technology kick in (or the policy makers have the support and the stamina that they need in order to push through legislation necessary for the technological innovations to take root). Here we describe the issues and questions that we think, if answered, will ultimately lead to the largest reductions in crashes. We organize the discussion chronologically, starting at a time before a teen ever thinks of driving. The important issues and questions fall upon different points along this timeline. But there are also several questions that transcend the timeline, and those we address at the end of the chapter.

30.2.1 Predriver Education

A battery of safety-related behaviors should be in place even before teens get behind the wheel. Of those, there are two that are relatively underresearched as they relate to teens. First, consider seat belts. Seat belt use is one of the great traffic safety battles, and it was fought on many different fronts. The regulatory battle is now largely won. Seat belt use is mandatory in all states in the United States but one, New Hampshire. The scientific evidence that seat belts save lives was never in doubt. But the battle has been a long one. And it is still being waged with teens, the frequency of seat belt use being smallest among this age group (National Highway Traffic Safety Administration, 2016). Much is known about how to increase use of seat belts among teens, but more clearly needs to be learned (Chapter 10).

Second, consider distractions. The evidence for the effects of distractions on driving is much more nuanced in the general population. In some cases, there is no disagreement. Long glances away from the forward roadway are dangerous, period (Klauer et al., 2014; Simmons et al., 2016). Texting is an activity that invariably entails such long glances (Caird et al., 2014), and consequently, it is banned in more and more countries and states (WHO, 2015). However, the evidence for the effect of cognitive distractions, such as cell phone conversations (not dialing), is mixed. Epidemiological studies generally find a higher crash risk for cell phone use (Elvik, 2011), whereas naturalistic studies do not seem to show a relationship to safety-critical event risk (Simmons et al., 2016). A meta-analysis of simulator and on-road studies finds reliable evidence for the effect of cell phone conversations on driving performance (Caird et al., 2008; Chapter 12). The Strategic Highway Research Program 2 (SHRP2), a naturalistic study, revealed an odds ratio of 2.0 for cell phone conversations (Kidd & McCartt, 2015). The cost of a cognitive conversation is less than texting but greater than driving without distractions.

There is no doubt that distractions play a larger role in crashes for novice teen drivers than they do for adults (Chapter 12). Moreover, the studies showing an effect of countermeasures on reductions in distracting behavior are few and far between. Several studies have shown that teens can be trained to reduce the duration of especially long glances inside the car (Divekar et al., 2013; Pradhan et al., 2011), but it is not known whether this actually just encourages them to take more glances. And one study showed that teens can be trained to reduce their engagement in secondary activities in the presence of latent hazards (Krishnan et al., 2015), but this study has not been replicated. Effective ways to reduce distracted driving remain elusive (e.g., Caird et al., 2014; McCartt et al., 2014). Distractions remain a huge issue as technologies increasingly enter vehicles though smartphones, and teen and young drivers are frequently the adopters of these social technologies. Finally, the impact of teen driver distraction on injuries and fatalities in the developing world has had almost no research or intervention work (WHO, 2015).

30.2.2 Driver Education

As noted throughout the handbook, standard driver education programs have had little, if any, effect on the crash rates of novice teen drivers (Nichols, 2003). Indeed, it would appear that they have had little impact on the behaviors that influence crashes. All this will possibly change. We now know the behaviors that decrease crash risk, both the cognitive and the personal behaviors (Chapters 9, 10, and 18) that are critical to this decrease. We know how to design training programs that influence these critical behaviors (Chapter 18). And recently it was determined that just 17 minutes of hazard anticipation training can decrease significantly the crashes of novice teen males by 43% and 35% in, respectively, 17 and 18 year olds (Thomas et al., 2016). But we do not know whether the many other training programs actually decrease crashes and, if so, by how much. Nor do we know whether these other training programs are scalable from the training context to real-world driving. The various cognitive skills training programs typically require only half an hour to administer and can be downloaded or accessed anytime on the web. They are easily scalable if they are engaging or of interest to the teen. The personal and social skills training programs typically often require at least an entire day of participation along with adult instructors (Senserrick et al., 2009). It is less clear how easily they would scale.

Not only are the cognitive skills training programs likely to be easy to scale, but also, they could benefit greatly from the recent and rapid advances in simulation technology (Chapter 25), including the head-mounted displays that are now taking the gaming world by storm (e.g., Oculus Rift), displays that can often track the eyes as well. In general, the closer is the actual situation in which a driver is asked to apply a skill, the more likely there is to be generalization to the real world of driving. Moreover, these new head-mounted displays are now more likely to be widely deployed because they are so much less expensive than equivalent technologies currently being employed in training.

30.2.3 Supervised Driving Phase of GDL

In Chapter 29, the authors note that after the learner's permit period ends, novice teen drivers still show a dramatic decrease in their crash risk. GDL sets the stage for the interactions between parents and their teens. But, in general, parents have not been apprised of the combination of cognitive and personal skills that their teen drivers need in order to reduce their crash risk (Chapter 19). Yet time and again, it has been shown that parental involvement is a key to the success of an intervention, whether it be involvement in new technology or an advance in parent–teen communication (Chapters 10, 19, and 20). We need to learn how to harness the driving expertise of parents. And with that, we need a greater emphasis on reaching underserved populations in order to maximize the influence of parents when it comes to the multiple domains related to learning to drive that depend critically on the culture (skills, attitudes, behaviors, self-regulation).

30.2.4 Driver Licensing

Driver licensing requirements have seen real changes in Europe and Australia (Chapters 17 and 28), in terms both of the time of licensing (GDL) and of the content of the licensing exam (hazard perception testing). The advances in license phasing are now incorporated into the various jurisdictions in North America and have produced major reductions in the per capita crash rate among novice teen drivers (Shope, 2007). These reductions are seen primarily in 16-year-olds, largely because they have a much more limited exposure under GDL programs (Chapter 17). It is not clear that the age at which licensing first occurs will ever move higher than 18 years of age. However, there are still a large number of states where the age at which drivers can first receive their restricted license could be raised, by either lengthening the period of supervised driving (a period during which there is little exposure to risk) or increasing the current age at which a driver can receive a learner's permit.

Unfortunately, the incorporation of content into the licensing exams in North America that has some evidence-based relation to crashes is still almost entirely absent. A look at the substance of Chapter 22 suggests that it will take more than just research to lead to changes in the content of the licensing exam. It will also take a rare blend of individuals from the worlds of research (academics, usually, though not always) and application (program administrators and agency officials at every level). Furthermore, for this initiative to succeed, the social climate has to be right; there should be no other issues competing for attention in the same arena. In some countries, the required constellation of events has occurred, and hazard anticipation is incorporated into the licensing exam. In other countries, like the United States, advances in technology (e.g., automated vehicles) have occupied the public's attention in transportation safety, and a focus on novice teen drivers has been put on the back burner. The situation is even more dire in developing countries (Chapter 24). However, this focus, too, will change. In the not-too-distant future, the time may be ripe for the introduction of change into licensing exams, making them more reflective of content with a predictive relation to crashes.

30.2.5 Restricted License Phase of GDL

There are a number of real opportunities in the restricted license phase of the GDL program to make progress that will require advances in research and changes in policy. Perhaps the most influential are the restrictions on teen passengers and nighttime driving that are part of most GDL programs (Chapter 17). But these restrictions are not present in all jurisdictions, and there is clear evidence for their combined effect on the reduction in crash risk (Masten et al., 2013; Vanlaar et al., 2009). The progress here must be made in the legislative arena and through advocacy groups.

None of these restrictions are as effective without the involvement of parents as they are with the involvement of parents (Chapter 19). Various programs have been developed that attempt to increase parental involvement in the restricted license phase with positive results on compliance (Simons-Morton et al., 2005). Yet, there are still relatively few evidenced-based evaluations of the role that parental involvement could and should play in the restricted license phase (as well as other phases). The progress here depends on research.

Finally, there has been a real interest in in-vehicle data recorders (IVDRs) or smartphone-based apps that provide real-time and/or delayed feedback to teens and their parents on risky driving behaviors

(Chapter 20). These devices and related programs have been shown to reduce the high-g-force events that are known to be related to crash involvement (Simons-Morton et al., 2011). The reductions are largest, perhaps not surprisingly, when parents are involved. However, their effects appear not to persist over time (after the IVDR is removed from a teen's vehicle or even after the novelty has worn off, in a few months' time). This may be partly because the teens do not associate the high-g-force events with scenarios in which they are likely to crash (Tsippy Lotan, personal communication). However, PC-based training programs have now been developed that do make clear the relation between high-g-force events and crashes. When the data are evaluated, there is clear evidence of a reduction in risky high-g-force behaviors (Zhang et al., 2015). However, it remains to be seen whether such a training program, if coupled with the IVDRs, would lead to long-term decreases in high-g-force events after an IVDR were removed from a teen's vehicle.

Insurance companies may solve the problem for us. Some are now offering reductions in premiums to drivers who install such devices in their vehicle. When the price of these devices comes down enough, insurance companies may be able to offer large discounts on insurance policies for teen drivers that they would not offer otherwise, perhaps extending throughout the restricted license phase and into the early part of the unrestricted license phase (until, say, age 21). The discounts may be too attractive to ignore for the majority of teen drivers. That said, the issues described in the previous paragraph remain.

30.2.6 Melding of Naturalistic and Simulator Studies

Finally, there have been great advances in our ability to study the behavior of novice teen drivers, including naturalistic studies (Chapter 26) and simulator studies (Chapter 25). These studies have made major contributions to our knowledge about why novice drivers crash and of what sorts of countermeasures can reduce those crashes. However, more recently, researchers are discussing how one might combine the two types of approaches and, synergistically, reduce the weaknesses of either approach separately. From the naturalistic studies, we can observe actual examples of crashes and near crashes and can infer—or directly observe—the causal factors. But in the final analysis, all we have are correlations: associations between environmental conditions, actions, and consequences. Although simulator studies do not have these limitations, they have different, equally concerning problems. In particular, in simulator studies, we can find evidence for or against a given hypothesis, but there is always the question of whether the results will generalize to the open road.

Given the increase in the detail available from naturalistic studies and the increase in the power of simulators, it is theoretically possible to replicate the environmental conditions in which drivers crash in a driving simulator. Consequently, a driving simulator could be used to evaluate countermeasures, that is, to determine whether a given countermeasure decreased the probability of the risky behaviors associated with the crash. Because the virtual world is almost identical in appearance and in all other relevant ways (e.g., traffic), the likelihood that the results will generalize to the real world is thus increased (Caird & Horrey, 2011).

30.3 The Future

Finally, we want to speak to what the future holds for developed countries on the one hand, and low- and middle-income countries on the other.

30.3.1 Developed Countries

It is clear that there is much that can be done in the next decade. But, 10 years from now, it is likely that advances in technology will radically change the landscape, even more so than the developments described previously. In fact, we, the editors, remain optimistic that we can reach near zero deaths among novice teen drivers in developed countries in the not-too-distant future. It may be a character flaw among us, a Panglossian bent that blinds us to the reality of the difficulty of making change despite what Heraclitus might have to say. Still, if you stick with us for just the next two pages, maybe you, too, will see some real light at the end of the tunnel, just as we do.

First, consider alcohol. It is estimated that alcohol is involved in a quarter of all crashes among teens (National Highway Traffic Safety Administration, 2013a). Ignition interlock devices can prevent the engine from starting based on blood alcohol concentration (BAC) levels. They have proven effective for first-time and repeat offenders in several different countries, so much so that the National Highway Traffic Safety Administration (2013b) has developed model guidelines for their implementation at the state level. From a technical standpoint, the question to date has been whether the devices are both reliable and fast enough to be useful. The quality of the evaluation of breathalyzers was greatly enhanced recently and makes possible much more rigorous testing of them in the Alcohol Countermeasures Laboratory at the Volpe National Transportation Systems Center in Cambridge, MA. Current estimates are that they cost $100 to install and $50–100 per month to maintain. All the evidence suggests that these costs will come down.

Not unrelated to alcohol, in our opinion, there will be a dramatic increase in the ability to determine other aspects of driver state, including instances of driver distraction as well as fatigue, drowsiness, or conversely, alertness. This will reduce markedly the large percentage of crashes in which distraction and fatigue are factors (e.g., Åkerstedt & Kecklund, 2001; Carney et al., 2015). The increase will come about through a combination of improvements in our ability to sense driver behaviors (eye movements, for example) along with advances in the mathematical models used to infer driver state.

In recent years, there has been much discussion of the future of automobiles. Advanced automation stands to fundamentally change the nature and function of cars and, more importantly, the role of the driver. With partially or fully automated vehicles, drivers could become relegated to a role of a systems' monitor or one in which they assume control of the vehicle only in very select situations. While the many inherent issues have been considered in a number of papers (e.g., Cunningham & Regan, 2015; Fisher et al., in press), there are also important implications for our future novice and teen driver. For example, as the needs and demands of drivers stand to change over time, so too will the approaches by which teen and novice drivers will learn to drive.

Consider also V2X communications (V2V, vehicle-to-vehicle; V2I, vehicle-to-infrastructure; V2B, vehicle-to-bicycle; and so on)—one dimension in the discussion of advanced automation and telematics. Currently, it is estimated that V2V communications will contribute to a dramatic reduction for intersection crashes, one of the more prevalent crash types among novice drivers (Department of Transportation, 2015). This is because drivers (or their vehicles) can now literally see around obstacles that prior types of collision warning devices were not able to do. V2I communications will reduce run-off-road crashes, another major type of crash in which novice drivers are overinvolved. And V2V communications along with automatic emergency braking should do away with most rear-end crashes, at least those that occur at low velocities. If the same reduction were to hold across only these two additional major crash types, the advance in V2V communications alone would lead to a 50% reduction in teen crashes.

Finally, there is the relentless advance of safety features that reduce injuries when a crash occurs. How much further this can go is unclear, but it can only get better.

30.3.2 Low- and Middle-Income Countries

The chapter on low- and middle-income countries makes clear the challenges that these countries face (Chapter 24; WHO, 2015). Fortunately, for their solution, these challenges don't simply depend on large sums of money. Instead, the development of safety-related policies on seat belts, GDL programs, and licensing reform requires political will. And there is evidence that the political will is emerging. For example, the Third Traffic Safety Forum was held in Saudi Arabia in November of 2015, attended mostly by representatives from the surrounding Arab countries. In total, there were close to 1000 attendees, including all of the necessary players—government officials, the police, academicians and other researchers, private industry and foundations, the Saudi Arabian equivalent of Mothers Against Drunk Driving (MADD) in the United States, and, perhaps most importantly, teens themselves.

Additional resources have been provided by international organizations. For example, WHO (2015) has provided resources to transportation and public health officials in low- and middle-income countries that help them improve the safety of all drivers, including resources used to: reduce speed; increase motorcycle helmet use; increase seat belt use and child restraints; and decrease drunk-driving, drugs, and distractions. All of these areas of focus are covered by our handbook.

The more technical advances should come down in cost as well, making much more affordable enforceable drunk driving countermeasures and the safety-critical advances in V2X communications. In short, there is every indication that novice driver deaths may approach zero in both developed and low- and middle-income countries in the not too distant future.

Acknowledgments

We want to acknowledge one final time the contributions of the many different authors to this handbook. The ideas here, at least the good ones, are theirs. Their passion for their research and for the purposes that it can serve is everywhere evident. That passion has kept us going when our candles were burning not only at both ends but also in the middle as the deadline to finish the handbook drew ever closer. Thank you.

References

Åkerstedt, T., & Kecklund, G. (2001). Age, gender and early morning highway accidents. *Journal of Sleep Research, 10*(2), 105–110.

Caird, J.K., & Horrey, W. (2011). Twelve practical and useful questions about driving simulation. In D.L. Fisher, M. Rizzo, J.K. Caird, & J.D. Lee (Eds.), *Handbook of Driving Simulation for Engineering, Medicine, and Psychology* (pp. 5-1–5-18). Boca Raton, FL: CRC Press.

Caird, J., Willness, C., Steel, P., & Scialfa, C. (2008). A meta-analysis of the effects of cell phones on driver performance. *Accident Analysis & Prevention, 40*, 1282–1293.

Caird, J.K., Johnston, K., Willness, C., Asbridge, M., & Steel, P. (2014). A meta-analysis of the effects of texting on driving. *Accident Analysis & Prevention, 71*, 311–318.

Carney, C., McGehee, D., Harland, K., Weiss, M., & Raby, M. (2015). *Using Naturalistic Driving Data to Assess the Prevalence of Environmental Factors and Driver Behaviors in Teen Driver Crashes.* Washington, DC: AAA Foundation for Traffic Safety.

Cunningham, M., & Regan, M.A. (2015). Autonomous vehicles: Human factors issues and future research. *Proceedings of the 2015 Australasian Road Safety Conference.* Gold Coast, Australia: Australasian College of Road Safety.

Department of Transportation. (2015). *Vehicle-to-Vehicle Communication Technology.* Accessed January 5, 2015, retrieved from file:///C:/Users/Donald.Fisher/Downloads/V2V_Fact_Sheet_101414_v2a.pdf.

Divekar, G., Pradhan, A., Masserang, K., Pollatsek, A., & Fisher, D. (2013). A simulator evaluation of attention maintenance training on glance distributions of younger novice drivers. *Transportation Research F: Traffic Psychology and Behaviour, 20*, 154–159.

Elvik, R. (2011). Effects of mobile phone use on accident risk: Problems of meta-analysis when studies are few and bad. *Transportation Research Record, 2236*, 20–26.

Fisher, D.L., Lohrenz, M., Moore, D., Nadler, E.D., & Pollard, J.K. (in press). Humans and intelligent vehicles: The hope, the help and the harm. *IEEE Transactions on Intelligent Vehicles.*

Kidd, D.G., & McCartt, A.T. (2015). The relevance of crash type and severity when estimating crash risk using the SHRP2 naturalistic driving data. *The 4th International Driver Distraction and Inattention Conference*, Sydney, Australia.

Klauer, S., Guo, F., Simons-Morton, B., Ouimet, M., Lee, S., & Dingus, T. (2014). Distracted driving and risk of road crashes among novice and experienced drivers. *New England Journal of Medicine, 370*, 54–59.

Krishnan, A., Samuel, S., Dundar, C., & Fisher, D. (2015). Evaluation of a training program (STRAP) designed to decrease young drivers secondary task engagement in high risk scenarios: A strategic approach to training hazard anticipation in young drivers. *Proceedings of the Transportation Research Board 94th Annual Meeting.* Washington, DC: National Academies Press.

Masten, S., Foss, R., & Marshall, S. (2013). Graduated driver licensing program component calibrations and their association with fatal crash involvement. *Accident Analysis & Prevention, 57,* 105–113.

McCartt, A.T., Kidd, D.G., & Teoh, E.R. (2014). Driver cellphone and texting bans in the United States: Evidence of effectiveness. *Annals of Advances in Automotive Medicine, 58,* 99–114.

National Highway Traffic Safety Administration. (2013a). *Model Guideline for State Ignition Interlock Programs (DOT HS 811 859).* Washington, DC: National Highway Traffic Safety Administration.

National Highway Traffic Safety Administration. (2013b). *Traffic Safety Facts 2011: Young Drivers.* Washington, DC: National Highway Traffic Safety Administration.

National Highway Traffic Safety Administration. (2016). *Teen Drivers—Seat Belt Use.* Accessed January 4, 2016, from Driving Safety, retrieved from http://www.nhtsa.gov/Driving+Safety/Teen+Drivers/Teen+Drivers+-+Seat+Belt+Use.

Nichols, J. (2003). *A Review of the History and Effectiveness of Driver Education and Training as a Traffic Safety Program.* Washington, DC: National Transportation Safety Board.

Pradhan, A., Divekar, G., Masserang, K., Romoser, M., Zafian, T., Blomberg, R.D., Thomas, F.D., Reagan, I., Knodler, M., Pollatsek, A., & Fisher, D.L. (2011). The effects of focused attention training (FOCAL) on the duration of novice drivers' glances inside the vehicle. *Ergonomics, 54,* 917–931.

Senserrick, T., Ivers, R., Boufous, S., Chen, H., Norton, R., Stenvenson, M., & Zask, A. (2009). Young driver education programs that build resilience have the potential to reduce road crashes. *Pediatrics, 124,* 1287–1292.

Shope, J. (2007). Graduated driver licensing: Review of evaluation results since 2002. *Journal of Safety Research, 38,* 165–175.

Simmons, S., Hicks, A., & Caird, J.K. (2016). Safety-critical events associated with cell phone tasks measured through naturalistic driving studies: A systematic review and meta-analysis. *Accident Analysis & Prevention, 87,* 161–169.

Simons-Morton, B., Hartos, J., Leaf, W., & Preusser, D. (2005). Persistence of effects of the Checkpoints program on parental restrictions of teen driving privileges. *American Journal of Public Health, 95,* 447.

Simons-Morton, B., Quimet, M., Zhang, Z., Klauer, S., Albert, P., & Dingus, T. (2011). Crash and risky driving involvement among novice adolescent drivers and their parents. *American Journal of Public Health, 101,* 2362–2367.

Thomas, F., Rilea, S., Blomberg, R., Peck, R., & Korbelak, K. (2016). *Evaluation of the Safety Benefits of the Risk Awareness and Perception Training Program for Novice Drivers (DOT HS 812 235).* Washington, DC: National Highway Traffic Safety Administration.

Vanlaar, W., Mayhew, D., Marcoux, K., Wets, G., Brijs, T., & Shope, J. (2009). An evaluation of graduated driver licensing programs in North America using a meta-analytic approach. *Accident Analysis & Prevention, 41,* 1104–1111.

World Health Organization (WHO). (2015). *Global Status Report on Road Safety 2015.* Geneva: World Health Organization.

Zhang, J., Romoser, M., & Fisher, D. (2015). Evaluation on a driving simulator of a training program designed to reduce risky behaviors associated with quick starts and quick stops: The LAG (Less Aggressive Goals) training program. *Proceedings of the Transportation Research Board Annual Meeting.* Washington, DC: National Academies of Science.

Index

Page numbers followed by f and t indicate figures and tables, respectively.

D

E

M

O

P

T

U

V